# Lecture Notes in Electrical Engineering

## Volume 471

"Lecture Notes in Electrical Engineering (LNEE)" is a book series which reports the latest research and developments in Electrical Engineering, namely:

- Communication, Networks, and Information Theory
- Computer Engineering
- Signal, Image, Speech and Information Processing
- Circuits and Systems
- Bioengineering

LNEE publishes authored monographs and contributed volumes which present cutting edge research information as well as new perspectives on classical fields, while maintaining Springer's high standards of academic excellence. Also considered for publication are lecture materials, proceedings, and other related materials of exceptionally high quality and interest. The subject matter should be original and timely, reporting the latest research and developments in all areas of electrical engineering.

The audience for the books in LNEE consists of advanced level students, researchers, and industry professionals working at the forefront of their fields. Much like Springer's other Lecture Notes series, LNEE will be distributed through Springer's print and electronic publishing channels.

More information about this series at http://www.springer.com/series/7818

Jaume Anguera · Suresh Chandra Satapathy
Vikrant Bhateja · K. V. N. Sunitha
Editors

# Microelectronics, Electromagnetics and Telecommunications

Proceedings of ICMEET 2017

 Springer

*Editors*
Jaume Anguera
Department of Electronics
   and Telecommunication
Ramon Llull University
Barcelona
Spain

Vikrant Bhateja
Department of Electronics
   and Communication Engineering
SRMGPC
Lucknow, Uttar Pradesh
India

Suresh Chandra Satapathy
Department of Computer Science
   and Engineering
PVP Siddhartha Institute of Technology
Vijayawada, Andhra Pradesh
India

K. V. N. Sunitha
BVRIT Hyderabad College
   of Engineering for Women
Hyderabad, Andhra Pradesh
India

ISSN 1876-1100          ISSN 1876-1119   (electronic)
Lecture Notes in Electrical Engineering
ISBN 978-981-10-7328-1          ISBN 978-981-10-7329-8   (eBook)
https://doi.org/10.1007/978-981-10-7329-8

Library of Congress Control Number: 2017960919

Printed on acid-free paper

This Springer imprint is published by Springer Nature
The registered company is Springer Nature Singapore Pte Ltd.
The registered company address is: 152 Beach Road, #21-01/04 Gateway East, Singapore 189721, Singapore

# Preface

This volume contains high-quality research papers presented at the third International Conference on Microelectronics, Electromagnetics and Telecommunications (ICMEET) held in Departments of Electronics and Communication Engineering of BVRIT Hyderabad College of Engineering for Women, Hyderabad, Telangana, India, during September 9–10, 2017. ICMEET aims to bring together academic scientists, researchers, and research scholars to discuss the recent developments and future trends in the fields of microelectronics, electromagnetics, and telecommunication. Previous editions of the conference were held in GITAM University (2015), Visakhapatnam, and Raghu Institute of Technology (2016), Visakhapatnam. ICMEET received a total of 325 submissions for LNEE series of the conference. Each paper was peer-reviewed by at least two members of the Program Committee. Finally, a total of 94 papers were accepted for publication in this proceeding. ICMEET was technically supported by IETE student forum and IE student chapter. Several special sessions were offered by eminent professors in many cutting-edge technologies. Several eminent researchers and academicians delivered talks addressing the participants in their respective field of proficiency. Topic on evolutionary optimized power control methodologies and mitigation of $H_2S$ from hydrothermal power plant has been discussed by Naeem M. S. Hannoon, Professor, Technology of University, Mara, Malaysia. A thorough discussion on MEMS and NEMS was made by Prof. Sanket Goel from BITS Pilani, Hyderabad campus.

We would like to express our appreciation to the members of the Program Committee for their support and cooperation in this publication. We are also thankful to the team from Springer for providing a meticulous service for timely production of this volume. Our heartfelt thanks to our Chairman Sri Vishnu Raju K. V. garu and other management members of Sri Vishnu Educational Society for extending wholehearted support to host this in their campus. Special thanks to all guests who have honored us by their presence on the inaugural day of the conference. Our thanks are due to all special session chairs, track managers, and reviewers for their excellent support. Last, but certainly not the least, our special thanks go to all the authors who submitted papers and all the attendees for their contributions and fruitful discussions that made this conference a great success.

| | |
|---|---|
| Barcelona, Spain | Jaume Anguera |
| Vijayawada, India | Suresh Chandra Satapathy |
| Lucknow, India | Vikrant Bhateja |
| Hyderabad, India | K. V. N. Sunitha |
| September 2017 | |

# Organizing Committee

## Chief Patrons

Sri K. V. Vishnu Raju, Chairman, Sri Vishnu Educational Society
Sri Ravichandran Rajagopal, Vice-Chairman, Sri Vishnu Educational Society

## Patrons

Dr. S. Ramakumar, Director, Sri Vishnu Educational Society
Dr. Srinivasan Sundarrajan, Director, Sri Vishnu Educational Society
Dr. K. V. N. Sunitha, Principal, BVRIT Hyderabad College of Engineering for Women

## Program Coordinator

Dr. J. Naga Vishnu Vardhan, HOD-ECE, BVRIT Hyderabad College of Engineering for Women

## Special Session Chairs

Dr. Sudheer Kumar Terlapu, Vishnu Group, Bhimavaram
Dr. P. S. R. Chowdary, RIT, Vizag
Dr. V. V. S. S. Sameer Chakravarthy, RIT, Vizag

## Conveners

Dr. Seetaiah Kilaru, BVRIT Hyderabad College of Engineering for Women

## Co-Conveners

Dr. Anwar Bhasha Pattan, BVRIT Hyderabad College of Engineering for Women
Mrs. Praveena M., BVRIT Hyderabad College of Engineering for Women

## Proceedings Committee

Dr. M. Parvathi, Professor, BVRIT Hyderabad College of Engineering for Women
V. Hindumathi, Assistant Professor, BVRIT Hyderabad College of Engineering for
Women
A. Radha, Assistant Professor, BVRIT Hyderabad College of Engineering for
Women

## Technical Program Committee

S. Madhavi, Assistant Professor, BVRIT Hyderabad College of Engineering for
Women
R. Shylaja, Assistant Professor, BVRIT Hyderabad College of Engineering for
Women
P. Prashanth, Assistant Professor, BVRIT Hyderabad College of Engineering for
Women
G. Siva Sankar Varma, Assistant Professor, BVRIT Hyderabad College of Engi-
neering for Women

## Organizing Committee

R. Priyakanth, Assistant Professor, BVRIT Hyderabad College of Engineering for
Women
K. Mahesh Babu, Assistant Professor, BVRIT Hyderabad College of Engineering
for Women
R. Sridevi, Assistant Professor, BVRIT Hyderabad College of Engineering for
Women
G. Rama Lakshmi, Assistant Professor, BVRIT Hyderabad College of Engineering
for Women

## Publicity Committee

K. Suneela, Assistant Professor, BVRIT Hyderabad College of Engineering for
Women

K. Brunda Devi, Assistant Professor, BVRIT Hyderabad College of Engineering for Women

Mr. P. Raj Bhagath, Assistant Professor, Raghu Institute of Technology

## Finance Committee

A. Radha, Assistant Professor, BVRIT Hyderabad College of Engineering for Women

V. Hindumathi, Assistant Professor, BVRIT Hyderabad College of Engineering for Women

Siva S. Sinthura, Assistant Professor, BVRIT Hyderabad College of Engineering for Women

G. Drakshayani, Assistant Professor, BVRIT Hyderabad College of Engineering for Women

## Hospitality Committee

Y. Anil Kumar, Assistant Professor, BVRIT Hyderabad College of Engineering for Women

N. M. Saikrishna, Assistant Professor, BVRIT Hyderabad College of Engineering for Women

P. Rajesh Kumar, Assistant Professor, BVRIT Hyderabad College of Engineering for Women

D. Venkata Siva Prasad, Assistant Professor, BVRIT Hyderabad College of Engineering for Women

M. Venkatesh, Assistant Professor, BVRIT Hyderabad College of Engineering for Women

## Transportation Committee

Y. Anil Kumar, Assistant Professor, BVRIT Hyderabad College of Engineering for Women

K. Mahesh Babu, Assistant Professor, BVRIT Hyderabad College of Engineering for Women

N. M. Sai Krishna, Assistant Professor, BVRIT Hyderabad College of Engineering for Women

G. Siva Sankar Varma, Assistant Professor, BVRIT Hyderabad College of Engineering for Women

# International And National Advisory Committee

Prof. Jaume Anguera, Universitat Ramon Llull, Spain
Dr. Ashwin Kumar, Carnegie Mellon University
Dr. Vannet Agarwal, Purdue University
Dr. Krishnaswami Hari Srihari, State University of New York
Dr. Alison Carrington, Staffordshire University
Dr. A. K. Raviteka, Drenden Institute of Technology, Germany
Dr. Narasimhasarma N. V. S., NIT Warangal
Dr. Adithya K. Jagannatham, IIT Kanpur
Dr. Syed Nazeemuddin, IIT Hyderabad
Dr. M. B. Srinivas, BITS Pilani, Hyderabad
Dr. L. Pratap Reddy, JNTU Hyderabad
Dr. M. Madhavi Latha, JNTU Hyderabad
Dr. D. Srinivasa Rao, JNTU Hyderabad
Dr. M. Asha Rani, JNTU Hyderabad
Dr. B. N. Bhandari, JNTU Hyderabad
Dr. T. Satya Savithri, JNTU Hyderabad
Dr. B. Lakshmi, NIT Warangal
Dr. P. Naveen Kumar, Osmania University
D. Rama Krishna, Osmania University
Dr. B. Rajendra Naik, Osmania University
Dr. R. Hemalatha, Osmania University
Dr. Sai Dhiraj Aruma, Samsung R&D Institute, Bangalore
Dr. G. Sasibhushana Rao, Andhra University
Dr. P. G. Krishna Mohan, JNTU Hyderabad (Retd)
Dr. K. Padmapriya, JNTU Kakinada
Dr. Zinka Sreenivasa Rao, BITS Pilani
Dr. I. A. Pasha, BVRIT Narsapur, Hyderabad
Dr. Amit Acharya, IIT Hyderabad
Mrs. N. Mangala Gowri, JNTU Hyderabad

# Panel of Reviewers

Prof. Vikranth Bhateja
Prof. Suresh Chandra Satapathy
Dr. Elsa Macias Lopez
Dr. Abdul Ella Hassanien
Prof. P. Mallikarjuna Rao
Dr. Santoshi Ganala
Prof. P. Satish Rama Chowdary

Dr. V. V. S. S. Sameer Chakravarthy
Dr. Ashish Singh
Dr. Sudheer Kumar Terlapu
Dr. T. Venkateswara Rao

# Contents

# About the Editors

**Dr. Jaume Anguera** is a Professor in the Department of Electronics and Telecommunication, Universitat Ramon Llull. He is also an R&D Manager at Fractus, Barcelona, Spain. He is directing projects in the areas of multi-frequency and miniature monopoles; dual-frequency dual-polarized microcell antennas; monoband dual-polarized microstrip patch array; dual-band dual-polarized microstrip patch arrays; network feeding architectures for GSM-UMTS arrays; broadband matching networks; high cross-polarization techniques, array pattern synthesis with genetic algorithms, handset antennas, automotion antennas, and antennas for military applications among others. He obtained the degrees of Technical Engineer (Electronic Systems) and Electronic Engineer from Ramon Llull University, Barcelona, and degree of Telecommunication Engineer and Ph.D. (Doctor Ing.) from Polytechnic University of Catalunya (UPC), Barcelona. His research interests are miniature, wideband, multi-band, and high directivity printed antennas and arrays; fractal-shaped antennas genetic algorithms; handset antennas: internal and external multi-band antennas; base station antennas: multi-band cross-polarized antennas; low cross-polarization and high-isolation techniques for BTS antennas; automotive antennas: integration of antennas; short-range wireless and UWB antennas; and educational methods for electromagnetics. He has over 50 publications in journals and 100 papers in conference proceedings.

**Dr. Suresh Chandra Satapathy** is currently working as Professor and Head, Department of Computer Science and Engineering, PVP Siddhartha Institute of Technology, Andhra Pradesh, India. He obtained his Ph.D. in Computer Science and Engineering from JNTU Hyderabad and M.Tech. in CSE from NIT Rourkela, Odisha, India. He has 26 years of teaching experience. His research interests are data mining, machine intelligence, and swarm intelligence. He has acted as program chair of many international conferences and edited six volumes of proceedings from Springer LNCS and AISC series. He is currently guiding eight Ph.D. scholars. He is also a senior member of IEEE.

**Prof. Vikrant Bhateja** is Associate Professor, Department of Electronics and Communication Engineering, Shri Ramswaroop Memorial Group of Professional Colleges (SRMGPC), Lucknow, and also the Head (Academics & Quality Control) in the same college. His areas of research include digital image and video processing, computer vision, medical imaging, machine learning, pattern analysis and recognition, neural networks, soft computing, and bio-inspired computing techniques. He has over 90 quality publications in various international journals and conference proceedings. He has been on TPC and chaired various sessions from the above domain in international conferences of IEEE and Springer. He has been the track chair and served in the core-technical/editorial teams for international conferences: FICTA 2014, CSI 2014, and INDIA 2015 under Springer-ASIC Series and INDIACom-2015, ICACCI-2015 under IEEE. He is Associate Editor in International Journal of Convergence Computing (IJConvC) and also serving in the editorial board of International Journal of Image Mining (IJIM) under Inderscience Publishers. At present, he is a guest editor for two special issues floated in International Journal of Rough Sets and Data Analysis (IJRSDA) and International Journal of System Dynamics Applications (IJSDA) under IGI Global Publications.

**Dr. K. V. N. Sunitha** completed her B.Tech. ECE from Nagarjuna University, M.Tech. Computer Science from REC Warangal. She was awarded Ph.D. from JNTU, Hyderabad, in 2006. She has 24 years of teaching experience. She has been working as Founder Principal, BVRIT Hyderabad College of Engineering for Women, Nizampet, Hyderabad, since August 2012. She received "Academic Excellence Award" by G. Narayanamma Institute of Technology & Science on September 18, 2005, "Best Computer Science Engineering Teacher Award for the year 2007" by Indian Society for Technical Education (ISTE) on February 23, 2008, "Best Faculty" award in Academic Brilliance Awards-2013 by Indian Education Expo on April 2013 at New Delhi, "Distinguished Scientist Award in NLP" by Venus International on December 3, 2016, "Distinguished Principal Award" by CSI Mumbai on January 15, 2017 at IIT Bombay. She has been recognized and invited by AICTE as NBA Expert Evaluator. Her autobiography was included in "Marquis Who's Who in the World," 28th edition, 2011. She has authored four textbooks published by NOVA publishers, USA. Her areas of research include natural language processing, speech processing, network and Web security. She has guided six Ph.D. students and currently guiding eight research scholars. She received a research grant from CSI for her R&D project in March 2011 for 1 year and also received 38.5 lakhs of funding for DST-TIDE project. She has published over 125 papers in international and national journals and conferences. She is a reviewer for many national and international journals. She is a fellow of Institute of Engineers, senior member for IEEE & International Association CSIT, and life member of many technical associations like CSI and ACM.

# A DCT-CS Watermarking Method for Monochrome and Color Image

D. Susmitha and S. M. Renuka Devi

**Abstract** In order to overcome the threat of data transmission over Internet, watermarking techniques have been developed. Watermarking can be implemented in spatial or in frequency domain. In this paper, watermarking is implemented in frequency domain using DCT transform and compressive sensing technique. The paper deals with a proposed watermarking scheme that can be applied to binary, monochrome, and color image. Here, the features of compressive sampling are considered in order to overcome the drawback of insecure data transmission. A random Gaussian matrix is used as the secret key for encrypting the watermark in the process of compressive sensing. This encrypted watermark is then embedded into the mid-frequency DCT coefficients of the host image. Experimental results depict that the performance of our algorithm is better than basic DCT watermarking in terms of robustness. It is observed that the security level and embedding capacity are also improved by the usage of compressive sampling.

**Keywords** Discrete cosine transform · Watermarking · Compressive sampling

## 1 Introduction

In the current years, the growth of technology has increased vastly in the concern of transfer of information over Internet. As Internet is an open source, the security of the digital media is put under risk. This threat can be faced by providing some authentication information along with the data. Embedding of authentication information into the digital media is referred as watermarking. Watermarking can be applied to video, audio, images, and other digital media also. Research in this

D. Susmitha (✉) · S. M. Renuka Devi
G. Narayanamma Institute of Technology and Science (for women),
Hyderabad, India
e-mail: susmitha.daparty18@gmail.com

S. M. Renuka Devi
e-mail: renuka.devi.sm@gmail.com

© Springer Nature Singapore Pte Ltd. 2018
J. Anguera et al. (eds.), *Microelectronics, Electromagnetics
and Telecommunications*, Lecture Notes in Electrical Engineering 471,
https://doi.org/10.1007/978-981-10-7329-8_1

1

area had been developed enormous from the past decade onward [1–3]. The resulting image after embedding the ownership or signature information is the watermarked image. Properties of HVS (human visual system) are used for embedding the data into the cover image imperceptibly.

Many authors [4–7] have discussed the watermarking performance evaluation by testing the algorithm's robustness against various attacks, embedding capacity and transparency, i.e., visual degradation of host and extracted watermark image. It is required to keep a trade-off between all these parameters.

Some of the authors [8–10] have already used watermarking based on compressive sampling and DCT. The main contribution of our work is in using compressive sampling and discrete cosine transform for providing better security by using a measurement matrix in compressive sampling and a full reference image watermarking scheme is used. Also, this paper explores the use of DCT-CS method for color images, which is the first of its kind.

This paper is organized as follows: Sect. 2 is about basics of compressive sampling and orthogonal matching pursuit algorithm; Sect. 3 discusses the proposed method of watermarking technique; and Sect. 4 concludes with the results by comparing the proposed method with basic DCT watermarking technique.

## 2  Compressive Sampling

Bandwidth is of major concern when transmitting information over channel and so the data needs to be compressed as per the Shannon–Nyquist sampling theorem before transmitting [11, 12]. In order to satisfy this theorem, it requires acquiring large number of samples (i.e., double the maximum frequency of input signal), which is expensive in applications like radar and medical imaging [11]. Compressive sampling deals with this problem by acquiring the signal directly in the compressed domain. This is possible by representing the signal in an appropriate basis [11, 12]. For example, consider a 1-D signal X (of finite length N) that can be expressed in an appropriate basis $\Psi$ using

$$X = \sum_{i=1}^{N} S_i \Psi_i \text{ or } X = \Psi S, \tag{1}$$

where S is the N × 1 column vector. X is similar to S except that their domain representation is different. If the number of samples in S that is nonzero, far less than the total number of samples, then S is said to be sparse representation of X, i.e., X is sparse in $\Psi$ domain. Now, the signal X is in compressed form.

But recovering the signal X at the receiver, without having any knowledge about the location of nonzero coefficients is computationally complex problem [12]. This is overcome by computing M(M < N) inner products between X and $\phi_j$ where $j = 1$ to M as $y_j = \langle x, \phi_j \rangle$.

$$Y = \phi X = \phi \Psi S = \Theta S \qquad (2)$$

$$\Theta = \phi \Psi \qquad (3)$$

$\Theta$ is an M × N matrix. $\Theta$ can also be referred as dimensionally reduced space of signal X. The signal can be recovered from this dimensionally reduced matrix by using the recovery algorithms like OMP, $l_1$-minimization [11, 13]. In our context, OMP is used since its performance is better as compared to $l_1$-minimization and also OMP is a generalized algorithm and suits all kinds of applications [14].

## 2.1 OMP (Orthogonal Matching Pursuit)

The OMP algorithm is an iterative process [14]. It makes use of the orthogonal projections of measurement matrix Y onto the random Gaussian matrix A. The algorithm for implementing the OMP is given below.

1. Initially, measurements $Y_{M \times 1}$ and the measurement matrix $A_{M \times N} = [a_1, a_2, ..., a_N] \in R^M$ are given, where M is the number of measurements and N is the number of samples.
2. Calculate the angles $\Theta_i$ for i = 1 to N, by orthogonally projecting Y onto the atoms, $a_i$, $\Theta_i$ = angle (Y, $a_i$).
3. Find the inner product, $\lambda_i$, between Y and $a_i$ and preserve the maximum value. This will be the first nonzero coefficient.
4. Calculate the index $I_1$ = argmin ($\Theta$i), for i = 1 to N.
5. Calculate the residue Y1 = Y − $\lambda_{I_1} a_{I_1}$.
6. Now calculate the inner product between residue Y1 and atoms of A (exclude the $I_1$th atom as it is already perpendicular) and preserve the maximum value to be the second nonzero coefficient.
7. Repeat the procedure until the residue value tends to be less than a predefined threshold (this is the stopping criteria).

## 3 Proposed Method

### 3.1 Watermark Embedding

Figure 1 illustrates the watermarking scheme used. The cover image is taken to be a color image. Since modifying the three channels (R, G, B) is not recommendable,

**(a)**

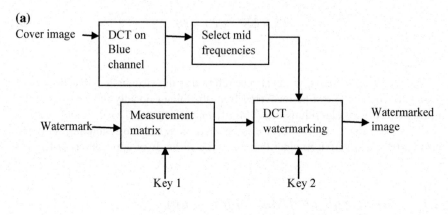

**(b)**

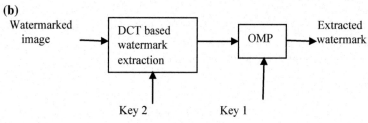

**Fig. 1  a** Watermark embedding **b** Watermark extraction

usually blue channel [4] is used for embedding the watermark. So, select the blue channel and apply 2-D DCT. For a monochrome watermark, apply compressive sampling principles. This can be done by generating a measurement matrix and calculating the measurement vector. These measurements are embedded into the mid-frequencies of B channel. The mid-frequencies are used, since in DCT the low frequencies carry most of the visual information and especially the DC component should not be disturbed; otherwise, this results in degradation of the cover image. The high-frequency components are more susceptible to noise and compression attacks. The measurement matrix acts as key1 and the equation used for embedding given below acts as key2:

$$I' = I + g*W, \tag{4}$$

where $I'$ is the watermarked image, g is the watermarking strength $(0 < g < 1)$ [7], and W is the watermark measurement vector. Watermarked image is produced by applying inverse DCT. In case of color cover image, combine the channels after applying inverse DCT to blue channel, for obtaining watermarked image.

## 3.2   Watermark Extraction

Extraction of watermark can be achieved only when the key is shared with the authorized party. Watermark can be extracted by applying DCT to the watermarked image and using the following formula:

$$W = \left(I' - I\right)/g \tag{5}$$

Now, the measurement vector is obtained, from which watermark can be produced by using the compressive sampling reconstruction algorithm OMP. Recovery is successful only when the same random matrix (key1) is used. In case of using a color cover image, it is recommended to separate the three channels initially and then extract the watermark.

## 4   Performance Measures and Results

Transform-domain watermarking provides less embedding capacity compared to the spatial-domain watermarking. For example, the DCT watermarking system offers less than half of the middle frequencies to be suitable for embedding without visual degradation [2]. This algorithm is carried out on $256 \times 256$ host color image and watermark of sizes $64 \times 64$ and $32 \times 32$. The extracted watermark quality is evaluated in terms of PSNR, SSIM, RMSE, and normalized correlation [2].

Figure 2 shows the results of the proposed system. Watermarks of size $32 \times 32$ and $64 \times 64$ are used for embedding the host image of size $256 \times 256$. Table 1 presents the performance evaluation of watermarked image for basic DCT method [5] and proposed DCT-CS method. Table 2 gives the quality measure of a $64 \times 64$ watermark considering one-third of total samples and half number of measurements.

**(a)**                         **(b)**                                    **(c)**
Original watermark       Watermarked image              Extracted watermark

**Fig. 2**  **a** $64 \times 64$ watermark; **b** Watermarked image of size $256 \times 256$; **c** Extracted watermark using DCT-CS method, obtained NC = 0.9652

**Table 1** Quality measure of watermarked image in DCT-CS watermarking method in comparison with the basic DCT [5]

| Watermark size (Text watermark) | PSNR (dB) | | SSIM | | RMSE | | NC | |
|---|---|---|---|---|---|---|---|---|
| | Proposed DCT-CS | Basic DCT | Proposed DCT-CS | Basic DCT | Proposed DCT-CS | Basic DCT | Proposed DCT-CS | Basic DCT |
| 32 × 32 | **51.2432** | 38.4746 | **0.9955** | 0.9731 | **0.0333** | 0.0917 | **0.9999** | 0.9988 |
| 64 × 64 | **39.6751** | 34.3594 | **0.9430** | 0.9619 | **0.1220** | 0.1208 | **0.9991** | 0.9969 |

**Table 2** Quality measure of extracted watermark for DCT-CS watermarking method in comparison with basic DCT [5]

| Watermark size (Text watermark) | PSNR (dB) | | SSIM | | RMSE | | NC | |
|---|---|---|---|---|---|---|---|---|
| | Proposed DCT-CS | Basic DCT | Proposed DCT-CS | Basic DCT | Proposed DCT-CS | Basic DCT | Proposed DCT-CS | Basic DCT |
| 64 × 64 | **19.2819** | 11.5241 | **0.7675** | 0.2949 | **0.1178** | 0.2862 | **0.9652** | 0.8543 |
| 32 × 32 | **23.2600** | 16.9982 | **0.5613** | 0.7805 | **0.0741** | 0.1533 | **0.9866** | 0.9403 |

It can be observed from Tables 1 and 2 that the performance of our system is far better than the traditional approach of watermarking in DCT domain in terms of PSNR, SSIM, RMSE, and NC for both watermarked image and extracted watermark.

Robustness of the watermark is evaluated by subjecting the watermarked image to attacks like salt and pepper noise, Gaussian filtering, additive white Gaussian noise (AWGN) and measuring the PSNR, SSIM, RMSE, and NC of the extracted watermark. Table 3 summarizes the robustness of extracted watermark (64 × 64) under the presence of various attacks. The results are compared against the basic DCT system.

Figure 3 displays the effect of Gaussian noise on extracted watermark. It can be observed from the plot that the NC of the watermark extracted is at least 80% only when the SNR of the watermarked image (after the channel attacks) is greater than 35 dB.

**Table 3** Robustness evaluation of extracted watermark in proposed DCT-CS system compared with basic DCT system [5]

| Type of attack | PSNR (dB) | | SSIM | | RMSE | | NC | |
|---|---|---|---|---|---|---|---|---|
| | Proposed DCT-CS | Basic DCT | Proposed DCT-CS | Basic DCT | Proposed DCT-CS | Basic DCT | Proposed DCT-CS | Basic DCT |
| Salt and pepper (0.025) | **8.6410** | 3.156 | **0.2002** | 0.022 | **0.3989** | 0.750 | **0.6698** | 0.36 |
| Gaussian filter (5 × 5) | **23.0097** | 2.603 | **0.5532** | 0.040 | **0.0763** | 0.799 | **0.9862** | 0.47 |
| AWGN (SNR = 35 dB) | **11.0732** | 2.588 | **0.2430** | 0.041 | **0.3015** | 0.800 | **0.8070** | 0.47 |

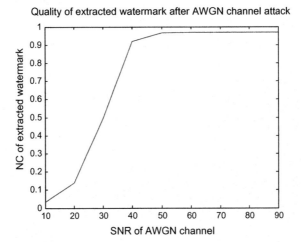

**Fig. 3** Performance of extracted watermark with Gaussian noise attack on channel

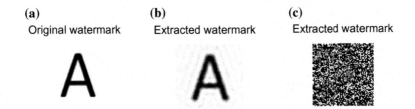

**Fig. 4 a** Embedded watermark of size 64 × 64; **b** Extracted watermark by using right key, obtained NC = 0.9652; **c** Extracted watermark by using wrong key, obtained NC = 0.0003

The degree of resistance of this system to intruders depends on watermarking key (key2) used for selecting the embedding location and the measurement matrix (key1) used for implementing compressing sensing. Figure 4 illustrates the importance of the keys in extracting the watermark.

Our method of DCT-CS is also suitable for color images, by embedding the watermark in one of the channels (R, G, B). The performance analysis by using a cover color image is similar to that of grayscale image since the same watermarking method is used except that the watermark is embedded in the blue channel. Figure 5 shows the performance of our algorithm for color image.

**(a)**                          **(b)**                          **(c)**
Watermarked image         Original watermark        Extracted watermark

**(d)**                                          **(e)**
Original watermark                        Extracted watermark

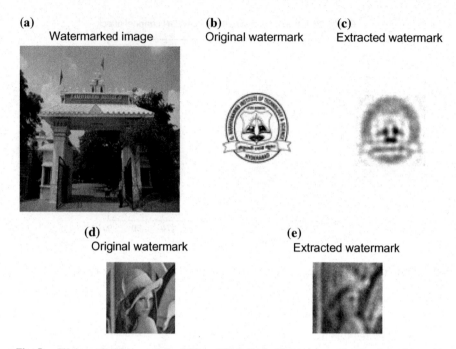

**Fig. 5** **a** Watermarked image of size 256 × 256; **b** Embedded logo watermark of size 64 × 64; **c** Extracted logo watermark, obtained NC = 0.7166; **d** Embedded Lena watermark of size 64 × 64; **e** Extracted Lena watermark, obtained NC = 0.7874

## 5   Conclusion

DCT-CS watermarking scheme proposed in this paper can be applied to binary, monochrome, and color images. This paper deals with the full reference watermarking where the original cover image is required for extracting the watermark. In this DCT-CS system with full reference watermarking, initially the watermark is compressively sensed and is embedded into the mid-band coefficients of DCT of the cover image. For improving the security and robustness, compressive sampling is considered. Since the compressive sampling reduces the total number of samples required for reconstruction of a signal, it therefore enhances the embedding capacity. In future, our algorithm can be stretched to other digital media like video and audio signals.

## References

1. Smita Pandey, Rohit Gupta, "A Comparative Analysis on Digital Watermarking with Techniques and Attacks", International Journal of Advanced Research in Computer Science and Software Engineering, Volume 6, Issue 6, June 2016.

2. Neetha K.K and Aneesh M. Koya, "A Compressive sensing approach to DCT watermarking system", International Conference on Control, Communication and Computing India (ICCC), IEEE conference publications, November 2015, pp. 495–500.
3. Potdar, V., S. Han and E. Chang, "A Survey of Digital Image Watermarking Techniques", in Proceedings of the IEEE International Conference on Industrial Informatics, Perth, Australia, 2005, pp. 709–716.
4. Verma, Bhupendra, et al., "A New color image watermarking scheme", INFOCOMP, Journal of computer science, Volume 5, Issue 3, 2006, pp. 37–42.
5. Barni, Mauro, et al. "A DCT-domain system for robust image watermarking", Signal processing, Volume 66, Issue 3, 1998, pp. 357–372.
6. J.J.K. O'Ruanaidh, F.M. Boland, W.J. Dowling, "Phase watermarking of digital images", Proceedings IEEE International. Conference on Image Processing (ICIP'96), Vol. III, Lausanne, Switzerland, 16–19 September 1996, pp. 239–242.
7. M.D. Swanson, B. Zhu, A.H. Tewfik, "Transparent robust image watermarking", Proceedings IEEE International Conference on Image Processing (ICIP'96), Vol. III, Lausanne, Switzerland, 16–19 September 1996, pp. 211–214.
8. P. K. Korrai and K. Deergha Rao, "Compressive Sensing and Wavelets Based Image Watermarking and Compression", TENCON 2014 - 2014 IEEE Region 10 Conference, IEEE Conference Publications, 2014, pp. 1–5.
9. Zhang, Xinpeng, et al. "Watermarking with flexible self-recovery quality based on compressive sensing and compositive reconstruction", IEEE Transactions on Information Forensics and Security, Volume 6, Issue 4, 2011, pp. 1223–1232.
10. Hsiang-Cheh Huang and Feng-Cheng Chang, "Robust Image Watermarking Based on Compressed Sensing Techniques," Journal of Information Hiding and Multimedia Signal Processing, Volume 5, Issue 2, April 2014, pp. 275–285.
11. Justin Romberg, "Imaging via compressive sampling," IEEE signal processing magazine, March 2008, pp. 14–20.
12. Richard G. Baraniuk, "Compressive sensing", IEEE signal processing magazine, July 2007, pp. 118–121.
13. V. Pavithra and S. M. Renuka Devi, "An image representation scheme by hybrid compressive sensing", 2013 IEEE Asia Pacific Conference on Postgraduate Research in Microelectronics and Electronics (Prime Asia), Visakhapatnam, 2013, pp. 114–119.
14. Tropp, Joel A., and Anna C. Gilbert, "Signal recovery from random measurements via orthogonal matching pursuit", IEEE Transactions on information theory, Volume 53, Issue 12, 2007, pp. 4655–4666.

# Intelligent Counter System for Generating Attendance

N. Edna Elizabeth, T. K. Gowthaman, J. Joannes Sam Mertens
and P. Likhitta Dugar

**Abstract** In the present time, in most educational institutes, proxy is witnessed as one of the most inexcusable violences of the rules and regulations. It is also often observed that the attendance in all the educational institutes is taken manually by the faculty in charge to avoid proxy, but ends up in reducing the productive time available in the classroom. The solution to the above-faced difficulties is to bring about an automated system to mark the student's presence in each hour. Thus, the aim of this project is to bring a two-factor authentication system for generating the attendance by integrating the radio frequency identity cards along with biometrics. The result shows the prototype of the system designed with security. Also, wireless communication through radio waves between the transmitter and the receiver is carried out in this work.

**Keywords** Two-factor authentication · Biometrics · Proxy avoidance
Automated system

N. Edna Elizabeth · J. Joannes Sam Mertens
Electronics and Communication Department, Sri Sivasubramaniya
Nadar College of Engineering, Old Mahabalipuram Road,
Kalavakkam 603110, Tamil Nadu, India
e-mail: ednaelizabethn@ssn.edu.in

J. Joannes Sam Mertens
e-mail: sam.mertens0007@gmail.com

T. K. Gowthaman · P. Likhitta Dugar (✉)
Sri Sivasubramaniya Nadar College of Engineering,
Old Mahabalipuram Road, Kalavakkam 603110, Tamil Nadu, India
e-mail: likhitta@dugar.in

T. K. Gowthaman
e-mail: gowthaman13029@ece.ssn.edu.in

© Springer Nature Singapore Pte Ltd. 2018
J. Anguera et al. (eds.), *Microelectronics, Electromagnetics
and Telecommunications*, Lecture Notes in Electrical Engineering 471,
https://doi.org/10.1007/978-981-10-7329-8_2

11

# 1 Introduction

The current scheme used for marking the attendance in the classroom is very time consuming and not entirely secure in terms of avoiding proxy. Hence, the idea to develop a handy terminal was implemented. The incapability of manipulating the biometric human characteristics was a major concern while implementing the project. It is used by assigning the fingerprint to their individual details, like name and unique register number in their passive Radio Frequency Identification (RFID) tags. The successful matching of both the authenticated factors helps in enlisting and creating a database for each time the attendance is generated.

This system is carried out by the integration of a fingerprint scanner along with an RFID reader and a Liquid Crystal Diode (LCD) screen. The LCD screen is used to display the acknowledgment and to avoid any mismatch of data. With the help of an Arduino UNO 3 using the transceiver, Serial Peripheral Interface (SPI) communication is the methodology used to communicate with the student terminal to the faculty terminal and vice versa. The faculty terminal is a complete integration of the transceiver with the database storage device as an individual module. As a device for local storage, a computer/laptop present in each classroom is used. The present local storage device is only accessible by the faculty in charge, of that particular hour. On the confirmation of the database observed on the Graphical User Interface (GUI), it is transferred to the faculty's Simple Storage Device (SSD) for anytime access. Our major concern is to access any particular hour's attendance by the faculty, within the premises, in case of any discrepancies. The faculty is the sole controller.

There are already available attendances marking systems but lack in some way or the other when taken into account the proxy or possible violations. The readily available systems are either wall mount or designed on only one-factor authentication.

When taken in the case of the wall mount biometric systems, there is a possibility of intentional time waste in doing the same. On the other hand, while analyzing the case of the available one-factor authentication system, i.e., RFID or biometric factor, there is no evidence of proxy or other malpractices.

In present time, at the end of each attendance slot allotted by the universities, it becomes a tedious job for faculties to create the database and get it reviewed by each and every student to avoid discrepancies, which lead us in creating this automated counter system.

# 2 Related Works

The article [1] has proposed a design using the emerging technology for identification (not proper format). Biometric refers to automatic identification of a person based on biological characters such as fingerprint, iris, facial recognition, etc. In this article, the main heart of the circuit is fingerprint module. This sends commands to the controller whenever fingerprint is matched.

Anif Jamaluddin et al. [2] have proposed a design for the improvement of assessment model using RFID technology which is implemented on Computer Based Test (CBT). But this system still needs human as a proctor for monitoring server.

G. V. Ambadkar and A. R. Karwankar [3] described the implementation techniques of RFID, integrated with biometric sensor to improve the security system. In this system, it uses both the techniques for registering proper attendance of student. No one can give proxy. It operates when the students swipe their RFID tag and also access the biometric, so that the attendance gets marked and the class door gets opened for the day. The limitation in the above paper is that the attendance was marked only at one instance in the whole day. This leads to high percentage of the students' attendance, for every subject even if the particular student was absent on particular class hours on several days.

Kong Shengli et al. [4] have explained the basic concept of the connection of RC522 reader to the Arduino platform for the acquisition of RFID card information. They have also given the optimum frequency of working as 13.5 MHz. They state that the usage of RC522 reader is for low power consumption, low cost, and low voltage requirements.

# 3 Proposed System Overview

## 3.1 Introduction

The system consists of a transmitter module and a receiver module. The microcontroller is common to all modules and it performs logical operations. The transmitter is placed on the students' desk, while the receiver is connected to the faculty accessible laptops. For communication between the transmitter and receiver, Serial Peripheral Communication (SPI) is used. This can be adapted easily by the windows operating software which is analyzed to be a common platform of operation in the present times. SPI communication was chosen as there are a common clock and other general signals. This type of communication is preferred as they communicate from one to many devices by just proving single and separate "chip enable" pin layouts for each device that is being connected and used.

## 3.2 Transmitter

The transmitter module consists of the Arduino UNO 3, fingerprint sensor module, the RFID reader, a 16 × 2 LCD screen, and the nRF24L01 transceiver. A communication link is established between the transmitter and the receiver, when the transmitter module receives the initializing signal transmitted from the module present in the faculty side.

Figure 1 describes the transmission module which contains the fingerprint sensor, Arduino, RFID reader, LCD, and transceiver. Once the initial signal is received, the Arduino supplies power to the LCD screen and the RFID reader. The LCD screen first displays, stating that the module is ready to accept and receive data. The students are asked to swipe the passive tags that are distributed individually consisting of unique RFID numbers.

Initially, a database is created in the faculty accessible laptop, when the students are simultaneously made to enroll one of their fingers as password along with the tag swipe, for the two factors of authentication. In the case of mismatching of the fingerprint and the RFID retrieved details, the error messages are displayed on the LCD screen. This ensures a high-level security in the attendance system.

## 3.3 Receiver

The message signals from the various transmitters present are received by the module that is present at the receiver end. This module is an integrated system of the Arduino UNO 3 with the transceiver.

Figure 2 describes the receiver module which contains an Arduino and a transceiver. This module is also used as the initialization system. When this receiver

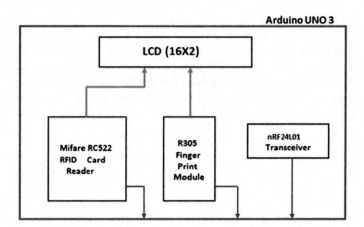

**Fig. 1** Transmission module

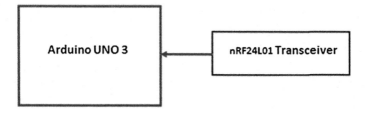

**Fig. 2** Receiver module

module is powered on by the faculty, through the method of SPI communication, it is observed that the receiver module sends the initial enabling signals to all the connected modules (transmitter) present in that classroom.

The signal is sent to each module and there is a particular time slot given to the students by the faculty within which they are asked to post in their attendance by passing through the two-factor authentication procedure. At the end of that time slot, the receiver module resends a signal which collects the generated list of attendance. On the successful reception of the data, the receiver sends the acknowledgement signal back to the students' module, stating the confirmation of the attendance received. On the contrary if there is any loss of data or errors observed, that particular unique ID number would not get an acknowledgement, by which they are required to repost their attendance.

# 4   Proposed System Execution

Implementing the process flow of the system is the major part that should be done carefully. Interfacing each component individually with Arduino UNO 3, and then integrating them completely, gives the final counter system. The student's terminal is the first step, which consists of an RFID reader and the fingerprint sensor with an LCD display connected to the Arduino UNO along with a transceiver. Next is the faculty terminal, which has an Arduino UNO 3 connected with a transceiver which receives the data sent from the students terminal and displays it on the laptop screen which the faculty accesses. The transceiver present at both the terminals is used for sending and receiving data. Further, there is a discussion about the interfacing steps for each component in the system.

In Fig. 3, the overall system process flow is shown. As seen in the process flow diagram, the proceedings of this counter system begin when the students' side terminal get powered on by the initializing signal received from the teacher's terminal. Once the student terminal is powered, the students are accessible to scan their tags along with matching their respective enrolled finger for marking their attendance.

On the event of both the factors of authentication, a temporary database is created and stored in the Arduino present in the student terminal. Once the faculty

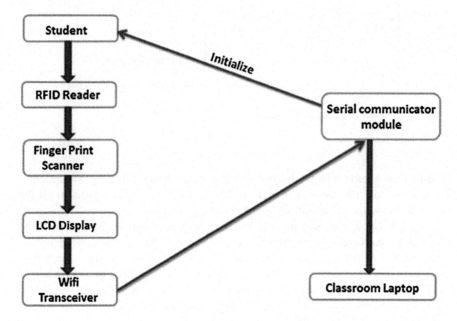

**Fig. 3** Attendance process flow

in charge clicks on the "STOP" button, there is a link created between the transceivers of the faculty and student terminals. Through SPI communication by the link, the temporary stored database is communicated and enlisted on the GUI screen on the faculty side terminal. The enlisted database created is then saved and used for further processing by the faculty.

### 4.1 Interfacing RFID with Arduino

This stage is the first factor for authentication, which is done using the RFID tag and reader. The passive RFID tag is made to swipe over the reader which uses its signals to capture the data from the tag swiped. The reader sends energy to an antenna which converts it into an RF wave and sends it into the read zone. Once the tag is read within the read zone, the RFID tag's internal antenna draws in energy from the RF waves. Mifare RFID reader is connected to the Arduino UNO R3 with its respective pin configurations. The LCD-connected pins, i.e., the secret slave pin and the reset pin of the RFID, are defined. Each card is read by the reader and the unique ID associated with each card is defined with its owner. The details of this owner (i.e., the name of the owner) of that particular card are made to display on the LCD screen as well as the serial monitor screen. When the faculty in the class is ready to take attendance, the faulty gives in the power signal from the terminal present there and once each unit receives the initial start signal, the LCD monitor

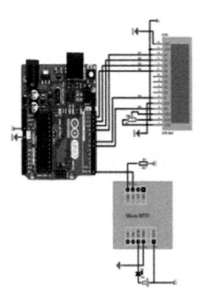

**Fig. 4** Interfacing RFID with Arduino

present in the students terminal displays the message that reads "SCAN CARD WHEN INSTRUCTED" as shown in Fig. 4. Once the student swipes his RFID tag on the reader, the LCD displays the student's name (owner's name).

In case of any unauthenticated card tapped on the reader, it will display an error message stating the detection of an invalid card.

## 4.2 Interfacing Biometric with Arduino

The second factor of authentication is carried out by using the biometric factor, i.e., the fingerprint sensor as shown in Fig. 5. Fingerprint authentication refers to the automated method of verifying a match between the saved fingerprint image and the fingerprint scanned at that moment. Fingerprint is one of many forms of biometrics which is said to be unique and is used to rightly identify individuals or verify their identity, addressed by the enrolment and matching of the fingerprint. The module is connected to Arduino UNO 3 and LCD with its respective pin configurations.

ENROLING—Here, the student's finger is placed twice for confirmation while creating the image and is then made to recognize by the Arduino. By the captured images, a database is created by the allocation of unique ID numbers (starting from 0).

**Fig. 5** Interfacing biometric with Arduino

MATCHING—The initial setup is kept the same as that in the enrolling step. The fingerprint sensor is programmed to blink until it identifies and captures a valid finger. Upon capturing, it is compared with the already existing images in the database. If a matched fingerprint is found, the associated ID is displayed on the LCD display.

Depending on the instructions displayed on the LCD, the student imprints and acquires the confirmation with the associated unique number. On the contrary, if it does not match, it displays the error message "NOT MATCHED".

### 4.3 Integrating RFID and Biometric

This step deals with the integration of the above two explained stages. Each RFID unique number is tagged not only with the owner's name but also with their respective biometric information. At first, the RFID is made to tap on the reader and once the first factor is authenticated successfully, it is then lead to the second level of authentication. At this stage, it captures the fingerprint and checks if it matches with the details pertaining to that of the unique ID of the matched RFID tag.

In Fig. 6, once the student "X" swipes the RFID tag and gets successfully recognized by the reader, the LCD displays as follows:

**Fig. 6** Integrating RFID and
biometric

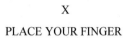

The student X then will place the enrolled finger on the fingerprint sensor, assuming it was enrolled with the number "1" and gets it verified as the second factor of authentication. If both, the RFID and fingerprint, match successfully, the LCD displays and the next continues the same process to mark his attendance. This is shown in Fig. 7.

**Fig. 7** Integrating RFID and
biometric

## 4.4   Transmission of Data

The nRF24L01 transceiver is connected on both ends to send and receive the student's data in each hour. The transmission/reception of data through the transceivers takes place only when the numbers of both the transceivers trying to connect and communicate are verified to be the same. In the case of identification of different numbers, there is no communication that takes place between those two transceivers.

## 4.5   Display of the Database

The final step of this counter system is the display of the database on the faculty accessible laptop screen. The data received through the transceiver is displayed on the GUI created for each laptop. The application (or) GUI created on the laptop is programmed such that the faculty has the control from the beginning to the end of attendance process. Once all the databases from each unit in the class are collected,

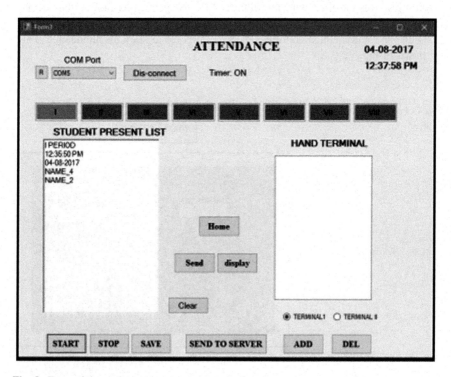

**Fig. 8** Form of the students present

the GUI gives the list of all the students who have marked their attendance in the classroom during that particular hour.

Figure 8 shows the created GUI which has a presentation screen on which the dates with the respective period numbers are mentioned. On the event of clicking upon a particular period, a list of the created database appears on a side tab.

The form is used for creating and storing the initial database of all the students in the educational institution. The form is filled when the students are enrolled in the educational institution at the time of admission. Alongside is also given a provision of viewing the personal information, such as their name, photo, register number, etc., upon selecting the student's database created.

## 5 Conclusion and Future Work

Thus, the attendance database generated by the intelligent counter system, through the two-factor authentication, is a secured system and does not involve any manipulation of data. Despite the existence of various methods which propose the generation of attendance using RFID solely or the generation of attendance by using only the biometrics factor, our proposed work stands high with the combination of both RFID system and the biometric for a secured way of authenticated attendance marking, thus avoiding proxy. The most common and efficient way of communication is purely wireless. So wireless communication through radio waves is what is carried out in this work. Interference or loss during the transmission is very less compared to the usual communication as SPI communication is the methodology used in this project. For reducing the burden of the faculty and by eliminating the large amount of time taken during marking the attendance, this intelligent counter system does justice.

Another aspect of work that can be done using the RFID tag is by replacing it with a smart card [5]. Each student having their own smart card can access it for multiple purposes in the college premises, for example, paying their fee, using it for paying bills in cafeteria, Xerox purposes, and also for transit application [5].

## References

1. "Biometric Attendance System Circuit" http://www.electronicshub.org/biometric-attendance-system-circuit/ (22 October 2015)
2. Anif Jamaluddin; Dewanto Harjunowibowo; M Akbar Rochim; Fajar Mahadmadi; HBulan Kakanita; Pringgo W. Laksono "Implementation of RFID on Computer Based Test (RF-CBT) system", Proceedings of the Joint International Conference on Electric Vehicular Technology and Industrial, Mechanical, Electrical and chemical Engineering (ICEVT & IMECE), pp. 153–156, https://doi.org/10.1109/ICEVTIMECE.2015.7496645, IEEE Conference Publications (2015)

3. G.V. Ambadkar, A.R. Karwankar, "RFID and Fingerprint Based Student Attendance System", International Journal of Industrial Electronics and Electrical Engineering, ISSN: 2347-6982, vol 3, (4 April 2015)
4. Kong Shengli, Zhao Jun, Shi Guang, Wu Chunhong, Zhao Wenpei, "The Design and Implementation of the Attendance Management System based on Radio Frequency Identification Technology", International Conference on Electronic Science and Automation Control, (ESAC 2015)
5. Edna Elizabeth. N, Nivetha S, "Design of a Two-factor Authentication ticketing system for Transit Applications" IEEE TENCON 2016-Technologies for Smart Nation, 22–25 November 2016, Marina Bay Sands, Singapore, pp 2498–2504 (IEEE Xplorer) https://doi.org/10.1109/TENCON.2016.7848483

# High-Throughput VLSI Architectures for CRC-16 Computation in VLSI Signal Processing

R. Ashok Chaitanya Varma and Y. V. Apparao

**Abstract** The intent of this paper is to design VLSI architectures for different CRC polynomial equations to achieve high throughput and low latency using DSP algorithms for signal processing. These architectures for CRC polynomials are designed using different techniques such a serial architecture, combined pipelining and parallelism, retiming technique, unfolding technique, and folding transformation. Linear Feedback Shift Register (LFSR) is an important component used in designing these architectures. A new formulation IIR filter-based design is proposed for designing these serial and parallel architectures using LFSR. In this paper, serial architectures, different levels of parallelism like one-level parallelism, two-level parallelisms, and three-level parallelisms are proposed for CRC-16 polynomial equation. Comparison is done between throughput and latency for different CRC polynomials for serial architectures and different levels of parallelism architectures. These architectures are designed and implemented in Verilog language and synthesized using Xilinx tool, cadence tool, etc.

**Keywords** Cyclic redundancy check · Linear feedback shift register
Throughput · Latency · Parallel architectures · IIR filter design
Look-ahead technique

R. Ashok Chaitanya Varma (✉)
Department of ECE, Shri Vishnu Engineering College for Women, Bhimavaram, India
e-mail: r.chaitanyavarma19@gmail.com

Y. V. Apparao
Department of ECE, GITAM University, Visakhapatnam, India
e-mail: yvarao@gitam.edu

© Springer Nature Singapore Pte Ltd. 2018
J. Anguera et al. (eds.), *Microelectronics, Electromagnetics
and Telecommunications*, Lecture Notes in Electrical Engineering 471,
https://doi.org/10.1007/978-981-10-7329-8_3

# 1 Introduction

To meet the real-world communication standards for high-throughput, low-latency Linear Feedback Shift Register (LFSR) is widely used in Digital Signal Processing (DSP) and communication systems, especially for Cyclic Redundancy Check (CRC) operations and BCH encoders. Recursive formulae have been performed for parallel CRC architecture computation, based on mathematical analysis. CRC is used to find errors in data communication during the receiving of data at the receiver using LFSR component to obtain data correctly for designing high-throughput Very Large-Scale Integration (VLSI) architecture [1, 2]. In the CRC technique, a few number of check bits, often called a checksum, are appended or added to the message being transmitted at the transmitter side. When the transmitted data is received at the receiver, the CRC checks the data and look-ahead technique is applied to the generator polynomial in correcting the errors [3]. For high-speed data transmissions, the normal serial architecture is designed and implemented which cannot meet the speed requirement because latency (number of clock cycles required to get the output) is high and throughput rate (number of instructions executing per a second) required is very less. Serial-to-parallel transformation is applied which is a very efficient technique to increase the throughput rate and reduce the latency of the clock cycles in serial architecture [4, 5]. A high-speed parallel architecture is introduced using a new formulation in the form of an IIR filter. Proposed high-speed architecture technique called combined pipelining and parallelism is applied to achieve good throughput rate [6, 7]. High-speed parallel architectures is based on the multiplication and division computations called look-ahead technique on CRC-16 generator polynomial equation in speeding up the parallel processing [8, 9]. The proposed design reduces the latency and increases throughput [10–12]. Single-level parallel, two-level parallel, and three-level parallel implementations are proposed to realize parallel processing [13, 14].

This paper is organized as follows. Section 2 describes a brief summary of mathematical analysis, algorithm for serial architecture; Sect. 3 describes the proposed designs of one-level parallel, two-level parallel, and three-level parallel architectures of CRC-16. Simulation waveforms of CRC-16, serial architecture, and three levels of parallel architectures, in comparison with the number of XOR gates, number of delay elements, latency of architecture, and throughput rate of the architecture of different parallel architectures, are presented in Sect. 4. Conclusion and remarks are presented in Sect. 5.

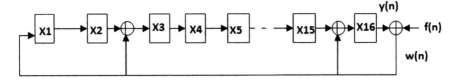

**Fig. 1** Serial architecture for CRC-16

# 2    Analysis of Serial Architectures of CRC-16

## 2.1    CRC-16 Serial Architecture

The generator polynomial for CRC-16 is

$$y(n) = y^{16} + y^{15} + y^2 + 1 \tag{1}$$

The output equation of CRC-16 generator polynomial is

$$y(n) = y(n-16) + y(n-14) + y(n-1) + f(n) \tag{2}$$

$$f(n) = u(n-16) + u(n-14) + u(n-1) \tag{3}$$

Figure 1 explains the CRC-16 serial architecture where output of the architecture is y(n), feedback of the architecture is w(n), and input to the architecture is u(n). Throughput achieved by serial architecture is 1. Latency is 16 clock cycles. The input for data is given as 1000000000000011. The output checksum bits are observed after 16 clock pulses and are observed as 1000000000000011. Since critical path is the loop with the maximum iteration bound, hence, critical path is $2T_{xor}$. Hence, latency is very high and throughput is very less; this architecture has disadvantage in digital signal processing. To overcome this throughput and latency, a high-speed technique called combined pipelining and parallelism is proposed.

# 3    Analysis of CRC-16 Parallel Architectures

## 3.1    CRC-16 Single-Level Parallel Architecture After Proposed Formulation

The proposed formulation (combined pipelining and parallelism technique) which is an IIR filter-based design architecture is introduced. The parallel algorithm technique processes an n-bit message in (n + k)/L clock cycles, where k is the order of the generator polynomial and L is the level of parallelism. n message bits can be processed in n/L clock cycles [7, 8].

**Fig. 2** Single-level parallel
architecture for CRC-16

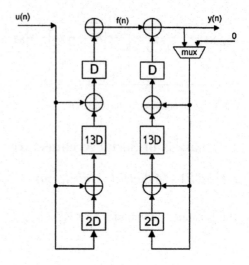

Block in one-level parallel architecture is delayed with block delay of 1 unit.
The output equation of CRC-16 generator polynomial is

$$y(n) = y(n-16) + y(n-14) + y(n-1) + f(n), \tag{4}$$

where

$$f(n) = u(n-16) + u(n-14) + u(n-1) \tag{5}$$

Figure 2 shows the single-level parallel architecture after which the proposed
formulation consists of both feedforward and feedback paths and multiplexer.
When selection input is "0", it will not allow output bit feedback to input. When
selection input is "1", it will allow output bit feedback to input. Latency and
throughput rate for CRC-16 single-level parallel architecture are 16 clock cycles
and 1, respectively.

## 3.2 CRC-16 Two-Level Parallel Architecture After Proposed Formulation

In Fig. 3, each block in two-level parallel architecture is delayed with block delay
by 2 units:

$$y(n) = y(n-16) + y(n-14) + y(n-1) + f(n), \tag{6}$$

**Fig. 3** Block delay of
two-level parallelism

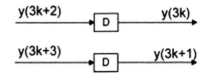

where

$$f(n) = u(n-16) + u(n-14) + u(n-1) \tag{7}$$

(With unit delay)

$$y(n-1) = y(n-17) + y(n-15) + y(n-2) + f(n-1) \tag{8}$$

After solving Eqs. (6), (7), and (8), the final equations are

$$y(3k+2) = y(3k-14) + y(3k-12) + y(3k+1) + f(3k+2) \tag{9}$$

$$\begin{aligned} y(3k+3) = {} & y(3k-14) + y(3k-13) + y(3k-12) + y(3k-11) \\ & + y(3k+1) + f(3k+2) + f(3k+3), \end{aligned} \tag{10}$$

where

$$f(3k) = u(3k-16) + u(3k-14) + u(3k-1) \tag{11}$$

$$f(3k+1) = u(3k-15) + u(3k-13) + u(3k) \tag{12}$$

In the above architecture, Fig. 4, two inputs (i.e., $u(3k+1), u(3k)$) are processed at a time and the corresponding output bits ($y(3k), y(3k+1)$) are computed. The

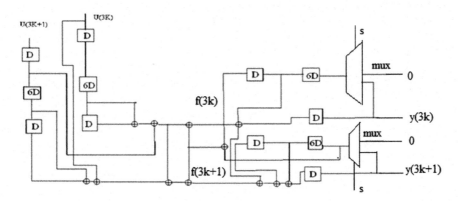

**Fig. 4** Two-level parallel architecture for CRC-16

critical path of the above architecture is 8Txor. Let n be the number of input sequence; then, latency for two-level parallel CRC-16 is $L = n/2$ clock pulses.

### 3.3 CRC-16 Three-Level Parallel Architecture After Proposed Formulation

In Fig. 5, each block in three-level parallel architecture is delayed with block delay by 3 units.

(with no delay)

$$y(n) = y(n-16) + y(n-14) + y(n-1) + f(n) \tag{13}$$

(with 1 unit delay)

$$y(n-1) = y(n-17) + y(n-15) + y(n-2) + f(n-1) \tag{14}$$

(with 2 units delay)

$$y(n-1) = y(n-17) + y(n-15) + y(n-2) + f(n-1) \tag{15}$$

After solving Eqs. (13)–(15), the final equations are

$$y(3k+3) = y(3k-13) + y(3k-11) + y(3k+2) + f(3k+3) \tag{16}$$

$$y(3k+4) = y(3k-12) + y(3k-10) + y(3k-13) + y(3k-11) \\ + y(3k+2) + f(3k+3) + f(3k+4) \tag{17}$$

$$y(3k+5) = y(3k-13) + y(3k-10) + y(3k-12) + y(3k-9) \\ + y(3k+2) + f(3k+3) + f(3k+4) + f(3k+5), \tag{18}$$

**Fig. 5** Block delay of three-level parallelism

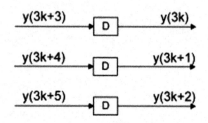

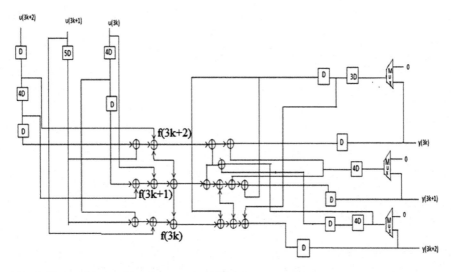

**Fig. 6** Three-level parallel architecture for CRC-16

where

$$f(3k+3) = u(3k-13) + u(3k-11) + u(3k+2) \tag{19}$$

$$f(3k+4) = u(3k-12) + u(3k-10) + u(3k+1) \tag{20}$$

$$f(3k+5) = u(3k-11) + u(3k-9) + u(3k) \tag{21}$$

In the above architecture, Fig. 6, three inputs (i.e., $u(3k+2), u(3k+1), u(3k)$) are processed at a time and the corresponding output bits $(y(3k), y(3k+1), y(3k+2))$ are computed. Latency for three-level parallel CRC-16 is $L = n/3$ clock pulses.

# 4   Simulation Results and Discussions

The results shown below are simulation results of the architecture simulated using Xilinx 12.2 i and the comparison table, for which the results have been obtained.

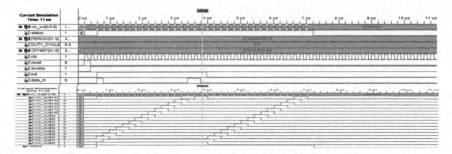

**Fig. 7** CRC-16 output waveform

## 4.1 CRC-16 Serial Architecture

The simulation waveform of the CRC-16 serial architecture is shown in Fig. 7.
Throughput obtained with serial architecture is 1 and latency obtained with serial
architecture is 16 clock cycles.

## 4.2 CRC-16 Single-Level Architecture After
## Proposed Formulation

In the proposed formulation technique, the simulation waveform of the CRC-16
single-level parallel architecture is shown in Fig. 8. Throughput obtained with serial
architecture is 1 and latency obtained with serial architecture is 16 clock cycles.

## 4.3 CRC-16 Two-Level Architecture After
## Proposed Formulation

The output waveform of the CRC-16 two-level parallel architecture is shown in
Fig. 9. Throughput obtained with serial architecture is 2 and latency obtained with
serial architecture is 8 clock cycles.

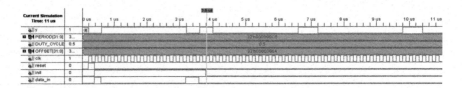

**Fig. 8** CRC-16 single-level parallel architecture output waveform

**Fig. 9** CRC-16 two-level parallel architecture output waveform

## 4.4 CRC-16 Three-Level Architecture After Proposed Formulation

The output waveform of the CRC-16 three-level parallel architecture is shown in Fig. 10. Throughput obtained with serial architecture is 3 and latency obtained with serial architecture is 5 clock cycles.

Table 1 describes the comparison in terms of number of checksum bits, number of XOR gates required, number of delay elements required, latency of the architectures, throughput rate of the architectures for CRC-16 serial architecture, single-level parallel architecture after proposed formulation, two-level parallel architecture, and three-level parallel architecture.

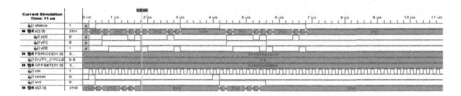

**Fig. 10** CRC-16 three-level parallel architecture output waveform

**Table 1** Results of the different architectures proposed for the CRC-12 generator polynomial

| Level of architectures | Number of checksum bits | Number of XOR gates | Number of delay elements | Critical path ($T_{XOR}$) | Throughput rate | Latency |
|---|---|---|---|---|---|---|
| Serial architecture | 16 | 3 | 16 | 2 | 1 | 16 |
| Single-level parallel architecture | 16 | 6 | 32 | 2 | 1 | 16 |
| Two-level parallel architecture | 16 | 12 | 32 | 8 | 2 | 8 |
| Three-level parallel architecture | 16 | 12 | 32 | 8 | 3 | 5 |

## 5    Conclusion

As the level of parallelism increases, the number of bits executing per a clock cycle also increases from single-level parallel architecture to three-level parallel architecture. In future, the further reduction of latency and increasing the throughput rate can be achieved by retiming technique, unfolding technique, and folding transformation DSP algorithms from parallel architectures.

## References

1. C. Cheng, K. K. Parhi, "High Speed VLSI Architecture for General Linear Feedback Shift Register (LFSR) Structures," Proc. of 43rd Asilomar Conf. on Signals, Systems, and Computers, Nov. 2009, Monterey, CA, pp. 713–717.
2. J. H. Derby, "High Speed CRC computation using state-space transformation," in Proc. Global Telecomm. Conf. 2001, GLOBECOM'01, vol. 1, pp. 166–170.
3. X. Zhang and K. K. Parhi, "High-speed architectures for parallel long BCH encoders," in Proc. ACM Great Lakes Symp. VLSI, Boston, MA, April 2004, pp. 1–6.
4. Manohar Ayinala, K. K. Parhi, "Efficient Parallel VLSI Architecture for Linear Feedback Shift Registers", IEEE Workshop on SiPS, pp. 52–57, Oct. 2010.
5. G. Campobello, G. Patane, and M. Russo, "Parallel CRC Realization," IEEE Trans. Computers, vol. 52, no. 10, pp. 1312–1319, Oct 2003.
6. K. K. Parhi, "Eliminating the fan-out bottleneck in parallel long BCH encoders", IEEE Transactions on Circuits and Systems I, Reg. Papers, vol. 51, no. 3, pp. 512–516, Mar. 2004.
7. C. Cheng, K. K. Parhi, "High Speed Parallel CRC Implementation based on Unfolding, Pipelining, Retiming," IEEE Transaction on Circuits and Systems II, Express Briefs, vol. 53, no. 10, pp. 1017–1021, Oct. 2006.
8. M. Surya Prakash, Rafi Ahamed Shaik, "Low-Area and High-Throughput Architecture for an Adaptive Filter using Distributed Arithmetic", IEEE Trans. Circuits and Systems II, Vol. 60, No. 11, pp. 781–785, Nov. 2013.
9. Y. Liu and K.K. Parhi, "Architectures for Recursive Digital Filters Using Stochastic Computing," IEEE Transactions on Signal Processing, 64(14), pp. 3705–3718, July 15, 2016.
10. M. Ayinala, M.J. Brown and K.K. Parhi, "Pipelined Parallel FFT Architectures via Folding Transformation", IEEE Trans. VLSI Systems, pp. 1068–1081, 20(6), June 2012.
11. M. Ayinala and K.K. Parhi, "High-Speed Parallel Architectures for Linear Feedback Shift Registers", IEEE Trans. Signal Processing, 59(9), pp. 4459–4469, Sept. 2011.
12. M. Garrido, K.K. Parhi, and J. Grajal, "A Pipelined FFT Architecture for Real-Valued Signals", IEEE Trans. Circuits and Systems-I: Regular Papers, 56(12), pp. 2634–2643, Dec. 2009.
13. C. Cheng and K.K. Parhi, "Hardware-Efficient Low-Latency Architecture for High-Throughput Rate Viterbi Decoders", IEEE Trans. Circuits and Systems-II: Express Briefs, 55(12), pp. 1254–1258, Dec. 2008.
14. C. Cheng and K.K. Parhi, "High-Throughput VLSI Architecture for FFT Computation", IEEE Trans. Circuits and Systems-II: Express Briefs, 54(10), pp. 863–867, Oct. 2007.

# Etch Time Optimization in Bulk Silicon MEMS Devices Using a Novel Compensation Structure

J. Grace Jency, M. Sekar and A. Ravi Sankar

**Abstract** The article aims to analyze the use of different compensation structures in MEMS micromachining technology to determine the minimum release time as fast as possible where undercutting is desirable. A high undercutting rate is advantageous for the formation of suspended structures. Thus, the implication of the present research includes analysis of corner undercutting behavior, their etching time, and etching characteristics using KOH and TMAH etchants. In this paper, the effective time and etchant concentration are studied using 33% KOH at 80 °C and 25% TMAH at 85 °C. It is found that wide bar with slit structure is the best compensation structure with minimum space competence.

**Keywords** Micromachining · Undercut · Etchants · Compensation structures
Etch rate · Processing time

## 1 Introduction

Micromachining is a common technique employed for developing MEMS (Microelectromechanical System) structures. It is a process of fabricating miniaturized devices and classified under surface and bulk micromachining. Surface

J. Grace Jency (✉)
Department of Electrical Sciences, Karunya University, Karunya Nagar,
Coimbatore 641114, Tamil Nadu, India
e-mail: gracegency@karunya.edu.com

M. Sekar
Department of Mechanical Sciences, Karunya University, Karunya Nagar,
Coimbatore 641114, Tamil Nadu, India
e-mail: sekar@karunya.edu

A. Ravi Sankar
School of Electronics Engineering, VIT University, Chennai Campus,
Chennai 600127, Tamil Nadu, India
e-mail: ravisankar.a@vit.ac.in

© Springer Nature Singapore Pte Ltd. 2018
J. Anguera et al. (eds.), *Microelectronics, Electromagnetics
and Telecommunications*, Lecture Notes in Electrical Engineering 471,
https://doi.org/10.1007/978-981-10-7329-8_4

micromachining structures are developed on top of the substrate by depositing sacrificial layers or by etching thin layers. Bulk micromachining etches material from the bulk substrate. The removal of substrate material is done either by dry etching or wet etching. Dry etching uses gas species or high-power laser to remove the substrate material and wet etching uses wet chemical species. Based on the etch rate, wet etching is either isotropic or anisotropic. If the etch rate is independent to the direction of crystallographic orientation, it is isotropic etching, and if the etch rate is dependent to the direction of crystallographic orientation, it is anisotropic etching [1]. Selection of micromachining etchants depends on factors like etch rate, anisotropy, handling, CMOS compatibility, and type of undercutting. Selection of micromachining etchants depends on factors like etch rate, anisotropy, handling, CMOS compatibility, and type of undercutting. Corner undercutting is a common phenomenon referred to as under-etching. It is the additional removal of materials below the substrate material due to fast etching in some planes. The effect of undercut is determined by undercut ratio which referred as the ratio undercut length to the etch depth. In the corner compensation method, extra structures called compensating structures are added at the convex corners in the mask layout design to eliminate the deformation at the convex corners during anisotropic wet chemical etching of silicon. A high accuracy of geometric pattern to crystal orientation is favorable to fabricate MEMS structures with controlled geometry. The shape and size of the etch profile are controlled by the alignment of mask edges along the crystallographic orientation direction. Typical MEMS piezoresistive device is considered under study by Ravi Sankar et al. to eliminate the deformation at the convex corners while fabricating the proof mass [2–5].

## 2 An Overview of Undercut Structures

The undercut structures both convex and concave are due to the intersection of {111} planes. However, undercut starts at the convex corner because in atomistic scale the break bond density is significantly higher in convex corners and there is no significant break bond density in concave undercut. The break bond density is not significant in concave undercut because all the concave corners are knotted by the neighboring atoms and hence they stay together by the intersection of {111} planes regardless of the etchant concentration, temperature, time, and etch depth. Conversely, in convex corners, though infinite planes pass through the tangent, only one plane is well noticed; the etching behavior at that tangent plane contains atoms at the convex corner that has higher etch rate than {111} planes [6–9]. Figure 1 shows the schematic of corner undercutting.

It is observed that corner undercutting affects the corners of the substrate and it causes corners to be etched away [4]. Undercutting results in loss of desired shape and functionality of the resultant microstructure. The various techniques to reduce corner undercutting are (a) by adding alcohols and surfactants with the etchants [10–12].

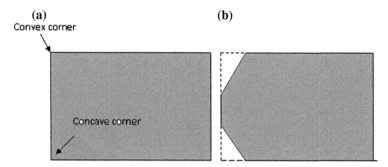

**Fig. 1** A schematic of convex corner undercutting; **a** perfect corners and **b** corners with undercutting

## 3 Analysis of Corner Undercutting

Corner undercutting depends on factors like type of the etchant, concentration of the etchant, the temperature of the etchant, its etch time, its geometry, and its typical etch planes [9–13].

The selection of etching solution is resolved by factors like etch rate, anisotropy, etched surface roughness, and selectivity of silicon dissolution with silicon dioxide. Though a simple KOH solution produces a clear etched surface, the surface is worse causing irregular shapes above 80 °C. In TMAH, the process is CMOS compatible and the advantage of non-toxicity is significant. For <100> and <110> surfaces, the etch rate increases as TMAH concentration increases with increase in temperature. Higher densities of pyramidal hillocks cover the surface at 5% concentration of TMAH. A very smooth surface is obtained when the TMAH concentration is greater than 22%. The etch rate for TMAH is constant for a longer etching time with a tolerable etch rate and it does not decompose below 130°. The (411) planes which emerge at the convex corners are found to be responsible for undercutting in KOH. Using wagon wheel experiment, it is observed that the etch rate of <111> plane is very slow using 25% of TMAH at 85°. The fast etching planes are <212> and <411> [6–9]. (Table 1 gives the etch rates of etchants KOH and TMAH for different etching planes.)

**Table 1** Etch rates of various etchants for different etching planes

| Etchant | {100} | {110} | {311} | {111} |
|---|---|---|---|---|
| KOH (33%, 80 °C) | 0.629 | 1.292 | 1.065 | 0.009 |
| TMAH (25%, 70 °C) | 0.272 | 0.532 | 0.576 | 0.009 |

**Table 2** A comparison of corner compensation methods

| Compensation structure | Upper edge | Bottom edge | Corner edge | Shape of the structure |
|---|---|---|---|---|
| Thin bar [7, 10] | Sharp | Sharp | Smooth | |
| Wide bar [11] | Sharp | Distorted | Smooth | |
| Superimposed squares [8, 9, 12] | Sharp | Sharp | Smooth | |
| <100> bar with narrow slit (proposed structure) | Sharp | Extruded | Uneven | |

# 4 Typical Corner Compensation Structures

Corner compensation structures retain the convex corner during micromachining. During etching, the etchant will attack the additional structures provided at the corners. By the time etchant reaches the corners, the desired portions are taken away and perfect convex corners are obtained. Some of the corner compensation structures are (1) Simple <100> oriented structure, (2) modified <100> oriented structure, (3) superimposed structure, and (4) slit-based structures.

In simple <100> oriented structure based on the boundary conditions for 33% KOH solution, the length of the structure should be equal or greater than the width of the structure. In modified <100> oriented structure, the width of the bar is preferred such that the lateral etching through (100) vertical planes stops at (410) sidewalls. The width of the beam is preferred twice the etching depth, and hence the length of the beam is greater than the width of the beam [7, 8]. The superimposed square structure has one large square with side length "a" and two short squares with side length "a/2" and are connected to the apex of the corner. Though the superimposed structure takes more time, a sharp corner is obtained occupying less amount of space [8, 9, 12, 13]. In the proposed design, slits are introduced at the convex corner for both thin bar <100> structure and wide bar <100> structure. The slit is made longer and it is twice the etch depth so that the <100> band is completely undercut. The length of the beam is concentration independent for the first half of etching. This structure consumes less space than other structures at the cost of leaving behind residue causing unevenness at the corner and it can be etched away. This is summarized in Table 2.

# 5 Results and Discussions

The above results use ACES (Anisotropic Crystalline Etch Simulator) to determine the etch rate for the various etchants at corresponding etchant concentration and temperature. The etchant concentration is inversely proportional to the etch rate and the temperature is directly proportional to the etch rate.

Figure 2 shows the accomplishment of perfect corners using corner compensation structures. It also determines the processing time required to attain perfect corners with different etchants at its corresponding etchant concentration and temperature. Etching is carried out for the proposed design using KOH etchant at a temperature of 80° with the concentration of 33% to ensure minimum surface roughness at optimum etching rate. More amount of KOH will inject the potassium ions into the substrate which will result in bad effects. The proposed structure is also carried out using TMAH etchant at a temperature of about 70 °C, as the best range of temperature. The etchant concentration is taken as 25% as usage of pure TMAH leads to severe undercutting; however, increase in TMAH concentration decreases the surface roughness.

**(i)**          **(ii)**

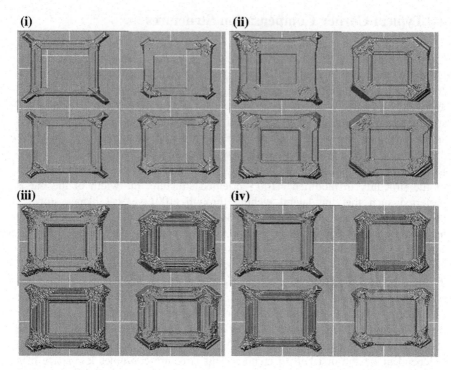

**(iii)**        **(iv)**

**Fig. 2** Using KOH etchant, (**i**) Thin bar with slit gets a perfect corner after 22 min. (**ii**) Wide bar with slit gets a perfect corner after 38 min. Using TMAH etchant, (**iii**) Thin bar with slit gets a perfect corner after 69 min. (**iv**) Wide bar with slit gets a perfect corner after 93 min.

**Table 3** A quantitative comparison of etch time consumption using KOH and TMAH

| Reference | Etch time consumption (min) | |
|---|---|---|
| | 33% KOH concentration at 80 °C temperature | 22% TMAH concentration at 70–90 °C temperature |
| Thin bar [7, 10] | 24 | 60 |
| Wide bar [11] | 57 | 121 |
| Thin bar with slit [13] | 22 | 69 |
| Wide bar with slit (Present work) | 38 | 93 |

Table 3 demonstrates the variation in the processing time of the etching process resulting in varied etching depth. As the processing time is increased, the etch rate increases for both KOH and TMAH etchants at its corresponding concentration and temperatures. Squares are obtained in the wafer at lesser processing time when KOH is used than that of TMAH and the etch rate table shows that the etch rate of KOH is higher than the etch rate of TMAH. However, though the etch time is less, perfect wafer squares are not obtained when the process involves KOH etchant. But the advantage of TMAH is that, even though it requires more processing time periods, the etched wafer squares are perfect.

# 6 Conclusions

The objective of this work is to analyze and design the different corner compensation structures, and to determine the etch rates for etchants like KOH and TMAH. The exact etch time period to etch the corner compensation structures and to obtain perfect wafer squares is determined. The above figure provides the optimized processing time period for various compensation structures and for both positive mask and negative mask. It is observed that the wide bar with slit structure is considered to be the best corner compensation structure because of its nature of space efficiency. There is a reduction of 2.3% in etch time when a wide bar with slit is used when compared to wide bar using TMAH. However, there is an increase of 1.76% when a wide bar with slit is used rather than narrow bar with slit using TMAH.

# References

1. M., Madou, Fundamentals of Microfabrication—The Science of Miniaturization. New York CRC Press (2002).
2. A., Ravi Sankar, S.K., Lahari, S., Das, Performance enhancement of a silicon MEMS piezoresistive single axis accelerometer with electroplated gold on proof mass. J. Micromechan. Microengg. 19 (2009) 1–9.
3. A., Ravi Sankar, J., Grace Jency, J., Ashwini, S., Das, Realization of silicon piezoresistive accelerometer with proof-mass-edge-aligned-flexures using wet anisotropic etching. Micro and Nano Letters. 7 (2012) 118–121.
4. A., Ravi Sankar, S., Das, A very-low cross axis sensitivity piezoresisitve accelerometer with an electroplated gold layer atop a thickness reduced proof mass. Sensors and Actuators. 189 (2013) 125–133.
5. A., Ravi Sankar, J., Grace Jency, S., Das, Design, fabrication and testing of a high performance silicon piezoresistive Z-axis accelerometer with proof mass-edge-aligned-flexures. Microsyst. Technol. 18 (2011) 9–23.
6. P., Pal, K., Sato, S., Chandra, Fabrication techniques of convex corners in a (100)-silicon wafer using bulk micromachining: a review. J. Micromech. Microeng., 17 (2007) 111–33.
7. W., Fan, D., Zhang, A simple approach to convex corner compensation in anisotropic KOH etching on a (100) silicon wafer. J. Micromech. Microeng., 16 (2006) 1951–1957.
8. H. K., Trieu, W., Mokwa, A generalized model describing corner undercutting by the experimental analysis of TMAH/IPA. J. Micromech. Microeng 8 (1998) 80–83.
9. K., Biswas, S., Das, S., Kal, Analysis and prevention of convex corner undercutting in bulk micro machined silicon microstructures. Microelectronics Journal, 37 (2006) 765–769.
10. Prem Pal, Kazuo Sato, A comprehensive review on convex and concave corners in silicon bulk micromachining based on anisotropic wet chemical etching. Micro and nano letters, (2015) 1–42.

11. Q., Zhang, L., Liu, Z., Li, A new approach to convex corner compensation for anisotropic etching of (100) Si in KOH. Sensors and Actuators, A 56 (1996) 251–254.
12. Chii-Rong Yang, Po-Ying Chen, Cheng-Hao Yanga, Yuang-Cherng Chioub, Rong-Tsong Lee, Effects of various ion typed surfactants on silicon anisotropic etching properties in KOH and TMAH solutions. Sensors and Actuators, A 119 (2005) 271–281.
13. J. W., Kwon, and E. S., Kim, Multi-level microfluidic channel routing with protected convex corners. Sensors and Actuators, A 97–98 (2002) 729–33.

# Eye Monitoring Based Motion Controlled Wheelchair for Quadriplegics

**Raju Veerati, E. Suresh, Adithya Chakilam and Sai Priya Ravula**

**Abstract** In today's world, people suffering from various disability problems is rising and more concernedly with quadriplegics (People, who are unable to walk in and around). To enhance their confidence and life independent, we have developed an effective alternative solution. The developed model uses the eye-tracking technique through circular Hough transform algorithm to control the movement of the wheelchair. The camera mounted aligns with the eye of the patient and captures continuous snapshots which are processed by image processing techniques in real time which, in turn, controls the direction of movement. Along with the control of motion of the wheelchair, this model also designed to detect the obstacles using ultrasonic sensors.

**Keywords** Image processing · Viola–Jones algorithm · Arduino
Wheelchair · Circular Hough transform

## 1 Introduction

Individuals utilizing wheelchairs are approximated 200 million around the world and in only 47% of the US population are suffering from spinal cord injuries and leading to quadriplegia problems. Some persons those unable to use the limbs are restricted and make their life difficult. Paralysis is another problem causing quadriplegia problems which can be of local, global, or tag on specific patterns. Most of the paralysis is almost identical nature and some of them vary such as periodic paralysis. Accidents and diseases causing the injuries to nervous system are regularly paralyzed duc to which the people are losing their ability to move.

For people with severe disabilities, there is a need for design and development of a semi-automatic wheelchair, where the motion of the wheelchair can be controlled

R. Veerati (✉) · E. Suresh · A. Chakilam · S. P. Ravula
Department of E.C.E, Kakatiya Institute of Technology and Science,
Warangal, Telangana, India
e-mail: rajureddyv@gmail.com

© Springer Nature Singapore Pte Ltd. 2018
J. Anguera et al. (eds.), *Microelectronics, Electromagnetics
and Telecommunications*, Lecture Notes in Electrical Engineering 471,
https://doi.org/10.1007/978-981-10-7329-8_5

41

by movements of the different organs of human body. Conventional wheelchair developed initially presumed that the user is capable of moving their hands. Semi-automatic wheelchairs had been developed which operates based on the movement of hands, fingers, or respiratory organs. Eye-controlled wheelchair movement offers alternative solutions to overcome such issues [1].

## 2 Literature Survey

The main reason to choose eye-tracking based technique can be understood from the disadvantages in various types of existing automated wheelchair technologies.

a. Head gesture-based wheelchair movement: It has two modes where different kinds of head movements are required to give the command to control. An EEG tool, Emotiv EPOC, is deployed to attain the head movement information from user which requires human effort and the devices used are expensive. It proves to be difficult for the people with deformities [2].
b. Voice-operated wheelchair: Even though it has less hardware requirement in similar to eye-tracking technique, there are drawbacks such as background noise and speaking style. It also provides less accurate results in accordance with speed of the speech and speaker variability [3].
c. Finger-based automated wheelchair: This type of wheelchair makes use of accelerometer and flex sensors. The value of those sensors is predictable to be within small range and few combinations can be made possible [4].
d. Motion-based method [5]: This depends on the effective movement of the human body organs to operate as computer input. Computer input is controlled by various organs like head, fingers, etc.; inadequacy of this method is that more human efforts are required to travel on a desired route using joystick, etc. and also not useful for a quadriplegics.

## 3 Block Diagram

The functionality of the system can be understood by architecture shown in Fig. 1. We have used Logitech C310 webcam for this prototype and operated it at a resolution of $240 \times 320$. It is mounted in front of the patient and the feed is set to MATLAB for further processing. Image processing toolbox of MATLAB is used extensively to take the continuous snapshots from the camera and performs morphological operations on them in real time. Depending upon the results from those operations, the direction which patient decides to move can be estimated. The commands estimated by MATLAB are transferred through an interface to the Arduino microcontroller at a baud rate of 9600 and suitable parity. Arduino UNO R3 microcontroller which is interfaced with L293D motor driver is programmed to

**Fig. 1** Basic architecture of automated eye-tracking system

act according to the commands and sends signals to motors of wheelchair. To detect the obstacles in the path of the wheelchair, an ultrasonic sensor is employed which calculates the distance between wheelchair and obstacle as a precautious [6].

# 4   Methodology

The camera is connected to a computer with the help of image acquisition toolbox shown in Fig. 2; continuous snapshots are taken and they are given as input to MATLAB software, where the further processing is done through image processing toolbox. Viola–Jones algorithm is used to detect the presence of face in the snapshot taken. The next step is to process the image by using some morphological operations to detect the direction of motion. Another technique is simultaneously employed to detect the state of eye (open/close). Based on the information obtained from the circle detection, the direction of the wheelchair is determined. Before the movement of wheelchair, obstacle detection will be done using ultrasonic sensors.

## 4.1   Viola–Jones Algorithm

Viola–Jones object detection algorithm is used to detect the human eye in the camera-captured images. Actually, this algorithm was developed for face detection though, but in recent times it is used in most of the applications for detecting all sorts of objects. This algorithm mainly works by taking aggregate of pixels of a rectangular array. Initially, the face detection is done by Viola–Jones algorithm from the smaller rectangles also called as fundamentally feature points; and if face detection is not identified, then it will detect from the large rectangles. These large

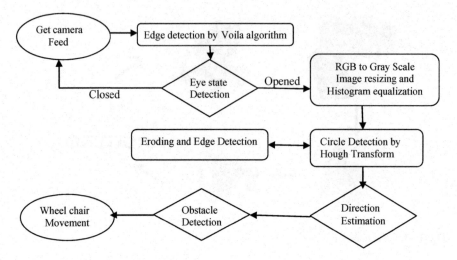

**Fig. 2** Proposed model methodology flow chart

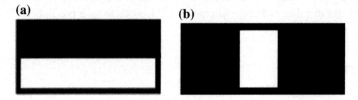

**Fig. 3** Haar features (Left-3 (**a**) Right-3 (**b**))

rectangles are constructed from many such smaller rectangles. For instance, upper cheeks is less darker than the eye region (Fig. 3a) and nose bridge region is brighter than the eyes (Fig. 3b). These features are termed as Haar features.

The "cascadeobjectdetector" on MATLAB utilizes this algorithm to detect the eyes of the person. We then extract the region of interest of detected eye by cropping the image at the appropriate position of the eye [7].

## 4.2 Morphological Operations

The image then undergoes processing as described below:

(1) Image resizing: The obtained image is resized into 256 × 256 resolutions so that it can be easily processed without large delays. This is done by using imresize().

(2) Color to black and white conversion: Now, the image is converted to grayscale image using the rgb2gray() function as shown in Fig. 4 because we do not need the color information for extracting the eye feature points [8].

**Fig. 4** RGB to gray conversion

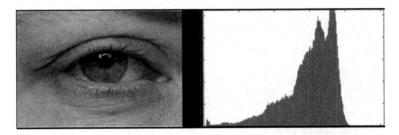

**Fig. 5** Before histogram equalization (Image and intensity v/s number of pixel plot)

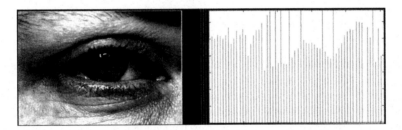

**Fig. 6** After histogram equalization (Image and intensity v/s number of pixel plot)

(3) Histogram equalization: Using this technique, the intensities are better distributed on the histogram. In this region, the lower local contrast is transformed into higher contrast. The most significant intensity values are effectively spread out by the Histogram equalization. The command histeq is used for this equalization and the images before and after Histogram equalization are shown in Figs. 5 and 6.

(4) Edge detection: An edge can be detected where there is a drastic change in the intensities of two adjacent pixels shown in Fig. 7. The image is eroded using a suitable structural element and resultant image is subtracted from the original image to obtain the boundary around the eye. The resultant image is binary.

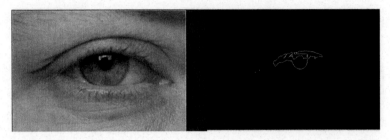

**Fig. 7** Comparison between original and edge detected image

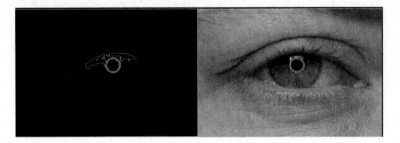

**Fig. 8** Mapping of detected circle on original image

## 4.3 Direction Estimation and Control

After the edge detection process, we obtain a binary image where a boundary is obtained around the iris of the eye. The boundary is in the shape of partial circle. To detect that circle and identify its coordinates, we use circular Hough transform. This algorithm will try to detect all possible circles in the given image by storing the votes of each pixel in an accumulator array [9] (Fig. 8).

The Hough transform returns the center of the eye with its x, y coordinates and radius. Since we normalized the resolution of every snapshot that has been taken by camera, decision can be made by comparing the x-coordinate with established thresholds (85 and 120). We can also change the thresholds dynamically using the graphical user interface.

We must stop the movement of the wheelchair if the eye is closed. To differentiate between the open and closed eye, we have calculated the column-wise mean of the image and plot it as shown in Fig. 9. Since most of the regions are black in an open eye, mean calculated is low when compared to the closed eye.

User interface is created by using GUIDE of MATLAB and with the help of deploy tool, it is developed into an executable file which can be installed and used on any personal computer [10]. It also provides the ability to change thresholds dynamically for decision making. Developed user interface is shown in Fig. 10a, b.

The direction commands taken by the MATLAB are sent to Arduino serially, which has been pre-programmed to act accordingly. The Arduino is also connected

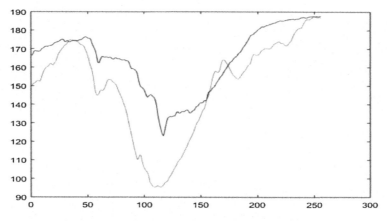

**Fig. 9** Column-wise mean plot (Green-open eye, red-closed eye)

**(a)**

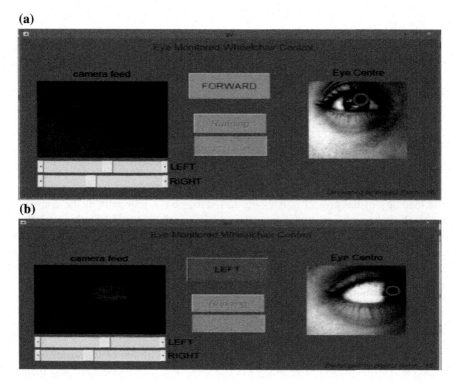

**(b)**

**Fig. 10 a** Developed user interface indicating forward directions, **b** developed user interface indicating forward and left directions

to an ultrasonic sensor which detects the obstacles in its path and supervises the wheels of wheelchair by using a motor driver circuit [11].

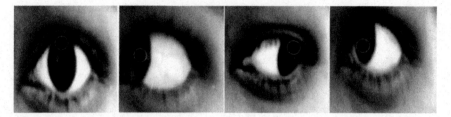

**Fig. 11** Results from real experiments

**Table 1** Test results for developed wheelchair prototype

| Frame capturing speed (of camera) | 20 fps |
| --- | --- |
| Wheelchair maximum speed | 4 km/h |
| Braking distance | 1 feet |
| Radius of rotation | 1–2 feet |

## 5  Results

The developed hardware prototype for eye-controlled wheelchair is able to detect the eye movement accurately even in the low-level lighting conditions with use of IR illuminators as shown in Fig. 11. Table 1 provides the test date of the developed wheelchair specifying the different parameter values.

## 6  Conclusion

The ideology of the project is not only to overcome the problems of the conventional wheelchair systems but also to take a supplementary step in bringing back the independent life of the disabled persons. The developed wheelchair model is easy to operate by the user. The user needs to position his eye towards the camera mounted on the wheelchair to move in the desired direction for small duration. The developed model movements are successfully verified in real time. In future, faster image processing techniques can be employed for improving the speed of the wheelchair as well as the detection of blinking nature of the eye in various lighting conditions.

Statement of consent:
The proposed work prototype is successfully operated with iris of the student who is the part of this project. The images containing the iris of the student and interfaces developed for indicating the direction are provided in the result section. The student is also the co-author of this work. In this regard, I request the editorial to accept this manuscript as it was.

# References

1. Rascanu, George Constantin, and Razvan Solea. "Electric wheelchair control for people with locomotor disabilities using eye movements." In *System Theory, Control, and Computing (ICSTCC), 2011 15th International Conference on*, pp. 1–5. IEEE, 2011.
2. Tameemsultana, S., and N. Kali Saranya. "Implementation of head and finger movement based automatic wheel chair." *Bonfring International Journal of Power Systems and Integrated Circuits* 1 (2011): 48.
3. Kokate, Jayesh K., and A. M. Agarkar. "Voice Operated Wheel Chair." *International Journal of Research in Engineering and Technology* 3, no. 2 (2014): 269–271.
4. Pande, Vishal V., Nikita S. Ubale, Darshana P. Masurkar, Nikita R. Ingole, and Pragati P. Mane. "Hand gesture based wheelchair movement control for disabled person using MEMS." *International Journal of Engineering Research and Applications* 4, no. 4 (2014): 152–8.
5. Kerbyson, D. J., and T. J. Atherton. "Circle detection using Hough transform filters." (1995): 370–374.
6. Arduino Ultrasonic sensor: http://playground.arduino.cc/Main/UltrasonicSensor.
7. Viola, Paul, and Michael Jones. "Rapid object detection using a boosted cascade of simple features." In *Computer Vision and Pattern Recognition, 2001. CVPR 2001. Proceedings of the 2001 IEEE Computer Society Conference on*, vol. 1, pp. I-I. IEEE, 2001. 1.
8. Get-rid-of-red-eye. http://howrid.com/health/get-rid-of-red-eye.
9. Purwanto, Djoko, Ronny Mardiyanto, and Kohei Arai. "Electric wheelchair control with gaze direction and eye blinking." *Artificial Life and Robotics* 14, no. 3 (2009): 397–400.
10. Creating GUI with Guide: https://www.mathworks.com/videos/creating-a-gui-with-guide-68979.html.
11. Patel, Shyam Narayan, and V. Prakash. "Autonomous camera based eye controlled wheelchair system using raspberry-pi." In Innovations in Information, Embedded and Communication Systems (ICIIECS), 2015 International Conference on, pp. 1–6. IEEE, 2015.

# Design and Analysis of Spherical Inverted-F Antenna Cavity Model

Parisa Jwalitha and G. Sambasiva Rao

**Abstract** The spherical inverted-f antenna having a spherically conformal rectangular patch antenna terminates with a quarter section of a metallic sphere. The spherical inverted-f antenna cavity model will be analyzed by using a new coordinate system which is custom curvilinear coordinate system. The conventional cavity method procedures can be applied by which the spherical structure is converted to an equivalent rectangular topology by mapping of coordinate transformation. By using this transformation, the wave equation is solved and also for predicting the input impedance, model characteristics, and antenna radiation parameter. Analytically simulated models are obtained for SIFA cavity model for different parameters of its fabrication.

**Keywords** Patch antenna · Spherical inverted-f antenna · Cavity model
Conformal antenna

## 1 Introduction

An antenna having ability to transmit the guided electromagnetic waves into free space which is a key component of modern electrical systems requires the free space communication. The antennas were used in many applications like commercial, industrial, military for the purpose of sensing, and also for communication. The spherical geometry is mostly used in mobile and aerodynamic, and also for biomedical devices, microsatellite systems, and mobile ad hoc networks (MANETs).

P. Jwalitha (✉)
Department of Electronics & Communication, KKR & KSR Institute
of Engineering & Sciences, Guntur, India
e-mail: jwalithagoud@gmail.com

G. S. Rao
Department of Electronics and Communication, Tirumula Engineering College,
Narasaraopet, India
e-mail: Sambasivacet@gmail.com

© Springer Nature Singapore Pte Ltd. 2018
J. Anguera et al. (eds.), *Microelectronics, Electromagnetics
and Telecommunications*, Lecture Notes in Electrical Engineering 471,
https://doi.org/10.1007/978-981-10-7329-8_6

The planar microstrip patch antennas cavity model was applied to different ranges of conformal antennas with different geometrics like cylindrical, spherical [1, 2, 3], and conical. The spherical geometry improves antenna visibility in the direction of back radiation, in contrast to planar microstrip antenna technologies [4], where a larger ground plane can limit radiation to just one hemisphere. Analysis of the cavity model provides the operation of different types of antennas and enables radiation behavior, input impedance, and accurate prediction of model characteristics, input impedance, and radiation behavior. To apply the cavity model, the geometry of the spherical inverted-f antenna [3] is suitable because of its structure. In its general form, the SIFA is defined by a planar inverted-f antenna that is conformed onto a sphere quadrant and inserted into its volume [5, 6]. Hence, it is taken into the category of microstrip antennas [7] which are conformal. By using any standard coordinate system, the topology of patch does not get any desired solution for the wave equation. The analysis of the cavity which is proposed mitigates the geometric complexity which is through the custom coordinate transformation provides a desired solution for wave equation.

In Sect. 2, the basic model of spherical inverted-f antenna is described, by using custom curvilinear coordinate system which is newly proposed to describe the spherical inverted-f antenna cavity model in Sect. 3. Based on the new coordinate system, the wave equation and its parameters are computed. Finally, in Sect. 4, the simulated results of cavity model are described. In Sect. 5, the conclusion is viewed.

## 2   Simulated Model of SIFA

A simulated model of spherical inverted-f antenna [3] is shown in Fig. 1. The structure of the simulated model has a metallic patch and metallic ground plane. The metallic patch is conformed to an outer sphere and the other one which is a metallic ground plane is conformed to a smaller inner sphere. The structure of the patch takes the area less than one-fourth of the sphere surface area. By using a small metallic shorting strip, the connection between the patch and the ground plane will be made. The SIFA [3] which is taken in this is feed by the coaxial probe which is radially radiated. For simplification, analysis of the cavity which is developed here is applied at which the design satisfies $\emptyset_s = \emptyset_w$. Table 1 gives information about different design parameters of canonical SIFA which are the specific dimensions and substrate parameters.

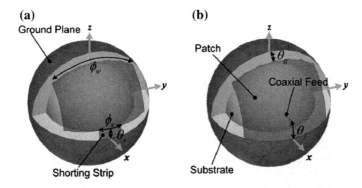

**Fig. 1** The SIFA-simulated model of patch geometry, **a** general, **b** canonical

**Table 1** Design parameters of the SIFA cavity

| Values | Name of the parameter | Design values |
|--------|----------------------|---------------|
| $\varepsilon_r$ | Dielectric constant | 1 |
| $\tan \delta$ | Loss tangent | 0 |
| $R_o$ | Outer radius (mm) | 180 |
| $R_i$ | Inner radius (mm) | 170 |
| $\varnothing_w$ | Patch width | 90° |
| $\theta_g$ | Length of the gap | 15° |
| $\theta_f$ | Angle for the feed | 20° |
| $f_r$ | Resonant frequency (MHz) | 380 |

# 3 SIFA Cavity Model

## 3.1 Boundary Conditions for SIFA Cavity

The geometry proposed for the SIFA cavity is shown in Fig. 2. The cavity geometry can be achieved by extruding the structure of two-dimensional patch on the outer sphere which is radially included in the ground plane. There are two boundary conditions of PEC: one is metallic patch and the another is ground plane. A radially directed shorting wall which has the PEC boundary conditions is applied between the patch and ground plane. PMC surfaces are taken here as three radiating slots.

## 3.2 Custom Curvilinear Coordinate System

To describe the geometry more easily and to achieve the desired solution for wave equation, the custom curvilinear coordinate system is used because of the peculiar

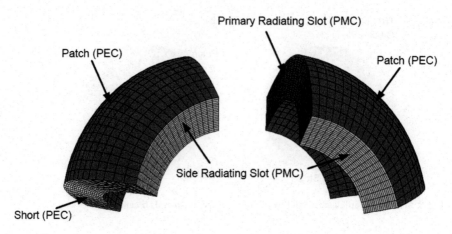

**Fig. 2** Boundary conditions for the geometry of the SIFA cavity

nature of the SIFA [3]; it does not allow the spherical coordinate system. The coordinates used by curvilinear coordinate system are $u$, $v$, and $w$, which is defined by (1), (2), and (3) in relations with the coordinates of the Cartesian which is in the standard form:

$$x = u \, \cos(w) \sin(v) \tag{1}$$

$$y = u \sin(w) \tag{2}$$

$$z = u \, \cos(w) \cos(v) \tag{3}$$

Here, $u$ is a radial quantity and $v$ *and* $w$ are the angular quantities. The custom coordinate system $u$, $v$, $w$ is shown in Fig. 3 graphically. In space, $u$, $v$, $w$, the spherically inverted-f antenna is defined from the Eqs. (4)–(6). The mapping is done from Cartesian space to custom coordinate system; the spherical inverted antenna is represented as a rectangular prism and it is symmetrical to *PIFA* cavity. This provides a desired solution to the wave equation and gives a clarification on the application of cavity method.

$$R_i \leq u \leq R_o \tag{4}$$

$$\theta_g \leq v \leq \frac{\pi}{2} - \theta_s \tag{5}$$

$$-\frac{\varnothing_w}{2} \leq w \leq \frac{\varnothing_w}{2} \tag{6}$$

**Fig. 3** The coordinate
system for the custom
curvilinear, $P_1$ is $(u_1,\ v_1,\ w_1)$

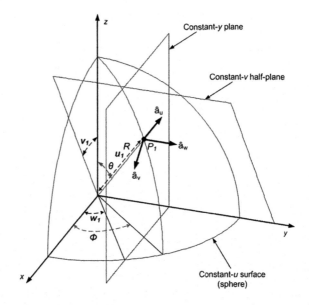

Helmholtz wave equation is given by

$$\nabla^2\varphi + k^2 = -\frac{uJ_u}{u} \tag{7}$$

The scalar function is

$$\varphi = \frac{A_u}{u}$$

Helmholtz wave equation which is in homogeneous form is a distinct form of
(8). By applying the boundary conditions of spherical inverted-f antenna cavity
model, the solutions for the separated equations are obtained by using the technique
[8] and are given by (9)–(11).

$$\psi_{\mu vk} = A(u)\ B(v)\ C(w) \tag{8}$$

$$A(u) = n'_{v_r}(k_dR_i)\,j_{v_r}(k_du) - j'_{v_r}(k_dR_i)\,n_{v_r}(k_du) \tag{9}$$

$$B(v) = \cos\left(\mu_r\left[v - \theta_g\right]\right) \tag{10}$$

$$C(w) = Q^{\mu_r}_{v_r}\left(\sin\frac{\varnothing_w}{2}\right)P^{\mu_r}_{v_r}(\sin w) - P'^{\mu_r}_{v_r}\left(\sin\frac{\varnothing_w}{2}\right)Q^{\mu_r}_{v_r}(\sin w) \tag{11}$$

For satisfying all boundary conditions, two transcendental Eqs. (12), (13) are must required for getting the parameters $v_r$ and $k_d$. Another parameter, $\mu_r$, is represented by Eq. (14). $\mu_r$, $v_r$ and $k_d$ are represented as discrete model quantities which are based on integer index. Every mode has its corresponding resonant frequency represented by Eq. (15) which is defined from $k_d$ and the model wavenumber is given by

$$Q_{v_r}^{\mu_r'}\left(\sin\frac{\emptyset_w}{2}\right)P_{v_r}^{\mu_r'}\left(\sin\left[-\frac{\emptyset_w}{2}\right]\right) = P_{v_r}^{\mu_r'}\left(\sin\frac{\emptyset_w}{2}\right)Q_{v_r}^{\mu_r'}\left(\sin\left[-\frac{\emptyset_w}{2}\right]\right) \qquad (12)$$

$$n_{v_r}'(k_dR_i)j_{v_r}'(k_dR_o) = j_{v_r}'(k_dR_i)n_{v_r}'(k_dR_o) \qquad (13)$$

$$\mu_r = \frac{m\pi}{2\left(\frac{\pi}{2}-\theta_g\right)} \qquad (14)$$

$$f_r = \frac{k_d}{2\pi\sqrt{\mu\varepsilon}} \qquad (15)$$

Table 2 gives the list of computed model quantities for the design with respective resonant frequencies of the computed model quantities for design. In the model, computations were performed with numerically.

## 3.3 Fields for Cavity Model

The magnetic vector potential is allowed by the total solution ψ and also in the cavity electric, magnetic fields are found. Extensively, the quantities of the field in the custom coordinate *uvw* components is given by Eqs. (16)–(21).

$$E_u = \frac{1}{j\omega\mu\varepsilon}\frac{\partial^2(\varphi u)}{\partial u^2} - j\omega(\varphi u) \qquad (16)$$

$$E_v = -\frac{1}{j\omega\mu\varepsilon u \cos w}\frac{\partial^2(\varphi u)}{\partial u\,\partial\theta} \qquad (17)$$

**Table 2** *SIFA* cavity model parameters

| $\mu_r$ | $v_r$ | $k_d$ | $f_r$ (MHz) cavity | $f_r$ (MHz) simulated | Error |
|---------|-------|-------|--------------------|------------------------|-------|
| 1.18 | 0.925 | 14.67 | 188.22 | 192.46 | 2.18 |
| 1.18 | 2.107 | 28.78 | 364.48 | 371.32 | 1.85 |
| 3.47 | 3.438 | 43.85 | 555.46 | 567.56 | 2.13 |
| 1.14 | 3.846 | 48.22 | 610.91 | 620.83 | 1.61 |
| 3.54 | 4.258 | 53.03 | 672.03 | 686.27 | 2.06 |

$$E_w = \frac{1}{j\omega\mu\varepsilon u} \frac{\partial^2(\varphi u)}{\partial u \, \partial w} \tag{18}$$

$$H_u = 0 \tag{19}$$

$$H_v = \frac{1}{\mu u} \frac{\partial(\varphi u)}{\partial w} \tag{20}$$

$$H_v = -\frac{1}{\mu u \cos w} \frac{\partial(\varphi u)}{\partial v} \tag{21}$$

## 3.4  Slot Radiation

The SIFA cavity fields are used to derive the radiation from primary radiating slot and also for the two side radiating slots. In Eq. (22), $\cos \alpha$ is the cosine angle between far-field point and any point on the slot, R is the distance from origin to a point which is in the far field, **n** be the normal vector to the surface, and $S'$ is the slot surface. In $uvw$ coordinates, $\cos \alpha$ and **n** are taking too long for the exact expressions of it. By using the electric vector potential, the radiated electric field is computed and through the standard techniques the radiated power is found to be

$$F = -\frac{2 \in E_o e^{-jkR}}{4\pi R} \iint E_u(u, v, w)(n\hat{a}_u)e^{jku \, |\cos \alpha|} \, ds \tag{22}$$

## 3.5  Probe Model and the Effective Loss

By using effective loss tangent [7], the energy loss from the cavity because of the radiation parameter is to be accounted [9]. The effective loss tangent (23) is derived as inverse of cavity quality factor. The stored electric energy is used to compute the quality factor (24) $(W_e)$ at resonance in cavity model.

$$\tan \delta_{eff} = \frac{1}{Q} \qquad Q = \frac{2\omega W_e}{P_{rad}} \tag{23}$$

$$W_e = \frac{\varepsilon}{4} \iiint |E_u| \, u^2 \cos w \, du \, dw \, dv \tag{24}$$

The coaxial feed probe which causes the current density $J_u$ is designed from standard convention to planar cavity model [10]. The approximation of two-dimensional sheet having current density (25) is stretching radially from ground plane to patch. The arc length is taken as five times the probe feed diameter by chosen angular width $w_f$ of the strip. The corresponding scalar current (27) is unity by choosing the magnitude (26) of

$$J_f = \begin{cases} J_u \delta\left(v - \left(\frac{\pi}{2} - \theta_f\right)\right)\hat{a}_u & -\frac{w_f}{2} \leq w \leq \frac{w_f}{2} \\ 0 & otherwise \end{cases} \tag{25}$$

$$J_u = 1/2u^2 \, \sin\left(\frac{w_f}{2}\right) \tag{26}$$

$$I_f = \iint J_f dA = \int\limits_{-w_f/2}^{w_f/2} \int\limits_{}^{} J_u u^2 \cos w dw = 1 \tag{27}$$

## 3.6 Input Impedance

Equation (28) is the input impedance which is defined by the ratio of feed voltage to the feed current. Equation (27) defines the feed current and the feed voltage is defined by the average voltage between the patch and ground plane. Here, the average is taken by the width of present feed strip.

$$z_{in} = \frac{V_f}{I_f} = -\frac{1}{I_f w_f} \int\limits_{-w_f/2}^{w_f/2} \int\limits_{R_i}^{R_o} E_u \, du \, dw \tag{28}$$

Equation (28) describes the ideal cavity which is given by input impedance. Practically, due to the capacitance between the ground plane and the patch, the impedance is affected and also coaxial probe adds the inductive component to that impedance. For planar microstrip antennas [10], inductive reactance effected by the probe is determined and is given in Eq. (29). Equation (30) describes by using parallel plate capacitance the estimation of capacitive reactance $X_c$ between the ground plane and the patch, where **A** represents the cross-sectional area of the spherical inverted-f antenna patch, computed at the average radius between $R_i$ and $R_o$. The input reactance is represented by the addition of two reactances which defines from Eq. (28)

$$X_{probe} = \frac{\eta k h}{2\pi} \left[ ln\left(\frac{4}{kd} - 0.577\right) \right] \tag{29}$$

$$X_c = \frac{h}{w \varepsilon_{eff} \varepsilon_o A} \tag{30}$$

## 4   Results

The wave equation can be derived by using custom curvilinear coordinate system. Figures show the different parameters of SIFA (spherical inverted-f antenna) cavity model. Figure 4 describes the VSWR versus frequency, and Figs. 5 and 6 show the input impedance versus frequency for both real and imaginary. The fields of SIFA cavity describe the radiation parameter of both the primary and side radiating slots which are shown in Figs. 7 and 8.

**Fig. 4** VSWR versus frequency

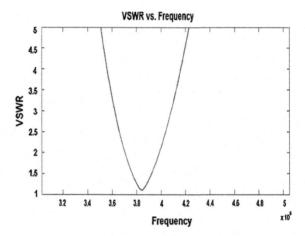

**Fig. 5** Real impedance versus frequency

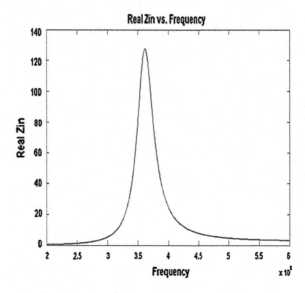

**Fig. 6** Imaginary impedance
versus frequency

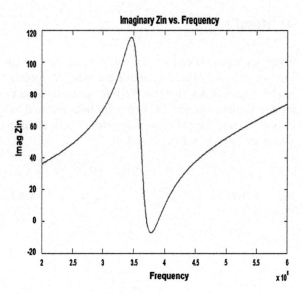

**Fig. 7** Side slot resistance
versus frequency

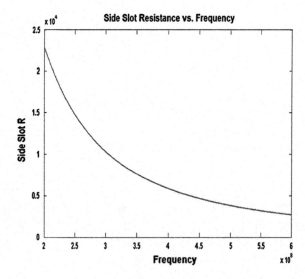

**Fig. 8** Primary side slot
resistance versus frequency

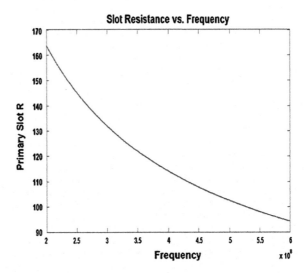

## 5  Conclusion

The custom curvilinear coordinate system is analyzed by the spherical inverted-f antenna cavity model. In this, the curvilinear coordinate system mapped to the Cartesian coordinate system so that it allows a desired solution to the wave equation which can easily understand the cavity method. Transformation of the system solves the wave equation and also analyzes different parameters like input impedance, slot radiation, and primary slot radiation of spherical inverted-f antenna cavity model.

## References

1. L.A. Costa, O.M.C Pereira-Filho, and F.J.S. Moreira, "Analysis of spherical-rectangular microstrip antennas," in *Proc. SBMO/IEEE MTT-S Int.Conf.Microw. and Optoelectronics*, Brasilia, Brazil, Jul. 2005, pp. 279–282.
2. A.C. de C. Lima, J.R. Descardeci, and A.J. Giarola, "Microstrip antenna on a spherical surface," in *Proc. Antennas Propag. Society Int. Symp.*, Jun 1991, vol. 2, pp. 820–823.
3. G.H. Huff and J.J. McDonald, "A spherical inverted-f antenna(SIFA)," *IEEE Antennas Wireless Propag. Lett.*, vol. 8, pp. 649–652, May 2009.
4. G Sambasiva Rao, K. Rama Krishna, "Design And Simulation Of Dual Band Planar Inverted F Antenna (Pifa) For Mobile Handset Applications," International Journal of Antennas(JANT)., vol 1, no. 1, pp. 37–47, Sep 2015.
5. K-Y.Wu and J.F. Kauffman, "Radiation pattern computations for cylindrical-rectangular microstrip antennas." in *Proc. Antennas Propag. society Int. Symp.*, Houston, TX, USA, May 1983, vol. 21, pp. 39–42.

6. K-Y. Wu and J.F. Kauffman, "Radiation pattern computations for spherical-rectangular microstrip antennas," in *Proc. Antennas Propag. society Int. Symp.*, Houston, TX, USA, May 1983, vol. 21, pp. 43–46.

7. W.F. Richards,Y.T. Lo, and D.D. Harrison, "An improved theory for microstrip antennas and applications," *IEEE Trans.Antennas Propag.,vol.* AP-29, no. 1, pp. 38–46, Jan. 1981.

8. K-M. Luk and W-Y. Tam, "Patch antennas on a spherical body," *Inst. Elect. Eng. Proc. H-Microw., Antennas Propag.,* vol. 138, no. 1, pp. 103–108, Feb 1991.

9. Matlab Student ver.7.1, Math Works. Natick, MA, USA, 01760.

10. J.R. James, *Handbook of Microstrip Antennas*. London, U.K.:Peter Peregrinus, 1989, by vol. 2.

# Performance of Ternary Sequences Using Adaptive Filter

**K. Renu and P. Rajesh Kumar**

**Abstract** The wide application of pulse compression is mainly because of its ability to reduce peak transmission power that causes improvement in SNR value. Pulse compression is a standard signal processing technique which can be used to achieve desired range resolution and detection range. Better range resolution can be achieved by using matched filter. In this paper, chaotic maps are used to generate ternary chaotic sequences of any length. A new approach is used to achieve superior performances in range detection and range resolution. The method used in this paper is adaptive filtering technique. The performance parameter is peak sidelobe ratio which has been estimated with and without adaptive filtering technique. The simulation results show significant improvement in the performance. Least mean square algorithm is one of the well-known algorithms due to ease of implementation in various applications. In this paper, the convergence property of this algorithm is measured in terms of mean square error which is clearly investigated.

**Keywords** Pulse compression · Chaotic maps · Autocorrelation sidelobe peak PSLR · Adaptive filtering · LMS algorithm · Mean square error

## 1 Introduction

In general, most of the radars transmit long duration pulses to achieve large energy for good range detection. But good range resolution can be achieved with short pulses. Pulse compression is a technique that differentiates nearby targets by combining both the high energy and resolution [1]. Various waveforms have been used for this purpose such as linear FM, nonlinear FM, bi-phase codes, etc. [2].

K. Renu (✉)
Department of ECE, GITAM University, Visakhapatnam, AP, India
e-mail: renuengg12@gmail.com

P. R. Kumar
Department of ECE, Andhra University, Visakhapatnam, AP, India
e-mail: rajeshauce@gmail.com

© Springer Nature Singapore Pte Ltd. 2018     63
J. Anguera et al. (eds.), *Microelectronics, Electromagnetics and Telecommunications*, Lecture Notes in Electrical Engineering 471,
https://doi.org/10.1007/978-981-10-7329-8_7

The code indicates the phases of sub-pulses of a phase-coded waveform. In pulse compression, these sequences have much importance.

The selection of a radar signal depends on the performance of range resolution and energy efficiency. This range resolution is determined from the autocorrelation pattern of the coded waveform which is the output of the matched filter. Pulse compression is a standard signal processing technique used to reduce peak transmission power and maximize SNR. For more better range resolution, matched filter is used [3]. The drawback of binary sequence appears for longer length codes where low peak sidelobe ratio is required. Boehmer, Linder, Rao, and Reddy proved that longer length binary sequences violating the condition of barker code [4–6]. Hence, the necessity of multilevel code came to picture. The elements of multilevel sequence are of unequal magnitude. Some examples of multilevel sequences are ternary code having elements 0, +1, and −1, whereas quinquenary sequences have 0, ±1, and ±2 as elements. Moharir proved that the ternary barker sequences exist for longer length [7, 8].

An exclusive set of radar signal which is used in spread spectrum communications is ternary chaotic sequence. The correlation properties of these ternary chaotic codes are surveyed by Bateni and MCGillen [9]. The generation of good ternary sequences and their performance are compared using different chaotic maps [10, 11]. The performance measure considered is peak sidelobe ratio which is the reciprocal discrimination factor. The peak sidelobe ratio is obtained by finding the ratio between sidelobe peak to peak of the mainlobe. This PSLR value must be as low as possible. This can be achieved in this paper by using adaptive filtering technique. Adaptive filtering has been widely used in various signal processing applications such as sonar, radar biomedical applications [12, 13]. However, least mean square adaptive filter is quite popular due to its simplicity. In this technique, an adaptive filter is placed after the matched filter whose coefficients are updated in every iteration. This results in reduced error in the output. The technique discussed here has the advantage over the previous method and low peak sidelobe ratio is achieved.

This paper is organized as follows. Section 2 reports the chaotic ternary sequence generation. Section 3 reports the adaptive filtering technique. The design implementation of the proposed method is explained in this section. The simulation and comparison results are presented in Sect. 4.

## 2  Sequence Generation

The chaotic signals in radar are easy to generate. Chaotic sequences are generated using various types of chaotic maps such as logistic map, improved logistic map, cubic map, tent map, and quadratic map. Logistic equation is a key example that represents the behavior like stable, chaotic, and periodic depending on the parameter called bifurcation factor $\mu$ in the following logistic map equation:

$$X_{n+1} = \mu X_n (1 - X_n) \tag{1}$$

The logistic map equation in Eq. (1) is very simple and fairly innocuous. The range of $\mu$ is (0, 4) and x is (0, 1). It exhibits chaotic behavior when the value of $\mu$ is 4. Logistic map is the second-order polynomial mapping considered as a prototype model. It represents how complex and chaotic behavior can arise from a nonlinear dynamical equation. The system exhibits chaotic state when $\mu$ is between 3.57 and 4, and beyond 4 the value of x diverges for all initial value by leaving the interval (0, 1). A small difference in initial value leads to different waveforms after some iterations. Cubic map equation can be written as

$$X_{n+1} = 4(X_n)^2 - 3X_n \tag{2}$$

The chaotic behavior exhibits for $x_n \in (-1, 1)$. If n is infinity, $x_0$ is larger than 1 and at the same time $x_n$ also tends to infinity. Similarly, improved logistic map is described by the equation

$$X_{n+1} = 1 - 2(X_n)^2 \tag{3}$$

Here, x ranges from $-1$ to 1. This equation exhibits chaotic behavior for $x_0$ between 0 and 1. If $x_0 > 1$ when n tends to infinity, $x_n$ also tends to infinity.

The method of generation of sequence is described below:

1. The bifurcation parameter is chosen as 4 so that map is in the chaotic region.
2. These maps are sensitive enough on initial values. So select the initial value $x_n$ between the proper ranges.
3. Using the map equations, generate various raw sequences depending on different initial values.
4. Ternary sequences are generated by quantization of raw sequences into three defined levels depending on the threshold levels a and b.

For logistic map, threshold levels are chosen by taking the mean of the raw sequence, i.e., 0.5 and the levels a and b are chosen above and below this mean value of the raw sequence.

$$
\begin{aligned}
S &= 1 \quad \text{for } x_n < a \\
&= 0 \quad \text{for } a < x_n < b \\
&= -1 \text{ for } x_n > b
\end{aligned}
\tag{4}
$$

A completely uncorrelated infinite sequence can be obtained by varying the initial condition $x_0$ for a given length of the sequence. Good sequences were filtered with smaller value of peak sidelobe ratio as defined in Eq. (6). Various lengths of sequences were generated based on threshold levels and maps. For logistic map, the values of a and b were 0.7 and 0.3, respectively. Similarly, for improved logistic map, a and b values were 0.7 and $-0.7$, respectively, and for cubic map 0.6 and

−0.4. Sequence generation using these maps is very simple and fast. The output of the matched filter is the aperiodic autocorrelation of the sequence which is given by

$$r(k) = \sum S_i S_{i+k}, \qquad (5)$$

where "i" ranges from 0 to N −1 − k and k = 0,1, 2, … N − 1. The performance in terms of peak sidelobe ratio is obtained from the autocorrelation pattern r(k) which is defined as the ratio of peak sidelobe amplitude to the main lobe peak.

$$PSLR = 20 * \log_{10}(\max(r(k))/r(0)) \text{ where } k \neq 0 \qquad (6)$$

It is expressed in decibels. A sequence having low PSLR is said to be best sequence. These PSLR values for good sequences for different lengths were tabulated using all map equations. The reciprocal of PSLR is discrimination factor.

## 3 Adaptive Filtering Technique

An adaptive filter consists of two components. They are transversal filter and an adaptive algorithm. The output of the transversal filter in response to the input signal is continuously compared with the desired reference signal. The adaptive algorithm is responsible for adjusting the coefficients of the transversal filter. The block diagram of adaptive filter including matched filter is shown in Fig. 1.

### 3.1 LMS Algorithm

The wide use of least mean square algorithm lies in its computational simplicity and the convergence characteristics. It is one of the classes of adaptive filter which imitates the desired filter by finding its coefficients. These coefficients produce least mean square error signal that is the difference between desired signal and actual signal. Initially, small weights, in most of the cases zero weights, are assumed and then in each step these are updated according to the algorithm equations written

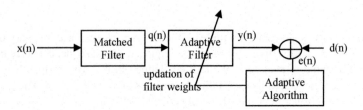

**Fig. 1** Block diagram of adaptive filter with matched filter

below. The advantage of using LMS algorithm is that it is stable and gives robust performance for different signal conditions. LMS algorithm is widely used in radar systems because of its simplicity in implementation and slow convergence. The step-size parameter is critical to the performance of the LMS and determines how fast the algorithm converges along the error performance surface.

*Algorithm:*
Parameters:      N = filter order
               $\mu$ = step size
               Initialization: $\hat{h}\,(0) = 0$
Computation:     For n = 0, 1, 2... N
               $X(n) = [x(n), x(n-1), .... x(n-N+1)]^{T}$
               $e(n) = d(n) - y(n)$
               $y(n) = W^{T}(n)\, x(n)$
               $W(n+1) = W(n)+2\mu e(n)x(n)$           (7)

The basic concept behind LMS algorithm is to approach the optimum filter weights by updating the weights. The value of the step size "$\mu$" is properly chosen to avoid mislead in convergence of mean. The filter weights never reach optimum weights but the convergence in mean can be possible. The adaptive filter calculates the output signal y(n) by using the following equation:

$$y(n) = X^{T}(n) \cdot W(n),$$

where W(n) is the filter coefficient vector. The adaptive algorithms adjust the filter coefficients to minimize the cost function. LMS algorithm has the advantage that it requires fewer computational resources and memory than other adaptive algorithms. In this paper, the output of the matched filter, autocorrelation function, is applied to adaptive filter whose weights are adjusted and updated based on the LMS algorithm. The peak sidelobe ratio is measured and compared with that obtained after matched filter for different lengths of the sequence. This procedure is repeated for different orders of the filter to achieve minimum peak sidelobe ratio.

## 3.2 MSE Behavior

It is one of the properties of LMS algorithm that depends on the step-size parameter. In LMS algorithm, the filter weights never attain optimal value. Hence, there is a possibility of convergence in mean. But the value of the optimal weights changes even with a small change in filter weights. The main cause of this problem arises if step size will not be chosen properly. This parameter controls the convergence speed of the algorithm. Selection of right constant value for this parameter is important and also difficult task. A very low value of step size is desired for this algorithm that results in minimum mean square error value. In this paper, the maximum value of mean square error is obtained and tabulated for various lengths of the sequence. It is observed that the maximum value of mean square error decreases with increase in length of the sequence for fixed number of iterations.

# 4 Simulation Results

Simulation is performed with the help of MATLAB software. Figures 2 and 3 show the autocorrelation pattern of the best sequences that were generated using logistic map equations without and with LMS algorithm for length 5000, respectively. The peak sidelobe ratios obtained using Eq. (6) before and after least mean square algorithm, which is given in Eq. (7), for good ternary sequences at different lengths for the logistic map are tabulated in Table 1.

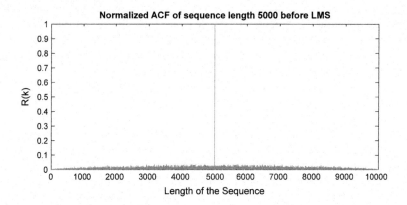

**Fig. 2** ACF pattern of logistic sequence of length 5000

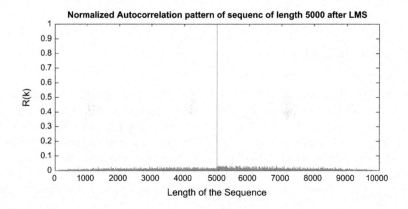

**Fig. 3** ACF pattern of logistic sequence of length 5000 with LMS

**Table 1** Comparison of PSLR of logistic map sequence without and with LMS algorithm

| Length of the sequence | Logistic map | | | | |
|---|---|---|---|---|---|
| | Without LMS | | With LMS | | Maximum value of MSE |
| | PSLR | ASP | PSLR | ASP | |
| 20 | −17.5012 | 0.1333 | −25.7712 | 0.0515 | 1.0975 |
| 30 | −17.6921 | 0.1304 | −20.8364 | 0.0908 | 0.8205 |
| 50 | −18.8402 | 0.1143 | −22.2532 | 0.0772 | 0.4778 |
| 70 | −18.5884 | 0.1176 | −18.7603 | 0.1153 | 0.4394 |
| 90 | −18.4597 | 0.1194 | −20.6247 | 0.0931 | 0.3383 |
| 100 | −18.5314 | 0.1184 | −20.8866 | 0.0903 | 0.3250 |
| 200 | −20.2848 | 0.0968 | −23.2037 | 0.0692 | 0.1521 |
| 300 | −20.5993 | 0.0933 | −22.0398 | 0.0791 | 0.1077 |
| 500 | −21.8451 | 0.0809 | −23.2414 | 0.0689 | 0.0631 |
| 700 | −22.7244 | 0.0731 | −24.4511 | 0.0599 | 0.0459 |
| 900 | −23.4572 | 0.0672 | −23.7869 | 0.0647 | 0.0353 |
| 1000 | −23.5336 | 0.0666 | −25.3883 | 0.0538 | 0.0329 |
| 2000 | −26.0917 | 0.0496 | −27.4281 | 0.0425 | 0.0163 |
| 3000 | −27.3804 | 0.0428 | −28.2946 | 0.0385 | 0.0111 |
| 4000 | −27.9588 | 0.0400 | −28.7488 | 0.0365 | 0.0084 |
| 5000 | −28.9338 | 0.0358 | −29.5468 | 0.0333 | 0.0066 |

Table 1 shows how the peak sidelobe ratio of ternary logistic sequence varied with respect to length. The PSLR has measured from the output of the matched filter. The input to adaptive filter is the output of the matched filter.

A plot of PSLR versus length of the ternary logistic sequence without and with least mean square algorithm is shown in Fig. 4. With the investigation, it is also observed that, as $\mu$ increases from 0 to 1 but not equal to 1, the maximum value of MSE decreases. The maximum value of mean square error is investigated for fixed step size and included in the table. The number of iterations chosen is 50 and step size is 0.05.

Similarly, the performance of peak sidelobe ratio for improved logistic sequence and cubic sequence before after employing LMS algorithm is tabulated in Tables 2 and 3, respectively. With analysis, it is also concluded that the maximum value of MSE increases with order of the filter.

It is clearly observed from the above analysis from the tables that the value of PSLR is increasing and the maximum value of MSE is going on decreasing with length for minimum value of step size, which is required for LMS algorithm to satisfy. Figure 5 represents the behavior of mean square error for 50 iterations with sequence length 5000.

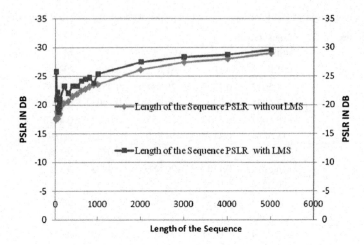

**Fig. 4** Comparison of PSLR of logistic map without and with LMS

**Table 2** Comparison of PSLR of improved logistic sequence without and with LMS algorithm

| Length of the sequence | Improved logistic map | | | | |
|---|---|---|---|---|---|
| | Without LMS | | With LMS | | Maximum value of MSE |
| | PSLR | ASP | PSLR | ASP | |
| 20 | −20.8279 | 0.0909 | −25.6701 | 0.0521 | 1.0892 |
| 30 | −19.0849 | 0.1111 | −21.9554 | 0.0798 | 0.9712 |
| 50 | −18.4164 | 0.1200 | −19.9979 | 0.1000 | 0.5677 |
| 70 | −18.5884 | 0.1176 | −20.0732 | 0.0992 | 0.4322 |
| 90 | −18.6900 | 0.1163 | −20.5679 | 0.0937 | 0.3349 |
| 100 | −19.0849 | 0.1111 | −20.0810 | 0.0991 | 0.2972 |
| 200 | −20.7485 | 0.0917 | −22.8375 | 0.0721 | 0.1507 |
| 300 | −20.9399 | 0.0897 | −21.7370 | 0.0819 | 0.1040 |
| 500 | −22.1371 | 0.0782 | −22.9685 | 0.0711 | 0.0674 |
| 700 | −22.8024 | 0.0724 | −24.1078 | 0.0623 | 0.0469 |
| 900 | −23.4804 | 0.0670 | −24.1169 | 0.0623 | 0.0370 |
| 1000 | −24.4089 | 0.0602 | −26.4352 | 0.0477 | 0.0321 |
| 2000 | −26.1176 | 0.0494 | −27.0794 | 0.0443 | 0.0165 |
| 3000 | −27.2231 | 0.0435 | −29.0352 | 0.0353 | 0.0112 |
| 4000 | −28.1748 | 0.0390 | −28.9364 | 0.0357 | 0.0084 |
| 5000 | −29.0712 | 0.0352 | −29.6382 | 0.0330 | 0.0067 |

**Table 3** Comparison of PSLR of cubic map sequence without and with LMS algorithm

| Length of the sequence | Cubic map | | | | |
| | Without LMS | | With LMS | | Maximum value of MSE |
| | PSLR | ASP | PSLR | ASP | |
|---|---|---|---|---|---|
| 20 | −20.8279 | 0.0909 | −30.9226 | 0.0284 | 1.1653 |
| 30 | −19.0849 | 0.1111 | −19.8496 | 0.1017 | 0.8525 |
| 50 | −20.0000 | 0.1000 | −21.7463 | 0.0818 | 0.5000 |
| 70 | −18.2763 | 0.1220 | −19.3356 | 0.1079 | 0.4195 |
| 90 | −18.7570 | 0.1154 | −19.7470 | 0.1030 | 0.3260 |
| 100 | −18.7152 | 0.1159 | −21.2510 | 0.0866 | 0.3026 |
| 200 | −20.0000 | 0.1000 | −21.0831 | 0.0883 | 0.1604 |
| 300 | −21.3079 | 0.0860 | −24.2402 | 0.0614 | 0.1044 |
| 500 | −22.1039 | 0.0785 | −24.3158 | 0.0608 | 0.0635 |
| 700 | −22.7568 | 0.0728 | −25.0277 | 0.0561 | 0.0456 |
| 900 | −23.5810 | 0.0662 | −25.1020 | 0.0556 | 0.0361 |
| 1000 | −23.8407 | 0.0643 | −25.1673 | 0.0552 | 0.0333 |
| 2000 | −25.6739 | 0.0520 | −26.0207 | 0.0500 | 0.0166 |
| 3000 | −27.0651 | 0.0443 | −27.3001 | 0.0432 | 0.0112 |
| 4000 | −27.8147 | 0.0407 | −28.1074 | 0.0393 | 0.0085 |
| 5000 | −28.3786 | 0.0381 | −28.4439 | 0.0378 | 0.0068 |

**Fig. 5** Mean square error for the logistic sequence of length 5000

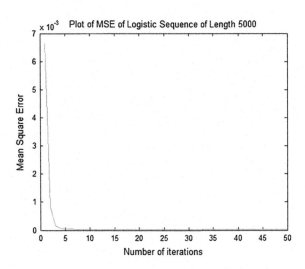

# 5 Conclusion

In this paper, PSLR of ternary chaotic sequence is optimized. The chaotic nature of these sequence helps for secure communication. A significant improvement in PSLR performance is obtained by using adaptive filtering with the implementation of LMS algorithm. The advantage of using this technique is that the maximum value of mean square error decreases with increase in sequence length. This indicates the good convergence property of the ternary chaotic sequence using LMS algorithm. The performance of least mean square algorithm is measured in terms of its convergence characteristics which depend on the minimum mean square error (MSE). The MSE is sensitive to the choice of scaling factor or step size $\mu$. By using LMS algorithm, the maximum value of mean square error decreases by varying the value of '$\mu$' between 0 and 1. This work can be extended to obtain better results for different phase sequences of any length.

**Acknowledgements** I would like to express my sincere gratitude to my advisor Prof. P. Rajesh Kumar for the continuous support of my Ph.D study and related research, for his patience, motivation, and immense knowledge. His guidance helped me in all the time of research and writing of the thesis. I could not have imagined having a better advisor and mentor for my Ph.D study.

# References

1. Levanon, N. and Mozeson, E. 2004. Radar Signals.
2. Merrill I. Skolnik 2002, Introduction to Radar System. (3rd ed.) (2002).
3. COOK, C. E., BERNFIELD, M. Radar Signals - An Introduction to Theory and Application. New York: Academic press (1967).
4. BOEHMER, A. M. Binary Pulse Compression Codes. IEEE Trans. IT-13, April 1967, no. 2, pp. 156–157 (1967).
5. LINDER, J. Binary Sequences up to Length 40 with Best Possible Autocorrelation Function. Electronic Letters, Oct 1975, vol. 11, no. 21, pp. 507–508 (1975).
6. VEERABHADRA RAO, K., UMAPATHY REDDY Biphase Sequences with Low Sidelobe Autocorrelation Function. IEEE Transactions on Aerospace and Electronic Systems, March 1986, vol. 22, no. 2, pp. 128–133 (1986).
7. MOHARIR, P. S. Signal Design. Journal of IETE, Oct. 1976, vol. 41, pp. 381–398 (1976).
8. MOHARIR, P. S. Ternary Barker Codes. Electronics Letters, Oct. 1974, vol. 10, pp. 460–461 (1974).
9. HEIDARI-BATENI, G., MCGILLEN, C. D. Chaotic Sequences for Spread Spectrum: An alternative to PN- sequences. IEEE, ICSTWC, 1992, pp. 437–440 (1992).
10. Xin Wu, Weixian Liu, Lei Zhao and Jeffrey, S., 2001, "Chaotic Phase code for Radar Pulse Compression", Proceedings of IEEE National Radar Conference, Atlanta (USA), pp. 279–283 (2001).
11. Seventline, J.B., Elizabeth Rani, D., RajaRajeswari.K, 'Ternary Chaotic Pulse Compression Sequences' Journal of Radio Engineering, Vol. 19, No: 3, pp. 415–420, September 2010 (2010).

12. S. Haykin, "Adaptive Filter Theory", 4th edition, Pearson Education Asia, 2002 (2002).
13. J.S. Fu, and Xin Wu, "Sidelobe Suppression Using Adaptive Filtering Techniques", in Proc. CIE International Conference on Radar, pp. 788–791, Oct. 2001 (2001).

# Effects of Square- and Rectangular-Shaped Slots Kept Over the Microstrip Antenna

Parsha Manivara Kumar, Nalam Ramesh Babu
and Lam Ravi Chandra

**Abstract** This paper clearly explains the slot loading effects of microstrip antenna. The effect of a series of square-shaped slots of different sizes is studied by using parametric analysis, and radiation characteristics are observed. Slot dimensions and position are chosen to get the better results. The variation in parameters like 10 db bandwidth, resonant frequency, return loss, radiation pattern, and VSWR with different slot sizes is presented. A rectangular slot width is chosen by using parametric analysis and it is positioned properly over the patch and simulated to obtain dual-band operation antenna which resonates at two different frequencies of 6.96 GHz in C-band and 9.66 GHz in X-band.

**Keywords** Square slots · Microstrip antenna · Dual band · Return loss
Parametric analysis

## 1  Introduction

Usage of a plain microstrip antenna is limited to few applications because of low bandwidth and poor gain. Lot of research has been carried out on this antenna for the past decade [1]. Slotted microstrip antenna gives high radiation efficiency if the slot is properly kept over the patch. Different techniques to achieve high isolation and bandwidth are adopted [2]. Extended dual-slot designs are being used to solve

P. M. Kumar (✉) · L. R. Chandra
Department of ECE, Chalapathi Institute of Engineering and Technology,
522034 Guntur, India
e-mail: manijw80@gmail.com

L. R. Chandra
e-mail: ravic.ciet@gmail.com

N. R. Babu
Department of ECE, DRK College of Engineering and Technology,
500043 Hyderabad, India
e-mail: nalam.rameshbabu@gmail.com

© Springer Nature Singapore Pte Ltd. 2018
J. Anguera et al. (eds.), *Microelectronics, Electromagnetics
and Telecommunications*, Lecture Notes in Electrical Engineering 471,
https://doi.org/10.1007/978-981-10-7329-8_8

beam tilt problems [3]. Slot loading is often used to control the frequency at which the antenna resonates [4]. Miniaturized structures of loop, slit, and strip loading techniques are considered and the results achieved are almost similar to the unloaded antennas [5]. Bandwidth of the microstrip antenna can be increased using slots over the patch [6].

This paper gives a more elaborate study and analysis of the effects of different slots kept over the patch. Our design starts with a basic line feed microstrip antenna which is simulated using HFSS tool. Same antenna is used to study the characteristics of the slot-loaded microstrip antenna. Parametric analysis is used to obtain the maximum dimensions of the square-shaped slots. Position of the slot is carefully adjusted to obtain good radiation characteristics. Radiation pattern and VSWR are observed continuously to check whether any undesirable radiation is present. Parameters like return loss, resonant frequency, and bandwidth are measured for each slot size. After this, the paper shows a geometry of a dual-band microstrip antenna with a rectangular slot which works better for C and X bands. Parametric analysis is done to get the optimum width of the rectangular slot for a specified length.

## 2 Basic Microstrip Design

For our basic design, we have chosen the operating frequency 7.5 GHz. Rogers RT/Duroid 5880 is suitable for the frequencies from 5 to 10 GHz band; hence, it is chosen as substrate. Its dielectric constant is $\epsilon_r = 2.20$.

Width of the microstrip antenna can be determined by using

$$W = \frac{1}{2f_r\sqrt{\mu_0\mu\epsilon_0}}\sqrt{\frac{2}{\epsilon_r+1}} \tag{1}$$

If $f_r = 7.5$ GHz, we obtain the W = 12.45 mm.
Length can be calculated from

$$L = \frac{1}{2f_r\sqrt{\epsilon_{\text{reff}}}\sqrt{\mu_0\epsilon_0}} - 2\Delta L, \tag{2}$$

where

$$\Delta L = 0.412h \times \frac{(\epsilon_{\text{reff}}+0.3)\left[\frac{W}{h}+0.264\right]}{(\epsilon_{\text{reff}}-0.258)\left[\frac{W}{h}+0.8\right]}, \tag{3}$$

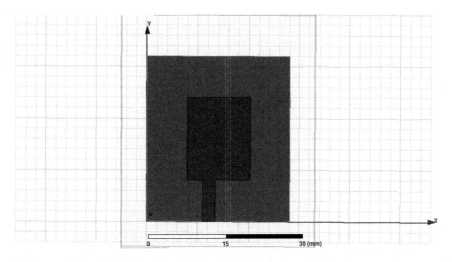

**Fig. 1** Microstrip without slot

where h is the thickness of the substrate

$$\epsilon_{\text{reff}} = \frac{\epsilon_r + 1}{2} + \frac{\epsilon_r - 1}{2}\left[1 + \frac{12h}{W}\right]^{\frac{1}{2}}$$ (4)

By substituting $\epsilon_r = 2.20$ in (4), we obtain $\epsilon_{reff}$.
Then, we substitute (4) in (3) and (3) in (2) to obtain the length of the patch

$$L = 16\,\text{mm}$$

Ground plane and substrate dimensions are chosen as 28.1 and 32 mm and height of substrate is 0.794 mm. Microstrip antenna geometry with these dimensions is shown in Fig. 1.

We used perfectly electric conductor for the patch area, feed line, and ground plane. Substrate is Rogers RT/Duroid 5880, which is suitable for 7.5 GHz. The microstrip antenna geometry with these dimensions is designed in HFSS and it is shown in Fig. 1. Simulation results of S11 for this basic design are shown in Fig. 2.

Figure 3a shows the radiation pattern and Fig. 3b shows the 3D polar plot of the basic microstrip. These figures indicate the radiation characteristics of basic microstrip antenna.

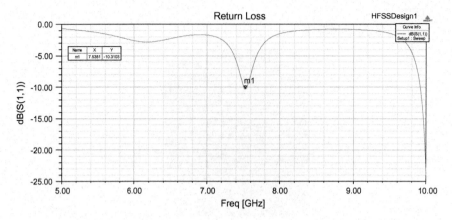

**Fig. 2** Return loss versus frequency graph of the designed microstrip

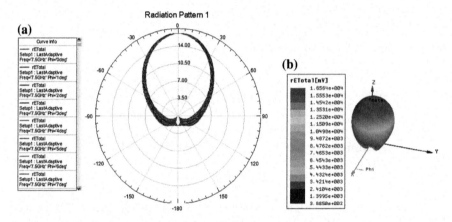

**Fig. 3** **a** Radiation pattern and **b** 3D polar plot of the microstrip antenna

## 3 Slot-Loaded Microstrip

Slot is kept over the microstrip, just by subtracting a portion of the specified patch area. Figure 4 shows the slot-loaded microstrip with slot dimension of 2 mm 2 mm. Size and shape of the slot decide operating frequency of the antenna. As the slot area increases from 1 to 16 mm$^2$, the resonant frequency shifts from 7.5 to 6.8 GHz and slight change in the radiation characteristics is observed without affecting the basic performance of the antenna.

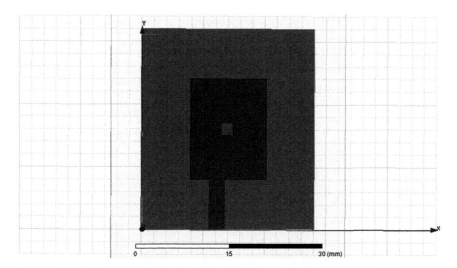

**Fig. 4** Slot-loaded microstrip antenna with 2 mm × 2 mm slot

Figure 5 shows the simulation results of 2 mm × 2 mm slot-loaded patch. These results are similar to the results shown in Fig. 3, which belongs to without slot. Only the resonant frequency shifts a bit to 7.45 GHz. Different dimensions of the square-shaped slot are chosen with the help of parametric analysis shown in Fig. 6. This figure gives the plots of S11 versus frequency for 16 different slot dimensions. After studying these 16 graphs, only seven slots are chosen. All these seven antennas with properly positioned slots are constructed and simulated separately without the use of parametric analysis. The S11 results are depicted in Fig. 7 and the values of return loss, 10 db bandwidth, and VSWR are indicated in Table 1. Numbers in the yellow color boxes indicate dimension of the square slot. We can easily notice the change in 10 db bandwidth and return loss as shown in Fig. 7.

VSWR curves for all the seven square slots and rectangular slot antennas are shown in Fig. 8. Yellow color labeling in Fig. 8 is same as in Fig. 7. VSWR values at resonant frequencies are ranging from 1.0099 to 1.87 as shown in Table 1. The dimension of the slot has no considerable effect on the VSWR of the antenna. Resonant frequency is a dependent quantity on slot dimension. As the slot size increases from 1 to 16 mm$^2$, the resonance frequency decreases from 7.5 to 6.8 GHz.

Return loss varies with the slot size. Therefore, more radiation is expected with the increase in slot size. Same is the case with 10 db bandwidth. Radiation characteristics remain unaltered from 1 to 3.7 mm slot. Radiation pattern slightly changes for the 3.8 and 4 mm slots.

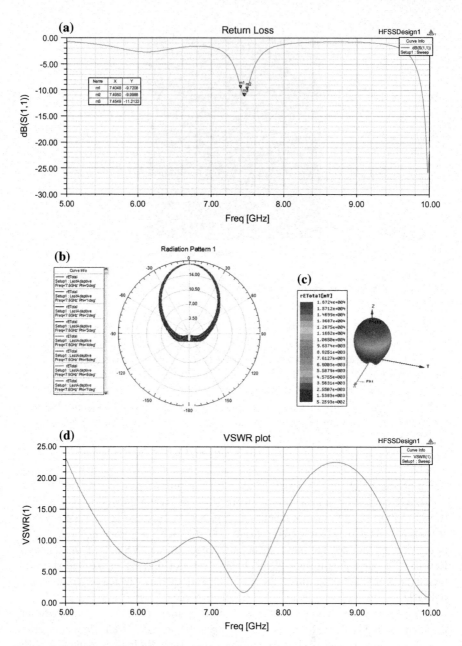

**Fig. 5** **a** S11 (Return loss), **b** radiation pattern, **c** 3D polar plot, **d** VSWR curve of slot-loaded microstrip antenna with 2 mm × 2 mm slot

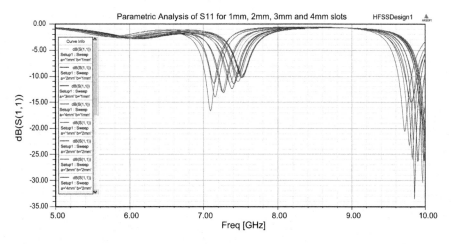

**Fig. 6** S11 results for microstrip antenna with 16 different slots

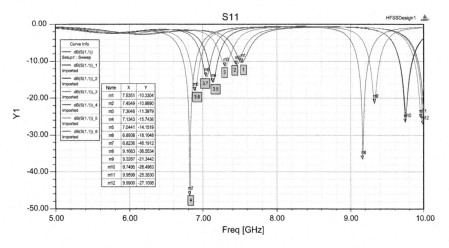

**Fig. 7** S11 of all the seven antennas with different square-shaped slots (Color figure online)

# 4 Dual-Band Behavior with Rectangular Slot

Dual-band operation can be achieved by keeping a rectangular slot as shown in Fig. 9. The length of the slot is 6 mm and the width is chosen based on the parametric analysis. Return loss at the two frequencies is almost same when the width is between 2.25 and 2.5 mm as shown in Fig. 10. Hence, the width is chosen as 2.282 mm and the simulations are carried again. S11 analysis for this width is shown in Fig. 11. We observe almost same return loss at the two frequencies and it is indicated in Table 2. Radiation pattern and VSWR for this rectangular

**Table 1** Simulation results for different slot sizes

| Antenna | Dimensions of square slot (mm) | Frequency (in GHz) | Return loss in db | 10 db bandwidth (MHz) | Shape of radiation pattern | VSWR |
|---------|-------------------------------|--------------------|-------------------|------------------------|----------------------------|------|
| Ant 1 | 1 × 1 | 7.5351 | 10.33 | 20 | No change | 1.87 |
| Ant 2 | 2 × 2 | 7.45 | 10.9 | 30 | No change | 1.81 |
| Ant 3 | 3 × 3 | 7.30 | 11.39 | 55 | No change | 1.73 |
| Ant 4 | 3.5 × 3.5 | 7.1343 | 15.74 | 130 | No change | 1.39 |
| Ant 5 | 3.7 × 3.7 | 7.044 | 14.15 | 110 | No change | 1.48 |
| Ant 6 | 3.8 × 3.8 | 6.8938 | 18.10 | 140 | Slight change in RP | 1.28 |
| Ant 7 | 4 × 4 | 6.8236 | 46.19 | 160 | Slight change in RP | 1.0099 |

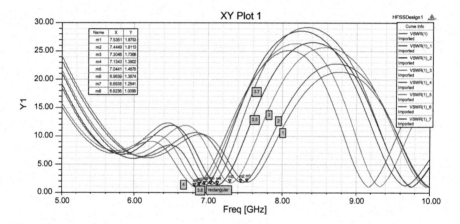

**Fig. 8** VSWR curves of all the seven antennas with different slots (Color figure online)

slot-loaded antenna is shown in Fig. 12. We can also observe the bandwidth, VSWR, in the table. The slot is kept at the right extreme of the patch as shown. If it is kept along the feed line, its radiation characteristics are undesirable and affect the antenna performance. This antenna operates at two resonant frequencies 6.9639 and 9.6693 GHz and the S11 is −16.8088 and −16.1823 db, respectively. S11 of this antenna is compared to the 1-mm-square and 4-mm-square loaded microstrips as shown in Fig. 13. Radiation pattern and VSWR curve of this rectangular slot-loaded microstrip antenna remain same when compared to the unloaded microstrip antenna shown in Fig. 1.

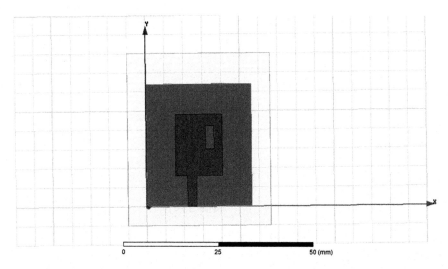

**Fig. 9** Rectangular slot-loaded microstrip antenna

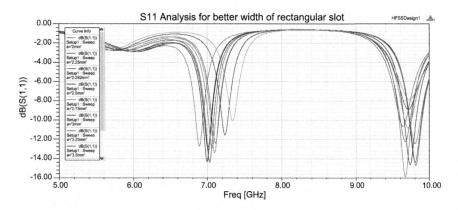

**Fig. 10** S11 analysis to find the width of the slot

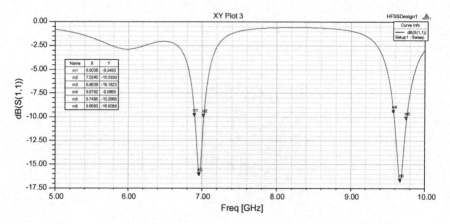

**Fig. 11** S11 of the slot-loaded microstrip antenna

**Table 2** Simulation results of rectangular slot-loaded antenna at two frequencies

| Dimension of rectangular slot | Frequency (in GHz) | Return loss in db | 10 db bandwidth (MHz) | Shape of radiation pattern | VSWR |
|---|---|---|---|---|---|
| 2.282 mm × 6 mm rectangular | 9.6693 | 16.8088 | 180 | No change | 1.3375 |
| 2.282 mm × 6 mm rectangular | 6.9639 | 16.1823 | 120 | No change | 1.3674 |

## 5   Conclusion

Basic microstrip antenna is designed and its simulation results were shown. Then, square-shaped slots are loaded on the microstrip and the radiation behavior is observed. We have used different sizes of square-shaped slots. No deviation is observed in radiation patterns, VSWR, and 3D polar plots for a slot size of up to 3.7 × 3.7 mm. However, a change in radiation pattern is observed for the slot sizes beyond 3.7 × 3.7 mm. Other parameters like return loss, resonant frequency, and −10 db bandwidth are tabulated for each slot size. Slot-loaded radiation characteristics are found similar except for a shift in frequency when compared to the unloaded antenna. Dimensions of the rectangular slot are chosen based on parametric analysis. Dual-band microstrip antenna which works better at 6.9639–9.6693 GHz is presented. Rectangular slot S11 results are compared with square slots results.

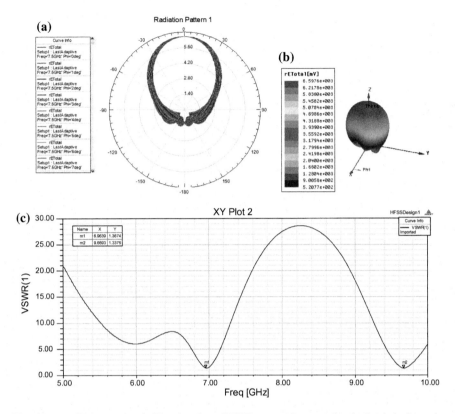

**Fig. 12 a** Radiation pattern, **b** 3D polar plot, **c** VSWR curve of rectangular slot-loaded microstrip antenna

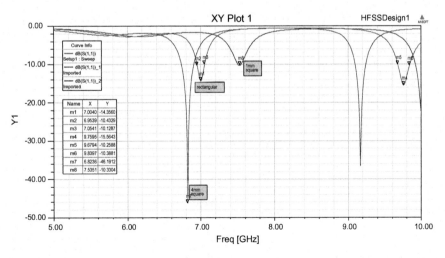

**Fig. 13** Comparison of S11 plot of microstrip antenna with 1-mm-square slot, 4-mm-square slot, and rectangular slot

# References

1. S. Maci, B. Biffi Gentili, P. Piazzessi, C. Salvador: Dual-band Slot-loaded patch antenna, IEE proceedings—Microwave, Antennas and Propagation Vol.142 Issue 3, June1995 pp 225–232.
2. Prashant K. Mishra, Dhananjay R Jahgirdar, Girish Kumar: A Review of Broadband Dual Linearly Polarised Microstrip antenna designs with High Isolation, IEEE Antennas and Propagation Vol 56, Issue 6, Dec 2014.
3. Kang Ding, Cheng Gao, Tongbin Yu, Dexin Qu: Wideband CP slot antenna with backed FSS reflector, IET Microwaves Antennas &Propagation Vol. 11, Issue 7, June 2017 pp 1045–1050.
4. Sara Mahmoud, W. Swelam, Mohamed Hassan: Parametric Study of Slotted Ground Microstrip patch antenna, IOSR Journal of Electronics and Communication Engineering Vol 11 Issue 1 ver III(Jan–Feb 2016) pp 1–8.
5. Sk. Moinul Haque, Khan Masood Parvez: Slot Antenna Miniaturization Using Slit, Strip, and Loop Loading Techniques, IEEE Transactions on Antennas and Propagation Vol. 65, Issue 5, May 2017 pp 2215–2221.
6. Sudheer Kumar Terlapu, Ch Jaya and GRLVN Srinivasa Raju: On the Notch Band Characteristics of Koch Fractal Antenna for UWB Applications, International Journal of Control Theory and Applications, Vol. 10, No. 06, pp. 701–707, 2017.

# Quality Monitoring of Water Through Electromagnetic Sensor

**Sheetal Mapare and G. G. Sarate**

**Abstract** The proposed sensor is designed on the principle of electromagnetic. The sensor structure is capable of generating electromagnetic field; this field is used for sensing contaminations in water. This electromagnetic sensor is a hexagonal-shaped coil surrounding the interdigital capacitor, implemented for monitoring the quality of water. The modelling and simulation results are experimentally tested. The experiments were performed to find out the behaviour characteristics of sensor and also to observe the response of sensor to materials involving air and distilled water. The sensor is capable of differentiating the presence of material. The sensor is tested for phosphate detection. The ammonium dihydrogen phosphate is added to distilled water so that solution of varying phosphate concentration is prepared. The sensor is immersed in the solution and its varying characteristics were obtained. The experimental results and future improvements that will be considered are also incorporated in the paper.

**Keywords** Water quality monitoring · Phosphate · Electromagnetic sensors

## 1 Introduction

The most vital element that provides essential nourishment for the perpetuation of life for human being and animal life is the water. The two main sources of water are the surface- and groundwater. The unsafe substances, in general, that cause the pollution in water resources are possibly generated from chemical waste of industries and solutions that are used for metal plating, surplus fertilizers and

S. Mapare (✉) · G. G. Sarate
Sant Gadge Baba Amravati University, Amravati, India
e-mail: smapare@gmail.com

S. Mapare
Vidyalankar Institute of Technology, Wadala, Mumbai, India

G. G. Sarate
Government Polytechnic, Amravati, Amravati, India

© Springer Nature Singapore Pte Ltd. 2018
J. Anguera et al. (eds.), *Microelectronics, Electromagnetics and Telecommunications*, Lecture Notes in Electrical Engineering 471,
https://doi.org/10.1007/978-981-10-7329-8_9

pesticides from the farming land and feedlot of animals wastes [1]. On the other hand, livestock and poultry farms contribute to nitrates contamination waste, harmful bacteria growth, biotic materials, phosphates and the solids that are not completely decomposed [2]. The nitrates and phosphates are the most worried pollution-causing chemicals in the water supplies [3]. Existing methods for monitoring water supplies involve spectrophotometric [4], electrochemical detection [5], ion-exchange chromatography combined with spectrometric, biosensors [6] and reaction due to different chemicals [7]. The high cost of material required for detection, difficulty in handling the elements of sensors, huge devices that make it bulky structures and reagents used for detection of impurities may be harmful to the environment and also involve number of steps for measurement of impurities so the current technologies are not favourable for revealing in situ measurement. Besides this, most of the devices technologies cannot be made transportable. Therefore, all current technologies drawbacks motivate the development of a sensor that will be build in as a cost-efficient, easy to use conveniently and suitable for making measurement as and when required system for quality monitoring of natural water sources. Planar electromagnetic sensors for testing the conductivity and magnetic property of materials have been reported in [8]. For numerous applications, the presence of moisture in the pulp has been measured using interdigital sensors [9]; the growth of immobilized bacteria is monitored by measuring the change in impedance [10], humidity measurement by sensors [11], inspection of food for human health safety [12] and also dielectric properties of the material that can be estimated [13]. The designed electromagnetics is a successive turn of coil and interdigitally placed elements are introduced. The simple printed circuit board (PCB) fabrication technology is used for fabrication of the sensor.

## 2 Modelling and Simulation of Sensor

The 3D modelling and simulation view of the sensor is shown in Fig. 1. The front view of sensor has hexagonal spiral inductor coil enclosing interdigital capacitor. The back side of sensor has ground plate as shown in Fig. 2. The sensor impedance and the characteristics of the sensor are evaluated that show the combination of electric and magnetic fields as electromagnetic field. The three-layer properties of the sensor are listed in the table (Fig. 3, Table 1).

The above figure shows the simulated impedance characteristics of the sensor. The sensor is simulated for frequency ranging from 100 to 10 MHz. The characteristics in Fig. 2c show the capacitive behaviour at lower frequencies and inductive at higher frequencies above 400 kHz. For intermediate frequencies, the field effect produced by the sensor is electromagnetic field. Magnetic lines are generated around the coil, and electric field lines are produced by the centrally placed interdigital capacitor. These two, magnetic and electric, fields combine to form an electromagnetic wave. This electromagnetic wave will be passed throughout the water sample to be tested that varies the sensor impedance.

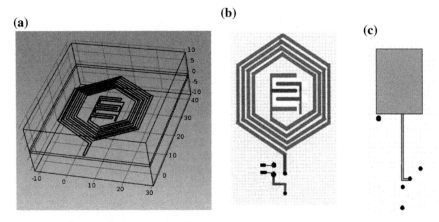

**Fig. 1** Modelling of sensor. **a** 3D view of sensor. **b** Front view of sensor. **c** Sensor with back plate

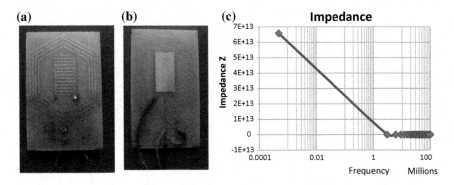

**Fig. 2** **a** Hexagonal sensor HS2 top layer. **b** Hexagonal sensor HS bottom ground. **c** Simulation characteristics of sensor

The impedance characteristics of sensor are measured in air as shown in Fig. 4. The sensor reference is obtained by immersing it in pure water that is the reference point for measurement. The characteristics of the sensor immersed in pure water show a drastic change in the impedance that is initially reduced to lower frequencies and then for a long range of frequencies it remains almost constant. This means the movement of ions in the pure water is almost constant at lower to intermediate frequencies; at higher frequencies, the electromagnetic field intensity increases that causes the free ions and thus displacement of those free ions in turn causing the impedance reduction slightly at higher frequencies.

The sensor is then tested for phosphate measurement. A sample of ammonium dihydrogen phosphate is mixed with distilled water prepared and the sensor is immersed in this solution. The sensor characteristics are shown in Fig. 5. This difference in the impedance value of the solution under test and distilled water

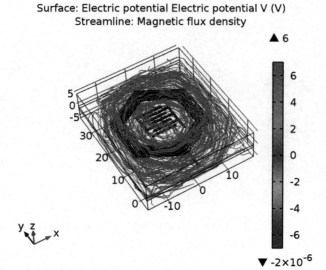

**Fig. 3** Magnetic flux density and electric potential

**Table 1** Layer properties

| Properties | Layer 1 air | Layer 2 copper | Layer 3 dielectric |
|---|---|---|---|
| σ | – | $5.998 \times 10^7$ | – |
| ε−o | $8.854 * 10^{-12}$ | 1 | 1 |
| ε−r | 1 | – | 4.3 |
| μ₀ | $4\pi * 10^7$ | – | – |
| μᵣ | 1 | 0 | 1 |

**Fig. 4** Air characteristics of sensor

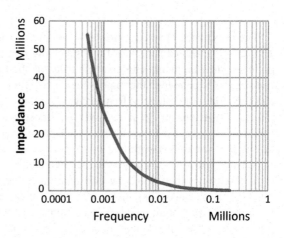

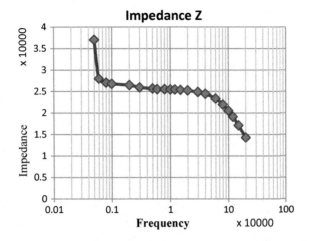

**Fig. 5** Impedance characteristics of sensor with distilled water

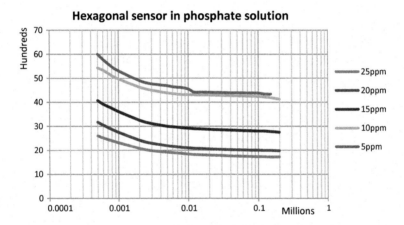

**Fig. 6** Sensor response at various concentrations

yields the detection of phosphate by the sensor. The impedance variation with frequency is noted. The ion conduction of the sample due to the presence of phosphate increases thereby decreasing the impedance with change in frequency. The conduction in distilled water is negligible due to less number of freely available ions in it. The characteristic difference shows the detection capability of the sensor.

The sensor response for various concentrations is shown in Fig. 6. As the phosphate concentration dissolved in distilled water increases, the impedance exponentially falls. For highest concentration, the impedance value is lowest. Due to increase in phosphate concentration, there is an increase in conduction phosphate ions that cause conduction in solution and reduction in dielectric of the solution.

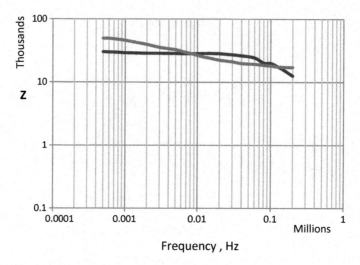

**Fig. 7** Comparison of impedance

These ions are generated due to the electromagnetic field produced by the sensor that interacts with the solution containing distilled water and phosphate ions. But these ions are deposited on the surface of the sensor that reduces the dielectric between the solution and the sensor. This results in reduction of overall impedance of the sensor.

The comparison of impedance measured, with sensor immersed in distilled water to that of solution containing ammonium dihydrogen phosphate dissolved in distilled water. From Fig. 7, the difference in the impedance values measured is equivalent to the phosphate contained in the solution of distilled water with phosphate ions.

## 3   Conclusion and Future work

The primary behaviour of the sensor shows distinctive difference between the characteristics in air and in pure water. Also, the phosphate detection is possible with sensor. This proves the sensing capabilities of the sensor that will be used for testing the water samples from the identified natural sources. A cost-effective microcontroller-based system will be well thought-out for the relative change in electrical quantity due to the excess material present in the water solution.

# References

1. E. M. Stanley, Fundamental of Environmental Chemistry, 3rd ed. Boca Raton: CRC Press, Taylor and Francis Group, 2009.
2. L. P. Donald and D. F. Charles, "Water Quality for Livestock Drinking," MU Extension Publication, University of Missouri-Columbia, vol. EQ 381, pp. 1–4, 2001.
3. N. F. Metcalf, W. K. Metcalf, and X. Wang, "The Differing Sensitivities of the Hemoglobin of Fetal and Adult Red-Cells to Oxidation by Nitrites in Man - the Role of Plasma," Journal of Physiology-London, vol. 407, pp. P44–P44, Dec 1988.
4. M. A. Ferree and R. D. Shannon, "Evaluation of a second derivative UV/visible spectroscopy technique for nitrate and total nitrogen analysis of waste water samples", Water Research, vol. 35, pp. 327–332, 2001.
5. S.-J. Cho, S. Sasaki, K. Ikebukuro, and I. Karube, "A simple nitrate sensor system using titanium trichloride and an ammonium electrode, "Sensors and Actuators B: Chemical, vol. 85, pp. 120–125, 2002.
6. T. Kjær, L. Hauer Larsen, and N. P. Revsbech, "Sensitivity control of ion-selective biosensors by electrophoretically mediated analyte transport," Analytica Chimica Acta, vol. 391, pp. 57–63, 1999.
7. N. J. Goldfine and D. Clark, "Near-surface material property profiling for determination of SCC susceptibility," in 4th EPRI Balance-of-Plant Heat Exchanger NDE Symp, 1996.
8. Y. Sheiretov, D. Grundy, V. Zilberstein, N. Goldfine, and S. Maley, "MWM-Array Sensors for In Situ Monitoring of High-Temperature Components in Power Plants," IEEE Sensors Journal, vol. 9, pp. 1527–1536, Nov 2009.
9. K. Sundara-Rajan, L. Byrd, and A. V. Mamishev, "Moisture content estimation in paper pulp using fringing field impedance Spectroscopy," IEEE Sensors Journal, vol. 4, pp. 378–383, Jun 2004.
10. S. M. Radke and E. C. Alocilja, "Design and fabrication of a microimpedance biosensor for bacterial detection," IEEE Sensors Journal, vol. 4, pp. 434–440, Aug 2004.
11. P. Furjes, A. Kovacs, C. Ducso, M. Adam,B. Muller, and U. Mescheder, "Poroussilicon-based humidity sensor with interdigital electrodes and internal heaters, "Sensors and Actuators B-Chemical, vol. 95, pp. 140–144, Oct 15 2003.
12. S. M. Radke and E. C. Alocilja, "Amicrofabricated biosensor for detecting food borne bioterrorism agents," IEEE Sensors Journal, vol. 5, pp. 744–750, 2005.
13. S. C. Mukhopadhyay, C. P. Gooneratne, S. Demidenko, and G. S. Gupta, "Low Cost Sensing System for Dairy Products Quality Monitoring, "Proceedings of the IEEE Instrumentation and Measurement Technology Conference, 2005. IMTC 2005.vol. 1, pp. 244–249, May 2005.

# Design and Implementation of Low-Power Memory-Less Crosstalk Avoidance Codes Using Bit-Stuffing Algorithms

Battari Obulesu and P. Sudhakara Rao

**Abstract** The crosstalk problems of interconnects are one of the main problems in DSM of switching network high-speed buses. To avoid the problem of crosstalk, we provided the crosstalk avoidance codes (CACs) to avoid the crosstalk problem. In this chapter, we will traverse and then produce FTC that should not have opposed directions of transitions, any direction of n number of neighboring wires in the channel. In this, we proposed a new method called a low-power algorithm for sequential and parallel bit stuffing. The low-power algorithm is for sequential and parallel bit stuffing by just inserting inverters (NOT gate) by avoiding the opposite transitions in the channel. We show the results of both algorithms (serial and parallel) of bit-stuffing (bus encoding) simulations and bit-removing (bus decoding) simulations using Verilog HDLs and synthesis and implement in FPGA. Compared to sequential bit stuffing, algorithms are somewhat more rapidly fast than the bit stuffing. And also we are finding the coding rate of both algorithms. The algorithms achieved not only higher coding rates but also lower power. Finally, we can extend the bit stuffing encoding system for generating forbidden transition codes (FTC) that avoid the two transition patterns, "$01 \rightarrow 10$" and "$10 \rightarrow 01$", on any four adjacent wires.

**Keywords** CACs · Bit stuffing · FTC · Bus encoding and bus decoding
Switching networks · Inverters

B. Obulesu (✉)
E.C.E Department, G. Pullaih College of Engineering and Technology, Kurnool,
Andhrapradesh, India
e-mail: obulesub@gmail.com

P. Sudhakara Rao
E.C.E Department, Vignan's Institute of Management and Technology for Women,
Hyderabad, Telangana, India
e-mail: sparvatha@gmail.com

© Springer Nature Singapore Pte Ltd. 2018                                        95
J. Anguera et al. (eds.), *Microelectronics, Electromagnetics
and Telecommunications*, Lecture Notes in Electrical Engineering 471,
https://doi.org/10.1007/978-981-10-7329-8_10

# 1   Introduction

The VLSI IC technology in DSM is an advanced technology, which is possible to group a large number of wires into on-chip buses and connect more I/Os. Nevertheless, the propagation delay through elongated on-chip buses have more problems in (DSM) designs [1], the problem crosstalk noise, the transmission of bit streams were damaging to be data were transmitted. To reduce the crosstalk noise effect, it is suggested to use above patterns on wires [1], and for this purpose, several bus encoding and bus decoding schemes are proposed. Among them, "adjacent opposite transitions" are preferred on n number [2–14]. The simplest way to avoid opposite or reverse transitions in two adjacent wires is to have only odd number of wires for data transmission, and in even-numbered wires, transmit 0 in any all time is called "ground shielding" in [15]. However, no explicit constructions of such codes and the associated encoding and decoding were given in [6].

We proposed the lower power algorithms for sequential and parallel bit stuffing, and also for achieving a code rate, implemented by the $2 \times 2$ bar state and $2 \times 2$ crossbar state switches with the help of above-mentioned algorithms using Verilog code, which is simulated with the help of modelism hardware tool.

# 2   Channels for FTP

In this system, let us consider that the sampling time should be ts = 1, 2, 3 .... This forbidden transition channel is defined and there is no differing transition in any two nearby wires. We can transmit the data with n binary sequences via n parallel wire by considering forbidden transition channel with n parallel wires, i.e., from 1 to n, any two nearby wires. The code word $c_i(t)$, where i = 1, 2, 3, ... and ts = 1, 2, 3, ..., be the bit transmitted on the wire ith time ts in n parallel wires. Then, for i $\geq$ 2 and t $\geq$ 2, let

$$\begin{bmatrix} b_i(t) \\ b_{i-1}(t) \end{bmatrix} = \begin{bmatrix} 1 \\ 0 \end{bmatrix} \tag{1}$$

Let us consider the NOT gate operation, i.e., if a = 0 and abar = 1, and also we can represent the sequence

$$c_{i-1}(t) = \overline{c_i}(t) \tag{2}$$

$$C_n = \frac{1}{n} \lim_{t \to \infty} \frac{\log_2 X_n(t)}{t} \, bits/wire/time \, unit. \tag{3}$$

The channel capacity, $C_n$, for determining the number of bits can transmit within the unit time. The adjacency channel matrix is defined by $(A_n)_{cc-1} = 1$ if there exists no $4 \leq i \leq n$ such that $\bar{b}_{i-1} = b'_{i-1} = b_i = b_i^{-1}$ and $(A_n)_{cc-1} = 0$; otherwise, where $b = (b_n, b_{n-1}, \ldots \ldots, b_1)$ and $b' = (b'_n, b'_{n-1}, \ldots \ldots, b'_1) \in \{1, 0\}^n$. In other words, if the bit is the previous unit delay, the output $= b$, and present bit $b(t) = b'$, then $(A_n)_{cc-1} = 1 \sum_{i=1}^{n} 2^{i-1} \in \{0, 1, \ldots 2^n - 1\}$.

The maximum eigenvalue $\lambda_{n,\,max}$ of the adjacency matrix An is expressed in terms of channel capacity $C_n$.

**Theorem 1** Forbidden transition channel with n parallel wires of channel capacity is given by

$$c_n = \frac{1}{n} \log_2 \lambda_{n,\,max} \tag{4}$$

## 3 Sequential Bit-Stuffing Algorithms

This sequential bit-stuffing algorithm can be used to send the data sequentially without loss of information and crosstalk occurrences in between the FTCs from sender and receiver (Fig. 1).

**The algorithm 1 for sequential bit-stuffing encoder**
   Given the bit stream $\{d_1, d_2, \ldots\}$ for input data.
   Initially, set $b_i = d_i$, for i = 1, 2, 3, ..., n.
   Generate the coded bit for every i $\geq$ 2, bi, i = 1, 2, 3, ...n; the bit-stuffing rules are given as follows:

(1)  Set the data bit stream for next input as $b_i$.
(2)  Every value is i = 2, 3, 4, ..., n.

   (a)  if $b_i$ is a stuffed bit, we set $b_i = b_{i-}$ (bit-stuffing condition).
   (b)  Otherwise, set $b_i =$ the data bit stream for next input.

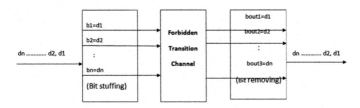

**Fig. 1** Bit-stuffing encoder and a bit-removing decoder for an n parallel transition channel

**The algorithm 2 for sequential bit-stuffing decoder**

Given received data bit stream $b_i$, $i = 1, 2, 3, \ldots, n$.

Initially, set $d_i = b_i$, for $i = 1, 2, 3, \ldots, n$.

For every value for $i \geq 2$, received coded bit is decoded, that is, $b_i$, $i = 1, 2, 3, \ldots n$ by the following bit-removing rules:

(1) Decoding the bit $b_i$ is the data bit stream for next data.
(2) Every value is $i = 1, 2, 3, \ldots, n$.

    (a) if $b_i = b_{i-1}$, the stuffed bit is $b_i$ and it is not needed (bit-removing condition).

    (b) Otherwise, decode the $b_i$ as the bit stream for data next.

## 4 Parallel Bit-Stuffing Algorithms

Only one bit stream data can be possible in the sequential bit-stuffing algorithm. When there is a parallel bit data stream, we will get a scalability problem. By using an algorithm for parallel bit stuffing, we can tackle the scalability problem.

**Algorithm:**

In encoder, let us consider the n data bit stream inputs $\{d_{i,1}\ d_{i,2}\ldots\}$, $i = 1, 2, 3, \ldots, n$, and convert them into codewords $(b_1(t), b_2(t), \ldots, b_n(t))$, $t = 1, 2, 3, \ldots, n$, so that

$$b_1(t) = d_{i,1} = \{d_{1,1}, d_{2,1}, d_{3,1}, \ldots\ldots\ldots d_{n,1}\}$$
$$b_2(t) = d_{i,2} = \{d_{1,2},\ d_{2,2},\ d_{3,2}, \ldots\ldots\ldots d_{n,2}\}$$
$$\vdots$$
$$b_n(t) = d_{i,n} = \{d_{1,n},\ d_{2,n},\ d_{3,n}, \ldots\ldots\ldots d_{n,n}\}$$

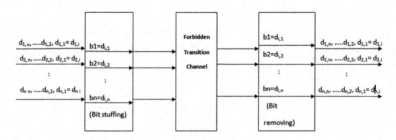

**Fig. 2** Encoding and decoding for parallel bit-stuffing and bit-removing algorithms

As shown in Fig. 2, among the n number of wires, there are no differing transitions in the two nearby wires for all values of time t. The rules for parallel bit-stuffing algorithm are represented as follows:

(1) (Odd wire) If i is odd, we can set the bit $b_i(t)$ to be ith data bit stream which is equal to the next data bit.
(2) (Internal even wire). If i is even and i < n, we can use the following three rules:

> R1: (for (i−1) wire bit-stuffing condition) if $b_i(t)$ bit is a stuffed, then we can set the bit $b_i(t) = b_i(t−1)$.
> R2: (for (i + 1) wire bit-stuffing condition) if $b_i(t)$ bit is a stuffed, then we can set the bit $b_i(t) = b_i(t−1)$.
> R3: Otherwise, the ith data bit stream is equal to the bit $b_i(t)$ that we set to be the next data bit.

## 5 Implementation and Simulation Results

1. **Sequential bit-stuffing algorithm**: In the sequential bit stuff, we are using the NOT gate or buffer-based inverter to avoid the 10 and 01 crosstalk combinations. Let us consider input 1 0 0 after FTC; again, the output will be 1 0 0 (Fig. 3).

   **Simulation result**:
   We can simulate the input and output, and simulation results will be obtained exactly as the same input 1 0 0 (Fig. 4).

2. **Parallel bit-stuffing algorithm**: In the parallel bit stuff, the information can be sent parallel through the FTC and the data without loss and crosstalk occurrences are obtained. This is very fast compared to sequential bit stuffing (Fig. 5).

   **Simulation result**:
   For the above simulation using parallel bit stuff, the input data can be sent 100,110,001 by doing this output also to replicate the same (Fig. 6).

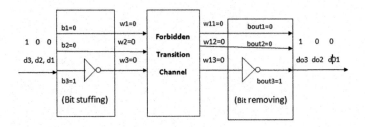

**Fig. 3** Encoding and decoding for sequential bit-stuffing and bit-removing algorithms

**Coding rate balance**:

1. **2 × 2 bar state switches for the crossbar**: By obtaining the high coding rate, we will use 2 × 2 bar state switches for the purpose of balancing the even and odd wire combination and to avoid the crosstalk occurrences between the input and output (Fig. 7).

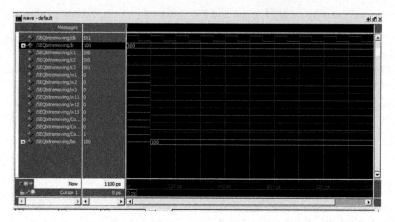

**Fig. 4** Simulation results for encoding and decoding for sequential bit-stuffing and bit-removing algorithms

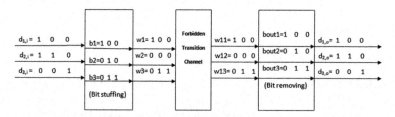

**Fig. 5** Encoding for bit stuffing and decoding for bit removing

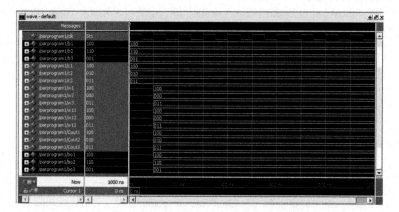

**Fig. 6** Simulation results for encoding bit stuffing and decoding bit removing

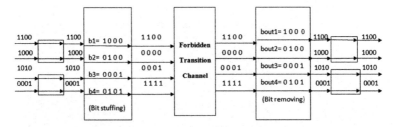

**Fig. 7** 2 × 2 bar state switches for the crossbar when the time t = 1, 3, for all values

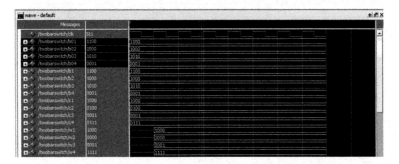

**Fig. 8** Simulation results for 2 × 2 bar state switches for the crossbar

**Encoder simulation results**:
The above encoder results state that the input will be considered as the data bits 1100, 1000, 1010, and 0001 by doing bit stuff, sent through the channel, and after bit removing, we can obtain the same input at the end (Fig. 8).

**Decoder simulation results**:
The above decoder results state that the input will be considered as the data bits 1100, 1000, 1010, and 0001 by doing bit stuff, sent through the channel, and after bit removing, we can obtain the same input at the end (Fig. 9).

2. **2 × 2 cross-state switches for the crossbar**: Even if any twisted pair or cross-state switches are there for sending the data, now we will use the 2 × 2 crossbar state switches (Fig. 10).

**Encoder simulation results**:
The above encoder simulation results state that the input will be considered as the data bits cross-coupled with 1100, 1000, 1010, and 0001 by doing bit stuff, sent through the channel, and after bit removing, we can obtain the same input at the end (Fig. 11).

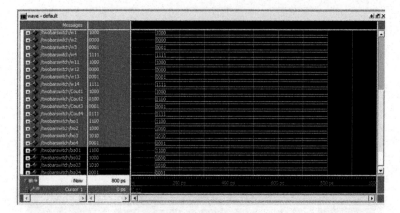

**Fig. 9** Simulation results for 2 × 2 bar state switches for the crossbar

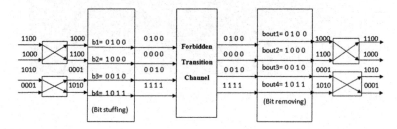

**Fig. 10** Simulation results for 2 × 2 cross-state switches for the crossbar when the time t = 1, 3, for all values

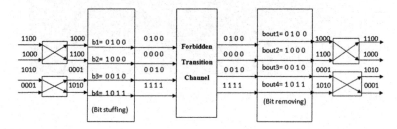

**Fig. 11** Simulation results for 2 × 2 cross-state switches for the crossbar when the time t = 1, 3, for all values

### Decoder simulation results:

The above decoder simulation results state that the input will be considered as the data bits cross-coupled with 1100, 1000, 1010, and 0001 by doing bit stuff, sent through the channel, and after bit removing, we can obtain the same input at the end (Fig. 12).

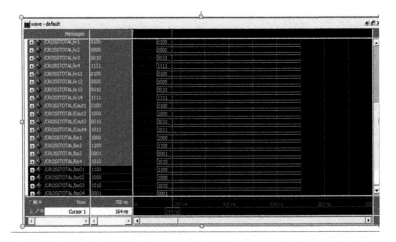

**Fig. 12** Simulation results for $2 \times 2$ cross-state switches for the crossbar when the time t = 1, 3, for all values

# 6   Conclusion

We implemented the algorithms for sequential and parallel bit stuffing that can produce prohibited switch codes in the channel for any two/three/four adjacent wires for avoiding the crosstalk. We achieved the high coding data rate in the channel using the bar switches, and thus we also achieved the lower power with both encoder and decoder. Further, we can also implement five/six/seven adjacent wires for avoiding the crosstalk.

# References

1. P. P. Sotiriadis, "Interconnect modeling and optimization in deep submicron technologies," Ph.D. dissertation, Dept. Elect. Eng. Comput. Sci., Massachusetts Inst. Technol., Cambridge, MA, USA, 2002.
2. B. Victor and K. Keutzer, "Bus encoding to prevent crosstalk delay," in Proc. IEEE/ACM Int. Conf. Comput.-Aided Design, SanJose, CA, USA, Nov. 4–8, 2001, pp. 57–63.
3. M. Mutyam, "Preventing crosstalk delay using Fibonacci representation," in Proc. 17th Int. Conf. VLSI Design, Mumbai, India, Jan. 5–9, 2004, pp. 685–688.
4. B. E. Moision, A. Orlitsky, and P. H. Siegel, "On codes that avoid specified difference," IEEE Trans. Inf. Theory, vol. 47, no. 1, pp. 433–442, Jan. 2001.
5. C. Duan, C. Zhu, and S. P. Khatri, "Forbidden transition free crosstalk avoidance CODEC design," in Proc. 45th Annu. Design Autom. Conf., Anaheim, CA, USA, Jun. 8–13, 2008, pp. 986–991.
6. X. Wu, Z. Yan, and Y. Xie, "Two-dimensional crosstalk avoidance codes," in Proc. IEEE Workshop Signal Process. Syst., Washington, DC, USA, Oct. 8–10, 2008, pp. 106–111.

7. S. R. Sridhara and N. R. Shanbhag, "Coding for system-on-chip networks: A unified framework," IEEE Trans. Very Large Scale Integr. Syst., vol. 13, no. 6, pp. 655–667, Jun. 2005.

8. W.-W. Hsieh, P.-Y. Chen, C.-Y. Wang, and T.-T. Hwang, "A busen coding scheme for crosstalk elimination in high-performance processor design," IEEE Trans. Comput.-Aided Design Integr. Circuits Syst., vol. 26, no. 12, pp. 2222–2227, Dec. 2007.

9. C.-S. Chang, J. Cheng, T.-K. Huang, and D.-S. Lee, "Explicit constructions of memoryless crosstalk avoidance codes via Ctransform," IEEE Trans. Very Large Scale Integr. Syst., vol. 22, no. 9, pp. 2030–2033, Sep. 2014.

10. L. L. Peterson and B. S. Davie, Computer Networks: A Systems Approach, 4th ed. San Francisco, CA, USA: Morgan Kaufmann, 2007.

11. R. M. Roth, P. H. Siegel, and J. K. Wolf, "Efficient coding schemes for the hard-square constraint," IEEE Trans. Inf. Theory, vol. 47, no. 9, pp. 1166–1176, Mar. 2001.

12. S. Halevy, J. Chen, R. M. Roth, P. H. Siegel, and J. K. Wolf, "Improved bit-stuffing bounds on two-dimensional constraints," IEEE Trans. Inf. Theory, vol. 50, no. 5, pp. 824–838, May 2004.

13. S. Aviran, P. H. Siegel, and J. K. Wolf, "An improvement to the bit stuffing algorithm," IEEE Trans. Inf. Theory, vol. 51, no. 8, pp. 2885–2891, Aug. 2005.

14. Cheng-Shang Chang, Jay Cheng, Tien-Ke Huang, Xuan-Chao Huang, Duan-Shin Lee and Chao-Yi Chen,"Bit-Stuffing Algorithms for Crosstalk Avoidance in High-Speed Switching" IEEE Transactions On Computers, Vol. 64, No. 12, December 2015.

15. J. D. Z. Ma and L. He, "Formulae and applications of interconnect estimation considering shield insertion and net ordering," in Proc. IEEE/ACM Int. Conf. Comput.-Aided Design, San Jose, CA, USA, Nov. 4–8, 2001, pp. 327–332.

# 180-nm 20 ps Resolution 0.29 LSB Single-Shot Precision Vernier Delay Line Based Time-to-Digital Converter

**R. S. S. M. R. Krishna, Debashis Jana, Sanjukta Mandal and Ashis Kumar Mal**

**Abstract** This article presents a Vernier Delay Line (VDL)-based Time-to-Digital Converter (TDC) in SCL 180-nm CMOS process technology. Delay elements are acknowledged through CMOS inverters and the transmission gate. Owing to vernier structure, the resolution of the TDC is that the distinction of delay lines instead of the delay of the single part. TSPC-based flip-flop uses the solitary clock and ensures to work at high frequencies with no skew. At supply voltage 1.8 V, the planned TDC demonstrates 20 ps resolution with most pessimistic scenario (PVT corners) DNL and INL of 0.3889 and 0.0032 LSB, respectively. This TDC indicates single-shot precision ($\sigma$) of 0.2903 LSB at an average power of 62.1936 $\mu$W.

**Keywords** Vernier delay line (VDL) · Time-to-digital converter (TDC) CMOS Inverter · Transmission gate · True single-phase clock (TSPC) · Resolution Single-shot precision · INL · DNL · Measurement range (MR)

## 1 Introduction

Digital building blocks take the total advantage of fast scaling of complementary metal oxide semiconductor (CMOS) technology regarding speed, size, and power wherever as analog building blocks degrade their attributes as far as intrinsic gain or output resistance since these circuits rely upon actual form of the transistor

R. S. S. M. R. Krishna (✉) · D. Jana · S. Mandal · A. K. Mal
Department of Electronics and Communication Engineering, National Institute
of Technology Durgapur, Durgapur 713209, India
e-mail: rssmrk.16ec1101@phd.nitdgp.ac.in

D. Jana
e-mail: debashisjana@ece.nitdgp.ac.in

S. Mandal
e-mail: sm.16ec1102b@phd.nitdgp.ac.in

A. K. Mal
e-mail: akmal@ece.nitdgp.ac.in

© Springer Nature Singapore Pte Ltd. 2018
J. Anguera et al. (eds.), *Microelectronics, Electromagnetics
and Telecommunications*, Lecture Notes in Electrical Engineering 471,
https://doi.org/10.1007/978-981-10-7329-8_11

characteristics instead of the quick transition between the states [1]. In time-domain circuits, wherever data is portrayed by the time contrast between the prevalence of digital events, opposition to the nodal voltages (voltage domain) or branch currents (current domain) of electric systems would take the full preferred standpoint of scaling [2]. Time-domain signal processing, thus Time-to-Digital Converters (TDC), is the unit projected for mainstream electronic circuits so as to bypass analog impairments in nanometre-scale CMOS technologies. Time-domain approaches have found a wide range of applications in time-of-flight (TOF) estimation, All Digital Phase-Locked Loops (ADPLL), Serializer/Deserializer (SerDes), Time-domain—Analog-to-Digital Converters, Software-Defined Radio (SDR), digital oscillators, and Space science instruments, to mention just a few.

Different structures and outline methods of time mode circuits have been developed as of late. TDC in light of parallel delay components using symmetric SAFF is given in [3]. In [4], TDC utilizing two-level change plans is shown. TDC used as a phase/frequency detector in an ADPLL is written in [5]. Curiously, TDC with customizable resolution for space instruments is appeared in [6]. Time-of-Flight (TOF) system-on-a-chip for spacecraft instruments is made public within [7]. Vernier Delay Line (VDL) alongside Delay-Locked Loop (DLL) is utilized as a part of [8]. A time interpolation circuit based on a DLL and a passive RC delay line is portrayed in [9]. Time digitizer with high-resolution based multi-stage delay line interpolation is employed in [10]. TDC by exploring local passive interpolation for on-chip portrayal is clarified in [11].

In this work, a simple way of designing of VDL-based TDC using True Single-Phase Clock (TSPC)-based flops and delay lines made up of inverters and transmission gate is shown. The methodology utilized here is that the resolution of the TDC is the distinction of delay lines instead of the delay of the single part. Attributable to symmetrical and straightforward design, this sort of TDCs would have higher linearity and precision.

The rest of this paper is structured as follows. Section 2 describes context and architectures of TDCs. Modified TSPC-based D flip-flop is bestowed in Sect. 3. Section 4 presents the simple way of designing delay elements. Section 5 discusses the performance matrices of the work and Sect. 6 presents the conclusions of the article.

## 2   TDC Architecture Description

TDC's resolution is restricted by technology to one delay unit ($\tau$). The delay unit ($\tau$) can be expressed as given in (1).

$$\tau = R_s * \square C_g \tag{1}$$

The sheet resistance $R_s$ and standard gate capacitance unit $\square C_g$ are specific to the technology. Many of the applications mentioned above demand higher resolution.

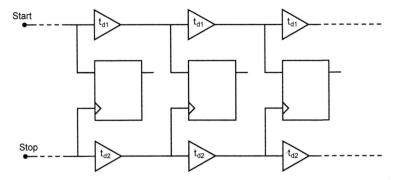

**Fig. 1** Vernier delay line based time-to-digital converter

So design strategies are essential that are not restricted by technology limits. The fundamental structures are parallel delay components and Vernier TDCs, modulating time variable techniques like pulse shrinking and stretching TDCs, TDC with interpolation, successive approximation TDC, and the Gated Ring Oscillator (GRO)-based TDC, to name a few. On a fundamental level, all structures specified above can accomplish a self-assertively high resolution. By and by, be that as it may, the resolution is restricted by process variations. TDCs, which make use of oversampling and noise shaping for accomplishing high resolution, are Gated Ring Oscillator (GRO)-based TDC. This article focused on Vernier TDC in light of its straightforward and symmetrical design.

The VDL-based TDC is best suitable for measuring time interims which are less than stage delay. As appeared in Fig. 1, it comprises two delay lines for start and stop pulses. The time interval to be measured is in between the start and stop signals. The delay blocks in start delay line have a postponement of $t_{d1}$, marginally bigger than the postponement of stop delay blocks $t_{d2}$. Amid estimation of the start pulse spreads through the start delay line and stop pulse happens just after the time input, yet the postponement offered here is littler. In this way, the stop pursues the start pulse. In every stage, stop pulse gears up by $T_{LSB}$.

$$T_{LSB} = t_{d1} - t_{d2} \tag{2}$$

The instant wherever each pulse in part is captured by flip-flops, sense amplifiers, arbiters, or just comparators. Toward the starting, these signals are unit skew by the menstruation time; however, each stage lessens the time contrast by $T_{LSB}$. For every single later stage, the initial order is turned around. The coarseness of the design, i.e., the resolution of TDC, is given by the delay distinction $T_{LSB}$ as appeared in Eq. (2). On a basic level, this distinction can be made self-assertively little, so the resolution not restricted by technology limits. Therefore, the Vernier TDC gives the design strategy to conquer the restrictions imposed by a specific technology.

## 3   Modified True Single-Phase Clock-Based D Flip-Flop

The True Single-Phase Clock (TSPC), proposed by Yuan and Svensson [12], utilizes a solitary clock, provided in here that is particularly unique to the conventional
dynamic CMOS circuit engineering in which it utilizes just a single-clock signal
which is never inverted since the inverted clock signal is not utilized in any place in
the framework; no clock skew issue exists. Therefore, higher clock frequencies can
be gone after element pipelined operation.

Modified positive edge-triggered TSPC-based D flip-flop with asynchronous reset
is shown in Fig. 2. The moment Clk = 0, the first stack of MOSFETs formed as
inverter with (M1–M3) and the altered contribution of input, will be given to node
X. The second stack of MOSFETs would be in the pre-charge mode, since M6 ener-
gizing the output node to supply voltage. The third stack of MOSFETs is within
the hold mode, because the input to M8 is Clk and the input to M9 is $V_{dd}$ both are
ineffective mode [13]. During negative edge of the clock, the input to the last static
inverter stage is decided by its previous value, so the output of the flop is stable. On
the positive edge trigger, dynamic inverter formed by M4–M6 is in evaluate phase,
so the output of the second stack of MOSFETs would be the inversion of its input,
i.e., X. During positive edge triggering, third stack of MOSFETs forms inverter with
M7–M8, so the input to this stack would be passed to the output node Q by means
of double inversion. One NMOS (M13) and another PMOS (M12) transistor are

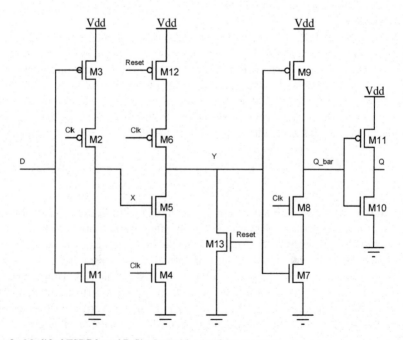

**Fig. 2**   Modified TSPC-based D flip-flop with asynchronous reset

delivered to the second inverter level to pull down the node Y to Gnd for reset action irrespective of Clk and X exchanges.

The propagation delay of flip-flop increases tremendously as the relative arrival of D and Clk approaches the point of setup/hold timings violations, this region of characteristics of the flop known as sampling window [14]. In spite of the fact that there are numerous meanings of the setup time, they are all identified with a similar key that the debasement of Clk-Q delay is attributable to an amendment within the relative time of arrival of D and clock signals. The Clk-Q versus D-Clk characteristics of the composed flip-flop can be used for extraction of setup time and is shown in Fig. 3.

There is no such equivocalness in the meaning of the hold time; this is just in light of the fact that the D-Q delay does not catch the hold time violation. Rather, the hold time is acquired from the Clk-Q versus D-Clk qualities [14]. It is commonly resolved to be the D-Clk offset like some such increment in Clk-Q delay from its value as associate degree illustrated in Fig. 4.

**Fig. 3** Clk-Q versus D-Clk for setup time

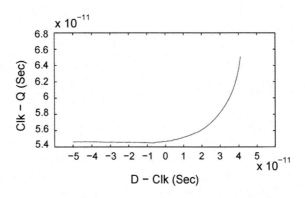

**Fig. 4** Clk-Q versus D-Clk for hold time

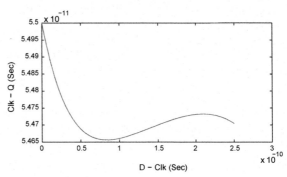

**Fig. 5** Stop delay element

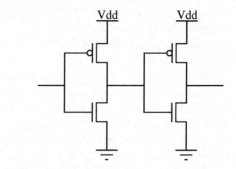

**Fig. 6** Start delay element

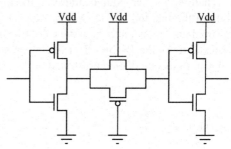

## 4 Delay Lines

CMOS buffer is framed by falling two CMOS inverters consecutively. The operation of CMOS inverter is to alter the signal from its rationale level [15]. In this way, a combo of inverters will convey back the signal to the initial rationale level after a certain delay. This property of CMOS buffer is exploited here for realizing stop delay line as shown in Fig. 5.

Stop delay ought to be not as much as start delay element in light of the fact that the stop pulse pursues the start pulse as clarified in Sect. 2. However, the prime enthusiasm here is to realize the buffer with slightly more delay than the stop delay elements, to accomplish the same as CMOS transmission gate [16] is cushioned by the CMOS inverters as appeared in Fig. 6.

## 5 Performance Analysis

The designs are synthesized at SCL 180 nm 1.8 V 1P4M CMOS process on Cadence Virtuoso Analog Design Environment using Synopsys's HSPICE simulator. The transfer characteristics of designed TDC are shown in Fig. 7. The resolution of the design is 20 ps ascertained by the slope of the transfer characteristics. The general execution is outlined in Table 1. The most extreme Differential Nonlinearity (DNL) is 0.3889 LSB, whereas the utmost Integral Nonlinearity (INL) is 0.0032 LSB.

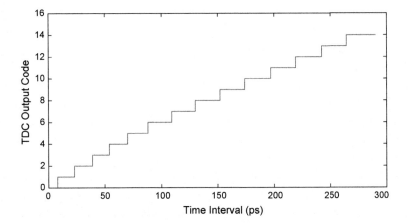

**Fig. 7** TDC Transfer characteristics

**Table 1** Performance summary

| Parameter | Value |
|---|---|
| Technology | SCL180 1P4M |
| Supply voltage | 1.8 V |
| Power | 62.1936 μW |
| Resolution | 20 ps |
| Measurement range (MR) | 320 ps |
| DNL(max) | 0.3889 LSB |
| INL | 0.0032 LSB |
| Single-shot precision (avg) | 0.2903 LSB |
| ENOB | 3.9919 |

These nonlinearities are produced by criss-cross loads to the flop because of different contributions of TDC. As per simulations, the average value of the current used by the design is around 34.552 μA.

Measurement Range (MR) of TDC is distinction amongst most extreme and least quantifiable time interims. For ideal n-bit TDC, MR is $2^n$ time of LSB so the MR of designed TDC is 320 ps. Single-shot precision ($\sigma$) can be defined as the standard deviation of the estimation that comes about appropriation when a solitary time interim is measured over and again, under repeatability conditions. Just the most pessimistic scenario esteem is generally given. The designed TDC is subjected to single-shot precision ($\sigma$) test and the result is given in Fig. 8. In many applications, Effective Number of Bits (ENOB) can be communicated as follows:

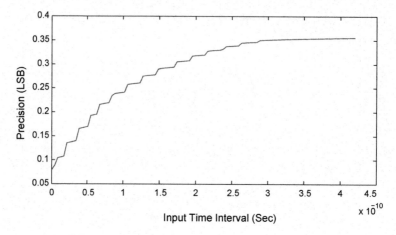

**Fig. 8** Single-shot precision of the TDC

$$
\begin{aligned}
ENOB &= \log_2\left(\frac{MR}{\sqrt{12}\,noiser\,ms}\right) \\
&= \log_2\left(\frac{MR}{\sqrt{12(\sigma^2+d^2)}}\right) \\
&\simeq \log_2\left(\frac{MR}{\sqrt{12}\sigma}\right),
\end{aligned}
\tag{3}
$$

where $d$ models the impact of systematic deviations, whereas $\sigma$ considers absolutely random variations and the root-mean-square value of noise is given by noise rms. As declared in [17], because of accumulative jitter, $\sigma$ will increase as square root of MR. Accordingly, the most pessimistic scenario $\sigma$ is typically registered for time interims beside the maximum MR. In these cases, $\sigma$ is normally substantially more prominent than $d$, hence, supporting the above estimation.

Table 2 shows the designed TDC of earlier reported work. This TDC operates at 20 ps resolution within the MR of 320 ps. The total power utilization is not specifically tantamount on the grounds that the outcomes from alternate works are compared to various measurement ranges. Nonetheless, despite everything it shows that this TDC expends low power owing to its compact design.

The anticipated design is put through PVT corners. The DNL of composed TDC is simulated at numerous corners of the process and the results are shown in Fig. 9, while INL variety is within 0.0032 LSB. The DNL of planned TDC is measured at various corners of supply voltage and furthermore the outcomes are shown in Fig. 10, whereas there is no important variation in INL. The DNL of planned TDC is measured at various temperature corners and the similar can be found in Fig. 11, while there is no critical variety in INL. The PVT examination graphs clear that the anticipated configuration has the minimal deviation (i.e., most pessimistic scenario for DNL and INL are 0.3889 and 0.0032 LSB, respectively.) and subsequently it is the sensible practicability.

**Table 2** Performance summary

| Work | Technique | CMOS [μm] | $V_{DD}$ [V] | Power [mW] | LSB [ps] | INL/DNL [LSB] |
|------|-----------|-----------|--------------|------------|----------|---------------|
| This | Vernier | 0.18 | 1.8 | 0.0622 | 20 | 0.0032/0.3889 |
| [4] | 2-level DL | 0.35 | 3 | 50 | 24 | −1.5/0.55 |
| [5] | Pseudo-diff DL | 0.09 | 1.3 | 6.9 | 17 | 0.7/0.7 |
| [6] | Pulse shrinking | 0.8 | 3 | 10 | 50 | − |
| [7] | Pulse shrinking | 0.8 | 5 | 20 | 180 | − |

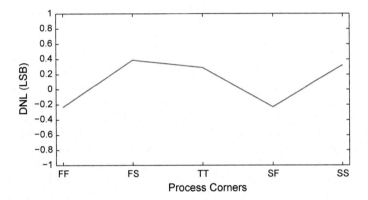

**Fig. 9** DNL versus process corners

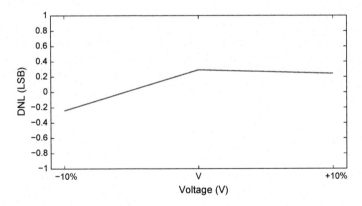

**Fig. 10** DNL versus voltage corners

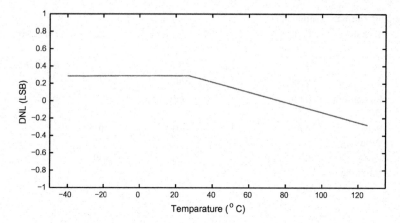

**Fig. 11** DNL versus temperature corners

## 6 Conclusion

In this paper, Vernier Delay Line (VDL)-based Time-to-Digital Converter (TDC) has been conferred ranging from delay elements to architectural design in SCL 180 nm CMOS process technology. Outlining of delay lines utilizing inverters and transmission gate has been illustrated. TSPC-based DFF has been incontestable to satisfy timing requirements of the designed TDC. With supply voltage 1.8 V, the composed TDC has accomplished 20 ps resolution with most pessimistic scenario (PVT corners) DNL and INL of 0.3889 and 0.0032 LSB, respectively. This TDC has indicated single-shot precision ($\sigma$) of 0.2903 LSB at an average power of 62.1936 $\mu$W.

**Acknowledgements** The authors would like to extend their heartfelt thanks to Dr.P.V.Ananda Mohan (Technology Advisor, CDAC, Bangalore), Prof. P. Mandal (IIT Kharagpur) and V.B. Chandratre (Electronics Division, BARC, Mumbai) for their valuable suggestions and comments. Harshil Jada (NIT Kurukshetra), Pritis Kumar Sahu (Asiczen Technologies), and Sumalya Ghosh are also acknowledged.

## References

1. Henzler, S.: Time-to-digital converters, vol. 29. Springer Science & Business Media (2010)
2. Yuan, F.: CMOS Time-Mode Circuits and Systems: Fundamentals and Applications, vol. 53. CRC Press (2015)
3. Yao, C., Jonsson, F., Chen, J., Zheng, L.R.: A high-resolution time-to-digital converter based on parallel delay elements. In: Circuits and Systems (ISCAS), 2012 IEEE International Symposium on. pp. 3158–3161. IEEE (2012)
4. Hwang, C.S., Chen, P., Tsao, H.W.: A high-precision time-to-digital converter using a two-level conversion scheme. IEEE Transactions on Nuclear Science 51(4), 1349–1352 (2004)

5. Staszewski, R.B., Vemulapalli, S., Vallur, P., Wallberg, J., Balsara, P.T.: 1.3 v 20 ps time-to-digital converter for frequency synthesis in 90-nm cmos. IEEE Transactions on Circuits and Systems II: Express Briefs 53(3), 220–224 (2006)
6. Karadamoglou, K., Paschalidis, N.P., Sarris, E., Stamatopoulos, N., Kottaras, G., Paschalidis, V.: An 11-bit high-resolution and adjustable-range cmos time-to-digital converter for space science instruments. IEEE Journal of Solid-State Circuits 39(1), 214–222 (2004)
7. Paschalidis, N., Stamatopoulos, N., Karadamoglou, K., Kottaras, G., Paschalidis, V., Sarris, E., McNutt, R., Mitchell, D., McEntire, R.: A cmos time-of-flight system-on-a-chip for spacecraft instruments. IEEE Transactions on Nuclear Science 49(3), 1156–1163 (2002)
8. Dudek, P., Szczepanski, S., Hatfield, J.V.: A high-resolution cmos time-to-digital converter utilizing a vernier delay line. IEEE Journal of Solid-State Circuits 35(2), 240–247 (2000)
9. Mota, M., Christiansen, J.: A high-resolution time interpolator based on a delay locked loop and an rc delay line. IEEE journal of solid-state circuits 34(10), 1360–1366 (1999)
10. Mantyniemi, A., Rahkonen, T., Kostamovaara, J.: An integrated 9-channel time digitizer with 30 ps resolution. In: Solid-State Circuits Conference, 2002. Digest of Technical Papers. ISSCC. 2002 IEEE International. vol. 1, pp. 266–465. IEEE (2002)
11. Henzler, S., Koeppe, S., Kamp, W., Mulatz, H., Schmitt-Landsiedel, D.: 90nm 4.7 ps-resolution 0.7-lsb single-shot precision and 19pj-per-shot local passive interpolation time-to-digital converter with on-chip characterization. In: Solid-State Circuits Conference, 2008. ISSCC 2008. Digest of Technical Papers. IEEE International. pp. 548–635. IEEE (2008)
12. Yuan, J., Svensson, C.: High-speed cmos circuit technique. IEEE Journal of Solid-State Circuits 24(1), 62–70 (1989)
13. Rabaey, J.M., Chandrakasan, A.P., Nikolic, B.: Digital integrated circuits, vol. 2. Prentice hall Englewood Cliffs (2002)
14. Oklobdzija, V.G., Stojanovic, V.M., Markovic, D.M., Nedovic, N.M.: Digital system clocking: high-performance and low-power aspects. John Wiley & Sons (2005)
15. Saini, S.: Low power interconnect design. Springer (2015)
16. Krishna, R.S.S.M.R., Madhumati, G.L., Mal, A.K.: Design of substantial delay block using voltage scaled cmos inverter and transmission gate blend. In: 2016 International Conference on Microelectronics, Computing and Communications (MicroCom). pp. 1–6 (Jan 2016)
17. Napolitano, P., Moschitta, A., Carbone, P.: A survey on time interval measurement techniques and testing methods. In: Instrumentation and Measurement Technology Conference (I2MTC), 2010 IEEE. pp. 181–186. IEEE (2010)

# Low-Cost Portable Gas Pollutants Detection System for People with Olfactory Impairment

Pushpa Kotipalli, Jyothi Chinta and M. Mohan Varma

**Abstract** Olfactory impairment refers to complete loss of normal ability to smell. Problems in nose, nervous system, or brain can lead to olfactory impairment. People with olfactory impairment need assistive devices for detecting the presence of harmful gases so that they can take necessary steps to protect themselves. Here, we propose a gas pollutants detection system. It is designed for four harmful gases methane, carbon monoxide, LPG, and air pollutants. It displays the amount of gases using LCD, provides audio output about safe or dangerous levels in a particular area by comparing the measured values with threshold levels and to save the information in the SD card. It is a low-cost portable system.

**Keywords** Olfaction deficiency · Harmful gases · Gas detection
LPG · Methane · Carbon monoxide · Air pollutants

## 1 Introduction

According to the 2014 WHO report, gas pollutants caused the deaths of around seven million people worldwide in 2012. Neurological disorders like Alzheimer's disease, tumors in the brain, head injuries, dementia, over the use of nasal decongestant can cause olfactory impairment. People with olfactory impairment face several problems in day-to-day life. If LPG leak occurs in the kitchen, the

P. Kotipalli (✉) · J. Chinta
Shri Vishnu Engineering College for Women Vishnupur, West Godavari District,
Bhimavaram 534202, Andhra Pradesh, India
e-mail: pushpak@svecw.edu.in

J. Chinta
e-mail: chintajyothi21@gmail.com

M. Mohan Varma
Shri Vishnu College of Pharmacy Vishnupur, West Godavari District,
Bhimavaram 534202, Andhra Pradesh, India
e-mail: mohan@svcp.edu.in

© Springer Nature Singapore Pte Ltd. 2018                                    117
J. Anguera et al. (eds.), *Microelectronics, Electromagnetics
and Telecommunications*, Lecture Notes in Electrical Engineering 471,
https://doi.org/10.1007/978-981-10-7329-8_12

person with olfactory impairment will be unable to recognize the toxic odor of the LPG and may face fire accidents leading to losing a life. Persons with olfactory impairment cannot detect the odor of any items [1]. As long as those items belong to foods, fruits, flowers, and scents, persons with olfactory impairment cannot get odor related satisfaction. But the toxic gases may severely affect the health, i.e., they may affect the lungs, heart, and even brain of those people with olfactory impairment. They need an assistive device which works similar to their olfactory system.

The biological olfactory system is as shown in Fig. 1a. The inhaled air through two nasal cavities carries volatile materials such as gases present in the air. Olfactory receptors act as a variety of chemical sensors. The volatile compounds, also known as odorants, stimulate olfaction receptors located in the olfactory epithelium, and produce signals. The olfactory nerves in the mucous membrane carry signals related to odor to olfactory bulbs located in the forebrain. Olfactory bulbs further transport the signals through olfactory tract to the olfactory cortex. Olfactory cortex processes the signals and identifies the odor of the gas. The electronic gas detection system, as shown in Fig. 1b, consists of the sensor array. Each sensor is sensitive to all volatile molecules but each in their specific way. They produce responses in the form of analog signals whenever gases are present in the surrounding air. These signals are preprocessed by reducing noise and converting it to digital form. In the digital domain, pattern recognition algorithms are applied and odor of the gas is identified.

Several toxic gases in the surrounding environment cause health hazards to the people with olfactory impairment. Some of the toxic gases are discussed below.

- **Carbon monoxide (CO)**: CO is released when natural gas, wood, or coal is a colorless, odorless, toxic yet non-irritating gas. It is a product of incomplete combustion of fuel, such as natural gas, coal, or wood. Vehicular exhaust is a major source of carbon monoxide.

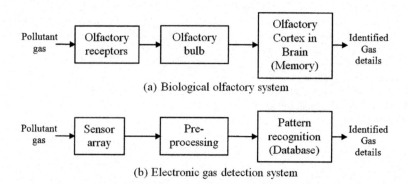

(a) Biological olfactory system

(b) Electronic gas detection system

**Fig. 1** Basic diagram showing the analogy between biological and electronic gas detection systems

- **Ammonia (NH₃)**: Ammonia gas is emitted by agricultural processes. It is pungent in odor and it is both caustic and hazardous. In the atmosphere, ammonia reacts with oxides of nitrogen and sulfur to form secondary particles.
- **Methane (CH₄)**: Methane is generated by the waste deposition in landfills. Due to its high flammable nature, it forms explosive mixtures with air. As methane displaces oxygen content, the people in enclosed space get suffocation when oxygen content falls below 19.5%.
- **Liquefied Petroleum Gas (LPG)**: LPG is generated by mixing commercial butane and propane gases having both saturated and unsaturated hydrocarbons. LPG vapor accumulated in the low lying area in the eventuality of the leakage or spillage may lead to an explosion. The combustion of LPG produces carbon dioxide ($CO_2$) and water vapors, but sufficient air must be available.

In [2, 3], authors suggested a system using a set of sensors, such as $CO_2$, CO, VOCs, $H_2$, temperature, and relative humidity and a Bluetooth link. The acquired data from Sensors is transmitted to the smartphone through the Bluetooth link. In [4], authors developed a system similar to the one suggested in [2, 3] but added an additional feature Wi-Fi. Bluetooth link transmits data from sensors to a smartphone. The smartphone forwards the received data from Bluetooth link to the server using a built-in Wi-Fi module. Wi-Fi module is also utilized for localization. These systems consist of Bluetooth, Wi-Fi modules, and smartphone which make the cost of the system very high. People with olfactory impairment need a low-cost portable system which they can carry everywhere and protect themselves whenever gas pollutants are detected. Chandrasekaran et al. [5] simulated the air pollutants detection for motor vehicles but practical implementation was not done.

Keeping this in mind, we propose a low-cost portable gas pollutants detection system with the following objectives:

(1) To find the amount of gases present in the atmosphere which are responsible for air pollution, such as $NO_2$, $SO_2$, CO, and $CH_4$.
(2) To give the indication about safe or dangerous levels, and to produce the audio output.

Remaining paper is organized in the following way. Section 2 proposes gas pollutants detection system. Section 3 discusses hardware design. Section 4 describes the software design of the proposed system. Section 5 discusses the hardware results as well as software results and concluding remarks.

## 2 Proposed Gas Pollutants Detection System

As shown in Fig. 2, the gas pollutants detection system consists of an array of sensors, ATmega 2560 microcontroller, and audio amplifier. The microcontroller ATmega 2560 gets the data from the array of sensors. The output of the sensors

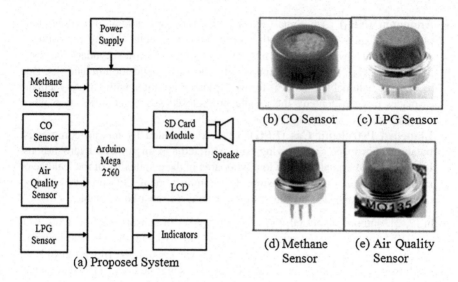

**Fig. 2** Proposed gas pollutants detection system and sensors used in it

from Arduino can be used in many ways. This output is printed on the LCD and can be stored in the micro SD card. Then, it is compared with the threshold and if that output value crosses the threshold then the red LED will glow and the buzzer starts sounding otherwise it will be OFF. Different audio outputs can also be obtained depending on the output value. In order to get the audio output, first, the audio files are to be converted into.wav files and to be stored in the SD card. These files can be retrieved by the Arduino depending on the output value from the sensors. The audio output will have a very low voice and is amplified using the LM386 audio amplifier module.

## 2.1 Sensors Array

The device that is used for measuring a physical quantity is called sensor. The sensor produces an electrical signal corresponding to the physical quantity measured. A sensor's sensitivity indicates how much the sensor's output changes when the input quantity being measured changes. In the proposed system MQ series sensors are used. These sensors have low conductivity in the fresh air. Whenever the gas is exposed to the MQ series sensors the conductivity of the sensing material, here $SnO2$, increases thereby, the analog output increases indicating the change in the concentration of the gas. As shown in Fig. 2, (a) CO sensor (MQ-7), (b) LPG sensor (MQ-5), (c) Methane sensor (MQ-4), and (d) Air quality sensor (MQ-135), are used in the proposed gas pollutant detection system.

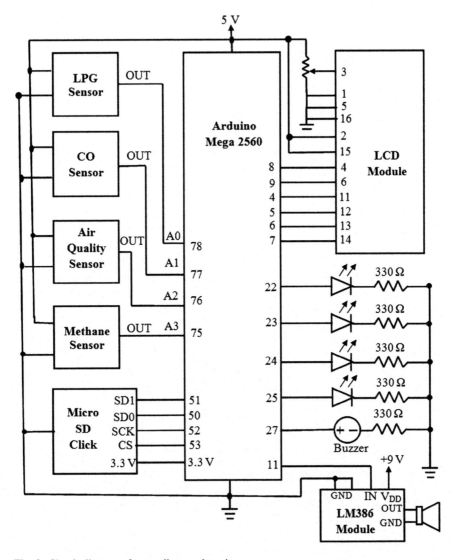

**Fig. 3** Circuit diagram of gas pollutants detection system

## 3 Hardware Implementation

Figure 3 shows how the Air Pollution Monitoring System is implemented using different components, such as Arduino, LCD, micro SD card, sensors, speaker, etc. ATmega2560 microcontroller is used in Arduino Mega board. ATmega2560 microcontroller consists of 100 pins out of which 54 digital input/output pins, 16 analog input pins. 15 PWM pins are part of digital input/output pins. The clock is

provided by 16 MHz crystal oscillator. Other features of Arduino board include
USB connection, ICSP header, power jack, and a reset button. It operates at 5 V.
Preferable input supply voltage falls in the range 7 to 12 V.

This circuit diagram shows the connections of the sensors to the Arduino pins
and the interfacing of the micro SD click, LCD module, and the light-emitting
diodes for indicating the dangerous levels when the threshold value is crossed.

The threshold level for each pollutant gas is fixed based on the threshold limit
values given by the government [5]. The threshold value is nothing but the safe
limit up to which the person can expose to that gas. And the buzzer is also connected similarly to the Arduino. The speaker is also connected to the Arduino for
the audio output. In this LM386 is used to amplify the audio output.

## 3.1  Flow Chart

Flowchart for implementation of the air pollution monitoring system project is
given in Fig. 4. The steps for implementing the proposed system, as shown in the
flowchart, are as follows:

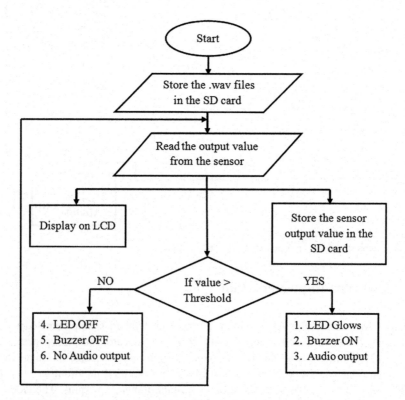

**Fig. 4** Flowchart of implementation of gas pollutants detection system

(1) Initially, store the audio files on the SD card. These audio files must be converted into the wav files.
(2) Read the analog output from the sensor. Print the amount of gases on the LCD and Store the values into the SD card.
(3) Compare the threshold with the obtained output value. If the Threshold is greater than the obtained value, then the LED will glow for 10 s and the buzzer will ring, otherwise, the LED will be off.
(4) The audio output will be according to the sensor output values.
(5) For continuous monitoring, step 2 to step 8 are repeated.
(6) In the proposed system, four different sensors are used for detecting different pollutants present in the atmosphere. Hence, the above procedure should be done for four sensors simultaneously.

# 4 Results

Complete implemented gas pollutants detection system is shown in Fig. 5. Whenever the Arduino mega board is powered it supplies power to all the external devices connected to it. Depending on the program written in the Arduino software the results will be obtained. The sensor output pins are connected to the Arduino and it is interfaced with the LCD and micro SD click. Then, the output value of each sensor is displayed on the LCD. If the threshold of any of the sensor is crossed then the respective LED glows for indicating the dangerous level and the audio output will also be obtained from the speaker.

The CO sensor value is displayed on the LCD when it is exposed to camphor. The obtained value is in milli Volts (mV). This value has to be converted into the ppm in order to compare it with the thresholds given the Permissible Exposure Limits for Chemical Contaminants [6].

A prototype of the gas pollutants detection system is as shown in Fig. 6a. LCD display is a display of CO reading 770 mV. The ppm conversion formula for this sensor is given by

$$ppm = 3.027 * e^\wedge (1.0698 * VRL) \text{where } VRL = \text{analogValue} * (5V\ /4095)$$

Similar conversion formulae are available for other sensors for converting analog reading in mV to ppm.

The settling time of each sensor is determined by exposing with their corresponding gases and readings on LCD display on the prototype are noted down. Readings are taken for every 5 s time. This raw data of sensors is plotted with respect to time as shown in Fig. 6b. It is observed from Fig. 6b that minimum 1 minute is required for sensor readings to settle down and negligible variation afterwards.

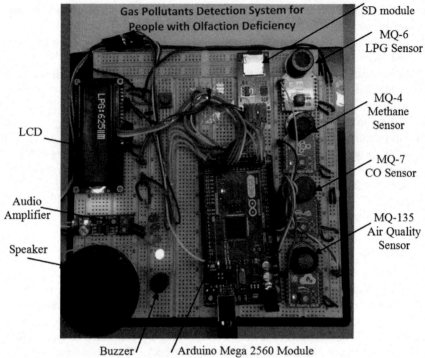

**Fig. 5** Circuit of implemented system

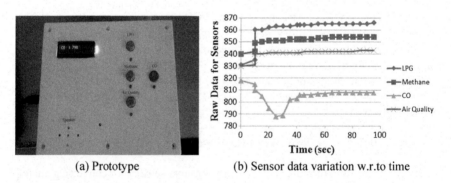

(a) Prototype                              (b) Sensor data variation w.r.to time

**Fig. 6** Settling period of sensors tested using prototype of proposed system

## 5   Conclusions

A portable gas pollutants detection system for the people with olfactory impairment is developed. It detects harmful gases like CO, methane, LPG, and air quality and provides audio messages. With the help of these messages, the people with olfactory impairment can take necessary actions to get least affected by the harmful gases. Total cost required to develop this system is INR 1500 which is far less than that of existing systems. The proposed system is portable and can be carried by individuals for protection from pollutant gases. Sensor working capability slowly decays with temperature and time. Therefore, it needs to be calibrated at regular intervals of time and this calibration process is complex. More concentration of gases may lead to damage to sensors. The proposed system can be used at chemical industries, traffic junctions, and crowded places to identify pollutant gases and save the people before facing health hazards.

**Acknowledgements** This work was supported by Department of Science and Technology under CSRI with order no. 4036/(IFD)/2015-16 dated 20.11.2015. Authors would like to acknowledge the encouragement given by management of Sri Vishnu Education Society and contributions made by colleagues and students.

# References

1. Miwa T1, Furukawa M, Tsukatani T, Costanzo RM, "Impact of olfactory impairment on quality of life and disability", NCBI. 2001 May; 127(5): 497–503.
2. Mendez, D.; Perez, A.J.; Labrador, M.A.; Marron, J.J. P-Sense, "A Participatory Sensing System for Air Pollution Monitoring and Control", Proc. IEEE Intl, Conf. on Pervasive Computing and Comms. Workshops, Seattle, USA, 21–25 Mar. 2011; Pp. 344–347.
3. Tsow, F.; Forzani, E.; Rai, A.; Wang, R.; Tsui, R.; Mastroianni, S.; Knobbe, C.; Gandolfi, A.J.; Tao, N.J. "A Wearable And Wireless Sensor System For Real-Time Monitoring of Toxic Environmental Volatile Organic Compounds", IEEE Sens. J. 2009, 9, 1734–1740.
4. Jiang, Y.; Li, K.; Tian, L.; Piedrahita, R.; Yun, X.; Mansata, O.; Lv, Q.; Dick, R.P.; "A Personalized Mobile Sensing System for Indoor Air Quality Monitoring", Proc. of 13th Intl. Conf. Ubicomp'11, Beijing, China, 17–21 Sept. 2011; PP. 271–280.
5. S.S. Chandrasekaran, S. Muthukumar and S. Rajendran, "Automated Control System for Air Pollution Detection in Vehicles", Intl. Conf. Intelligent Systems Modeling & Simulation, 29–31 Jan, 2013.
6. "Threshold Limit Values & Biological Exposure Indices", the American Conference of Governmental Industrial Hygienists (ACGIH) in 2005.

# Seismic Signal Processing by Using Root-MUSIC Algorithm

G. Pradeep Kamal, B. L. Prakash and S. Koteswara Rao

**Abstract** The overall purpose of this study is to obtain a high-resolution seismic signal from a raw seismic data by avoiding all the noise. The earthquake data, which is received from an earthquake station, contains noise. In order to remove that noise, we implement signal processing techniques by using Root-MUSIC algorithm, which is an improvement to the actual MUSIC algorithm. In this paper, better results are obtained by using Root-MUSIC algorithm.

**Keywords** Seismic signal · Power spectral density · Complex exponentials Eigendecomposition

## 1 Introduction

Earthquake is an unpredictable natural phenomenon; the seismic signal acquired from the earthquake monitoring stations contains enormous data which has to be processed in order to analyze the magnitudes of the seismic events; many techniques are there to predict an earthquake and its pattern. The only problem with seismic signal processing is the lack of resolution to clearly identify a peak of the magnitude of the earthquake event. The resolution can be increased by applying some algorithms to the actual process of the seismic signal. In this paper, Root-MUSIC algorithm is implemented. The algorithm's basics are clearly described below for the better understanding of the algorithm and its mechanism.

G. P. Kamal (✉) · B. L. Prakash · S. Koteswara Rao
Department of ECE, K L University, Green Fields, Guntur DT 522502, AP, India
e-mail: gollamudipradeep@gmail.com

B. L. Prakash
e-mail: prakashklu4368@kluniversity.in

S. Koteswara Rao
e-mail: rao.sk9@gmail.com

© Springer Nature Singapore Pte Ltd. 2018
J. Anguera et al. (eds.), *Microelectronics, Electromagnetics and Telecommunications*, Lecture Notes in Electrical Engineering 471,
https://doi.org/10.1007/978-981-10-7329-8_13

## 2  Frequency Estimation

A complex sinusoidal signal's frequency estimation in noise is one of the crucial problems in the signal processing field, because estimation of frequency is used extensively in several areas such as communications, radar, image analysis, and sonar [1]. The pole-zero types presume a system which is linear time-invariant excited by white noise. However, in several implementations, the complex exponentials are signals of interest which are lying in white noise for which a sinusoidal or harmonic model is more appropriate. Signals which have complex exponentials are identified as formant frequencies in radar's moving targets, speech processing, and signals which propagate spatially in array processing. In real signals, complex conjugate pair (sinusoids) is made by complex exponentials, but for complex signals, they may happen at only one frequency. For complex exponentials found in noise, the parameters of interest are the frequencies of the signals. Hence, the primary aim is the estimation of these frequencies from the data, by calculating the power spectrum and by utilizing the nonparametric techniques or the minimum-variance spectral estimate. The complex exponentials frequency estimates are hence the frequencies at which peaks appear in the spectrum. Certainly, the usage of these non-parametric techniques appears appropriate for signals of complex exponentials as they do not make assumptions about the basic procedure. As some of these techniques can get very good resolution, none of these techniques accounts for the basic model of complex exponentials in the noise, beginning by harmonic signal model description, deriving the model from a vector notation, and looking at the eigendecomposition of the correlation matrix of complex exponentials in noise. Then, we describe frequency estimation methods based on the harmonic model: Root-MUSIC. This method has the ability to resolve complex exponentials closely spaced in frequency and has led to the name *superresolution* commonly being associated with them [2]. The extreme level of performance in the means of resolution is acquired by supposing a foundational model of the data. By considering all the other parametric techniques, the performance of these techniques depends upon how nearly this mathematical technique corresponds to the actual physical technique that generates the signals. Deviations from this assumption result in model mismatch and will generate frequency estimates for a signal that may not have been generated by complex exponentials.

## 3  Root-MUSIC Algorithm

This algorithm has been proposed as an improvement over the MUSIC algorithm which is proposed by Barabell. Root-MUSIC indicates that the MUSIC algorithm is simplified to detecting polynomial roots as contrary to solely plotting the pseudospectrum or looking for peaks in the pseudospectrum [3]. The eigenanalysis method presumes that the received data can be decomposed into two mutually orthogonal subspaces. One is the signal plus noise subspace, and the another is the noise-only subspace. Consider the sample of a signal that contains P complex exponentials present in noise.

$$x(n) = \sum_{p=1}^{P} \alpha_p e^{j2\pi n f_p} + \omega(n) \tag{1}$$

The discrete-time, normalized frequency of the $p$th component is where $\omega_p$ is the discrete-time frequency which has the units in radians, $p$th complex exponential's actual frequency is $F_p$, and sampling frequency is denoted by $F_s$ [2]. Basically, the goal is to estimate not only the frequencies but also the amplitudes of these signals. Take into account that the amplitude contains phase of each complex exponential, which is

$$\alpha_p = |\alpha_p| \, e^{j\psi_p} \tag{2}$$

Here, the phases $\psi_p$ are nothing but some random variables which are uncorrelated and uniformly distributed in the range of $[0, 2\pi]$. Here, the deterministic quantities are magnitude $|\alpha_P|$ and the frequency $f_P$. By considering the harmonic process spectrum, it can be seen that at white noise's power of a constant level of a background, it contains an impulse set.

$$\sigma_w^2 = E\{|\omega(n)|^2\} \tag{3}$$

Hence, the complex exponentials spectrum of power is generally related to a spectrum of line, and based on some length M of time window, some matrix methods are put into use, and it is helpful to typify the model of the signal as an arrangement of a vector above the time span consisting of sample delays of signal [4]. Consider a signal x(n) from Eq. (1) in its initial and upcoming values of M − 1.

Equation (3) represents a frequency vector which has a time span. At frequency $f$, $v(f)$ is casually a discrete Fourier transform vector of length M, by differentiating between the signal s(n) which has the summation of noise component w(n) and complex exponentials.

Consider the time window form of vector which has the summation of noise complex exponentials in (5). This model's autocorrelation matrix is represented as the summation of autocorrelation matrices of noise and signal, respectively

$$A = \begin{bmatrix} |\alpha_1|^2 & 0 & \cdots & 0 \\ 0 & |\alpha_2|^2 & \ddots & \vdots \\ \vdots & \ddots & \ddots & 0 \\ 0 & \cdots & 0 & |\alpha_P|^2 \end{bmatrix}$$

The above diagonal matrix has the exponents of every single one of particular complex exponentials. The white noise's autocorrelation matrix is

$$R_w = \sigma_w^2 I, \tag{4}$$

which is of complete rank, as against to $R_s$ is a $P < M$ rank-lacking. Basically, select the number of P complex exponentials in which the time window $M$ length is greater. In the same way of its eigendecomposition, the autocorrelation matrix can also be represented

$$R_x = \sum_{m=1}^{M} \lambda_m q_m q_m^H = Q \wedge Q^H \tag{5}$$

Here, eigenvalues $\lambda_m$ appear in decreasing order, which is $\lambda_1 \geq \lambda_2 \geq \ldots \geq \lambda_M$, and $q_m$ are the equivalent eigenvectors of them. By utilizing the sample correlation matrix's eigenvectors and eigenvalues, estimate the subspaces of noise and the signal. Note that for notational expedience we will not differentiate between eigenvectors and eigenvalues of the true and sample correlation matrices [5]. However, the sample correlation matrix's eigendecomposition is what must be used for implementation. Consider that the use of an estimate rather than the true correlation matrix will result in a degradation in performance.

## 4    Mathematical Modeling

Like the Pisarenko harmonic decomposition, by using the eigenvectors of the correlation matrix, the space of $M$-dimension is split into components of noise and signal from

$$Q_s = [q_1 q_2, \ldots, q_P] \quad Q_w = [q_{P+1}, \ldots, q_M] \tag{6}$$

However, rather than limit the time window's length to $M = P+1$, which is a unity 1 more than the amount of complex exponentials, consider the time window's size to be $M > P+1$. Therefore, the subspace of noise has a dimension more than a unity 1. Using this dimension which is greater allows for averaging over the noise subspace, providing an improved, more robust frequency estimation method than Pisarenko harmonic decomposition [3], from

$$P_s v(f_p) = v(f_p) \, P_w v(f_p) = 0 \tag{7}$$

Hence each eigenvector ($P < m \leq M$), to every complex exponential, the $P$ frequencies $f_p$. Pseudospectrum for every eigenvector of noise is calculated as

$$\overline{R}_m\left(e^{j2\pi f}\right) = \frac{1}{\left|V^H(f)q_m\right|^2} = \frac{1}{\left|Q_m(e^{j2\pi f})\right|^2} \tag{8}$$

The $M - 1$ are the roots contained in the polynomial $Q_m\left(e^{j2\pi f}\right)$, among those $P$ roots relate to the complex exponentials frequencies. Inside the pseudospectrum, $P$ peaks are produced by these roots through (9). Take into account that all $M - P$ noise eigenvector's pseudospectra distribute these roots that are because of the subspace of the signal [3]. The noise eigenvector's leftover roots, anyway, appear at frequencies which are unique. On these roots' locations, there exist no restrictions, so that some may produce extra peaks in the pseudospectrum by being near to the unit circle. A necessity of decreasing these false peak levels in the pseudospectrum is to calculate the average of the separate noise eigenvectors of the $M - P$ pseudospectra.

$$\overline{R}_{music}\left(e^{j2\pi f}\right) = \frac{1}{\displaystyle\sum_{m=P+1}^{M} \left|v^H(f)q_m\right|^2} = \frac{1}{\displaystyle\sum_{m=P+1}^{M} \left|Q_m(e^{j2\pi f})\right|^2} \tag{9}$$

which is called as the MUSIC pseudospectrum. The $P$ complex exponentials' frequency estimates are then considered as the pseudospectrum's $P$ peaks. The name *pseudospectrum* is utilized due to which the amount in (10) has no data regarding the complex exponentials powers or the surrounding level of noise. Note that for $M = P + 1$, the MUSIC method is equivalent to Pisarenko harmonic decomposition [4]. The implicit assumption in the MUSIC pseudospectrum is that as the noise is white Gaussian, the eigenvalues of noise have the same power $\lambda_m = \sigma_w^2$. In general, the eigenvalues of noise will not be same, if an actual correlation matrix is replaced by an estimate. When the estimation of correlation matrix is done from a tiny amount of information samples, the differences will be more noticeable. Thus, the *eigenvector method,* a slight variation on the MUSIC algorithm [6], was proposed to account for the possibly different eigenvalues of noise. In this technique, the pseudospectrum is

$$\overline{R}_{ev}\left(e^{j\omega}\right) = \frac{1}{\displaystyle\sum_{m=P+1}^{M} \frac{1}{\lambda_m}\left|V^H(f)q_m\right|^2} = \frac{1}{\displaystyle\sum_{k=P+1}^{M} \frac{1}{\lambda_m}\left|Q_m(e^{j2\Pi f})\right|^2} \tag{10}$$

Relating to the eigenvector $q_m$, $\lambda_m$ is an eigenvalue. By the corresponding eigenvalue, the pseudospectrum of all eigenvectors is normalized [6]. The eigenvector and MUSIC techniques are similar when it comes to the case of eigenvalues $(\lambda_m = \sigma_\omega^2)$ of equal noises for $P + 1 \leq m \leq M$. The peaks in the MUSIC pseudospectrum correspond to the frequencies at which the denominator in (10)

$\sum_{m=P+1}^{M} \left| Q_m\left(e^{j2\pi f}\right) \right|^2$ approaches zero. Hence, by considering the denominator's z-transform,

$$\overline{P}_{music}(z) = \sum_{m=P+1}^{M} Q_m(z) Q_m^* \left(\frac{1}{z^*}\right) \tag{11}$$

That denominator is nothing but the pseudospectrum's sum of the z-transforms because of every eigenvector of noise. $M - 1$ pairs of roots are present in this $(2M - 1)$th-order polynomial with one in and one out of the unit circle. The complex exponentials corresponding roots should be present on the unit circle as they are assumed to be not damped. The unit circle's $P$ closest roots will relate to the complex exponentials by finding the $M - 1$ roots of 12. Finally, these frequency estimates are nothing but the phases of these roots. This technique of polynomial rooting relating to pseudospectrum of MUSIC is called as Root-MUSIC [7]. Consider that in several scenarios, calculating a pseudospectrum at a very good resolution of frequency which probably involves a very big FFT is not so efficient when compared to a rooting method [8].

## 5  Simulation and Results

As an initial step, a synthetic signal is analyzed at first, Root-MUSIC algorithm is applied, and the power spectral density of the synthetic signal is produced; this synthetic signal is generated by using a mathematical function; this mathematical function simply generates a signal which has the characteristics of an earthquake signal. Two types of synthetic signals are taken here, one is a synthetic signal with noise and the another one is without noise. Later, a raw seismic signal is taken and detrended. Detrending is a technique which involves the removal of some noisy characteristics present in a statistical data. The data utilized for the observation is acquired from Book_Seismic_Data.mat of east Texas landline which is the file name. We have taken the source as a dynamite blast which took place at a depth of around 100 ft; one trace has 1501 samples of 0.002 s sampling interval. The normalized frequencies are $0.1\pi$ and $0.3\pi$, respectively. In Fig. 1, we can observe that two types of synthetic signals are shown, where one has noise and the another does not have noise. In Fig. 2, we can see the power spectral density obtained for the respective synthetic signal, which has two normalized frequencies at $0.1\pi$ and $0.3\pi$. Root-MUSIC algorithm is applied on detrended seismic signal and the PSD obtained is shown in Fig. 5. The max peak is at 0.958 normalized frequency (Figs. 3 and 4)

**Fig. 1** Synthetic signal

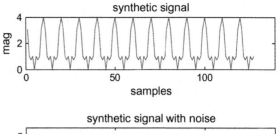

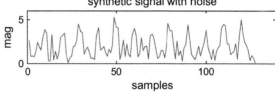

**Fig. 2** PSD of the synthetic signal

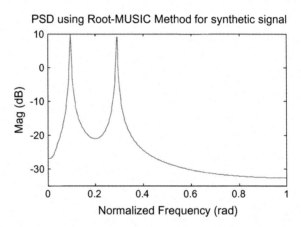

**Fig. 3** Raw seismic signal

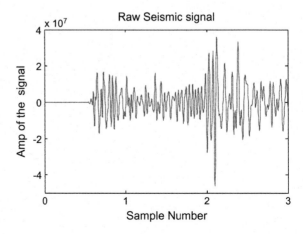

**Fig. 4** Detrended raw
seismic signal

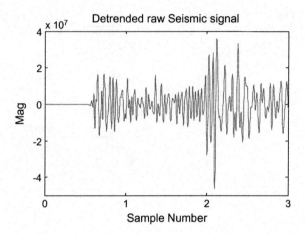

**Fig. 5** Frequency spectrum
of FIR bandpass filter

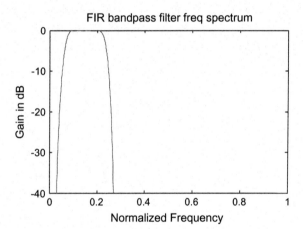

$$w = \frac{2\pi f}{f_s} = 0.0958\pi$$

$$= \frac{2\pi f}{f_s} = 0.0958\pi$$

$$f = \frac{500}{2} * 0.0958$$

$$= 23.950 \, \text{Hz}$$

In the reference book, it is written that the data is bandpass filtered in the range [15, 60 Hz]. For ensuring purpose, a BP filter with FIR order 8 is realized.

**Fig. 6** FIR bandpass-filtered signal

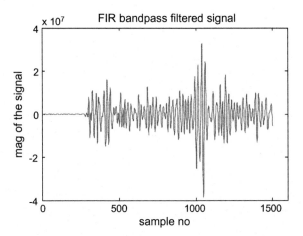

**Fig. 7** PSD using Root-MUSIC

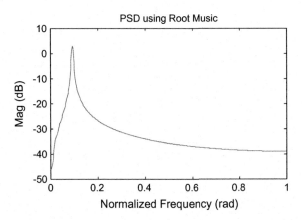

The transfer function of the same is shown in Fig. 6. The detrended seismic signal is convolved with FIR BPF and the output is shown in Fig. 6. The same PSD, as shown in Fig. 7, is obtained. So, the seismic signal tonal is 23.95 Hz. Another insignificant tonal is

$$w_1 \frac{2\Pi f}{f_s} = 0.1\pi$$

$$f = \frac{0.1\pi * 500}{2\pi}$$

$$\text{i.e.} f = 250 * 0.1 = 25 \, \text{Hz}$$

# 6  Conclusion

In this paper, a time series data of a seismic signal has been taken and processed by applying Root-MUSIC algorithm; the actual signal strengths in the seismic signal from time series data taken are clearly estimated with peaks of high resolution; power spectral density for both synthetic and raw seismic signals is obtained. The step-by-step process of the seismic signal analysis has been perfectly presented in the results section. By using this algorithm, computational efforts have been greatly reduced.

# References

1. Wail A. Mousa, Abdullatif A. Al-Shuhail, "Processing of Seisemic Reflection data using Matlab", Synthesis Lectures on Signal Processing., Morgon & Claypool Publishers.
2. Monson H. Hayes, "Statiscal Digital Signal Processing and Modeling", John Wiley & Sons, Inc, 1996.
3. G. Manolakis and Vinay k. Ingle "Statistical and Adaptive Signal Processing" McGraw-Hill, 2000.
4. Michael J Grimble and Emeritus Michael A. Johnson, "Power Spectral Density Analysis", Advanced Textbooks in Control and Signal Processing, 2005.
5. Andy Vesa, "Direction of Arrival Estimation Using MUSIC and Root-MUSIC Algorithm", 18th Telecommunications Forum TELFOR 2010, Serbia, Belgrade, November 23–25, 2010.
6. Yanping Liao and Aya Abouzaid, "Resolution Improvement for MUSIC and ROOT MUSIC Algorithms" *Journal of Information Hiding and Multimedia Signal Processing,* Volume 6, Number 2, March 2015, pp 189–197.
7. N Purnachandra Rao, "Earthquakes", Editor of publications-Amaravati popular science series, 2016.
8. Petre Stoica and Randolph Moses, "Spectral Analysis of Signals", Prentice Hall, Inc, 2005.

# Low-Power and Area-Efficient FIR Filter Implementation Using CSLA with BEC

**M. Sumalatha, P. V. Naganjaneyulu and K. Satya Prasad**

**Abstract** Carry Select Adder (CSLA) is the best and effective adder utilized in digital signal processing to implement high-speed arithmetic applications. CSLA adder will solve fast arithmetic functions in multiple data processing methods. CSLA method is mainly used to diminish the power and area instead of using normal adder. This adder is influenced by many system structures to avoid the carry delay. The main intention of this paper is to use Binary to Excess-1 Converter (BEC) instead of Ripple Carry Adder (RCA) with Cin = 1 in the normal CSLA to get high-speed operations, small area, and low power utilization. Here, binary excess converter will become the number of minor logic gates when compared to n bit Full Adder (FA) structure. According to this deliberation, the delay of time also will be reduced. In this paper, the proposed BEC method will give the significant results with regard to reducing power and area. The CMOS process technology is implemented on 0.18 m custom design and layout.

**Keywords** CSLA · BEC · RCA

## 1 Introduction

In VLSI design process, the performance of the system, efficient area, and low power are considerable in recent research, which are used in many applications like robots, embedded systems, communication systems—Software-defined radios and biomedical instrumentation [1]. In digital signal processing, a multiplier and a adder

M. Sumalatha (✉) · K. Satya Prasad
Department of ECE, JNTUK, Kakinada, India
e-mail: suma.sekhar4@gmail.com

K. Satya Prasad
e-mail: prasad_kodati@yahoo.co.in

P. V. Naganjaneyulu
MVR College of Engineering & Technology, Paritala, Vijayawada, India
e-mail: pvnaganjaneyulu@gmail.com

© Springer Nature Singapore Pte Ltd. 2018
J. Anguera et al. (eds.), *Microelectronics, Electromagnetics and Telecommunications*, Lecture Notes in Electrical Engineering 471,
https://doi.org/10.1007/978-981-10-7329-8_14

play vital role in many applications. Hence, multipliers are moderately multifaceted circuits, which are usually operated at a high system clock rate. Multiplications are highly expensive and the overall operation will be slowed. Multiplication process requires single two-input adder to perform the arithmetic operation, which is having an arithmetic unit. Here, the sum of each one-bit position is added and this carry is transferred to the subsequently place. Finally, the transferred carry diminishes the adder speed.

A perfect adder design really increases the performance of a multifaceted digital signal processing system. A Ripple Carry Adder (RCA) will be used for simple design and Carry Propagation Delay (CPD) is the major concern in this adder. Carry look-ahead and Carry Select (CS) mechanisms are recommended to diminish the CPD of adders. The CSLA will be used to mitigate this problem [2]. CSA is one of the best adders having small area and low power utilization. It generates fractional sum and carry by conceding for carry input $Cin = 0$ and $Cin = 1$, and then the multiplexers will choose the last sum and carry. The main drawback of CSLA is larger area. It will be rectified by modified CSLA. The proposed structure of 16-bit regular SQRT (Square Root) CSLA has five distinct groups size of RCA with $Cin = 1$. Each one group contains Binary to Excess-1 Converter (BEC) and multiplexer [3]. The area evolutions have done by counting the total number of AOI gates. Here, the delay will be obtained by adding more number of gates in the highest lane of logic block.

## 2  Basic Structure of BEC Logic

A modified CSLA will use BEC. In this circuit, add1 is applied to the input numbers and the four-bit binary excess converter is shown in Fig. 1. The main theme of the proposed method is to get better addition speed with the minimum number of logic gates when compared to the Full Adder (FA) structure. The main significance of the BEC logic reduces the huge silicon area with large number of bits which are used in CSLA design [4–10].

BEC contains four inputs and the output is achieved by adding "1" with each of it. The prime intention of this exertion is to use binary excess converter instead of the RCA with $Cin = 1$ in order to diminish the area and power utility of the normal CSLA. To restore the n-bit RCA, an n+1 bit binary excess converter is needed. Figure 2 shows how the basic function of the CSLA is obtained by using 4-bit BEC collectively with the multiplexer. The four-bit input (B3, B2, B1, and B0) is applied to the input of 8:4 multiplexer and an extra input of the multiplexer is the BEC output.

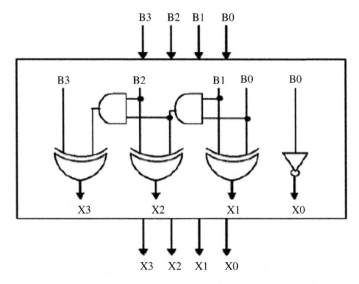

**Fig. 1** BEC-1 convertor with four-bit

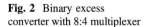

**Fig. 2** Binary excess converter with 8:4 multiplexer

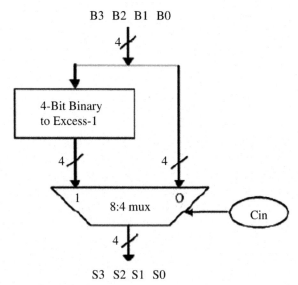

# 3   Area Estimation of CSLA Using BEC

The overall architecture of FIR filter is shown in Fig. 3, which consists of the coefficient ROM, data RAM, input data reader, clock generator, and a filter. The clock signal can be generated from the clock generator. The coefficient value can be stored in the coefficient ROM. The input data can be moved to the data RAM for

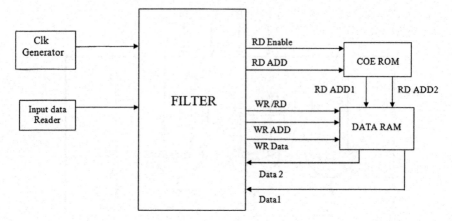

**Fig. 3** Proposed FIR filter architecture

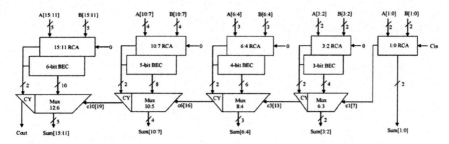

**Fig. 4** Modified 16-bit CSLA

storing purposes. To perform the filtering operation, we need coefficient ROM and data RAM.

Here, the data unit can arrange the input data, and the following operation can be done in the above architecture. In the first stage, calculate the memory address and enable data RAM to store the input data. The second stage enables the coefficient ROM to perform the read coefficient one by one to get the filter results. Finally, the data RAM is enabled to get the filter coefficients. In that FIR filter, only digital adder can be caused to get the result. It occupies less area and power. The proposed structure uses 16-bit square root CSA with binary excess converter for RCA with Cin = 1 to reduce the area and power which is shown in Fig. 4.

In the proposed system, the two-bit RCA has one FA and one-half adder for Cin = 1, where three-bit binary excess converter is utilized which enhances one to the output from two-bit RCA. According to this deliberation, the time delay has been reduced. The output of the MUX is depending on the input of the MUX and BEC. The input arrival time is lesser than the multiplexer selection input arrival time. By selecting the BEC output or the straight inputs, there are two possibilities such as parallel and multiplexer rendering to the regulate signal Cin. While designing CSLA, the area will be reduced based on the logic gates of RCA with BEC and multiplexer selection output, which leads to get less power consumption.

## 4  Results and Discussion

Table 1 and Fig. 5 show the comparison of the existing as well as proposed method by considering the parameters such as area, power, and delay. The performance of CSLA with BEC is implemented by using 0.18 m CMOS technology. In this table, we can observe that all the parameters are reduced when compared with the existing methods.

**Table 1** Comparison of area, power, and delay for existing and proposed method for 180 nm CMOS technology

| Design | Area (square meters) | Power (nw) | Delay (ps) |
|---|---|---|---|
| CSLA with BEC | 159.040 | 5843.963 | 2188 |
| SQRT | 195.200 | 6783.889 | 2188 |
| CSLA RCA | 151.012 | 6883.889 | 2195 |

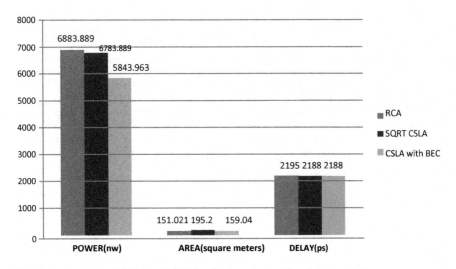

**Fig. 5** Performance of CSLA with BEC and comparison with existing methods

# 5 Conclusion

In this paper, we have proposed new FIR filter architecture with the help of a clock generator, coefficient ROM, data RAM, and CSA with binary excess-1 converter to diminish the area, power, and also to improve the speed and performance of the VLSI system. The main approach is to reduce the area and power by using proposed modified CSLA architecture with less number of gates which makes it efficient VLSI implementation. Finally, the major parameter of VLSI implementation like area, power, and delay is reduced in the proposed system with the help of CSLA adder with BEC.

**Acknowledgements** The author likes to express sincere thanks to the management and principal of Sai Spurthi Institute of Technology, B. Gangaram for providing the necessary infrastructure.

# References

1. K. K. Parhi, *VLSI Digital Signal Processing*. New York, NY, USA:Wiley,1998.
2. T. Y. Ceiang and M. J. Hsiao, —Carry-select adder using single ripple carry adder,|| Electron. Lett., vol. 37, no. 10, pp. 614–615, May 2001.
3. Y. He, C. H. Chang, and J. Gu, "An area efficient 64-bit square root carry-select adder for lowpower applications," in Proc. IEEE Int. Symp. Circuits Syst., 2005, vol. 4, pp. 4082–4085.
4. Y. He, C. H. Chang, and J. Gu, "An area efficient 64-bit square root carry-select adder for low power applications," in Proc. IEEE Int. Symp. Circuits Syst., 2005, vol. 4, pp. 4082–4085.
5. Kuldeep Rawat, Tarek Darwish and magdy Bayoumi||A low power and reduced area carry select adder||.
6. J. M. Rabaey, Digital Integrated Circuits—A Design Perspective. Upper Saddle River, NJ: Prentice-Hall, 2001.
7. Y. Kim and L.-S. Kim, "64-bit carry-select adder with reduced area," Electron. Lett., vol. 37, no. 10, pp. 614–615, May 2001.
8. Youngjoon Kim and Lee-Sup Kim, "A low power carry select adder with reduced area", IEEE International Symposium on Circuits and Systems, vol. 4, pp. 218–221, May 2001.
9. B. Ramkumar and H.M. Kittur, "Low-power and area-efficient carry-select adder," IEEE Trans. Very Large Scale Integer. (VLSI) Syst., vol. 20, no. 2, pp. 371–375, Feb. 2012.
10. S. Manju and V. Sornagopal, "An efficient SQRT architecture of carry select adder design by common Boolean logic," in Proc. VLSI ICEVENT, 2013, pp. 1–5.

# Design and FPGA Implementation of TPFT-Based Channelization for SDR Applications

P. Sri Lekha and K. Pushpa

**Abstract** A particular channel selection can be done in wideband communication receivers that are performed by using channelization. Generally, channelization performs down conversion of signal to baseband and filtering of channel, since the features of channelization generally affected by means of software, which is much advantageous to implement channelization performance as much as possible with digital signal processing. Number of channels must be received simultaneously in base stations. Thus, each channel must contain independent channelizer, which can be achieved by Tunable Pipelined Frequency Transform (TPFT)-based channelization. So this paper deals with the design as well as the implementation of the TPFT-based channelization for any of the applications that are of software-defined radio. The coding is completed with a Verilog Hardware Description (HDL) and its simulation is completed in a Xilinx ISE 14.5 environment.

**Keywords** Cognitive radio · Coarse channelization · Interleaver
Processing element (PE) · Software-defined radio (SDR) · Tunable pipeline
frequency transform (TPFT)

## 1 Introduction

Cognitive Radio (CR) is considered as a technology which is emerging and used for a Dynamic Spectrum Access (DSA) system. Joseph Mitola [1] was the first to introduce the cognitive radio concept. A cognitive cycle in the CR contains four resource management functions, which are the spectrum sensing, the spectrum deciding, the spectrum share, and the mobility of the spectrum. Any Cognitive

P. Sri Lekha (✉) · K. Pushpa
Department of ECE, Shri Vishnu Engineering College for Women,
Vishnupur, Bhimavaram 534202, Andhra Pradesh, India
e-mail: ece.srilekha@gmail.com

K. Pushpa
e-mail: pushpak@svecw.edu.in

© Springer Nature Singapore Pte Ltd. 2018       143
J. Anguera et al. (eds.), *Microelectronics, Electromagnetics
and Telecommunications*, Lecture Notes in Electrical Engineering 471,
https://doi.org/10.1007/978-981-10-7329-8_15

Radio Network (CRN) is one containing the cognitive radios or Secondary Users (SUs). Owing to continuous changes in the available spectrum and different Quality of Services (QoS), the needs of different applications, the CRNs, impact some unique challenges. All the SUs in the CRNs adapt best methods dynamically to transmit like the transmission channel, the rate, and the transmit power. This leads to an enhancement in the parameters of the QoS. On seeing an increase in the users of multimedia, the CR may prove to be an optimum solution for the transmissions of multimedia. The features of spectral efficiency of the systems of multicarrier may be utilized properly for addressing the transmissions of multimedia over the CRN.

Software-defined radio (SDR) is that wireless communication system is re-configured by means of software reprogramming to be operated on various frequencies and protocols. It is normally executed by the SDR platform. This idea was first used by Mitola [1] stating "The point where the wireless personal digital assistants and the related networks are sufficiently computationally intelligent about radio resources and linked computer to computer communications to identify the needs of communicating user as a function of use context, and providing radio resources and wireless services much related to those needs."

## 1.1 Software-Defined Radio (SDR)

The Software-Defined Radio (SDR) elevates the scope of digital reconfigurability. It is particularly known to encompass the baseband digital processing and its functions like demodulation and modulation, coding, time recovery, equalization, and channelization, but the RF functions are usually completed by analog front-end components [2].

Generally, a classic approach for building the systems of SDR is divided into three phases: the analog front-end, the digital front-end, and the digital back-end, as shown in Fig. 1. The analog front-end interfaces the antenna and the DSP hardware. This is a multi-standard front-end that allows receiving along with transmitting the bandwidths and arbitrary frequencies. At the following stage, digital front-end groups and the Analog-to-Digital Converter (ADC) along with the channelizer that

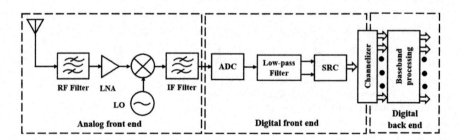

**Fig. 1** Practical SDR receiver implementation [2]

is before an independent channel and its operation done using digital back-end. These works are also done in various platforms.

## 1.2  Channelization

The word channelization originally refers to the technology of sharing communication median with different users, so that many conversations can be submitted simultaneously on a signal band, with each conversation being on a separate channel. By this means, each channel is used to transfer different information simultaneously. The commonly known multiplexing methods including TDMA (Time-Division Multiplexing), FDMA (Frequency-Division Multiplexing), and CDMA (Code-Division Multiplexing) belong to the channelization concepts. In recent years, especially after the development of the UWB communication, the word channelization refers specifically to the frequency channel separation of a wide band. Signals are transferred to different channels of a predefined frequency band. In the UWB communication systems, the entire UWB frequency range is split into several bands, and each band has many channels.

A major factor in SDR systems is a real-time configurable digital channelizing that is needed to receive and resample the radio signal promptly, and is also important to keep track of the entire spectrum in order that the inactive bands are identified. Compatibility in various standards of communication desires the channelization to be reconfigurable dynamically. Designing the resource efficiently is one more basic need to implement the channelization. Obtaining a good solution to balance targets, like performance and resource, is tedious in case of digital channelizing. Many attributes like rate of sampling, bandwidth, resolution in frequency, and dynamic configuring affect the designing and implementation of the channelization. Hence, determining the cost/performance trade-offs forms a significant problem in digital channelizing [3].

## 1.3  Overview of Existing Channelization Techniques

Digital channelization is achieved by making use of any of the techniques given below: (1) the multichannel Digital Down Converter (DDC); (2) the Fast Fourier Transform (FFT); (3) the Polyphase Discrete Fourier Transform (DFT) filter bank; (4) the Goertzel Filter bank; and (5) the Tree-structured filter bank.

The DDC normally includes a Numerically Controlled Oscillator (NCO) and also a Sampling Rate Converting filter (SRC). These are made use normally for channelization of single channel but do not suit multichannel channelization as it has low utilization of resources. The FFT-depended channelizing consists of a simple built and a high resources usage, and has a filtering performance that so bad and can be enhanced by a technique known as window functioning [4, 5]. The polyphase DFT filter bank contains high usage of resources and also a better filtering performance, but proves a uniform division of channels [6]. A Goertzel filter bank, however, solves a problem in fixing center frequency that is related to a polyphase DFT and its filter bank, but will not be able to extract un-uniform bandwidth channels [6]. The tree-structured filter bank approach by the RFEL Ltd. has brought about a method for channelization known as the Pipelined Frequency Transform (PFT). The main problem in the power-of-two channel stacking is solved by means of Tunable Pipelined Frequency Transform (TPFT) technique [7]. This can locate accurately the radio signals and ensure proper reception.

The paper deals with both the design and the implementation of the TPFT-based channelization for various SDR applications. Section 1 deals with the introduction of software, digital channelization, and also an overview of the techniques of channelization. Section 2 talks about the features and the architecture of the Tuneable Pipelined Frequency Transform (TPFT) that is based on channelization. Section 3 deals with results of simulation with the TPFT. Conclusions drawn are presented in Sect. 4.

## 2  Tuneable Pipelined Frequency Transform (TPFT)-Based Channelization

In simple terms, the PFT produces frequency bins that are equally spaced. For overcoming this limitation, a form that is derived and called "Tuneable PFT" (TPFT) is made use which permits independent tuning of middle frequencies of all bins and independent filters for every bin. As there are different frequency resolutions, the result is flexible like DDC approach with better efficiency which is needed for large channel numbers.

We notice a significant improvement of making usage of the PFT cascade structure in which in-between outputs are present. By modifying the PFT architecture and for extracting the frequency bands and further giving them any frequency, this is completed and this tunability is got by two stages: first, signals are tuned coarsely inside the PFT stages, and fine-tuned by using a complex converter with a local oscillator (LO) and a numerical control oscillator (NCO) that is driven using a routing engine as shown in Fig. 2. The tuning range for each stage is decreased by any two factors and the DDC needs to be finely tuned for the entire bandwidth of the input. Overall, this structure is very ideal for replacing multiple DDCs in applications like multi-standard base stations, intelligent antenna systems, and the satellite communications.

This TPFT-based coarse channelizing structure contains two parallel TPFT structures along with two parallel inter-leavers as shown in Fig. 3. For every future stage of this, the output signal will be decimated by two. The factor of maximum decimation of a TPFT block will be 512, so every block will be composed of nine processing elements (PE) that are cascaded and used for processing of one complex

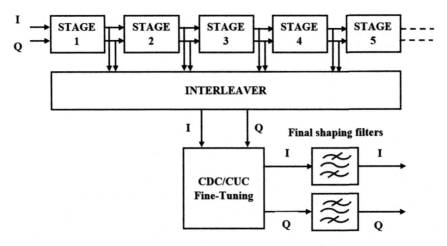

**Fig. 2** Schematic of tuneable PFT architecture

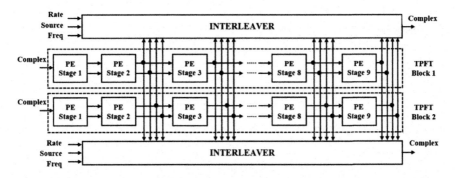

**Fig. 3** Structure of the TPFT-based channelization block [3]

input data. This TPFT channelization block will be scalable for the input signal number and is easily adaptable for applications.

## 3 Simulation Results

The TPFT-based channelization structure shown in Fig. 3 was designed by using TPFT-based channelization block as shown in Fig. 3 that had been designed by making use of Verilog HDL. The synthesis and simulation results were produced making use of Xilinx ISE 14.5 for the Spartan 3E family device. In the results of simulation, the technology view designates a top block that indicates an input and output set. The Register Transfer Logic (RTL) view depicts internal architecture

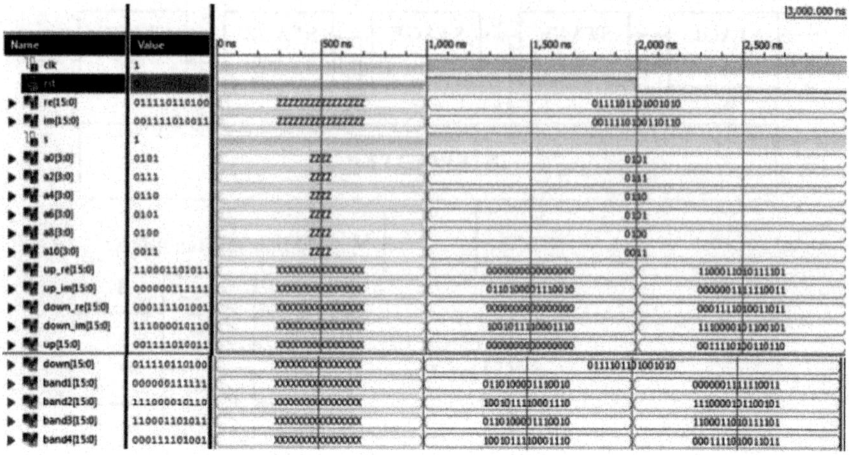

**Fig. 4** Simulation result for Processing Element 1

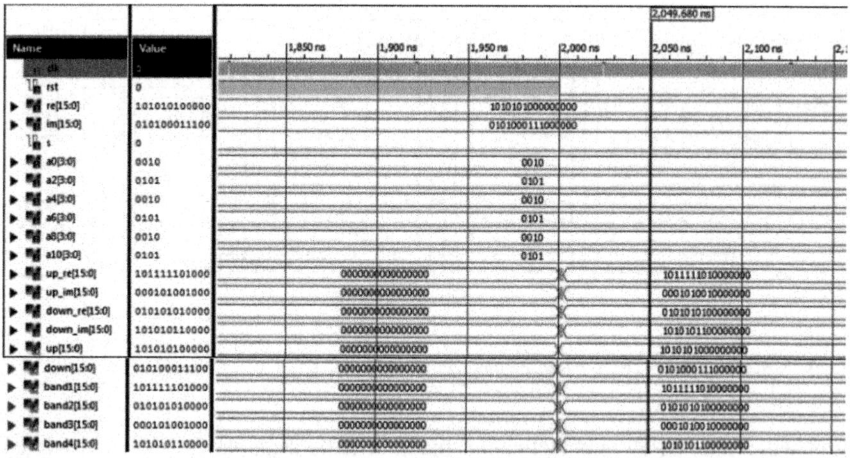

**Fig. 5** Simulation result for Processing Element 2

blocks along with the links that are in-between the output and the input terminals. Simulation results are produced by using a writing test bench program for each design. This program contains an input test vectors' set used in designing. The simulation outcomes of the Processing Element 1(PE1), Processing Element 2 (PE2), and TPFT-based coarse channelization structure are shown in Figs. 4, 5, and 6. The complex inputs for coarse channelization structure are the rate, the source, and the frequency. Figures 7, 8, and 9 depict the RTL view of Processing Element 1 (PE1), Processing Element 2 (PE2), and TPFT-based coarse channelization block.

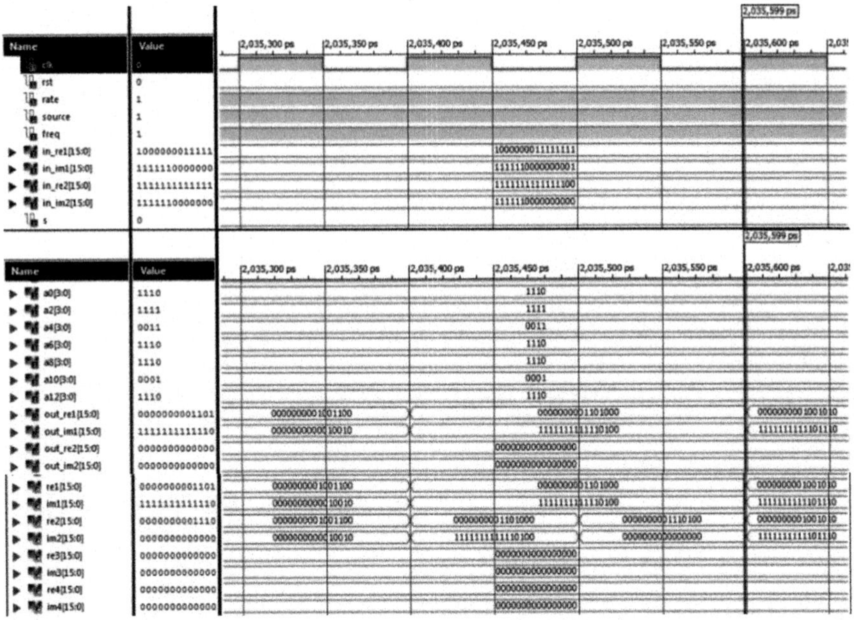

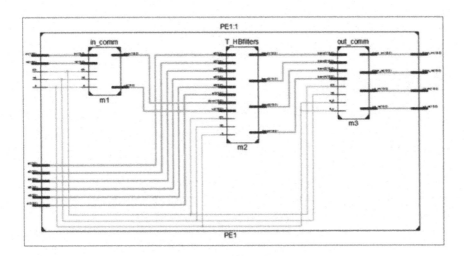

**Fig. 6** Simulation result for TPFT coarse channelization block

**Fig. 7** RTL schematic of Processing Element 1

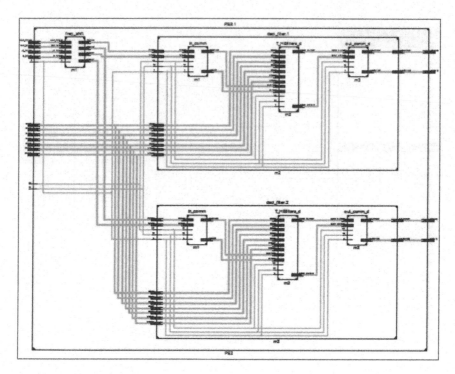

**Fig. 8** RTL schematic of Processing Element 2

Synthesis report (device utilization summary) for the TPFT-based coarse channelization block is shown in Fig. 10 and a time delay of 114.968 ns is observed for
the implemented architecture.

## 4   Conclusions

In this paper, the area and speed efficient channelization that is TPFT-based has
been proposed for the application of the SDR. The design has been tested and also
verified by a Verilog HDL, and coding and simulation has been completed by
Xilinx ISE 14.5. This gives freedom to the user in specifying channels using
bandwidth and frequency. Further, there is also the possibility of direct application
of masks that shape spectrum to the outputs inside the very architecture. It further
provides a highly efficient and elegant method that can channelize wideband
signals.

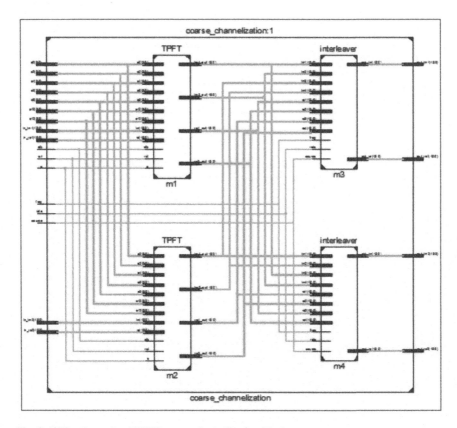

**Fig. 9** RTL schematic of TPFT coarse channelization block

| Device Utilization Summary (estimated values) | | | | [-] |
|---|---|---|---|---|
| Logic Utilization | Used | Available | Utilization | |
| Number of Slices | 22783 | 4656 | | 489% |
| Number of Slice Flip Flops | 19739 | 9312 | | 211% |
| Number of 4 input LUTs | 42359 | 9312 | | 454% |
| Number of bonded IOBs | 162 | 232 | | 69% |
| Number of MULT18X18SIOs | 12 | 20 | | 60% |
| Number of GCLKs | 1 | 24 | | 4% |

**Fig. 10** Synthesis report for TPFT coarse channelization block

# References

1. J. Mitola.: Cognitive Radio: An Integrated Agent Architecture for Software Defined Radio. PhD Thesis, Royal Institute of Technology, KTH, (2000).
2. Alvaro Palomo Navarro.: Channelization for multi standard software-defined radio base stations. PhD Thesis, National University of Ireland, (2011).

3. Xue Liu, Ze-Ke Wang, Qing-Xu Deng.: Design and FPGA implementation of a reconfigurable 1024 channel channelization architecture for SDR application. IEEE Transactions on Very Large Scale Integration (VLSI) Systems. 24(7) (2016) 2449–2461.
4. G. Lopez-Risueno, J. Grajal and A. Sanz-Osorio.: Digital Channelized receiver based on time-frequency analysis for signal interception. IEEE Trans. Aerosp. Electron. Syst. 41(3) (2005) 879–898.
5. Z. Wang, X. Liu, B. He and F. Yu.: A combined SDC-SDF architecture for normal I/O piplined radix-2 FFT. IEEE Transactions on Very Large Scale Integration (VLSI) Systems. 23(5) (2015) 973–977.
6. T. Hentschel.: Channelization for software defined base-stations. Ann. Telecommun. 57(5–6) (2002) 386–420.
7. J. Lillington.: Flexible channelization architectures for software defined radio front ends using the tuneable piplined frequency transform. in Proc. IEE Colloq. DSP Enabled Radio. (2003) 1–13.

# Frequency Estimation Using Minimum Norm Algorithm on Seismic Data

Ch. Namitha, V. Uma Mahesh, M. Anusha, S. Koteswara Rao
and T. Vaishnavi Chandra

**Abstract** Seismic signals are generally produced due to disturbances in the earth or an explosion that propagate through earth layers. The signal-to-noise ratio is very low for the generated signal. In order to increase its SNR and reduce the noise, various preprocessing techniques using FIR-based bandpass filter are considered. Minimum norm algorithm is implemented in the frequency domain to analyse the spectrum of the seismic signal in addition to analyse the tonals of the signal.

**Keywords** Seismology · Applied statistics · Stochastic signal processing
Adaptive signal processing

## 1 Introduction

### 1.1 Seismology

Seismology is the scientific study of earthquakes, tsunamis, etc. Earthquakes are caused due to the sudden transference of the tectonic plates which results in devastation of mankind. The signals are generated using a source and a seismogram is used to predict the seismic signals. As these signals are contaminated with noise,

Ch. Namitha (✉) · V. Uma Mahesh · M. Anusha · S. Koteswara Rao
T. Vaishnavi Chandra
Department of ECE, K L University, Green Fields, Guntur DT 522502, AP, India
e-mail: namithachsr@gmail.com

V. Uma Mahesh
e-mail: vannemahesh@gmail.com

M. Anusha
e-mail: anureddy.anusha@gmail.com

S. Koteswara Rao
e-mail: rao.sk9@gmail.com

T. Vaishnavi Chandra
e-mail: vaishuchandra.dalu999@gmail.com

© Springer Nature Singapore Pte Ltd. 2018
J. Anguera et al. (eds.), *Microelectronics, Electromagnetics
and Telecommunications*, Lecture Notes in Electrical Engineering 471,
https://doi.org/10.1007/978-981-10-7329-8_16

various signal processing techniques are used. Signal processing is an applied science that is used to implement and transform the information obtained. So, signal processing is adapted for this purpose [1].

Seismic waves are elastic waves and they are of various types like body waves, surface waves, and normal waves [2]. Many seismic stations are present across the world to collect the data from the seismogram for further analysis and predictions for the future. Generally, surface waves are slower than the body waves but stronger in nature as they are captured by the layers of the earth surface. At present, spectrograms are also used to record seismic signals but the data is collected through the satellites where the pictures are represented and further analysis is performed through various image processing techniques [3].

## 1.2 Seismology in Signal Processing

Seismic reflection data is collected during the destruction; various signal processing techniques are applied to analyse the geological features of the earth. The data collected is contaminated with large amount of noise of various types. Random noise is a result of lack of phase coherency among adjacent traces and the main sources of this type of noise are wind, rain, instruments used for the process, etc. Coherent noise is generated by the source of the explosion. Some examples of the coherent noise are multiple reflections of surface waves like ground roll and airwaves, etc. Generally, seismic data is composed of ground roll noise. These noise waves are of low velocity, that is, few thousands of metres and are of low frequency. They are mainly affected by the variation in the frequencies, thus creating various modes of the noise. Due to these noises, the SNR of the seismic trace is very low. Thus, various preprocessing techniques are adapted to increase the SNR of the traces. Considering various filtering and Fourier transform methods, the SNR is improved in time–frequency domain [4, 5]. Seismic signal resolution also decreases and various preprocessing techniques are to be included to make the signal more ideal. For this, several convolution and post-stack techniques have to be involved to increase the resolution and attenuate the noise [6].

Power spectrum analysis is useful in order to examine the seismic traces obtained [7]. The different processing techniques as well as noise constraints can be applied to seismic wavelets with slight changes in the basic procedure like finding the earth's reflectivity function, etc. and the below-discussed algorithm is also applicable in the estimation of seismic wavelets [8]. For this, various algorithms of both parametric and nonparametric types are taken into account. These algorithms require time-to-frequency conversions, whereas in frequency estimation algorithms, the synthetic signal is directly considered in frequency domain. In this paper, minimum norm a frequency estimation algorithm is examined for a seismic signal.

In frequency estimation algorithms, signal is considered as a sum of complex exponentials. Hence, frequencies are estimated directly from the spectrum. In these methods, the signal is considered as a combination of signal and noise subspaces and they are computed based on eigendecomposition forms of autocorrelation matrix.

## 1.3 Minimum Norm Frequency Estimation

Eigenvector is a vector which is a nonzero whose value does not change on linear transformation. Eigenvalue is the root of the eigenvector obtained and in its determination characteristic polynomial of the matrix is very essential. These parameters are of vital importance in finding roots of a matrix as well as many parameters like whether the matrix is invertible or not and how sensitive to errors and so on. The matrix illustration is possible with the eigenvalues and vectors, and this is known as eigenvalue decomposition.

## 1.4 Autocorrelation Matrix

An autocorrelation matrix is a significant characteristic of a statistical second-order discrete-time random process. Random process is generally considered as an event based on probability conditions [9]. Let us consider a matrix x to analyse the autocorrelation.

$$x = [x(0), x(1), x(2), \ldots, x(p)]^{\mathrm{T}} \tag{1}$$

There are $p + 1$ components of the vector $x(n)$ and the outer product is given as

$$xx^H = \begin{bmatrix} x(0)x^*(0) & x(0)x^*(1) & x(0)x^*(2)\ldots\ldots & x(0)x^*(p) \\ x(1)x^*(0) & x(1)x^*(1) & x(1)x^*(2)\ldots\ldots & x(1)x^*(p) \\ \ldots & \ldots & \ldots & \ldots \\ \ldots & \ldots & \ldots & \ldots \\ x(p)x^*(0) & x(p)x^*(1) & x(p)x^*(2)\ldots\ldots & x(p)x^*(p) \end{bmatrix} \tag{2}$$

$x(n)$ matrix is of the order $(p + 1)$ and $x(n)$ is WSS [10]. Hermitian symmetry of the autocorrelation signal is used and the autocorrelation matrix $R_x$ is given as follows:

$$R_x = E\{xx^H\} = \begin{bmatrix} r_x(0) & r_x^*(0) & r_x^*(2) & \ldots & r_x^*(p) \\ r_x(1) & r_x(0) & r_x^*(1) & \ldots & r_x^*(p-1) \\ r_x(2) & r_x(1) & r_x(0) & \ldots & r_x^*(p-2) \\ . & . & . & \ldots & . \\ r_x(p) & r_x(p-1) & r_x(p-2) & \ldots & r_x(0) \end{bmatrix} \tag{3}$$

Slight modification in the $R_x$ matrix leads to the autocovariance matrix. The $R_x$ matrix has various properties like the eigenvalues are real values and nonnegative and the values in the two diagonals are equal. The most important characteristic of this matrix is that it is a Hermitian matrix for wide-sense stationary random process. Random process is modelling from the signal space onto discrete-time signals.

There are many types of frequency estimation algorithms depending on eigen-decomposition methods like eigendecomposition autocorrelation matrix method in which autocorrelated matrix should be known exactly, Pisarenko harmonic decomposition in which minimum value of the autocorrelation matrix is to be taken into account, and MUSIC [11] is an improvement of the Pisarenko method which uses only one noise eigenvector to determine the peaks at different frequencies and so on.

Section 2 deals with mathematical modelling related to minimum norm algorithm, Sect. 3 deals with methodology and the results obtained, and the paper is concluded in Sect. 4.

## 2 Mathematical Modelling

Minimum norm algorithm uses single eigenvector in the noise sub-spectrum to determine the peaks. Here, let us consider a single vector that lies in the noise subspace as **a**. The samples received from the target are considered as an n-element array.

The signal is corrupted with additive Gaussian noise with zero mean. x (n) from the receiver is a combination of p complex exponentials result of a random process consisting of Gaussian noise [12, 13]. Vector x (n) is given by

$$X(n) = \sum_{i=1}^{p} A_i e^{j(n\omega i \sin \theta i + \varphi i)} + W(n), \qquad (4)$$

where

$\omega_i$     —ith target frequency,
$\theta_i$     —ith target angle at which it is located,
$A_i$     —ith target amplitude of the exponential,
w (n)   —ith target associated Gaussian noise,
p      —Count of the targets present, and
$\phi_i$     —Random variable which is uncorrelated and uniformly distributed with zero mean in the interval [−pi, pi]

It is predicted that the magnitude $|A_i|$, frequency $\omega_i$, and angle $\theta_i$ are unknown and random in nature.

The spectrum of the "p" impulses and the power spectrum of w (n) combine to form the power spectrum of the signal x (n). This method is generally carried out in two subspaces, that is, noise and signal subspace.

First, the subspaces are predicted to find the frequencies accurately followed by the eigendecomposition of the autocorrelation matrix. Let $R_x$ be the autocorrelation matrix of N*N with eigenvalues as $\lambda_1, \lambda_2, \lambda_3, ..., \lambda_N$ and the eigenvectors as $v_1, v_2, v_3, ..., v_N$ where N > p+1 in the descending order. These are further divided into all p values having large eigenvalues as one group and the N-p eigenvectors with eigenvalues as $\sigma_w^2$ as another group.

The estimation function of frequency from which the frequencies can be estimated is

$$P_{MN}\left(e^{j\omega}\right) = \frac{1}{\left|e^H a\right|^2} \tag{5}$$

Null frequencies are present in $|e^H a|^2$ corresponding to each complex exponential term. The coefficient of **a** factored in the z-transform may be represented as

$$\sum_{k=0}^{p} a(k)z^{-k} = \prod_{k=1}^{p} \left(1 - e^{j\omega k}z^{-1}\right) \prod_{k=p+1}^{M-1} \left(1 - z_k z^{-1}\right) \tag{6}$$

In the above equation, $z_k$ are the roots that do not exist on unit circle where k = p + 1,..., M − 1. The main problem lies in the estimation of the exact noise vector that is present in the noise subspace along with reducing the effect of spurious zeroes on the peaks of the frequency estimation function.

The vector "a" in this algorithm should satisfy these three conditions:

1. "a" should be present in the noise subspace.
2. It should have minimum norm.
3. "a" should have the first element as unity.

The first condition implies that the p roots are present in the unit circle of a (z). The second condition makes that the spurious roots exist inside the unit circle of a (z) that is $|z_k|$ < 1. The last condition shows that the key of minimum norm is not a zero vector.

The vector "a" that lies in the noise subspace is expressed as the following to reduce the minimisation problem:

$$a = P_n v \tag{7}$$

Here, the projection matrix $P_n$ projects onto the vector **v** which is arbitrary expressed as $P_n = V_n V_n^H$. Thus, the last condition is written as

$$a^H u_1 = 1 \tag{8}$$

$u_1 = [1000...0]^T$ in the last expression. The above condition plus the condition $a = P_n v$ is given as

$$v^H\left(P_n^H u_1\right) = 1 \tag{9}$$

From the equation $a = P_n V$, the vector **a** in its norm form is

$$\|a\|^2 = \|P_n v\|^2 = v^H\left(P_n^H P_n\right)v \tag{10}$$

The Hermitian matrix of $P_n$ is $P_n = P_n^H$, and the idempotent is $P_n^2 = P_n$, which leads to the expression

$$\|a\|^2 = v^H P_n v \tag{11}$$

Estimating the vector **v** that reduces the quadratic form of $v^H P_n v$ is similar to defining the minimum norm of the vector "a". Using the optimization theory, the condition of minimization problem is resolved and obtained as

$$\min v^H P_n v \text{ subject to } v^H\left(P_n^H u_1\right) = 1 \tag{12}$$

As the solution to the above expression is obtained, then by extending the vector **v** onto the noise subspace, the result of minimum norm is found.

By the optimisation theory [14], the reduced form of Eq. (12) can be

$$V = \lambda P_n^{-1}\left(P_n^H u_1\right) = \lambda u_1 \tag{13}$$

The eigenvalue $\lambda$ is given as

$$\lambda = \frac{1}{u_1^H P_n u_1} \tag{14}$$

The final expression of minimum norm is

$$a = P_n v = \lambda P_n u_1 = \frac{P_n u_1}{u_1^H P_n u_1} \tag{15}$$

This is nothing but the extension of the unit vector onto the noise subspace and the first coefficient is made equal to one by normalisation. From the eigenvectors of the autocorrelation matrix, the solution to the minimum norm is

$$a = \frac{\left(V_n V_n^H\right)u_1}{u_1^H\left(V_n V_n^H\right)u_1} \tag{16}$$

## 3   Simulation and Results

Step 1: From the file Book_Seismic_ Data.mat [15], the data used for analysis is retrieved which is a landline from east Texas. At 80–100-feet-deep holes, the dynamite source is placed. The seismic trace considered consists of 1501 samples and sampling interval of 0.002 s.

Step 2: To estimate the tonals of the seismic signal, initially the minimum norm algorithm is applied to the synthetic signal to assess the algorithm performance.

Step 3: Frequencies of the synthetic signal are assumed as 0.2 π, 0.3 π, 0.8 π and 1.2 π and are represented in the form of complex exponentials as shown in Fig. 1.

Step 4: The input synthetic signal is buried with noise; Monte Carlo algorithm is applied for 10 times and obtained average of the signal using minimum norm method.

Step 5: The assumed frequencies are obtained in Fig. 2 after the application of the algorithm. Thus, proving that the algorithm is correct.

Step 6: The data regarding the raw seismic signal is shown in Fig. 3 which is obtained from [15] consists of one shot loaded.

Step 7: The seismic signal is determined as shown in Fig. 4 in order to remove bias.

Step 8: The minimum norm algorithm is applied to the seismic signal and the normalised frequency is obtained as 20.75 Hz as shown in Fig. 5.

$$w = \frac{2\pi f}{fs} = 0.083\,\pi$$

$$= \frac{2\pi f}{500}$$

$$= \frac{2\pi}{fs}f = 0.083\,\pi$$

$$Total\ frequency f = \frac{500}{2} * 0.083 = 250 * 0.083 = 20.75\,Hz$$

**Fig. 1** Frequency estimation using minimum norm algorithm

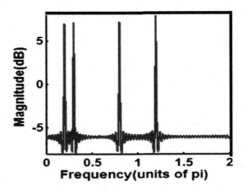

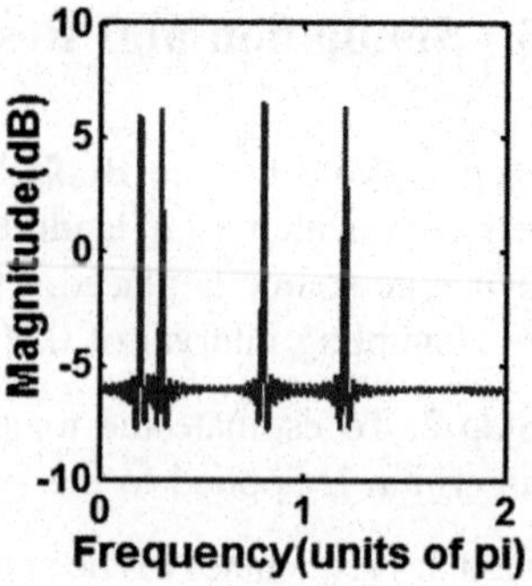

**Fig. 2** Average of frequency
estimation using minimum
norm algorithm

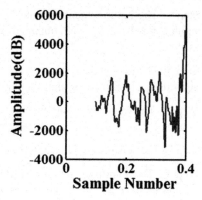

**Fig. 3** Raw signal

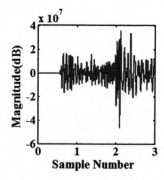

**Fig. 4** Detrended raw
seismic signal

Step 9: Earthquake signals are generally in the frequency range of 15–60 Hz. So, a bandpass filter in the same frequency range is considered [15].

Step 10: A bandpass filter with FIR [16] of order 8 is realised and used for smoothening the signal. The filter frequency spectrum is shown in Fig. 6.

Step 11: The FIR bandpass-filtered signal of seismic data is shown in Fig. 7, and normalised frequency versus magnitude in decibels is represented in Fig. 8.

**Fig. 5** Frequency estimation
using minimum norm method
for raw seismic signal

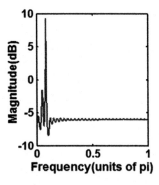

**Fig. 6** FIR bandpass-filtered
spectrum

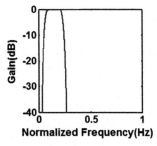

**Fig. 7** FIR bandpass-filtered
signal

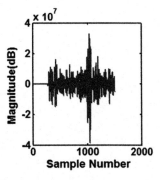

Step 12: The minimum algorithm is applied to find frequency estimation of the
seismic signal after bandpass filtering is shown in Fig. 9 and the normalised frequency can be derived using the same formula as in step 7 and obtained as
21.75 Hz.

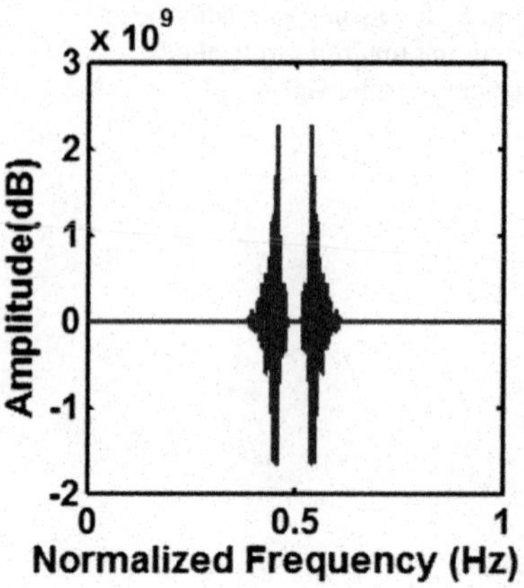

**Fig. 8** FFT spectrum seismic
after BPF in norm frequency
versus magnitude in dB
representation

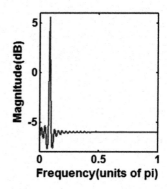

**Fig. 9** After BPF frequency
estimation using minimum
norm method

## 4    Conclusion

The harmonics of the seismic signal are estimated from the results. Accurate results
are obtained as minimum norm algorithm is considered and the spurious peaks
which lead to uncertainty in the estimation of exact harmonics are also reduced. As
the signal is considered directly in the frequency domain, the conversions are
avoided and exact harmonics are obtained from the data making the algorithm most
efficient. Thus, this algorithm can be utilised in frequency estimation of seismic
data.

## References

1. Naihao Liu, Jinghuai Gao, Xiudi Jiang, "Seismic Time–Frequency Analysis via STFT-Based
   Concentration of Frequency and Time", IEEE Geoscience and Remote Sensing Letters,
   Volume: 14, Issue: 1, IEEE, 9 December 2016, p.p. 127 – 131.

2. Dr. N. Purnachandra Rao, "Earthquakes", Amaravathi Popular Science Series.
3. Zhong-lai Huang, Jianzhong Zhang, Tie-hu Zhao, "Synchrosqueezing S-Transform and Its Application in Seismic Spectral Decomposition", IEEE Transactions on Geoscience and Remote Sensing, Volume: 54, Issue: 2, 28 August 2015, p.p. 817–825.
4. Yan Zhao, Yang Liu, Xuxuan Li, Nansen Jiang, "Time–frequency domain SNR estimation and its application in seismic data processing", Journal of Applied Geophysics, Volume 107, August 2014, p.p. 25–35.
5. Wei Liu, Siyuan Cao, Yangkang Chen. "Seismic Time–Frequency Analysis via Empirical Wavelet Transform", IEEE Geoscience and Remote Sensing Letters, Volume: 13, Issue: 1, 06 November 2015, p.p. 28–32.
6. Yangkang Chen, Zhaoyu Jin, "Simultaneously Removing Noise and Increasing Resolution of Seismic Data Using Waveform Shaping", IEEE Geoscience and Remote Sensing Letters, Volume: 13, Issue: 1, January 2016, pp. 102–104.
7. G. Manolakis and Vinay K. Ingle, "Statistical and Adaptive Signal Processing", Mc Graw-Hill, 2000.
8. Shibin Wang, Xuefeng Chen, Chaowei Tong, Zhibin Zhao, "Matching Synchrosqueezing Wavelet Transform and Application to Aero engine Vibration Monitoring", IEEE Transactions on Instrumentation and Measurement, Volume: 66, Issue: 2, IEEE, 2 December 2016, p. p. 360–3.
9. Athanasio Papoulis, "Probability, random variables and Stochastic Processes", Third Edition.
10. Dr. S. Koteswara Rao, K. Bramaramba, Dr. K. Raja Rajeswari, "Performance evaluation of noise subspace methods of frequency estimation techniques", Springer (2012).
11. S. Koteswara Rao, "Application of MUSIC algorithm to power quality analysis: detecting interharmonic components", DMIAPR-2011.
12. S. Koteswara Rao, "Minimum-Norm with an application of estimation of bearing angles passive under water multi target scenario", Proceedings of ICISIP-2005, pp. 371–373.
13. PetreStoice and Randolph Moses, "Spectral Analysis of Signals", Prentice Hall, Inc., 2005.
14. Monson Hayes, "Statistical Digital Signal Processing and Modelling", John Wiley & Sons, INC.
15. Wail A. Mousa, Abdullatif A. Al-Shuhail, "Processing of Seismic Reflection data using Matlab", Synthesis Lectures and Signal Processing. Morgon & Claypool Publishers.
16. Sanjit K. Mitra, "Digital Signal Processing a computer based approach", Second Edition, McGraw-Hill publications.

# Processing of Seismic Signal Using Minimum Variance Algorithm

Md. Basha Saheb, U. Neeraj Kumar, S. Koteswara Rao
and V. Lakshmi Bharathi

**Abstract** Raw seismic signals contain noise which corrupts the real seismic data. To overcome this type of interference in the seismic data, preprocessing is done using the FIR bandpass filter. A new method is proposed in this paper for non-parametric estimation of seismic signals. Minimum variance spectral estimation is an eminent spectrum analysis process that offers a high-frequency resolution in comparison with remaining nonparametric methods. Here, an assured band of frequencies is allowed for processing from supplied data to nullify the unwanted signals. Minimum variance algorithm is used to find out the spectrum of the seismic signal and to improve the resolution of the signals.

**Keywords** Stochastic signal processing · Adaptive signal processing
Seismology · Applied statistics

## 1 Introduction

The process of sharing any data of intent among individuals is termed as communication. The transfer of such signals from one place to another may vary in distance through a diverse means of media. The signals that can be construed mathematically are called deterministic signals [1]. Noise, being one such random signal, is ubiquitous and has many forms. Spectral estimation is the process in

Md. Basha Saheb (✉) · U. Neeraj Kumar · S. Koteswara Rao · V. Lakshmi Bharathi
Department of ECE, K L University, Green Fields, Guntur DT 522502, AP, India
e-mail: bashasaheb38@gmail.com

U. Neeraj Kumar
e-mail: neerajkumar.uppalapati@gmail.com

S. Koteswara Rao
e-mail: rao.sk9@gmail.com

V. Lakshmi Bharathi
e-mail: valluri.bharathi@gmail.com

© Springer Nature Singapore Pte Ltd. 2018
J. Anguera et al. (eds.), *Microelectronics, Electromagnetics
and Telecommunications*, Lecture Notes in Electrical Engineering 471,
https://doi.org/10.1007/978-981-10-7329-8_17

which frequency of a signal is defined cardinally. On filtering a process with a group of cramped bandpass filters, the power spectrum is resolved.

Spectral estimation essentially converges on assessing noise with great resolution [2]. To accomplish this obligation, there is a necessity of enhancing the signal detection and attaining a consistent measure of the signal with resolution [3]. Seismic waves are used to determine topology of the subsurface layers for better identification of their boundaries [4]. Due to the asymmetry in the layers of the earth, seismic waves are reflected in distinct directions initiating multipath propagation. Earthquake's origin is determined by the seismic data of that earthquake recorded from at least three diverse receiver positions [5, 6].

## 1.1 Seismic Signal Processing

Enhancement in the raw seismic source by nullifying the noise improves the authenticity of the seismic signals replicating the seismic event parameters [7]. At the end of any seismic propagation, the seismic waves have minute energy that can be lost at the reception end due to the noise intervention. Considering the random noise as additive white Gaussian noise, it can be attenuated easily through seismic data processing methods. Here, stacking overcomes most of the random noise, thereby improving the SNR by a factor of $\sqrt{Q}$, where Q is equivalent to the number of stacked traces.

Coherent noise is mainly caused by ground roll, consistently scattered waves, etc. The seismic data recorded contains ground roll noise. As a part of surface waves, they have high amplitudes. Implementing the bandpass filters in this perspective improves the SNR by reducing this noise [5]. In succeeding section, nonparametric minimum variance spectral estimation process is interpreted.

## 1.2 Minimum Variance

The minimum variance spectral estimation is the modification of maximum likelihood technique proposed by Capon to interpret two-dimensional power spectral density [8]. Capon's estimator can be interpreted as a set of filters which are optimized to reduce their response of frequency outside the circle of interest and the width of each filter depends on the information [2, 9]. Thus, the pattern of the filter depends on opted frequency range and information adaptiveness [10].

Here, assumptions are not made unlike the parametric spectral estimation methods [11]. During the propagation of the signal, additive noise dominates the information at low-energy components ensuing into mislaid features. The key concept of minimum variance is to restrict entire output filter's energy [12].

By calculating the filter bank from its own signal, capon's method determines the signal power.

Periodogram has high sidelobes rising ambiguity in the amplitude response of the obtained signal [13]. Minimum variance method is used to reduce the sidelobes in turn enlightening resolution and diminishing variance over periodogram. During the signal propagation, there is a chance of system degradation due to several factors intruding the user signal [10].

The minimum variance spectrum estimation technique includes these subsequent steps as follows:

1. Create a set of bandpass filters $g_i(n)$, in order to discard the maximal extent of power outside the confined band and thereby achieve distortion less propagation at a given frequency $\omega_i$.
2. Measure power for every output process $y_i(n)$ by filtering $x(n)$ with all filters available in the given set.
3. Initiate $\widehat{p}_x(e^{j\omega_i})$ equivalent to the power estimated in the second step divided by filter bandwidth.

The minimum variance spectrum approximation for the given signal is $\widehat{p}_{MV} = \frac{p+1}{e^H R_x^{-1} e}$, where $R_x$ is the $p \times p$ autocorrelation matrix.

Depending on the filter length (p), the resolution and variance of the minimum variance method vary accordingly. For better resolution, bandwidth of the filter should be small that can be attained only when p is large [3]. In the second section, mathematical modeling of minimum variance algorithm is described. Later in the third section, simulation and results of minimum variance method are explained. In the final section, the paper is concluded by summarizing the minimum variance technique.

## 2  Mathematical Modeling

Let $x(n)$ be a wide-sense zero-mean immobile arbitrary mode with $P_x(e^{j\omega})$ as power spectrum and let $g_i(n)$ be a perfect bandpass filter by center frequency $\omega_i$ and bandwidth $\Delta$,

$$|G_i(e^{j\omega})| = \begin{cases} 1 & ; & |\omega - \omega_i| < \Delta/2 \\ 0 & ; & otherwise \end{cases} \tag{1}$$

In the output $y_i(n)$ by filtering $x(n)$ with $g_i(n)$, the power spectrum is determined to be

$$P_i(e^{j\omega}) = P_x(e^{j\omega})|G_i(e^{j\omega})|^2 \tag{2}$$

and the power is specified as

$$E\left\{|y_i(n)|^2\right\} = \frac{1}{2\pi}\int\limits_{-\pi}^{\pi} P_i(e^{j\omega})d\omega = \frac{1}{2\pi}\int\limits_{-\pi}^{\pi} P_x(e^{j\omega})|G_i(e^{j\omega})|^2 d\omega$$

$$= \frac{1}{2\pi}\int\limits_{\omega_i-\Delta/2}^{\omega_i+\Delta/2} P_x(e^{j\omega})d\omega \tag{3}$$

If $\Delta$ is small enough so that $P_x(e^{j\omega})$ is almost steady throughout the filter's passband, then power in output process is approximately

$$E\left\{|y_i(n)|^2\right\} \approx P_x(e^{j\omega_i})\frac{\Delta}{2\pi} \tag{4}$$

Therefore, it is possible to estimate the power spectral density at the given condition $\omega = \omega_i$ from the noise removal scheme by assessing the power in (n) and distributing the spectrum through the controlled frequency limit ranging about $\Delta/2\pi$,

$$\widehat{p}_x(e^{j\omega_i}) = \frac{E\left\{|y_i(n)|^2\right\}}{\Delta/2\pi} \tag{5}$$

The periodogram produces an estimate of the power spectrum in a similar fashion. Specifically, $x(n)$ is made noise resistant through a set of bandpass filters, $h_i(n)$, where

$$|H_i(e^{j\omega})| = \frac{\sin[N(\omega-\omega_i)/2]}{N\sin[(\omega-\omega_i)/2]} \tag{6}$$

and the power in every filtered signal is calculated through a single-point model average,

$$\widehat{E}\left\{|y_i(n)|^2\right\} = |y_i(N-1)|^2 \tag{7}$$

On separating the power approximation with the means of the controlled frequency limit range $\Delta = 2\pi/N$, the periodogram is designed.

In this technique, all the noise-reducing filters available remain identical, altering in terms of the middle frequency. So, they are known for noncontingent nature to the given information. All the random signals have a nonuniform representation all over the path of propagation. So, they are present in the sidelobes of the bandpass filter as well. As the sidelobes are not a desired feature for an efficient technique, their presence is not entertained to overcome the power seepage glitches that occur falsehood in the power approximations. Adaptiveness to the filter removes the sidelobe signals by using the circle of interest for accepting the spectrum of the confined band. This feature makes the complete finest design of the spectral estimation technique.

To estimate the power spectral density of input signal at frequency, $\omega_i$, let $g_i(n)$ be a compound p order FIR bandpass filter. To safeguard the changes in the input power at the given frequency $\omega_i$, $G_i(e^{j\omega})$ is forced to attain a unity gain at the condition $\omega = \omega_i$,

$$G_i(e^{j\omega_i}) = \sum_{n=0}^{p} g_i(n)e^{-jn\omega_i} = 1 \tag{8}$$

Let $g_i$ be the vector of filter coefficients $g_i(n)$,

$$g_i = [g_i(0), g_i(1), \ldots, g_i(p)]^T \tag{9}$$

and let $e_i$ be the vector of complex exponentials $e^{jk\omega_i}$,

$$e_i = \left[1, e^{j\omega_i}, \ldots, e^{jp\omega_i}\right]^T \tag{10}$$

The constraint on the frequency response given in Eq. (8) may be written in vector form as follows:

$$g_i^H e_i = e_i^H g_i = 1 \tag{11}$$

Now, for the power spectrum of $x(n)$ at frequency $\omega_i$ to be measured as accurate as possible, the set of filters must deny the power outside the circle of interest. Therefore, criterion is used for designing the bandpass filter for minimizing the power with respect to the linear constraints of the output process as given in Eq. (11). The power in $y_i(n)$ may be indicated through the autocorrelation matrix $R_x$ by

$$E\left\{|y_i(n)|^2\right\} = g_i^H R_x g_i \tag{12}$$

The approach for designing the apt filter has got some challenging limitations. To overcome these problems, minimizing Eq. (12) satisfies the condition with respect to the linear constraints given in Eq. (11). The key for this complication is

$$g_i = \frac{R_x^{-1} e_i}{e_i^H R_x^{-1} e_i}, \tag{13}$$

where the smallest amount of $E\left\{|y_i(n)|^2\right\}$ is equivalent to

$$\underset{g_i}{min}\, E\left\{|y_i(n)|^2\right\} = \frac{1}{e_i^H R_x^{-1} e_i} \tag{14}$$

Thus, Eq. (13) defines the best filter to approximate the input power at frequency $\omega_i$, and Eq. (14) gives power in $y_i(n)$, which is used as the estimate, $\hat{\sigma}_x^2(\omega_i)$, of the input power at frequency $\omega_i$. However, the above equations are derived at a fixed frequency $\omega_i$, although these equations were derived for a specific frequency $\omega_i$; since this frequency was arbitrary, then these equations are valid for all $\omega$ [14]. Thus, the desired filter to approximate the input power at frequency $\omega$ is

$$g = \frac{R_x^{-1}e}{e^H R_x^{-1}e} \tag{15}$$

whereas power estimate is given by

$$\hat{\sigma}_x^2(\omega) = \frac{1}{e^H R_x^{-1}e} \; ; \; e = \left[1, e^{j\omega}, \ldots, e^{jp\omega}\right]^T \tag{16}$$

Having designed the bandpass filter bank and estimated the distribution of power in $x(n)$ as a function of frequency, we may now approximate the power spectrum by separating the power approximate by the confined frequency set. Even if distinct conditions are present to describe a range of frequencies, using the appropriate value for $\Delta$ generates exact white noise power spectral density [14]. Since the minimum variance approximation of power in white noise is $E\left\{|y_i(n)|^2\right\} = \sigma_x^2/(p+1)$, it follows from Eq. (5) that the spectrum estimate is

$$\hat{P}_x(e^{j\omega_i}) = \frac{E\left\{|y_i(n)|^2\right\}}{\Delta/2\pi} = \frac{\sigma_x^2}{p+1}\frac{2\pi}{\Delta} \tag{17}$$

Therefore, if the bandwidth is given as

$$\Delta = \frac{2\pi}{p+1} \tag{18}$$

the resultant $\hat{P}_x(e^{j\omega}) = \sigma_x^2$. Using Eq. (18) as the bandwidth of the filter g(n), the power spectrum estimate becomes, in general,

$$\hat{P}_{MV}(e^{j\omega}) = \frac{p+1}{e^H R_x^{-1}e} \tag{19}$$

which is the minimum variance spectrum estimate. Note that $\hat{P}_{MV}(e^{j\omega})$ is determined through the autocorrelation matrix $R_x$ of the input signal [3].

# 3  Simulation and Results

Step 1: The seismic signals used in this paper are attained from MATLAB file present in [5], Book_Seismic_Data.mat through a geophone array in Southern United States. This reference data is recorded from the man-made seismic waves to idealistically represent the real-time earthquake scenario. Here, source is observed to be a high-explosive material filled in about 100 feet below the earth surface by making holes. There are 33 traces, each divided individually into 1500 samples with sampling interval of 0.002 s. To analyze this spectrum, one of the traces is considered among the supplied traces.

Step 2: The accuracy of the code written is assessed on comparing with the known synthetic signal. Minimum variance algorithm is then determined to estimate the harmonics of the earthquake recorded seismic data.

Step 3: Consider the input signal to be a tonal sinusoidal signal $0.98xe^{\pm j0.3\pi}$. The produced signal is represented in Fig. 1 with a normalized frequency of $0.4\pi$. The considered signal is corrupted with white noise of variance 1.

Step 4: Power Spectral Density (PSD) of the synthetic signal is shown in Fig. 2. The peak occurs at 0.4 normalized frequency as in Fig. 2. This makes it vivid that the code taken is accurate and functioning well.

Step 5: In Fig. 3, trace 5 of the seismic signal data is shown.

**Fig. 1** Synthetic signal

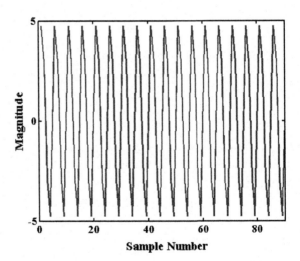

**Fig. 2** Minimum variance
spectrum of synthetic signal

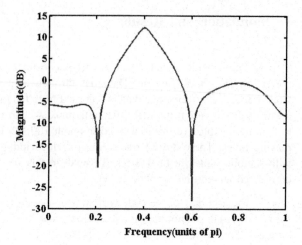

**Fig. 3** Raw seismic signal

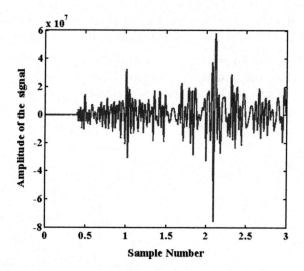

Step 6: The bias observed in the given input signal is withdrawn by subtracting the mean of the raw seismic signal to determine the detrended seismic signal as shown in Fig. 4.

Step 7: The prescribed minimum variance algorithm is applied on the detrended seismic signal and the PSD achieved out of it is represented in Fig. 5. From this figure, the maximum peak is determined to be at $0.09375\pi$ normalized frequency. The sampling frequency is calculated by using the sampling interval 0.002 s as

**Fig. 4** Detrended seismic signal

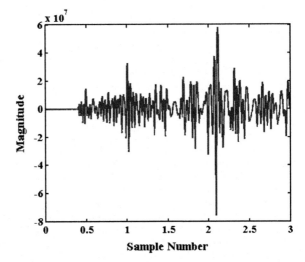

**Fig. 5** Minimum variance spectrum of detrended synthetic signal

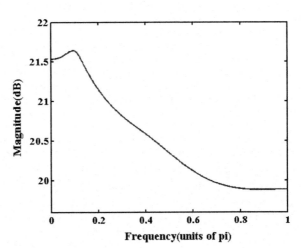

$$fs = 1/0.002 = 500\,\mathrm{Hz}$$
$$w = \frac{2\pi f}{fs} = 0.09375\pi$$
$$= \frac{2\pi f}{500}$$
$$= \frac{2\pi}{fs}f = 0.09375\pi,$$

**Fig. 6** FIR bandpass-filtered
spectrum

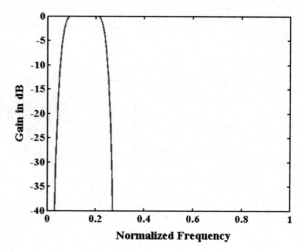

where f is the frequency of the signal. It is given by,

$$\text{Tonal frequency} f = \frac{500}{2} * 0.09375$$
$$= 250 * 0.09375$$
$$= 25 * 0.9375 = 23.4375 \, \text{Hz}$$

Step 8: The considered seismic data has earthquake signal frequency ranging between 15 and 60 Hz. For realization of bandpass filter, Finite Impulse Response (FIR) order is taken as 8. Its frequency spectrum estimation is shown in Fig. 6.

**Fig. 7** FIR bandpass-filtered
detrended synthetic signal

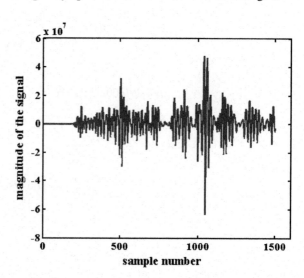

**Fig. 8** FFT of
bandpass-filtered seismic
signal

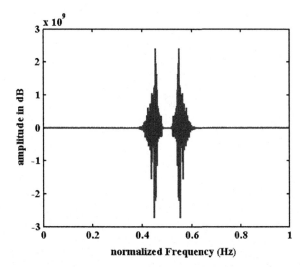

**Fig. 9** Minimum variance
spectrum of the BPF
detrended seismic signal

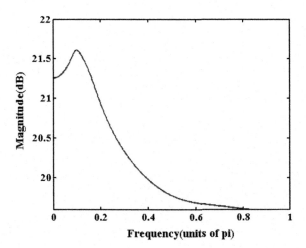

Step 9: The convoluted output of the detrended seismic signal with the FIR bandpass filter is shown in Fig. 7. This eliminates the unwanted signals existing outside the order of the bandpass filter. Fast Fourier transform of bandpass-filtered spectrum is represented in Fig. 8.

Step 10: PSD of the bandpass-filtered signal is determined using minimum variance and it is shown in Fig. 9. The maximum peak is obtained at $0.09961\pi$. By repeating the process like step 7, the frequency of the signal is 24.9025 Hz.

# 4   Conclusion

Spectral estimation technique offers good resolution and minimum variance. It overcomes sidelobe leakage by making the filters' data adaptive and avoid the signals out of band. These features make minimum variance method efficient over periodogram. By this technique, SNR is improved by reducing the noises like ground roll noise using FIR bandpass filters and in turn retaining the desired seismic data.

# References

 1. Yanwei Wang, Jian Li and Petre Stoica, "Spectral analysis of signals", Morgan and Claypool publishers.
 2. Dimitris G Manolakis, Vinay K Ingle and Stephen M Kogon, "Statistical and adaptive signal processing", Artech house, Inc.
 3. Monson H. Hayes, "Statistical digital signal processing and modeling", John Wiley and sons Inc., publishers.
 4. Pantelis Soupios and Dimitrios Ntarlagiannis, "Characterization and Monitoring of Solid Waste Disposal Sites Using Geophysical Methods: Current Applications and Novel Trends", "Modelling Trends in Solid and Hazardous Waste Management" pp. 75–103.
 5. Wail A. Mousa and Abdullatif A. Al-Shuhail, "Processing seismic reflection data using MATLAB", Morgan and Claypool publishers.
 6. N. Purnachandra Rao, "Earthquakes", Editor of publications-Amaravati popular science series, 2016.
 7. Robert Hossa, Ryszard Makowski and Radoslaw Zimroz, "Automatic segmentation of seismic signal with support of innovative filtering", International Journal of Rock Mechanics & Mining Sciences 91 (2017), pp. 29–39.
 8. Umar Mujahid, Jameel Ahmed, Abdur Rehman, Umair Shahid, Abbass and Mudassir, "Performance Analysis of Spectral Estimation for Smart Antenna System", 2013, pp. 1695–1705.
 9. V U Reddy and K Maheswara Reddy, "Eigen structure based spatial spectrum estimation: Problems encountered in practice and signal processing solutions", July and September 2001, pp. 405–438.
10. Tsung-Ching Liu and Barry D. Van Veen, "Multiple window based minimum variance spectrum estimation for multidimensional random fields", IEEE transactions on Signal Processing, Vol. 40, No. 3, March 1992, pp. 578–589.
11. M. Durnerin and N. Martin, "Minimum variance filters and mixed spectrum estimation", Signal Processing 80, 2000, pp. 2597–2608.
12. J. Naga Vishnu Vardhan and Dr. P. G. Krishna Mohan, "Minimum variance power spectral estimation of noisy signals with improved SNR", pp. 341–349.
13. Jian Li and Petre stoica, "An adaptive filtering approach to spectral estimation and SAR Imaging", IEEE transactions on Signal Processing, Vol. 44, No. 6, June 1996, pp. 1469–1484.
14. Michael J. Grimble and Emeritus Michael A. Johnson, "Power Spectral Density Analysis", Advanced Textbooks in Control and Signal Processing, 2005.

# IoT-Based Patient Health Monitoring System

**Akash Vaibhav and Imtiaz Ahmad**

**Abstract** This paper presents the design and implementation of a wireless biomedical parameters monitoring system using various sensors and Arduino UNO as the MCU (Master Control Unit). The system can be used to continuously monitor the biomedical parameters of a patient wiz. The body temperature and pulse rate from anywhere on the globe using IoT (Internet Of Things). IoT is implemented using an ESP WiFi module, which allows the various signals to be transmitted seamlessly over the internet. The device is portable and can be powered by a 5 V DC source.

**Keywords** Arduino · ESP8266 WiFi module · DS18B20 temperature sensor Heartbeat sensor · Patient health monitoring system

## 1 Introduction

In the contemporary world, electronics have become an inevitable part of human existence. Every aspect of human life has been enriched by the electronic technology, and the medical field is no exception. With the increasing rush in their day-to-day lives, people find it difficult to take out time for their loved ones. In this modern cut-throat competition, illness of near and dear ones can become a hindrance in the path of success [1–3]. Today, it is next to impossible for a student or working professional to stay beside an ill relative round the clock. Although, advanced healthcare facilities are now available which guarantee 24 × 7 monitoring of patients in facilities like ICU (Intensive Care Unit) by expert doctors, such services are costly and are beyond the reach of the common man in case of small ailments like cough and cold, mild fever, etc. The proposed system has been

A. Vaibhav · I. Ahmad (✉)
ECE Department, B.I.T Sindri, Dhanbad, Jharkhand, India
e-mail: imtiazahmadbitsindri@gmail.com

A. Vaibhav
e-mail: akashvbhv@gmail.com

© Springer Nature Singapore Pte Ltd. 2018                                        177
J. Anguera et al. (eds.), *Microelectronics, Electromagnetics
and Telecommunications*, Lecture Notes in Electrical Engineering 471,
https://doi.org/10.1007/978-981-10-7329-8_18

designed keeping in mind the expectations and limitations of the common people. This system allows a person to continuously monitor the vital body parameters of a human subject like body temperature and pulse rate from anywhere across the globe while allowing them to perform their regular duties unhindered. Thus, the person can continuously monitor the health of an ill relative and if need be, seek immediate medical attention [4–9]. This system can also be employed to monitor the health of elderly people, who tend to be vulnerable to sudden change in health parameters and the correct treatment at the right time might add years to their lives. The important enabling factor of IOT is in medical and health care. IOT devices are used to collect, monitor, evaluate, and notify the patient of the information.

## 2   Proposed System

In this system, we aim to continuously monitor the health parameters of a human subject like the temperature and pulse rate of the subject using various sensors. An Arduino UNO is used as the MCU which processes the data from various sensors and displays the readings on the interfaced LCD. The readings are also transmitted to the internet using the WiFi module (Fig. 1).

## 3   Components

### 3.1   DS18B20 Temperature Sensor

The DS18B20 is basically a digital thermometer which can provide 9-bit to 12-bit Celsius temperature measurements. It is designed to communicate over a 1-Wire

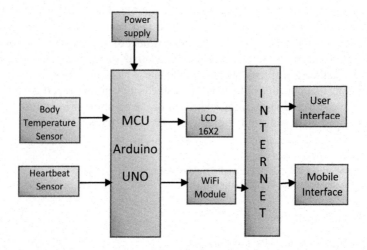

**Fig. 1** Block diagram of the proposed system

bus which requires only one data line (and ground) for communicating with a central microprocessor. It runs on "parasite power," i.e., it derives power from the data line, thus eliminating the need for an external power supply.

## 3.2 Heartbeat Sensor

The heartbeat detection module combines a phototransistor and IR LED. When a finger is placed between the IR LED and the phototransistor, blood pumped during a heart pulse provides a varying signal. Reading this analog signal, we can interpret a change in signal as a heartbeat

## 3.3 Arduino Uno

The Arduino UNO is a microcontroller board ("computer-on-a-chip"). It based on the Atmel's ATmega328 and runs an 8-bit architecture. It has a total of 28 pins of which, 14 pins are digital I/O pins and 6 are analog I/O pins. It also has other power pins, such as GND, VCC. It can provide two voltage outputs of 5 and 3.3 V to external peripherals. It has an onboard 16 MHz ceramic resonator. Other components include a USB socket, a power jack, an ICSP header, and a reset button. The onboard ATmega328 has2 Kb of SRAM, an EEPROM of 1 KB, and Flash memory of 32 kb. The Arduino UNO is an open source hardware development board which supports a wide variety of open-source libraries in order to interface various external components.

## 3.4 ESP8266 Module

The ESP82668266 is basically a low-cost WiFi Module which can be easily interfaced with Arduino. It runs on 3.3 V (derived from the Arduino board) and can be programed by the end user, though, each ESP8266 comes preprogramed with a ready-to-use AT command set firmware. The ESP8266 is a self-contained System on Chip with integrated TCP/IP protocol stack which provides any microcontroller access to WiFi network. The ESP82668266 can either host an application or offload all WiFi networking functions from another application processor.

## 3.5 LCD Module

LCD stands for Liquid Crystal Display. The LCD module basically comprises a controller, driver, LCD panel and backlight (optional). The LCD module can be

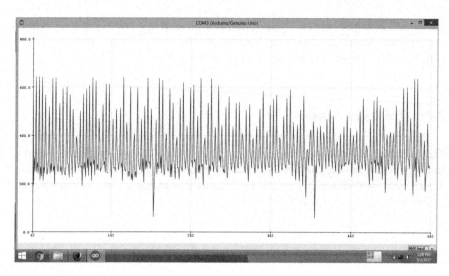

**Fig. 2** Screenshot of heartbeat plot

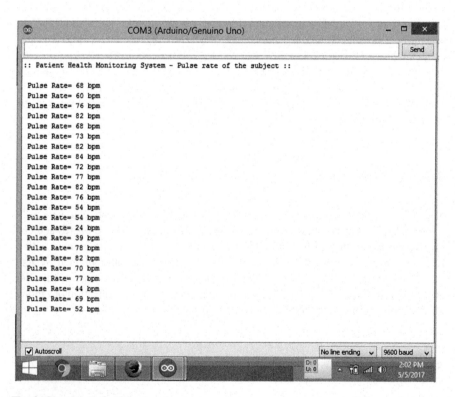

**Fig. 3** Screenshot of pulse rate values

directly hooked to the Arduino UNO board and can be used as its output device. It can be easily programed using the open-source library "LiquidCrystal.h." It is the most popular output component and is used in a variety of appliances and projects.

## 4 Software Description

The MCU, i.e., the Arduino Uno, can be programed directly via the Arduino IDE (Integrated Development Environment), without the need for any additional software. The Arduino Uno uses an ATmega328 IC at its core, which has a bootloader already burned into it so that new codes can be burned into it without the need for an external hardware programer. The Arduino board communicates with the computer using the original STK500 protocol. The Arduino IDE software has a serial monitor which can be used to monitor simple textual data from the board and allows data to be sent to it. It also includes a serial plotter which can be used to monitor the data graphically. Arduino programs are called "Sketches". The Arduino Software (IDE) can be extended through the use of libraries, to provide extra functionality to sketches. The Sketch for the proposed system requires the use of various libraries wiz. LiquidCrystal, SoftwareSerial, StdLib, OneWire, DallasTemperature, etc. for controlling the functionality of various components.

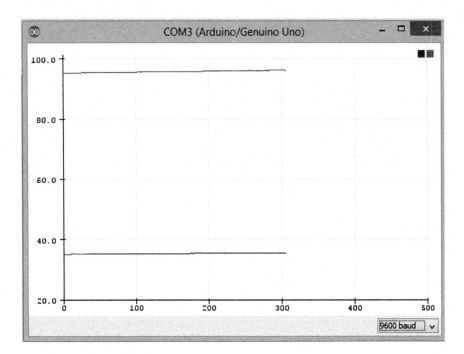

**Fig. 4** Screenshot of body temperature values

The Arduino Uno has the facility to read and write digital as well as analog data. Analog to Digital conversions are done using software programing and no external analog to digital conversion is necessary. The proposed system uses an ESP8266 module to communicate over WiFi. The IoT (Internet Of Things) implementation exploits the services of various websites which allow the streaming of data over an IP. These can also be used to monitor or analyze online and even though mobile apps, thus making it easy to access and manipulate data anywhere anytime.

## 5  Results

The patient health monitoring system takes the data input from the temperature sensor and the heartbeat sensor every 500 ms. The output from the DS18B20 temperature sensor is serially read in the MCU. The analog output from the heartbeat sensor module is first converted into digital equivalent using software programing and is then analyzed to determine the pulse rate of the subject. The

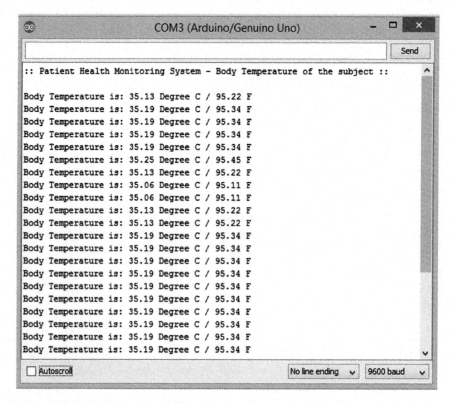

**Fig. 5** Screenshot of body temperature values

body temperature and the pulse rate is displayed continuously on the LCD, the serial monitor on the Arduino IDE and is also transmitted to the internet channel using the ESP8266 WiFi module. The transmitted data can be viewed by logging into the public channel using a browser and a mobile application (Figs. 2, 3, 4 and 5).

# 6 Conclusion and Future Work

This system based on the various biosensors, Arduino Microcontroller and ESP8266 WiFi module to transmit data over the Internet, can be of prime importance in the field of health and patient monitoring. The data can be collected on the internet and can be analyzed to determine the improvement or deterioration in the subject's health. The system consumes very low power and can be easily moved and assembled. This system is highly cost-efficient and can revolutionaries the way a common man looks after their ill near and dear ones.

More modules can be interfaced to this system in order to expand its horizon, for example, the blood pressure module, ECG Module, etc. Moreover, user end experience can be enhanced by adding various analytical and communication programs, which can be used to analyze the improvement or deterioration in a subject's health and can even contact the hospital automatically in case of an emergency.

# References

1. ArunaDevi. S, Godfrey Winster. S, Godfrey Winster. S: Patient Health Monitoring System using IOT Devices, International Journal of Computer Science & Engineering Technology (IJCSET), Vol. 7 No. 03 Mar 2016
2. Harshavardhan B. Patil, Prof. V.M. Umale,: Arduino Based Wireless Biomedical Parameter Monitoring System Using Zigbee, International Journal of Engineering Trends and Technology (IJETT) – Volume 28 Number 7 - October 2015
3. ManishaShelar, Jaykaran Singh, Mukesh Tiwari: Wireless Patient Health Monitoring System, International Journal of Computer Applications (0975–8887), Volume 62, No. 6, January-2013
4. Haux R, Medical informatics: past, present, future, International journal of Medical Informatics, 2010, 79(9):599–610
5. Kaplan B. Health IT Success and failure: Recommendations from literature and an AMTA workshop. Journal of American Medical Informatics Association, 2009 16(3):291–299
6. Lang T. Advancing global health research through digital technology and sharing data science, 2011, 331(6018): 714–717
7. utte A. J. Translational Bioinformatics: Coming of Age. Journal of American Medical Informatics, 2008, 15(6): 709–714
8. Varner H. R. Olasted C. M., Rutherford B. D., HELP-A program for medical Decision Making, Computers and Biomedical Research, 1972, 5(1):65–74
9. Alan G. Smith: Introduction to Arduino

# A Synoptic Review on Dielectric Resonator Antennas

G. Divya, K. Jagadeesh Babu and R. Madhu

**Abstract** A brief review of research on dielectric resonator antenna (DRA) is presented in this paper. Basic characteristics of DRA and its comparison with microstrip patch antenna are discussed. Different types of DRAs, excitation mechanisms, various bandwidth enhancement schemes, and isolation improvement techniques are also discussed in this paper. The recent inventions associated with DRA are also included.

**Keywords** Dielectric resonator antenna · Isolation · Excitation
Glass DRA

## 1 Introduction

Antennas are the essential communication link in the present wireless world. As the new wireless products surface every day, the need for sophisticated antennas is much more demanded in day-to-day life.

In today's wireless era, a low cost, more gain, highly efficient, broadband antennas are a major challenge for the antenna designers. Over the last three decades, an extensive research shows that microstrip antenna (MSA) and dielectric resonator antenna (DRA) are suitable for modern microwave and wireless communications because of lightweight, low profile, inexpensive, and compatibility with integrated circuits. In contrast to MSA, DRA is considered as the viable solution to the traditional conductor antennas at millimeter-wave frequencies.

G. Divya (✉) · R. Madhu
Department of ECE, Jawaharlal Nehru Technological University, Kakinada, India
e-mail: divya.gudapaty@gmail.com

R. Madhu
e-mail: madhu_ramarkula@jntucek.ac.in

K. Jagadeesh Babu
St. Ann's College of Engineering & Technology, Chirala, India
e-mail: jagan_ec@yahoo.com

© Springer Nature Singapore Pte Ltd. 2018
J. Anguera et al. (eds.), *Microelectronics, Electromagnetics and Telecommunications*, Lecture Notes in Electrical Engineering 471,
https://doi.org/10.1007/978-981-10-7329-8_19

185

## 2 Literature Survey

The study of dielectric resonator antennas was started in the early 1980s by Stuart A
Long. But DRs [1] came into existence in 1939 by Robert Richtmyer. He
demonstrated that unmetallized dielectric objects could function as microwave
resonators and mentioned that if a DR is placed in free space, it radiates due to the
boundary conditions at the interface between the dielectric and the air. This provided the fundamental theory and later invented dielectric resonator antenna. Initially, DRs using high permittivity materials ($10 \leq \epsilon_r \leq 100$) with high-quality
factors are used as energy storage elements [2] in microwave circuits, such as filters,
amplifiers, oscillators. DRA [3, 4] was first designed and tested by Long et al. in
1982 by assuming a leaky waveguide model. Ever since, extensive research has
been carried out on analyzing DRA shapes, resonant modes, radiation characteristics, and excitation schemes [5, 6]. A historical review over the past three decades,
major research activities, and latest developments on DRAs are mentioned in [7].
The outcome of the research has highlighted the attractive features of DRAs.

## 3 Comparison of Characteristics of MSA and DRA

The characteristics of a microstrip antenna and dielectric resonator antenna are
given (Table 1).

**Table 1** Comparison of MSA and DRA

| Parameter | Microstrip antenna | Dielectric resonator antenna |
|---|---|---|
| Dimension | Resonant length, $L = 0.49\lambda_d$ | Maximum dimension, $D = \frac{\lambda_0}{\sqrt{\epsilon_r}}$ |
| Radiation efficiency | Poor due to surface waves | High (95%) as there are no conductor losses |
| Power handling | Low | High |
| Gain | High | Low |
| Bandwidth | Narrow bandwidth as it radiates through two narrow slots | Wide bandwidth as it radiates through whole DRA surface |
| Conductor losses | Do exist because of conducting patch | Minimum because of the absence of conductor |
| Ohmic losses | Large ohmic losses occur in the feed structure of the array | Do occur |
| Polarization purity | Difficult to achieve | Easily achieved |
| Waves | Surface waves are excited due to thicker dielectric substrates | Standing waves are formed due to abrupt change in permittivity |
| Resonant modes | TE, TM | TE, TM, HEM |

The comparison table shows that the characteristics of MSA and DRA are primarily dependent on the relative permittivity. DRA radiates through entire surface, so the impedance bandwidth of DRA is superior compared to MSA. Due to the presence of surface waves, the efficiency of MSA is lower than DRA. However, DRA suffers from low gain compared to MSA. Many snags that appear in MSA do not appear in DRA, making it more attractive for millimeter-wave applications.

## 4 Types of DRA

DRAs are volume devices, i.e., 3-D type antennas. These 3D structures excite various modes in the single antenna volume. So, a multifunction or diversity DRA using a single DR is to be designed. This reduces overall system size and cost. Some of the DRA shapes are shown in the following figure.

## 5 DRAs for Broad Banding

DRAs offer broad bandwidth due to their attractive features like lightweight, compact size, and low profile. Various broadband DRAs are presented in this section.

### 5.1 Mono DRA

It is also called as single DRA with a single dielectric material of any regular shape. A primary approach to improve the bandwidth of DRA is to cut and detach some portions of DRA geometries [8]. Re-moulding of DRA is not easy in fabrication because of the hardness of DRA materials. If a simple rectangular DRA is fed by a

**Fig. 1** Types of DRA

**Fig. 2** Rectangular DRA
with a notch

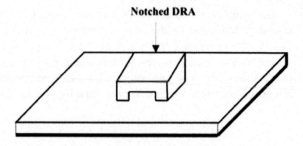

slot with a notch at the center shown in Fig. 2, bandwidth is increased up to 28%. By changing the dimensions of the notch, DRA can be used for broadband or dual band operation (Fig. 1).

## 5.2 Multi-DRA

It is also called as poly DRA with same or different dielectric materials with different sizes. In multi-DRAs different modes are excited in the resonator. These modes may be same or not.

**Fig. 3** Slot-coupled DRA

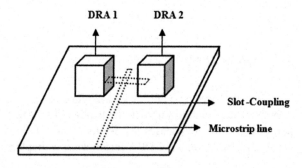

**Fig. 4** Multi-segment DRA

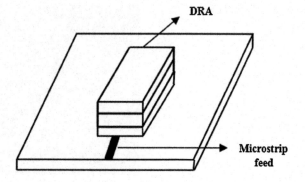

One of the examples of a multi-DRA is a pair of slot-coupled DRA [9], shown in Fig. 3. By employing such kind of geometry, the asset lies in tuning the DRA independently which is more flexible for designing. But the disadvantage is the size of the antenna is increased. An alternative approach is stacked DRA, i.e., combining two dielectric resonators as one shown in Fig. 5. A stacked DRA in which one resonator is loading the other is referred as stepped DRA [10, 11]. Embedded DRA [12] is obtained by inserting a smaller DRA into another larger resonator shown in Fig. 6 (Figs. 4 and 7).

**Fig. 5** Stacked DRA

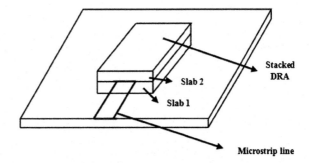

**Fig. 6** Embedded DRA

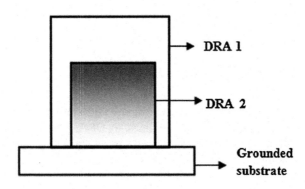

**Fig. 7** Coaxial probe feed

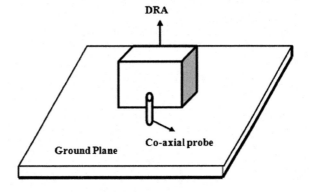

Even though these approaches give efficient results, reshaping of DRA is more difficult because they require a composite structure. Additionally, hybrid DRA [13, 14] and parasitic DRAs are also used.

# 6 Resonant Modes of DRA

Various DRA modes can be excited using different excitation techniques. The selection of the feed and its location both determine the number of modes excited within the DRA. The resonant modes that are realizable in DRA are Transverse Electric (TE) mode, Transverse Magnetic (TM) mode, and Hybrid Electro Magnetic (HEM) mode. The result of the modes is cross polarization. Depending on the requirement the possible resonant modes can be merged or separated or mixed.

# 7 Field Excitation Techniques

The antenna feed mechanisms play a prominent role in the overall performance of the antenna in terms of radiation efficiency. Different feeding methods are discussed hereunder.

## 7.1 Coaxial Probe Feed

This is one of the simplest techniques generally used to feed DRA. Coaxial probe is placed within the DRA or next to DRA. The main advantage of coaxial probe excitation is without using any matching network, a direct coupling into 50 $\Omega$ feeding system is possible. Excitation of modes inside DRA relies on the location of the probe. $TE_{11\delta}$ mode is excited if the probe is placed beside DRA. When this mode is excited, DRA radiate like a horizontal magnetic dipole. $TE_{011}$ mode is excited if the probe is placed at the center of cylindrical DRA. Here, DRA radiates like a vertical dipole.

## 7.2 Microstrip Line Feed

The microstrip line shown in the Fig. 8 can be placed side or underneath the DRA. By varying the size and shape of the microstrip line wide bandwidth can be achieved. If DRA is fed by using open microstrip line, it acts as an electric

**Fig. 8** Microstrip feed

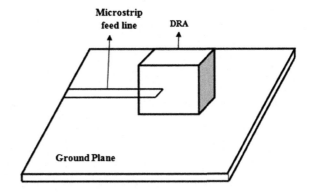

monopole. This feeding technique provides high impedance matching than coaxial probe excitation. Magnetic fields are excited in DRA with microstrip line feed. As the fabrication of microstrip feed is easier, this is one of the most commonly used techniques along with coaxial probe feed.

## 7.3 Aperture-Coupled Microstrip Feed

The aperture-coupled feeding mechanism is shown in Fig. 9. In this feeding method, an aperture in the ground plane excites DRA. An aperture is a slot cut in the ground plane. It is fed by a microstrip line placed beneath the ground plane. The slot aperture acts as a horizontal magnetic dipole. If rectangular DRA is used with a rectangular slot, $TE_{111}$ mode is excited. The challenge in this design is the length of the slot should be around $\lambda/2$. The main drawback of this feeding technique lies in the fabrication because of the presence of multiple layers.

**Fig. 9** Aperture-coupled feed

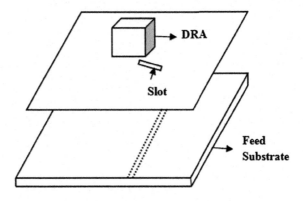

**Fig. 10** Coplanar waveguide

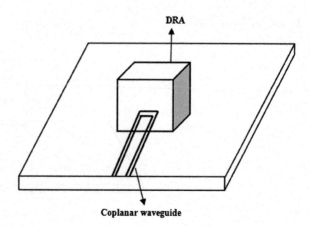

Coplanar waveguide

## 7.4 Coplanar Waveguide

Another common method for exciting DRA is coplanar waveguide. This technique is first proposed by R. A Karenburg et al. in [15] for a cylindrical DRA. Coplanar loop coupled to rectangular DRA is shown in Fig. 10. $HEM_{11\delta}$ mode is excited in cylindrical DRA if the coplanar loop is placed at the edge. $TM_{01\delta}$ mode is excited if the loop is placed in the middle.

## 7.5 Dielectric-Image Guide

This feed technique propounds various advantages compared to conventional microstrip line feed at millimeter-wave frequencies is dielectric-image guide feed mechanism, shown in Fig. 11. This model is best suited for an isolated DRA when placed in free space. When the dielectric image guide is operated near the cut-off

**Fig. 11** Dielectric image waveguide feed

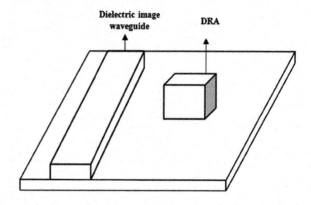

frequencies, the amount of coupling is increased. $TE_{10\delta}$ mode is excited in rectangular DRA using dielectric image waveguide feed. This method is used to feed series linear array.

# 8 Isolation Improvement Techniques

Mutual coupling is defined as the interaction of electromagnetic fields between the antenna elements in an array. Adjacently placed antennas of distance less than /4 cause high coupling between the elements. Even though DRA offers more advantages than MSA, DRA seriously suffers from mutual coupling than MSA. Some techniques for isolation improvement are presented hereunder.

## 8.1 Decoupling Networks

By providing negative coupling, the decoupling network decouples the input ports of the adjacent elements, so that the isolation is improved [16]. The decoupling network is composed of two directional couplers. It is constructed by using lumped elements along with distributed elements. Conventional decoupling networks suffer from narrow bandwidth. Broadband is achieved using parallel resonant circuit, which is suitable for mobile devices.

## 8.2 Parasitic Elements

Parasitic elements are passive elements, which are not directly coupled to the antenna. These elements are used to minimize the coupling effect by creating an opposite coupling fields. These parasitic elements are generally observed in Yagi-Uda antenna as reflector and directors. At millimeter-wave frequencies, parasitic elements can be of resonator type, floating type, or shorted stubs. Stubs improve the matching of the antenna and the slot within it reflects the radiation from the elements and improves the isolation.

## 8.3 Defected Ground Structures

The coupling between neighboring antenna elements can be minimized by modifying the ground plane. Defected ground structures (DGS) are used to enhance the antenna isolation by realizing a stop-band filter between the radiating elements. In [17], bent slits are introduced in the ground plane to enhance the isolation. Slits acts as band stop filter based on the parallel LC resonator and suppress surface waves, which further reduces mutual coupling.

### 8.4 Neutralization Lines

Neutralization lines enhance the isolation between the adjacent elements. Neutralization line is placed above the ground plane without any modifications in the ground plane. By choosing a suitable length for NL, the current is maximum and is phase reversed. This reversed current is fed to the nearby antenna to lessen the amount of coupled current. But it has the disadvantage of narrow banding.

### 8.5 Metamaterials

The materials that possess negative permittivity or permeability are metamaterials. Metamaterials also possess negative refractive index. Metamaterial antennas increase the performance of miniaturized antennas than traditional antennas.

## 9 Applications

The Dielectric Resonator Antenna covers several applications from our day-to-day life to important defense applications.

Wireless applications:   Digital cellular networks, Wireless sensor networks, Mobile Broadcasting system,
Satellite applications:   Direct Broadcast services, Doppler and other RADARs, GNS
Military applications:    Missiles and Telemetry, Surveillance systems.
Medical applications:    Biomedical radiators, Tumor detection

Newly developed technologies, detailed technical descriptions, and latest practical applications of antennas in all wireless systems ranging from communications to maritime applications are presented in [18].

# 10    Recent Advancements

The first transparent glass DRA of 2D and 3D type [19] was proposed by Lim and Leung.

Transparent hemispherical DRAs made of Borosilicate Crown Glass are proposed for optical applications. A dual function glass DRA using Omnidirectional hollow rectangular glass serves as a light cover is presented for the first time in [20].

A novel decoration DRA which is made out of crystal and glass wares is a Glass Swan antenna, proposed by K. W. Leung [21]. It is made out of K-9 glass. The presented DRA is an attractive candidate for wireless communications as well as for sensor designs and medical areas.

# 11    Conclusion

This paper aims to prove a brief review of dielectric resonator antenna. The comparison table summarizes the advantages of DRA over MSA. Different broad banding techniques, feeding mechanisms, and various isolation improvement techniques that are much useful for the researchers working on DRAs are discussed. Furthermore, novel glass and transparent DRAs are also mentioned in this paper.

# References

1. Richtmyer, R. D., "Dielectric resonator," J. *App. Phy.*, Vol. 10, 391–398, Jun. 1939.
2. D. Kajfez and P. Guillon, Eds., Dielectric Resonators. Norwood, MA: Artech House, 1986.
3. Long, S.; McAllister, M.; Shen, L. (1983), "The Resonant Cylindrical Dielectric Resonator Antenna," IEEE Transactions on Antennas and Propagation, 31: 406–412.
4. S. Long, M. McAllister, and L. Shen, "The resonant cylindrical dielectric cavity antenna," *IEEE Trans. Antennas Propag.*, Vol. 31, no. 3, pp. 406–412, May 1983.
5. A. Petosa, A. Ittipiboon, Y. Antar, D. Roscoe, and M. Cuhaci, "Recent Advances in dielectric-resonator antenna technology," *IEEE Trans. Antennas Propag.*, Vol. 40, no. 3, pp. 35–48, 1988.
6. K. M Luk and K.W Leung, *Dielectric Resonator Antennas*. Research Studies Press LTD. England, 2003.
7. Aldo Petosa and Apisak Ittipiboon, "Dielectric Resonator Antennas: A Historical Review and the Current State of the Art," *IEEE Antennas and Propagation Magazine,* Vol. 52, No. 5, pp. 91–116, October 2010.
8. Guillon, P. and Y. Garault, "Accurate resonant frequencies of dielectric resonators," *IEEE Transactions on Antennas and Propagation,* Vol. 25, No. 11, 916–922, Nov. 1977.
9. Fan, Z. and Y. M. M. Antar, "Slot-coupled DR antenna for dual-frequency operation," *IEEE Transactions on Antennas and Propagation,* Vol. 45, No. 2, 306–308, 1997.
10. Pliakostathis, K. and D. Mirshekar-Syahkal, "Stepped dielectric resonator antenna for wideband applications," *Proc. IEEE AP-S Int. Symp. Dig.*, Vol. 2, 1367–1370, USA, 2004.

11. Al Sharkawy, M. H., A. Z. Elsherbeni, and C. E. Smith, "Stacked elliptical dielectric resonator antennas for wideband applications," *Proc. IEEE AP-S Int. Symp. Dig.*, Vol. 2, 1371–1374, USA, 2004.

12. Kishk, A. A., "Experimental study of broadband embedded dielectric resonator antennas excited by a narrow slot," *IEEE Antennas and Wireless Propagation Letters*, Vol. 4, 79–81, 2005.

13. Rao, Q., T. A. Denidni, and R. H. Johnston, "A novel feed for a multi-frequency hybrid resonator antenna," *IEEE Microwave and Wireless Components Letters*, Vol. 15, No. 4, 238–240, 2005.

14. Esselle, K. P. and T. S. Bird, "A hybrid-resonator antenna: Experimental results," *IEEE Transactions on Antennas and Propagation*, Vol. 53, No. 2, 870–871, 2005.

15. R. A Kranenburg, S. A. Long and J. T. Williams, "Coplanar waveguide excitation of dielectric resonator antennas," *IEEE Transactions on Antennas and Propagation*, vol. 39, no. 1, pp. 119–122, 1991.

16. Chen W-J, Lin H-H. LTE700/WWAN MIMO antenna system integrated with decoupling structure for isola-tion improvement. Antennas and Propagation Society International Symposium (APSURSI); 2014. p. 689–90.

17. J. Li, Q. Chu, and T. Huang, "A Compact Wideband MIMO Antenna with Two Novel Bent Slits," *IEEE Transactions on Antennas and Propagation*, vol. 60, no. 2, pp. 482–489, 2012.

18. Petosa, Aldo, and Apisak Ittipiboon, "Handbook of Antenna Technologies", Springer Nature 2016.

19. E. H. Lim and K. W. Leung, "Transparent dielectric resonator antennas for optical applications," *IEEE Trans. Antennas Propag.*, vol. 58, no. 4, pp. 1054–1059, Apr. 2010.

20. Leung, K., X. Fang, Y. Pan, E. Lim, K. Luk, and H. Chan, "Dual-Function Radiating Glass for Antennas and Light Covers – Part II: Dual-Band Glass Dielectric Resonator Antennas," *IEEE Transactions on Antennas and Propagation*, Vol. 61, No. 2, pp. 587–597, August 2012.

21. K. W. Leung. E. H Lim and X. S. Fang, "Dielectric resonator antennas: From the basic to aesthetic," *Proceedings of the IEEE*, vol. 100, no. 7, pp. 2181–2193, Jul. 2012.

# DC Electric Field Analysis of Nomex, Kraft Paper, and PPLP Insulation Arrangement in Liquid Nitrogen by Using COMSOL Multiphysics

Vikram Singh and Shabana Urooj

**Abstract** For the supreme performance of the superconducting cable, insulation decides the ability of the cable to withstand the operating voltage or not. So insulation is the main focused area in the superconducting cable to decide its selectivity, preference, and performance over the other cable. The major aim of this simulation-based analysis is to analyze dielectric characteristics of Nomex, Kraft Paper, and PPLP Insulation arrangement in liquid Nitrogen. Dielectric characteristics this resulting combination is analyzed by using COMSOL Multiphysics software as a simulator. For determining the dielectric characteristics, a voltage of the style Ramp voltage and a step voltage have been provided and the behavior of Nomex insulation, Kraft Paper, and PPLP insulation is observed under these voltages.

**Keywords** Polypropylene laminated paper (PPLP) · Kraft paper
Nomex insulation · Liquid nitrogen · DC electric field

## 1 Introduction

Important of using DC superconducting cable in the power system network as it eliminates AC losses, as the losses reduce system efficiency by itself increase. In the application of transformers and high capacity cables Kraft paper was mainly used and it prominence for low-temperature applications as studies and investigated has been conducted for material implement [1]. The dielectrics properties of numerous materials, such as PPLP, 100HN, 100CR, and 150FCR019, Kraft paper, were performed in liquid nitrogen has been studied from many years with the purpose of estimate their capabilities as best electric insulation and optimizing the designs of insulation schemes for superconducting cables [2]. And most studies have aimed to develop HTS ac cables system mainly composed of power cable, termination

V. Singh · S. Urooj (✉)
Electrical Engineering Department, Gautam Buddha University, Greater Noida, India
e-mail: shabanaurooj@ieee.org

© Springer Nature Singapore Pte Ltd. 2018                                         197
J. Anguera et al. (eds.), *Microelectronics, Electromagnetics
and Telecommunications*, Lecture Notes in Electrical Engineering 471,
https://doi.org/10.1007/978-981-10-7329-8_20

bushing and cable joint [3]. For the enrichment of superconducting cables in a DC experience, losses in the superconducting cable can possibly reduced by proper insulation arrangement and selecting insulation material [4–7]. In insulation layer basically contains butt gaps these butt gap become the originator of partial discharge (P.D) and must of the breakdown takes place in this region only [8, 9]. The researcher also found in their work that, on increasing the pressure of liquid nitrogen no change in dielectric characteristic is observed [10, 11]. So it is important to study the new arrangement while wrapping. Nomex, Kraft paper, and PPLP contain butt gap between insulating PPLP layers and Kraft Paper become a source of partial discharge which may accelerate insulation failure. In this work, the voltage of the fashion ramp and step voltage is applied; Ramp voltage to generate the capacitive field, and step voltage for the transient field.

## 2 Simulation Model

DC Electric Field Analysis of Nomex, Kraft paper, and PPLP insulation arrangement in liquid Nitrogen using COMSOL simulation. Electric field analysis can be possible by using Finite element method. 2d symmetry Model of variety was built in COMSOL, semicircular electrodes of diameter 14 mm and its 10 mm flat surface

**Table 1** Dimensions of material used in simulation

| Material | Thickness (µm) | Length (mm) |
|---|---|---|
| Nomex insulation | 90 | 20 |
| Kraft paper | 20 | 20 |
| PPLP | 100 | 20 |
| Butt gap | 100 | 2.5 |

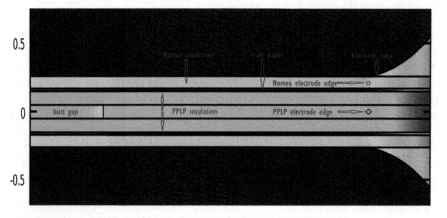

**Fig. 1** Measuring points and materials used in specimens with butt gaps

base cut to touching to the Nomex insulation sheets, and a material selected for this is aluminum and other dimensions used in the simulation are shown in Table 1.

One specimen model having Nomex insulation on the top and bottom then Kraft paper followed by three layer PPLP sheets absence of any butt gap and Fig. 1 shows the arrangement of the specimen, with single butt gap in the center insulation of PPLP.

Semicircular electrodes electrode and insulation sheets were covered in a box with a dimension 21 mm length and width of 8 mm. The materials used in the simulation are given in Table 2.

**Table 2** Electrical properties of materials use

| Material | Relative permittivity | Electrical conductivity [S/m] |
|---|---|---|
| Liquid nitrogen | 1.45 | $2.0 \times 10^{-14}$ |
| Aluminum | – | $3.77 \times 10^{7}$ |
| PPLP | 2.29 | $1.69 \times 10^{-16}$ |
| Kraft paper | 1.03 | $2.44 \times 10^{-14}$ |
| Nomex insulation | 1.7 | $1.66 \times 10^{-17}$ |

**Table 3** Step voltage generating style

| Step by step voltage providing sequence | Rate of voltage increase | Voltage achieved (kV) | For time duration (S) |
|---|---|---|---|
| Step 1 | 0.5 kV/s | 30 | 60 |
| Step 2 | Constant | 30 | 300 |
| Step 3 | 0.5 kV/s | 36 | 12 |
| Step 4 | Constant | 36 | 300 |
| Step 5 | 0.5 kV/s | 42 | 12 |
| Step 6 | Constant | 42 | 300 |
| Step 7 | 0.5 kV/s | 48 | 12 |
| Step 8 | Constant | 48 | 300 |
| Step 9 | 0.5 kV/s | 51 | 6 |
| Step 10 | Constant | 51 | 900 |
| Step 11 | 0.5 kV/s | 54 | 6 |
| Step 12 | Constant | 54 | 900 |
| Step 13 | 0.5 kV/s | 57 | 6 |
| Step 14 | Constant | 57 | 900 |
| Step 15 | 0.5 kV/s | 60 | 6 |
| Step 16 | Constant | 60 | 900 |
| Step 17 | 0.5 kV/s | 63 | 6 |
| Step 18 | Constant | 63 | 900 |

Upper semicircular electrode is made to be positive and the bottom electrode is Ground has been done in the simulation model. Ramp voltage with the rate of 0.5 kV/s was an increase on the positive electrode. Similarly, step voltage from time to time increasing the amount of voltage.

Step voltage generating method is given in the above Table 3. The simulation was performed for the duration of 6000 s so as to attain voltage exceeding 63 kV.

## 3   Simulation Result of the Ramp Voltage

With the increase in the rate of voltage 0.5 kV/s a Ramp voltage up to 70 kV is provided on the positive electrode in the simulation, carried out ramp voltage test.

The ramp voltage is provided to electrode, specimen containing single and without butt gap present. It is observed that electric field was linearly increasing as the ramp voltage is increasing and this was a capacitive electric field distribution. In Fig. 2 electric field strength at electrode edge without containing any butt gap or with containing follow the same path as shown in figure. The electric field inside butt gap and Nomex insulation without butt gap follow the same straight line and overlap each other. In Fig. 3 highest electric field is observed in the Kraft paper insulation in both specimens with butt gap present and without butt gap present. This may become the weakest point for the failure of insulation. At PPLP electrode edge, above butt gap and Nomex insulation near electrode edge electric field

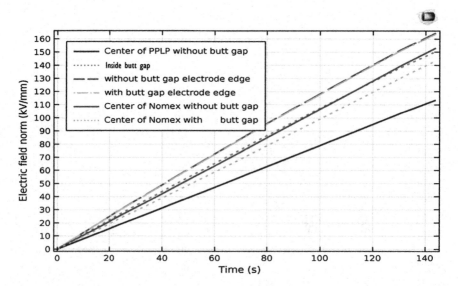

**Fig. 2** Electric field strength variation to the ramp voltage at PPLP electrode edge and inside PPLP insulation for all three specimen system without butt gap

observed is small. The electric field in Kraft paper insulation was very high compare to another region, this is due to capacitive fields is inversely proportional to the permittivity.

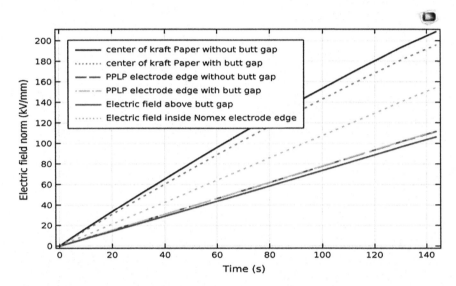

**Fig. 3** Electric field strength variation to the ramp voltage at Kraft paper, PPLP electrode edge, above butt gap, and Nomex electrode edge for two specimen system

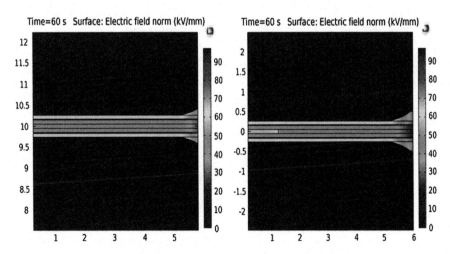

**Fig. 4** Electric field strength at 60 s. Consists of two specimens left: without butt gap, right: with butt gap

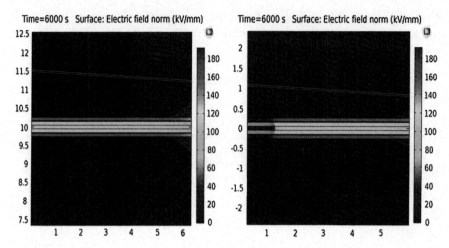

**Fig. 5** Electric field strength at 6000 s. Consists of two specimens left: without butt gap, right: with butt gap

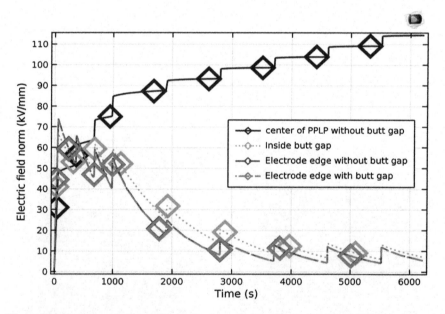

**Fig. 6** Electric field strength variation to the step voltage at Electrode edges, inside butt gap, and center of PPLP insulation without butt gap for two specimen system

# 4 Step Voltage Simulation Result

Step voltage was applied as shown in the table above, the sequence of generating this step voltage so as to attain maximum voltage level above of 63 kV in the simulation model.

Figures 4 and 5 describe the electric field distribution at 60 s and 6000 s, respectively. Figure 6 shows the measured electric field strength at electrode edge, inside butt gap where ramp voltage reaches the first step at 60 s has sharp increase to highest capacitive electric field distribution and after 997 s electric field start decreasing till the voltage reached 63 kV in both specimens. In Fig. 7 Kraft paper shows the same characteristic as that of inside at the electrode edge electric field and butt gap varies as voltage changes in steps. Kraft paper is the region with highest electric field strength in the first step of (60 s) step voltage.

Figure 7 shows electric field in Nomex in increases as the step voltage increase in both specimens without butt gap and with butt gap and slightly high-electric field with containing butt gap. Figure 8 shows electric field inside PPLP insulation is the lowest in other region and an electric field is increased as we move measuring point from the center of PPLP toward the electrode edge.

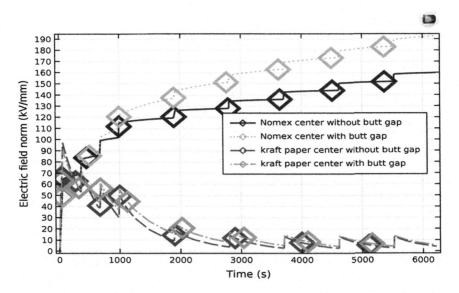

**Fig. 7** Electric field strength variation to the step voltage at Nomex center and Kraft paper for two specimen system

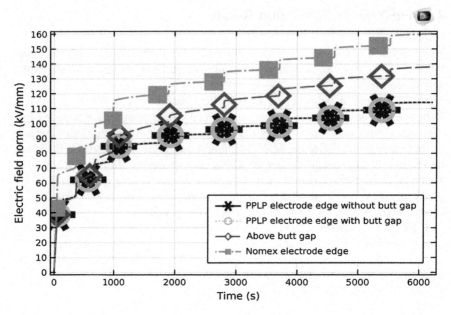

**Fig. 8** Electric field strength variation to the step voltage at PPLP electrode edge without and with butt gap, above butt gap, and Nomex electrode edge for two specimen system

## 5 Conclusion

The influence of Kraft paper, Nomex insulation, and PPLP composite insulation in absence of butt gap and with butt gap in $LN_2$ is analyzed to determine breakdown characteristics. Kraft paper shows the similar electric field characteristics as shown in the butt gap and electrode edge point. From these results, it was clear that region of Kraft paper is experiencing the highest electric field strength in this composite system without containing any butt gap or with containing butt gap. It is also susceptible to breakdown due to high-electric field intensity. Presence of Kraft paper decreased the breakdown voltage during ramp and step voltage tests, respectively. It can be estimated that most of the breakdowns may take place in the Kraft paper insulation first and then followed by electrode edge and inside butt gap of the insulation composite system. Consequently, it was deduced from the obtained results that selection of insulation materials in $LN_2$ would be a challenging and significant feature in determining complex insulating performance for superconductor power cable.

# References

1. Bin Wei, Zhi-Kai Liu, Ming Qiu, Wei-Guo Li, Xing-Jun Gao, Chao Gao,Yong-Qing Zhao, Jing-Zhou Hou, and Pan-Pan Chen "A Study of the Composite Insulation Breakdown Properties for Wrapping Cables in Liquid Nitrogen" IEEE Transactions on Applied Superconductivity, Volume: 25, Issue: 1, Feb. 2015
2. H. J. Kim, D. S. Kwag, J. W. Cho, K. C. Seong, K. D. Sim, and S. H. Kim "Insulation studies and experimental results for high Tc superconducting power cable" IEEE Transactions on Applied Superconductivity, Volume: 15, Issue: 2, June 2005
3. Hae-Jong Kim, Jeon-Wook Cho, Woo-Jin Kim, Yeon Suk Choi, And Sang-Hyun Kim. "The Basic Dielectric Characteristics of Insulating Materials for HTS DC Cable System" IEEE Transactions on Applied Superconductivity, Volume: 26, Issue: 3, April 2016
4. D. S. Kwag, V. D. Nguyen, S. M. Baek, H. J. Kim, J. W. Cho, and S. H. Kim "A study on the composite dielectric properties for an HTS cable" IEEE Transactions on Applied Superconductivity, Volume: 15, Issue: 2, June 2005
5. A. Masood, M. U. Zuberi and E. Husain "Breakdown strength of solid dielectrics in liquid nitrogen" IEEE Transactions on Dielectrics and Electrical Insulation, Volume: 15, Issue: 4, August 2008
6. Jonggi Hong, Jeong Il Heo, Seokho Nam, and Hyoungku Kang "Analysis on the Dielectric Characteristics of Solid Insulation Materials in $LN_2$ for Development of High Voltage Magnet Applications" IEEE Transactions on Applied Superconductivity, Volume: 24, Issue: 3, June 2014
7. J. -W. Choi; H. -G. Cheon; J. -H Choi; H. -J. Kim; J. -W. Cho; S. -H. Kim."A Study on Insulation Characteristics of Laminated Polypropylene Paper for an HTS Cable" IEEE Transactions on Applied Superconductivity, Volume: 20, Issue: 3, Year: 2010
8. Jae-Kyu Seong, Won Choi, Umer A. Khan, Bang-Wook Lee "Correlation between DC Electric Field Intensity and Electrical Breakdown of Butt Gap in $LN_2$/PPLP Composite Insulation System" IEEE Trans. on Dielectrics and Electrical Insulation Vol. 22, No. 1; February 2015
9. J. K. Seong, W. Choi, W. J. Shin, J. S. Hwang, B. W. Lee, "Experimental and analytical study on DC breakdown characteristics of butt gap condition in $LN_2$/PPLP composite system," IEEE Trans. Appl. Supercond., Vol. 23, No. 3, Article No. 5401604, 2013.
10. Wei-Guo Li, Zhi-Kai Liu, Bin Wei, Ming Qiu, Xing-Jun Gao, Chao Gao,Yong-Qing Zhao, Song Li, Jing-Zhou Hou, and Pan-Pan Chen"Comparison Between the DC and AC Breakdown Characteristics of Dielectric Sheets in Liquid Nitrogen" IEEE Trans. Appl. Supercond., Vol. 24, NO. 6, December 2014
11. M. Runde, N. Magnusson; O. Lillevik; G. Balog; F. Schmidt "Comparative tests of tape dielectrics impregnated with liquid nitrogen" IEEE Transactions on Dielectrics and Electrical InsulationYear: 2007, Volume: 14, Issue: 2

# Power and Area Efficient Opamp for Biomedical Applications Using 20 nm-TFET

**Bellamkonda Saidulu and Arun Manoharan**

**Abstract** This paper presents an ultralow power and area efficient TFET-based opamp for portable and wearable IoT devices in smart health monitoring and recording applications. The two-stage operational amplifier is designed with 20 nm Tunnel Field Effect Transistor (TFET). The unique features of TFET transistor would helpful to meet requirements in analog circuit designs where more demand in the area and low-voltage operation. This work shows better improvement in area and power consumption. Area optimized with an absence of miller capacitance ($C_{Miller}$) is the novelty of this opamp design as compared to conventional. The Opamp is designed and simulation carried for 1–10 kHz bandwidth in Cadence environment. The simulation results show a gain of 46 dB, Phase is 68° and power consumption is 1.5 μW with a supply of 0.5 V.

**Keywords** Two-stage opamp · Tunnel FET (InAs) · Biomedical · Ultralow power · Miller capacitance ($C_{Miller}$)

## 1 Introduction

Medical and prosthetic gadgets have risen as a promising possibility for treatment of patients with neurological issues ranging from epilepsy, Parkinson's, and Alzimerith diseases. This portable medical gadgets and instruments should operate with low power for the long-time monitoring of neurons activities without interruptions to avoid battery charging cycles. Bio-potentials are collected as electrical signals from

B. Saidulu (✉) · A. Manoharan
School of Electronics Engineering, VIT University, Vellore 632014, Tamilnadu, India
e-mail: bellamkonda.saidulu@gmail.com

A. Manoharan
e-mail: arunm@vit.ac.in

© Springer Nature Singapore Pte Ltd. 2018
J. Anguera et al. (eds.), *Microelectronics, Electromagnetics and Telecommunications*, Lecture Notes in Electrical Engineering 471, https://doi.org/10.1007/978-981-10-7329-8_21

**Fig. 1** Classification of
bioelectrical signals from
human body

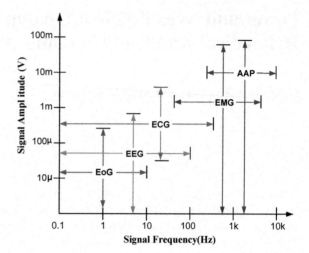

thousands of electrodes which are placed on the scalp. Typically, Bio-potentials are
covered bandwidth of Hz–10 kHz with the amplitude of μV to hundreds of mV range
as shown in Fig. 1.

These weak signals are needed to be processed for faithful signal acquisition by
the amplification with high gain and filtered by the rejection of noise. This process is
continuous and important for all signals more increases the power consumption may
raise the temperature which leads to tissue damage. This work mainly focused on
low power, low noise, and small area are the key challenges of opamp which decides
the overall performance of neural recording system. The outline of opamp design
constraints is power and area with nanoscale devices which operate in low-supply
voltage like. CMOS-SOI, FinFETs, Tunnel FETs. This design performs recording
of Neural signal from brain and interface to the digital system for further diagnoses
process in medical applications. The main focus of this work is to design low power
and small area operational amplifier which suites fulfillment of challenges in biomedical applications. In this recording system, the opamp is collected the micro volt's
range signals from electrodes and amplified to the milli volt's range. The amplified
signals are fed to analog/digital systems for further processing which results in more
power consumption. This Opamp design helps to minimize the overall power consumption of the system. In [1], presented a low power and low noise neural amplifier of two-stage opamp structure with the current buffer. This design used the small
value of compensation capacitor (Cc) for the improvement of bandwidth of OTA
and results show gain of 46 dB, bandwidth of 0.9–13.8 kHz, input referred noise
is 5 $\mu V_{rms}$ and power consumption of 2.4 μW with the supply of 1.2 V in CMOS
180 nm technology. In [2], a neural amplifier with another topology is Folded cascode OTA with source degeneration achieves 40 dB gain, the power dissipation of
2.2 μW, and|vadjust input referred noise is 4.3 $\mu V_{rms}$ in the range of 1–10 kHz is

presented [3]. Presents a neural recording amplifier array with low power and low noise operation. This work improves power supply rejection with reference sharing structure in two-stage amplifier topology [4]. Describes the compact neural amplifier with high-input impedance in folded cascode differential amplifier with current feedback. This feedback connection separates the input signal and improves the input impedance of preamplifier by avoiding the CMFB circuitry and clumsy capacitors [5]. Presents Inverter-based analog front-end amplifiers as low-power topologies for medical imaging applications [6]. Presented low-power fully differential CMOS OTA architecture with dual input cascode inverters at the differential input stage [7]. Presents low power and low noise CMOS telescopic OTA for biomedical applications. In this work, the $g_m/I_D$ methodology is adopted to optimize size for each transistor having a benefit of low power and low noise [9]. Presented a TFET-based 14 nm folded cascode OTA with reduced power consumption and increased gain of the amplifier. From the previous works, the neural amplifier design still there is scope to improve in terms of area, noise, and power consumption.

This work proposes an area efficient and low-power nanoscaled TFET-based two-stage opamp. This two-stage opamp topology is implemented and simulated with Universal TFET (20 nm) without miller capacitor ($C_{miller}$) compensation. And, the advanced features of TFET model helps to operate with low voltage for the benefit of low-power consumption. This paper is organized as follows. Section 2 introduces about Universal Tunnel FET principle model, and I–V characteristics. Section 3 describes neural amplifier analysis and simulation results. Last Sect. 4 is conclusion and acknowledgment.

## 2    Universal Tunnel Field Effect Transistor (Uni-TFET)

Author [8] developed a Dual-Metal-Gate (DMG) InAs TFET with the supply of 0.5 V. Benchmarks of ITRS (2020) specifications for multi-gate technology versus DMG-TFET has met the $I_{ON}/I_{OFF}$ ratio and got doubled ($I_{ON}$ = 1.322 mA/μm, $I_{OFF}$ = 0.01 nA/ μm). The subthreshold slope (SS) is less than 60mV/dec is the advanced feature of TFET based on the Band-To-Band Tunneling (BTBT), which is more useful for analog low-power applications. A lower value of SS, which enhance the energy efficiency and lower parasitic capacitances as compared to CMOS devices in digital as well as analog applications. Transconductance efficiency ($g_m/I_D$) of TFET is providing higher value in subthreshold region. The I–V characteristics of N-TFET are identical to P-TFET except capacitances due to different electron and hole density-of-states in this material (InAs). The output characteristics of TFET determines the saturation region can be approximated by [10],

$$V_{DSAT} = V_{gs} - V_{th} + V_{Dth} \tag{1}$$

$V_{Dth}$ is the drain threshold or super linear threshold voltage results in the last onset of saturation region. For analog circuit designs, overdrive voltage ($V_{ov}$) less than 1V,

**Fig. 2** AC model for
Uni-Tunnel FET (Uni-TFET)

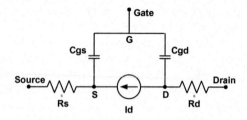

$V_{Dth}$ is calculated by $\lambda_D(V_{gs} - V_{off})$ and constants are $\lambda_D$, $V_{off}$ with values of 0.4, $V_{th}$, respectively [11]. InAs (III–V group) homojunction TFET is one of popular TFET devices.

## 2.1   Small-Signal Model

Transconductance ($g_m$) and output resistance ($r_o$) are the important small signal model Fig. 2 parameters of the device for analog applications. The transconductance in subthreshold region which as relation with subthreshold Swing (SS) is given by

$$g_m = \frac{ln(10)I_{DS}}{SS} \tag{2}$$

Subthreshold swing is sensitive to as increase the current in TFET [12]. For large current densities, SS is constant for MOSFET. In the saturation region, the drain current depends on band–band tunneling near source terminal side of the channel less impact of $V_D$. TFET does not have channel length modulation due to the absence of p–n junction at the drain terminal. The output resistance of TFET is higher than CMOS. The small signal model of TFET has two capacitances ($C_{gs}$, $C_{gd}$) along with parasitics. For the TFET Cgs is low and $C_{gd}$ is large compared to CMOS. The value of $C_{gd}$ raises only when TFET entering into the linear region and $V_{gs}$ is much larger than $V_{ds}$. This is the advantage of TFET to operate in a saturation region, not in the linear region for analog applications. Small signal resistance ($r_{on}$) of TFET is playing important in the linear region when used as a switch. Ron will observe for N-type transistor when $V_g = V_{DD}$, for P-type $V_g$ = GND. To find the $V_{th}$ for this device simulation with transfer characteristics ($I_{ds}$-$V_{gs}$ plot) is shown in Fig. 3 and observed $V_{th}$ is 0.17 V. Table 1 shows the aspect ratio of each device used in the proposed circuit (Figs. 4 and 5).

**Fig. 3** Transfer
characteristics of TFET

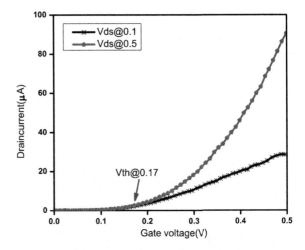

**Table 1** Sizing of TFET
devices in OPAMP circuit

| Device | (W/L) $\mu$m |
|---|---|
| M0 (Tail current device) | 1.05/0.06 |
| M1, M2 (Input pair of first stage) | 1.05/0.02 |
| M3, M4 (diode connected load) | 1.0/0.02 |
| M5 (load device of second stage) | 1.76/0.02 |
| M6 (Input device of second stage) | 1.0/0.02 |

**Fig. 4** Drain characteristics
of TFET

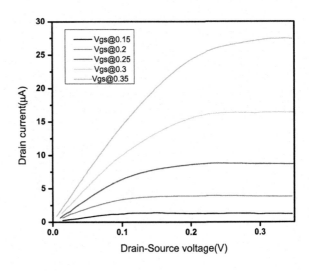

**Fig. 5** Transconductance
efficiency versus gate voltage

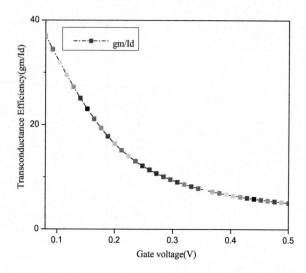

## 3 Proposed Two-Stage Opamp

A novel TFET-based two-stage opamp is designed and simulated in cadence environment. This opamp shows improved results as compared to the conventional structure. And, the novelty of proposed opamp is there is no need for miller compensation
capacitor, intern which is saving the on-chip area. The unity gain frequency is 3 MHz
having two poles at output node and diode connected node. The gain is 46 dB, phase
69° and ultra level power consumption of 1.5 µW with low operating voltage 0.5 V.
The proposed schematic of two-stage opamp is shown in Fig. 6

$$A_v = G_m R_{out} \tag{3}$$

where $G_m = g_m$ is the transconductance of opamp and $R_{out} = (r_{op}//r_{on})$ is the output
resistance of second stage.

$$I_{out} = g_m v_{gs,in} \tag{4}$$

$$\omega_{p1} = \frac{1}{R_{out} C_{out}} = \frac{1}{R_{out}(C_L + C_{gdpTFET} + C_{gdnTFET})} \tag{5}$$

$$f_T = \frac{g_m}{2\pi(C_{gd} + C_{gd})} \tag{6}$$

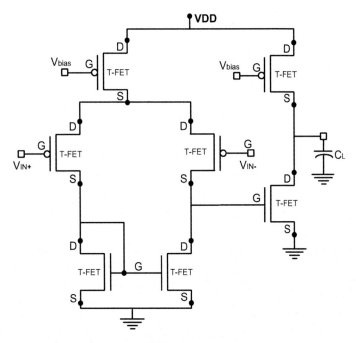

**Fig. 6** Proposed two-stage opamp with TFETs

## 4 Conclusion

This paper proposes a novel TFET-based two-stage operational amplifier in Cadence environment. The proposed topology is highly efficient as compared to other two-stage operational amplifiers in terms of area and power. In the absence of miller capacitor ($C_{miller}$), it minimizes the area consumption which is the novelty of this work. The TFET has lower threshold voltage is the added benefit for the low-voltage operation. The TFET (20 nm)-based opamp implemented with the bias current of nanoAmp level and supply voltage of 0.5 V. The simulation results shown in Table 2 of proposed design shows the gain of 46 dB, Phase is 68° with power consumed is 1.5 $\mu$W having a $V_{DD}$ of 0.5 V. The proposed topology shows the benefits, as the usage for ultralow-power applications such as biomedical applications (Fig. 7).

**Table 2** Simulation results of proposed two-stage OPAMP

| Parameter | Value |
| --- | --- |
| Technology (μm) | 0.02 |
| Supply voltage (V) | 0.5 |
| Total bias current (μA) | 3.1 |
| Gain (dB) | 46 |
| Phase (deg) | 68 |
| Power consumption (μW) | 1.5 |
| CMRR (dB) | 66 |
| Bandwidth (Hz) | 1–10 k |
| UGF (MHz) | 3 |

**Fig. 7** Simulation results of proposed opamp gain and phase

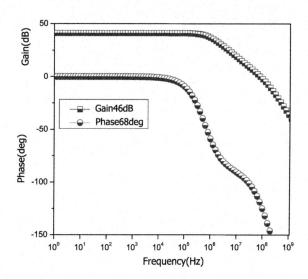

# References

1. Kim, H.S., Cha, H.K.: A low-power, low-noise neural recording amplifier for implantable bio-medical devices. In International SoC Design Conference (ISOCC), (2016) 275–276
2. Kim, H.S., Cha, H.K.: A low-power capacitive-feedback CMOS neural recording amplifier for biomedical applications. In International SoC Design Conference (ISOCC), (2016) 279–280
3. Yang, T., Holleman, J.: An ultralow-power low-noise CMOS biopotential amplifier for neural recording. IEEE Transactions on Circuits and Systems II: Express Briefs, 62 (2015) 927–931
4. Liu, J., Zhang, X., Li, B., Liu, M., Chen, H.: A compact neural amplifier with high input impedance. In Advanced Information Technology, Electronic and Automation Control Conference (IAEAC), (2015) 233–237
5. Reddy, C.L., Singh, T., Ytterdal, T.: Inverter-based 1 V analog front-end amplifiers in 90 nm CMOS for medical ultrasound imaging. Analog Integrated Circuits and Signal Processing, 67 (2011) 73–83

6. Tanimoto, H., Yazawa, K., Haraguchi, M.: A fully differential OTA based on CMOS cascode inverters operating from a 1-V power supply. Analog Integrated Circuits and Signal Processing, 78 (2014) 23–31
7. Saidulu, B., Manoharan, A., Sundaram, K.: Low Noise Low Power CMOS Telescopic-OTA for Bio-Medical Applications. Computers, 5 (2016) 25
8. Beneventi, G.B., Gnani, E., Gnudi, A., Reggiani, S., Baccarani, G.: InAs TFET optimized by means of TCAD to meet all the ITRS specs at VDD = 0.5 V. In Proc. Int. Semicond. Device Res. Symp. (ISDRS), (2013) 1–2
9. Sedighi, B., Hu, X.S., Liu, H., Nahas, J.J., Niemier, M.: Analog circuit design using tunnel-FETs. IEEE Transactions on Circuits and Systems I: Regular Papers, 62 (2015) 39–48
10. Bhushan, B., Nayak, K., Rao, V.R.: DC compact model for SOI tunnel field-effect transistors. IEEE Transactions on Electron Devices, 59 (2012) 2635–2642
11. Lu, H., Kim, J.W., Esseni, D., Seabaugh, A.: Continuous semi-empirical model for the current-voltage characteristics of tunnel FETs. In 15th International Conference on Ultimate Integration on Silicon (ULIS), (2014) 25–28
12. Saurabh, S., Jagadeesh kumar, M.: Fundamentals of Tunnel Field-Effect Transistors, CRC, Florida, (2016)
13. A universal TFET compact model implemented in verilog-A, https://nanohub.org/publications/31/1

# Design of Ultralow Voltage-Hybrid Full Adder Circuit Using GLBB Scheme for Energy-Efficient Arithmetic Applications

Kishore Sanapala, L. Rekha Shree and R. Sakthivel

**Abstract** In recent years, ultra low voltage (ULV) operation is gaining more importance to achieve minimum energy consumption. In this paper, the performance of the gate level body biasing (GLBB) is evaluated in subject to the subthreshold hybrid full adder logic design which employs CMOS logic and Transmission Gate (TG) logic. The performance metrics—energy, power, area, delay, and EDP are calculated and compared with the conventional CMOS (C-CMOS) Full adder. The simulations are performed in cadence at ULV of 200 mV using 90 nm CMOS technology. The obtained results showed that the proposed subthreshold hybrid full adder circuit with GLBB scheme achieves more than 44% savings in delay, 20% savings in energy consumption, and 55% savings in EDP in comparison with the conventional CMOS configuration and other hybrid counterparts.

**Keywords** CMOS logic · Energy · GLBB · Full adder · Transmission gate logic · ULV

## 1 Introduction

With the technology trends scaling toward the submicron regime, and the growing demand for the need of portable battery operated devices like laptops, calculators, mobiles, wrist watches, and IOT devices, the energy consumption of digital circuits is becoming a major concern [1]. Scaling down the supply voltage ($V_{dd}$) is an

K. Sanapala (✉) · L. R. Shree · R. Sakthivel
Department of Micro & Nanoelectronics, School of Electronics Engineering,
VIT University, Vellore, India
e-mail: kishore.technova@gmail.com

L. R. Shree
e-mail: rekha181295@gmail.com

R. Sakthivel
e-mail: rsakthivel@vit.ac.in

© Springer Nature Singapore Pte Ltd. 2018
J. Anguera et al. (eds.), *Microelectronics, Electromagnetics and Telecommunications*, Lecture Notes in Electrical Engineering 471,
https://doi.org/10.1007/978-981-10-7329-8_22

effective knob to minimize the energy and power consumption for ultralow power (ULP) applications. To achieve ULP with acceptable performance, operating digital circuits in the subthreshold region ($V_{dd}$ is less than the transistor threshold voltage ($V_T$)) is one of the solutions [2]. The MOS transistor's subthreshold current is given by Eq. 1 [1].

$$I_{Sub} = I_0 \exp\left(\frac{V_{GS} - V_T}{nV_{th}}\right), \tag{1}$$

and $I_0 = \mu_0 C_{ox} W / L (n-1) V_{th}^2$,

where

| | |
|---|---|
| $V_T$ | Threshold voltage |
| $V_{th}$ | Thermal voltage $(KT/q = 26\,mV)$ |
| $V_{GS}$ | Gate to Source voltage |
| n | subthreshold slope factor $(1 + C_{depletion}/C_{oxide})$ |
| W/L | Effective channel width to the length ratio |
| $\mu_0$ | zero bias mobility |
| $C_{depletion}$ | depletion capacitance |
| $C_{oxide}$ | oxide capacitance |

Arithmetic operations play a crucial role in most of the computing applications. Frequently applied arithmetic operations are add, subtract, multiply, and accumulate, where the fundamental unit for these operations is 1-bit full adder circuit. Hence, evaluating the performance of 1-bit full adder circuit is necessary to enhance the overall system performance.

Many full adders circuit designs employing different logic styles have been reported earlier in the literature. C-CMOS [3], pass transistor logic (PTL) [4], and TG [5, 6] logic styles are the most conventional logic designs. Each of the logic designs is having its own advantages and drawbacks with one of the performance parameters—power, delay, and area. However, these designs show good functionality and performance in the above threshold operation ($V_{dd} > V_T$), but the case differs when comes to the subthreshold operation ($V_{dd} < V_T$). The performance of the circuit degrades because of the exponential increase in delay with the supply voltage scaling [1]. Also because of the reduced output swing, the logic styles like CPL may lead to functionality failure for some input test cases. The C-CMOS logic is the most optimal design style for subthreshold operation, as it provides full output swing and more robust against PVT variations than the other logic designs [1, 2] but it requires more number of transistors which lead to more power consumption and long carry propagation paths in the design of full adder circuits.

In this paper, a new ULV energy-efficient hybrid full adder circuit, employing multiple logic styles (C-CMOS and TG) along with GLBB scheme has been designed for energy-efficient arithmetic applications. The remaining of the paper is structured as below. The proposed full adder's design methodology is mentioned in Sect. 2 followed by results and comparative study in the third section. Lastly in the fourth section conclusions were made.

## 2 Proposed Hybrid Design

The proposed hybrid design employs static CMOS logic [3], Transmission gate logic (TG) [5, 6] with GLBB scheme [7] as shown in the Fig. 1. The basic logic design structure of the proposed design is similar to the hybrid design proposed in [8] which uses static CMOS and TG logic. Two identical XNOR gates and transmission gates were used to generate the output sum and the output carry ($C_{out}$), respectively. Since the sum block consumes more power, the XNOR gates used are designed with minimum transistor count to reduce power consumption and swing restoring transistors (M5 and M6) are used to ensure full output swing. Full logic swing is ensured for generating $C_{out}$ by using the transmission gates. The carry propagation delay is greatly reduced, as only one transmission gate (M15 and M16) is used for propagation of input carry ($C_{in}$). As supply voltage scales down, the performance of the circuit degrades due to the exponential increase in delay. If this supply voltage scaling continues toward the subthreshold regime, the degradation in the circuit performance may results in the huge energy consumption [1]. Here, GLBB scheme is employed with this design to improve the energy efficiency while operating in the subthreshold region. We have employed GLBB scheme, since it is the most robust against the process and temperature variation [9].

In general, body biasing involves connecting transistor body terminal to a bias network in the circuit, instead of $V_{dd}$ or gnd. The body bias can be supplied from an external (off-chip) or an internal (on-chip) source. In GLBB scheme, an external body biasing generator (BBG) is used for each logic blocks or gates in the circuit. The BBG's as mentioned in Fig. 1 controls the body voltage of all the transistors in the corresponding logic blocks. The BBG is a simple push–pull amplifier. It acts as a voltage follower for the output voltage ($V_{out}$) while decoupling large body capacitances from the output node [7]. When $V_{out}$ is high (low), the BBG makes the NMOS (PMOS) transistors in the logic block to switch faster by transferring a high (low) voltage value on the $V_B$ net. Since the MOSFET'S in the logic block (either PMOS or NMOS gates) will be forward biased in prior to the arrival of gate inputs, the driving current capability of the output is greatly improved. The primary advantage of using GLBB scheme in comparison with the other ULV body biasing schemes like Dynamic Threshold (DTMOS) [10] and forward body biasing (FBB) [11] is that the parasitic delay in charging the body terminal of the transistors does not affect the speed of logic gates. Unlike, in DTMOS logic scheme, GLBB reduces the unnecessary switching of the body capacitances in the case, where the input signals switch without any change in the output of the logic gate.

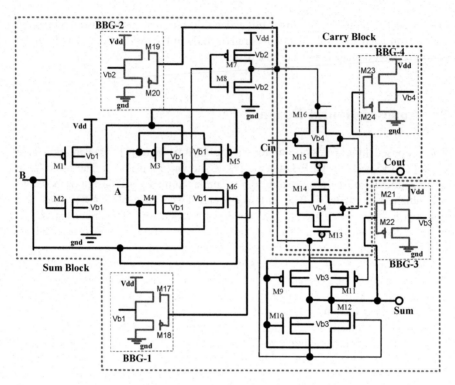

**Fig. 1** Proposed hybrid full adder circuit

# 3   Results and Discussion

The proposed subthreshold hybrid full adder circuit is simulated and the results are analyzed in comparison with the conventional full adder logic designs. All the simulations are performed in cadence at ULV of 200 mV and operational frequency of 20 kHz using 90 nm CMOS technology. Table 1 shows the comparison of the performance metrics: power, delay, energy, and EDP obtained from the simulations. The comparisons clearly infer that the proposed design consumes less energy and EDP than the other designs.

**Table 1** Simulation results for different ULV full adder designs

| Design | Average power (pW) | Delay (ns) | Energy (aJ) | EDP (yJs) |
|--------|--------------------|------------|-------------|-----------|
| C-CMOS | 943.4 | 50.9 | 48.019 | 2.44 |
| GLBB-CMOS | 1043.4 | 43.22 | 42.379 | 1.827 |
| Hybrid FA | 1018.2 | 42.23 | 42.99 | 1.811 |
| Proposed | 1494.3 | 23.32 | 34.847 | 0.812 |

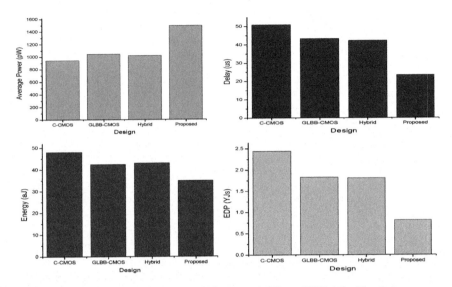

**Fig. 2** Power, delay, energy, and EDP comparisons of different ULV full adder designs

The comparison plots of performance metrics for different full adder designs are shown in Fig. 2. The layout of the proposed subthreshold hybrid full adder design is shown in Fig. 3. Figure 4 shows the obtained timing waveform after post-layout functional simulation of the proposed full adder circuit. It can be noticed that the

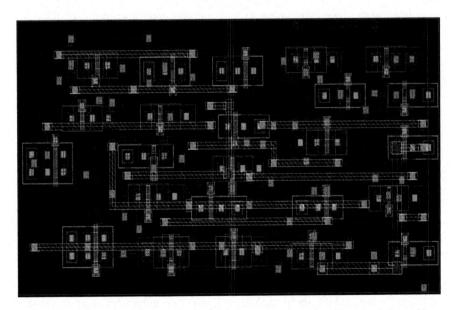

**Fig. 3** Layout design of the proposed subthreshold hybrid full adder circuit in cadence 90 nm technology

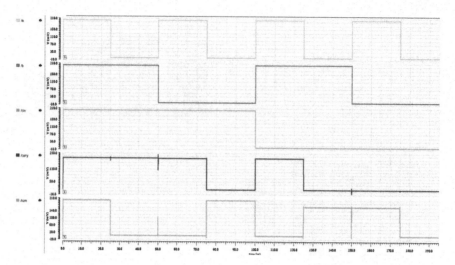

**Fig. 4** Post-layout functional simulation waveform of the proposed subthreshold hybrid full adder circuit in cadence 90 nm technology

proposed design with the layout area of 37.6 μm², achieved more than 44% savings in delay, 20% savings in energy consumption and 55% savings in EDP in comparison with the conventional CMOS configuration and other hybrid counterparts. The proposed design manages to consume only 34.847aJ of energy, which is 30% and 20% lesser than the C-CMOS and hybrid designs, respectively, despite suffering from more power consumption (58% and 46% more than the C-CMOS and hybrid designs, respectively). This is because of the constant forward body biasing mechanism provided by the GLBB technique used in the proposed design which results in huge delay savings (more than 54% and 45% than the C-CMOS and hybrid designs, respectively).

## 4 Conclusions

In this paper, a new subthreshold hybrid full adder circuit which employs CMOS and TG logic with Gate Level Body Biasing (GLBB) scheme is designed to operate in the subthreshold regime for achieving minimum energy consumption. The simulations of the circuits have done using cadence 90 nm technology with a supply voltage of 200 mV. The obtained simulation results showed that the proposed design outperforms the other designs (CMOS, GLBB-CMOS, Hybrid-FA) by achieving more than 44% delay savings, 20% energy savings, and 55% EDP savings. Hence, the proposed full adder circuit can be used as one of the substitutes instead of many subthreshold adders designed for energy-efficient arithmetic applications.

# References

1. Alice Wang, B.H. Calhoun, and A. Chandrakasan. Subthreshold design for ultra low-power systems. 1st edition. Springer: 2006.
2. H. Soeleman and K. Roy. Ultra low power Subthreshold digital logic circuits. Proceedings of IEEE conference on Low power electronics and design, 1999, 94–96.
3. Neil. H. Weste and David Harris. CMOS VLSI design-A circuits and systems perspective. 3rdEdition. Addison Wesley: 2004.
4. Zimmermann R, Fichtner W. Low-power logic styles: CMOS versus pass-transistor logic. IEEE Journal of Solid-State Circuits, 1997, 32(7):1079–1090.
5. Alioto M, Cataldo GD, Palumbo G. Mixed full adder topologies for high-performance low-power arithmetic circuits. Microelectronics Journal, 2007, 38(1):130–139.
6. Shams AM, Darwish TK, Bayoumi MA. Performance analysis of low-power 1-bit cmos full adder cells. IEEE Transactions on VLSI Systems, 2002, 10(1):20–29.
7. M. Lanuzza, R. Taco and D. Albano, Dynamic gate-level body biasing for subthreshold digital design, 2014 IEEE 5th Latin American Symposium on Circuits and Systems, Santiago, 2014, pp. 1–4.
8. Partha B, Bijoy K, Sovan G and Vinay K. Performance Analysis of a Low-Power High Speed Hybrid 1-bit Full Adder Circuit. IEEE Transactions on VLSI Systems, 2015, 23(10): 2001–2008.
9. R. Taco, M. Lanuzza, and D. Albano, "Ultra-Low-Voltage Self Body Biasing Scheme and its application to basic Arithmetic Circuits". J. VLSI Design, 2015, pp. 1–10.
10. Assaderaghi F, D. Sinitsky, S. Parke, J. Bokor, P. K. Ko, and C. Hu, A dynamic threshold voltage MOSFET (DTMOS) for ultra-low voltage operation. IEDM Tech. Dig., 1994, pp. 809–812.
11. Shih-Fen Huang et.al, Scalability and biasing strategy for CMOS with active well bias. 2001 Symposium on VLSI Technology, pp. 107–108.

# Closed-Loop Blood Glucose Control for Type I Diabetes Patients Using PID Controller

Bharat Singh, Shabana Urooj and Ravi Sharma

**Abstract** In this paper, we have designed a closed-loop controller for continuous infusion for diabetic therapy, while the typical way of insulin infusion is discrete manner, based upon long-term interval measurement. We proposed an optimal PID (proportional integral derivative). An automated system integrated with biosensor used for measuring blood glucose concentration and infusion pump mimics the action of natural insulin secretion of a healthy person. Using controllers, we maintained blood glucose concentration in a certain range. The overall control strategy based upon feedback to overcome glucose variations patient body. The result of this PID control strategy reveals that controller can maintain blood glucose concentration under certain range. The results of Proportional Integral Derivative shows us stable response with presence uncertainty of parameters that can vary from patient to patient.

**Keywords** PID controller · Type I diabetes patient · Glucose management

## 1 Introduction

Approx. 177 million people in the world are facing the diabetic problems this number is going to be twice at the end of 2030. Based upon 2002 death certificate data of US, diabetes was the sixth largest cause of death in the US [1]. Almost 73 249 death certificates issued in which diabetes was the underlying cause [1]. There are plenty of natural feedback systems in the human body to maintain the proper level of hormones in the body. A dysfunction of any of the natural feedback loop in the human body caused illness, which may give short-term and long-term effect. Diabetes mellitus is a metabolic disorder in which insulin a hormone which

B. Singh
Bharati Vidyapeeth's College of Engineering, New Delhi, India

S. Urooj (✉) · R. Sharma
School of Engineering, Gautam Buddha University, Greater Noida, UP, India
e-mail: shabanaurooj@ieee.org

© Springer Nature Singapore Pte Ltd. 2018
J. Anguera et al. (eds.), *Microelectronics, Electromagnetics and Telecommunications*, Lecture Notes in Electrical Engineering 471,
https://doi.org/10.1007/978-981-10-7329-8_23

promotes glucose into cell cannot perform its role properly. In the pancreas, β cells are responsible for producing insulin which makes the level down of blood glucose concentration, α cells of the pancreas which produce glucagon this further break-down glycogen to glucose and increase the blood glucose concentration. The interplay of these hormones maintains the blood glucose concentration within certain range.

For type I diabetes patient, the β cells of the pancreas are completely destroyed. So, for this types of diabetes patient completely depends upon external sources of insulin. According to Diabetes Control and Complications Trial (DCCT) [2] the normal range of blood glucose concentration is 60–120 mg/dL. If the excess level of insulin supplied blood glucose concentration is falling under 60 mg/dL this state is unknown as hypoglycemia or in another case if the supplied level of insulin is not sufficient the blood glucose concentration rises above the normal value of 130 mg/dL is a state is known as hyperglycemia [3]. As it is shown that both cases are different and have different effects like hypoglycemia has short-term effect organ failure or diabetic coma, hyperglycemia has long-term effect like nephropathy, retinopathy, and tissue damage.

In current fashion, the patient has to measure its blood glucose concentration and according to concentration, they should have injected the appropriate amount of insulin to maintain blood glucose concentration. As such treatment lacks a factor of reliability because of there is no system for continuous monitoring of blood glucose concentration. So using this system cannot deal with patients having more fluctuations in blood glucose concentration. Hence, there is always a need of a system that continuously monitors the blood glucose concentration and releases insulin accordingly. Development of such system is a revolutionary improvement in the treatment of diabetes, in some special cases, where there is no medical supervision is possible and patient has lack medical knowledge this continuous system will decrease errors during injection of insulin in the patient body [4] (Fig. 1).

A continuous closed-loop system for controlling glucose concentration of human body is consists of three elements: a glucose sensing element, a control algorithm, and insulin infusion pump. Glucose sensor has been developed for continuous monitoring of blood glucose concentration these sensors may be needle type, extracorporeal sensors, etc. The pump mechanism has already studied and a lot of

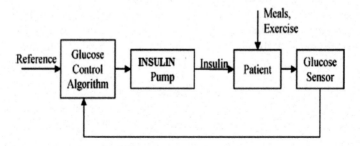

**Fig. 1** Block diagram of closed-loop controller

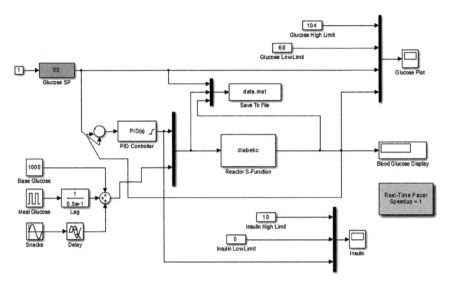

**Fig. 2** Simulation diagram of closed-loop PID control for diabetes patient

work has been done on pump mechanism. The most important thing is to design a control algorithm for controlling insulin infusion rate with taking care of blood glucose concentration at the real time. And our main focus in this paper is to design a control strategy.

There many control algorithms for controlling blood glucose concentration. PID is one of the traditional controller in such type of control strategies. The PID controller has been used in many researches as there shown that it can function well with human body parameters. The disadvantage if using PID controller that it does not consider the changes in parameters and that is the main factor in the biological system that parameters are variable from patient or patient or even for the same patient.

## 2 Insulin—Glucose System Model

Bergmenn minimal model is most widely used for blood glucose concentration control. To reduce the nonlinearity of the system this linear system was proposed bergmenn. Three equations represent the three different compartments [5].

$$\dot{G}(t) = -\left[p_1 + \dot{X}(t)\right]G(t) + p_1 G_b + m(t) \tag{1}$$

$$\dot{X}(t) = -p_2 X(t) + p_3 I(t) \tag{2}$$

$$\dot{I} = \tau \cdot u(t) - n \cdot I(t) \tag{3}$$

G(t) is known as plasma glucose concentration (mg/dl), I(t) is known as plasma insulin concentration (mg/dl), X(t) is known as effective insulin response(1/dl). p1, p2, p3, n, $\tau$ are the different biological parameter of bergmenn minimal model. Here Gb (mg/dl), Ib (mU/dl) is the basal value of glucose and insulin. m(t) exogenic rate of insulin infusion.

## 2.1 Transfer Function

The transfer function of the framework is the Laplace of Output that is glucose concentration of the in blood to the Laplace of information which is insulin level that given to the body.

With a specific end goal to rearrange the nonlinearity, the possibility of linearity of Bergman model is proposed in, so we take the more disentangled condition of above.

$$\dot{G}(t) = [p_1 X(t)]G(t) - \overline{G} \cdot X(t) + \overline{GX} + p_1 G_b + m(t) \tag{4}$$

$$\dot{X}(t) = -p_2 X(t) + p_3 I(t) \tag{5}$$

$$\dot{I} = \tau \cdot u(t) - n \cdot I(t) \tag{6}$$

After taking Laplace of Eqs. (4)–(6) we have to apply values of following:

Function f(t) is drawn closer by a high-order binomial. The model of insulin to glucose, by applying Laplace for (4)–(6) the transfer function from insulin to glucose is given the parameters said above.

$$G(s) = \frac{0.00003}{s^3 + 0.274s^2 + 0.013656s + 0.00018} \tag{7}$$

The fundamental aggravation brought on by three suppers can be delineated. Under the unsettling influence, we need to design a controller to meet the necessities of typical individual's glucose concentration of around 60–100 mg/dL before the feast and under 140 mg/dL after dinner. As the feast unsettling influence is significantly higher than ordinary glucose level, it requests a controller with great execution of aggravation dismissal. Considering the conceivable unmolded nonlinearities, one control procedure is acquainted with configuration comparing controllers, the PID controller.

# 3  PID Controller

Real plants are probably not going to have an immaculate first-order lag, yet this estimate is sensible to depict the frequency response roll off in a greater part of cases. Higher arrange posts will present an additional phase shift, in any case. Regardless of the possibility that they do not influence the state of the gain roll-off, phase shift matters a great deal to circle solidness. You cannot rely on a single "lag" shaft to coordinate both the adequacy roll off and the stage move precisely.

Ziegler–Nichols display presumes an extra anecdotal phase adjustment that does not misshape the expected size roll off. At the dependability edge, there is a 180° phase shift around the input circle (Nyquist's stability criterion). First-order lag can be close to 90° of the phase shift. Whatever remains of the watched phase shift must be secured by the simulated phase modification. Phase modification is ventured to be a straight line in the vicinity of zero and the basic recurrence where 180° of phase shift happens.

## 3.1  PID Controller Theory

For given PID controller, we have used Ziglar–Nicholas method for tuning the parameters. Ziglar Nicholas is well-known PID tuning rule which gives best values of PID parameters.

Applying simple rules of Ziegler–Nichols, we have tuned our parameters for getting the desired result in terms blood glucose concentration.

## 3.2  Simulation

In Fig. 3 we can show how meal glucose error and disturbance works. The base value of glucose is taken 1000 which is in addition to other values of glucose like meal glucose and snacks. Here, meal glucose has been adding at the meal time which is after every 6–8 h and snacks which depend on person or patient can be taken in a random manner which has no particular time whenever patient wants to take some extra food either then its meal that counts in snacks. Snacks also matter a lot hence, it also increases the blood glucose level. For example, in night patient feels hungry it has eaten an apple that apple increases his or her blood glucose concentration but you have not designed your system according to it. So, for creating a biological system you to take each and every side system in your account so that you can according to it. As you can see above that meal is added to the system at every 6 h which works as breakfast, lunch, and dinner. When patient taken its meal glucose concentration of patient also changes because patient takes

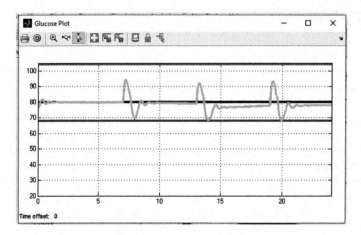

**Fig. 3** Blood glucose output

some amount of glucose with the food it takes and rest amount of glucose enters in the body when digestive enzymes work at that time body releases glucose hormone in i the body in larger scale and for type I diabetes patient there is internal insulin generation, hence there is sudden rise body glucose concentration. When designing a system meal glucose also taken into account better functioning of the system and your system can deal with adverse effects.

Both delay and lag element work in similar way lag element we have used simple lag equation and for random errors, we have used available delay system block in MATLAB with inherent or default values. These blocks basically used when we designed any biological system it helps our system to makes it more controllable and we get a system that we can relate to our real system. So, we when we take some food there is some delay time our body took for releasing glucose or we have random snacks that time we defined using these two blocks so that we can make our simulation realistic and more usable (Fig. 2).

The output of PID controller feed to the diabetic function which includes the patient transfer function and patient transfer function depends on many variables like:—Blood Insulin response, Blood glucose sensitivity, Blood insulin sensitivity, and Insulin response time. The output of PID is in addition to the meal disturbances. After feeding these outputs to the diabetes function we get the result which blood glucose concentration.

# 4 Result

As, above Fig. 4 shows that blood glucose concentration is under certain range that is 70–100 mg/dl. Despite of meal disturbances, our blood glucose level does not cross limited range, hence it can be bearable for humans. When blood glucose

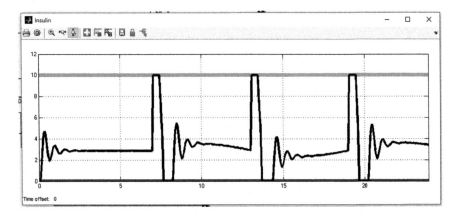

**Fig. 4** Output of PID controller

concentration rises pump releases the insulin accordingly and blood glucose concentration comes below. After reaching a minimal value of 80 mg/dl insulin infusion stops so blood glucose concentration maintain the basal value that is 80 mg/dl. The output of the controller is actually the insulin infusion to the patient.

## 5 Conclusion

In the end, we conclude that despite meal disturbances acting on our system using PID controller we maintained under blood glucose concentration under limited range that can be bearable to the patient. Our system works well for disturbance and gives fruitful results, Using Fuzzy can provide us more stable results on disturbances and on our system function in better form.

## References

1. Jiming Chen, Kejie Cao, Youxian Sun, Yang Xiao, and (Kevin) Su "Continuous Drug Infusion for Diabetes Therapy: A Closed-Loop Control System Design" EURASIP Journal on Wireless Communications and Networking Volume 2008.
2. Pinky Dua, Francis J. Doyle, III, and Efstratios N. Pistikopoulos "Model-Based Blood Glucose Control for Type 1 Diabetes via Parametric Programming" IEEE TRANSACTIONS ON BIOMEDICAL ENGINEERING, VOL. 53, NO. 8, AUGUST 2006.
3. Camelia Owens, Howard Zisser, Lois Jovanovic, Bala Srinivasan, Dominique Bonvin, and Francis J. Doyle, "Run-to-Run Control of Blood Glucose Concentrations for People With Type 1 Diabetes Mellitus" III IEEE TRANSACTIONS ON BIOMEDICAL ENGINEERING, VOL. 53, NO. 6, JUNE 2006.

4. D. U. Campos-Delgado*, M. Hernández-Ordoñez, R. Femat, and A. Gordillo-Moscoso "Fuzzy-Based Controller for Glucose Regulation in Type-1 Diabetic Patients by Subcutaneous Route" IEEE TRANSACTIONS ON BIOMEDICAL ENGINEERING, VOL. 53, NO. 11, NOVEMBER 2006
5. Mohamed Al-Fandi, Mohammad A. Jaradat, Yousef Sardahi " Optimal PID-Fuzzy Logic Controller for Type 1 Diabetic Patients" Jordan University EURASIP Journal on Wireless Communications and Networking Volume 2010.

# FPGA-Based Implementation of AES Algorithm Using MIX Column

S. Neelima and R. Brindha

**Abstract** This article deals with the clear analysis and experimental simulation results of the modified AES-128-bit algorithm which can be personalized. To improve this technique, we introduced the high-level increased parallelism scheme which will reflect even in Mi columns of the AES architecture. By using this technique, we can increase the throughput efficiency and is implemented on Quartus of FPGA device. With this technique, usage can increase the stack usage for 5% more with a minimum reduction of 30% area.

**Keywords** AES · MIX Column · FPGA

## 1 Introduction

Nowaday technological evolution has rapidly been increasing but on the flipside, fraudulent practices to have been increasing which in turn has become a challenge to the matters concerning security and confidentiality. Cryptography is used for sending a secret message to someone else when the scope for open access is presumed to be huge. It plays a vital role in this regard to provide security and file integrity. So, to improve security and to enhance the performance, we need to perform algorithms to encipher and decipher the text. To implement these algorithms, the processor will take lots of CPU memory, and the coding complexity will increase. So, the embedded systems cannot yield satisfactory and speedy output.

S. Neelima
Gandhiji Institute of Science and Technology, Jaggayyapet, Bhimavaram, India

R. Brindha (✉)
Faculty of Engineering, Avinashilingam Institute for Home Science
and Higher Education for Women, Coimbatore 641043, India
e-mail: brin1kalai@yahoo.co.in

© Springer Nature Singapore Pte Ltd. 2018
J. Anguera et al. (eds.), *Microelectronics, Electromagnetics
and Telecommunications*, Lecture Notes in Electrical Engineering 471,
https://doi.org/10.1007/978-981-10-7329-8_24

The encryption standards are classified into two categories; They are symmetric key algorithms and asymmetric cipher algorithms. The symmetric algorithm uses one key whereas asymmetric algorithm uses two different keys. These were faster than public key algorithms because the CPU cycles needed for secret key encipherment and some are for asymmetric encipherment. Encryption Methodology (AES), Decryption Methodology, RC2 and RC6, and Blowfish are some of the symmetric algorithms. Remote Secure Access is an asymmetric cipher algorithm.

The AES works on 128 bits of data. It can encrypt and decrypt the blocks by using cipher keys. The key size may vary to suit the application in hand. For instance, that may be of 128, 192, or 256 bits.

Initially, it was AES silicon and gave a flow rate of 2.29 Gbps via a pipeline architecture no [1]. Improved system performance to AES pipeline, which has a flow rate of 8 Gbps [2]. The first implementation of AES at a rate greater than 10 Gbps by the implementation of Recommendation T box, which includes a combination of sub-Bytes Shift rows and columns, includes AES [3] algorithm. Processor pipeline is fully AES more complex operations and a flow rate of 30 Gbps and 70 [4]. A powerful AES installation with an architecture step includes ten pipeline system performance up to 1.85 Gbps limited by using data in [5] memory. The implementation of AES full pipeline processor on FPGAs indicated an indication of 21.54 Gbps, but took a large area of 5177 segments and showed a latency of 31 cycles given in [6]. AES algorithm uses partial dynamic reconfiguration performed by the new technique. This uses pipeline processing technology and at the same time under the partial dynamic reconfiguration and provides a flow rate of 24922 Gbps but 3576 slices and uses a larger range [7]. A parallelism applied to blocks to cause more delays causing the flow of 68.82 Gbps Mix Columns with Many-Core 167 [8] In June 2003, the National Security Agency (NSA) AES-128 could be used for classified SECRET and AES 192/256 Level TOP SECRET documents.

## 2 Background Methodology

AES (Advanced Encryption Standard) is the algorithm for the encryption of electronic data from the various sources. The algorithm converts the data into an encrypted format with help of key; the encrypted data cannot be read by the third person. The receiver can read the encrypted data after the decryption is performed. For both encryption and decryption, the algorithm uses the same key. That is why this algorithm is also known as a symmetric key algorithm. The steps followed in the algorithm are given below,

|  | Algorithm steps for AES encryption standard |
|---|---|
| Initial round | Add round key |
| First round | Sub-bytes |
|  | Shift rows |
|  | Mix column |
|  | Add round key |
| Final round (no mix columns) | Sub-bytes |
|  | Shift rows |
|  | Add round key |

In the process of encryption and decryption, the algorithm follows same steps repeatedly. Each step is considered as round. The number of rounds depends on the block size of input data to be modified. The size input block varies based on the user requirement. For example, AES-128 consists of 10 rounds, AES-192 consists of 12 rounds, and AES-256 consists of 14 rounds. The iterative process of such steps is to improve the integrity of encryption.

### Add round key:

Add round key is the step which is followed through all rounds of encryption and decryption process, where the input and key generated from key generation algorithm are combined together. For that, the algorithm performs bitwise XOR operation on input and key.

### Sub-bytes:

Input bytes are illustrated in the form of $4 \times 4$ matrixes. The input matrix also called state matrix. In sub-bytes the algorithm holds a predefined matrix. This is stored in terms of the lookup table. The lookup table is named as S-BOX (Substitution Box). In the above-mentioned step each byte in input matrix is replaced by each byte from the lookup table. This is shown in Fig. 1.

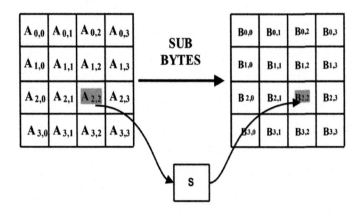

**Fig. 1** Process diagram performing substitution bytes

*Shift rows*:

In this step, each byte in the state matrix is shifted in a certain method, the shifting method is explained in Fig. 2.

In the Fig. 2 it can be observed that the first row remains the same during the entire process. The second row undergoes complete byte shift by one subblock. Similarly, the third and fourth row undergoes twice and thrice shifts, respectively.

*Mix column*:

During mix column operation each column from input state matrix is multiplied with the predefined constant matrix. The resulting column is illustrated as the corresponding column. So each input matrix will affect each column of resulting matrix (Fig. 3).

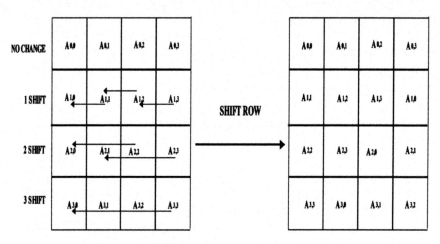

**Fig. 2** Process diagram performing shifting of rows

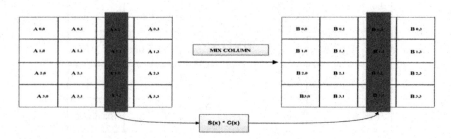

**Fig. 3** Process diagram performing mix column

## Four Stage Parallelisms

### In Mix Column

The Mix Columns transformation of AES functions on a four-column data block is shown in Fig. 4. It works on each column and it is independent. The introduction of the parallelism in MIX Column block would lessen 60% latency of the mix column execution delay in a distinct loop. Consequently, the throughput of the Mix Columns performance is increased. Every Mix Column block computes only one column at a stretch rather than a whole data block in four stage parallelism.

### Eight Stage Parallelisms in Mix Column

By modifying the fourth stage to eighth stage parallelism, whose procedure called "Increased Parallelism technique." This technique will increase the efficiency of throughput. The clear analysis is explained in the below figure. So the simulation/execution of these blocks will be done in single cycle, where eight stages of parallelism include and divide each clock period. In the matrix, two elements are executed with the 1/8th clock period. By this reduction leads to reduce the process of latency and optimizes the area with an increase in speed.

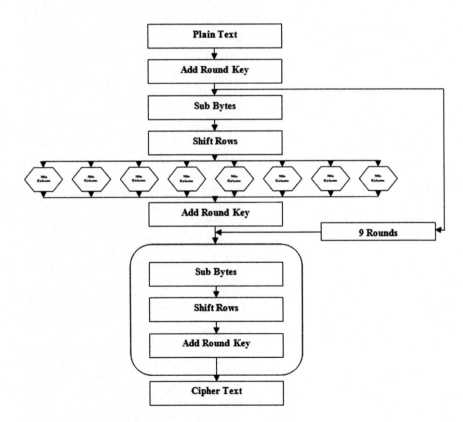

**Fig. 4** Parallel MIX Column

## Simulation Results

The algorithms presented in this paper are implemented in different devices. The metrics like power, delay, and LUT are measured and tabulated. Table 1 shows the

**Table 1** Comparision of results

| Parameter | Cyclone 2 | Stratix 3 |
|---|---|---|
| Power | 601.80 mw | 441.21 mw |
| Delay | 14.605 ns | 12.663 ns |
| Number of LUTs | 256 | 221 |

**Fig. 5** RTL view 1

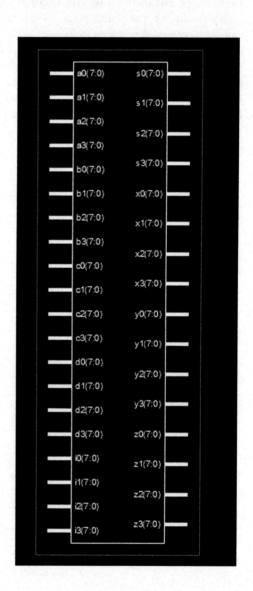

performance of the AES algorithm. The power in stratix 3 is improved by about 47% in cyclone 2. Similarly, the performance toward delay is also improved by about 16%. When power is considered the stratix II kit gives a 20% improvement in power when compared to Cyclone II. The delay is reduced by about 50%.

**RTL View**:

To view the pictorial representation of implemented design RTL Schematic view is shown in Figs. 5 and 6. Here, we have implemented the parallel mix column algorithm. The Fig. 4 shows the RTL Schematic view of implemented algorithm.

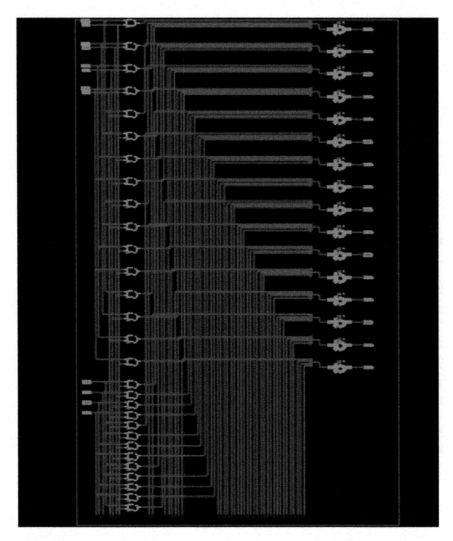

**Fig. 6** RTL view 2

The Fig. 4 represents about a number of inputs and outputs. Also, provide information about the bit details of each input and output. More details about the blocks and inner blocks available in the implemented design are shown in Fig. 6. This includes all detailed information about input and output.

## 3 Simulation Output

The simulation output part consists of two parts representing the function of input and output. All values of input and output shown in hexadecimal values. Based on the user convenience it can be changed to binary or decimal values. In the output window a, b, c, d, and i denotes inputs on the other hand s, x, y, and z denotes the outputs (Figs. 7 and 8).

**Gate Count**:

Figure 9 represents the design summary of implemented design. It includes information about the number of LUTs, number of slices and number of gates occupied by the implemented design. Also, it displays utilization percentage. It compares the available and utilized components from which we can identify utilization percentage.

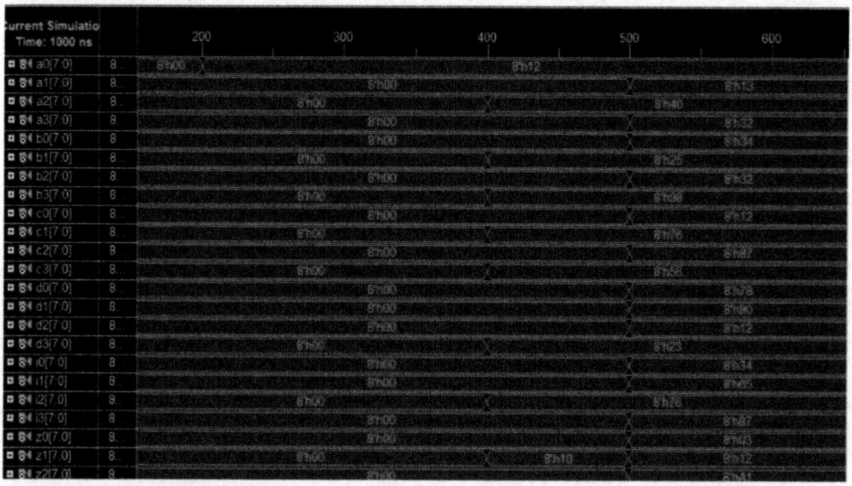

**Fig. 7** Simulation output 1

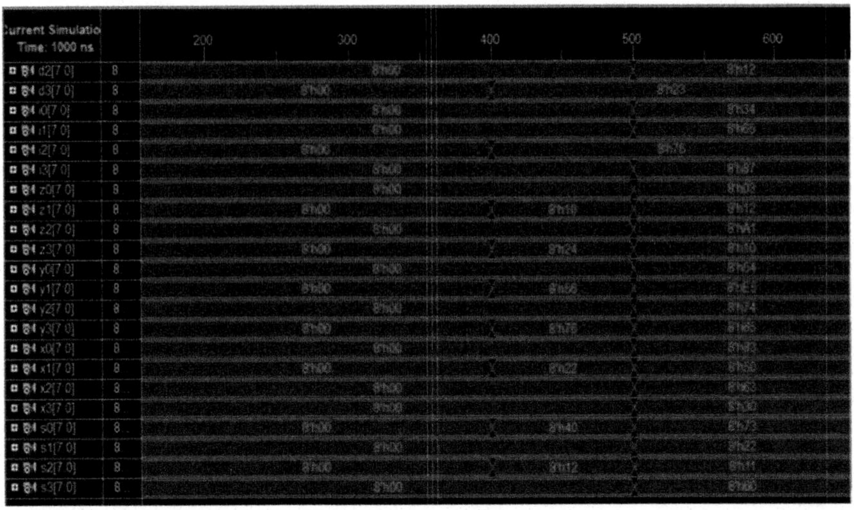

**Fig. 8** Simulation output 2

| AES Project Status | | | |
|---|---|---|---|
| Project File: | AES.ise | Current State: | Mapped |
| Module Name: | matm | • Errors: | 0 Error |
| Target Device: | xc3s100e-4cp132 | • Warnings: | 3 Warnings (3 new, 0 filtered) |
| Product Version: | ISE 9.2i | • Updated: | Sat Mar 25 13:16:15 2017 |

| AES Partition Summary |
|---|
| No partition information was found. |

| Device Utilization Summary | | | | |
|---|---|---|---|---|
| **Logic Utilization** | **Used** | **Available** | **Utilization** | **Note(s)** |
| Number of 4 input LUTs | 384 | 1,920 | 20% | |
| **Logic Distribution** | | | | |
| Number of occupied Slices | 192 | 960 | 20% | |
| Number of Slices containing only related logic | 192 | 192 | 100% | |
| Number of Slices containing unrelated logic | 0 | 192 | 0% | |
| **Total Number of 4 input LUTs** | 384 | 1,920 | 20% | |
| Number of bonded IOBs | 288 | 83 | 346% | OVERMAPPED |
| Number of bonded Out/Bidir IOBs | 128 | 72 | 177% | OVERMAPPED |
| Number of bonded Input IBUFs | 160 | 11 | 2666% | |
| **Total equivalent gate count for design** | 2,304 | | | |
| Additional JTAG gate count for IOBs | 13,824 | | | |

Detailed Reports

**Fig. 9** Gate count

## Delay Result:

The important parameter in FPGA design is a delay. Here, the delay is calculated from design summary. In that design summary window to calculate delay should go with synthesis report. The calculated delay is shown in Fig. 10.

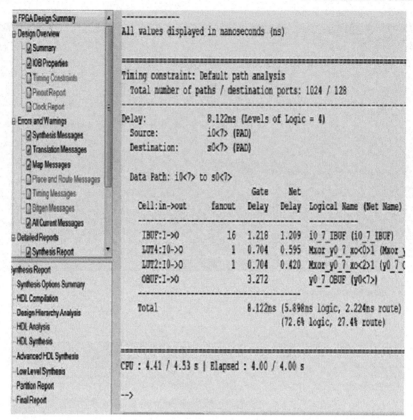

**Fig. 10** Delay result report

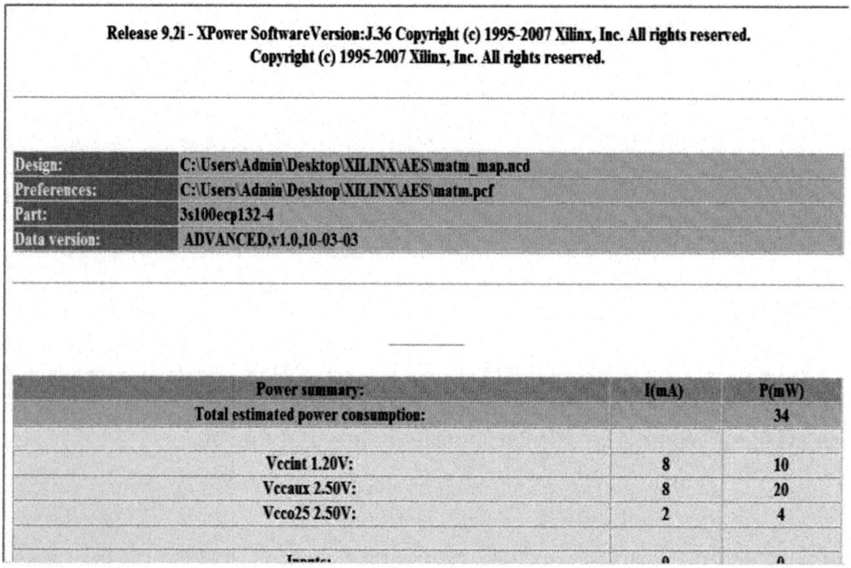

**Fig. 11** Power analysis 1

**Power Analysis**:

For calculating the power the separate simulation tool XPower is used here. Even though it is separate it can calculate the power of the supporting files generated from Xilinx 9.2i. We need to attach the files generated during the implementation of the proposed design in Xilinx 9.2i. That attached file consists of all the parameter details about our implemented design. Figures 11 and 12 show the HTML view of power report. The Fig. 13 shows power summary report.

| Inputs: | 0 | 0 |
|---|---|---|
| Logic: | 0 | 0 |
| Outputs: | | |
| Vcco25 | 0 | 0 |
| Signals: | 0 | 0 |
| | | |
| Quiescent Vccint 1.20V: | 8 | 10 |
| Quiescent Vccaux 2.50V: | 8 | 20 |
| Quiescent Vcco25 2.50V: | 2 | 4 |

| Thermal summary: | |
|---|---|
| Estimated junction temperature: | 27C |
| Ambient temp: | 25C |
| Case temp: | 26C |
| Theta J-A: | 62C/W |

**Fig. 12** Power analysis 2

**Power Summary**:

```
Part:            3s100ecp132-4
Data version:  ADVANCED,v1.0,10-03-03

Power summary:                              I (mA)    P (mW)
----------------------------------------------------------------
Total estimated power consumption:                     34
                                   ---
                        Vccint 1.20V:        8         10
                        Vccaux 2.50V:        8         20
                        Vcco25 2.50V:        2          4
                                   ---
                        Inputs:              0          0
                        Logic:               0          0
                        Outputs:
                           Vcco25            0          0
                        Signals:             0          0
                                   ---
            Quiescent Vccint   1.20V:        8         10
            Quiescent Vccaux   2.50V:        8         20
            Quiescent Vcco25   2.50V:        2          4

Thermal summary:
----------------------------------------------------------------
    Estimated junction temperature:                    27C
                    Ambient temp:    25C
                       Case temp:    26C
                       Theta J-A:    62C/W

Analysis completed: Sat Mar 25 13:31:00 2017
----------------------------------------------------------------
```

**Fig. 13** Power summary

# 4 Conclusion

The Advanced Encryption Scheme has a very high throughput which has been implemented in this paper. A new method was introduced along with 128-bit AES which will modify its architecture in Mix Column round. This feature will provide an efficient throughput in the overall performance of AES algorithm. To implement this method device used is Quartus FPGA device which will be most helpful in power consumption. The power consumed for existing Cyclone 2 is about 600 mW and for Stratix 3 is 440 mW. It is about 26% improvement. The delay and area are reduced as observed in the table below.

# References

1. Verbauwhede, P., Schaumont, and Kuo, H.: Design and Performance Testing of a 2.29 gb/s Rijndael Processor. IEEE J. Solid-State Circuits, vol. 38, no. 3, (2003) 569–572
2. Mukhopadhyay, D., RoyChowdhury, D.: An Efficient end to End Design of Rijndael Cryptosystem in 0:18 m CMOS. Proc. 18th Int'l Conf. VLSI Design, (2005) 405–410
3. Morioka, S., Satoh, A.: A 10-gbps full-AES Crypto Design with a Twisted BDD s-Box Architecture. IEEE Trans. Very Large Scale Integration Systems, vol. 12, no. 7, (2004) 686–691
4. Hodjat, A., Verbauwhede, I.: Area-Throughput Trade-Offs for Fully Pipelined 30 to 70 Gbits/s AES Processors IEEE Trans. Computers, vol. 55, no. 4, (2006) 366–372
5. Biglari, M., Qasemi, E., Pourmohseni, B.: Maestro: A high performance AES encryption/decryption system, Computer Architecture and Digital Systems (CADS). 17th CSI International Symposium, (2013) 145–148, 30–31

6. Hodjat. A., Verbauwhede, I.: A 21.54 gbits/s Fully Pipelined AES Processor on FPGA. Proc. IEEE 12th Ann. Symp. Field Programmable Custom Computing Machines, (2004) 308–309
7. Granado-Criado, J., Vega Rodriguez, M., Sanchez Perez, J., Gomez Pulido, j.: A New Methodology to Implement the AES Algorithm Using Partial and Dynamic Reconfiguration. Integration, the VLSI J., vol. 43, no. 1, (2010) 72–80
8. Bin Liu, Bevan M. Baas.: Parallel AES Encryption Engines for Many-Core Processor Arrays. IEEE transactions on computers, vol. 62, no. 3, (2013)

# Impact of High Geomagnetic Activity on Global Positioning System Satellite Signal (L-Band) Delay and Klobuchar Algorithm Performance Over Low Latitudinal Region

K. C. T. Swamy

**Abstract** The Klobuchar algorithm is currently being used by the single frequency Global Positioning System (GPS) user to compute the ionospheric time delay at anywhere in the world. The aim of this paper is a preliminary assessment of Klobuchar algorithm performance by using ionospheric time delay estimated with the data provided by International Global Navigation Satellite System (GNSS) Service (IGS) network on a geomagnetic storm day of high solar activity year, 2016. This work is carried out at various IGS stations, namely PBRI, IISC, HYDE, LCK4 and LHAZ. Klobuchar model mean results agreement with experimental data is acceptable at the considered stations. This is the preliminary work done in the process of improvement of the Klobuchar algorithm for low latitude regions.

**Keywords** GPS · IGS · Ionospheric time delay

## 1 Introduction

Day to day the dependence on Global Navigation Satellite System (GNSS) is increasing in high precision applications such as air transportation, marine communication, missile tracking and guidance civil aviation along with timing applications. But, the radio signals of GNSS satellites are being affected by the different phenomena that occur in space between the Earth and the Sun. The effect is potentially severe on GNSS signals during the geomagnetic storm [1]. Geomagnetic storms have the impact on ionosphere behaviour and produce disturbances in the density of free electrons present in the F2-region. The understanding of ionospheric storms presented in [2–5]. One of the major effects on GNSS signal is refraction, it

K. C. T. Swamy (✉)
Department of Electronics and Communication Engineering, G. Pullaiah College of Engineering and Technology, Kurnool 518452, AP, India
e-mail: kctswamyece@gpcet.ac.in

© Springer Nature Singapore Pte Ltd. 2018
J. Anguera et al. (eds.), *Microelectronics, Electromagnetics and Telecommunications*, Lecture Notes in Electrical Engineering 471,
https://doi.org/10.1007/978-981-10-7329-8_25

causes additional travel time of the signal, and the amount of additional time can be expressed as,

$$\Delta t = \left(\frac{40.3}{cL_1^2}\right) . TEC \quad (\text{Seconds}),$$

where,

'c'  is velocity of light (i.e. $3 \times 10^8$ m/s)
'$L_1$'  is carrier signal frequency in Hz

Total Electron Content (TEC) gives the number of free electrons present along the signal path and it is expressed in TECu (1TECu = $10^{16}$ electrons/m$^2$). Using dual frequency GPS receiver pseudoranges (P1 and P2), Total Electron Content (TEC) along the signal path can be estimated as [6],

$$TEC = \frac{1}{40.3} \frac{L_1^2 . L_2^2}{(L_1^2 - L_2^2)} (P1 - P2) \quad (TECu)$$

where,

$L_1$ (1.575 GHz) and $L_2$ (1.2747 GHz) are the carrier frequencies of GPS.

In this regard, GPS proposed and being used an algorithm known as Klobuchar algorithm [7]. Later, for Galileo single frequency users NeQuick model was proposed [8, 9]. Some significant work was done on klobuchar algorithm performance evaluation over low latitude region during low solar activity year [10]. Since the ionospheric time delay is highly variable in low latitude regions, the continuous study of it is needed. Hence, this paper provides an investigation of geomagnetic activity impact on GPS signal delay variation and evaluation of Klobuchar algorithm with respect to the IGS data over the low latitude region.


## 2   Data Analysis

To accomplish time delay variations of GPS signals and performance evaluation of Klobuchar algorithm in low latitude regions three successive days (7th, 8th and 9th May 2016) data is considered. For the considered days, the ionosphere activity parameters Sun Spot Number (SSN), Kp-index and Ap-index which are measured using a network of data are presented in Table 1. Dual frequency GPS receivers data obtained from IGS network (http://sopac.ucsd.edu/dataBrowser.shtml) is measured on two frequencies (L1 = 1.575 GHz and L2 = 1.227 GHz) at five low latitude stations (Table 2). GPS data obtained from IGS network is in hatanaka format. By using CRX2RNX tool data was converted into compact RINEX observation file format [11, 12]. The ionospheric time delay of GPS signals transmitted on 1.575 GHz is computed by extracting pseudoranges from rinex data.

**Table 1** Solar activity of selected days

| Date | Kp-index | SSN | Ap-index |
|------|----------|-----|----------|
| 7th May, 2016 | $1 \leq Kp \leq 4$ | 34 | 09 |
| 8th May, 2016 | $5 < Kp \leq 6$ | 44 | 70 |
| 9th May, 2016 | $3 \leq Kp \leq 6$ | 52 | 30 |

**Table 2** Coordinates of selected low latitude IGS stations

| S.No | Station ID | Latitude | Longitude |
|------|-----------|----------|-----------|
| 1 | PBRI | 11.63°N | 92.71°E |
| 2 | IISC | 13.02°N | 77.57°E |
| 3 | HYDE | 17.41°N | 78.55°E |
| 4 | LCK4 | 26.91°N | 80.95°E |
| 5 | LHAZ | 29.65°N | 91.10°E |

# 3   Results and Discussion

Though the GPS is being affected by many errors, dual frequency users are able to get required accuracy. Single frequency users accuracy is limited by irremovable ionospheric time delay error. For reducing time delay error an efficient model needs to be developed. In this process, preliminary work is carried out and results presented in this paper. First, the ionospheric time delay variations during geomagnetic storm conditions at low latitudes are analysed. Later, GPS single frequency ionospheric time delay model performance is evaluated.

GPS data for the period 7–9 May 2016 is archived from the IGS network website. The resolution of raw data provided is 30 s and satellites with elevation angle greater than $20^0$ are used to derive Slant TEC which is the quantity used in the computation of ionospheric time delay. At each epoch, the receiver is able to acquire a minimum of 8 GPS satellite signals from different directions. In STEC computation, all the available satellite signals are considered and plotted corresponding ionospheric time delay variations of $L_1$ signal at the selected IGS stations on 8th May 2016 (Fig. 1). Further, results compared with Klobuchar model estimated ionospheric time delay for the same satellite signals. In the figures, red colour corresponds to Klobuchar model and blue colour corresponds to experimental data.

The intensity of ionospheric time delay impact at selected five receiver stations is different because of geographical separation. Since the receiver station is stationary and GPS satellites orbital period is 11 h 58 min, it is possible to acquire signals only for a few h/day from each satellite. This is reflected as a discontinuity in the time delay profile. From the above figures, it can be observed that the Klobuchar model estimated peak delay is greater than 20 ns for PBRI, IISC and HYDE. Whereas at other two stations LCK4 and LHAZ it is nearly close to 20 ns. During daytime (5:00-15:00 UTC) experimental data is greater than the Klobuchar model estimated delay. Statistics of mean, maximum and minimum delays of Klobuchar model and experimental data are presented in Tables 3, 4 and 5. At all the stations, mean and maximum delays have a higher value for experimental data and

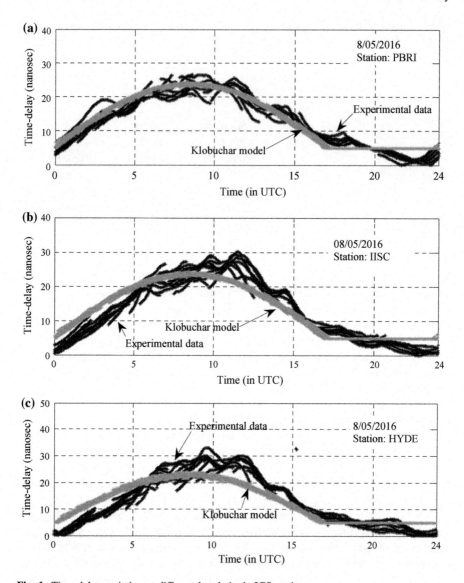

**Fig. 1** Time delay variation at different low latitude IGS stations

Klobuchar model on the day which is having high Kp-index ($5 \leq$ Kp $\leq$ 6) and Ap–index (70) values, i.e. 8th may 2016. For an instant, consider a day, i.e. 8th may 2016, the mean delay for HYDE, IISC, PBRI, LCK4 and LHAZ is 13.10 ns, 13.17 ns, 12.86 ns, 11.04 ns and 10.18 ns, respectively. Corresponding values due to Klobuchar model are 12.94 ns, 13.56 ns, 13.41 ns, 8.32 ns, and 7.79, respectively.

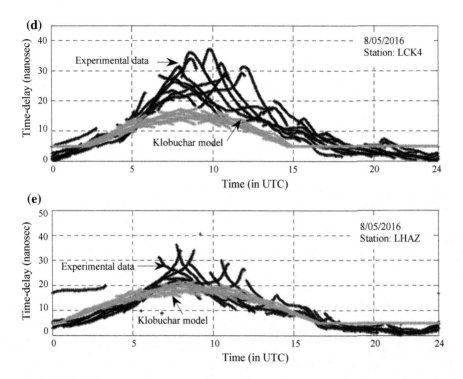

**Fig. 1** (continued)

**Table 3** Ionospheric time delay (in nano sec). Variation of GPS Signals on 7th May 2016

| Station ID | Experimental data | | | Klobuchar model | | |
|---|---|---|---|---|---|---|
| | Mean | Max | Min | Mean | Max | Min |
| PBRI | 12.0654 | 26.3238 | 0.0054 | 13.1695 | 24.6619 | 5 |
| IISC | 11.5092 | 26.2696 | 0.0921 | 13.2701 | 24.5518 | 5 |
| HYDE | 11.7582 | 28.5549 | 0.0108 | 12.6095 | 23.8650 | 5 |
| LCK4 | 8.1907 | 32.2210 | 0 | 7.8124 | 16.0748 | 5 |
| LHAZ | 7.2241 | 27.9417 | 0.0921 | 10.2536 | 20.0219 | 5 |

**Table 4** Ionospheric time delay (in nano sec). Variation of GPS signals on 8th May 2016

| Station ID | Experimental data | | | Klobuchar model | | |
|---|---|---|---|---|---|---|
| | Mean | Max | Min | Mean | Max | Min |
| PBRI | 10.1887 | 26.1684 | 0.1895 | 7.7795 | 16.4671 | 5 |
| IISC | 13.1789 | 30.0333 | 0 | 13.5623 | 24.1576 | 5 |
| HYDE | 13.1058 | 33.1308 | 0 | 12.9488 | 23.9997 | 5 |
| LCK4 | 12.8670 | 37.6920 | 0 | 13.4122 | 24.1574 | 5 |
| LHAZ | 11.0448 | 36.1002 | 0.0054 | 8.3288 | 17.3304 | 5 |

**Table 5** Ionospheric time delay (in nano sec). Variation of GPS signals on 9th May 2016

| Station ID | Experimental data | | | Klobuchar model | | |
|---|---|---|---|---|---|---|
| | Mean | Max | Min | Mean | Max | Min |
| PBRI | 6.9887 | 10.4630 | 0.0054 | 6.1164 | 13.1639 | 5 |
| IISC | 12.7142 | 30.7806 | 0.0054 | 12.4629 | 23.0926 | 5 |
| HYDE | 11.9594 | 32.3131 | 0.0054 | 11.6318 | 22.4315 | 5 |
| LCK4 | 13.0700 | 30.3094 | 0.7581 | 12.4160 | 23.2085 | 5 |
| LHAZ | 7.7051 | 25.8310 | 0.3682 | 6.7401 | 14.4845 | 5 |

Moreover, the performance of Klobuchar model over these locations is evaluated by computing a parameter called deviation. Mean deviation is maximum at LCK4.

Therefore, the results presented revealed that the Klobuchar model performance is not good at LCK4 compared to other stations. Moreover, the model performance is good in mean delay estimations rather than an epoch.

# 4 Conclusions

In India, there is a requirement for regional ionospheric time delay model because of great increment in the number of GNSS users for important applications. To develop a new model, first, we need to do a rigorous investigation of existing models. So, in this paper, GPS single frequency ionospheric time delay model was considered and investigated to find out inability conditions during high solar activity days. There is a marked deviation between the experimental data and Klobuchar model during noon time and night times. The results are suggesting that the model needs improvement to get better results over the Indian region.

**Acknowledgements** The research work presented in this paper has been carried out under the project entitled 'Investigation of Indian Ionosphere Irregularities Correlation with Space Weather Parameters for Navigation Applications' funded by SAC (ISRO), NavIC-GAGAN Utilization Programme, Project ID: NGP-12, Dated: 23 January, 2017.

# References

1. Cannon P et al: Extreme space weather: impacts on engineered systems and infrastructure. Royal Academy of Engineering, London, (2013) 1–68v.
2. Prolss, G. W.: Ionospheric F-region storms, in handbook of Atmo-spheric electrodynamics, CRC press, Boca Ration, Fla, (1995) 195.
3. Buonsanto, M. J., Mendillo, M., and Klobuchar, J. A.: The ionosphere at L = 4: Average behavior and the response to geomagnetic storms. Ann. Geophys., 35, (1979) 15–26.
4. Danilov, A. D. and Lastovicka, J.: Effects of Geomagnetic storms on the ionosphere and atmosphere, Int. J. Geomagn. Aero, 2(2001) 209–224.

5. Mendillo, M.: Storms in the ionosphere: Patterns and processes for total electron content. Vol. 44, Rev. Geophys., (2006) RG4001.
6. Misra, P., and Per Enge: Global Positioning System: Signals, Measurements, and Performance, Ganga-Jamuna press, 2nd Edition, New York (2006).
7. Klobuchar, J.A.: Ionospheric time-delay algorithm for single-frequency GPS users. Vol. 23 (3), IEEE Trans Aerosp Electron Syst. (1987) 325–331.
8. Hochegger, G., B. Nava, S. M. Radicella, and R. Leitinger: A family of ionospheric models for different uses. Vol 25(4) Phys. Chem. Earth, (2000) 307–310.
9. Radicella, S. M., and R. Leitinger: The evolution of the DGR approach to model electron density profiles, Vol. 27, Adv. Space Res., (2001) 35–40.
10. Swamy K.C.T, Sarma A.D, Srinivas V.S, Naveen Kumar P, and Somasekhar Rao P.V.D: Accuracy evaluation of estimated ionospheric delay of GPS signals based on Klobuchar and IRI-2007 models in low latitude region, Vol. 10, IEEE Geosci. Remote Sens. Lett., (2013) 1557–1561.
11. Hatanaka, Y.: Compact RINEX Format and Tools (beta-test version), proceeding of 1996 Analysis Center Workshop of IGS, March 19–21, (1996), 121–129.
12. Hatanaka, Y.: A RINEX Compression Format and Tools, Proceedings of ION GPS-96, September 17–20, (1996), 177–183.

# Scalable Recursive Convolution Algorithm for the Development of Parallel FIR Filter Architectures

Anitha Arumalla and Madhavi Latha Makkena

**Abstract** The paper presents 2-parallel and 3-parallel scalable recursive short convolution algorithms. The performance of these two short convolution algorithms is verified for different order filters. Hardware complexity is reduced by a factor of 3/4 in 2-parallel and 2/3 in 3-parallel filter implementation over conventional convolution. Synthesis results with 45 nm nangate library show that the scalable recursive convolution method has similar frequency performance for 2-parallel and 3-parallel implementations while the area and power are increasing for higher order filters.

**Keywords** Parallel FIR filter · Fast FIR algorithm · Successive recursive convolution

## 1 Introduction

Higher order signal and image processing algorithms require area and efficient algorithms to support portable devices. Parallel/block signal processing has gained importance for high-speed signal processing hardware design. Many parallel filter architectures for both FIR [1–9] and IIR [2, 10] filters are developed in the literature. N output samples every cycle can be processed using n-parallel convolution technique while duplicating the hardware for n times. But consecutive sample processing can provide a scope in resource optimization. Therefore, parallel processing is aimed at reducing hardware cost while designing high-speed signal processing applications.

A. Arumalla (✉)
Velagapudi Ramakrishna Siddhartha Engineering College, Vijayawada, India
e-mail: anithaarumalla83@gmail.com

M. L. Makkena
Jawaharlal Nehru Technological University Hyderabad, Hyderabad, India
e-mail: mlmakkena@yahoo.com

© Springer Nature Singapore Pte Ltd. 2018
J. Anguera et al. (eds.), *Microelectronics, Electromagnetics and Telecommunications*, Lecture Notes in Electrical Engineering 471, https://doi.org/10.1007/978-981-10-7329-8_26

A modular Walsh–Hadamard decomposition is used for the realization of block FIR filter structure in [1]. This implementation is found to have reduced latency than conventional design. The regular and modular filter implementation ensures high-frequency operation, but requires hardware resources increasing linearly with parallel block size. Also, the complex representations of the block filter, could not attract much attention. Two-dimensional digital FIR and IIR filter [2] implementations are derived from the parallel implementation of one-dimensional digital filters. Three diverse algorithms are discussed for two-dimensional digital filter implementations.

hardware efficient block FIR filter in [3, 4] has reported reduced area than conventional filter up to 45%. While the conventional implementation of filters has area increase by O[n] with the raise in order of parallelism, a low power, and low area solution is offered in Fast FIR Algorithm (FFA). Quantized filter coefficients are encoded using Canonical Signed Digit (CSD) recoding for implementing multiplier less filter. Further reduction in the filter hardware resources is achieved using subexpression elimination and adjacent coefficient sharing. This work has presented high area efficient filter design for high sampling rate operation. However, the considerable manual effort is necessary for implementing subexpression elimination and coefficient sharing for every modification in filter specification.

In [5], frequency characteristics of FIR filter are used for proper selection of FFA algorithm. Also, filter coefficients quantized using block quantization algorithm to achieve reduced the binary adders in the filter architecture. Depending upon the significant filter coefficients, a narrow band filter is implemented using difference FFA architecture subfilters. Similarly, FFA with addition subfilters is used for wide band filter. Fast two-dimensional block matching algorithm [6] is proposed using FFA. A 2 × 2 block convolution is implemented using two parallel one-dimensional FFA decomposition. Block overlap redundancy is considered in the high-speed block matching algorithm. The two-dimensional filter implementation is done using nine subfilters. Higher order filter realization is difficult with this algorithm.

A polyphase decomposition [8] is used for a non-separable filter 2D filter bank implementation. The filter bank implementation is carried out with time multiplexed hardware resource sharing while throughput is not compromised. The area efficiency of polyphase implementation is a factor of number of filter banks than non-polyphase implementation, i.e., high area saving for large filter banks. But not all filters can be implemented using non-separable implementation. A 256-tap filter [9] is implemented using split-based distributed arithmetic (DA) lookup tables (LUT). But the memory requirement increases tremendously with increase in data-width and order of the filter.

A systolic parallel bit-level architecture [10] is incorporated in two-dimensional IIR filter design without feedback. This is a block processing algorithm with minimum hardware requirement and maximum sampling frequency. The feedback in IIR filter avoided by using systolic architecture array, and hence can produce one output sample every clock cycle. But systolic architecture needs larger area resources for block processing since no area optimization is considered.

In this paper, scalable recursive convolution algorithm is presented that eliminates the maximum possible redundant hardware in parallel filter architecture while demonstrating a simple representation and implementation.

The rest of the paper is organized as follows: two parallel and three parallel implementations of conventional FIR filter is presented in Sect. 2, scalable recursive convolution algorithm is detailed in Sect. 3. The performance analysis of scalable recursive convolution algorithm is presented in Sect. 4 and conclusion is drawn in Sect. 5.

## 2 Parallel FIR Filter

If $y_n$, $x_n$ are infinite output, input samples, and $h_n$ is the filter impulse response, then the time domain and frequency domain representations of the filter is given by (1) and (2)

$$y_n = \sum_{k=1}^{N} h_k x_{n-k}, n = 0, 1, 2, \ldots \infty \tag{1}$$

$$Y_n = HX_n \tag{2}$$

where $Y_n$, $X_n$ are frequency domain representation of output, input signals, and $H$ is the frequency response of the filter.

To generate, two parallel output samples $y_n$, $y_{n+1}$ for each clock, two signal samples $x_n$, $x_{n+1}$ are given as input to the filter, i.e., alternative samples fed to each input. The frequency domain representation of two parallel filters is shown in (3)

$$\begin{aligned}
Y_{2n} &= X_{2n}H_0 + Z^{-2}X_{2n+1}H_1 \\
Y_{2n+1} &= X_{2n}H_1 + X_{2n+1}H_0
\end{aligned} \tag{3}$$

where $X_{2n}$, $X_{2n+1}$, $Y_{2n}$, $Y_{2n+1}$ and $H_0$, $H_1$ are transforms of $x_{2n}$, $x_{2n+1}$, $y_{2n}$, $y_{2n+1}$ and $h_0$, $h_1$, respectively. $x_{2n}$, $x_{2n+1}$, $y_{2n}$, $y_{2n+1}$, and $h_0$, $h_1$ are alternately decimated vectors of $x_n$, $y_n$, and $h_n$, respectively. From the equations, it is evident that the number of multipliers required is $2N$ and the number of adders required is $2(N-1)$ for an $N$ tap filter.

Similarly, a three parallel filter can process three decimated inputs and generate three output samples per clock as represented in Eq. (4) with decimated bythree coefficient vectors $h_0$, $h_1$, and $h_2$.

$$Y_{3n} = X_{3n}H_0 + Z^{-3}X_{3n+1}H_2 + Z^{-3}X_{3n+2}H_1$$
$$Y_{3n+1} = X_{3n}H_1 + X_{3n+1}H_0 + Z^{-3}X_{3n+2}H_2 \tag{4}$$
$$Y_{3n+2} = X_{3n}H_2 + X_{3n+1}H_1 + X_{3n+2}H_0$$

Each subfilter $H_0$, $H_1$, and $H_2$ is a $N/3$ tap filter and the number of multiplication and addition operations required are $3N$ and $3(N-1)$.

# 3    Scalable Recursive Convolution Algorithm (SRCA)

The SRCA is a regular architecture which is closely inspired by FFA eliminating delay elements within the structure. The algorithm can be effortlessly scaled and used recursively for developing higher order parallel filters.

## 3.1    Two Parallel SRCA (2PSRCA)

The 2PSRCA decomposes the two parallel filter equations to exploit redundancy between the equations in (3). Equation (5) represents the 2PSRCA decomposition with a subfilter shared for the computation of two outputs. The matrix representation and symbolic representation of the 2PSRCA are given in (6) and (7), respectively. $P_{2S}$ is a preprocessing matrix for processing difference input samples for the subfilters, $Q_{2S}$ is a post-processing matrix for compiling output from the subfilter outputs and $H_{2S}$ is a precomputed coefficient matrix representing the subfilters required to process the inputs. $X_2$ and $Y_2$ are two parallel input and output matrices. Block diagram representation is shown in Fig. 1 with post-processing, preprocessing, and coefficient matrices.

$$Y_{2n} = (X_{2n-1} - X_{2n})H_1 + X_{2n}(H_0 + H_1)$$
$$Y_{2n+1} = X_{2n}(H_0 + H_1) + (X_{2n+1} - X_{2n})H_0 \tag{5}$$

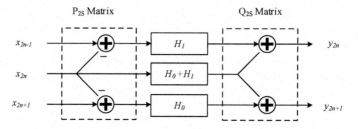

**Fig. 1** Two parallel successive recursive convolution algorithm

$$\begin{bmatrix} Y_{2n} \\ Y_{2n+1} \end{bmatrix} = \begin{bmatrix} 1 & 1 & 0 \\ 0 & 1 & 1 \end{bmatrix} diag \begin{bmatrix} H_1 \\ H_0 + H_1 \\ H_0 \end{bmatrix} \begin{bmatrix} 1 & -1 & 0 \\ 0 & 1 & 0 \\ 0 & -1 & 1 \end{bmatrix} \begin{bmatrix} X_{2n-1} \\ X_{2n} \\ X_{2n+1} \end{bmatrix} \quad (6)$$

$$Y_2 = Q_{2S}H_{2S}P_{2S}X_2 \quad (7)$$

The 2PSRCA architecture requires $3N/2$ multipliers, $3N/2 + 1$ adders which is approximately 3/4 times the number of multipliers and adders required for the conventional filter.

## 3.2 Three Parallel SRCA (3PSRCA)

The 3PSRCA can be derived from the two-stage decomposition of 2PSRCA applied on Eq. (4). The optimal redundant equation decomposition is presented in Eq. (8). The matrix representation and symbolic representation of the 3PSRCA are given in (9) and (10), respectively, where $P_{3S}$ is a preprocessing matrix, $Q_{3S}$ is a post-processing matrix and $H_{3S}$ is a precomputed coefficient matrix for 3PSRCA.

$$Y_{3n} = X_{3n}(H_0 + H_2) + X_{3n-1}(H_1 + H_2) + H_2(X_{3n-2} - X_{3n-1} - X_{3n})$$
$$Y_{3n+1} = X_{3n+1}(H_0 + H_1) + X_{3n-1}(H_1 + H_2) + H_1(-X_{3n-1} + X_{3n} - X_{3n+1}) \quad (8)$$
$$Y_{3n+2} = X_{3n}(H_0 + H_2) + X_{3n+1}(H_0 + H_1) + H_0(-X_{3n} - X_{3n+1} + X_{3n+2})$$

$$\begin{bmatrix} Y_{3n} \\ Y_{3n+1} \\ Y_{3n+2} \end{bmatrix} = \begin{bmatrix} 1 & 0 & 0 & 1 & 1 & 0 \\ 0 & 1 & 0 & 1 & 0 & 1 \\ 0 & 0 & 1 & 0 & 1 & 1 \end{bmatrix} diag \begin{bmatrix} H_2 \\ H_1 \\ H_0 \\ H_2 + H_1 \\ H_2 + H_0 \\ H_1 + H_0 \end{bmatrix} \begin{bmatrix} 1 & -1 & -1 & 0 & 0 \\ 0 & -1 & 1 & -1 & 0 \\ 0 & 0 & -1 & -1 & 1 \\ 0 & 1 & 0 & 0 & 0 \\ 0 & 0 & 1 & 0 & 0 \\ 0 & 0 & 0 & 1 & 0 \end{bmatrix} \begin{bmatrix} X_{3n-2} \\ X_{3n-1} \\ X_{3n} \\ X_{3n+1} \\ X_{3n+2} \end{bmatrix}$$
$$\quad (9)$$

$$Y_3 = Q_{3S}H_{3S}P_{3S}X_3 \quad (10)$$

Block diagram of 3PSRCA is shown in Fig. 2. The number of multipliers and adders required by 3PSRCA are $2N$ and $2N + 6$, respectively, which are approximately 2/3 times the requirement of conventional 3 parallel filter design. The signal power and computational effort highly depend on the proper selection of pre and post-processing matrices. Most of the times, the coefficient matrix is preloaded values so as to avoid additional area overhead in the architecture.

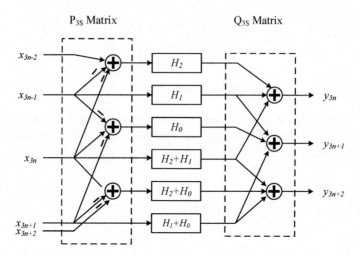

**Fig. 2** Three parallel successive recursive convolution algorithm

# 4 Results and Discussions

Three parallel and two parallel SRCA architectures for parallel FIR filter are designed using Verilog HDL and synthesized with 45 nm nangate library using Cadence RTL Compiler. The designs are validated on Virtex 6 FPGA with system generator interface.

**Table 1** Area, frequency, and power performance of 2PSRCA and 3PSRCA

| Filter order | Bit width | Area ($\mu m^2$) | | Maximum frequency (MHz) | | Power (nW) | |
|---|---|---|---|---|---|---|---|
| | | 2PSRCA | 3PSRCA | 2PSRCA | 3PSRCA | 2PSRCA | 3PSRCA |
| 12 | 8 | 9783 | 13337 | 182.38 | 189.538 | 90584 | 122814 |
| | 12 | 18695 | 25340 | 141.6 | 146.456 | 173900 | 234437 |
| | 16 | 30353 | 41046 | 116.4 | 119.303 | 264338 | 349301 |
| 24 | 8 | 15311 | 26456 | 174.52 | 172.087 | 154304 | 219417 |
| | 12 | 28954 | 50392 | 137.04 | 135.538 | 270103 | 407075 |
| | 16 | 46522 | 81691 | 112.55 | 112.082 | 436436 | 641995 |
| 48 | 8 | 30592 | 52140 | 152.16 | 147.341 | 291766 | 420482 |
| | 12 | 57548 | 99821 | 124.24 | 119.048 | 538273 | 802365 |
| | 16 | 92932 | 161822 | 104.14 | 103.961 | 811320 | 1233612 |
| 96 | 8 | 61007 | 104139 | 138.08 | 141.764 | 582894 | 826757 |
| | 12 | 115275 | 198927 | 113.47 | 116.036 | 1058095 | 1538743 |
| | 16 | 185928 | 323132 | 96.35 | 98.2415 | 1612447 | 2395709 |
| 192 | 8 | 122124 | 207178 | 129.17 | 128.816 | 1148345 | 1604839 |
| | 12 | 232586 | 396587 | 101.3 | 106.758 | 1868326 | 3058816 |
| | 16 | 358946 | 646328 | 89.9 | 93.1706 | 2952419 | 4642530 |

**Fig. 3** Plots representing
power analysis for 3PSRCA
and 2PSRCA

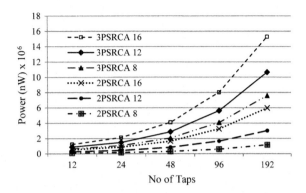

A low pass filter is designed for different filter lengths 12, 24, 48, 96, and 192
with multiple bit widths 8, 12, and 16. The designs are evaluated for the area,
power, and maximum frequency of operation. The results of two parallel and three
parallel are presented in Table 1. For a conventional filter, a 33% resource and the
power increase is expected with an increase in parallelism from 2 parallel to 3
parallel filter. From the table below, the maximum operating frequency is nearly
equal to both 2PSRCA and 3PSRCA. The power dissipation for 8-bit, 12-bit, and
16-bit filters has increased with level of parallelism between 35–44, 34–63, and
52%, i.e., larger bit widths exhibit a huge increase in power dissipation. The plots
representing power dissipation for various designs are shown in Fig. 3. Similarly,
the area requirement for 8-bit, 12-bit, and 16-bit filters has increased with level of
parallelism between 36–72, 35–74, and 35–80%. The plots representing area
requirements for various designs are shown in Fig. 4. From the results, it is
indicative that area requirements are enormously increasing for higher order and
large bit width filters.

**Fig. 4** Plots representing
area analysis for 3PSRCA and
2PSRCA

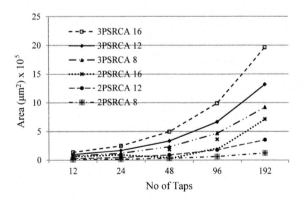

## 5 Conclusion

In this paper, a parallel FIR filter architecture design is presented and performance of 2PSRCA and 3PSRCA is evaluated in terms of area, power, and maximum operating frequency. The SRCA architecture has simple architecture and flexible mathematical representation that can be recursively used for scaling the architecture to achieve a high level of parallelism. The SRCA algorithm has optimized the operator level resources by appropriate decomposition of parallel output expressions and a mere theoretical analysis has shown significant saving in hardware resources.

## References

1. Mertzios, B.G., Venetsanopoulos, A.N.: Fast Block Implementation of Two-Dimensional FIR Digital Filters via the Walsh–Hadamard Decomposition. Int. J. Electron. 68, 991–1004 (1990).
2. Mitra, S.K., Gnanasekaran, R.: Block implementation of two-dimensional digital filters. J. Franklin Inst. 316, 299–316 (1983).
3. Parker, D.A., Parhi, K.K.: Low-Area/Power Parallel FIR Digital Filter Implementations. J. VLSI Signal Process. Syst. Signal Image. Video Technol. 17, 75–92 (1997).
4. Parhi, K.: VLSI Digital Signal Processing Systems: Design and Implementation. John Wiley & Sons (2007).
5. Chung, J.G., Parhi, K.K.: Frequency Spectrum Based Low-Area Low-Power Parallel FIR Filter Design. EURASIP J. Adv. Signal Process. 2002, 944–953 (2002).
6. Naito, Y., Miyazaki, T., Kuroda, I.: A Fast Full-Search Motion Estimation Method for Programmable Processors with a Multiply-Accumulator. In: Acoustics, Speech, and Signal Processing, ICASSP-96. pp. 3221–3224. IEEE (1996).
7. Parhami, B., Kwai, D.M.: Parallel Architectures and Adaptation Algorithms for Programmable FIR Digital Filters with Fully Pipelined Data and Control Flows. J. Inf. Sci. Eng. 19, 59–74 (2003).
8. Mohanty, B.K., Al-Maadeed, S., Amira, A.: Systolic Architecture for Hardware Implementation of Two-Dimensional Non-Separable Filter-Bank. In: 2013 8th IEEE Design and Test Symposium. pp. 1–6. IEEE (2013).
9. Nandal, A., Vigneswarn, T., Rana, A.K., Dhaka, A.: An Efficient 256-Tap Parallel FIR Digital Filter Implementation Using Distributed Arithmetic Architecture. In: Procedia Computer Science (2015).
10. Zhijian Hu, Gaston, F.: A Bit-level Systolic 2D-IIR Digital Filter without Feedback. In: Proc. of the 13th Asilomar Conf. on Signals, Systems and Computers. pp. 1063–1066. IEEE Comput. Soc. Press (2002).

# IoT-Based Green Environment for Smart Cities

Naveen Kishore Gattim, M. Gopi Krishna, B. Raveendra Nadh,
N. Madhu and C. Lokanath Reddy

**Abstract** In the present scenario, severe actions are required to manage waste from its formation to its allocation. Monitoring the waste is essential for proper recycling. The smart container garbage bins indicate the level of municipal waste. When the bins are full it gives an indication to the authority to clean the bin. This is considered to minimize harmful effects of waste on health and environment. Bad smell spreads to the surrounding areas creating disorders. So, the solid waste generated from the residents is indicated using "IOT based green environment for smart cities". A smart waste collection management using smart containers providing intelligence to the containers along with the IoT sensors which can read, transmit and collect data on the Internet. In the proposed system, we use different garbage bins to monitor the solid waste gathered in the garbage bins which can be indicated by a web page and SMS. This system is also used to monitor the level of hazardous gases produced in garbage bins and it is indicated through buzzer. In this, a truck is used to collect the garbage from the colonies and few minutes before the truck arrival the message will be sent to locality people so that they can come and dump their dust.

**Keywords** IoT · Smart container · Message · Buzzer · Truck
Smart city

N. K. Gattim (✉) · M. G. Krishna · B. R. Nadh · N. Madhu · C. L. Reddy
Department of ECE, Vardhaman College of Engineering, Hyderabad, India
e-mail: naveengattim@live.com

M. G. Krishna
e-mail: m.gopikrishna@vardhaman.org

B. R. Nadh
e-mail: bravindra64@hotmail.com

N. Madhu
e-mail: n.madhu@vardhaman.org

C. L. Reddy
e-mail: clreddy@vardhaman.org

© Springer Nature Singapore Pte Ltd. 2018
J. Anguera et al. (eds.), *Microelectronics, Electromagnetics
and Telecommunications*, Lecture Notes in Electrical Engineering 471,
https://doi.org/10.1007/978-981-10-7329-8_27

# 1  Introduction

Things (Embedded devices) that are connected to Internet and sometimes these devices can be controlled from the Internet which are used in hospitals, agriculture, homes, locate the changes in environment and many advantages with the usage of internet of things is commonly called as Internet of Things. It is forecasted that 41 billion IoT devices are utilized in 2020 [1]. The rapid increase in the number of IoT devices has opened new horizons in cloud services. The devices build the data using the Internet. The response time requirement, data safety, and privacy is possible by using edge computing [2]. Waste is produced during the removal of underdone materials utilized for consumption or other human activities [3, 4]. Our implementation is on smart waste container which is continuously monitored and connected to the Internet for getting the level indication of waste filled in the smart container. This project IOT-based green environment for smart cities is a novel system for indicating the level of waste produced and to keep the cities clean. The smart waste containers placed at different places are observed by the system which gives an indication of the level to the concerned person via a web page and an alert message is sent to them. The level of the smart container can be determined by placing the ultrasonic sensors above the smart container to observe the garbage level which is measured per the total depth of the smart container. The system is assembled using Arduino UNO, 12 V transformer for power supply. It helps the locality people to know the arrival of truck by using an RF sensor. It also has a unique performance where it monitors the amount of gas present in the garbage bins and if the amount of hazardous gases exceeds a certain limit. Then, the buzzer gets activated immediately.

## 1.1  Related Work

The GSM (Global System for Mobile Communication) and Arduino are used to form the integrated system to monitor the waste bins remotely [5]. The smart containers are placed at different public places with sensors placed on top of the smart container. The sensor identifies the level of the garbage which is indicated to the Arduino controller. Through the controller, an alert message is sent to the GHMC people through SMS using GSM, so that they collect the garbage bin which is filled and it indicates through a web page also. So, that the authorities can come to that place where the smart container is filled with the garbage. Some of the waste collection methods must be practised and disposed as per the category of waste. So, that the environment achieves zero waste. In this system, we also implemented the detection of hazardous gases present in the smart container and it will be indicated through a buzzer. This chapter's objective is to track the arrival of GHMC truck at the colonies when the smart container is filled.

This paper highlights the disposal of waste through the internet of things where the municipal authorities can track the status of the smart containers. Thereby reducing the environmental consequences.

## 2   System Architecture

The system architecture includes the hardware section of Arduino which is a microcontroller board based on ATmega328P. The technical specification is as follows: The operating voltage is 5 V, clock speed is 16 MHz, the input recommended voltage is 7–12 V. The Arduino has 14 digital I/O pins, 6 PWM digital I/O pins, 6 Analog input pins, 20 mA DC current per I/O pin, 50 mA DC current for 3.3 V pin, 2 kb SRAM, 1 kb EEPROM, 13 LED builtin are the key technical specifications.

### 2.1   GSM Modem

GSM module is interfaced to Arduino which requires only three connections between Arduino and the GSM module. A GSM module (SIM 300) works on frequencies EGSM 900 MHz. The interface of GSM modem is made through a 60-pin connector. It includes a keypad and SPI, LCD interface, two serial ports, two Arduino channels which include two microphones inputs and two speaker outputs that can be configured by AT commands [6, 7].

### 2.2   AT Commands

Modems are determined by AT commands. There are basic AT commands which are used to test the modem. To send and receive SMS using GSM module AT commands are used to communicate from Arduino. There are many commands to execute separate tasks using GSM module [8, 9]. There are AT commands library to interpret the functionality of GSM module. Some of the AT commands are send message () —"AT + CMGF = 1" Receive message ()—"AT + CNMI = 2, 2, 0, 0, 0".

### 2.3   Liquid Crystal Display (LCD)

Liquid crystal display operates on the light modulating property of liquid crystals. It uses reflector to construct images in color. LCDs can also be obtained with low information content. The LCD interface is made up of the following pins: a. Register select (RS), b. Read/Write pin, and c. Enable pin and 8 data pins (D0–D7). This LCD module is operated in two modes: 4-bit and 8-bit. The 4-bit mode involves 7 I/O pins from the Arduino. The 8-bit mode involves 11 pins. A 4-bit mode is sufficient to control a 2 × 16 LCD.

## 2.4  Ultrasonic Sensor

The ultrasonic sensor is used to assess the distance from objects. The ultrasonic pulse generated is 40 kHz which waits for the pulse to echo back computing the time taken in microseconds. The ultrasonic sensor has 4 pins namely VCC, trigger, echo, and GND used with 5 V supply. Ultrasonic sensor converts AC to sound and vice versa. It has many applications in medicine, automated factories, parking sensors, and cleaning devices.

## 2.5  MQ3 Gas Sensor

MQ3 sensor module functionality is to detect gas leakage, i.e., $H_2$, LPG, CO, $CH_4$, smoke, and alcohol. It has high sensitivity and fast response time. The gas sensor has VCC, GND, and signal pin. The concentration of gas increases which is identified by the increase in the output voltage from the gas sensor. The potentiometer is used to adjust the sensitivity.

## 2.6  Buzzer

Buzzers are two types which are passive and active, active buzzer is required only electric signal to give a sound. Its operating frequency is 2 kHz. Come to passive buzzers it required sound signal to generate tone, giving the PWM to buzzer or on and off the buzzer with different frequencies (1.5–2.5 kHz). Buzzers consists of 3 pins which are VCC, GND, and signal pin connect the VCC to VCC of the Arduino GND pin to GND pin of Arduino and signal pin to any digital pins (D0–D13) of Arduino.

## 2.7   WiFi Module

ESP8266 module consists of 8 pins. Those are VCC, Tx, Rx, CH_PD, and GPIO 0, GPIO 2 last two pins (GPIO 0, GPIO 2) are not used to interface with the Arduino. But these pins are useful for updating the software of the ESP8266 board, while updating process connect GPIO pin to ground. To interface the WiFi module with Uno make sure of connections giving the VCC to 3.3 V or not because WiFi module is not compatible with the 5 V, Tx pin of the WiFi module given to the Rx pin of the UNO board, Rx pin of the WiFi module given to the Tx pin of the Uno board. GND to GND of UNO. After connecting WiFi module adjust the baud rate of the module which is 115200 or 9600 in serial monitor.

## 2.8   Arduino Pro Mini

Arduino Uno is used to dump the code into Arduino pro mini without using the USB adaptor. To dump a code into Arduino pro mini first takeoff the microcontroller of the Uno and connect the Uno pin 5 V to mini main VCC pin
Uno's pin GND to Pro GND
Uno's pin Rx to mini's Rx1
Uno's pin Tx to mini's Tx0
Uno's pin Reset to mini's reset, and then upload the code into pro mini without pressing any button.

## 2.9   RF Module

Tx the data without using wires are cables RF module is useful. RF module Tx the data within the range of 100 m or more depending on the antenna which is using, RF module is present varies operating frequencies like 433, 315 MHz. It will Tx up to 600 bps to 10 Kbps, it uses the ASK modulation to Tx data, for Tx and Rx purpose two kinds of modules are used.

**Tx Module**
This module consisting of 4 pins those are VCC, GND, Serial data input pin, and Antenna for Tx purpose, connect the serial data input pin of the RF Tx to any digital pins (D0–D13) of the Uno.

**Rx Module**
This module consists of 4 pins those are VCC, GND, Serial data output, and Antenna. For Rx purpose, connect the serial data output pin of the RF Rx to any digital pins (D0–D13) of the Uno.

## 3  Architecture and Software Design of Smart Container

The following Fig. 1 shows the architecture of Arduino which is interfaced to various sensors.

### 3.1  Arduino IDE

The Arduino Integrated Development Environment [10] is used for writing the source code and then uploading to the board where all the hardware sensors are connected to the specific pins of the Arduino. The following is the module of detecting the level of smart container.

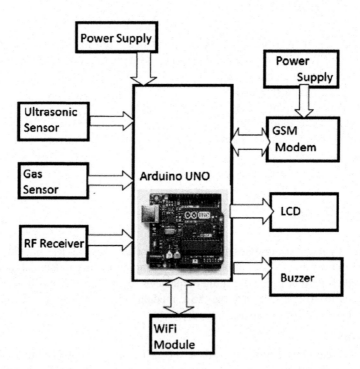

**Fig. 1** Architecture of Arduino

**Fig. 2** Arduino file with
distance identification

```
Serial.println(distance);
if (vw_get_message(buf, &buflen))
{
  Serial.print("Got: ");
  s=buf;
}
if(distance<15)
{
  lcd.clear();
  lcd.setCursor(0,0);
  lcd.print("DUSTBIN FULL");
  lcd.setCursor(0,1);
  lcd.print("SENDING SMS");
  sendsms("9618013940","DUSTBIN IS FULL");
  lcd.clear();
}
Serial.println(analogRead(gas));
if(analogRead(gas)>600)
{
  digitalWrite(buzz,LOW);
}
if(analogRead(gas)<600)
{
  digitalWrite(buzz,HIGH);
}
lcd.setCursor(0,0);
lcd.print("Distance:");
```

## 4 Results

The complete implementation is shown in Fig. 2 connecting different sensors to the
Arduino Uno board. A sequence of alert message is given to the authorities when
the smart container is filled with household waste and harmful gases. The smart
containers level of garbage can be monitored in the webpage (Figs. 3, 4, and 5).

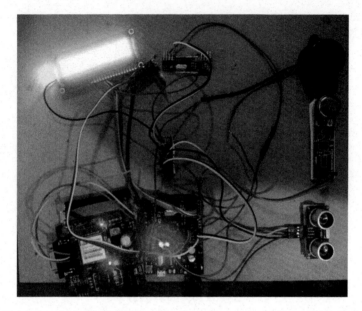

**Fig. 3** Connections of Arduino, GSM module, ultrasonic, gas sensor, buzzer, LCD display, WiFi module, Arduino pro mini, transmitter, and receiver

**Fig. 4** **a** An alert message indicating that smart container is full and **b** Indication of arrival of truck

## 5 Conclusion

IOT-based green environment for smart cities using Arduino UNO is a collaboration of software and hardware. This system can easily avoid the overflow of garbage bins and can also detect the hazardous gases present in the garbage bins which

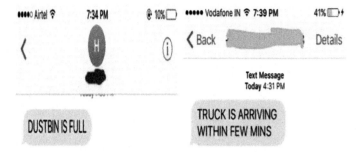

**Fig. 5** Alerts communication to the authorities to collect the waste

are harmful to the human life. This system also indicates the arrival of trucks in colonies so that there is no overflow of trash in our houses. Further, the work can be extended to develop application for municipal staff, city administrators, and recycling factories.

# References

1. Weiwei Fang, "On the throughput-energy tradeoff for data transmission in IoT devices", Journal of Information Sciences, Elsevier, Vol. 283, No. 1, pp. 79–93, 2014.
2. Alexey Medvedev, Pert Fedchenkov, A Arkady Zaslavsky, "Waste Management as an IoT Enabled Service in Smart Cities", Conference on Smart Spaces, Springer, pp. 104–115, 2012.
3. Lingling H., Haifeng L., Xu X., Jian L., "An Intelligent Vehicle Monitoring System, Based on Internet of Things", IEEE 7th International Conference on Computational Intelligence and Security (CIS), pp. 231–233, Hainan, December, 2011.
4. Nuortio T., Kytojoki J., Niska H., Braysy O., "Improved route planning and scheduling of waste collection and transport", Journal of Expert Systems with Applications, vol. 30, No. 2, pp. 223–232, 2006.
5. Li J. Q., Borenstein D., Mirchandani P. B., "Truck scheduling for solid waste collection in the City of Porto Alegre, Brazil", Journal of Omega, vol. 36, No. 6, pp. 1133–1149, December, 2008.
6. Sauro Longhi, Davide Marzioni Waste Management Architecture Using Wireless Sensor Network Technology", International Journal of Engineering Research & Technology, Vol. 3, No. 4, June, 2014.
7. Guerrero L. A., Maas G., Hogland W., "Solid waste management challenges for cities in developing countries", Journal of Waste Management, vol. 33, No. 1, pp. 220–232, January, 2013.
8. Marshall R. E., Farahbakhsh K., "Systems approaches to integrated solid waste management in developing countries", Journal of Waste Management, vol. 33, No. 4, pp. 988–1003, April, 2013.
9. Caniato Marco, Mentore Vaccari, Chettiyappan Visvanathan, Christian Zurbrügg, "Using social network and stakeholder analysis to help evaluate infectious waste management: A step towards a holistic assessment", Management, vol. 34, no. 5, pp. 938–951, 2014.
10. http://www.arduino.cc/.

# Design of a Robust Estimator for Submarine Tracking in Complex Environments

D. V. A. N. Ravi Kumar, S. Koteswara Rao and K. Padma Raju

**Abstract** A new filter which could satisfy most of the requirements of a modern-day tracking estimator was developed in this chapter. This estimator is designed particularly for the purpose of the undersea tracking purpose, where the erroneous bearing information is the only data available. The technique involved is to combine the various important techniques available in the literature to develop a less complex more robust estimator. First of all, the weighted expectation of the current and the previously received sensor measurements are computed. These named as the preprocessed measurements contain less variance of noise are in turn applied to the various nonlinear algorithms such as UKF. The consolidation of all the UKF outputs is done in such a way that the least mean square error is obtained. This is possible with the least squares estimation filter (WLSE). The output of WLSE is a much superior one because of performing the refining procedure in two different steps. Monte Carlo simulations are performed to verify the robustness of the introduced mechanism. The algorithm is tested for tracking a submarine located in various quadrants, various ranges, and various noise levels.

**Keywords** Estimator · Submarine tracking · Estimation error
Divergence issues · Monte Carlo

D. V. A. N. Ravi Kumar (✉)
Department of ECE, GVPCEW, Madhurawada, Visakhapatnam, India
e-mail: ravikumardwarapu@gvpcew.ac.in

S. Koteswara Rao
Department of ECE, KL University, Guntur, India

K. Padma Raju
Department of ECE, JNT University, Kakinada, India

© Springer Nature Singapore Pte Ltd. 2018
J. Anguera et al. (eds.), *Microelectronics, Electromagnetics
and Telecommunications*, Lecture Notes in Electrical Engineering 471,
https://doi.org/10.1007/978-981-10-7329-8_28

# 1  Introduction

Tracking enemy's vehicles like torpedoes, submarines is important for field requirements to remain safe or to demolish the enemy. This deals with finding the location of the enemy in the coming instants. This critical task is assisted by the advanced electronic equipment called RADAR and SONAR. Unfortunately, RADAR cannot handle the issue of tracking in the underwater environments because of the incapability of the Electromagnetic waves produced by the RADAR to propagate through the water. So, the best-suited equipment for undersea tracking issue is the SONAR which works with sound waves instead of the EM waves. SONARs are broadly classified into two categories namely active and passive. Active SONAR believes in the principle of the traditional RADAR system, which uses the mechanism of echo processing to localize the enemy. On the other hand, passive SONAR concentrates to listen to the enemy generated noise to compute the foe's future position. In this chapter, the passive SONARs measurements are processed to estimate the enemy's velocity and position.

The measurements processed by any of the equipment whether it be a RADAR or a SONAR is polluted by noise. The extraction of targets position from the contaminated sensor measurements is an important task performed by the digital signal processing unit. Various available algorithms runs in the processor to serve the purpose are Particle Filter (PF), Unscented Kalman filter (UKF), Extended Kalman filter (EKF), or a simple Kalman filter (KF). Not only these but many of their extensions play a key role in the present day tracking issues. KF and its various forms are available in [1–3], EKF is modified in various ways to improve its applications [4–6], similarly UKF takes various shapes to fit itself for a variety of applications [7–9, 12]. PF is a recent extension of KF available in forms of [10, 11] to ease the estimation problems. Each of the algorithms mentioned presents some supporting points and some criticizing points. So in this chapter some algorithms available in the literature are fused to develop a robust estimator with better performance. The robust estimator fuses the UKF, integration technique, and preprocessing technique to bring the estimation error to a lesser value compared to any of them individually. Not only this but the introduced method offers better superiority in the unfavorable conditions such as that of far submarine tracking and dealing with the heavily polluted measurements. The superiority of attaining least estimation error is due to the integration technique, the superiority of tough condition tracking is due to the preprocessing mechanism and also the algorithm is a nondiverging one because of the usage of the UKF. These all features certainly support the importance of the novel algorithm.

Section 2 deals with integration technique which shows how the superiority in estimation is achieved by a combination of different algorithms, Sect. 3 handles the topic of preprocessing where the procedure of conditioning of measurements is shown, Sect. 4 explains about the proposed RUKF where the methods of integration technique and preprocessing are used together to achieve superior results, and Sect. 5 talks about mathematical representation of the targets moments and the sensor measurements, Sect. 6 deals with results and chapter concludes in Sect. 7.

## 2 Integration Technique

The principle of integration technique is to attain a better estimate by fusing the individual filter estimates with the support of a combiner like WLSE. The block diagram is shown in the Fig. 1 and the expressions related to them is given in the current Sect. 2. This is implemented in [7, 12].

$$\widehat{M}_{(i)}(t/t) = \left(\sum_{j=1}^{p} \frac{1}{p_{ji}^2}\right)^{-1} \left(\sum_{j=1}^{p} \frac{Mj_{(i)}(t/t)}{p_{ji}^2}\right), \quad P(t/t) = \frac{1}{p}\sum_{i=1}^{p} Pi(t/t) \quad (1)$$

$\widehat{M}(t/t) = \left[\widehat{M}_{(1)}(t/t)\ \widehat{M}_{(2)}(t/t)\ \ldots\ \widehat{M}_{(n)}(t/t)\right]^T$ $\widehat{M}_{(i)}(t/t)$ is the consolidated RUKF estimate, $Mj_{(i)}(t/t)$ is the estimate provided by jth filter, p is the no of filters. $P(t/t)$ is the consolidated covariance, $Pi(t/t)$ is the ith filter covariance. $p_{ji}$ is the jth, ith element of the covariance matrix.

**Fig. 1** Integration technique

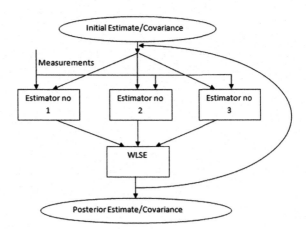

## 3   Preprocessing Technique

The principle involved in the preprocessing technique is to smoothen the measurements prior to usage by averaging the sensor-generated data in time with the support of predictors. Averaging results in drastic reduction of noise variance in measurements. The block diagram is displayed in Fig. 2 and the mathematical expressions are given in equations of the Sect. 3. This technique is available in [9].

$$ppmeas(1) = meas(1), \quad ppmeas(2) = \frac{1}{2}[meas(2) + predmeas(1)]$$

$$ppmeas(3) = \frac{1}{2}[meas(3) + predmeas(2)] \quad \text{are preprocessed readings.} \quad (2)$$

$$predmeas(i) = \frac{1}{s} \sum_{i=2-s}^{1} [meas(i) - meas(i-1)] \text{ is predicted reading from s data}$$

$$(3)$$

## 4   Robust Estimator Design

The block diagram of the proposed robust estimator is shown in Fig. 3. First of all the weighted expectation of the current and the previously received sensor measurements are computed. These named as the preprocessed measurements contain less variance of noise are in turn applied to the various nonlinear algorithms such as UKF. The consolidation of all the UKF outputs is done in such a way that the least mean square error is obtained. This is possible with the least squares estimation filter (WLSE). The output of WLSE is a much superior one because of performing the refining procedure in two different steps.

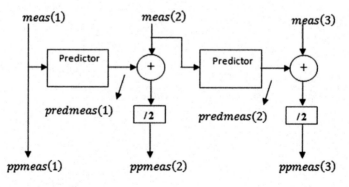

**Fig. 2** Preprocessing technique

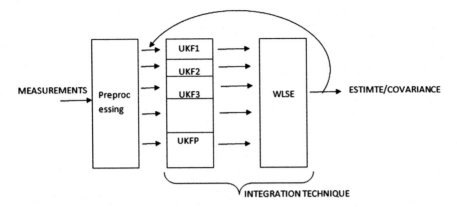

**Fig. 3** RUKF mechanism

# 5 Modeling of the Issue in Mathematical Form

## 5.1 Modeling of Submarine Trajectory

If the state vector $M(t)$ noted at time k is made up of four elements, out of which two are related to location $(lx, ly)$ and two related to speed $(sx, sy)$ then the state equation for a non-maneuvering submarine takes shape as follows.

$$M(t+1) = TrM(k) + Qi(k) \qquad (4)$$

$Tr = \begin{bmatrix} 1 & 0 & 1 & 0 \\ 0 & 1 & 0 & 1 \\ 0 & 0 & 1 & 0 \\ 0 & 0 & 0 & 1 \end{bmatrix}$ $Qi(k)$ is corruption in dynamics resulted from the

uncertain factors like waves.

## 5.2 Modeling of Corrupted Bearings

A couple of angle measurements are required to compute the location of the submarine. Let these be noted by $ps1(t), ps2(t)$. These are expressed in terms of the submarine location as follows from the simple coordinate geometry.

$$ps1(t) = arctan\left(\frac{ly(t) - S1(2)}{lx(t) - S1(1)}\right), \quad ps2(t) = arctan\left(\frac{ly(t) - S2(2)}{lx(t) - S2(1)}\right) \qquad (5)$$

The pure bearings $ps1(t), ps2(t)$ are turned out to extrinsic form $psm1(t)$, $psm2(t)$ by contaminations $psn1(t), psn2(t)$. s1 and s2 above denote sensor locations.

$$psm1(t) = ps1(t) + psn1(t), \quad psm2(t) = ps2(t) + psn2(t) \tag{6}$$

Expressing the above in the matrix language results in the measurement equation as follows:

$$meas(k) = T(x(t), t) + v1(t) \tag{7}$$

$$meas(t) = \begin{bmatrix} psm1(t) \\ psm2(t) \end{bmatrix}, \quad T(x(t), t) = \begin{bmatrix} arctan\left(\frac{ly(t) - S1(2)}{lx(t) - S1(1)}\right) \\ arctan\left(\frac{ly(t) - S2(2)}{lx(t) - S2(1)}\right) \end{bmatrix}, \quad v1(t) = \begin{bmatrix} psn1(t) \\ psn2(t) \end{bmatrix}$$

## 5.3   Comparison Metrics

### 5.3.1   RMS Error (M)

It is the physical separation between the true and estimated location of submarine. It is a Location error.

### 5.3.2   Missile Firing Time (S)

It is the time, the observer takes before releasing the missile. The missile can be released only if the estimation error is believed to be less than a preset value. It is denoted by Tm.

# 6   Results Obtained from Simulation

## 6.1   Tracking Environment Simulation

Observer at origin having three sensors (500 m apart) lying along the y axis tries to track a submarine positioned 14000 m away and $-60°$ from x-axis. The submarine is in the moving state with a speed of 40 kmph and a course $+90°$. The observer's sensors pick the impure bearing measurements from the submarine after a span of 1 s. The impurity belongs to Gaussian family with bias of zero and a variance of 0.015 rad. These polluted measurements are applied to UKF, preprocessed UKF, Integrated UKF and RUKF and the estimation errors in position are plotted in

**Fig. 4** Location errors for different estimators

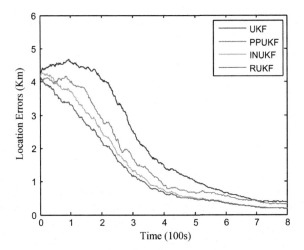

Time (100s)

**Table 1** Performance of different filters

| Parameter \estimator | UKF | PPUKF | INUKF | RUKF |
|---|---|---|---|---|
| RMS error (m) | 394 | 328 | 192 | 186 |
| Tm | 407 | 347 | 297 | 281 |
| % Superiority | 0 | 17 | 51 | 53 |

Fig. 4. All the algorithms are initialized with an error of ±3000 in location coordinates and ±5 m/s in speed.

In Table 1 and Fig. 4 it is clear that the estimation error produced by all the individual estimation algorithms namely UKF, PPUKF, INUKF are larger than the combination of them (RUKF). This means the combination of the existing algorithms produces better results with a negligible increase in the complexity. This is clear from the least value of the missile releasing time (281 s) associated with RUKF compared with the 407, 347, and 297 s associated with UKF, PPUKF, and INUKF.

## 6.2 Performance of Robust Estimator in Different Target Trajectories

For the scenario given in Sect. 1 except the bearing and course of the target are changed and the algorithms are tested. The used bearing and course are Case 1 B = 315, CR = 135, Case 2 B = 45, CR = 225, Case 3 B = 135, CR = 315, and Case 4 B = 225, CR = 45.

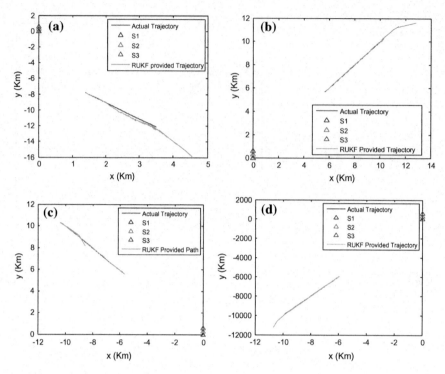

**Fig. 5** Tracking submarine in different scenarios

In all the cases Fig. 5a and d the novel algorithm produces excellent results and does not diverge. The trajectory produced by the introduced RUKF is superimposed on the true trajectory of the target justifies that the algorithm proposed works well in all the possible environments.

## 6.3 Performance of Robust Estimator for Different Submarine Ranges

The tracking of the submarine is performed in the same situation as given in the Sect. 1 but the range was changed and the performance is observed. The different range cases considered are 15, 20, 25, 30, 35, and 40 k. The increasing spread between the curves in Fig. 6 indicates that the proposed RUKF's superiority tends to get better and better with the raising range.

**Fig. 6** RUKF at different
submarines ranges

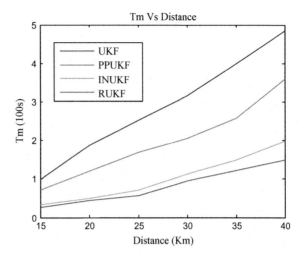

## 6.4 Performance of Robust Estimator for Different Sensor Noise Levels

The tracking of the submarine is performed in the same situation as given in the Sect. 1 but the noise level associated with the measurements was changed and the performance is observed. The different levels of noise considered are 0.01, 0.02, 0.03, 0.04, and 0.05 rad. The increasing spread between the curves in Fig. 7 indicates that the proposed RUKF's superiority tends to get better and better with the raising measurement corruption levels (NLev).

**Fig. 7** RUKF at different
sensor noise levels

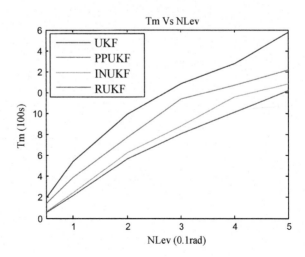

## 7  Conclusions

The reduced estimation errors, excellent superiority in the tougher environments certainly makes a key point that this robust algorithm can serve the purpose of the most advanced estimation algorithm for undersea issues of tracking. The superior performance exhibition of the current algorithm is because of the combination of the advantages of existing UKF, Integration technique, and the preprocessing mechanism with which the RUKF is made up of. The scope of this work is that more algorithms with different combinations of the existing algorithms can be developed to exhibit better performance than the individual ones.

## References

1. Kalman, R.E.: A New Approach to Linear Filtering and Prediction Problems, Transactions of the ASME-Journal of Basic Engineering, Vol. 82, series D, pp. 35–45.
2. Lerro, D. and Shalom, Y.B.: Tracking With Debiased Consistent Converted Measurements Versus EKF, IEEE transactions on Aerospace and Electronic systems, Vol. 29, no. 3, (1993), pp. 1015–1022.
3. Zhou, G., Pelletier, M., Kirubarajan, T., and Quan, T.P: Statically Fused Converted Position and Doppler Measurement Kalman Filters, Aerospace and Electronic Systems, IEEE Transactions on, Vol. 50,no. 1, (2014), pp. 300 – 318.
4. Rao, S.K., Kumar, D.V.A.N. Ravi Kumar, and Raju, K.P.: Combination of Pseudo Linear Estimator and modified gain bearings-only extended Kalman filter for passive target tracking in Abnormal conditions, Ocean Electronics (SYMPOL), (2013), pp. 3–8.
5. Aidala, V.J. and Hammel, S.E.: Utilization of modified polar coordinates for bearings only tracking, IEEE Trans. Automatic Control Vol. AC-28, no. 3, (1983), pp. 283–94.
6. Song, T.L., Speyer, J.L.: A Stochastic Analysis of a Modified Gain Extended Kalman Filter with Applications to estimation with Bearings-only measurements, IEEE Trans. Automat. Contr, Vol. AC-30, no. 10, (1985), pp. 940–9.
7. Rao, S.K., Kumar, D.V.A.N. Ravi Kumar, and Raju, K.P.: Integrated Unscented Kalman Filter for underwater passive target tracking with towed array measurements, Optik, Vol. 127, no. 5, (2016), pp. 2840–2847.
8. Julier, S. and Uhlmann, J.: Unscented Filtering and Nonlinear Estimation, Proceedings of the IEEE, Vol. 92, no. 3, (2004), pp. 401–422.
9. Rao, S.K., Kumar, D.V.A.N. Ravi Kumar, and Raju, K.P.: "A Novel Stochastic Estimator using Pre-Processing Technique for Long Range Target Tracking in heavy Noise Environment", Optik, Vol. 127, no. 10, (2016), pp. 4520–4530.
10. Simon, D.: Optimal State Estimation: Kalman, $H_\infty$,and Nonlinear Approaches, Wiley-Interscience, (2006), New Jersey.
11. Ristick, B., Arulampalam. S., and Gordon, N.: Beyond the Kalman Filter–Particle Filters for Tracking Applications, Artech House, DSTO, (2004), London.
12. Rao, S.K., Kumar, D.V.A.N. Ravi Kumar, and Raju, K.P.: "Underwater Bearings-Only Passive Target Tracking Using Estimate Fusion Technique", Advances in military technology, Vol. 10, no. 2, (2015), pp. 31–44.

# Geometry Scaling Impact on Leakage Currents in FinFET Standard Cells Based on a Logic-Level Leakage Estimation Technique

Zia Abbas, Andleeb Zahra, Mauro Olivieri and Antonio Mastrandrea

**Abstract** Static power consumption is one of the most critical issues in CMOS digital circuits, and FinFET technology is being recognized as a valid solution for the problem. In this chapter, we utilize a logic-level leakage current estimation technique relying on an internal node voltage-based model. The model is implemented in the form of VHDL packages. By utilizing the capability of the model, the behavior of major leakage component has been analyzed separately for FinFET technology scaling over single- and multi-stage digital standard cells.

**Keywords** FinFET · VHDL · Leakage · Digital · Standard cell

## 1 Introduction

As per International Technology Roadmap for Semiconductors (ITRS) for the trend of power dissipation with respect to technology progress, static power dissipation is expected to exceed dynamic power dissipations in the total power budget [1]. CMOS technology scaling is the apparent reason for the enormous increase in static power dissipation (i.e., leakage) and facing severe challenges scaling beyond 22 nm technology node. Despite adopting some strategies like the introduction of High-K dielectric in order to control excessive gate oxide leakage current [2–4], yet the

Z. Abbas (✉)
Center for VLSI and Embedded System Technology, IIIT Hyderabad,
Hyderabad, India
e-mail: zia.abbas@iiit.ac.in

A. Zahra · M. Olivieri · A. Mastrandrea
DIET, Spaienza University of Rome, Rome, Italy
e-mail: andleeb.zahra@uniroma1.it

M. Olivieri
e-mail: mauro.olivieri@uniroma1.it

A. Mastrandrea
e-mail: antonio.mastrandrea@uniroma1.it

© Springer Nature Singapore Pte Ltd. 2018
J. Anguera et al. (eds.), *Microelectronics, Electromagnetics
and Telecommunications*, Lecture Notes in Electrical Engineering 471,
https://doi.org/10.1007/978-981-10-7329-8_29

283

sub-threshold leakage is the dominant source of leakage power [5]. Therefore, designers are adopting new emerging device structures like fin-type field-effect transistor (FinFET), which is scalable well below 22 nm regime and has better electrostatic control over the channel and hence less leakage power dissipation [6].

In this context, an early and accurate estimation of the leakage currents in the design flow is valuable [7, 8]. Further, it is of great interest to have a clear assessment of the impact of technology scaling on leakage power behavior especially on major leakage components separately. The separate estimation would allow a more clear definition of countermeasure trade-offs [5]. SPICE-level leakage estimation promises the most accurate values at the cost of very long computation time and therefore are not feasible means of estimating leakage currents in integrated circuits (IC) of medium–high complexity. Second, the SPICE-level simulation does not allow a straightforward distinction among the contributions of different physical sources of leakage in a complex IC. On the other hand, logic-level calculation models are inherently faster and therefore have the computational speed advantage over SPICE simulations. This work utilizes the implementation of the logic-level model of leakage currents to characterize a set of FinFET standard cells and apply the same methodology to analyze the behavior of multi-stage complex cell design cases. Advantage of the logic-level characterization approach is the possibility of discriminating among the different leakage components within the circuit under analysis. The final results show a quantitative assessment of the trend in leakage components when geometry scaling is applied to FinFET standard cells.

## 2 Logic-Level Leakage Estimation Model

The proposed logic-level leakage estimation technique is based on the offline SPICE characterization of (i) voltage profiles at the internal nodes of the stacks inside the cells and (ii) currents in a single FinFET device. The library characterization data are encoded in a hardware description language (VHDL) to support logic-level simulation of the leakage power consumption. This is accomplished by the implementation of the following VHDL packages:

***Single_FinFET_leakage_current.vhd***: This package consists of data arrays of extracted leakage currents through drain, gate (front and back gates shorted), source and body terminals for the corresponding voltages at each terminal. Separate matrices of extracted currents values have been created for four FinFET device operating modes, namely nFinFET-*on*, nFinFET-*off*, pFinFET-*on*, and pFinFET-*off*. All the leakage current, values in the single FinFET device have been characterized for the following configurations:

- Number of fins 'NFIN' ranging from the minimum to eight times the minimum in the respective technology (similar to multiple of minimum width in CMOS technology).

**Fig. 1** Simulation setup for
static current extraction
FinFET

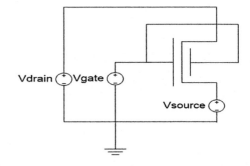

- Three FinFET technology nodes, i.e., 20, 16, and 10 nm.

- Terminal voltages ranging from 0 to 1.2 V with a step size of 0.01 V, therefore, simulations at 120 values of voltage.

The simulation setup for static current characterization in a single FinFET device is shown in Fig. 1 (for n-type FinFET-conducting case). Three VHDL functions have been created in the package in order to extract the three different leakage currents at the logic level. The three VHDL functions are *Isub, Igate,* and *Ibody* corresponding to sub-threshold, gate, and junction leakage respectively. The function for *Isub* along with their parameter lists is shown below. Other functions are considered similarly.

```
Function Isub( Lmin:   in integer;        -- gate length [nm],
               Wmin:   in integer;        -- number of fins
               Vds:    in real;           -- Vds
               Temperature: in integer    -- temperature [°C]
               )
```

*Intermediate_node_voltage.vhd*: This package contains the data arrays of all the internal node voltages in pull-up and pull-down stacks in the matrix form. The internal node voltage characterization is required in the circuit structures (stacked transistors) normally present in real cells; otherwise evaluation of leakage current values would be inaccurate. Figure 2 shows the n-type and p-type FinFET basic structures needed to characterize all the standard cells in the target library. When two or more transistors in the pull-down or pull-up network of a logic cell are off, it is difficult to predict the voltages at the nodes connecting the transistors based on simple models of transistor behavior. As an example, in 2-input NAND cell which contains pull-down *(A and B)n* stack, for input AB = *"00"*, INV in Fig. 2b is not actively driven by any device (may be referred to high-impedance state) and its voltage is only determined by nonlinear voltage–current characteristic of the connected devices in off [9]. Note that the same pull-up and pull-down stacks often exist in different standard cells and have the same internal node voltage values. For example the same *(A and B)n* structure also exists in XOR2, AND2, AO12, Full Adder, AO112, and AO22 circuit.

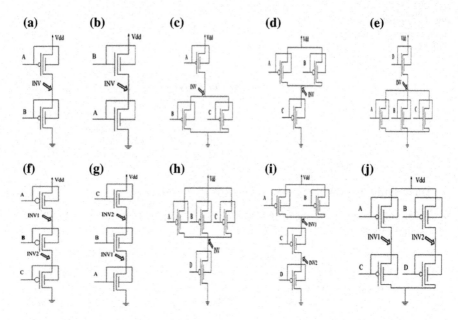

**Fig. 2** Internal node voltage stacks of n-type FinFET and p-type FinFET used (single and two internal nodes)

Interestingly, as a result of the voltage characterization, we found the voltages at the internal nodes of the "off" pull-down or pull-up network of a digital cell do not depend on the resize factor of the cell. Another relevant observation is that the internal node voltages do not depend on technology scaling. Table 1 reports the evidence of such result referring to the 2-input NAND cell which has pull-down *(A and B)n* stack at Vdd = 0.9 V.

***Multi cell_leakage.vhd***: This optional package is used to simply characterize the leakage in multi-stage composite cells from single-stage basic standard cells based. However, this can be done by adding all the VHDL equations of corresponding single-stage cells.

In the proposed work, all the SPICE-level characterizations are performed with the HSPICE simulator, using Berkeley Short-channel IGFET Model (BSIM-CMG) [10] for 20, 16, and 10 nm common multi-gates (CMG) FinFETs, based on

| **Table 1** Observations in internal node voltage | Resize factor (shown for 20 nm) | | | Technology scaling (at min size) | | |
|---|---|---|---|---|---|---|
| | Comb | Min size | Resize (×10) | 20 nm | 16 nm | 10 nm |
| | 00 | 0.657 | 0.657 | 0.657 | 0.664 | 0.684 |
| | 01 | 0.837 | 0.837 | 0.837 | 0.836 | 0.833 |
| | 10 | 0.0 | 0.0 | 0.0 | 0.0 | 0.0 |

predictive technology model parameters (PTM) [11]. The method supports separate calculation of gate leakage, sub-threshold leakage, and junction leakage, including input pattern dependence, stacking and loading effects.

# 3 Model Accuracy Verification and Separate Estimation of Major Leakage Components

Sixteen combinational and sequential standard cells of different fan-in and seven multi-stage complex cells are considered to test the accuracy of the model for three technologies, i.e., 20, 16, and 10 nm bulk FinFET at 0.9 V. We also employed the capability of the proposed model and estimated sub-threshold, gate leakage and junction body leakage separately in the three-scaled FinFET technology nodes in order to investigate the impact and behavior of each leakage component separately and analyses the dominant leakage component with scaling in technology. Naturally, summing up the three leakage components for a particular input combination in the respective technology, returns the total leakage current ($Itot = Isub + Igate + Isub$), as in tables.

Table 2 reports all leakage currents for single-input and two-input cells estimated by the VHDL model and by HSPICE, along with error percentage for all input combinations. The error percentage between the VHDL model and HSPICE is below 1%. Table 3 report the leakage current values of three input, four input, and sequential cells. While each cell was characterized and verified for each of the all possible input combinations, due to space limitations the Table reports the combinations corresponding to the minimum and maximum leakage, and the average leakage computed among all the input combinations. Figure 3 shows the trend for average sub-threshold and gate leakage over the three-scaled technologies. As per our analysis, we observe that on average there is approximately a 2.5 and 4.5% increase in sub-threshold leakage when scaling from 20 nm to 16 nm and from 16 nm to 10 nm, respectively, at $V_{DD} = 0.9$ V. An interesting outcome of our analysis, for a scaling factor of around 0.7 in geometry downsizing (0.7 times the preceding technology node), we notice an approximately double percentage increase in sub-threshold leakage currents in the second scaling step than in the first scaling step. As a whole, however, the increase in sub-threshold leakage with geometry scaling is weak. Again as a general trend from Fig. 3, the percentage increase in gate leakage is enormous in the second scaling step, i.e., scaling from 16 nm to 10 nm. On average in two-input standard cells at $V_{DD} = 0.9$ V, the percentage increase in gate leakage currents are 17.5 and 175% in the first and second scaling steps, respectively. Even if the contribution to the total leakage is

**Table 2** Input pattern-dependent $I_{sub}$ in [nA]; $I_{gate}$ and $I_{body}$ in [pA]; $I_{tot}$ [nA] through model and spice [nA] in 2-input cells

| Std. Cell | Sig Pat | 20nm FinFET | | | | | | 16nm FinFET | | | | | | 10nm FinFET | | | | | |
|---|---|---|---|---|---|---|---|---|---|---|---|---|---|---|---|---|---|---|---|
| | | Model | | | | SPICE | Err % | Model | | | | SPICE | Err % | Model | | | | SPICE | Err % |
| | | Isub | Ibody | Igate | Itot | | | Isub | Ibody | Igate | Itot | | | Isub | Ibody | Igate | Itot | | |
| NOT | 0 | 7.287 | 0.653 | 0.024 | 7.288 | 7.286 | 0.02 | 7.413 | 0.096 | 0.035 | 7.413 | 7.412 | 0.01 | 7.450 | 0.013 | 0.201 | 7.450 | 7.450 | 0.01 |
| | 1 | 7.196 | 0.373 | 36.320 | 7.233 | 7.232 | 0.01 | 7.412 | 0.064 | 42.822 | 7.455 | 7.454 | 0.01 | 7.931 | 0.007 | 118.004 | 8.049 | 8.048 | 0.01 |
| | Avg | 7.242 | 0.513 | 18.172 | 7.261 | 7.259 | -0.02 | 7.413 | 0.080 | 21.429 | 7.434 | 7.433 | -0.01 | 7.691 | 0.010 | 59.103 | 7.750 | 7.749 | -0.01 |
| NAND2 | 00 | 1.550 | 1.306 | 0.048 | 1.552 | 1.554 | 0.17 | 1.620 | 0.192 | 0.069 | 1.620 | 1.620 | 0.00 | 1.606 | 0.025 | 0.401 | 1.606 | 1.617 | 0.65 |
| | 01 | 12.625 | 1.398 | 0.024 | 12.627 | 12.611 | 0.12 | 12.810 | 0.224 | 0.035 | 12.811 | 12.809 | 0.02 | 12.837 | 0.027 | 0.201 | 12.837 | 12.753 | 0.66 |
| | 10 | 14.569 | 1.397 | 72.659 | 14.643 | 14.628 | 0.10 | 14.824 | 0.222 | 85.676 | 14.910 | 14.897 | 0.08 | 14.899 | 0.026 | 236.187 | 15.135 | 15.125 | 0.07 |
| | 11 | 28.780 | 1.489 | 145.280 | 28.927 | 28.919 | 0.03 | 29.648 | 0.254 | 171.292 | 29.820 | 29.811 | 0.03 | 31.722 | 0.027 | 472.000 | 32.194 | 32.186 | 0.02 |
| | Avg | 14.381 | 1.398 | 54.503 | 14.437 | 14.428 | -0.06 | 14.726 | 0.223 | 64.268 | 14.790 | 14.784 | -0.04 | 15.266 | 0.026 | 177.197 | 15.443 | 15.420 | -0.15 |
| NOR2 | 00 | 14.574 | 2.610 | 0.088 | 14.577 | 14.574 | 0.02 | 14.825 | 0.382 | 0.131 | 14.826 | 14.825 | 0.01 | 14.900 | 0.047 | 0.795 | 14.901 | 14.900 | 0.01 |
| | 01 | 28.780 | 1.677 | 36.364 | 28.818 | 28.767 | 0.18 | 29.648 | 0.254 | 42.888 | 29.691 | 29.650 | 0.14 | 31.723 | 0.030 | 118.402 | 31.841 | 31.806 | 0.11 |
| | 10 | 23.914 | 0.372 | 36.320 | 23.951 | 23.908 | 0.18 | 24.222 | 0.063 | 42.822 | 24.265 | 24.262 | 0.01 | 25.843 | 0.007 | 118.004 | 25.961 | 25.762 | 0.77 |
| | 11 | 2.294 | 0.744 | 72.640 | 2.367 | 2.372 | 0.22 | 2.254 | 0.127 | 85.644 | 2.340 | 2.340 | 0.01 | 2.405 | 0.014 | 236.008 | 2.641 | 2.658 | 0.64 |
| | Avg | 17.391 | 1.351 | 36.353 | 17.428 | 17.405 | -0.13 | 17.737 | 0.207 | 42.871 | 17.781 | 17.769 | -0.06 | 18.718 | 0.025 | 118.302 | 18.836 | 18.782 | -0.29 |
| AND2 | 00 | 15.940 | 1.678 | 36.368 | 15.978 | 15.981 | 0.02 | 16.444 | 0.256 | 42.891 | 16.487 | 16.486 | 0.01 | 17.467 | 0.032 | 118.405 | 17.585 | 17.595 | 0.05 |
| | 01 | 27.015 | 1.771 | 36.344 | 27.053 | 27.054 | 0.00 | 27.634 | 0.288 | 42.857 | 27.678 | 27.686 | 0.03 | 28.698 | 0.034 | 118.205 | 28.816 | 28.740 | 0.26 |
| | 10 | 28.959 | 1.769 | 108.984 | 29.070 | 29.073 | 0.01 | 29.648 | 0.286 | 128.503 | 29.777 | 29.777 | 0.00 | 30.760 | 0.033 | 354.205 | 31.114 | 31.114 | 0.00 |
| | 11 | 36.067 | 2.142 | 145.304 | 36.214 | 36.238 | 0.07 | 37.061 | 0.350 | 171.327 | 37.232 | 37.247 | 0.04 | 39.172 | 0.040 | 472.201 | 39.644 | 39.657 | 0.03 |
| | Avg | 26.995 | 1.840 | 81.750 | 27.079 | 27.087 | 0.03 | 27.697 | 0.295 | 96.395 | 27.794 | 27.799 | 0.02 | 29.024 | 0.035 | 265.754 | 29.290 | 29.277 | -0.05 |
| XOR2 | 00 | 67.268 | 5.403 | 145.372 | 67.419 | 67.348 | 0.10 | 68.695 | 0.826 | 171.427 | 68.867 | 68.842 | 0.04 | 72.466 | 0.098 | 472.799 | 72.939 | 72.718 | 0.30 |
| | 01 | 48.871 | 4.469 | 109.067 | 48.985 | 48.984 | 0.00 | 49.871 | 0.697 | 128.629 | 50.001 | 50.006 | 0.01 | 51.047 | 0.080 | 354.986 | 51.402 | 51.326 | 0.15 |
| | 10 | 48.871 | 4.377 | 109.067 | 48.984 | 48.977 | 0.01 | 49.871 | 0.666 | 128.629 | 50.001 | 50.001 | 0.00 | 51.047 | 0.078 | 354.986 | 51.402 | 51.322 | 0.16 |
| | 11 | 81.474 | 4.843 | 217.964 | 81.697 | 81.586 | 0.14 | 83.518 | 0.763 | 257.002 | 83.776 | 83.717 | 0.07 | 89.288 | 0.089 | 708.406 | 89.996 | 89.748 | 0.28 |
| | Avg | 61.621 | 4.773 | 145.368 | 61.771 | 61.724 | -0.08 | 62.989 | 0.738 | 171.422 | 63.161 | 63.142 | -0.03 | 65.962 | 0.086 | 472.794 | 66.435 | 66.279 | -0.24 |

negligible. Finally, body leakage currents are definitely negligible in total leakage budget, as from Tables, show almost a saturated behavior.

As a result, in FinFET, the percent increase in sub-threshold leakage currents with technology scaling, are comparatively very less with bulk CMOS technology [5], and hence provides the better room for further scaling and confirming that FinFET cells are a promising alternative to CMOS if low static power design is a primary concern.

**Table 3** $I_{sub}$ in [nA]; $I_{gate}$ and $I_{body}$ in [pA]; $I_{tot}$ [nA] through model and spice [nA] in 3-input, 4-input, and sequential cells (Max, Min, Avg)

| Std. cell | Sig pat | 20 nm FinFET | | | | | | 16 nm FinFET | | | | | | 10 nm FinFET | | | | | |
|---|---|---|---|---|---|---|---|---|---|---|---|---|---|---|---|---|---|---|---|
| | | Model | | | Itot | Spice | Err % | Model | | | Itot | Spice | Err % | Model | | | Itot | Spice | Err % |
| | | Isub | Ibody | Igate | | | | Isub | Ibody | Igate | | | | Isub | Ibody | Igate | | | |
| NAND3 | Min | 1.408 | 1.958 | 0.072 | 1.411 | 1.417 | 0.43 | 1.548 | 0.287 | 0.104 | 1.549 | 1.562 | 0.85 | 1.612 | 0.037 | 0.602 | 1.613 | 1.609 | 0.24 |
| | Max | 43.17 | 3.348 | 326.82 | 43.5 | 43.478 | 0.05 | 44.472 | 0.569 | 385.404 | 44.858 | 44.844 | 0.03 | 47.583 | 0.06 | 1062.06 | 48.645 | 48.622 | 0.05 |
| | Avg | 13.836 | 2.572 | 95.353 | 13.934 | 13.919 | 0.1 | 14.178 | 0.417 | 112.457 | 14.291 | 14.284 | 0.05 | 14.589 | 0.047 | 310.056 | 14.899 | 14.871 | 0.19 |
| NOR3 | Max | 43.174 | 4.283 | 36.445 | 43.215 | 43.067 | 0.34 | 44.472 | 0.633 | 43.015 | 44.516 | 44.4 | 0.26 | 47.584 | 0.075 | 119.19 | 47.7 | 47.6 | 0.22 |
| | Min | 1.985 | 1.115 | 108.948 | 2.095 | 2.093 | 0.09 | 2.053 | 0.19 | 128.466 | 2.181 | 2.17 | 0.52 | 2.28 | 0.02 | 354.01 | 2.63 | 2.612 | 0.84 |
| | Avg | 18.442 | 3.985 | 54.536 | 18.501 | 18.473 | 0.15 | 18.769 | 0.594 | 64.326 | 18.834 | 18.812 | 0.11 | 19.879 | 0.07 | 177.576 | 20.057 | 19.977 | 0.4 |
| AND3 | Min | 15.644 | 2.331 | 36.392 | 15.683 | 15.844 | 1.02 | 16.313 | 0.351 | 42.926 | 16.357 | 16.422 | 0.4 | 17.418 | 0.044 | 118.61 | 17.537 | 17.589 | 0.3 |
| | Max | 50.457 | 4.001 | 326.844 | 50.788 | 50.811 | 0.05 | 51.885 | 0.665 | 385.439 | 52.271 | 52.289 | 0.03 | 55.033 | 0.073 | 1062.26 | 56.096 | 56.111 | 0.03 |
| | Avg | 27.319 | 2.98 | 127.136 | 27.449 | 27.467 | 0.07 | 28.068 | 0.485 | 149.931 | 28.219 | 28.228 | 0.03 | 29.392 | 0.055 | 413.335 | 29.805 | 29.793 | 0.04 |
| AO12 | Max | 52.901 | 3.447 | 145.304 | 53.05 | 53.033 | 0.03 | 53.499 | 0.541 | 171.327 | 53.671 | 53.699 | 0.05 | 56.588 | 0.064 | 472.201 | 57.06 | 56.622 | 0.77 |
| | Min | 11.544 | 2.513 | 181.624 | 11.728 | 11.744 | 0.14 | 11.615 | 0.412 | 214.149 | 11.83 | 11.831 | 0.01 | 11.946 | 0.046 | 590.205 | 12.536 | 12.561 | 0.2 |
| | Avg | 33.303 | 3.388 | 95.415 | 33.402 | 33.4 | 0 | 34.033 | 0.52 | 112.521 | 34.146 | 34.148 | 0.01 | 35.666 | 0.061 | 310.424 | 35.977 | 35.915 | 0.17 |
| MUX | Min | 28.964 | 1.683 | 109.008 | 29.075 | 29.078 | 0.01 | 29.649 | 0.26 | 128.537 | 29.778 | 29.778 | 0 | 30.761 | 0.036 | 354.405 | 31.116 | 31.122 | 0.02 |
| | Max | 65.027 | 2.517 | 145.304 | 65.175 | 65.2 | 0.04 | 66.71 | 0.416 | 171.325 | 66.881 | 66.889 | 0.01 | 69.933 | 0.05 | 472.209 | 70.405 | 70.411 | 0.01 |
| | Avg | 46.996 | 2.478 | 90.848 | 47.089 | 47.104 | 0.03 | 48.179 | 0.396 | 107.125 | 48.287 | 48.294 | 0.02 | 50.347 | 0.05 | 295.407 | 50.643 | 50.65 | 0.01 |
| FA | Min | 98.014 | 17.232 | 200.33 | 98.232 | 99.111 | 0.89 | 101.362 | 2.569 | 236.388 | 101.601 | 101.678 | 0.08 | 105.573 | 0.307 | 654.271 | 106.227 | 105.856 | 0.35 |
| | Max | 143.409 | 19.09 | 272.845 | 143.701 | 141.444 | 1.6 | 147.197 | 2.883 | 321.852 | 147.521 | 145.222 | 1.58 | 154.528 | 0.339 | 889.167 | 155.417 | 153 | 1.58 |
| | Avg | 125.851 | 15.466 | 460.014 | 126.327 | 126.403 | 0.06 | 128.831 | 2.394 | 542.78 | 129.376 | 129.182 | 0.15 | 134.812 | 0.275 | 1497.062 | 136.31 | 135.843 | 0.34 |
| Latch | Max | 65.253 | 2.008 | 36.378 | 65.253 | 65.167 | 0.13 | 66.754 | 0.307 | 42.91 | 66.754 | 66.778 | 0.04 | 69.091 | 0.042 | 118.507 | 69.091 | 68.711 | 0.55 |
| | Min | 36.289 | 2.008 | 36.378 | 36.289 | 36.256 | 0.09 | 37.105 | 0.307 | 42.91 | 37.105 | 37.156 | 0.14 | 38.33 | 0.042 | 118.507 | 38.33 | 38.867 | 1.38 |
| | Avg | 48.309 | 2.294 | 78.735 | 48.309 | 48.317 | 0.02 | 49.508 | 0.362 | 92.854 | 49.508 | 49.531 | 0.05 | 51.765 | 0.047 | 256.093 | 51.765 | 51.744 | 0.04 |
| Flip Flop | Max | 108.539 | 4.196 | 145.335 | 108.539 | 108.589 | 0.05 | 111.355 | 0.673 | 171.399 | 111.355 | 111.444 | 0.08 | 117.03 | 0.083 | 472.605 | 117.03 | 117.111 | 0.07 |
| | Min | 72.379 | 4.521 | 145.344 | 72.379 | 72.422 | 0.06 | 74.294 | 0.719 | 171.409 | 74.294 | 74.322 | 0.04 | 78.338 | 0.087 | 472.708 | 78.338 | 78.367 | 0.04 |

(continued)

**Table 3** (continued)

| Std. cell | Sig pat | 20 nm FinFET | | | | | | 16 nm FinFET | | | | | | 10 nm FinFET | | | | | |
|---|---|---|---|---|---|---|---|---|---|---|---|---|---|---|---|---|---|---|---|
| | | Model | | | Itot | Spice | Err % | Model | | | Itot | Spice | Err % | Model | | | Itot | Spice | Err % |
| | | Isub | Ibody | Igate | | | | Isub | Ibody | Igate | | | | Isub | Ibody | Igate | | | |
| | Avg | 86.93 | 4.633 | 159.87 | 86.93 | 86.758 | 0.2 | 89.136 | 0.732 | 188.539 | 89.136 | 89.1 | 0.04 | 93.477 | 0.09 | 519.907 | 93.477 | 93.576 | 0.11 |
| AO112 | Max | 72.72 | 4.095 | 145.285 | 72.869 | 72.933 | 0.09 | 73.703 | 0.632 | 171.327 | 73.875 | 73.911 | 0.05 | 78.093 | 0.072 | 472.201 | 78.565 | 77.967 | 0.77 |
| | Min | 9.778 | 2.885 | 217.916 | 9.999 | 10.049 | 0.5 | 9.909 | 0.475 | 256.971 | 10.166 | 10.253 | 0.85 | 10.278 | 0.052 | 708.209 | 10.986 | 10.996 | 0.09 |
| | Avg | 33.643 | 5.421 | 109.068 | 33.757 | 33.755 | 0.01 | 34.278 | 0.818 | 128.649 | 34.407 | 34.407 | 0 | 35.952 | 0.096 | 355.11 | 36.307 | 36.219 | 0.24 |
| AO22 | Max | 64.853 | 4.187 | 217.936 | 65.075 | 64.922 | 0.23 | 66.709 | 0.664 | 257.005 | 66.966 | 66.856 | 0.17 | 70.896 | 0.074 | 708.399 | 71.605 | 71.5 | 0.15 |
| | Min | 11.883 | 3.629 | 290.544 | 12.177 | 12.178 | 0 | 11.92 | 0.602 | 342.619 | 12.263 | 12.267 | 0.03 | 12.261 | 0.066 | 944.201 | 13.205 | 13.244 | 0.3 |
| | Avg | 41.766 | 4.429 | 138.54 | 41.909 | 41.953 | 0.1 | 42.692 | 0.688 | 163.393 | 42.856 | 42.897 | 0.1 | 44.702 | 0.08 | 450.672 | 45.152 | 45.089 | 0.14 |
| AO31 | Max | 75.621 | 5.303 | 326.862 | 75.954 | 74.144 | 2.44 | 76.543 | 0.853 | 385.439 | 76.929 | 74.967 | 2.62 | 79.157 | 0.094 | 1062.261 | 80.219 | 79.344 | 1.1 |
| | Min | 13.684 | 4.371 | 363.178 | 14.051 | 13.733 | 2.32 | 13.717 | 0.726 | 428.261 | 14.146 | 13.867 | 2.01 | 14.194 | 0.078 | 1180.265 | 15.374 | 15.122 | 1.67 |
| | Avg | 35.061 | 4.818 | 160.111 | 35.226 | 35.101 | 0.36 | 35.89 | 0.749 | 176.789 | 36.067 | 36.098 | 0.09 | 37.489 | 0.086 | 487.657 | 37.976 | 37.943 | 0.09 |

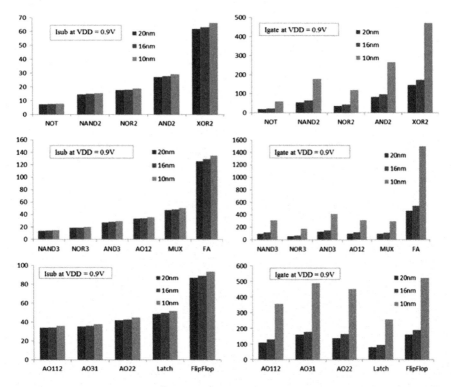

**Fig. 3** Impact of technology scaling over sub-threshold and gate leakage in single cells (average values)

## 4 Multi-stage Complex Cells

Table 4 reports the estimated leakage currents in complex cells through the logic-level model and HSPICE, along with the relative error; all cases show error percentages below 1%. Table 4 also reports separately estimated major leakage currents. The trends for sub-threshold and gate leakage with technology scaling are shown in Fig. 3 for single cells and Fig. 4 for multi-cell benchmark circuits. The result perfectly matching the behavior found in single cells can be ascribed to a negligible impact of loading effects among the cells.

**Table 4** $I_{sub}$ and $I_{gate}$ in [nA]; $I_{body}$ in [pA]; $I_{tot}$ [nA] through model and spice [nA] in multi-stage complex cells. (Avg)

| Multi-stage Std. cell | 20 nm FinFET | | | | | | 16 nm FinFET | | | | | | 10 nm FinFET | | | | | |
|---|---|---|---|---|---|---|---|---|---|---|---|---|---|---|---|---|---|---|
| | Model | | | Itot | Spice | Err % | Model | | | Itot | Spice | Err % | Model | | | Itot | Spice | Err % |
| | Isub | Ibody | Igate | | | | Isub | Ibody | Igate | | | | Isub | Ibody | Igate | | | |
| 4-bit magnitude comparator | 154.137 | 12.854 | 0.509 | 154.65 | 153.34 | 0.86 | 160.658 | 2.035 | 0.600 | 161.26 | 160.34 | 0.58 | 167.144 | 0.239 | 1.654 | 168. 8 | 168.44 | 0.21 |
| 18-input combinational unit | 256.117 | 31.105 | 0.872 | 257.02 | 258.71 | 0.65 | 263.943 | 4.758 | 1.029 | 264.97 | 265.89 | 0.34 | 283.238 | 0.560 | 2.838 | 286.08 | 284.11 | 0.69 |
| 8-bit parity checker | 430.747 | 34.785 | 0.972 | 431.75 | 432.10 | 0.08 | 440.217 | 5.364 | 1.146 | 441.36 | 441.61 | 0.05 | 460.976 | 0.630 | 3.162 | 464.14 | 463.28 | 0.19 |
| Interrupt controller | 494.50 | 41.26 | 1.635 | 496.18 | 500.11 | 0.79 | 513.830 | 6.516 | 1.928 | 515.77 | 512.33 | 0.67 | 536.982 | 0.763 | 5.317 | 542.3 | 538.33 | 0.74 |
| 4-bit ripple carry adder | 498.935 | 66.958 | 1.128 | 500.13 | 497.88 | 0.45 | 509.738 | 10.110 | 1.333 | 511.08 | 508.67 | 0.47 | 533.021 | 1.193 | 3.682 | 536.71 | 533.5 | 0.60 |
| Simple 8-bit ALU | 1156.35 | 107.08 | 3.64 | 1160.1 | 1160.0 | 0.01 | 1182.42 | 16.68 | 4.29 | 1186.72 | 1186.00 | 0.06 | 1239.87 | 1.93 | 11.82 | 1251.7 | 1247. 7 | 0.31 |
| 4-bit multiplier | 1440.649 | 233.632 | 2.986 | 1443.86 | 1446.0 | 0.15 | 1479.454 | 34.914 | 3.523 | 1483.01 | 1484.66 | 0.11 | 1546.341 | 4.199 | 9.746 | 1556.1 | 1552. 4 | 0.23 |

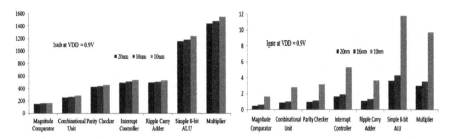

**Fig. 4** Impact of technology scaling over Isub and Igate in multi-stage cells (average values)

# 5 Conclusion

We have presented a characterization of leakage currents in FinFET standard cells with respect to geometry scaling based on logic-level leakage estimation technique. The characterization methodology allowed to investigate dominant leakage component, i.e., sub-threshold leakage as expected with FinFET technology scaling.

# References

1. International Technology Roadmap for Semiconductors. International SEMATECH, Austin, TX. http://public.itrs.net.
2. R. Chau et al. "Application of High-k Gate Dielectric and Metal Gate Electrodes to enable Silicon and Non-Silicon Logic Nanotechnology", Microelectron Engineering, Vol. 80, pp. 1–6, June 2005.
3. Z. Abbas, V. Genua, and M. Olivieri, "A novel logic level calculation model for leakage currents in digital nano-CMOS circuits," in Proc. IEEE 7th Conf. PRIME, Jul. 2011, pp. 221–224.
4. Zia Abbas, Mauro Olivieri, Andreas Ripp "Yield-Driven Power-Delay-Optimal CMOS Full Adder Design complying with Automotive Product Specifications of PVT Variations and NBTI Degradations" Journal of Computational Electronics (Springer), Vol. 15, Issue. 4, pp. 1424–1439, 2016.
5. Zia Abbas, Mauro Olivieri, "Impact of Technology Scaling on Leakage Power in nano-scale bulk CMOS Digital Standard Cell Library" Elsevier Microelectronics Journal, Vol. 45, Issue. 2, pp. 179–195, Feb. 2014.
6. T.-J. King, "FinFETs for nanoscale CMOS digital integrated circuits," in Proc. Int. Conf. Computer-Aided Design, Nov. 2005, pp. 207–210.
7. M. Olivieri, G. Scotti and A. Trifiletti, "A Novel Yield Optimization Technique for Digital CMOS Circuits Design by Means of Process Parameters Run-Time Estimation and Body Bias Active Control", IEEE transactions on Very Large Scale Integration (VLSI) Systems, Vol. 13, No. 5, May 2005.
8. S. Mukhopadhyay, S. Bhunia, K. Roy, "Modeling and Analysis of Loading Effect in Leakage of Nano-Scaled Bulk- CMOS Logic Circuits," IEEE Trans. on Computer Aided Design of Electronic Circuits and Systems, vol. 25, n. 8, Aug. 2006.

9. Z. Abbas, A. Mastrandrea and M. Olivieri, "A Voltage-Based Leakage Current Calculation Scheme and its Application to Nanoscale MOSFET and FinFET Standard-Cell Designs", IEEE transactions on Very Large Scale Integration (VLSI) Systems, Vol. 22, No. 12, Dec 2014.

10. Venugopalan, S, Karim, M. A., Lu, D. D., Niknejad, A. M., Hu C., "Compact models for real device effects in FinFETs," Proc. of International Conference on Simulation of Semiconductor Processes and Devices (SISPAD), Denver, CO, 2012.

11. Predictive Technology Model, http://ptm.asu.edu/.

# Blind Recognition of Error-Correcting BCH Codes Using GFFT

**A. Abhishek Nath and V. Navya**

**Abstract** In this chapter, a novel method Galois Field Fourier Transform is proposed for extracting the primitive BCH Code, which is intercepted with the noise. First, the intercepted bit stream is divided into sequences of same length and GFFT is performed on all the sequences, from which spectral components are recorded. Then based on the locations of common null spectral components, the code length and the roots of respective generator polynomial are found. Finally, the performance of the proposed technique is measured by computing the false alarm and miss detection probabilities for both roots and non-roots of the Generator Polynomial. The code recognition of the proposed method is plotted for different code lengths and also the proposed method is compared with previous techniques.

**Keywords** BCH codes · Galois field Fourier transform (GFFT)
Blind recognition · Binary symmetric channel · Generator polynomial
Optimal threshold

## 1 Introduction

In digital communication, if information is transmitted over a channel, noise may be added to the required signal. So such errors should be corrected, which is done by using coding theory. Coding theory is the study of error correction codes, it is used to solve the problem of transmission of data over a defective communication channel. Typically, if $N$ is transmitted through a channel it may be recovered as $N'$, which is little different from the original. The purpose of coding theory is to find a way to recover $N$ from $N'$. There are two major kinds of coding schemes; linear block codes and convolutional codes. BCH Codes comes under linear block codes which have

A. Abhishek Nath (✉) · V. Navya
Sree Vidyanikethan Engineering College, Tirupati, India
e-mail: rockabhishek62@gmail.com

V. Navya
e-mail: vuppalapatinavya@gmail.com

© Springer Nature Singapore Pte Ltd. 2018
J. Anguera et al. (eds.), *Microelectronics, Electromagnetics
and Telecommunications*, Lecture Notes in Electrical Engineering 471,
https://doi.org/10.1007/978-981-10-7329-8_30

better recognition probabilities in coding theory. BCH Code have the capability of correcting $h$ error combinations in a block length of $n = 2^m - 1$. This code is also known as $h$-error-correcting BCH Code. The specification of this code is done in terms of generator polynomial roots from the Galois Field $GF(p^m)$ [1]. Blind recognition of error-correcting code is a research area for non-cooperative contexts such as military interruption and coding AMC, which is depicted as the receiver to distinguish the encoder parameters that utilized by the transmitter from the information captured from the channel. Because of the absence of earlier data and the presence of noise, this issue turns out to be hard now and again, this issue turns out to be hard sometimes. To overcome this problem some methods have been proposed in recent years, Channel Coding Techniques for Wireless Communications [2] explains basic construction of BCH Codes. Statistical recognition method of binary BCH code is proposed based on the frame length [3] recognition performance is not much degraded for higher Bit Error Rates, but it is not suitable to long code lengths. The method [4] of Reconstruction of BCH Codes using Probability Compensation technique is based on cyclic properties and finally probability compensation method is used to progress the performance. In, [5], a method based on the Information dispersion entropy-based blind recognition of binary BCH Codes in soft decision situations are proposed but in [4, 5] the authors do not give a reasonable threshold to achieve Optimal recognition. In [6] Blind recognition of BCH codes based on Galois Field Fourier Transform, method mainly focused on 2-ary-one-error correction of BCH Code at different lengths and shown Optimal Threshold value to distinguish the root and non-roots which belongs to generator polynomial. The proposed method of focal point is initiated from [6]. Where the proposed method explains, 2-ary-2-error correction of BCH code and 2-ary-3-error correction of BCH code.

## 2  Galois Field Fourier Transform

Assume $c(x) = c_0 + c_1 + \cdots + c_{n-1}x^{n-1}$ is a polynomial over $GF(q)$, where $n = q - 1$ and $q = 2^m$, the degree of the primitive polynomial for $GF(q)$ is $m$, then, $\alpha^n = 1$ which is the root of $x^n - 1$. The defined polynomial from Galois Field Fourier Transform (GFFT) of c(x) is

$$C(x) = C_0 + C_1 + \cdots + C_{n-1}x^{n-1} \tag{1}$$

The $j$th spectral component of $C(x)$ is $C_j$ coefficient which is called as spectral polynomial of $c(x)$ as per this paper,

$$C_j = \sum_{i=0}^{n-1} c_j \alpha^{ij}, 0 \leq j < n \tag{2}$$

$F_j = ((\alpha^j)^0, (\alpha^j)^1, (\alpha^j)^2, \ldots, (\alpha^j)^{n-1}, Fb_j$ is the binary form of $F_j$, then the binary matrix form of Eq. (2) is

$$C_j = Fb_j \times (c_0, c_1, \ldots, c_{n-2}, c_{n-1})^T \tag{3}$$

The $C(x)$ and $c(x)$ have the following properties:

1. The $j$th spectral component of $C_j$ is zero if and only if $\alpha^j$ is a root of $c(x)$.
2. The $i$th spectral component of $c(x)$ is zero if and only if $\alpha^{-j}$ is a root of $C(x)$.

## 3 Proposed Method

Consider 2 -ary-$h$-error-correcting BCH Code which has $n$ as its length,

$$n = 2^m - 1 \tag{4}$$

The proposed method will calculate the length of the BCH code up to 255. The range of $m$ value is $m_{min} = 3$ to $m_{max} = 8$. Because of polynomial $c(x)$ degree $n-1$ or less over $GF(2)$ is a code polynomial that is possible when $c(x)$ is divisible by $g(x)$ [1]. That means $c(x)$ is a code polynomial and $c(x)$ has roots $\alpha, \alpha^2, \alpha^3 \ldots, \alpha^{2t}$. So the expected work is to find the roots of $c(x)$ from the intercepted bit stream and the generator polynomials $g(x)$ recovering is performed based on Eq. (5).

$$g(x) = (x - \alpha_1)(x - \alpha_2)(x - \alpha_3), \ldots, (x - \alpha_r) \tag{5}$$

Therefore, these roots can be determined in two ways. By the properties of GFFT transform pair $c(x)$ and $C(X)$, we can find the roots of the polynomials by the positions of zero spectral components. Assuming that $m$ possible range is known synchronization is perfect, then intercepted bit stream $l$ is split into M sequences of length $n_0 = 2^{m_0} - 1$. where $m_0$ is the estimated value of $m$ subsequently $n_0$ is the estimated value of $n$ in Eq. (4). By preforming GFFT calculation based on the transform pair of $c(x)$ and $C(x)$ we can find the roots of all the $M$ code polynomials. As in the form of analysis matrix $M \times n$ where, $M = \left[\frac{l}{n_0}\right]$. In that if common roots exits then those common roots will be the roots of $M$ polynomial of $g(x)$.

If $m_0 = m$, the intercepted bitstream is composed of $M$ code polynomials in the form of analysis matrix $M \times n$, then the $i$th code polynomial is $c_i(x) = c_{i,0} + c_{i,1}x + \cdots + c_{i,n-1}x^{n-1}$ and that can be written in GFFT as $C_i(X) = C_{i,0} + C_{i,1}X + \cdots + C_{i,n-1}X^{n-1}$, where $i = 1, 2, \ldots, M$. $S_j$ is calculated to those M spectral polynomials with the cumulative spectral component of $j$th order.

$$S_j = \sum_{i=1}^{M} C_{i,j} \tag{6}$$

If root of generator polynomial $g(x)$ is $\alpha^j$ then $v_i(\alpha^j) = v_{i,j} = 0$. If we have $S_j = 0$ and $\alpha^j$ will not be a root of $g(x)$, $c_i(\alpha^j) = 0$ to the code polynomials $c_i(x)$, then $S_j = 0$ in this condition it is not possible for all M polynomials.

Since the common zero spectral component positions of all M polynomial GFFT can be found by the positions of zeros in number of groups cumulative spectral components of all those M polynomials GFFT.

Let us cumulate all the number of group spectral components as $S$.

$$S = (S_0, S_1, \ldots, S_{n-1}) \tag{7}$$

When $m_0 \neq m$, we considered M sequences as series of random data, and the common roots of all the M polynomials do not exist. Since common roots are the roots of generator polynomials. If $S_j = 0$ the common roots do not exist.

## 3.1   GFFT Calculation on Binary Symmetric Channel

In noisy condition, we considered Binary symmetric channel. BSC channel is best memory less channel to reduce above difficulty. In this, zeros may not exist in vector S even when $m_0 = m$. So it reduces the above toughness. We assume that bitstream is intercepted after BSC channel with the cross-over probability $t$. Below equation replaces the Eq. (2) to analyze the problem more convenient.

$$C_j = F(\alpha^j) \times (c_0, c_1, \ldots, c_{n-2}, c_{n-1})^T$$
$$F(\alpha^j) = \left( (\alpha^j)^0, (\alpha^j)^1, \ldots, (\alpha^j)^{n-2}, (\alpha^j)^{n-1} \right) \tag{8}$$

Substitute $\alpha^j$ with binary vector of length $m$ because $\alpha^j$ is an element in $GF(2^m)$. Now $F(\alpha^j)$ can be written in GF(2) as $Fb(\alpha^j)$ in the form of binary matrix of size $m \times n$.

If $m_0 \neq m$: When $m_0 \neq m$, in this case intercepted bitstream could be measured as M random sequences of code length $n_0 = 2^{m_0} - 1$. If $\alpha^j$ will be the nonzero element of $j$th spectral component in $GF(2^{m_0})$, for a random polynomial $c(x)$ of degree $n_0 - 1$, the corresponding $j$th spectral component $C_j$, we have $j = 0, 1, 2, \ldots, n_0 - 1, k = 1, 2, \ldots, 2^{rk_j} - 1$.

$$\Pr[C_j = 0] = \Pr[C_j = l_k] = \frac{1}{2^{rk_j}} \tag{9}$$

where $l_k = k$-th nonzero element of $C_j$ and $rk_j = \text{rank } (Fb(\alpha^j))$.

The mean and variance of $C_j$ can be deduced as follows:

$$E(C_j) = \sum_{k=1}^{2^{rk_j}-1} l_k \Pr[C_j = l_k] = \frac{1}{2^{rk_j}} \sum_{k=1}^{2^{rk_j}-1} l_k$$

$$D(C_j) = \frac{1}{2^{rk_j}} \sum_{k=1}^{2^{rk_j}-1} l_k^2 - E^2(C_J) \tag{10}$$

Due to M intercepted sequences are independent the Fourier transform of those M polynomials can be considered as independent repeated process at M times. Then, the cumulative spectral component $S_j$ have mean $E_0 = M \times E(C_j)$ and variance $D_0 = M \times D(C_j)$.

If $m_0 = m$: When $m_0 = m$, in this case the intercepted bitstream can be measured as M noisy codewords with in the code length of $n = 2^m - 1$. If $\alpha^j$ is the root of generator polynomial $g(x)$ with respect to each noisy code polynomials. For the $j$ th spectral component of $C_j$ of the GFFT, we have

$$\Pr[C_j = 0] = (1-\tau)^n + \frac{1}{2^{rk_j}}(1 - (1-\tau)^n)$$

$$\Pr[C_j = l_k] = \frac{1}{2^{rk_j}-1}(1 - \Pr[C_j = 0]) = \frac{1 - (1-\tau)^n}{2^{rk_j}} \tag{11}$$

Mean of $C_j$ is and variance of $C_j$ is

$$E(C_j) = \sum_{k=1}^{2^{rk_j}-1} l_k \Pr[C_j = l_k] = \frac{1 - (1-\tau)^n}{2^{rk_j}} \sum_{k=1}^{2^{rk_j}-1} l_k$$

$$D(C_j) = \frac{1 - (1-\tau)^n}{2^{rk_j}} \sum_{k=1}^{2^{rk_j}-1} l_k^2 - E^2(C_J) \tag{12}$$

For all noisy code polynomials mean and variance of spectral component $S_j$ are $E_1 = M \times E(C_j)$ and $D_1 = M \times D(C_j)$ respectively.

## 3.2 Analytical Expression of the Optimal Threshold

The proposed recognition technique is based on the false alarm and miss detection probabilities. In which false alarm $P_{fa}$ and miss detection probabilities $P_{md}$ can be deduced from

$$P_{fa} = Pr[S_j \leq T/H_0] = \int_0^T \frac{1}{\sqrt{2\pi D_0}} exp\left(\frac{-(x-E_0)^2}{2D_0}\right) dx$$

$$P_{md} = Pr[S_j \leq T/H_1] = \int_\tau^\infty \frac{1}{\sqrt{2\pi D_1}} exp\left(\frac{-(x-E_1)^2}{2D_1}\right) dx \tag{13}$$

The problem we have faced with a binary hypothesis testing is the null hypothesis $H_0$, which corresponds to that case where the $\alpha^j$ is not the root of generator polynomial $g(x)$ and corresponds to alternated the hypothesis $H_1$, where $\alpha^j$ is the root of $g(x)$.

$P_t = P_{fa} + P_{md}, P_t$ be the minimal as the threshold. It is equivalent to find the value of threshold $T$, where as $\frac{\partial p_t}{\partial T} = 0$, we get

$$\frac{\partial p_t}{\partial T} = 0 = aT^2 + 2bT + c = 0 \tag{14}$$

The optimal threshold $T$ can be described as

$$T^* = \frac{-b + \sqrt{b^2 - ac}}{a} \tag{15}$$

where

$$a = D_1 - D_0,$$
$$b = D_0 E_1 - D_1 E_0,$$
$$c = D_1 E_0^2 - D_1 E_1^2 - 2D_0 D_1 \ln \sqrt{D_1/D_0}.$$

## 4  Simulation Results

In this section, results for recognition of BCH Codes are simulated. For all the simulations m-range should be $m_{min} = 3$ to $m_{max} = 8$ and number of intercepted code words are M = 1000. Finally, results will be obtained after 1000 simulations. In [4] author explained with BCH (63, 39), in which this method is not suitable for other code lengths. Figure 1 explained about the comparison between the proposed method and the Reconstruction of BCH Codes are done using Probability Compensation [4]. Below comparison says that proposed method superior to the method in [4].

In [5], the author explained about transmitted channel which is effected by Additive white Gaussian noise (AWGN) with the variance of $\alpha^2 = N_0/2$, with BPSK modulation mode. Figure 2 explained about the comparison between the proposed method and method [5]. Below comparison says that proposed method is superior to the method in [5] at lower false recognition probabilities.

Figures 3, 4, and 5 show the probability recognition of BCH codes at different lengths.

Based on the primitive elements of $h$-error BCH codes Figs. 3, 4 and 5 have been plotted. With proposed method, we can recognition of BCH Code can be perfectly done when the code is not too long.

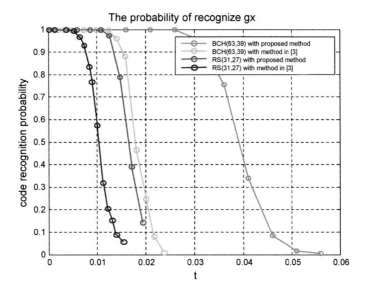

**Fig. 1** Comparison of the code recognition performance of proposed method and the method in [4]

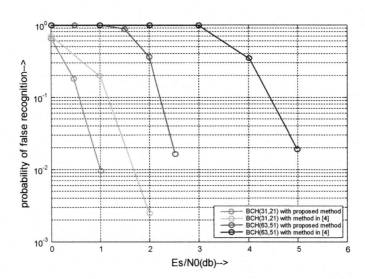

**Fig. 2** Comparison of the code recognition performance of proposed method and the method in [5]

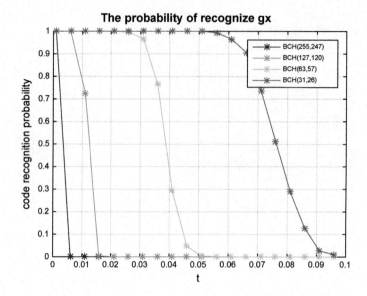

**Fig. 3** Recognition performance of 2-ary-1-error-correcting BCH code with different lengths

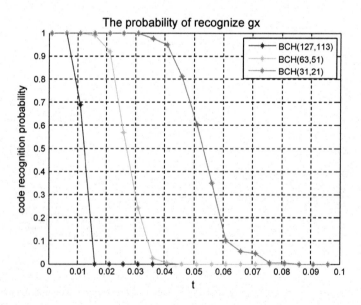

**Fig. 4** Recognition performance of 2-ary-2-error-correcting BCH code with different lengths

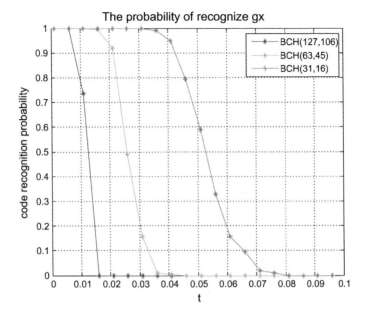

**Fig. 5** Recognition performance of 2-ary-3-error-correcting BCH code with different lengths

For example, in Fig. 3 we can see that recognition performance of

BCH (255, 247) < BCH (127, 120) < BCH (63, 57) < BCH (31, 26).

In Fig. 4 recognition performance of

BCH (127, 113) < BCH (63, 51) < BCH (31, 21).

In Fig. 5 recognition performance of

BCH (127, 106) < BCH (63, 45) < BCH (31, 16).

That means probability of recognition is decreasing with increasing of code lengths. To solve this problem, more intercepted code words and a soft decision situation may be considered in the feature work.


# 5 Conclusion

In this chapter, a novel method is proposed to recognize BCH codes under noisy environment. This method is based on Galois Fast Fourier Transform which is applied to each of the splitted intercepted bit stream and the spectral components are calculated at the end. Finally, based on positions of zero spectral components the correct codes can be estimated.

# References

1. S. Lin and D. Costello, "Error Control Coding", 2nd ed. Upper Saddle River, NJ, USA: Prentice-Hall, Inc., 2004.
2. CTT_K. Deergha Rao, "Channel Coding Techniques for Wireless Communications" Springer India-2015.
3. Jiafeng Wang, Yang Yue, Jun Yao, "Statistical Recognition Method of Binary BCH Code" Communications and Network, Vol. 3 No. 1 (2011), Article ID:3988.
4. H.Lee, C.Park, J.Lee, and Y.Song, "Reconstruction of BCH codes using probability compensation," in Proceedings of IEEE APCC, Jeju Island, Korea, October 2012, pp. 591–594.
5. Z. Jing, H. Zhiping, L. Chunwu, S. Shaojing, and Z. Yimeng, "Information-Dispersion-Entropy-Based blind recognition of binary BCH codes in soft decision situations", Entropy, vol. 15, no. 5, pp. 1705–1725, 2013.
6. Gang Wu, Bangning Zhang, Xiang Wen and Daoxing Guo, "Blind Recognition of BCH Code Based on Galois Field Fourier Transform," in Proc. WCSP2015, 2015.

# Simultaneous Power Quality Disturbances Analysis Using Modified S-Transform and Evolutionary Approach

G. Sahu and A. Choubey

**Abstract**  This chapter proposes a hybrid technique based on modified S-transform (ST) and Differential Evolution (DE) algorithm for the visual detection and pattern classification of different nonstationary power quality (PQ) events. The presence of Gaussian window in ST provides a high time resolution in low-frequency bands. The modified Gaussian window in modified ST is capable of depicting a high-resolution time–frequency representation (TFR) for different simultaneous PQ disturbance signals. Further, the modified ST is used for extraction of relevant features from the available PQ disturbance waveforms. Then, the features obtained by modified ST are clustered by using a fuzzy C-mean (FCM)-based DE algorithm. The analysis and experimental results show that the proposed hybrid technique provides a considerable improvement in PQ detection and classification.

**Keywords**  Power quality disturbances · Time–frequency representation
Modified S-transform · Fuzzy C-means clustering · Differential evolution
algorithm

G. Sahu (✉) · A. Choubey
Department of Electronics and Communication Engineering, National Institute
of Technology, Jamshedpur 831014, India
e-mail: gupteswar.sahu@gmail.com

A. Choubey
e-mail: achoubey.ece@nitjsr.ac.in

G. Sahu
Department of Electronics and Communication Engineering, Raghu Engineering College,
Visakhapatnam 531162, India

© Springer Nature Singapore Pte Ltd. 2018
J. Anguera et al. (eds.), *Microelectronics, Electromagnetics
and Telecommunications*, Lecture Notes in Electrical Engineering 471,
https://doi.org/10.1007/978-981-10-7329-8_31

# 1 Introduction

The PQ monitoring for academic and industrial fields has become a key issue. In general, a PQ disturbance is caused due to any change in electric service current or voltage from the standard sinusoidal waveform [1]. The occurrence of PQ disturbances in a power system not only mortifies the performance of the device, but also reduces its overall lifetime. Hence, the improvement of power quality and supply of clean power is very much crucial. In order to take a precise mitigation action and to control the malfunction in electrical devices, an accurate identification and classification of PQ disturbances are a prerequisite. A realistic approach to achieve the objective of identification and classification of different PQ events consists of three major steps. They are: (i) signal analysis; (ii) feature extraction and selection; (iii) disturbance classification by using an appropriate classifier.

Many signal processing techniques have been proposed by various researchers for the analysis of PQ events in a power system. This includes visual recognition, feature extraction, and classification of PQ disturbances. Most commonly used approach to deal with the time series data is Fourier transform (FT). In FT, the signal is assumed to be periodic, due to which it loses the information in the time axis. For this reason, the Short-time Fourier transforms (STFT) is adopted as an alternative tool for time–frequency analysis of PQ disturbance waveforms. However, due to use of a fixed window length, it is not effective to describe the signal disturbances adequately. For this reason, wavelet transforms (WT)-based approaches are introduced and used as an efficient technique for nonstationary signal analysis [2–5]. WT is very much capable of providing good time resolution and frequency resolution for high-frequency events and low-frequency events respectively. But in WT, it is observed that the time–frequency resolution depends on the chosen wavelet. The ST used as an effective tool for time–frequency representation of nonstationary signals. Due to use of a fixed Gaussian window, the original ST is not capable of providing an effective TFR for all types of PQ disturbances [6]. In order to circumvent this problem, in literature different schemes are proposed to modify the original Gaussian window function. In [7] introduced adjustable parameters to control the width of the window function. During the past decade, many hybrid approaches have been proposed for automatic detection and classifications of PQ events. P.K. Dash et al. [8] have classified different PQ events by using the fuzzy expert system along with Fourier linear combiner. Probabilistic neural-based classifiers with ST [9] were used for identification and classification of a wide range of PQ events. In [10], B. Biswal et al. has reported a hybrid approach for the identification and classification of eight types of PQ disturbances by adopting Hilbert–Huang transform with a balanced neural tree. A combination of Fuzzy C-means with adaptive particle swarm optimization technique has been presented in [11] for automatic classification of nonstationary PQ events. In this chapter, a new approach for automatic PQ detection and classification based on modified S-transform and DE algorithm is presented.

## 2 Modified S-Transform

The ST of $y(t)$ is given as [12] follows:

$$s(t,f) = \int_{-\infty}^{+\infty} y(\tau) * w(t - \tau, f) * e^{-2i\pi f \tau} d\tau \tag{1}$$

$$s(t,f) = \int_{-\infty}^{+\infty} y(\tau) * \frac{1}{\sigma(f)\sqrt{2\pi}} * e^{\frac{-(t-\tau)^2}{2\sigma(f)^2}} * e^{-2i\pi f \tau} d\tau \tag{2}$$

The standard deviation $\sigma(f)$ selected for the modified Gaussian window is:

$$\sigma(f) = \frac{k}{p + q\sqrt{f}} \tag{3}$$

where $f$ is the signal frequency, $p$ and $q$ are positive constants, and $k \le \sqrt{p^2 + q^2}$. The window $w$ selected in Eq. (1) is the Gaussian one. In [11], a new Gaussian function is proposed by varying the stretch of the initial Gaussian function along with frequency. The modified version of the window function is represented as

$$w(t,f) = \frac{p + q\sqrt{|f|}}{k\sqrt{2\pi}} * exp^{-\frac{\left(p + q\sqrt{|f|}\right)^2 t^2}{2k^2}}, \ k > 0 \tag{4}$$

In which t and $\tau$ represents the time variables. Whereas $k$ and $q$ are the scaling factors. Which are basically used to control the count of oscillations in the window. With the increased value of k, it is observed that the frequency resolution is increased in the frequency domain. Whereas the window broadens in the time domain. Thus, the generalized ST of a continuous time signal with modified Gaussian window is given by

$$S(\tau,f) = \int_{-\infty}^{\infty} Y(\alpha + f) * exp^{\frac{\left(-2\pi^2\alpha^2k^2\right)}{\left(p + q\sqrt{|f|}\right)^2}} * exp^{2i\alpha\pi\tau} d\alpha \tag{5}$$

The generalized ST of a signal in discrete domain is given as

$$S[j,n] = \sum_{m=0}^{N-1} Y[m + n] * exp^{\frac{\left(-2\pi^2 m^2 k^2\right)}{\left(p + q\sqrt{|f|}\right)^2}} * exp^{\frac{i2\pi mj}{N}} \tag{6}$$

where $Y[m]$ is given by

$$Y[m] = \frac{1}{N} * \sum_{k=0}^{N-1} Y(k) * exp^{-j\frac{2\pi mk}{N}} \tag{7}$$

# 3 Time–Frequency Analysis Using Modified S-Transform

For testing purpose, eight types of simultaneous power signal problems are investigated. All PQ disturbance signals are simulated using MATLAB software. The PQ disturbance signals are sampled at a rate of 3.84 kHz. In this section, for simplicity, only two power quality disturbances results are shown.

The TFRs generated by using modified S-transform are plotted in the time–frequency co-ordinate system. Figure (1) shows the swell and flicker voltage waveform and its corresponding modified ST-based TFR plot. Figure (2) shows sag with harmonics in a voltage signal and its modified ST-based TFR plot. The Modified ST provides clear information about the presence of the disturbance and the nature of the type of disturbance. The simulation results indicate the effectiveness of modified ST in terms of detection and visual localization of the PQ disturbances. Further, the modified ST used to extract different time domain and frequency domain feature vectors. Among all the extracted features, normalized value and variance are found to be the best set of two features. These two features are chosen as input parameters for pattern classification.

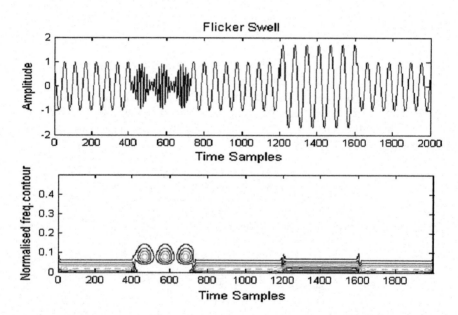

**Fig. 1** Swell plus flicker in a voltage waveform and TFR using modified S-transform

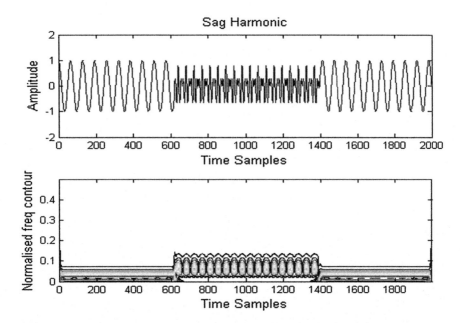

**Fig. 2** Sag with harmonics in a voltage signal and TFR using modified S-transform

# 4 Fuzzy C-Means Clustering

The FCM is a fuzzy-based clustering technique that permits a single piece of data to fit into more than one cluster. In 1971, the FCM was originally proposed by Dunn and in 1981; the FCM clustering technique was further improved by Bezdek. It is used as an important tool to solve pattern classification of different data sets. The minimization of the objective function is one of the important criteria in FCM. Several parameters such as energy, mean, standard deviation, variance, autocorrelation, and normalized values are extracted from the power quality signals. Normalized value and variance are chosen for clustering of various nonstationary power signals [11]. In this chapter, one hundred of each disturbance (Feature vectors) is considered as input parameters to the FCM. In FCM, The objective function $H_m$ is to be minimized in order to determine the values of the cluster center $C_j$ and membership matrix $U$. The objective function is represented as

$$H_m = \sum_{i=1}^{N} \sum_{j=1}^{C} v_{ij}^{m} \left\| y_i - c_j \right\|^2 \tag{8}$$

where N represents the number of data points, m represents the weighting exponent, C is the number of clusters, $v_{ij}$, denotes the degree of membership of $y_i$ in the cluster

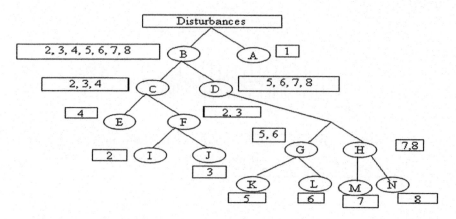

**Fig. 3** Fuzzy C-means tree

$j$. $y_i$ is the $i_{th}$ of n-dimensional measured data and $c_j$ is the n-dimensional center of the cluster.

$$c_j = \frac{\sum_{i=1}^{N} V_{ij}^m y_i}{\sum_{i=1}^{N} V_{ij}^m} \qquad (9)$$

$$V_{ij} = \sum_{k=1}^{c} \left[ \frac{\|y_i - c_j\|}{\|y_i - c_k\|} \right]^{\frac{-2}{m-1}} \qquad (10)$$

The following eight types of power quality disturbances are considered for clustering. (1) voltage flicker with swell; (2) voltage swell with harmonics; (3) voltage sag with flicker; (4) voltage flicker with harmonic; (5) momentary interruption with flicker; 6) momentary interruption with harmonic; (7) voltage sag with harmonic. (8) momentary interruption with transient. A fuzzy C-means tree is used for clustering process as shown in Fig. 3.

## 5 Differential Evolution (DE)

The DE algorithm introduced by Storrn and Price [13], has appeared as one of the high-speed, dynamic, efficient, and powerful population-based global optimization technique. It has been effectively applied to various fields of science and engineering. In comparison to most of the proposed clustering methods, the DE algorithm decides the optimal number of partitions of the data on the run. However, it does not require any past knowledge of the data to be classified [14]. Three important operators in DE are mutation, crossover, and selection. For a population of $D$-dimensional vectors space, let us assume $Y_{i,G}$ $i = 1, 2, 3 \dots N$, correspond to the population of each generation $G$. $V_{i,G+1}$ denote the mutant vector in $(G+1)$th generation and $U_{i,G+1}$ is the trial vector in $(G+1)$th generation. Three operators associated with DE are described as

1. Mutation: DE generates new parameter vectors by adding the weighted difference between two parameter vectors to a third vector. For each target vector, a mutant vector is generated according to

$$V_{i,G+1} = Y_{r1,G} + F(Y_{r2,G} - Y_{r3,G}) \quad i = 1, 2, 3, \ldots, N$$

where the values of $r_1, r_2$ and $r_3$ are mutually exclusive random numbers in the interval [0, 1]. These values are integers and different from the index $i$. $F$ is the amplification controlling factor of the differential variation, which is termed as mutation scale factor. The initial value of F chosen is 0.8.

2. Crossover: The crossover operation generally increases the diversity of the targeted vector. A trial vector is generated by mixing the mutated vector with the parameters of another predetermined vector.

$$U_{i,G+1} = (U_{1i,G+1}, U_{2i,G+1}, \ldots, U_{Di,G+1})$$

where

$$U_{ji,G+1} = \begin{cases} V_{ji,G+1} & if \ (ran[0,1] \leq CR) \ or \ j=k \\ Y_{ji,G} & if \ (ran[0,1] \geq CR) \ and \ j \neq k \end{cases} \quad j = 1, 2, 3, \ldots, D$$

$CR$ is the crossover constant and $CR \in [0, 1]$. $k$ is a random integer in $[1, D]$. The initial value of $CR$ chosen is 0.9.

3. Selection: In DE's selection scheme, the next generation, $G+1$, is selected from the child population based on objective function value.

$$Y_{i,G+1} = \begin{cases} U_{i,G+1} & if f(U_{i,G+1}) < f(Y_{i,G}) \\ Y_{i,G} & otherwise \end{cases}$$

where $f(Y_{i,G+1})$ is the objective function.

From the experimental results, it is found that fuzzy C-means algorithm based clustering technique provides very low percentage of classification accuracy. In order to enhance the classification accuracy, the fuzzy C-means algorithm is extended using DE algorithm. The suitable selection of centers plays a key role in improving the classification accuracy. The estimated centers attained from fuzzy C-means clustering are used as the initial parameter for DE.

## 6    Results and Discussion

The complete clustering procedure is carried out by using the fuzzy C-means algorithm. Some of the clustering results are shown in Figs. 4, 5, and 6. The percentage of classification accuracy obtained using fuzzy C-means algorithm and the proposed DE-based algorithm are presented in Table 1. The results show the superior performance of the proposed algorithm in terms of classification accuracy.

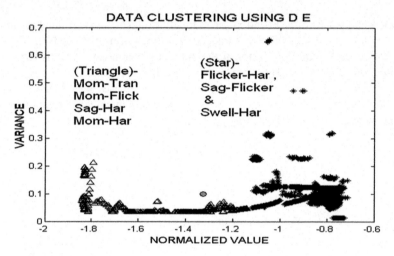

**Fig. 4** Classification of cluster B into C and D using DEA

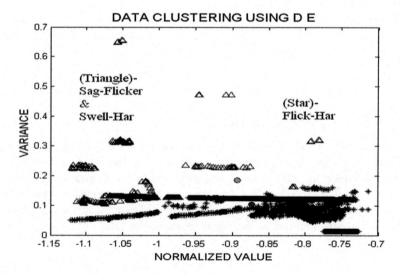

**Fig. 5** Classification of cluster C into E and F using DEA

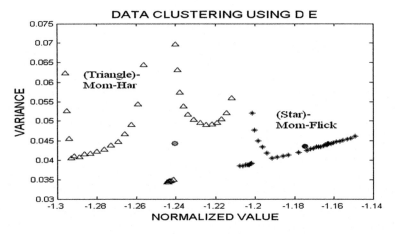

**Fig. 6** Classification of cluster G into K and L using DEA

**Table 1** Classification results in percentage using FCA and DEA

| Sl. No. | Disturbances | Accuracy in (%) | | |
|---|---|---|---|---|
| | | Input features | FCA | DEA |
| 1 | Swell-flicker | 100 | 90 | 100 |
| 2 | Sag-flicker | 100 | 88 | 98 |
| 3 | Swell-harmonic | 100 | 87 | 97 |
| 4 | Flicker-harmonic | 100 | 85 | 98 |
| 5 | Momentary interruption-flicker | 100 | 92 | 99 |
| 6 | Momentary interruption-harmonic | 100 | 89 | 97 |
| 7 | Momentary interruption-transient | 100 | 91 | 99 |
| 8 | Sag-harmonic | 100 | 87 | 98 |
| | % Average accuracy | | 88.625 | 98.25 |

## 7    Conclusions

In this chapter, a hybrid scheme is presented for the visual recognition and pattern classification of simultaneous PQ disturbances present in a power signal. The modified ST is used to extract the features from the available PQ disturbance signals. The modified ST is proven to be very effective in improving the time–frequency resolution of the PQ disturbance signals and it is observed that the extracted features are very simple and effective. Finally, a FCM and DE-based algorithm is implemented to improve the classification accuracy. From the experimental results, it is shown that the proposed method has a very high classification accuracy compared to the traditional Fuzzy C-means algorithm.

# References

1. Fuchs E.F., Masoum M.A.S.: Power Quality in Power Systems and Electric Machines. 1st edn. Academic Press, (2008).
2. Santoso S., Powers E. J., Grady W. M.: Power quality disturbance data compression using wavelet transform methods. IEEE Trans. Power Del., vol. 12, pp. 1250–1257, (1997).
3. Gaouda A. M., Salama M. M. A., Sultan M. R., Chikhani A. Y.: Power quality detection and classification using wavelet-multiresolution signal decomposition. IEEE Trans. Power Del., vol. 14, pp. 1469–1476, (1999).
4. Karimi M., Mokhtari H., Iravani M. R.: Wavelet based on-line disturbance detection for power quality applications. IEEE Trans. Power Del., vol. 15, pp. 1212–1220, (2000).
5. Santoso S., Grady W. M., Powers E. J., Lamoree J., Bhatt S. S.: Characterization of distribution power quality events with Fourier and wavelet transforms. IEEE Trans. Power Del., vol. 15, pp. 247–254, (2000).
6. Dash P. K., Panigrahi B. K., Panda G.: Power Quality analysis using S-Transform. IEEE Trans. Power Del., vol. 18, pp. 406–411, (2003).
7. Sejdic E., Djurovic I., Jiang J.: A window width optimized S-transform, EURASIP, J. Adv. Signal Process. 672941, (2008).
8. Dash P. K., Mishra S., Salama M. A., Liew A. C.: Classification of power system disturbance using a fuzzy expert system and a Fourier linear combiner. IEEE Trans. Power Del., vol. 15, pp. 472–477, (2000).
9. Mishra M. Bhende C. N. Panigrahi B. K.: Detection and Classification of Power Quality Disturbances Using S-Transform and Probabilistic Neural Network, IEEE Trans. Power Del., vol 23, pp. 280–287, (2008).
10. Biswal B., Biswal M., Mishra S., Jalaja R.: Automatic Classification of Power Quality Events Using Balanced Neural Tree. IEEE Trans. Ind. Elect., vol. 61, pp. 521–530, (2014).
11. Biswal B., Dash P. K., Panigrahi B. K.: Power quality disturbance classification using fuzzy C-means algorithm and adaptive particle swarm optimization. IEEE Trans. on Ind. Elec., vol. 56, pp. 212–220, (2009).
12. Stockwell R. G., Mansinha L, Lowe R. P,: Localization of the complex spectrum: The S-transform. IEEE Trans. Signal Process., vol. 44, no. 4, pp. 998–1001. (1996).
13. Storn R., Price K..: Differential Evolution – A Simple and efficient Heuristic for global Optimization for Continuous spaces. Journal of Global Optimization, vol. 11, pp. 341–359, (1997).
14. Das S., Abraham A., Konar A.: Automatic Clustering Using an Improved Differential Evolution Algorithm, IEEE Trans. Sys. Man. Cyber., vol. 38, pp. 218–237, (2008).

# Efficient Integration of Zoom ADC with Temperature Sensors for System on Chip Applications—A Perspective

**Rafath Unnisa and P. Trinatha Rao**

**Abstract** For different types of measurements, various sensors are readily available in the market. These sensors have their output signal which is less than the optimal. Low-cost versions of the sensors introduce quite a large gain and offset variation in a temperature sensor. Addition of amplification ICs and minimum change in amplification settings before integrating sensing element to ADC helps in overcoming such situations. As an alternative, more sensitive and better-controlled sensors can be considered but at an extra cost. Another way to overcome the gain and offset variation is to use Zoom ADC which has the advantages of oversampled ADCs with extra amplification stage for measuring sensing elements with higher resolution. Efficient integration of Zoom ADC with temperature sensor in System on Chip helps in improving performance of the SoC. By the usage of an advanced algorithm for the development of zoom ADC has been implemented in this research, which further improves the performance and efficiency of System on Chip.

**Keywords** Temperature sensor · Zoom ADC · Radio frequency Programmable gain amplifier · SoC

## 1 Introduction

Voltage delivered by sensors can be much lesser compared to supply voltage which has a significant offset. Considering this signal with a regular Analog-to-Digital Converter those results for complete loss of resolution or a need for several external components that should be integrated to the sensor with the help of complex

R. Unnisa (✉)
Muffakham Jah College of Engineering and Technology, Hyderabad, Telangana, India
e-mail: unnisa.rafath@gmail.com

P. Trinatha Rao
School of Technology, GITAM University, Hyderabad, Telangana, India
e-mail: trinath@gitam.in

© Springer Nature Singapore Pte Ltd. 2018
J. Anguera et al. (eds.), *Microelectronics, Electromagnetics and Telecommunications*, Lecture Notes in Electrical Engineering 471,
https://doi.org/10.1007/978-981-10-7329-8_32

calibration techniques or algorithms. The Zoom ADC helps in acquiring such signals without any loss in resolution and without any external amplifier intervention or compensation in offset settings. This paper focuses on the capability of Zoom ADC to improve the efficiency and performance which has been integrated in a System on Chip (SoC).

Advanced research on Radio Frequency (RF) focuses on higher integration for meeting expectations of the client in low power with low budget.

Smaller feature size helps in less power consumption. The performance of the device is highly affected as we move towards smaller feature size. Generally Analog-to-Digital Converters with high speed, less noise and less offset voltage are required for applications such as PDA or mobiles. Gain amplifiers and comparators cater a major role in performance improvement of the ADC. Oversampling is very important in analog-to-digital conversion, which has become very popular due to its efficiency increment and better performance. A sampling rate that uses a higher bandwidth compared to the signal of consideration is considered in oversampling. Next filtering and down sampling is introduced. Zoom ADC has low-bandwidth, low-power, very high resolution. It is a low-cost ADC that is used in different SoCs for various signal processing and wireless applications.

Zoom ADC has become a preferable choice for audio and high-resolution industrial measurement applications currently.

## 1.1  Analog-to-Digital Conversion

Conversion of signals from analog-to-digital domain can be processed by using digital signal processing. Therefore, analog-to-digital conversion is of very high importance. Analog-to-digital conversion process based on sigma-delta modulation implied in Zoom ADC is the main alternative for high-resolution converter, which can be used as an integrated part in System on Chip. Sigma-delta modulators have good importance in digital VLSI technologies. Implementations of converters are critical as they need to have analog components in their filters which are not easy to implement. Huge possibility exists for conversion circuits to get susceptible to noise and interference as well.

## 1.2  Zoom ADC

Analog-to-Digital Converter (ADC) can be classified into two parts based on variations in sampling rate. When signals are sampled at Nyquist rate, i.e., $f_N = 2F$ ($f_N$ is the sampling rate and F is the bandwidth of the input signal). In second case, the signal is sampled at a much higher sampling rate than the signal bandwidth which is known as oversampling. High performance with better resolution and efficiency are the outcomes expected from this kind of converters. The fundamental

**Fig. 1** Sigma-delta ADC
block diagram

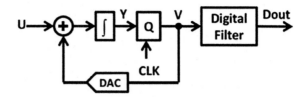

theory of sigma-delta data converters is to sample the signal many more times in order to decrease the quantization noise and for better signal representation (Fig. 1).

With some extra offset and gain stages, the Sigma-Delta ADC works as Zoom ADC.

Zoom ADC amplifies the signal up to 1000 times with its three amplification stages. Its architecture is composed of three amplifiers with two offset cancelation stages before addition of the sigma-delta modulator. The offset cancelation stages permits the signal for re-center allowing non-saturation of amplifier. In Zoom ADC, the part of the signal which carries the information gets selected, amplified and is converted to digital with full ADC resolution. The key for the Zoom ADC's extreme signal magnification capability is the precise settings together with the strong integration of pre-amplifiers and ADC [1].

The Zoom ADC makes it possible to compensate for large offsets on smaller signals without resolution or saturation losses by sharing the offset and gain compensation over three stages. Zoom ADC has an extra feature to capture single and differential signals. It is based on the differential signal path, accommodating single ended signals along with the controlled offset.

### 1.3  Architecture Model

Zoom ADC is generally used in many sensing applications nowadays. Including it with two more additional stages along with the sigma-delta modulator out of which one stage is a multiplexer integrated with a programmable gain amplifier (PGA) before the sigma-delta modulator stage results in a Zoom ADC for high-end applications (Fig. 2).

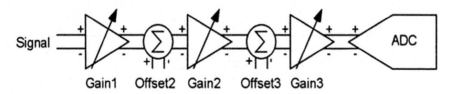

**Fig. 2** Zoom ADC block diagram [1]

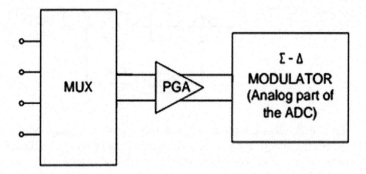

**Fig. 3** Block diagram of Zoom ADC for high-end applications

ADC of 1 bit is used for basic modulation which is a comparator as well as a DAC of 1 bit. There are several ADCs with multiple bits which produces the advantage of efficient linearity.

1-bit stream output is generated from the modulator because of negative feedback around the integrator. If input voltage is zero equal number of 0 and 1 can be observed as output data. When input becomes more positive, number of 1s gets increased subsequently decreasing the number of 0s as well. Similarly, as input signal becomes more negative, number of 1s decreases, subsequently increasing the count of number of 0s. Count of 1 in output data stream to the total number of samples present in that duration is proportional to input DC value (Fig. 3).

## 1.4 Linearity

Linearity is one of the most important factors when sensors are used because perfection inaccuracy of the measurement can be determined by a good linearity (Fig. 4).

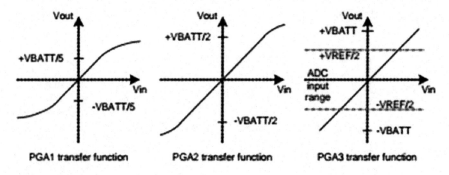

**Fig. 4** PGA transfer function

To make linearity optimized amplification, signal coming out of the initial stages PGA1, PGA2 should be much lesser of ±VBATT/5 and ±VBATT/2 respectively. The strongest gain can be considered at the stage closest to ADC. In that case, initially the gain is set for PGA3 and PGA2 stages and at last for PGA1 if it is really required. A total gain of 50 is achieved with PGA3 = 10, PGA2 = 5 when PGA1 is kept disabled.

By implementation of the three-stage amplification process, Zoom ADC can amplify the signal coming as input up to a factor of 1000. There are several advantages for this huge amplification gain.

Three amplifiers and two offset cancellation stages before addition of sigma-delta architecture constitute Zoom ADC [2].

Achieving a huge gain is not impossible even though the signal comprises of DC component due to the additional offset stages which help in offset cancellation and restructure the signal. PGA1 is used for obtaining full-scale signal, PGA2 is mainly integrated for offset cancellation whereas PGA3 is linear throughout all the ranges and its output range is equal to ±Vref/2 i.e. ADC input range.

## 1.5  Noise in Zoom ADC

The quantization noise in ADC arises because of rounding error between ADC's input signal which is in analog form and the output which is in digitized form. Generally, noise depends on a signal which varies with the resolution of ADC.

Due to thermal agitation of electrons, thermal noise is generated which changes with temperature. A circuit consisting of a signal of narrow band generates lower noise compared to a circuit present in a hot environment with large band signal. Noise generated in the Zoom ADC is generally random which follows a curve similar to Gaussian distribution. Standard deviation of the data set can be presented by root mean square or RMS noise as per measurement [3].

When a PGA stage is enabled, it adds a certain amount of noise in input signal transmission in Zoom ADC. The noise which arrives in the end of PGA3 is the sum of noise generated in three different stages as per implementation. Generally, noise signals are random and not correlated. The equivalent root mean square value is usually the square root of the squares sum result. While PGA1 is active, the generated noise of PGA1 stage output is amplified by the stages PGA2 and PGA3. This has been designed carefully to minimize the noise (volt rms) as it is one of the most dominant noise source. Input referred noise is also important which is generated when output noise is returned back to the input by dividing with the gain factor. The noise can be reduced if bigger gains can be set in the initial stages.

A gain of 1000 is set by setting PGA1 = 20, PGA2 = 10 with PGA3 = 5 for noise minimization when the output voltage of amplifier stages should be in the appropriate linear range. As per the need for required application, a trade-off should be achieved for the noise as well as linearity factors.

## 2 Zoom ADC Temperature Sensor

### 2.1 Temperature Sensors

The temperature sensors have important usage in different commercial applications from industrial and domestic appliances to environmental monitoring usage.

Below is the architecture of temperature sensor. The circuit design has bias current generation with Ibias, the bipolar front-end, and Zoom ADC circuit together with bipolar core bias current selection availability [4].

Bias Circuit is integrated with front-end and timing and control unit to fed into the decimation unit for the temperature output [5] (Fig. 5).

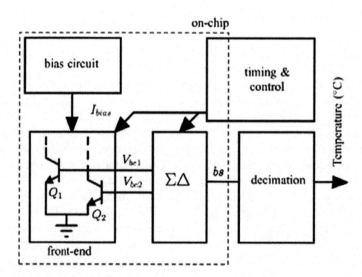

**Fig. 5** Block diagram of temperature sensor

## 2.2  Algorithm

Algorithmic steps for Zoom ADC model is as shown below

1. Start by declaring function zoom_adc with input-output parameters say wavein—waveout and for no. of samples used declare sample and loop control variable i.
2. Initialize zeroth sample of input waveform with some variable, i.e., integrator.
3. Declare a for loop.
4. If the loop control variable is greater than zero, update the integrator value with the difference between current input signal value and previous output signal value.
5. If the value of integrator is greater than zero, output of ADC is 1 otherwise -1.
6. Repeat steps 5 and 6 for all values of input samples and at each sample display the integrator value.
7. End the program.

## 3   Results

It is implemented by a function that takes three parameters, input waveform, output waveform, and several samples in input data. It is started by setting the integrator value to the wavein's first element. The integrator accumulates the history. We set it to the first sample since we do not have any history. With the help of printf statement, we can see the value of integrator in action (Fig. 6).

Moving on to the signals, if we put sine wave into one. What exactly happens is Next, digitized the sine wave with Zoom ADC converter (Fig. 7).

The spectrum of the Zoom ADC digitized sine wave consisting of five cycles has been presented (Fig. 8).

When we will integrate Zoom ADC in our targeted SoC with other efficiency improvement techniques for power reduction and performance improvement, it can

**Fig. 6** Five-cycle sine wave

**Fig. 7** Digitized sine wave by Zoom ADC converter

**Fig. 8** Zoom ADC sine wave spectrum

give approximately around 15–18% efficiency improvement compared to SoCs used for different applications having different types of ADCs and few performance improvement techniques.

## 4 Conclusion

Sampling rate, resolution and power supply voltage are the most important factors to be considered while selecting the specific ADC for targeted application. *Data acquisition, industrial measurement, voice band audio*, high-speed applications are the four general classifications of application categories of the ADC. Voice band and audio segment are highly controlled by Zoom ADC with other segments as well up to a certain extent.

Modern Zoom ADCs can replace the integrating-type ADCs used to a greater extent such as dual-slope ADC, triple-slope ADC, in applications that need higher resolution which can vary from 16 to 24 bits and effective sampling rates. High resolution combined with amplifiers having on-chip programmable gain can help in the conversion of small output voltages of sensors to get digitized. Efficient bandwidth and sample rate choice helps in rejection of 50–60 Hz ranged frequencies. By sharing the gain and offset compensation over three stages, the Zooming ADC makes it possible to compensate for huge offsets for small signals without any saturation or loss of resolution. In addition, Zoom ADC has the capability to capture single as well as differential signals. The Zoom ADC is based out of a differential signal path, and accommodates the specific signals which are single ended with controlled offset and gains which prove far better compared to other available ADCs.

Due to the integration of Zoom ADC with the MCU in the family of SOCs, the MCU can be used in extending the acquisition system's capabilities. Major applications of MCUs are calibration (on demand or automatic), communication, signal processing, and power supply management.

Zoom ADCs can be considered as one of the most attractive options for different applications like controlling motors, conditioning of sensors, monitoring energy with resolutions of 12–24 bits.

# References

1. SX8724 datasheet, XE8801A datasheet: www.semtech.com.
2. N.P. Pendharkar: Design, Development and performance investigations of Sigma-Delta ADC using CMOS technology, Published in International Journal of Advanced Engineering & Application issue, (Jan 2011).
3. G.C. Meijer: Thermal sensors based on transistors, Sensors and Actuators, vol. 1, no. 2, (Aug, 2001) 98–108.
4. R.J. Baker, H.W. Li, D.E.: CMOS Circuit Design, Layout, and simulations. New York, IEEE, (1997) 637.
5. Fabio Sebastiano, Kofi A.A. Makinwa, D.: A 1.2-V 10-μW NPN-Based Temperature Sensor in 65-nm CMOS With an Inaccuracy of 0.2 C (3σ) From 70 °C to 125 °C, IEEE Journal of Solid-State Circuits, Vol. 45, No. 12, (Oct. 2010) 2591–2601.

# A Compact Pattern Reconfigurable Antenna for WiMAX Application

Devi Perla and Rajya Lakshmi Valluri

**Abstract** The chapter describes the design of a Compact Radiation Pattern Reconfigurable Antenna (RPRA). The RPRA can reconfigure its radiation pattern without changing its polarization and operating frequency. The antenna is placed on FR4 substrate of thickness 1.6 mm and switch technique is being used. The antenna can change its radiation pattern based on the state of the switch which changes current distribution on the antenna. The antenna resonates at 3.3 GHz frequency and is useful for WiMAX application. Simulations were done with HFSS.

**Keywords** Reconfigurable antenna · PIN diode · Return loss
Radiation pattern · VSWR

## 1 Introduction

In MIMO systems, satellite communications, radars, navigation, and in remote sensing applications, if an antenna wants to cover five areas, at present 5 antennas or smart antenna array is being used. As the technology has been increasing day to day, the need to reduce the number of antennas used in various applications is increasing in order to minimize the cost and volume requirement. The radiation pattern reconfigurable antenna is the best candidate for these kinds of applications as it is a single antenna whereas smart antennas are an array of antennas; it can change its radiation pattern based on the application.

The reconfigurable antennas can reconfigure its any one of the parameters (radiation Pattern, frequency, and polarization) based on the application. One of the advantages of the Radiation Pattern Reconfigurable Antenna is that it can effectively suppress the signals from the interfering sources and increases the signal

D. Perla (✉) · R. L. Valluri
Anil Neerukonda Institute of Engineering and Technology, Visakhapatnam, India
e-mail: deviperla5893@gmail.com

© Springer Nature Singapore Pte Ltd. 2018
J. Anguera et al. (eds.), *Microelectronics, Electromagnetics*
*and Telecommunications*, Lecture Notes in Electrical Engineering 471,
https://doi.org/10.1007/978-981-10-7329-8_33

strength in the intended direction by changing its radiation pattern toward the desired direction.

The main aim of this antenna is to change the current distributions on the antenna structure because the current distribution of the antenna directly determines the radiation pattern. The radiation pattern reconfigurability can be achieved by using any one of the two techniques such as switch technique and modification of material characteristics [1–3].

3 × 3 electrically small square-shaped metallic pixels are considered as an antenna, the adjacent pixels are connected by PIN diodes. When the state of the switch is changed, the geometry of the parasitic surface is changed, which in turn changes the radiation pattern of an antenna [4]. Self-oscillating pattern reconfigurable antenna consists of two monopoles which are connected to semi-ring radiator via PIN diodes and the varactor diode is placed into the semi-ring radiator. When the state of the diodes is changed, the radiation pattern of an antenna is also changed because current distribution of antenna is changed [5]. PIN diodes can be used to change the radiation pattern [6, 7].

Antenna structure consists of a driven element surrounded by parasitic elements and they act either as director or reflector depending on the switching arrangement [8]. The antenna consists of an array of two microstrip dipoles and the length of the dipole-arm strip is changed by using PIN diode switches, which lead to different radiation patterns [9]. The design consists of two-element dipole array loaded with varactor diode and pattern reconfiguration is achieved by tuning the varactor diode [10].

The proposed design achieves beam reconfiguration by using only 4 PIN diodes. RPRA can change its radiation pattern to different directions by changing the state of the switches without any change in its operating frequency and polarization. The proposed design does not use any parasitic elements or array structure and works with very few switching elements thus minimizes the cost and volume requirement.

## 2 Antenna Design

The geometry of the RPRA is shown in Fig. 1, the design has one vertical patch and two bracket-shaped patches. These three patches are connected by using 4 PIN diodes P1, P2, P3, and P4 and the dimensions of an antenna is shown in Table 1.

In this design, the shape and direction of the radiation pattern can be changed by changing the states of the PIN diode. When the PIN diode is in ON state, the switch (PIN diode) acts like a short circuit and when the PIN diode is in OFF state, the switch acts like an open circuit.

The design is simulated in HFSS. After optimization, when the diode is in ON state, the equivalent resistance value is chosen as 3 Ω and when the diode is in OFF

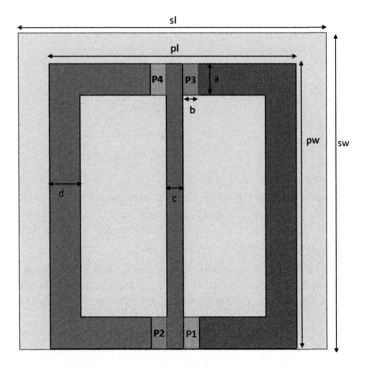

**Fig. 1** Geometry of radiation pattern reconfigurable antenna

**Table 1** Dimensions of the proposed antenna

| S. no | Parameter | Description | Dimension (mm) |
|---|---|---|---|
| 1 | sw | Width of the substrate | 20 |
| 2 | sl | Length of the substrate | 20 |
| 3 | pw | Width of the patch | 18 |
| 4 | pl | Length of the patch | 16 |
| 5 | a | Width of the switches | 2 |
| 6 | b | Length of the switches | 1 |
| 7 | c | Length of the vertical patch | 1.1 |
| 8 | d | Length of the bracket-shaped patch | 2 |

state, it has the resistance and capacitance values are chosen as 20 kΩ and 0.04 pF respectively in parallel combination. Figure 2 shows the linear circuit model of the PIN diode in both ON and OFF states.

The proposed antenna has been simulated in 8 switching configuration.

**Fig. 2** The linear circuit
model of the PIN diode

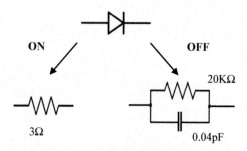

### Case 1: P1 is ON

When the diode P1 is in ON state, then the vertical patch is connected to the bottom part of the right bracket shaped patch. In this case, the antenna resonates at a frequency of 3.3 GHz and the radiation pattern has a main lobe in −30° direction.

### Case 2: P4 is ON

When the diode P4 is in ON state, then the vertical patch is connected to the top part of the left bracket shaped patch. In this case, the antenna resonates at a frequency of 3.3 GHz and the radiation pattern has two main lobes in direction of −140° and +40°.

### Case 3: P1 and P4 are ON

When the diodes P1 and P4 are in ON state, then the vertical patch is connected to the bottom part of the right bracket shaped patch and the top part of the left bracket shaped patch. In this case, the antenna resonates at a frequency of 3.3 GHz and the radiation pattern has two main lobes, they are in the direction of +50° and −150°.

### Case 4: P2 and P3 are ON

When the diodes P2 and P3 are in ON state, then the vertical patch is connected to the bottom part of the right bracket-shaped patch and the top part of the left bracket-shaped patch. In this case the antenna resonates at a frequency of 3.3 GHz and the radiation pattern has two main lobes, they are in the direction of 0° and 180°.

### Case 5: P2, P3, and P4 are ON

When the diodes P2, P3, and P4 are in ON state, then the vertical patch is connected to the left bracket shaped patch and the top part of the right bracket shaped patch. In this case, the antenna resonates at a frequency of 3.3 GHz and the radiation pattern has two main lobes in the direction of −160° and +25°.

### Case 6: P1, P3, and P4 are ON

When the diodes P1, P3, and P4 are in ON state, then the vertical patch is connected to the right bracket shaped patch and top part of the left bracket shaped patch. In this case the antenna resonates at a frequency of 3.3 GHz and the radiation pattern has two main lobes in the direction of −40° and +120°.

### Case 7: P1, P2, P4 are ON

When the diodes P1, P2, and P4 are in ON state, then the vertical patch is connected to the left bracket shaped patch and bottom part of the right bracket shaped patch. In this case the antenna resonates at a frequency of 3.3 GHz and the radiation pattern has a main lobe in −45° direction.

### Case 8: P1, P2, P3, and P4 are ON

When the diodes P1, P2, P3, and P4 are in ON state, then the vertical patch is connected to both left bracket shaped patch and right bracket shaped patch completely. In this case the antenna resonates at a frequency of 3.3 GHz and the radiation pattern has two main lobes in the direction of −145° and +45°. This antenna can change its radiation pattern to different directions without changing its polarization and operating frequency.


## 3  Results

Simulation results are described using eight switching configurations as follows:

### Case1:

When the switch P1 is in ON state, the antenna resonates at a frequency of 3.33 GHz with a VSWR of 1.01, return loss of −25.21 dB and an axial ratio of 39 dB.

### Case2:

When the switch P4 is in ON state, the antenna resonates at a frequency of 3.33 GHz with a VSWR of 1.05, return loss of −42.17 dB and an axial ratio of 33 dB.

### Case3:

When the switch P1 and P4 are in ON state, the antenna resonates at a frequency of 3.33 GHz with a VSWR of 1.05, return loss of −34.19 dB, and axial ratio of 44.3 dB.

### Case4:

When the switch P2 and P3 are in ON state, the antenna resonates at a frequency of 3.33 GHz with a VSWR of 1.05, return loss of −31.75 dB and axial ratio of 36 dB.

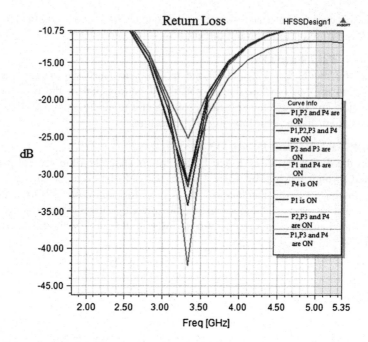

**Fig. 3** Return loss plot in all cases

**Case5**:

When the switch P2, P3, and P4 are in ON state, the antenna resonates at a frequency of 3.33 GHz with a VSWR of 1.05, return loss of −31.40 dB and axial ratio of 34 dB.

**Case6**:

When the switch P1, P3, and P4 are in ON state, the antenna resonates at a frequency of 3.33 GHz with a VSWR of 1.11, return loss of −31.06 and 36 dB.

**Case7**:

When the switch P1, P2, and P4 are in ON state, the antenna resonates at a frequency of 3.33 GHz with a VSWR of 1.05, return loss of −31.02 dB and axial ratio of 32 dB.

**Case8**:

When the switch P1, P2, P3, and P4 are in ON state, the antenna resonates at a frequency of 3.33 GHz with a VSWR of 1.11, return loss of −31.75 dB and axial ratio of 44 dB.

Figures 3 and 4 show the plots of return loss and radiation pattern plots in all the cases respectively.

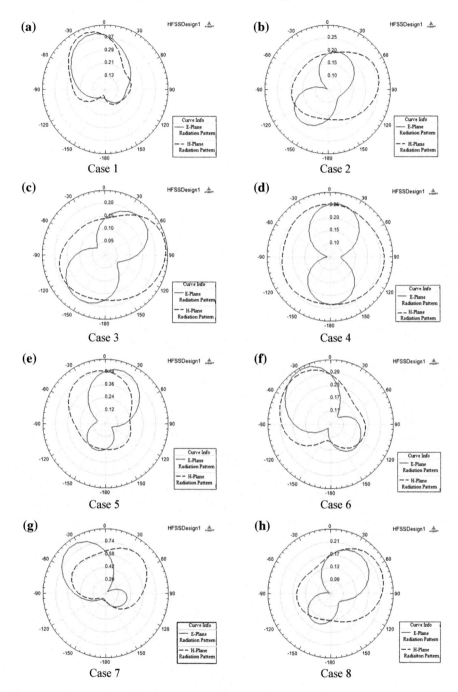

**Fig. 4** E-Plane and H-Plane radiation patterns in all cases

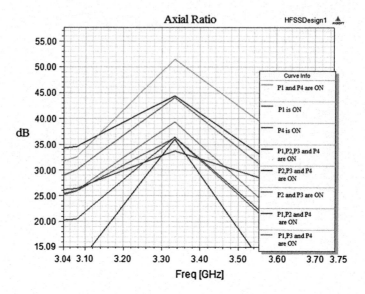

**Fig. 5** Axial ratio plot in all cases

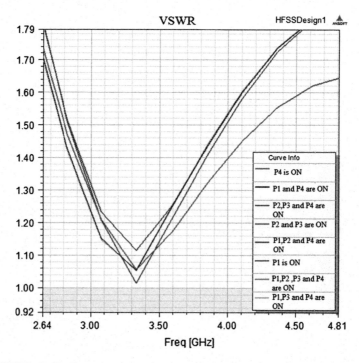

**Fig. 6** VSWR plot in all cases

**Table 2** Summary of results

| Switching combinations | Resonant frequency (GHz) | Return loss (dB) | VSWR | Radiation pattern direction | Polarization |
|---|---|---|---|---|---|
| Case 1 | 3.3 | −25.21 | 1.01 | −30° | Linear |
| Case 2 | 3.3 | −42.17 | 1.05 | −140° and +40° | Linear |
| Case 3 | 3.3 | −34.19 | 1.05 | +50° and −150° | Linear |
| Case 4 | 3.3 | −31.75 | 1.05 | 0° and 180° | Linear |
| Case 5 | 3.3 | −31.40 | 1.05 | −160° and +25° | Linear |
| Case 6 | 3.3 | −31.06 | 1.11 | −40° and +120° | Linear |
| Case 7 | 3.3 | −31.02 | 1.05 | −45° | Linear |
| Case 8 | 3.3 | −31.75 | 1.11 | −145° and +45° | Linear |

The antenna is exhibiting linear polarization in all the cases. Figure 5 shows the plot of axial ration in all cases and the VSWR plot is shown in Fig. 6.

The results are summarized in Table 2.

# 4 Conclusion

The proposed compact pattern reconfigurable antenna can change its direction of radiation pattern to different directions by changing the configuration of the switches. The antenna resonates at 3.3 GHz frequency with minimum −25 dB return loss and VSWR of 1.1 in all configurations. It uses very less number of switches and it does not use any array structure or parasitic elements for beam steering. Hence, it occupies less space and cost required is also less. The antenna is useful for WiMAX application.

# References

1. J.T. Bernhard. Reconfigurable Antenna. Sam rafeal. CA, USA, Morgan & claypool, 2007.
2. CG. Christodoulou. Y. Tawk, S.A. Lane, and S.R. Erwin "Reconfigurable antenna for wireless & space applications", Proc. IEEE, vol.100, no. 7, pp 2250–2261, Jul. 2012.
3. R.L. Haupt and M. Lanagan "Reconfigurable antennas" IEEE Antennas & Prop. Mag., vol. 55, no. 1, pp 49–61, Feb 2013.
4. Pei-Yaun Qin, Y. Jay Guo, Andrew R. Weily and Chang-Hong Liang, "A Pattern Reconfigurable U-slot Antenna and its applications in MIMO systems", IEEE Transactions on Antenna and Propagations, Vol. 60, No. 2, Feb 2012.
5. Cheng-Hsun Wu and Tzyh-Ghuang Ma, "Pattern-Reconfigurable Self-Oscillating Active Integrated Antenna With Frequency Agility", IEEE Transactions on Antenna and Propagations, Vol. 62, No. 12, Dec 2014.
6. Chih-Hsiang Ko, I-Young Tarn and Shyh-John Chung, "A Compact Dual-Band Pattern Diversity Antenna By Dual Band Reconfigurable Frequency-Selective Reflectors With A

Minimum Number Of Switches", IEEE Transactions on Antenna and Propagations, Vol. 61, No. 2, Feb 2013.

7. George D. Sworo, Kapil R. Dandekar and Moshe Kam, "Design and Analysis of Reconfigurable Antennas For WiMAX Applications", IEEE Transactions on Antenna and Propagations, 2013.

8. Yang-Ying, Shaoqiu Xiao, Changrong Liu, Xiang Shuai and Bing-Zhong Wang, " Design of Pattern Reconfigurable Antennas Based on a Two-Element Dipole Array Model", IEEE Transactions on Antenna and Propagations, Vol. 61, No. 9, Sept 2013.

9. W. Kang, K.H. Ko and K. Kim, "A Compact Beam Reconfigurable Antenna For Symmetric Beam Switching" Progress In Electromagnetic Research, Vol. 129, 1–16, 2012.

10. Muzammil Jusoh, Tamer Aboufoul, Thennarasan Sabapathy, Akrram Alomainy and Muhammad Ramlee Kamarudin, "Pattern Reconfigurable Microstrip Patch Antenna With Multidirectional Beam For WiMAX Applications", IEEE Transactions on Antenna and Propagations, Vol. 13, 2014.

# Low-Power Adiabatic Logic—Design and Implementation in 32-Nanometer Multigate Technology

Suresh Kumar Pittala and A. Jhansi Rani

**Abstract** A new FinFET-based adiabatic NAND logic circuit with Self-Adjustment of Rail Potential (SARP) is proposed. The proposed logic provides reduced power consumption when compared to conventional CMOS and adiabatic circuits. A new FinFET-based adiabatic logic is implemented based on Complementary Energy Path structure. The proposed design reduces the second-order effects, short-Channel effects occurring in Conventional CMOS circuits. The performance of the proposed SARP-FinFET-based adiabatic NAND gate is dominant when compared to the SARP-CMOS-based adiabatic NAND gate. The proposed adiabatic circuits are designed using double gate FinFET using predictive technology models (PTM) in 32 nm Technology using Synopsis HSPICE. The experimental results for the proposed adiabatic FinFET design demonstrate their effectiveness with energy consumption and power optimization.

**Keywords** FinFET · Adiabatic logic · Adder · NAND gate
Shorted gate · Energy efficient · Power optimization

## 1 Introduction

The advancements in electronics design technology led to the innovation of portable and battery-operated devices, consequently, the power consumption and speed of operation has become the primary concern. In high-performance handheld devices, the power consumption is the fundamental constraint. In the literature, several adiabatic circuits are presented and the results show their dominance in energy-saving capability. Even though the complexity in design is high when compared to the conventional circuits. Yong moon [1] presented the efficient charge

S. K. Pittala (✉)
Acharya Nagarjuna University, Nagarjuna Nagar, Guntur 522510, Andhra Pradesh, India
e-mail: dr.sureshkumarpittala@gmail.com

A. J. Rani
Velagapudi Ramakrishna Siddhartha Engineering College, Vijayawada, India

© Springer Nature Singapore Pte Ltd. 2018                                                    335
J. Anguera et al. (eds.), *Microelectronics, Electromagnetics
and Telecommunications*, Lecture Notes in Electrical Engineering 471,
https://doi.org/10.1007/978-981-10-7329-8_34

recovery logic (ECRL) which performs precharge and evaluation phase simultaneously. An adiabatic differential cascade voltage switch with a complementary pass-transistor logic tree (ADCPL) is presented by Chun-Keung Lo and Philip C H Chan [2]. The presented work lowers the gate complexity of operating the circuit from a two-phase nonoverlapping supply clock. Matthew Morrison [3] in their chapter presented an algorithm for minimization of Boolean functions by correlating the horizontal offsets in the permutation matrix instead of library. The presented algorithm frames an adiabatic s-box structure which reduces the energy imbalance compared to previous benchmarks. The algorithm does the forward encryption and reverse decryption with minimal overhead.

F. Liu and K. T. Lau [4] presented a pass-transistor-based low-power adiabatic logic with NMOS pull-down configuration. Using the implementation of a multiplexer working at 20 MHz, a power saving of 80% is achieved when compared to 2N2N2P Multiplexer. In [5] Dragan Maksimovic presented a clocked CMOS Adiabatic Logic operating with Single-Phase Power Clock Supply. The presented low energy logic incorporates the design of the power control unit within the chip itself. In another work [6] a Pass-Transistor-based single power clock supply Adiabatic Logic is presented. The energy saving happens till 160 MHz for a 1.2 μm technology when applied to a PAL structure. A two-phase clocking dual-rail adiabatic logic known as 2PCDAL is presented in [7] which is based on 2N2N2P structure. The presented circuit uses a two-phase clocking scheme instead of a four-phase clocking scheme used in conventional 2N2N2P. A two-phase clocked adiabatic static CMOS logic (2PASCL) is presented in [8] which uses the adiabatic switching and energy recovery logic as principles. Suhwan Kim and Marios C. Papaefthymiou have presented a True Single-Phase Adiabatic logic instead of complex Dynamic logic [9] which exhibit increased energy consumption and low-performance high-speed design. The work is based on energy recovering logic operating using single-phase sinusoidal clock. Cihun-Siyong Alex Gong et al. presented [10] an irreversible energy recovery logic, which inherits the advantages of Quasi-Static Energy Recovery Logic (QSERL). When compared with QSERL, the circuit excludes the hold phase avoiding the use of feedback keeper. The work states the advantage of the circuit toward the reduction of area and power overheads. The throughput is twice when compared to the counterpart and the power clock used is similar.

In this chapter, we have presented a new adiabatic logic based on FinFET. The proposed design incorporates the best performance efficiency of FinFET and adiabatic in a single platform. The SARP-level circuit structure improves the efficiency of the design by reducing the leakage current. The rest of the chapter is organized in this manner. In the next section, the FinFET-based NAND structure is presented, followed by the proposed leakage reduction technique for adiabatic logic in Sect. 3. Section 4 presents the proposed design and implementation of the methodology. Section 5 discusses the experimental setup and results obtained followed by the conclusion and reference.

## 2   FinFET-Based Adiabatic NAND Structure

The improvement in technology aggressively scaled down the geometric parameters of the MOS devices to achieve the power consumption reduction, increased speed of operation and larger integration density. The scaling down of geometric parameters has limited to 45 nm due to the performance degradation of the MOSFET devices. The leakage current is the main bottleneck below submicron level scaling. The power consumption can be reduced further by reducing the power supply voltage level, but the leakage current increases exponentially. At the same time, voltage scaling reduces the energy-per-operation during switching instance due to the relation between energy and supply voltage. In multicore devices, the energy is saved using dynamic voltage scaling. The other issues in CMOS devices are the increased sensitivity to process variation and Short-Channel Effects (SCE). Previous studies show that the multigate transistor [11] reduces subthreshold and the gate tunneling leakage current. The device increases the driving current when compared to the standard single-gate MOSFETs below submicron technologies [12, 13]. The promising multigate device, which emerged to replace MOSFET device, is the FinFET. Due to its thin silicon body and dual electrically coupled gate, it suppresses the SCE which reduces the subthreshold leakage current. This device shows better parameter variations in immunity than the conventional single-gate device.

Figure 1 illustrates the FinFET-based adiabatic NAND gate structures based on Shorted Gate (SG) mode FinFETs for conventional NAND gate and 2N2N2P NAND gate.

Since FinFET being a nonplanar device and can work in different modes of operation like SG mode or independent gate mode it is efficient in logic implementation. In this chapter, the 32 nm FinFET PTM is used for HSPICE simulation. The physical characteristics of the PTM FinFET are given in the Table 1.

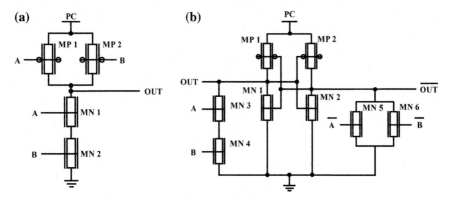

**Fig. 1** FinFET-based adiabatic NAND gate structures based on Shorted Gate (SG) mode FinFETs for **a** conventional NAND gate and **b** 2N2N2P NAND gate

**Table 1** Primary parameters in PTM

| Parameters | Value n-type FinFET | Value p-type FinFET |
|---|---|---|
| Channel length (nm) | 32 | 32 |
| Fin height (nm) | 40 | 50 |
| Gate oxide thickness (nm) | 1.4 | 1.4 |
| Threshold voltage (V) | 0.29 | −0.25 |
| Fin thickness (nm) | 8.6 | 8.6 |
| $V_{DD}$ (V) | 1 | 1 |

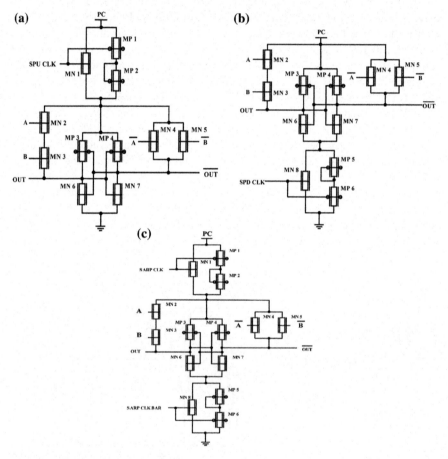

**Fig. 2** Various leakage reduction technique structures **a** SPU **b** SPD **c** SARP

# 3  Proposed Leakage Reduction Techniques for the Adiabatic Logic

Several leakage reduction techniques are available for CMOS circuits used in memory devices [14–16] like SRAM. A leakage reduction technique SARP is designed for the proposed adiabatic circuits using FinFET. A SARP circuit supplies maximum DC voltage to an active-load circuit on request. This makes the load circuit to work quickly. When the load circuit is in active mode, the SARP circuit provides a maximum supply voltage and minimum ground voltage. The three structures of SARP technique the SARP circuit for pull-up, pull-down SARP circuit and combined SARP circuit for pull-up and pull-down is designed with the FinFET-PFAL NAND as shown in Fig. 2. The NAND gate is the load circuit with the lower SARP circuit consists of a single p-type switch while the n-type switches are connected in series.

# 4  Proposed Technique

A new adiabatic logic using FinFET-based Complementary Energy Path Adiabatic Logic (CEPAL) is proposed. The proposed design is been restructured by SG mode FinFET. The proposed circuits outperform the previous adiabatic logic families in terms of energy efficiency and operating speed. The proposed circuit is based on two clock signals working with four stages namely WAIT, EVALUATE, HOLD, and RECOVER. Two clock signals are used for the SARP and power supply. The phase difference between any two adjacent stages is a quarter of a period. The typical time sequence of the logic with respect to the four-stage operations are the WAIT state in which the power supply is zero, the input is valid and the pre-evaluated result is generated by the evaluation logic and the outputs keep low voltage. In EVALUATE phase, the rise time of the power supply starts from zero and increasing, the input becomes stable. The output follows the power supply and is valid. To keep the output valid (HOLD), the power supply stays in high state, providing a constant input signal to next stage. In RECOVER stage, the power supply falls to zero shutting down the current access to ground and allows the charge stored in the capacitor to recover through the cross-coupled FinFETs. The PFAL circuit eliminates the charge stored in the output through the positive feedback. The proposed CEPAL circuit inherits three advantages when compared to all the previous circuits presented earlier. The circuit operation eliminates the hold phase by which the throughput is increased. This feature is similar to QSERL where the complexity is reduced. The proposed CEPAL circuit consists of two charging transistors (MP1 and MP2), a pull-up network, two discharging transistors (MN1 and MN2), a pull-down network and the SARP block for three configurations. The CEPAL has two paths apart from the charging and discharging paths. The number of devices is more so implementation cost is higher. The adiabatic logic

**Table 2** Performance analysis of different FinFET-based adiabatic circuits for PCL = 1 MHz

| Switching frequency | 10 kHz | 20 kHz | 40 kHz | 80 kHz | 100 kHz | 250 kHz | 500 kHz | 1 MHz | 10 MHz |
|---|---|---|---|---|---|---|---|---|---|
| Device name | Energy in fJ | | | | | | | | |
| Conventional FinFET | 172.4 | 218 | 1460 | 8150 | 12600 | 21400 | 10700 | 10600 | 10400 |
| ECRL | 108 | 135 | 315 | 1910 | 979 | 14500 | 41600 | 9810 | 5330 |
| 2N2N2P | 134000 | 134010 | 134060 | 134870 | 984 | 142000 | 139000 | 72600 | 72600 |
| PFAL | 3240 | 3240 | 3240 | 3240 | 592 | 3230 | 3230 | 3240 | 3200 |

**Table 3** Performance analysis of different SARP-FinFET-based adiabatic circuits for various power clock frequencies

| Device name | CEPAL NAND | | SARP CEPAL | | SPU CEPAL | | SPD CEPAL | |
|---|---|---|---|---|---|---|---|---|
| Power clock frequency | Iavg | E in fJ | Iavg | E in fJ | Iavg | E in fJ | Iavg | E in fJ |
| 10 MHz | 4.18E-07 | 418 | 1.74E-06 | 1741.76 | 4.03E-06 | 4035.36 | 1.08E-07 | 108.76 |
| 500 MHz | 5.71E-07 | 571 | 1.69E-07 | 169.59 | 4.41E-06 | 4406.89 | 1.69E-07 | 169.59 |
| 1 GHZ | 3.59E-07 | 359 | 1.99E-06 | 1990.11 | 4.48E-06 | 4485.31 | 1.33E-07 | 133.09 |
| 2 GHZ | 4.77E-07 | 477 | 2.29E-06 | 2288.01 | 4.81E-06 | 4814.78 | 1.33E-07 | 133.09 |

operating speed depends on the pull-up transistors since the charging and discharging operation is performed through those transistors. The pull-down transistors maintain the voltage at certain nodes for operation. The leakage current and power consumption is controlled by the pull-down transistors. Therefore, in the proposed design the SARP circuit reduces the leakage current and power consumption.

## 5 Simulation Results

To demonstrate the effect of switching activity on the energy consumption and average current, vast experiments and simulations were conducted for different adiabatic logics designed using CMOS and FinFET. Table 2 shows the performance of different devices measured with different parameters. The switching frequency is varied from 10 kHz to 10 MHz. From Table 3, it is observed that the SPD CEPAL circuit offers reduced energy consumption when compared to the SARP and SPU circuits. The CMOS circuits show the steady-state response, but more energy dissipation while the FinFET-based design even though have peaks and trough the energy consumed is very less. Thus, the proposed designs have efficient performance when compared to the existing methods.

## 6 Conclusions

We have shown in this chapter the effectiveness of the proposed SARP-FinFET-based CEPAL adiabatic NAND gate. The existing CMOS-based adiabatic design is noticeably improved using our proposed SARP-FinFET-based adiabatic structures. The proposed circuits have been simulated and analyzed using HSPICE with Predictive Technology Models. The results show that the CMOS adiabatic energy dissipation are reduced on average for PFAL and CEPAL. In future, the design will be utilized for the design of the arithmetic unit and power clock circuit.

# References

1. Yong Moon,: An Efficient Charge Recovery Logic Circuit. IEEE Journal of Solid-State Circuits. 31 (1996) 514–522.
2. Chun-Keung Lo, Philip C H Chan,: An Adiabatic Differential Logic for Low-Power Digital Systems. IEEE Transactions on Circuits and Systems II: Analog and Digital Signal Processing. 46 (1999) 1245–1250.
3. Matthew Morrison,: Synthesis of Dual Rail Adiabatic Logic for Low Power Security Applications. IEEE Transactions on Computer-Aided Design of Integrated Circuits and Systems. 33 (2014) 975–988.
4. Liu, F., Lau, K. T.,: Pass-transistor adiabatic logic with NMOS pull-down configuration. Electronics Letters. 34 (1998) 739–741.
5. Dragan Maksimovic, Vojin G Oklobdzija, Borivoje Nikolic, Wayne Current, K.,: Clocked CMOS Adiabatic Logic with Integrated Single-Phase Power-Clock Supply. IEEE Transactions on Very Large Scale Integration (VLSI) Systems. 8 (2000) 460–463.
6. Vojin G. Oklobdzija,: Pass-Transistor Adiabatic Logic Using Single Power-Clock Supply. IEEE Transactions on Circuits and Systems II: Analog and Digital Signal Processing. 44 (1997) 842–846.
7. Yasuhiro Takahashi, Zhongyu Luo, Nazrul Anuar Nayan, Michio Yokoyama,: 2PCDAL Two-Phase Clocking Dual-Rail Adiabatic Logic. IEEE Asia Pacific Conference on Circuits and Systems (APCCAS), 1 (2012) 124–127.
8. Nazrul Anuar, Yasuhiro Takahashi, Toshikazu Sekine,: Two Phase Clocked Adiabatic Static CMOS Logic and its Logic Family. Journal of Semiconductor Technology and Science, 10 (2010) 1–10.
9. Suhwan Kim, Marios C Papaefthymiou,: True Single-Phase Adiabatic Circuitry. IEEE Transactions on Very Large Scale Integration (VLSI) Systems, 9 (2001) 52–63.
10. Cihun-Siyong Alex Gong, Muh-Tian Shiue, Ci-Tong Hong, Kai-Wen Yao,: Analysis and Design of an Efficient Irreversible Energy Recovery Logic in 0.18-μm CMOS. IEEE Transactions on Circuits and Systems I: Regular Papers. 55 (2008) 2595–2607.
11. Prateek Mishra, Anish Muttreja, Niraj K Jha,: FinFET Circuit Design. Nanoelectronic Circuit Design, 1 (2011) 23–33.
12. Dhruva Ghai, Saraju P Mohanty, Garima Thakra,: Comparative Analysis of Double Gate FinFET Configurations for Analog Circuit Design. IEEE 56th International Midwest Symposium on Circuits and Systems, 1 (2013) 809–812.
13. Matteo Agostinelli, Massimo Alioto, David Esseni, Luca Selmi,: Leakage–Delay Tradeoff in FinFET Logic Circuits: A Comparative Analysis With Bulk Technology. IEEE Transactions on very large scale integration (VLSI) systems, 18 (2010) 232–245.
14. Manorama, Saurabh Khandelwal, Shyam Akashe,: Design of a FinFET Based Inverter Using MTCMOS and SVL Leakage Reduction Technique. Students Conference on Engineering and Systems (SCES), 1 (2013) 1–6.
15. Mindaugas Drazdziulis, Per Larsson-Edefors,: A Gate Leakage Reduction Strategy for Future CMOS Circuits. Proceedings of the 29th European Solid-State Circuits Conference, 1 (2003) 317–320.
16. Akashe, Shyam, Meenakshi Mishra, Sanjay Sharma,: Self-controllable voltage level circuit for low power, high speed 7T SRAM cell at 45 nm technology. Students Conference on Engineering and Systems, 1 (2012) 1–5.

# Design of Low-Power Binary Content Addressable Memory for Future Nanotechnologies

G. Surekha, N. Balaji and Y. Padma Sai

**Abstract** In today's industrial situation, there is a vast demand for devices with low power consumption. Therefore, the demand for reducing the power consumption in memory elements become vital as it occupies a significant portion of chip area. Content Addressable Memory is a kind of memory element used for search applications. The foremost CAM design requirement is to decrease power consumption connected with the huge amount of parallel active circuitry. In this work, a low-power Binary Content Addressable Memory (BCAM) design is implemented. The proposed CAM is simulated using Cadence Virtuoso simulator in 45, 90, and 180 nm technology. The proposed technique can be applied to nanotechnologies to reduce the power consumption without affecting the original functionality of the memory cell.

**Keywords** Low-power design · Binary content addressable memory
Leakage power and nanotechnology

## 1 Introduction

Memory is a basic element used in all the electronic system architectures for storing programs and data. Memory elements are generally classified into Random Access Memory (RAM), Serial Access Memory and Content Addressable Memory (CAM). Content Addressable Memory addresses the memory based on the input

G. Surekha (✉)
Department of ECE, GRIET, Bachupally, Hyderabad, Telangana, India
e-mail: yandamurisurekha@gmail.com

N. Balaji
Department of ECE, JNTUK, Vizianagaram, Andhra Pradesh, India

Y. Padma Sai
Department of ECE, VNR VJIET, Bachupally, Hyderabad, Telangana, India

© Springer Nature Singapore Pte Ltd. 2018
J. Anguera et al. (eds.), *Microelectronics, Electromagnetics
and Telecommunications*, Lecture Notes in Electrical Engineering 471,
https://doi.org/10.1007/978-981-10-7329-8_35

343

data rather than the address (physical) location [1]. CAM compares the stored data with the input variable. CAMs are widely used in many applications because of its multiple features. These features include translation look-aside buffers for virtual memory systems, tag directories for associated cache organizations, collision detection processor in intelligent automobiles, the interconnection of different routers, reconfigurable CAM, and other uses in the fields of image processing and artificial intelligence [2]. CAM can also be used in applications where the requirement is such that only specific address ranges are to be searched. Functions of CAM are write, read, and match. The write operation describes the stored data into the memory. The read operation describes the stored data retrieval and refreshes purposes and the match operation compares the new data with the previously stored data and indicates if both are same. Types of memory CAMs are Binary CAM, Ternary CAM, and Memory-resistor CAM.

In this chapter, Sect. 2 describes the need for low power, existing Binary CAM, and proposed Binary CAM are discussed in Sects. 3–5 explains Simulation Results and Sect. 6 is for Conclusion.

## 2  Need for Low Power

In earlier days, less number of transistors is integrated per a chip and power consumption is more because of the long channel length [3]. Moore's law states that for every year the number of transistors integrated per chip is doubled [4]. Based on this law, the number of transistors integrated on a single chip is increased and the power supply is decreased, hence power consumption is low as the length of the channel is decreased, the area will be reduced which will further reduces the inter connection wire length between logical blocks. In CMOS circuits, there are three sources of power dissipation [5]. First is due to signal transitions between to logical levels, the capacitances associated with these logical level nodes get charged and discharged called dynamic power dissipation. Second, power dissipation comes from short circuit currents which flow directly from the supply voltage and ground called the short circuit power dissipation and the last is because of leakage currents, which flow when the inputs and outputs are not changing and is called static power dissipation [4]. The dynamic power dissipation is proportional to the supply voltage. As the technology is scaling down the power supply required is also reduced hence reducing dynamic power but this increases the leakage currents because of low threshold voltages of CMOS transistors. This low threshold voltage of CMOS leads to increased leakage currents results in increased static power dissipation. Even after scaling down the technology power dissipation is not reduced because of existing static power. So, the power dissipation should be less as the low power is the main consideration factor for all the Electronic systems.

**Fig. 1** Binary CAM cell

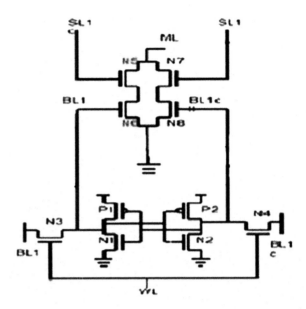

## 3   Existing Binary CAM

Existing Binary CAM cell is shown in Fig. 1. To store the data in 1's and 0's form Binary CAM cell is used. PMOS transistors (P1, P2) and NMOS transistors (N1, N2, N3, N4) are used to design the basic SRAM cell. Activating the word line (WL) as high data can be written in the SRAM memory cell. After storing the data in SRAM cell word line (WL) should be low and before checking the input data matched line (ML) should be pre-charged to $V_{DD}$. If the given data is matched with stored data then matched line (ML) is remains at $V_{DD}$ else matched line (ML) is discharged.

## 4   Proposed Binary CAM

Proposed Binary CAM cell is shown in Fig. 2. The proposed Binary CAM cell uses an extra NMOS transistor N1 is used for reducing the power dissipation of the cell without changing the supply voltage. This proposed Binary CAM cell is also used to store data in 1's and 0's form but in improved performance of power. This extra NMOS transistor is connected to match cell circuitry in association with SRAM cell and is used to store the data. Apart from SRAM cell the upper circuit is used to verify the stored data by taking the inputs.

**Fig. 2** Proposed binary
CAM cell

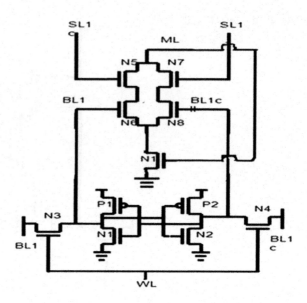

## 4.1   Leakage Power Control in Proposed Binary CAM

Subthreshold leakage current is reduced by connecting transistors in series called as
"stacking effect" [5]. NMOS transistor N1 is connected in series with N7 and N8 in
mismatch condition as shown in Fig. 2. Assuming SRAM stored values are "0" and
"1" for BL1, BL1c and the input values given to SL1 and SL1c are "1" and "0" as it
is in mismatch condition. So, in mismatch condition ML discharges through N7,
N8, and N1. As the node voltage between N8 and N1 is positive because of small
drain current flowing through N8 and N1 which are in off conditions. In match
condition also N8 and N1 are in series, due to this positive potential gate to source
voltage of N8 becomes less than zero hence reduced subthreshold current.

## 5   Simulation Results

All the simulations are obtained using Cadence Tool Virtuoso simulator 45, 90, and
180 nm technology shown in Table 1.

Table 1 shows the power comparisons of the existing Binary CAM and the
proposed Binary CAM. For all the three technologies, power consumption is more
in 180 nm. As the technology is reducing power consumption is greatly reduced
using the proposed Binary CAM cell. Power dissipation is reduced in both matched

**Table 1** Power comparison of the existing binary CAM and the proposed binary CAM in 45, 90, and 180 nm technology

| Technology (nm) | Binary CAM power dissipation | | Proposed binary CAM power dissipation | |
|---|---|---|---|---|
| | Matched condition (nW) | Un-matched condition (nW) | Matched condition (nW) | Un-matched condition (nW) |
| 45 | 0.353 | 1.799 | 0.317 | 0.6623 |
| 90 | 2.267 | 92.49 | 1.842 | 86.2 |
| 180 | 137.2 | 184 | 122.6 | 179 |

**Fig. 3** Power dissipation of the existing binary CAM and the proposed binary CAM for matched condition

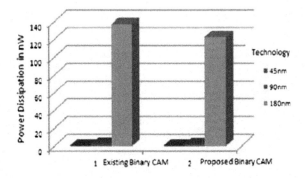

condition and unmatched conditions for all the used technologies. Power dissipation of existing Binary CAM and proposed Binary CAM for the matched condition is shown in Fig. 3.

Power dissipation of the existing Binary CAM and the proposed Binary CAM for the unmatched condition is shown in Fig. 4. In this diagram technology is taken on X-axis and power dissipation on Y-axis.

**Fig. 4** Power dissipation of the existing binary CAM and the proposed binary CAM for unmatched condition

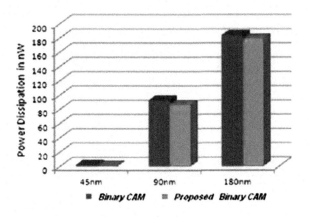

## 6 Conclusion

This paper presents new Binary CAM which will be differing from the existing Binary CAM. The proposed Binary CAM decreases power of 11, 19 and 12 for matched condition and 63%, 7%, and 3%, respectively, for unmatched condition in 45, 90, and 180 nm technology. Results are obtained using Cadence Virtuoso simulator. It is observed from the results that proposed Binary CAM reduces more power dissipation for 45 nm technology. Thus, the proposed technology can be used to reduce the power in short channel CMOS transistor circuits.

## References

1. Master–Slave Match line Design for Low-Power Content-Addressable Memory, yen-jen chang and tung-chi wu, IEEE VLSI, Vol. 23, No. 9, September 2015, pp (1740–1749).
2. Decoupled Dynamic Ternary Content Addressable Memories José g. Delgado-Frias, Jabulani Nyathi and Suryanarayana b. Tatapudi, IEEE CS, Vol. 52, No. 10, October 2005, pp (2139–2147).
3. J. M. Rabaey, Digital Integrated Circuits-Prentice-Hall, 1996.
4. Kiat-Seng Yeo, Kaushik Roy-Low Voltage Low Power VLSI Subsysyems. Tata McGraw Hill, 2009.
5. Lin, Ming-Bo. (2011). Introduction to VLSI Systems: A Logic, Circuit, and System Perspective. CRC Press.

# Cross Talk Delay Reduction in System on Chip

**R. Sridevi, P. Chandra Sekhar and B. K. Madhavi**

**Abstract** Integrated circuit design has undergone immense progress in the past few years. Semiconductor applications had a wide range of growth in technological inventions. Effects on SOC timing and functionality with cross talk are one of the most important aspects of this chapter. An efficient analysis has been carried out and a systematic flow is developed and the efficiency has been compared with other existing methods. Synopsis prime time tool is used here to evaluate the delay analysis. Prime time runs more quickly than other cross talk analysis techniques. Among the existing techniques, some are considered for addressing the issue of cross talk in system on chip design.

**Keywords** SOC · Cross talk delay · Cross talk noise · Performance

## 1 Introduction

CMOS scaling to ultra-deep sub-micron technology has increased the sensitivity of CMOS technology to various noises such as cross talk noise, leakage noise, supply noise, etc.

Reducing switching activity at the interconnect nodes is one of the low power coding methods used for off chip buses, where the coupled capacitance is the major source of noise where as inter-wire capacitance is the major factor in on chip buses and should be minimized to reduce energy consumption. Two main factors, which cause crosstalk in VLSI, are mutual inductance and inter-wire capacitance.

R. Sridevi (✉)
BVRITH College of Engineering for Women, Hyderabad, India
e-mail: rsrideviprasad@yahoo.co.in

P. Chandra Sekhar
University College of Engineering, Osmania University, Hyderabad, India

B. K. Madhavi
Sridevi Women's Engineering College, Hyderabad, India

© Springer Nature Singapore Pte Ltd. 2018
J. Anguera et al. (eds.), *Microelectronics, Electromagnetics and Telecommunications*, Lecture Notes in Electrical Engineering 471,
https://doi.org/10.1007/978-981-10-7329-8_36

Cross talk can be considered as an interaction between two different signals on two electrical wires. The creator of cross talk effect is known as an "aggressor", while the one that receives the effect is called a "victim". A single wire can be an aggressor or a victim. Two major cross talk classifications are as follows [1].

## 1.1  Inductive Cross Talk

A magnetic field can be generated by electrical current flowing in a loop and if this field changes, it can either radiate energy or can couple two adjacent loops which causes inductive cross talk. Two conductors are referred to as magnetically or inductively coupled, when the change in current through a wire induces a voltage in other wire through electromagnetic induction. Generally, inductive coupling between two conductors is calculated by their mutual inductance. Shielding technique, increasing metal-to-metal separation, buffer insertion technique are some of the techniques used to reduce inductive coupling.

## 1.2  Electrostatic Cross talk (Capacitive Coupling)

An electric field creates an electrical voltage on a line, which on changing couples capacitive to adjacent lines. When geometries shrink below 0.25 μm electrostatic cross talk is becoming significant. Avoiding floating nodes, the increment in rise and fall times, usage of cascaded buffers, driving large capacitances, usage of shielding wires and shielding layers help in reduction of cross talk generated by capacitive coupling.

Winding of wires into coils and placing them closer to the common axis, induces the magnetic field of one coil into another increases the coupling between wires. Usage of materials like iron or ferrite in coils increases magnetic flux and coupling. Intentionally or unintentionally coupling may occur. Unintentional coupling in signals of one circuit to other circuit leads to electromagnetic interference.

When compared to wire capacitance, gate capacitance is negligible in small geometries. Due to this, delays in timing paths are dominated by interconnect delays than delays in cells. From 0.7 to 0.09 μm technology the delay due to capacitive coupling increases by ten times. Interconnect delay can be reduced by changing metal in wires from aluminum to copper.

Interconnect delay dominates with the reduced geometries in latest technologies when compared to the gate delays. The interconnect delays do not scale as in case of intrinsic delays.

When high-speed digital designs are considered, the mutual inductance plays a predominant role in interconnects when compared with the inter-wire capacitance between bus lines. Mutual inductance becomes dominating when the spacing between adjacent lines is very less. Current is induced from aggressor line to victim line due to mutual inductance, which causes cross talk.

## 1.3 Setup and Hold Violations are Caused by Cross talk in System on Chip Designs

### 1.3.1 Setup Violations

Setup violations occur when setup time requirements are not meet the data inputs in the sequential element. During cross talk delay analysis for the system on chip setup violations are observed, though less in number.

### 1.3.2 Hold Violations

Hold violations in SOC occur when the data input does not satisfy the minimum required hold timing duration. Generally, clock networks are very much susceptible to cross talk issue as these are widely spread throughout the system on chip in every sequential element present in the SOC.

The effect of cross talk is mostly due to hold violations. This primarily occurs when any of clock networks in the chip becomes a victim of a very fast aggressor. The clock network is in coupling with other wire, which is generally driving a buffer of high driving strength so the clock network becomes the victim of the aggressor.

### 1.3.3 Bus Violations

Bus violations are one of the possible effects of cross talk delay can be seen on bus signals. Interconnect between two far placed blocks is done by long stretched bus signals. Performance requirements are generally for these bus signals and need multiple repeaters in the path. For reduction of skew destination, the routing of bus signals is done together with minimum space allowed between them.

## 2 SOC for Consideration of Cross Talk and Noise Analysis

For SOCs with large size and complexity, plenty of violations can be encountered the first time a cross talk analysis is done [2]. The main challenging task is to find the timing reports generated by the cross talk delay and static timing analysis to identify the possible violations. Physical routing in the concerned paths causes cross talk-induced timing violations. It is very crucial to have routing details of the path to find the violations. Paths that meet timing should be untouched in fixing process [3].

In SOC, there are several peripherals present with processor and memories (ROM, RAM, and Cache) [4].

Case 1: Individual protocol functionality is checked which is known as block-level functionality check in SOC.

Case 2: Once block-level functionality check for all peripherals is done, SOC-level functionality check is done where some of the tests which have an overall flow in SOC are considered.

Processor— > DMA— > Peripheral, etc.

The SOC designed is for mobile communication applications. The internal blocks of this SOC are as follows:

It can support both synchronous and asynchronous data transmission with different speed requirements. It consists of

ARM11 which is most suitable for phone and embedded applications and its frequency is 350–1 GHz. In this pipelining Technique can be used and more faster than ARM9

Video processor is used for transmission of image and video transfer
I2C, I2S, UART, GPIO, and SPI are low speed peripherals
DMA for accessing the memory
Ethernet is medium speed bus for communication through optical fibers
Interconnect buses used here are AXI, AHB, and APB
MMC is a controller used for controlling media transfer between different blocks

Inputs for different peripherals depend on specification or as per user need.

Generally, 1 or 2 bytes of data is transferred depending on protocols which consist of – > preamble, start bit, R/W bit, address, payload, stop bit, CRC, etc. (General format, varies with protocols)

I2C: 1 or 2 bytes (diff. check)
I2S: 4 bytes in LC/RC: 8 bytes
UART: 2 bytes
GPIO: 2 bytes
SPI: 2 bytes

1 byte of data is enough to check the basic functionality for I2C, GPIO, UART, SPI, etc.

When running tests using processor, the C code is compiled and converted into a Hex file and loaded into the Boot Medium (ROM, System Memory or External Flash). The processor starts start execution of Hex Code from Boot Medium. The interrupts will be serviced according to the ISR-Interrupt ID mapping specified in the Interrupt Handler routine [5] (Fig. 1).

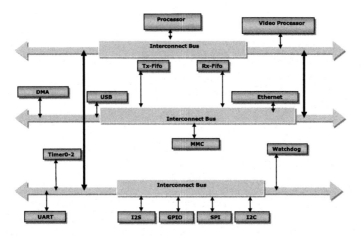

**Fig. 1** SOC block diagram

## 2.1 I2c

With SDA low and SCL high the START bit(S) initiates data transfer. When SCL is low, SDA sets the 1st data bit. When SCL rises (green) for the first bit (B1), data is sampled. With both SDA and SCL high STOP bit (P) is signaled. The transition of SDA with SCL low data is read.

False marker detection can be avoided when SDA is changed on the SCL falling edge and is sampled and captured on the rising edge of SCL.

## 2.2 I2s

Data may be sent either with the most significant bit first or the least significant bit first. The timing diagram shows that the left channel is sent before the right channel for a stereo signal.

## 2.3 UART

Here communication may be in one direction only, i.e., simplex, with no provision for the receiver device to send back information to the transmitting device), full duplex (both sender and receiver at the same time), or half duplex (devices do transmitting and receiving in turns).

## 2.4  SPI

SPI output data changes at the clock edge which is the transmit edge,

The sampling edge is the clock edge at which the sampling of the SPI input data takes place. The sampling edge and the transmit edge are normally the opposite of each other.

## 2.5  Bus Bridge

VALID and READY should be active for clock cycle to complete the transaction. VALID must not depend on READY for the assertion (this is to prevent deadlock because READY can depend on VALID for assertion). Once asserted, VALID cannot de-assert until READY accepts the transaction. In other words, de-assertion of VALID can depend on READY.

AMBA is a bus protocol generally targeted for high-performance system design. A core in SOC which has both master and slave interfaces, the AXI compliant signals of the IP blocks are packetized by another interface. The network interface injects or absorbs the packets that leave or arrives in functional blocks of the design and also helps in packetizing/de-packetizing the signals that come or reaches AXI compatible cores in packet form. As per the consideration, generally, data is coded at the source interface and is decoded in destination interface [2, 3, 6].

## 2.6  Ethernet

Ethernet is a medium speed bus. It uses a broadcast topology with baseband signaling and a control method called CSMA/CD to transmit data. Physical layer in Ethernet is for LAN cabling and data link layer. MAC consists of physical hardware address of source and destination.

Payload is actual data and is 46–1500 bytes. It consists of preamble, source, and destination address, data, type. Analog blocks like AGC, FGC, FGA, DDFs are connected at the upper layer of Ethernet which are used for communicating with outside environment and these are the which cause crosstalk (how).

Result of experiment depends on when compared with other present research techniques [5]. Iterations required for timing closure and fixing the cross talk violations is one of the most important aspects while considering the effectiveness of current research technique implied. Results to fix the cross talk violations by different adopted techniques have been presented over here. For analysis of violations and fixing, those have been carefully encountered by implying various steps as mentioned earlier for each successive iteration.

The adopted methodology produces a good outcome in terms of timing closure while dealing with cross talk violations and number of iterations used during the experiment.

# 3    Methodology

Fixing techniques for cross talk violations:

Fixing violations without touching the clock tree is the best way. If the violation is caused due to a clock net which is aggressor, in that case the victim net should be re-routed with techniques like extra spacing or using repeaters can be inserted for reduction of the effect of coupling capacitance.

Reduction in number of iterations is also considered as a huge challenge in fixing cross talk violations. First setup and then hold violation fixing have been considered in this research as hold fixing is much easier compared to setup fixing by adding some extra buffers in the design [7].

Several iterations should be carried out for fixing the setup and hold violations for cross talk.

## 3.1    Cross Talk Coupling with STA

Cross talk coupling with static timing analysis is the first step for cross talk violation fixing from where a clear idea can be obtained regarding crosstalk violations.

## 3.2    Identify Aggressor and Victim

A script is generally used for parsing the reports, which contain all the crosstalk delay information timing generated by synopsis primetime as manual check is very difficult for huge number of cross talk violations.

To known the information about aggressor and victim nets for each violating timing path all the timing reports are considered which are automatically parsed by the script.

Coupling capacitance for the nets in the design is one of the most important inputs.

Delay effect caused by net coupling is useful for getting the details of aggressor and victim. Based on the clock domain, aggressors and victims are identified in each path, which are mainly responsible for multiple violations. Unique aggressors

and their victim nets are also identified which helps in reducing the number of nets
to be considered by a reasonable amount.

### 3.3   Filter Static Paths and Clock Isolation

Filter the static paths out of aggressors and victims. Static paths are fixed and its
value does not change. For each operating modes, static paths can be identified.
Reset signals and boot up configuration registers can cause static paths. Static paths
can safely be removed from aggressor list. Since during normal operation, static
paths do not alter. Therefore, there is no possibility of generation of cross talk with
respect to other paths. Fake cross talk violations can be filtered by identification of
static nets.

Minimization of existing clock tree and fixing the cross talk violations is one of the
most important aspects. The timing of the design is affected for any by small changes
in clock tree and might have to go through several iterations for timing closure. Cross
talk delay fixing is performed without any change in clock network [8].

The timing report review helps in isolating clock network nets from the aggressors
and victims. Prime time is used for clock network paths identification by using the
following command:

"report_ timing –from < launch > -to < capture > -path_ type full_ clock_
expanded"

Violating paths in expanded mode can be observed in this case to perform further
actions on them.

### 3.4   Aggressor and Victim Re-route Mechanism

Using increased spacing criteria routing software should take care of re-routing of
aggressor nets. This approach helps in reducing cross talk violations by fixing
aggressors and victims. The other nets should not be affected while doing this
incremental routing. This approach helps in reducing the violations to a greater
extent. Fixing the violations with extra routing area is one of the best solutions to
reduce the number of violations and hence reduces the number of iterations as well.

### 3.5   Victim Resizing and Splitting

Resizing the inverters and buffers present in the design which drive the victim nets
to allow sufficient driving strength for reduction of coupling effects from aggressor
nets. Cells that drive the aggressor paths can be considered for downsizing to

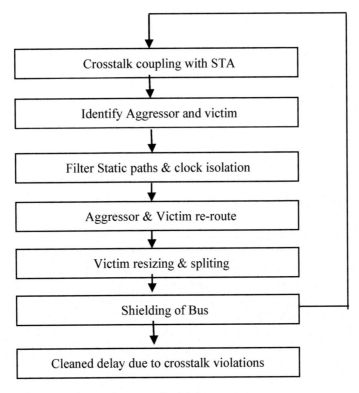

**Fig. 2** Flowchart with steps to fix cross talk violations

reduce their effect on the victim. If aggressor and the victim are mutually coupled, this approach cannot be used for such a scenario as it is.

Combination of spacing and insertion of repeater after nets breaking can be applied to tackle such situations. With the increased net size, the coupling capacitance also increases. Repeaters have a major role when longer interconnect nets are separated in smaller interconnects. Repeaters should not be the cause of the creation of new aggressors or victims on other nearby nets (Fig. 2).

## 3.6 Bus Shielding

Longer interconnect bus signals are to be shielded by placing Ground (VSS) signals on both the sides of the bus. Shielding mechanism helps in avoiding a huge amount of cross talk. For reducing bus violations, shielding has great importance. This approach seems to be a proven success for the experiments run on the design tested [7].

# 4  Result

For interconnect delay calculation

Process Technology 150, 130, 100, 70 nm
Intrinsic delay of device 48.1, 45.4, 39.0, 22.1 ps
1 mm wire 0.053, 0.041, 0.002, 0.041 ns.

## 4.1  Cross Talk Violations Comparison in SOC Between Traditional and Proposed Schemes

Simulation is the process of execution of a model in the software environment. Here simulation, synthesis, and static timing analysis for bridge interface have been carried out along with various protocols or peripherals such as I2C, I2S, UART, and SPI which are the integral parts of SOC [9].

Synopsys Primetime is used to take care of the cross talk analysis. It calculates the delay changes in the nets due to cross-coupling and generates the timing report incorporating these. Crosstalk timing report gives delay change value.

Number of iteration cycles can be declared to calculate the arrival window of the aggressor net. Setup and hold violations reduce with increment in number of iterations.

Fourteen consecutive checks have been performed in the traditional scheme for setup and hold violations (cross talk violation) and as can be seen from the values in

Table 1 Comparison between the proposed and traditional methods

| S. no. | Traditional scheme | | Proposed scheme | |
|---|---|---|---|---|
| | Setup | Hold | Setup | Hold |
| 1 | 410 | 1029 | 410 | 1029 |
| 2 | 317 | 978 | 139 | 615 |
| 3 | 231 | 522 | 120 | 230 |
| 4 | 211 | 425 | 111 | 190 |
| 5 | 150 | 300 | 65 | 98 |
| 6 | 129 | 90 | 16 | 26 |
| 7 | 79 | 76 | 4 | 15 |
| 8 | 52 | 75 | 1 | 4 |
| 9 | 41 | 61 | 0 | 0 |
| 10 | 30 | 42 | 0 | 0 |
| 11 | 3 | 7 | 0 | 0 |
| 12 | 0 | 0 | 0 | 0 |
| 13 | 0 | 0 | 0 | 0 |
| 14 | 0 | 0 | 0 | 0 |

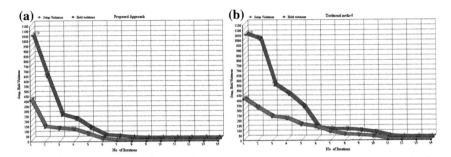

**Fig. 3** Comparison between the proposed **a** and the traditional **b** approaches for setup and hold violations

the table mentioned below the setup and hold violations are brought down to zero (Table 1), (Fig. 3).

Setup and Hold violations can be completely removed in less number of iterations by proposed schemes as shown in the graph where as number of iterations required to remove these violations is more in traditional scheme compared to the proposed scheme.

## 5 Conclusion

A very efficient technique has been presented here that encounters loss of energy and delay issues that are associated with crosstalk noise which is encountered in today's on chip buses implemented in System on chip (SOC). One of the most critical problems encountered in system on chip nowadays is signal integrity.

Crosstalk delay is the main cause of signal integrity issues in advanced process technology nodes. Fundamentals of the problem are charted out in the initial stage of this report. The effects of these crosstalk violations in the deep sub-micron designs have been explained in details as well.

The details about a complex system on chip design have been taken into consideration for the experiments are explained. Innovative encoding methodology and procedures to address the problem due to crosstalk violations have been considered here. Best ways of addressing the cross talk issue in system on chip are avoiding the possibilities of the cross talk and fixing the existing cross talk issues [10]. Both these methods are performed for this research with immense care by implying the proposed technique for cross talk and noise reduction in system on chip.

# References

1. Pranjal Patil, Pable, S.D., Crosstalk Delay Analysis in Very Deep Submicron VLSI Circuits, IRJET, 3(5), (2016), 1695–1698.
2. Osborne, S., Erdogan, A.T., Arslan, T., and Robinson, D.: 'Bus encoding architecture for low-power implementation of an AMBA-based SoC platform', IEE Proc., Comput. Digit. Tech., 149 (4), (2002), 152–156.
3. Sotiriadis, P.P., and Chandrakasan, A.: 'Bus energy minimization by transition pattern coding (TPC) in deep sub-micron technologies'. Computer Aided Design, ICCAD-2000, IEEE/ACM Int. Conf., San Jose, CA, USA, (2000), 322–327.
4. Khan,. Z., Arslan, T. and Erdogan, A.T., Low power system on chip bus encoding scheme with crosstalk noise reduction capability, IEE Proc. -Comput. Digit. Tech., Vol. 153(2), (2006).
5. Amitav Halder, Arunendra Tomar, Umesh Pratap, Arun Jain, Method for Booting ARM Based Multi-Core SOCs, (Freescale Semiconductor India Pvt. Ltd.).
6. Kim, K.-W., Baek, K.-H., Shanbhag, N., Liu, C.L., and Kang, S.-M.: 'Coupling-driven signal encoding scheme for low-power interface design'. Computer Aided Design, ICCAD-2000. IEEE/ACM Int. Conf., San Jose, CA, USA, (2000), 318–3216. Sotiriadis, P.P., and Chandrakasan, A.: 'Low power bus coding techniques considering inter-wire capacitances'. Proc. IEEE Custom Integrated Circuits Conf., CICC 2000, Orlando, FL, USA, (2000), 507–510.
7. Dennis Sylvester and Kurt Keutzer, Getting to the Bottom of Deep Submicron II: A Global Wiring Paradigm, SNUG Europe (2001).
8. Bijan Kiani and Anthony Hill, Static crosstalk analysis assures silicon success, EE Times, (2002).
9. Franzini, B., Forzan, C., Pandini, D., Scandolara, P. and Dal Fabbro, A. "Crosstalk Aware Static Timing Analysis Environment", ISQED, (2000).
10. Satyendra R. Datla, James Song, Stewart Shankel, Kaijian Shi and Yuanqiao zheng, "Overcoming challenges in STA for 3G wireless application", Comms Design Journal, May 2003, EE Times Journal, (2003).

# Design and Simulation of Boost Converter for Correction of Power Factor and THD Reduction

Renu Kadali and Srinivasa Rao Jalluri

**Abstract** In the present scenario, there is a steady increase in the usage of electronic equipment. During power conditioning, non-sinusoidal line currents are being drawn by these equipment, because of the presence of nonlinear elements. Non-sinusoidal currents produce harmonics and lead to distortion in line voltage waveform. Passive filters can be used, but they are not optimal in terms of cost, size and weight. In this chapter, Boost Converter with Active PFC is designed to obtain the power factor near to unity with less Total Harmonic Distortion (THD) and reduced output voltage ripple. Design is simulated and verified using MATLAB/Simulink.

**Keywords** Power factor · Power factor correction · Boost converter
Hysteresis current control · THD · Voltage ripple

## 1 Introduction

In modern power systems, there is a steady increase of nonlinear loads [1]. During power conditioning, an AC-to-DC converter draws a harmonic rich AC line current because it mainly consists of a passive diode bridge and large DC link capacitors [2]. High level of harmonics implies a low power factor, electromagnetic compatibility problems [3], lack of output voltage regulation [2]. Thus, making the line current to follow the line voltage provides higher efficiency with improved power factor and lower THD [4].

R. Kadali (✉) · S. R. Jalluri
Department of Electrical and Electronics Engineering, VNR VJIET, Hyderabad, India
e-mail: renukadali10@gmail.com

S. R. Jalluri
e-mail: srinivasarao_j@vnrvjiet.in

© Springer Nature Singapore Pte Ltd. 2018
J. Anguera et al. (eds.), *Microelectronics, Electromagnetics and Telecommunications*, Lecture Notes in Electrical Engineering 471,
https://doi.org/10.1007/978-981-10-7329-8_37

DC–DC converters can be used for PFC and to regulate the output voltage [1]. Of all DC–DC converters, Boost converters have gained significant importance, especially when they are used in Continuous Conduction Mode (CCM) [4] as it reduces the ripple in current and minimizes the losses. Also due to their simple structure with less number of power semiconductor devices, less passive components [2] they are being widely used for PFC. Hysteresis Current Control (HCC) is used as PWM technique. It is one of the widely used PWM techniques because of its simple implementation, increased system stability, fast response, and low distortion in supply current waveform. In this chapter, HCC technique is employed for shaping the input current waveform to be as sinusoidal as possible and controlling the output voltage of converter.

This chapter is organized as follows: Sect. 2 introduces Boost Converter with design of inductor and capacitor. Nonlinear loads and power factor correction techniques are presented in Sect. 3. Section 4 mentions about the PID controller. In Sect. 5, simulation results are presented and analyzed.

## 2 Boost Converter

Boost converter is one of the DC–DC converter topologies and it is used to increase the level of DC voltage. It consists of four elements namely diode, switch (electronic), inductor, and an output capacitor [5]. The design of inductor and capacitor plays a crucial role in the operation of boost converter. The boost converter is operable in two modes but the CCM is preferred in high and medium power applications [4]. The boost converter operation is explained with the concept of averaging. The output and input voltage of a boost converter are related by the following equation.

$$\frac{V_o}{V_i} = \frac{1}{1-D} \tag{1}$$

### 2.1 Design of Inductor

In order to operate the boost converter in CCM, the inductor value should be greater than that of the critical value as given below,

$$L_c \geq \frac{V_{in}*(V_{out} - V_{in})}{\Delta I_l * f_s * V_{out}} \tag{2}$$

where

L$_c$    Critical inductance
V$_{in}$    Input voltage
V$_{out}$    Output voltage
ΔI$_l$    Inductor current ripple
f$_s$    Switching frequency

Large inductance values increase the start-up time slightly whereas small inductance values allow the coil current to ramp up to higher levels before the switch turns off. Thus, inductors with a ferrite core or equivalent are recommended.

## 2.2 Design of Capacitor

$$C_b \geq \frac{V_{out} * D}{\Delta V_{out} * f_s * R} \tag{3}$$

where

C$_b$    Base capacitance
D    Duty cycle
R    Load resistance
ΔV$_{out}$    Output voltage ripple.

## 3 Nonlinear Loads and Power Factor Correction

Power factor is the utilization factor of the power from grid. Theoretically, it is the proportion of the real power to apparent power and is in the range of 0–1.

$$\text{Power Factor} = \frac{\text{Real power}}{\text{Apparent Power}} \tag{4}$$

For pure sinusoidal voltage and current waveforms

$$\text{Power Factor} = \cos \varnothing \tag{5}$$

where "cos $\varnothing$" is the displacement factor between voltage and current.

But for nonlinear loads (i.e.,) for sinusoidal line voltage and non-sinusoidal supply line current waveform the power factor can be described as [1].

$$\text{Power Factor} = \frac{V_{rms} * I_{1rms}}{V_{rms} * I_{rms}} \cos \emptyset$$

$$= \frac{I_{1rms}}{I_{rms}} \cos \emptyset \tag{6}$$

$$= DPF * \cos \emptyset$$

$$= \frac{I_{1rms}}{\sqrt{I_{1rms}^2 + I_{2rms}^2 + I_{3rms}^2 + \cdots + I_{nrms}^2}} \cos \emptyset$$

where DPF refers to distortion power factor. Here n represents nth order harmonic. Total Harmonic Distortion (THD) factor is

$$THD_i = \frac{\sqrt{I_{2rms}^2 + I_{3rms}^2 + \cdots + I_{nrms}^2}}{I_{1rms}^2} \tag{7}$$

Hence,

$$DPF = \frac{1}{\sqrt{1 + (THD_i)^2}} \tag{8}$$

Hence, a high power factor is possible with a reduced amount of harmonic content.

Thus, power factor improvement can be regarded as the reduction of the line current harmonics. The major objective of the thesis is to correct the power factor, i.e., maintaining a lesser angle difference between the input voltage and current with lesser THD level.

## 3.1 Power Factor Correction

To gain a high power factor, different power factor correction techniques are introduced [4] and they are broadly classified into two types (Fig. 1).

(1) Passive PFC
(2) Active PFC

Active switches in combination with passive element employed in "Active PFC Approach", help in improving the line current wave shape as well as obtain controlled output voltage. To achieve this, a DC–DC converter especially a Boost Converter is used and is made to work at high frequency to align the shape of the line current waveform to sinusoidal. This new topology is mainly used for research on AC–DC PFC pre-regulator system for qualitative improvement of power.

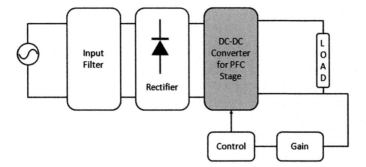

**Fig. 1** Schematic diagram of active PFC technique

## 3.2 Hysteresis Current Control Technique

For implementing the closed-loop control, few quantities such as input voltage, output voltage, input current of DC–DC converter are sensed. In outer most loops, the converter output voltage is compared with the reference voltage value. The difference of voltage, i.e., error is passed through a PI controller. Two sinusoidal current references are generated in HCC based on the minimum and maximum boundary limits of current (Fig. 2).

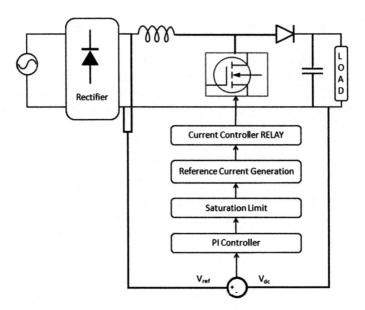

**Fig. 2** Block diagram of the proposed hysteresis current control scheme

## 4  PID Controller

Proportional–Integral–Derivative (PID) control is one of the commonly used control strategies in industry. In this chapter, PI control is adopted and its tuning is performed using trial and check method.

I term is set to zero initially and the P gain is increased gradually. Once we reach our response with some steady-state error, start increasing I term. The integral term is adjusted such that we achieve a minimal steady-state error.

## 5  Simulations and Results

### 5.1  Simulation Without Boost Converter

In this model, a diode rectifier is considered to be inducing the nonlinear properties into the R-load thus making the load nonlinear. The AC supply is given to the full bridge diode rectifier feeding R-load. Output capacitor is provided as a filter in order to ensure a ripple-free DC output voltage. If there is no capacitor present, then the voltage may fall to zero due to the rectification process and this can cause serious effects on the load side (Table 1).

### 5.2  Simulation with Boost Converter

The simulation of the diode bridge rectifier including boost converter and a suitable control mechanism is shown below. In this circuit, hysteresis current control mechanism is used to reduce the error between input and output. This method is one of the current mode control techniques where in simple design and more reliability can be promised (Tables 2 and 3).

### 5.3  Comparison of Results With and Without Control Technique

**Table 1**  Circuit element specifications

| S. no | Parameter | Value |
| --- | --- | --- |
| 1 | Supply voltage | 325(P-P) volts-50 Hz, AC supply |
| 2 | Capacitor | 400 μF |
| 3 | Resistor | 100 Ω |

**Table 2** Circuit element specifications

| S. no | Parameter | Value |
|-------|-----------|-------|
| 1 | Supply voltage | 32 V(P-P) volts-50 Hz, AC supply |
| 2 | Capacitor | 4000 μF |
| 3 | Resistor | 100 Ω |
| 4 | Boost inductor | 2 mH |
| 5 | Source impedance | 0.5 mH |

**Table 3** Various parameters used in HCC block

| Parameter | Value |
|-----------|-------|
| Proportional value | 3 |
| Integral value | 2 |
| Relay setting | Max = 10; Min = −10 |
| Saturation setting | Max = 0.9; Min = −0.9 |
| Gain | 1/325 |

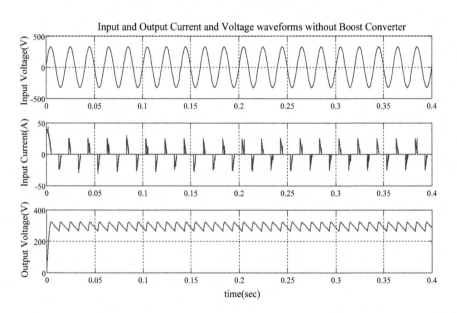

**Fig. 3** Input voltage, current and output voltage waveforms without boost converter

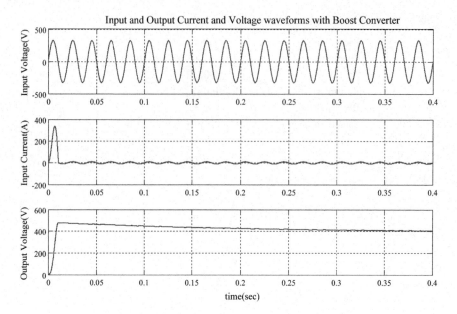

**Fig. 4** Input and output current and voltage waveforms with boost converter

# 6 Conclusions

For improving the utilization ratio of grid power, it is essential to minimize the line current harmonics. PFC strategy helps in reducing the effects of nonlinear loads/ elements that are responsible for making power factor less than one. For this, DC–DC Boost converter topology is operated at high frequency to align the shape the line current waveform to be as sinusoidal as possible and also simulation results depict the same (Figs. 3 and 4).

In "Active PFC approach", the Boost PFC converter is taken with suitable switching control strategy which improves the system stability irrespective of system loading conditions.

The circuit with boost converter topology employing hysteresis current mode control is simulated in MATLAB/Simulink (R2013a) and is compared with the simulation in which no control mechanism is utilized and the results are formulated in Table 4.

**Table 4** Comparison of various parameters

| Parameter | Without PFC | With active PFC |
|---|---|---|
| Power factor | 0.73 | 0.99987 |
| Total harmonic reduction (%) | 132 | 6.2 |
| Output ripple (volts) | 50 | 3 |

# References

1. Smita Rani Patra, Tanmoy Roy Choudhury and Byamakesh Nayak, "Comparative Analysis of Boost and Buck-Boost Converter for Power Factor Correction using Hysteresis Band Current Control", 1st IEEE International Conference on Power Electronics. Intelligent Control and Energy Systems (ICPEICES-2016)
2. Ayan Mallik, Bryan Faulkner and Alireza Khaligh, "Control of a single-stage three-phase boost power factor correction rectifier", Applied Power Electronics Conference and Exposition (APEC), 2016 IEEE
3. Arthur G. Bartsch, Christian J. Meirinho, Yales R. de Novaes, Mariana S. M. Cavalca and José de Oliveira, "Analysis of predictive control for boost converter in power factor correction application", 2016 12th IEEE International Conference on Industry Applications (INDUSCON)
4. Mukhzan Mobeen Ali, Sardar Shazali Sikander, Usman Ali and Arbab Waleed, "An active Power Factor Correction technique for bridgeless boost AC-DC converter", 2016 International Conference on Intelligent Systems Engineering (ICISE), 23 May 2016
5. B. M. Hasaneen and Adel A. Elbaset Mohammed "Design and Simulation of DC-DC Boost Converter", Power System Conference, 2008. MEPCON 2008. 12th International Middle-East, 15 July 2008

# Design and Simulation of High-Performance 2D Convolution Architecture

**V. S. Vishal and B. S. Kariyappa**

**Abstract** Two-dimensional 2-D convolution is always computationally intensive and memory-intensive process. There are many architectures to handle the computational load of 2D convolution constrained by its throughput, area requirement and memory bandwidth. All the architecture tried to optimize one of the three parameters keeping the others unbounded. This paper presents a new design of 2D convolution with new features to improve throughput, efficient data reuse to keep the area minimum. The design uses the principle of Double Data Rate register to use two clocks of the same period but opposite phase. Unlike legacy architecture using multiple window convolutions for throughput, the proposed design uses only 2 windows which triggers on the positive edge of two clocks, one after the other giving the throughput of 2 pixel/clock while keeping the area and memory bandwidth minimum.

**Keywords** Field-programmable gate array (FPGA) · Two-dimensional (2D) convolution · Shift-variant convolver · Moving window · Multi-window(MW) Double date rate (DDR) · Full buffering (FB) · Partial buffering (PB)

## 1 Introduction

Two-dimensional (2D) convolution is computationally intensive and memory-intensive process. For an R × S convolution filter (kernel) requires R × S multiplication with corresponding data elements, R × S − 1 additions with neighbors and R × S access to the input data from the memory. The main constraint vectors in designing a 2D convolution architecture are throughput, memory bandwidth, and area. Throughput is the number of convolved pixels/clk. Memory

V. S. Vishal (✉) · B. S. Kariyappa
R V College of Engineering, Bangalore 560059, Karnataka, India
e-mail: vishalsuryanarayanan@gmail.com

B. S. Kariyappa
e-mail: kariyappabs@rvce.edu.in

© Springer Nature Singapore Pte Ltd. 2018
J. Anguera et al. (eds.), *Microelectronics, Electromagnetics
and Telecommunications*, Lecture Notes in Electrical Engineering 471,
https://doi.org/10.1007/978-981-10-7329-8_38

bandwidth defines the number of input data elements (pixels) required for one-convolved output data. Area is the sum of the number of multipliers, adders, FIFO (First in First Out), and shift registers.

There are many architectures available for 2D convolutions but these architectures concentrate on optimizing only one of these three parameters. Delay line 2D convolution architecture [1–3] (also called Full buffering) uses a long FIFO and delay lines to hold input data elements giving a throughput of 1 pixel/clk with a constant memory bandwidth of one pixel, but fixed input data length prevents the design to process different data length. Large input length increases the size of the FIFO as well as the delay lines which reflect in a large area. Implementing 2D convolution using 1D convolution architectures requires large register count to hold the values of the partial product [4, 5]. When the size of the convolution kernel increases the number of partial products required also increases along with other requirements in implementing the large kernel. SWPB (Single Window Partial Buffering) [1, 6, 7] uses separate FIFO for each row of the convolution kernel giving a convolved output on every clock cycle. SWPB requires large memory bandwidth usage with respect to convolved pixel but the throughput is always one pixel/clock. The main disadvantage of SWPB is it's poor data management. MWPB (Multi-Window Partial Buffering) [1, 8, 9] uses multiple windows to store the input data values. For an R × S kernel, number of windows required is R + S − 1. MWPB reduces memory bandwidth by reducing the data access only one in every S cycles. R × S convolution kernel requires R + S − 1 SIPO (Serial In Parallel Out) shift register. FSM (Finite State Machine) select a group of SIPO in such way to process the data for a pixel. This architecture reduces the memory bandwidth by reading the data only once and stores in temporary space, but when the size of these filter increases, resources required to stores the temporary values of multiple windows is exponentially increased. So, implementation of area constraints is difficult.

This paper presents a new design evolved by adopting the pros of legacy architecture SWPB and MWPB. Section 1 gives an introduction to 2D convolution and architecture. Section 2 describes the proposed design, design challenges, and fixes. Section 3 contains performance analysis and comparison with legacy designs. Section 4 is conclusion and future scopes and possible modifications.

## 2  Proposed Design

### 2.1  Background

Mathematical relation of 2D convolution is given by Eq. 1, where i, j are the coordinates of the kernel in the input data matrix and $\mathbf{P}'_{ij}$ is the convolved pixel at coordinate (i, j) in the output matrix.

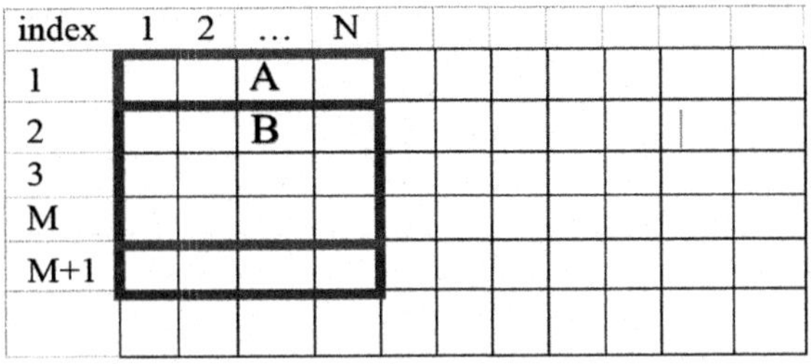

**Fig. 1** Convolution filter of size M × N

| index | 1 | 2 | ... | N | | | | | | |
|---|---|---|---|---|---|---|---|---|---|---|
| 1 | | | A | | | | | | | |
| 2 | | | B | | | | | | | |
| 3 | | | | | | | | | | |
| M | | | | | | | | | | |
| M+1 | | | | | | | | | | |
| | | | | | | | | | | |

$$P'_{ij} = \sum_{-R/2}^{R/2} \sum_{-S/2}^{S/2} Pij * Wij \tag{1}$$

For R × S convolution kernel moving in a horizontal direction (Column wise), requires only R new data values for a new convolved data. Remaining 80% of the input data required are already available to the convolution kernel from the FIFO. Similarly, when filter moving in the vertical direction (Row wise) only S new data values are required. This design moves in the horizontal direction to compute 2 convolved data. During the high level of the clock, design uses values belonging to M × N matrix (region A, i.e., red) and the low level uses values belonging to (M + 1) × N matrix (region B, i.e., blue). M and N are position of R × S kernel in the data matrix. In this way, on every clock period 2 convolved data are obtained giving the throughput of 2 pixel/clk. Figure 1 illustrates the process. During the positive half cycle, data values under region A (red) is convolved, while on the negative half cycle region B values (blue) are used. The overlapping data values of region A and region B is reused. Data reuse not only reduces memory bandwidth but also increases the throughput by twice.

## 2.2 Architecture

Figure 2 shows the convolution engine of the proposed design. For an R × S convolution kernel, the proposed design uses R + 1 SIPO (Serial In Parallel Out) and SISO (Serial In Serial Out) shift registers. SISO shift register act as a buffer between the memory and convolution engine. SISO shift register provides a constant data rate to the convolution engine. Size of the SISO is arbitrary, usually takes size as the column size of the filter and data width of FIFO is 8 bits. SIPO is implemented inside the convolution engine provides data to its corresponding multiplier depending upon the trigger signal. On the positive half cycle, SIPO(1) to SIPO(R) provides data to the multiplier circuit, while on the negative half data from SIPO(2) to SIPO(R + 1) is used. The output of SIPO is 128-bit width (16 Bytes) which is multiplexed with the adjacent SIPO for data reuse.

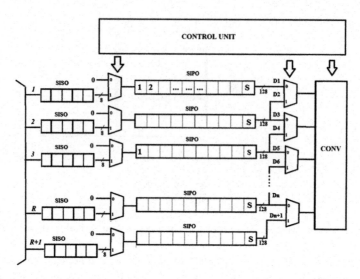

**Fig. 2** Convolution engine

On every positive edge of the clock all the shift registers (both SISO and SIPO) are pushed. The multiplier is designed in such a way to complete the process before the next data arrival. Control unit provides necessary control signals to the data path. Control unit can enable/disable the SISO data path depending upon the filter size. Increased throughput requires a multi-constant multiplier module which can multiply the data input with one of the 2 values selected by the clock signal using a 2:1 multiplexer. When the clk is HIGH multiplier uses the first constant, while on the LOW clk level uses the second constant value. If a multiplier is disabled, output of the multiplier will not be considered for convolution.

## 2.3 Design Challenges and Fixes

The proposed design gives a throughput of 2 pixel/clock. This is achieved by triggering the combination circuit on both positive and negative level of the clk. The data arrival time of this overlapping timing path (combinational logic) should be always less than the half the clock period of the new clock for flip-flop setup check and should be always greater than flip-flop hold time for hold check. To satisfy the required timing constraints, proper design of the combinational logic is crucial. Combinational logic is made of a number of multiplier blocks, adders, and multiplexers. To achieve minimum delay as possible, design uses Vedic multiplier based on Urdhva Tiryagbhyam [10–12] and Kogge stone adders. The Vedic multiplier is known for its fast computation with slightly large area compared to others. For a given throughput of 2 pixel/clock, use of Vedic multiplier can reduce the number of multipliers required to half the value of that required for the legacy

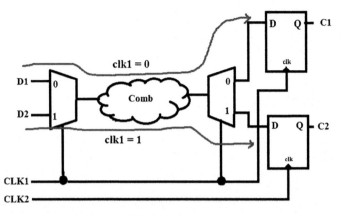

**Fig. 3** Data steering logic inside convolution engine

designs. Less area multiplier uses multiple stages to compute the result. This will degrade the throughput.

Convolution engine uses two new generated clocks clk1 and clk2. Clock clk1 and clk2 are related by clk2 = ~clk1 and clk2 is having same period but a phase shift of 90. Period of the clk1 and clk2 is constrained in such a way that no timing violation is present considering worst-case delay of the combinational circuit. In Fig. 3, data steering logic of the proposed design is presented. Data D1 and D2 are available at the input of the multiplexer (Inputs channel 0 and channel 1 are selected by select lines) on every positive edge of clk1 = clk. When clk1 is high, select lines of both mux selects the 1st channel. Combinational logic computes the data D2 and output is captured by the positive edge of clk2 at FFB (Flip-Flop B). Similarly, during the clk1 LOW 0th channel of both the multiplexer is activated. Data D1 already launched by the positive edge of clk1 is captured at the next positive edge of the same clk1 at FFA (Flip-Flop A), i.e., When clk1 = 0 (negative half), output of the combination logic will be captured by FFA. So, the operation can be summarized as on every positive edge of clk1 input data D1 and D2 are available. On the positive half cycle of the clock clk1, D2 is processed and captured by clk2 and on negative half of clk1 D1 is processed and captured by next positive edge of the clk1. This is possible because clk2 = ~clk1. The operation can be further simplified as data launched by the positive edge of clk1 is captured by the positive edge of clk2 and data launched by the positive edge of clk2 is captured by the positive edge of clk1. Path delay of the intermediate combinational circuit should be less the data required time of the both FFA (Tclk2 − Tsetup A) and FFB (Tclk2 − Tsetup B). Clocks clk1 and clk2 is having the same period with clock width large enough to accommodate Tcomb and Setup time of FF (Flip-Flop). Following equations should be passed for the design to be timing violation free

a) Tcomb < Tclk2 − Tsetup A & Tcomb < Tclk1 − Tsetup B

c) Tcomb > Thold A & Tcomb > Thold B

# 3 Performance Analysis

This section compares the proposed design with legacy architectures like DL, SWPB, and MWPB. When choosing the architecture for FPGA based 2D convolution, the cost is a function of different parameters such as throughput in terms of pixels, area utilization in terms of FF, and memory bandwidth. Table 1 gives a generalized comparison of various architectures. Throughput is the number of output pixel data available on one clock cycle. Conv register is the number of registers required for the convolution engine. These register stores the input data required for each filter values. Buffering registers are the temp storage of data between memory and convolution engine. Buffer bandwidth also called memory bandwidth is the number of input data pixels required for one output pixel. The comparison is done on a filter of size $R \times S$ on an input matrix of size $M \times N$ such that $R <= M$ and $S <= N$. $P = 8$ is taken as the depth of SISO.

Throughput of the proposed design is higher than any of the legacy designs. This is because, computation is done on both positive and negative level of the clock. The legacy design may have a higher frequency of operation. For a stipulated time, the net throughput of an architecture is given by Eq. 2.

$$\text{Throughput}_{\text{NET}} = \text{Throughput}_{\text{DESIGN}} \left[ \frac{Time\ Slot}{Clock\ Period} \right] \quad (2)$$

If the clock period of the legacy design is 10 ns and proposed design is 12 ns. In the time slot of 120 ns legacy design can give only 12 pixels. But the proposed

**Table 1** Specification of different 2D convolution architectures

| Design | Area utilization in FF | | Buffer bandwidth (pixels/ clock) | Case study | | Throughput (pixel/ clock) |
|---|---|---|---|---|---|---|
| | Conv register (window) | Buffering registers | | Area (ff count) | Buffer bandwidth (pixel/ clock) | |
| DL | $R \times S$ | $(R - 1) \times (N - S) + P$ | 1 | 16488 | 1 | 1 |
| SWPB | $R \times S$ | $R \times P$ | R | 520 | 5 | 1 |
| MWPB | $(R + S - 1) \times S$ | $(R + S - 1) \times P$ | $(R + S - 1)/S$ | 936 | 1.8 | 1 |
| Proposed | $(R + 1) \times S$ | $(R + 1) \times P$ | $(R + 1)/2$ | 624 | 3 | 2 |

design can give 20 pixels. In the worst case, if the time period of the proposed design is 20 ns, net throughput will be same as the legacy designs. So, the net throughput of the proposed design is always greater than or equal to the legacy designs. The area is measured in terms of no. of Flip-Flops (FF) used as FF covers a major part of all convolution architectures. Number of FF is the sum of the size of all shift registers multiplied by 8 (8-bit arithmetic is used). In Table 1, DL (Delayed Line) takes the largest area over any design and can be ignored in the comparison. Among the SWPB, MWPB and proposed design, SWPB takes lesser area than others. But Throughput of SWPB is lesser than the proposed design as only one computation is done on every cycle. Data reusability in the SWPB is the lowest among the other designs which helps in reducing the FF count. The number of conv registers is $R \times S$. In MWPB data reusability is maximum when compared with proposed design and SWPB. This is achieved by increasing the conv register count by $R + S - 1$. But when the size of the filter increases MWPB design takes a large amount of resources. The proposed design comes in between SWPB and MWPB taking only one additional shift register unlike MWPB having $R + S - 1$ additional shift registers. As the throughput of the proposed design is higher than SWPB for almost same area requirement proposed design can be preferred over others.

Memory bandwidth also called buffer bandwidth is the number of new input data pixels required for new convolved data. As shown in Table 1, DL has the smallest buffer bandwidth of 1. Since the throughput of DL is only 1, to achieve the throughput of 2 requires more resources than the old design. As area requirement for even one throughput is very much larger than other designs, DL design is not preferred. Moreover, the current data width of the I/O is greater than 8 bit wide. So, using a large data width bus to transfer only one-byte data per cycle is highly inefficient, because remaining data lines is left unused. Buffer bandwidth of SWPB for an $R \times S$ filter is R. That is, R new pixels are required for a new output data. As the size of the filter increases, SWPB requires R pixels for just a single output becomes inefficient. Data currently available in the SWPB registers is not reused efficiently. MWPB have less buffer bandwidth when compared to the SWPB and the proposed design. This is achieved by compromising with the area. In MWPB data is accessed only in every S clocks. This gives the buffer bandwidth as $(R + S - 1)/S$.

The proposed design gives a throughput of 2 pixel for every $(R + 1)$ input data. Proposed design can be preferred over SWPB as it gives 2 pixels for every $(R + 1)$ input pixels whereas SWPB can give only 1 pixel per R input pixels. Figure 4 shows the simulation results of the proposed design for convolution filter of size $7 \times 7$. The global clock clk generates two generated clock signals clk_1 and clk_2 of equal period but opposite phase. On every positive edge of the clk_1, convolution engine is triggered and data elements are multiplied by its corresponding filter values. All the multiplied values are added to give convolved pixel. In Fig. 4, red waveforms are the set of sums of all convolved pixel in a row. Similarly, on the positive edge of clk_2, another set of values are processed (Blue). In others words, positive edge of clk_1 compute one set of data(red) and the negative edge of clk_1 computes the second set of data elements. In this way, two pixels are processed in a

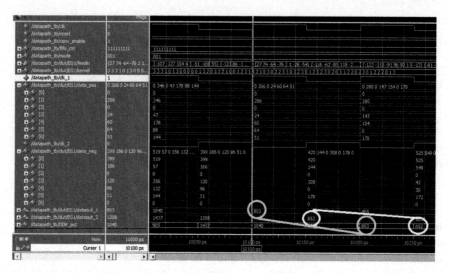

**Fig. 4** Simulation waveform of the convolution

single clock period. DDR (Double Date Rate) register output DDR_out combines both dataout_1 and dataout_2 to give 2 convolved pixels per period of clk. The two convolved pixels are highlighted in the Fig. 4. Since the throughput is increased to 2 pixels in a single clock period, proposed design outperforms the legacy designs.

For a case study, we consider an image of size $512 \times 512$ and Kernel size $5 \times 5$. Convolution IP will get data from the external memory and stored back to external memory after convolution. SISO (FIFO) of depth $P = 8$ is used for buffering the pixel data to the SIPO shift register. Size of the SIPO filter is 5. Number of SIPO is 6. First 5 SIPO, i.e., (SIPO(1) to SIPO(5)) is enabled on the positive edge. Next 5 SIPO, i.e., (SIPO(2) to SIPO(6)) is enabled, disabling first SIPO (1) is used for the negative edge. Shift registers and delay lines can be available IP blocks library. In this case, throughput of the proposed is higher than all legacy designs. Buffer bandwidth for proposed design is 3 because the proposed design uses 6 pixels to compute 2 output pixels, whereas SWPB uses 5 pixels to

**Table 2** Resource utilization and output data throughput comparison

| Filter size | DL | | SWPB | | MWPB | | Proposed design | |
|---|---|---|---|---|---|---|---|---|
| | Area (FF count) | Throughput (pixel/ clock) | Area (FF count) | Throughput (pixel/ clock) | Area (FF count) | Throughput (pixel/ clock) | Area (FF count) | Throughput (pixel/ clock) |
| $5 \times 5$ | 16488 | 1 | 520 | 1 | 936 | 1 | 624 | 2 |
| $7 \times 7$ | 24696 | 1 | 840 | 1 | 1560 | 1 | 960 | 2 |
| $11 \times 11$ | 41112 | 1 | 1672 | 1 | 3192 | 1 | 1824 | 2 |
| $13 \times 13$ | 49320 | 1 | 2184 | 1 | 4200 | 1 | 2352 | 2 |

compute only 1 output pixel. Table 2 presents the comparison of different architecture of different values of filter size.

# 4 Conclusion

Two-dimensional convolution is a powerful mathematical tool for all signal processing and analysis. 2D convolution will be used as the backbone of all machine learning algorithms to be used in the near future. Implementation of a high-performance architecture is very essential. In this brief, we have presented a new design for two-dimensional convolution giving twice the throughput of legacy designs keeping the area and memory bandwidth comparable. New buffering strategy and dual edge computation not only increases the efficiency but also keeps the FPGA resources minimum.

# References

1. Zhao-Bin Ma., Yang Yang., Yun-Xia Liu., and Anil Anthony.: Recurrently Decomposable 2-D convolvers for FPGA-Based Digital Image Processing: IEEE transactions on circuits and systems—Vol. 63. (2016)
2. F. J. Toledo-Moreo., J. J. Martínez-Alvarez., J. Garrigós-Guerrero., J. M. Ferrández-Vicente.: FPGA-based architecture for the real-time computation of 2-D convolution with large kernel size, J. Syst. Architect., vol. 58, 8. (2012) 277–285
3. M. Z. Zhang., V. K. Asari.: An efficient multiplier-less architecture for 2-D convolution with quadrant symmetric kernels: VLSI J. vol. 40, 4. (2007) 490–502; Bosi., G. Bois., Y. Savaria.: Reconfigurable pipelined 2-D convolvers for fast digital signal processing, IEEE Trans. VLSI Syst., vol. 7, 3. (1999) 299–308
4. F. Cardells-Tormo., P.L. Molinet.: Area-efficient 2-D shift-variant convolvers for FPGA-based digital image processing: IEEE Trans. Circuits Syst. II Exp. Briefs, vol. 53, 2. (2006)
5. Stefano Di Carlo., Giulio Gambardella., Marco Indaco., Paolo Prinetto.: An area-efficient 2-D convolution implementation on FPGA for space applications: Design and Test Workshop (IDT), IEEE 6th International (2011)
6. M. Z. Zhang, H. T. Ngo, V. K. Asari.: Multiplier-less VLSI architecture for real-time computation of multi-dimensional convolution, Microprocess. Microsyst. vol. 31, no. 1. (2007) 25–37
7. S. Perri, M. Lanuzza, P. Corsonello, G. Cocorullo.: A high-performance fully reconfigurable FPGA-based 2-D convolution processor, Microprocess. Microsyst, vol. 29, (2005) 381–39
8. Gian Domenico Licciardo.: Weighted Partitioning for Fast Multiplier-less Multiple-Constant Convolution Circuit: IEEE transactions on circuits and systems, vol. 64, 1. (2017)
9. H. Zhang., M. Xia., G. Hu.: A multiwindow partial buffering scheme for FPGA-based 2-D convolvers: IEEE Trans. Circuits Syst. II Exp. Briefs, vol. 54, no. 2. (2007)
10. R. Bhaskar, Ganapathi. Hegde, P.R. Vaya.: An efficient hardware model for RSA Encryption system using Vedic mathematics: Procedia Engineering, vol. 30. (2012) 124–128

11. Asmita Havelia.: FPGA implementation of a Vedic Convolution algorithm: International Journal of Engineering research and applications, vol. 2, 1. (2012) 678–884
12. L. Sriraman., T.N. Prabakar.: FPGA implementation of high performance multiplier using squarer: International Journal of Advanced Computer Engineering & Architecture, vol. 2, 2. (2012)

# CRC-Based Hardware Trojan Detection for Improved Hardware Security

N. Mohankumar, M. Jayakumar and M. Nirmala Devi

**Abstract** Several methodologies aim at tackling the issue of Hardware Trojans through the help of a "Golden Reference", which is not always available; thereby arises the need for an efficient method without a Golden Reference. This work involves the detection of Hardware Trojan in a circuit using an improved voting algorithm employing CRC. Two modifications to a conventional voting algorithm are proposed in this chapter along with CRC to improve the detection efficiency. This logic-based detection procedure avoids the requirements of complex pre-processing procedures like segmentation, fingerprinting, thermal imaging, etc., The following proposed modifications (i) incapacitates the bias toward 1s and incorporating CRC for comparison of bit streams and (ii) equal weight of 1 is given as initial weight to all CUTs, which gives better results in voting algorithm. Detection accuracy is found to be around 95.27% based on the detailed analysis with infected and non-infected ISCAS'85 and ISCAS'89 circuits.

**Keywords** Hardware security · Hardware Trojans · CRC · Voting algorithms
Malicious activity · Golden reference

N. Mohankumar (✉) · M. Jayakumar · M. Nirmala Devi
Department of Electronics and Communication Engineering, Amrita School
of Engineering Coimbatore, Amrita Vishwa Vidyapeetham, Amrita University,
Coimbatore 641112, India
e-mail: n_mohankumar@cb.amrita.edu

M. Jayakumar
e-mail: m_jayakumar@cb.amrita.edu

M. Nirmala Devi
e-mail: m_nirmala@cb.amrita.edu

© Springer Nature Singapore Pte Ltd. 2018
J. Anguera et al. (eds.), *Microelectronics, Electromagnetics
and Telecommunications*, Lecture Notes in Electrical Engineering 471,
https://doi.org/10.1007/978-981-10-7329-8_39

381

# 1 Introduction

Present day electronic chips are manufactured by complex, multi-level processes, out of which some processes are outsourced by the manufacturer and done by third-parties. Such practices are widely prevalent as it works out to be more profitable in terms of cost and effort to complete these processes offshore [1, 2] This comes with a potential risk of compromised security; a prominent and adverse example of which includes Hardware Trojan Insertion. A malicious attack on electronics hardware by hardware is generally termed as a Hardware Trojan Horse [3]. A Hardware Trojan is capable of affecting the functionality of the circuit along with causing other actions which may not manifest readily on the circuit, but will turn up at a later time. The design and fabrication stages are more susceptible to Hardware Trojan insertion. The presence of a Trojan and the effects of its attacks may or may not be observable thus making its detection a challenging task [2].

Some of the highly efficient methods employ the help of a "Golden Reference", which is a trustworthy sample of the considered chip. Our work aims at developing a Trojan detection method, which has improved accuracy and consumes less time period for execution without using a Golden Reference. The proposed algorithm uses the logic-based detection method for identifying infected circuits. It involves analyzing the output logic of the circuits. Most ICs are capable of giving digital outputs and analyzing the bit streams provides an avenue for Trojan detection [4]. Errors in bitstreams can be accounted for infections. Hence, it does away with the need for a golden reference. Comparisons of the bit streams obtained from a batch of ICs can be used to arrive at the infected circuit(s). Crafty algorithms which can isolate the infected circuits from the sample of circuits form the crux of this detection method.

Most of the techniques related to the proposed logic-based detection approach are formulated using multiple modular redundancy-based approaches [4–7]. Some of the existing methods employing logic-based detection are:

## 1.1 Simple Majority Voting

The simple voting algorithm polls the binary words for the number of ones and zeroes. The output bit is that which forms the majority and it is considered as the correct bit for that particular bit position. At the end of the polling, an output word, the same size as the input word is obtained [4]. Though the method is simple and fast, it favors the majority which might pose problems when the non-infected CUT/ IPs does not constitute the obvious majority of the samples considered.

## 1.2   Weighted Voting

This method assigns weights to the outputs based on a comparison between bit-streams and the circuit whose output has the highest weight is considered the non-infected one. Here, CUT/IPs are trusted gradually based on the output generated for different input patterns. One of the major advantages of weighted voting is that even when the majority (or all) of the circuits under test is infected, it is possible to identify the infected circuits [4, 6]. The major drawbacks are the algorithm biased towards 1's.

## 1.3   Weighted Voting

In hybrid voting, we compare power consumed and path delays among the ICs. We calculate total dynamic power and cell leakage power for every ICs and compare among them. The Trojan-affected IC will consume more power [6]. Likewise, we calculate net delay arc and cell delay arc for every IC and compare among them and the Trojan-affected IC will have more path delay [8].

## 2   Methodology

Always an odd number of CUT/IPs from different vendors; then outputs or logic of intermediate nodes are observed and validated on a bit-by-bit basis by doing effective voting to produce the correct output [8]. The outline of the approach is shown in Fig. 1.

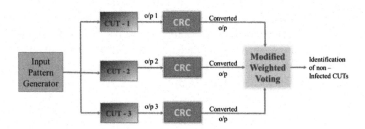

**Fig. 1** Outline of the scheme

## 2.1 Outline of Weighted Voting Algorithm

Voting algorithm selects the higher weighed CUT/IP bit. Each CUT/IP's initial weight counter is zero as mentioned in [4, 6], After each cycle, voting algorithm calculates the sum of weights for each IC cores resulting logical one and sum of weights for each CUT/IP resulting logical zero. Then it compares both the sums. The higher sum will be considered the correct result bit and the voted output, and the lower sum is the infected CUT/IP bit. Weight counters are increased by one for all CUT/IP which produces the same result as clear bit and the weight counters of disagreeing result CUT/IP's are right shifted [6]. It is done to simplify the hardware implementation voting circuit. This technique produces supreme results, especially when using a minimum number of smaller CUT/IPs units. The proposed Trojan detection method consists of the following.

## 2.2 Modified-Weighted Voting Algorithm

The proposed technique improves the existing methods by developing a Modified-Weighted Voting Algorithm which overcomes the bias towards 1s and incorporating Cyclic Redundancy Check (CRC) for comparison of bit streams. The following flowchart in Fig. 2 of modified-weighted voting algorithm clearly exhibits the operation procedure of the proposed scheme. It compares the output bit streams of a sample set of CUTs (Circuit under Threat) and identifies the infected circuit(s) based on a Modified Voting Technique and Cyclic Redundancy Check (CRC). In order to improve the accuracy of the existing weighted voting algorithm, the following modifications are made: equal initialization of weights and removing bias towards 1s by giving equal priority to both 1 and 0 in the bit stream.

Input Pattern Generator: It has certain predefined patterns which cover the critical input patterns. It makes sure every net in the circuit is toggled at least once from 0 to 1 and 1 to 0. Thus enabling us to find the infected circuit using very less number of input patterns. Evaluating the accuracy of the algorithm by processing the outputs of around 126 varieties of benchmark circuits ISCAS'85 and ISCAS'89 circuits with and without Trojan. To efficiently analyze the effect many variations were attempted in terms of Trojan type and location of the Trojan. The following Trojan circuits were designed and used

- NAND gate as Trojan
- Linear-Feedback Shift Register
- 3-bit asynchronous Counter.

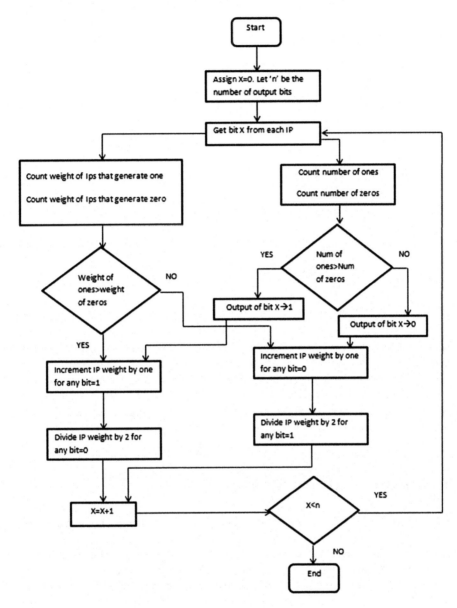

**Fig. 2** Flowchart of modified-weighted voting algorithm

## 2.3 Incorporating CRC in Trojan Detection

CRC is generally used in data transmission to check whether the data transmitted is received properly. It is based on the cyclic addition of redundant data which

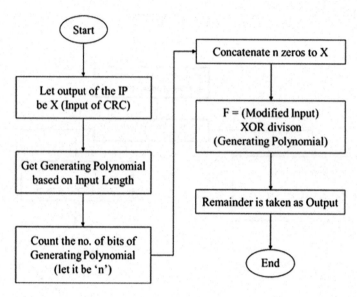

**Fig. 3** Simple outline of CRC

encodes the input to check for errors. It is one of the commonly used error detecting techniques which compares the transmitted and the received data. Here, we use CRC to convert the output from the CUT/IPs into simpler forms which enable easy and improved comparison. The generating polynomial used in CRC computation is selected based on the hamming distance between the outputs and the length of the outputs. A simple outline to understand CRC computation is shown in Fig. 3.

## 3   Implementation and Result Analysis

Each circuit is implemented in three ways: (1) The Direct implementation, (2) NAND implementation, and (3) NOR implementation. Then 4 locations in each circuit are chosen and different trojans are inserted in different locations. Here, we have identified and inserted 3 trojans and 4 locations. Thus, we have 12 types of infected circuits for each implementation. Readings are taken by selecting any one of the 12 infected circuits of one implementation and non-infected circuits of the other two implementations. So we have 36 cases. The resulting weight obtained from the algorithm will be lower for infected circuits.

The sample results in Figs. 4a, b and 5a, b highlights the difference in weights in case of the c6288 and s1488 circuits infected by Trojans in different locations of the circuit. A similar analysis is carried out for c17, c432, c499, c1908, s27, s298, and s1238. The detection accuracy of Modified-Weighted Voting (MWV) in each case is found to be better than Weighted Voting (WV). The Modified Voting algorithm

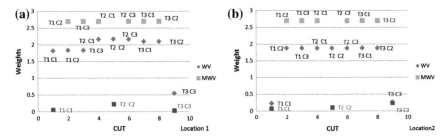

**Fig. 4** **a** Comparison chart showing variation of weight in C6288-location 1, **b** comparison chart showing variation of weight in C6288-location 2

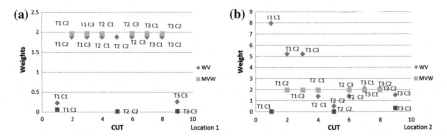

**Fig. 5** **a** Comparison chart showing variation of weight in S1488-location 1, **b** comparison chart showing variation of weight in S1488-location 2

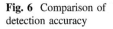

**Fig. 6** Comparison of detection accuracy

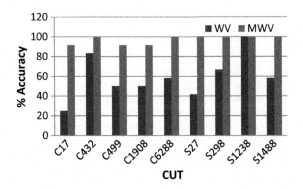

gives significantly higher accuracy for almost all the circuits, providing an average detection accuracy of 95.27% over the existing weighted Voting algorithm, which gives an average accuracy (among the tested circuit samples) of 58.21% following multiple modular redundancy approaches shown in Fig. 6.

Notations and terminologies used in the following tables:

WV—Weighted Voting (Existing); MWV—Modified-Weighted Voting (Proposed);

L1, L2, L3, L4—Locations of the circuit where Trojan is placed;

C1, C2, C3—Different implementations of a particular circuit; T1, T2, T3—Trojans.


# 4   Conclusion

It can be observed from the results that the modified-weighted voting algorithm gives a better discrimination between the infected and non-infected circuits. The better accuracy is achieved because of removing the bias toward 1s which was present in the existing voting algorithm.

Also, the modified usage of CRC method in our algorithm enables easy and encoded comparison. The inclusion of critical pattern generator module and toggling analysis provides time-saving results C6288. The accuracy of detection can be improved by feeding all the input patterns. But it becomes a time-consuming process for systems which have many inputs. If speed is preferred to accuracy then critical pattern feeding method can be used and accuracy is preferred to time then the entire input pattern feeding method can be used. Thus, we have been able to achieve a Trojan detection method which alleviates the need for a Golden Reference.


# References

1. M. Tehranipoor, F. Koushanjar, 'A Survey of Hardware Trojan Taxonomy and Detection', IEEE Design & Test, PP.1–1, 2013.
2. Tehranipoor M, Salmani H, Zhang X, Wang X, Karri R, Rajendran J, et al. Trustworthy hardware: Trojan detection and design for-trust challenges. Computer 2011;44(7):66–74.
3. Mohammad Tehranipoor and Cliff Wang, editors. Introduction to Hardware Security and Trust. Springer, 2012. ISBN 978-1-4419-8079-3.
4. H. Amin, Y. Alkabani and G. Selim, 'System-level protection and Hardware Trojan detection using weighted voting', Journal of Advanced Research, Vol.5, no.4, PP.499–505, 2014.
5. G. Latif-Shabgahi, J.M. Bass, S. Bennett, History-Based Weighed Average Voter: A Novel Software Voting Algorithm for Fault-Tolerant Computer Systems, Proc. PDP2001: 9th Euromicro Workshop on Parallel and Distributed Processing, February 7–9, Italy, (2001).
6. R. Bharath, G. Arun Sabari, Dhinesh Ravi Krishna, Arun Prasathe, K. Harish, N. Mohankumar, M. Nirmala Devi, "Malicious Circuit Detection for Improved Hardware Security", In Proc. of The Third International Symposium on Security in Computing and Communications (SSCC'15), Vol.:536, pp:464–472. 2015.

7. Venkata Raja et al., "HAPMAD: Hardware Based Authentication Platform for Malicious Activity Detection in Digital Circuits," 4th International Conference On Information System Design And Intelligent Applications, (India - 2017) 2017, Vietnam.

8. A. Al-Anwar, Y. Alkabani, M.W. El-Kharashi, and H. Bedour. Defeating hardware spyware in third party IPs. In Saudi Intl. Electronics, Communications and Photonics Conference (SIECPC), 2013.

# A EHO-Based Clustering and Routing Technique for Lifetime Enhancement of Wireless Sensor Networks

R. K. Krishna and B. Seetha Ramanjaneyulu

**Abstract** It is known that Clustering and Routing are two techniques that can be used for improving the lifetime of wireless sensor networks. Here, clustering has been implemented using the Elephant Herd optimization (EHO) technique. In this method, the behavior of the elephant groups has been taken into consideration. As it is known, elephants form groups consisting of the father elephant, mother elephant, and calves. As soon as the father elephant grows old, the mother elephant takes over the leadership and the older elephant moves away from the group and stays in isolation but is in contact with the group. An enhancing operator which selects the cluster head based on the highest energy and a separator operator used for separating the node with energy exhausted that is the father node is used for implementing the clustering technique. Routing technique has been implemented through the fittest cluster head method. While selecting the path the cluster heads in the neighboring clusters are compared on the basis of their energies and the cluster head with the higher energy is selected for the path.

**Keywords** Elephant herd optimization · Father elephant · Mother elephant
Enhancing operator · Separating operator

## 1 Introduction

In recent times, most of the work in wireless sensor networks is concentrated on saving energy and improving the lifetimes of wireless sensor networks. Recent studies indicate that metaheuristic algorithms are providing optimum solutions to many problems. An elaborate discussion on how metaheuristic algorithms such as PSO, GA, ACO are being used to find optimum solutions to these problems is

R. K. Krishna (✉) · B. Seetha Ramanjaneyulu
Department of ECE, VFSTR University, Vadlamudi, Guntur, India
e-mail: rkrishna40@rediffmail.com

B. Seetha Ramanjaneyulu
e-mail: ramanbs@gmail.com

© Springer Nature Singapore Pte Ltd. 2018
J. Anguera et al. (eds.), *Microelectronics, Electromagnetics
and Telecommunications*, Lecture Notes in Electrical Engineering 471,
https://doi.org/10.1007/978-981-10-7329-8_40

presented in [1]. New algorithms such as BFOA, ARM, EHO, RHESUS, GWO, AFSA, Firefly, and HABC are being increasingly used for providing optimization solution for many problems in wireless sensor networks. Our purpose here is to use these bioinspired techniques for clustering and routing techniques with the aim of improving the lifetime of wireless sensor networks. Clustering and routing are two techniques which substantially enhance the lifetime of wireless sensor networks. Our aim in this paper is to use some of the bioinspired techniques for clustering and routing techniques. Here, we use the EHO technique for clustering and fittest cluster head technique for Routing.

It is known that one of the largest animals is the elephant. Elephants have typical characteristics of behavior. They are social animals with a structure that consists of a group composed of male elephants and female elephants with calves. An elephant group consists of the male elephant which is the leader. Female elephants generally tend to live in groups, whereas male elephants prefer to live in isolation. Once the calves grow the male elephant moves away from the group and lives in isolation though it may remain in contact with the group by low-frequency vibrations. Then the female elephant or the mother elephant takes over the leadership of the group. In this chapter, this characteristic of elephant behavior has been applied to a wireless sensor network for clustering.

After clustering and selection of cluster heads is done with, the next task is data transmission using an appropriate routing technique. That is data is collected, aggregated, and sent to the base station after selecting a suitable route. This has been done using fittest cluster head selection method. While selecting the path from the source to destination the data is first transmitted to the cluster head. The cluster heads of the neighboring clusters are verified and the cluster head with the higher energy is selected and the data is passed through that cluster head. This process is continued till the base station is reached. This helps in data transmission through the fittest route. This method has been compared with the existing GA technique and this method has been found to give better results in terms of packet delivery ratio (PDR), delay, and energy.

## 2 Related Work

In [1], Chun Wei Tsai et al. have reviewed several metaheuristic algorithms that are attracting the researchers in providing optimum solutions to various problems for improving the lifetimes of wireless sensor networks. They have analyzed and discussed several lifetime problems of wireless sensor networks such as Number of Alive Nodes problem, Cluster Head Election Problem, Deployment Coverage Problem, Set-Cover Problem, and Data Routing problems. The discussions are divided into three parts: single-solution-based algorithms, evolutionary algorithms, and swarm intelligence. In the swarm-based techniques SA (Simulated Annealing), TA (Tabu Search), GA (Genetic Algorithm), ACO (Ant Colony Optimization), PSO (Particle Swarm Optimization) methods have been analyzed with respect to the above problems. Open issues such as representation, parameter setting, multi-objective, computation cost, and convergence have also been discussed.

As it is known, ACO is composed of three phases that are initialization, pheromone update, and solution construction. In [2] Wang et al. propose a new algorithm called AMR. In this method, the ants are able to communicate with each other indirectly. This is done by making changes in the thickness of pheromone. Each ant has a capability of searching paths themselves and they provide a feasible solution. In this technique, the authors use a halting state for the ants. In this algorithm, if the ants after returning to the depot are not in halting state, they are made to travel again in search of a new path. Each ant selects a new path using the random proportional rule. This iteration is repeated until a proper route is found. A further improvement has been suggested to the existing technique by providing a parameter to save distance. For this purpose, the authors propose that there should be two types of ants that are called ordinary ants and special ants. As far as ordinary ants are considered, they travel as per convention and use the random proportional rule and deposit pheromone for communication with other ants. The special ants construct routes in a similar manner as TSP ignoring the capacity constraint and they pick up the "short path" between two customers and deposit pheromone accordingly. Thus, the special ants have an advantage as they thus exploit the local optima.

Hajlaoui et al. [3] covers the survey of different metaheuristic methods that are used to solve the routing problems in VANETs. Further, they have also presented the existing trends. In this chapter, they have elaborately studied PSO (Particle swarm optimization), Genetic algorithm and Ant colony optimization techniques with respect to numerous parameters. How Particle Swarm Optimization (PSO) is used for Modified-Optimized Link State Routing Protocol Tuning (OLSR Routing) has been discussed. The method by which the PSO is used for Optimized DSDV is also analyzed. The other parameters discussed include Using Particle Swarm Optimization, Parameter Value Optimization of Ad Hoc on Demand, and Vector Routing for multipath distance, and Time Seed-Based Solution Distance Effect Routing Algorithm which is used for mobility analysis, simple forwarding over Trajectory, Geocasting Beacon Power Control Data Aggregation.

Lalwani and Sagnik Das [4] implements the BFO algorithm. The basic principle of this method is to mimic the foraging strategy of a swarm of E Coli. The basic objective of this method is to optimize multi-objective functions. The bacteria are always in search of nutrients to improve its energy. During this search process, the movement of the bacterium is by means of small steps in the environment. This process is called chemotaxis. This is the basic idea utilized by BFOA. In BFOA a virtual bacterium is the solution vector. It is used to solve the optimization problem. For this purpose, the chemotactic movement is performed in the problem search space. The aim is to get a global optimal solution vector. As it is known, cluster head selection and routing are techniques which are used widely for improving and enhancing the life of the wireless sensor networks (WSN). BFOA is used for implementing these two techniques. In the clustering technique, the cluster head is selected based on BFOA where the residual energy of nodes and intra-cluster distance are the parameters considered as fitness function for selection of cluster head whereas for the routing techniques residual energy of next hop and Euclidean distance between cluster head and next hop is considered for selecting the routing path.

In [5] Gai Ge Wang et al. have proposed a new metaheuristic optimization method, which is based on elephant herding behavior. As it is known elephants, though belonging to different clans live together. The matriarch is the leader of the group. The male elephants generally leave the family group after a few years. That is when the children grow up. The old elephant then lives separately away from the group but still is in contact with the family. The mother now becomes the leader of the group. This behavior of the elephants has been modeled into two operators. One is called clan updating operator and the other is named as separating operator. In EHO, when the old elephant leaves the group and its position is taken over by a young elephant then this process of replacement and updating is performed by the clan updating operator. The next step is the implementation of the separating operator. The purpose of this operator is to separate the old elephant from the group.

Zhou et al. [6] is based on the popular and widely used technique called the Particle Swarm Optimization (PSO). In this chapter, the authors have proposed an improved PSO algorithm. This has been done by adjusting the inertial weight so that the particles are not trapped in local optima. The purpose of the improved PSO algorithm is to increase the fitness function to the maximum level. This helps in the selection of cluster heads that are more suitable and better relay nodes. This results in the generation of an energy-efficient wireless sensor network.

Nugroho et al. [7] proposes a routing technique that is based on the behavior of an insect called Bombyxmori, a silk moth. This insect flies in a unique manner with the purpose of finding the source of pheromones. First, it flies in a straight manner and then it follows a zigzag path. The algorithm proposed, therefore, consists of two steps, the first step is called shortest path algorithm and the second step is named the zigzag path algorithm. The total process consists of three stages. The first stage is the node initialization stage. The second stage consists of the creation of dictionary manager and finally, the third stage is the forwarding of data in a zigzag manner. First, the function of each node is the initialization of nodes. Next, each node has to identify the adjacent nodes. Then the flow of traffic is recorded. The dictionary manager is the central controller. The dictionary manager's function is to find the best route from one node to another. The second algorithm is the zigzag algorithm. In the second stage, the data is transmitted in a zigzag manner. Thus, it can be said that data transmission consists of two parts. In the first stage, the shortest path algorithm is used for transmitting the data during the first $i/2$ hops. Thus, during the first stage, the shortest path algorithm is used for transmitting the data. This step is analogous to the straight line movement undertaken by the male silkmoth while flying to its destination. The second stage consists of the zigzag path algorithm. This path is used for data transmission during the remaining $i/2$ hops.

In [8], Miao et al. propose a Leach H protocol based on genetic algorithm, which is an improvement over the Leach protocol. This method improves the lifetime of the network. In the traditional leach method, cluster head election process is completely dependent on the random number and factors such as the current residual energy and location of the nodes are not considered resulting in a reduction of lifetime of the network. Also if the location is not decided then it may happen that in certain areas there may be more nodes and in some areas there may be a dearth of nodes resulting in uneven distribution affecting the data transfer. To overcome these problems, the authors have proposed a GA-based Leach protocol. In this method, energy and the position of nodes are taken into account to optimize this election mechanism. The improved algorithm introduces three parameters energy, the node's number of neighbors, the distance between node and base station to correct threshold. Then the genetic algorithm is used for clustering and selection of cluster heads by considering four parameters such as the size of population, cross probability, mutation probability, and weigh accuracy of influence factors. The genetic algorithm procedure consists of coding the chromosome according to the required accuracy. Then, the fitness value of each chromosome is calculated followed by selection, crossover and mutation and generation of new population and the selection of cluster head based on optimal weight values.

# 3  Proposed Work

As is known, elephants are social animals that tend to live in groups. They have a specific group structure consisting of many groups and each group is led by the oldest elephant. The rest of the group comprises of the female elephant and the calves. As soon as the elephant grows old, it tends to move away from the group and stay in isolation but is in contact with the group. The group is now led by the mother elephant.

This behavior has been idealized to form a general global optimization method. In order to make the herding behavior solve all the optimization problems, following idealized rules have been framed.

1. The total elephant populations consist of many groups and each group comprises of the father elephant, mother elephant, and its children.
2. The group is led by the father elephant and once it becomes old it will leave the group and stays in isolation but remains in contact with the group. Now the mother takes over the leadership.

This behavior is implemented using two operators (a) Clan updating operator. (b) Separating operator.
The equation for clan updating operator is given as

$$X_{newci,j} = X_{ci,j} + \alpha \times \left( X_{best,ci} - X_{ci,j} \right) \times r \tag{1}$$

where $x_{newci,j}$ is the newly updated position of elephant j in clan i and $x_{ci,j}$ is the old position for elephant j in clan ci.

$A \in [0, 1]$ is a scale factor that determines the influence of matriarch ci on $x_{ci,j}$. $X_{best,ci}$ represents matriarch ci, which is the fittest elephant individual in clan ci. $r \in$

[0, 1] is a kind of stochastic distribution and uniform distribution in the range [0, 1] is used in our current work.

The equation for separating operator is given by the formula as follows:

$$x_{worst, ci} = x_{min} + (x_{max} - x_{min} + 1)rand \tag{2}$$

where $x_{max}$ and $x_{min}$ are, respectively, upper and lower bound of the position of elephant individual. $X_{worst,ci}$ is the worst elephant individual in clan ci. rand $\in$ [0, 1] is a kind of stochastic distribution and uniform distribution in the range [0, 1].

This behavior of the elephant herding is applied to the wireless sensor network in this chapter. The nodes are idealized as elephants and the groups that are formed are analogous to the clusters and the leader is the cluster head. Here, we have formed the clusters with each cluster consisting six elephants. The node with the highest energy is selected as the cluster head. Then as soon as the energy of that cluster head reduces the mother cluster head takes over. This is implemented using the clan updating operator. Then the father elephant is isolated and moves away from the group. This is implemented using the separating operator.

Once clustering is done the next step is data transmission or routing. In routing, the selection of the path to be taken is generally based on the shortest path to reach the destination. But in this technique, while selecting the path the energies of the two neighborhoods cluster heads or the cluster heads in the neighboring clusters that form the path are compared. The cluster head with higher energy is selected and communication takes place through the selected cluster head. The remaining cluster heads are also selected in a similar way and these cluster heads form the path to the destination or base station. This results in improvement in the lifetime of the communication path and data transmission takes place for a longer amount of time thus enhancing the lifetime of the network. This scenario and methodology are simulated using NS2 simulator and the quantities of energy consumed, delay occurred and ratio of packet delivery are recorded.

## 4   Results

The simulation results for packet delivery ratio, delay, and energy are obtained for the proposed EHO-based clustering and routing techniques. These results are compared with the existing Genetic Algorithm (GA) technique. Performance comparisons are shown in Figs. 1, 2 and 3.

Figure 1 is the PDR comparison with the existing GA technique. It is found that the packet delivery ratio is 10–20% better than the GA technique with only small drop in packets.

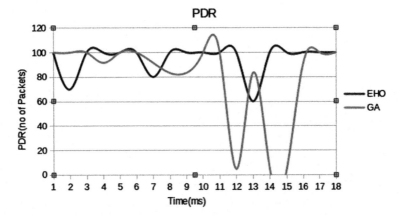

**Fig. 1** PDR comparison

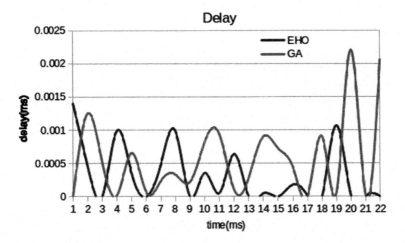

**Fig. 2** Delay comparison

Figure 2 shows the delay comparison of the proposed technique with the existing GA technique. The graph indicates that our proposed EHO method has a lesser delay of about 10% as compared to the existing GA technique.

Figure 3 shows the energy consumption comparison of the proposed technique with the existing technique. Energy consumed by the proposed technique is almost 40% lesser than the existing GA techniques. This shows that our method is far more efficient than the existing techniques.

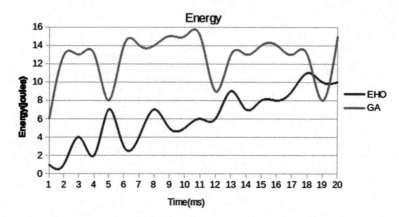

**Fig. 3** Energy consumption comparison

## 5   Conclusions

It is found that the proposed method that uses the EHO-based clustering and fittest cluster head routing technique gives a good improvement over the existing GA technique. The proposed technique not only yields better results by improving the PDR ratio and reducing the delay by about 10% but also reduces the energy consumption by almost 40% which helps in improving the lifetime of wireless sensor networks.

## References

1. Chun-Wei Tsai, Tzung Pei Hong, and Guo Neng Shiu.: Meta heuristics for the Lifetime of WSN: A Review, IEEE Sensors Journal, Vol. 16, No. 9, May 1, (2016), 2812–2831
2. Xinyu Wang, Tsan-Ming Choi, Haikuo Liu, and Xiaohang Yue.: Novel Ant Colony Optimization Methods for Simplifying Solution Construction in Vehicle Routing Problems, IEEE Transactions on Intelligent Transportation Systems, Vol. 17, No. 11, November, (2016), 3132–3141
3. Rejab Hajlaoui, Hervé Guyennet, and Tarek Moulahi.: A Survey on Heuristic-Based Routing Methods in Vehicular Ad-Hoc Network: Technical Challenges and Future Trends, IEEE Sensors Journal, Vol. 16, No. 17, September 1, (2016), 6782–6792
4. Praveen Lalwani I, SagnikDas.: Bacterial Foraging Optimization Algorithm for CH selection and Routing in Wireless Sensor Networks, 3rd International Conference on Recent Advances in Information Technology I RAIT (2016)
5. Gai-GeWang, Suash Deb, Leandrodos S. Coelho.: Elephant Herding Optimization, 3rd International Symposium on Computational and Business Intelligence, (2015) 1–5
6. Yuan Zhou, Ning Wang, and Wei Xiang.: Clustering Hierarchy Protocol in Wireless Sensor Networks Using an Improved PSO Algorithm, IEEE Access, (2017), 1–12
7. Dwi Agung Nugroho, Agi Prasetiadi, and Dong-Seong Kim.: Male Silk Inspired Routing Algorithm for Large-Scale Wireless Mesh Networks, Journal of Communications and Networks, Vol. 17, No. 4, August (2015) 384–393

8. Hongxia Miao, Xuanxuan Xiao, Bensheng Qi, Kang Wang.: Improvement and Application of Leach Protocol based on Genetic Algorithm for WSN2015 IEE 20$^{th}$ International workshop on computer aided modeling and design of communication links and Networks, (2015), 242–245

# Joint Multiview Video Plus Depth Coding with Saliency Approach

T. Manasa Veena, D. Satyanarayana and M. N. Giri Prasad

**Abstract** The main consideration of compression efficiency in multiview is to venture the temporal and interview analytical dependencies since all cameras capture the particular frame with variable viewpoints. The era of multiview coding (MVC) predictor selection statistics is used to compress MVD representation resulting in separate bitstreams of texture and video sequences. These coding schemes do not capture the similarities that arise in texture and depth video sequence, whereas joint multiview video plus depth coding (JMVDC) scheme employs the correlation co-efficiency of the motion in texture and depth sequence in rendering the object of interest in the scene. Large amount of data that is produced in this representation becomes a challenge for data storage and network transmission. In JMVDC, the structure enables interlayer motion prediction mechanism by representing the base and enhancement layers as texture and depth. The inter-dependency of motion texture and depth content of multiview is accomplished and achieved by employing different correlation coefficient methods. The proposed method utilizes depth map to enhance the salient region with the combination of local and global saliency information. The experimental results show that the JMVDC method in saliency map with enhancement achieves the reduction of computational complexity to detect the distinctive region.

**Keywords** MVD · Joint multiview video plus depth (JMVD)
Saliency method · Local saliency · Global saliency

T. Manasa Veena (✉) · D. Satyanarayana
RGMCET-Rajeev Gandhi Memorial College of Engineering & Technology, Nandyal, India
e-mail: manasaveenab4@gmail.com

D. Satyanarayana
e-mail: dsn2003@rediffmail.com

T. Manasa Veena · D. Satyanarayana · M. N. Giri Prasad
JNTUA-Jawaharlal Nehru Technological University Anantapur, Anantapur, India
e-mail: mahendragiri1960@gmail.com

© Springer Nature Singapore Pte Ltd. 2018
J. Anguera et al. (eds.), *Microelectronics, Electromagnetics
and Telecommunications*, Lecture Notes in Electrical Engineering 471,
https://doi.org/10.1007/978-981-10-7329-8_41

# 1 Introduction

In a three-dimensional (3D) video, view amalgamation is becoming a primary source to multiview video where texture plays a promising approach in acquiring pixel-based depth map used in MVD. Hence, a joint (JMVDC) structure is designed to accomplish the ideal coding productivity while concentrating on the regressive similarity of texture video coding of multiview in MVC. Multiview applications and answers for bolster generic multiview and additionally 3D services are presented. Including various operations of MVC [1] like interfacing, transmission and decoding capacity covers an extensive variety of prerequisites which are used for 3D video sequences are depicted by this approach. The portions which are acquainted in MVC with bolster these proposed arrangements incorporate reference pictures stamping, aided proficient view exchange, bitstream collaboration, the motion of view extensible in enhancing supplemental upgrade data and parallel decrypting. Merkle in [2] a review of multiview video plus depth representation (MVD) arrangement is exhibited. A navigational approach like 3D TV and free perspective video are empowered as 3D portrayal. Compression algorithms in video coding scheme for multiview misuses factual conditions of reference pictures for both temporal and between views related to shading and depth information.

A survey on the video top to bottom portrayal for multiview video game plans is shown in [3]. Compression efficiency and complexity are prominent parameters in the development of video coding standards and this serves as a typical approach for composite coding concepts utilizes motion compensation prior to previous coding structures. In [4] another new algorithm depicted for depth map which uses a prevalent video standard, H.264 in order to lessen encoding time fundamentally while keeping up high compression effectiveness. Rather than evaluating movement vectors straightforwardly in the depth map, create applicant movement modes by exploiting movement data of the comparing texture video. The test comes about demonstrate that the coding algorithm at low bit rates reduces the unpredictability of the past plan about 60% which utilizes two sequences independently and coding simulation is additionally enhanced up to 1 dB.

With the accession in [5] is broadly utilized for view amalgamation in 3D video applications. To effectively use restricted data transmission, novel standards like the Advanced Video Coding (AVC) standard also notable as H.264 extension of SVC, can be embraced 4:0:0 Chroma as a depth data. Jointly collaborating the surface video and its corresponding depth data which utilizes Scalable Video Coding (SVC), expansion of H.264 or AVC, a novel algorithm illustrated in DIBR model.

Recent presentations of the Advanced VC standards, used in 3D video critical upgrades have been exhibited in compression capacity. Collaboratively, The JVT of the ITU-T VCEG and the ISO/IEC MPEG has now likewise institutionalized a Scalable Video Coding (SVC) expansion of H.264 or AVC standard [6, 7]. SVC empowers the transmission and decoding of partial bitstreams to give video administrations bring down sensual along with time or structural measures or reduced constancy while holding a reproduction quality and inadequate bitstreams w.r.t rate phenomenon.

## 2  Problem Definition

As we are considering a depth data instead of texture, bitrate of depth data plays an important role and a minimum is required to encode a depth data rather than a texture due to its structure and entropy as depicted in [8]. In JMVDC structure the base layer and enhancement layer represent corresponding texture and depth of a scene. The per-pixel depth map is incurred in view interpolation with many advanced systems to access the depth bitstream in the enhancement layer. A large amount of data that is produced in this representation becomes a challenge for data storage and network transmission.

By partitioning the frame into texture and depth and analyzing depth data along with saliency mapping unveils the parameters like bitrate and complexity [9, 10] as saliency enables redundant data with spatiotemporal mapping. Previous methods are concentrated and capable to decode only texture video whereas JMVDC by achieving the preservation of bitrate is wider than MVC and MVD. Our Saliency approach enhances the texture and its corresponding depth map of multiview followed by MVC and depicted jointly MVD along with its depth reducing complexity and saving bitrate as possible. The JMVDC method in saliency map with enhancement achieves the reduction of computational complexity to detect the distinctive region.

## 3  Proposed Approach

In the present era of 3D content securing and advancement with the huge market potential to multiview video frameworks. Our present phenomenon, we consolidate multiview video coding and the forecasting macroblocks in a depth data outline from corresponding texture frame arrangement. The final depicted framework of proposed (JMVDC) structure is shown in Fig. 1, which is used to accomplish the ideal coding proficiency while keeping up the retrogressive similarity of the multiview texture to MVC along with saliency mapping. Tests demonstrate that the JMVDC structure beats the autonomous MVC of multiview depth map outline by 10–20% as far as bitrate sparing.

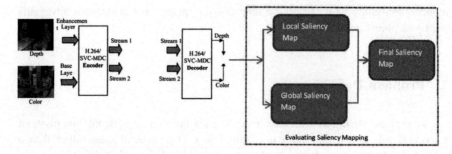

**Fig. 1** Block diagram of depicted framework

## 3.1 Depth Map

With the advancement of jointly coding texture and depth map enhances compression efficiency effectively beyond unconnectedly coding. In our approach, Fig. 2 depicted a flowchart to attain depth map of a video sequence. By considering

**Fig. 2** Flowchart of depth map creation

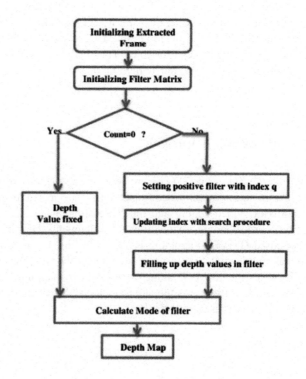

filter or blocks of the pixel region in which original is leaving aside i.e., working image. To identify filter result around the depth pixel an index is updated. Calculating depth value in a frame separating the entire pixel region with macroblocks of $3 \times 3$ or $5 \times 5$ or $8 \times 8$. Filling up all the depth values in particular MB so that we will have a surrounding zero value pixel. All nonzero entries and repeated values are carried out in mode process.

## 3.2 Saliency Approach

Human Visual System (HVS) plays a prominent role in visual aspects which is used to identify important regions in an image. To identify important region automatically in image saliency detection is the best approach [11]. A wide variety of image processing applications such as related compression, segmentation, enhancement and the valuation is extended up to video applications such as Video compression, Video Analysis. Many conventional algorithms are developed for saliency detection. Local Saliency is obtained by evaluating one of the feature extraction, i.e., an intensity which is a directional change and termed as image gradient [12]. To extract the information of an image intensity gradient is a permissible approach convolving with a filter. Global Saliency in this work is termed as the total gradient of DCT image [13] due to the visual attention to depth feature which is going to serve as salient for stereoscopic images.

## 3.3 Feature Extraction

For learning saliency approach, we extract the parameters related to color, intensity, texture, and depth. A compression domain saliency detection mapping which uses to extract the above features is implemented by DCT used in JPEG compression of $8 \times 8$ blocks. The color and texture related parameters used for object localization is formally extracted from coefficients of DCT. After obtaining a DCT block size image, enhancement technique is applied in order to improve quality of the image and also obtain highlighted features which are going to serve exact replication of saliency along with weighted features obtained from DCT block splitting.

## 4  Methodology

(a) *Multiview Video Coding*: MVC, referred to as both bury expectation and interview prediction to exploit the worldly and between view redundancies. The interview expectation is performed quite recently like the entomb prediction aside from that reference pictures in the forecast procedure are at no time in the future transiently neighboring pictures yet the photos of a similar testing moment in the nearby perspectives. It relies on upon the particular encoder setup, regardless of whether interview expectation is connected to non-anchor pictures.

(b) *Multiview Video Plus Depth Coding*: Due to the extended amount of data in a multiview video where each frame is independently coded and is a challenging task possible with any video standards along with H.264/AVC, MVDC utilizes combining texture and depth map deliberately. Moreover, the confinement of MVD extension to MVC was contrasted and independently coding of texture and depth map freely and alluded to MVD with simulcasting H.264 or AVC. The Practical implementation of both MVD structures is indistinguishable with the succession of Ballet whereas the proposed structure of MVD along with MVC outflanked the simulcasting MVD depth by 0.5 dB PSNR in Luma for Break-dancers dataset.

(c) *JMVDC SCHEME*: The pixel qualities are very extraordinary between a texture and its relative depth map delineate, developmental entities in a texture and the related profundity outline are typically comparable. A similar conclusion can likewise become to subjectively; the considered JMVDC plot employs the way that texture video and its profundity outline have comparative protest outlines and developments. It gives a sensible tradeoff between many-sided quality and compression productivity and facilitated the execution exertion of JMVDC. The prediction procedure of anchor picture is referred as dissimilarity of interlayer (alternative) and used to extract multiview depth map from texture and a clear exposition is depicted in Fig. 2, following the proposed approach.

## 5   Result Analysis

As delineated as above, the following result analysis is depicted in Fig. 3, where the consolidated MVC and the forecast depth map outline from corresponding texture frame. By evaluating the depth map, JMVDC, and its corresponding threshold JMVDC to get binary data from depth sequence. The cumulative histogram is analyzed for depth, JMVDC, and threshold in order to conquer the variations mathematically. In order to accomplish the ideal coding proficiency while keeping up the retrogressive similarity of multiview texture MVC along with saliency mapping is implemented separately combining with local and global mappings. The performance evaluations of all implemented saliency mapping conditions are as follows with Fig. 4. The overall performance evaluation of local, global along with combining saliency mapping analysis is measured in terms of precision, recall, and F-Measure with the correlation coefficients of all three mapping methods [14]. Based on correlation phenomenon of the depth map of each technique is compensated with the correlated sequences of user-defined parameters.

**Fig. 3** Interlayer view of
anchor pictures

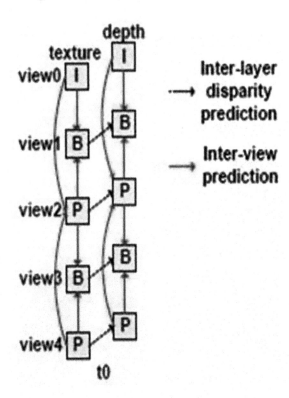

The objective measures of the proposed model are elevated in the calculation of signal to noise ratio and a bit per pixel comparison of proposed JMVDC and Saliency mapping methods. Our proposed JMVDC model with the combination of local and global saliency mapping contributes to compression ration based on bit per pixel of uncompressed to compressed formats which follows Table 1. In advance, the proposed method is used with a compression ratio of JMVDC method achieved is 28.9 whereas saliency mapping is 42.10 which is an added advantage. Simulation results are organized as follows:

- Reading the total number of frames from a video sequence as an input.
- Concatenating particular frame and extracted as an original frame as (a).
- Calculating Depth map for that particular texture as depicted in Sect. 3.1.
- A modulated intensity gradient and a total gradient is analyzed with the JMVDC.
- Evaluating proposed method starts with extracted frame and splitting into block-wise to get an enhanced image separating the color information by adjusting and preserving blocked image (h).
- The local map is obtained by normalizing the gradient estimate with filtering techniques. Global mapping is obtained by averaging all the coefficients.

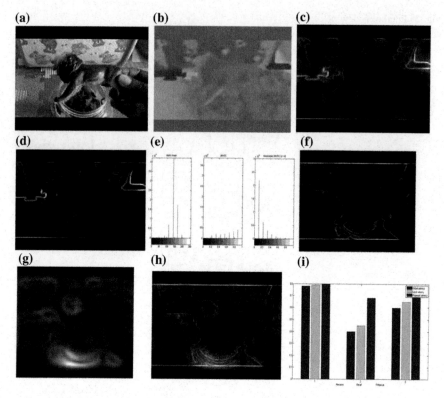

**Fig. 4 a** Original frame, **b** depth image, **c** JMVDC, **d** threshold JMVDC ($\tau = 4$), **e** histogram analysis of (**b**), (**c**), and (**d**), **f** local saliency mapping, **g** global saliency mapping, **h** final saliency mapping with enhancement, **i** overall performance evaluation for the local, global, and final saliency maps of the proposed model

**Table 1** Objective metrics of the proposed method

| Method type | PSNR (dB) | BPP |
|---|---|---|
| Using JMVDC | 32.65 | 0.83 |
| Saliency mapping | 72.22 | 0.57 |

## 6  Conclusion and Future Work

Our phenomenon is a productive joint coding plan, as JMVDC, an extension of MVD portrayal. The likeness of the mapping data in a texture and related Depth delineate and the interview relationship are employed through the interlayer motion expectation structure of SVC and MVC interview prediction structure in JMVDC. The proposed method utilizes depth map to enhance the salient region with the combination of local and global saliency information. The experimental result analysis depicted the JMVDC method in saliency map with enhancement achieves the reduction of computational complexity to detect the distinctive region. Future work may focus on extending with a novel transform analysis along with non-static texture and a corresponding depth map.

# References

1. Y. Chen, Y.-K. Wang, K. Ugur, M. M. Hannuksela, J. Lainema, and M. Gabbouj, "The emerging MVC standard for 3D video services," EURASIP Journal on Advances in Signal Processing, vol. 2009, ArticleID 78601513 pages 2009.
2. P. Merkle, A. Smolic, K. Müller, and T. Wiegand, "Multi-view video plus depth representation and coding," Proc. of IEEE International Conference on Image Processing, vol. 1, pp. 201–204, Oct. 2007.
3. T. Wiegand, G. J. Sullivan, G. Bjøntegaard and A. Luthra, "Overview of the H.264/AVC video coding standard," IEEE Transactions on Circuits and Systems for Video Technology, vol. 13, no. 7, Jul. 2003.
4. H. Oh, Y.-S. Ho, "H.264-based depth map sequence coding Using Motion Information of Corresponding Texture Video," Springer Berlin/Heidelberg, Advances in Image and Video Technology, vol. 4319, 2006.
5. C. Fehn, "Depth-image-based rendering (DIBR), compression, and transmission for a new approach on 3D-TV," Proc. SPIE, vol. 5291, 93 (2004). https://doi.org/10.1117/12.524762.
6. S. Tao, Y. Chen, M. M. Hannuksela, Y.-K. Wang, M. Gabbouj, and H. Li, "Joint texture and depth map video coding based on the scalable extension of H.264/AVC," Proc. of IEEE International Symposium on Circuits and Systems, pp. 2353–2356, May 2009.
7. H. Schwarz, D. Marpe, and T. Wiegand, "Overview of the scalable video coding extension of the H.264/AVC standard," IEEE Trans. Circuits Syst. Video Technol., vol. 17, no. 9, Sep. 2007.
8. Jun Zhang, Miska M. Hannuksela and Houqiang Li, "Joint Multiview Video Plus Depth Coding," IEEE International Conference on Image Processing, vol. 10, Sep 2010.
9. L. Itti, "Models of bottom-up and top-down visual attention," Ph.D. dissertation, Dept. Computat. Neur. Syst., California Inst. Technol, Pasadena, 2000.
10. L. Itti, C. Koch, and E. Niebur, "Model of saliency-based visual attention for rapid scene analysis," IEEE Trans. Pattern Anal. Mach. Intell., vol. 20, no. 11, pp. 1254–1259, Nov. 1998.
11. Jun Xu, Xiaoqiang Guo, Qin Tu, Cuiwei Li and Aidong Men, "A Novel Video Saliency Map Detection Model In Compressed Domain," Proc. Of IEEE Milcom 2015 in Communications, vol. 15, no. 9, Sep. 2015.
12. Wonjun Kim and J-J Han, "Video saliency detection using contrast of spatiotemporal directional coherence," Signal Processing Letters, IEEE, vol. 21, no. 10, pp. 1250–1254, 2014.
13. Yuming Fang, Zhenzhong Chen, WeisiLin, Chia-Wen Lin," Saliency Detection in the Compressed Domain for Adaptive Image Retargeting," IEEE TRANSACTIONS ON IMAGE PROCESSING, VOL. 21, NO. 9, SEPTEMBER 2012.
14. Imamoglu, N., W. Lin, and Y. Fang. "A Saliency Detection Model Using Low-Level Features Based on Wavelet Transform", IEEE Transactions on Multimedia, 2012.

# Two-Stage Enhancement of Dry Fingerprint Images Using Intensity Channel Division and Estimation of Local Ridge Orientation and Frequency

Ramagiri Priyakanth, Katta Mahesh Babu
and Nyshadam Sai Krishna Kumar

**Abstract** Quality of finger prints play a major role in justifying the performance of any automatic finger print identification or verification system specially in extracting minutiae. This chapter deals with the enhancement of dry fingerprints as it is very crucial in forensics. This enhancement algorithm improves the quality of dry fingerprints adaptively in two stages. First-stage enhancement is done using intensity channel division approach followed by the second, which is based on the presence of ridge regions in the image. The ridge regions recognized in the dry fingerprint image are normalized and hence, ridge orientations are determined. Finally, estimation of local ridge frequencies is carried out along with the application of relative filters with appropriate orientation and frequencies.

**Keywords** Dry finger prints · Intensity channel division · Ridge orientations
Ridge frequencies · Minutiae extraction

## 1 Introduction

Fingerprint analysis is one of the most prominent technologies under biometrics used to recognize suspects and unravel crimes, and it persists as an immensely valuable tool for the enforcement of law and criminal investigations [1]. This helps investigators to relate one crime scene to another connecting the same individual. Every person has his own fingerprints and do not match with that of the others. They are so unique, even if indistinguishable twins having synonymous DNA are

R. Priyakanth (✉) · K. Mahesh Babu · N. Sai Krishna Kumar
Department of Electronics and Communication, BVRIT Hyderabad
College of Engineering for Women, Hyderabad, India
e-mail: priyakanth.r@bvrithyderabad.edu.in

K. Mahesh Babu
e-mail: maheshbabu.k@bvrithyderabad.edu.in

N. Sai Krishna Kumar
e-mail: saikrishnakumar.n@bvrithyderabad.edu.in

© Springer Nature Singapore Pte Ltd. 2018
J. Anguera et al. (eds.), *Microelectronics, Electromagnetics
and Telecommunications*, Lecture Notes in Electrical Engineering 471,
https://doi.org/10.1007/978-981-10-7329-8_42

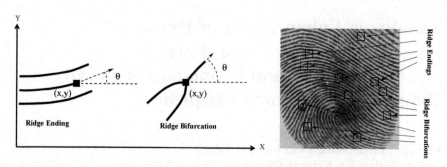

**Fig. 1** Ridge ending and ridge bifurcation in fingerprint images

considered. This exceptionality makes fingerprint identification significant in tracking the past record of the accused.

Presence of corrugations and ridges on the surface of tip of any finger represent a pattern of fingerprint. The relation between local ridges and their characteristics helps to determine the uniqueness of fingerprints [2]. The ridge alignment and their features in the impression acquainted are inspected locally and their distribution is observed aberrant. The two decisive ridge features named as minutiae used in the process of fingerprint matching are identified as ending and bifurcation of ridges. The point at which a ridge suddenly ends is termed as ridge ending and the point at which branch ridges are formed from a single ridge is understood as ridge bifurcation (Fig. 1 shows an example). Typically, the total number of minutiae in a best grade fingerprint impression image varies from 40 to 100. The fingerprint image quality will have a huge effect on the process of extraction of minutiae. Hence, dry fingerprint enhancement is required for quality extraction of minutiae [2].

Detection of ridges and locating minutiae in the case of an ideal binary dry fingerprint image is made simple, as the flow of ridges is in a constant direction locally and irregularities of ridges are termed as minutiae. But, in practice, because of disparity in impression conditions, the fingerprint can be dry with gaps in the ridges and will be of poor quality (Fig. 2 shows an example).

This leads to creation of false minutiae, ignoring a large percent of true minutiae. Hence, enhancement of the dry fingerprint image should be performed first before the extraction of minutiae.

**Fig. 2** Examples of dry fingerprint images with gaps in ridges

## 2   Fingerprint Enhancement

Dry fingerprint images contain discontinuities in ridges and their orientations. Local ridge orientations and ridge continuity are the significant visual evidences used by any fingerprint specialist to appropriately identify the minutiae. The enhancement algorithm presented in this chapter imposes these evidences to elevate clear ridges in dry finger print images only in specific regions which are recoverable. These recoverable regions contain ridges degraded by a little number of creases, spots, etc., but are still noticeable and adequate data about the true ridges is provided by the regions in neighborhood.

Enhancement of dry fingerprint image is needed before implementing the process of minutiae extraction and this is performed in two stages. First stage is implementation of Intensity Channel Division (ICD) and the second stage enhancement is connected with the process of estimating both the local ridge alignments and their respective frequencies. This fingerprint enhancement is conducted on gray images. The flowchart of two-stage enhancement algorithm for improving the quality of dry fingerprint image suitable for true minutiae extraction is represented in Fig. 3.

### 2.1   Intensity Channel Division

In Intensity Channel Division, the set of contrast pairs in the dry fingerprint image is formed. They are formed as two classes of contrast pairs. One is edge and the other one is smooth. Contrast pairs are termed as edge when the difference in the two intensities of sub-image is greater than the value of threshold (approximated to the value of 10), otherwise it is termed as smooth [3]. Edge pairs are clustered into different intensity channels and the curve of intensity transformation function designed is applied to the input dry fingerprint image. Texture regions are enhanced

**Fig. 3** Two-stage enhancement of dry fingerprint images

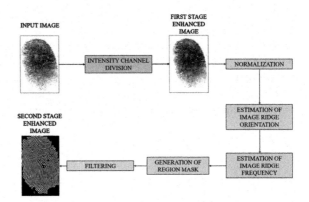

by intensity channels in the dry input fingerprint image maintaining flat regions avoiding artifacts.

The Local Contrast Indicator (LCI), $L^j(k)$ is framed for the specific intensity value $j$ as

$$L^j(k) = \sum_{s,t} \sum_{\lambda \in C_e^j(s,t)} \lambda(k) \tag{1}$$

where $L^j(k)$ is the $k$th location in the local contrast indicator function, $(s, t)$ are center pixel coordinates in the sub-image of dry fingerprint input image, $C_e^j(s,t)$ represent contrast pairs set for center pixel $(s, t)$ from its eight neighbors, such that the intensity $j$ is within those eight neighbor's pair intensities, and $\lambda(k)$ represent $k$th element in the edge contrast pair for the center pixel $(s, t)$. The range of values of $s$ and $t$ is [0, M], M being the maximum number of intensity levels. Definition of $C_e^j(s,t)$ for $(s, t)$ is illustrated by

$$C_e^j(s,t) = \left\{ \lambda_{I(s,t)}^{I(s',t')} \left| \begin{array}{c} (s',t') \in M(s,t) \\ I(s,t) - I(s',t') \geq 0 \\ \wedge (j = I(s,t \vee j) = I(s',t')) \end{array} \right. \right\} \tag{2}$$

where $I(s, t)$ and $I(s', t')$ are the intensities of pixels at $(s, t)$ and $(s', t')$, respectively, $M(s, t)$ represent any of the 8-neighbors of $(s, t)$, and $\psi$ is set as threshold (constant). Finally, the forcing function $F^j$ and hence the function of transformation for intensity channel $j$ are anticipated in (3) and (4) respectively.

$$F^j(n) = \frac{\sum_{k=0}^{n} L^j(k)}{\sum_{k=0}^{M} L^j(k)} \tag{3}$$

$$T^j(n) = \frac{I(n) + L^j(n)}{\max(I + L^j)} \quad 0 \leq n \leq M \tag{4}$$

Finally, the enhancement of dry fingerprint image is done by

$$I_{en}(s,t) = \zeta\{I(s,t)\} \tag{5}$$

where $I(s, t)$ represent pixel intensity at $(s, t)$ in the dry fingerprint input image, $\zeta$ is the Savitzky–Golay filtered intensity transformation. This filtering smoothens the transformation curve as a part of curve fitting and $I_{en}$ is the intensity replacing intensity $I$ at $(s, t)$ in dry fingerprint image [3]. This process gives an enhanced dry fingerprint image using ICD approach in first stage.

## 2.2  Estimation of Local Ridge Orientations and Frequencies

In the second stage of enhancement the output of Intensity Channel Division approach becomes the input. Let $I_1$ represent the grayscale dry fingerprint image after ICD defined as a matrix sized $N \times N$ and pixel intensity at coordinates $(s, t)$ is designated by $I_1(s, t)$.

1. As a first step of second stage enhancement, the process of normalizing the input is performed to minimize the deviations in the values of gray level all along the ridges. Let $G_{Norm}(s, t)$ refer to the normalized grayscale value at coordinates $(s, t)$ of any pixel and is defined as

$$G_{Norm}(s, t) = \begin{cases} MEAN_d + \sqrt{\frac{VARIANCE_d(I_1(s,t) - MEAN)^2}{VARIANCE}}, & if\ I_1(s, t) > MEAN \\ MEAN_d - \sqrt{\frac{VARIANCE_d(I_1(s,t) - MEAN)^2}{VARIANCE}}, & if\ I_1(s, t) \leq MEAN \end{cases} \qquad (6)$$

where $MEAN_d$ represent the value of expected mean and $VARIANCE_d$ is the expected variance. The mean and variance of ICD output image $I_1$ of size N x N can be computed as

$$MEAN(I_1) = \frac{1}{N^2} \sum_{s=0}^{N-1} \sum_{t=0}^{N-1} I_1(s, t) \qquad (7)$$

$$VARIANCE(I_1) = \frac{1}{N^2} \sum_{s=0}^{N-1} \sum_{t=0}^{N-1} \{I_1(s, t) - MEAN(I_1)\}^2 \qquad (8)$$

This process of normalization facilitates the successive processing steps.

2. After obtaining a normalized image $G_{norm}$, the following steps are involved using least mean squares method for estimating the local orientation image [4]:

   (i) Dividing normalized image $G_{norm}$ into square sized $m$ x $m$ blocks (16 × 16).

   (ii) Computing the gradients using Sobel operator, $\partial_x(s, t)$ and $\partial_y(s, t)$ at every pixel coordinate $(s, t)$ [5].

   (iii) Alignment of every block alignment using Sobel gradients at every pixel coordinate $(s, t)$ is estimated by

$$v_x(s, t) = \sum_{p=s-\frac{m}{2}}^{s+\frac{m}{2}} \sum_{q=t-\frac{m}{2}}^{t+\frac{m}{2}} 2\partial_x(p, q)\partial_y(p, q) \qquad (9)$$

$$v_y(s,t) = \sum_{p=s-\frac{m}{2}}^{s+\frac{m}{2}} \sum_{q=t-\frac{m}{2}}^{t+\frac{m}{2}} \left( \partial_x^2(p,q) - \partial_y^2(p,q) \right) \tag{10}$$

$$\Theta(s,t) = \frac{1}{2} tan^{-1} \left( \frac{v_y(s,t)}{v_x(s,t)} \right) \tag{11}$$

where $\Theta(s,t)$ represents correct orientation that is perpendicular to the effective direction of $m \times m$ window-sized Fourier spectrum and is the estimated local ridge orientation of any specific block using least mean squares algorithm.

The Consistency Level (CL) [2] of the ridge orientation in the 8-neighborhood of pixel (p, q) is computed by

$$CL(p,q) = \frac{1}{N} \sqrt{\sum_{(s,t) \in H} |\Theta(s,t) - \Theta(p,q)|^2} \tag{12}$$

where H, which is a $3 \times 3$ window, represents the 8-neighborhood around the pixel location (p, q); N represent Total number of pixels in H; $\Theta(s,t)$ and $\Theta(p,q)$ are ridge orientations at pixels (s, t) and (p, q) respectively.

This process of estimating ridge orientations locally is performed again at low resolution level of the dry fingerprint image, if the value of consistency level is less than a certain threshold.

3. Local ridge frequency is an additional essential property of a dry fingerprint image. This is because, the ridge gray levels can be shaped as a sinewave-oriented orthogonal to local ridge direction when minutiae do not appear locally. Figuring out the local ridge frequency [4] using the steps as under:

   (i)  Dividing $G_{norm}$ into $m \times m$ sized blocks (m = 16)
   (ii) Evaluation of an $n \times m$ sized (for n = 32, m = 16) orientation window should be done for every block which is centered at (s, t) (Fig. 4 Shows an example).

**Fig. 4** Oriented window and x-signature

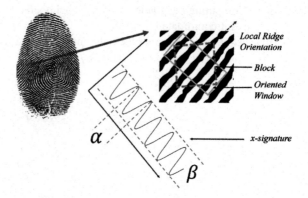

(iii) Compute $X[l]$, the $x$-signature for every block centered at the pixel coordinate $(s, t)$, for $l = 1, 2, 3....n-1$, inside the orientation window by,

$$X[l] = \frac{1}{m} \sum_{r=0}^{m-1} G_{norm}(p,q), l=0 \; to \; n-1 \tag{13}$$

where $u = s + \left(r - \frac{m}{2}\right) \cos\{\Theta(s,t)\} + \left(l - \frac{n}{2}\right) \sin\{\Theta(s,t)\}$

$$q = t + \left(r - \frac{m}{2}\right) \sin\{\Theta(s,t)\} + \left(\frac{n}{2} - l\right) \cos\{\Theta(s,t)\}$$

In the oriented window, if minutiae do not appear, sinusoidal shaped x-signature frequency and ridge frequency are equal. The ridge frequency $\varpi(s,t)$ is calculated as

$$\varpi(s,t) = \frac{1}{\daleth(s,t)} \tag{14}$$

where $\daleth(s,t)$ is the average number of pixels between peak-to-peak x-signature sinusoidal wave. Presence of minutiae in the oriented window corrupts the ridges. Hence for these blocks, the frequency values should be interpolated from the neighboring blocks whose frequencies are estimated.

4. The region mask is generated by dividing every block in the output of step-1 into two categories. They are named as recoverable or un-coverable. This can be attained by the evaluation of the sinusoidal shaped wave produced by ridges locally. The three important features related to each block are evaluated as

   (i)   $\alpha$ = Average peak-to-peak amplitude of the sinusoidal wave
   (ii)  $\beta = \frac{1}{\daleth(s,t)}$, where $\daleth(s,t)$ average number of pixels between peak-to-peak x-signature sinusoidal wave.
   (iii) $\gamma = \frac{1}{n}\sum_{r=1}^{n}\left(X[r] - \left(\frac{1}{n}\sum_{r=1}^{n}X[r]\right)\right)^2$, where $X[r]$ represent x-signature of the block

The input dry fingerprint image is accepted and filtering is applied only when the recoverable region percentage is greater than cutoff value (let it be 40%).

5. By tuning a parallelly realized Gabor filters to ridge frequency and orientation locally for efficient removal of undesired noise in step-1 output to produce enhanced dry fingerprint preserving true ridges. Gabor filter is used to filter every block with distinct directions. High-quality block orientation is strong represented by one or more filter outputs greater than that of other filters. Whereas in the case of poor quality block orientation, the filter outputs are similar. The quality of every block is designated as either good or poor based on the standard deviation of the filters responses [6].

## 3   Minutiae Extraction

Minutiae extraction is a process of marking or identifying ridge endings and ridge bifurcations. Before minutiae extraction an algorithm which gives thinned image needs to be performed by properly modifying deceptive ridges. Thinning is obtained by performing morphological operations on binary images [7]. As the result of two-stage enhancement is a binary image, specific morphological operation called *thinning* is applied which narrow the objects to points [8].

For extracting minutiae, crossing number method is used after thinning. This process searches for distinctive minutiae using pixels from $3 \times 3$ sub-images (Fig. 5 shows an example). The crossing number defined as one half times the sum of differences between pairs of 8-adjacent pixels is computed for a ridge pixel R by [9]:

$$Crossing\ Number = \frac{\sum_{i=1}^{8} |R_i - R_{i+1}|}{2} \tag{15}$$

where $R_i$ is the pixel value among the 8-neighbors of R scanned in counterclockwise direction.

Majority of the minutiae are visible to the naked eye of the expert individual, but by an automatic fingerprint identification system, rules stated by Kim Seonjoo et al. [10], must be used to eliminate false minutiae. During this process, the ridges at the end of the blocks which form the border regions of the image are ignored to avoid considered as artifacts due to thinning [11].

## 4   Evaluation of Enhancement Process Using Goodness Index

Quantitative assessment of two-stage enhancement of dry fingerprint image using intensity channel division and local ridge orientation and frequency estimation is performed by calculating the goodness index of the minutiae extraction with the help of which the performance of fingerprint identification machine can be quantified. Let $M_{DE}$ represent the set containing "n" values of minutiae marked by the above explained algorithm of minutiae extraction and $M_{EI}$ represent the set

**Fig. 5** 8-neighbors of centered pixel R

| $R_4$ | $R_3$ | $R_2$ |
|-------|-------|-------|
| $R_5$ | R     | $R_1$ |
| $R_6$ | $R_7$ | $R_8$ |

containing "m" values of minutiae marked by any trained individual in the dry fingerprint image. The goodness index [12] is computed using

$$Goodness\ Index = \frac{\sum_{r=1}^{R} QF_r(PM_r - MM_r - DM_r)}{\sum_{r=1}^{R} QF_r \cdot TM_r} \qquad (16)$$

where $R$ is the total number of $16 \times 16$ windows in the input dry fingerprint image, $PM_r$ is number of paired minutiae, $MM_r$ is number of missed minutiae, $DM_r$ is number of deceptive minutiae measured between $M_{DE}$ and $M_{EI}$ sets, $TM_r$ being the number of true minutiae in the $r$th window, $QF_r$ being the quality factor (defined with the values of 4 for Good, 2 for Medium and 1 for Poor) of the $r$th window.

The value of goodness index ranges from 0 to 1. Large value of goodness index indicates better performance of the minutiae extraction algorithm, which in turn depends on the quality of the input dry fingerprint image. Hence, goodness index can be used to evaluate the enhancement process quantitatively.

## 5  Simulation Results

Typical examples of dry fingerprint images are available in FVC2000 database. The process of image enhancement is performed in two stages, using intensity channel division followed by local ridge orientation and frequency estimation. As this enhancement process is crucial in every minutiae extraction algorithm executed by any automatic fingerprint recognition or identification system, minutiae extraction algorithm is performed for images which have undergone the two-stage enhancement process and the output images are recorded. For all the minutiae-extracted images, goodness index has been calculated before and after enhancement.

Table 1 shows the values of goodness index values measured for the three typical dry fingerprint images.

It has been observed that the goodness index is better only after two-stage enhancement of the dry fingerprint images. Table 2 shows the outputs obtained in the process of two-stage enhancement of three typical dry fingerprint images. Tables 3 and 4 show the images of the extracted minutiae without and after the enhancement process respectively. Red color markings indicate ridge endings and blue color markings are ridge bifurcations. Neglecting the minutiae marked at the

**Table 1** Goodness index values of three typical input dry fingerprint images

| Input image no. | Goodness index | |
|---|---|---|
| | Without enhancement | After enhancement |
| #1. | 0.38 | 0.59 |
| #2. | 0.29 | 0.52 |
| #3. | 0.31 | 0.53 |

**Table 2** Different stage outputs obtained in the process of applying the two-stage enhancement of three typical dry fingerprint images

| Input Dry Fingerprint Image No. #1,2,3→ | | | |
|---|---|---|---|
| **First Stage Enhancement:** ICD Output | | | |
| **Second Stage Enhancement** — Normalized Image | | | |
| Ridge Orientation Image | | | |
| Ridge Frequency Image | | | |
| Gabor Filtered Image | | | |
| Final Binary Enhanced Image | | | |

*Image    courtesy*    https://github.com/carandraug/PeterKovesiImage/tree/master/WWWImages, http://psychyogi.org/hall-player-2008-fingerprint-analysis/

**Table 3** Image outputs obtained in the minutiae extraction process of three typical dry fingerprint images without enhancement

| Binary Raw Input images No.#1,2,3 → | | | |
|---|---|---|---|
| Thinned Images | | | |
| Minutiae Extraction | | | |

boundaries of the fingerprint image, it is very important to note that spurious minutiae are marked in the minutiae extraction images without enhancement and are less in the case of minutiae extraction after enhancement.

**Table 4** Image outputs obtained in the minutiae extraction process of three typical dry fingerprint images after enhancement

| Binary Enhanced Input images No.#1,2,3 → | | | |
|---|---|---|---|
| Thinned Images | | | |
| Minutiae Extraction | | | |

# 6    Conclusion

"The distorted or less quality fingerprint images may have the missing and specious features, which may degrade the recognition performance of entire system" [13]. The two-stage dry fingerprint enhancement algorithm executed in this paper improves the quality of ridge structures using intensity channel division approach eliminating the presence of artifacts followed by the estimation of local ridge orientation and ridge frequency. Then the process of minutiae extraction is performed to extract the true ridge endings and bifurcations without enhancement and after enhancement process. This two-stage enhancement process is quantitatively evaluated using goodness index. This algorithm has neglected unrecoverable regions in the process and considered only recoverable regions making the enhancement algorithm best suitable for minutiae extraction.

This enhancement algorithm with some respectable changes can be extended for the enhancement of wet and wrinkled fingerprints which have more false acceptance rates and false rejection rates of fingerprint identification or matching systems.

# References

1. Lyle, D.P., M.D: Chapter 12: Fingerprints: A Handy Identification Tool, FORENSICS: A GUIDE FOR WRITERS (Howdunit), Writer's Digest Books, Cincinnati, Ohio, pp. 269–284 (2008)
2. Anil Jain, Lin Hong and Ruud Bolle: On-Line Fingerprint Verification, IEEE Transactions on Pattern Analysis and Machine Intelligence, Vol. 19, No. 4, pp. 302–314 (1997)

3. R. Priyakanth, Santhi Malladi, Radha Abburi, "Dark Image Enhancement through Intensity Channel Division and Region channels using Savitzky – Golay Filter", International Journal of Scientific and Research Publications (IJSRP), Vol. 3, No. 8, pp. 2050–2016 (2013)
4. Davide Maltoni, Dario Maio, Anil K. Jain, Salil Prabhakar: Handbook of Fingerprint Recognition, Second Edition, Springer-Verlag, London (2009)
5. Mikhail Yu. Kachay, Maxim Pasynkov: Theoretical approach to developing efficient algorithms of fingerprint enhancement, Analysis of Images, Social Networks and Texts, Procs. of 4th International Conference, Springer Yekaterinburg, Russia, pp. 83–95 (2015)
6. Fernando Alonso-Fernandez, Julian Fierrez, Javier Ortega-Garcia, Joaquin Gonzalez-Rodriguez, Hartwig Fronthaler, Klaus Kollreider, and Josef Bigun: A comparative study of fingerprint image-quality estimation methods, IEEE Transactions on Information Forensics and Security, Vol. 2, No. 4, pp. 734–743 (2007)
7. Rafael C. Gonzalez, Richard E. Woods, Steven L. Eddings: Digital Image Processing using MATLAB: 2nd Edition, Tata McGraw Hill, New Delhi (2009)
8. L. Lam, S. W. Lee, and C. Y. Suen: Thinning Methodologies-A Comprehensive Survey, IEEE Transactions on Pattern Analysis and Machine Intelligence, Vol. 14, No. 9, pp. 869–885 (1992)
9. Raymond Thai: Fingerprint Image Enhancement and Minutiae Extraction, Report, Honours Programme of the School of Computer Science and Software Engineering, The University of Western Australia, pp. 38–39 (2003)
10. Kim Seonjoo, Lee Dongjae, Kim Jaihie: "Algorithm for Detection and Elimination of False Minutiae in Fingerprint Images", Procs. of Third International Conference on Audio- and Video-based Biometric Person Authentication, Halmsted, Sweden, Lecture Notes in Computer Science, Vol. 2091. Springer-Verlag, Berlin Heidelberg New York pp. 235–240 (2001)
11. Fayadh. M. Abed, Adnan Maroof: Fingerprint Image Pre-Post Processing Methods for Minutiae Extraction, Raf. J. of Comp. & Math's, Vol. 6, No. 1, pp. 97–110 (2009)
12. Lin Hong, Yifei Wan, A. Jain: Fingerprint Image Enhancement: Algorithm and Performance Evaluation, IEEE Transactions on Pattern Analysis and Machine Intelligence, Vol. 20, No. 8, pp. 777–789 (1998)
13. Anand V. Telore: Study of Distortion Detection and Enhancement Methods for Fingerprint Images, IEEE International Conference on Computational Intelligence and Computing Research (2016)

# Virtual Instrumentation-Based Malicious Circuit Detection Using Weighted Average Voting

**G. Aishwarya, Hitha Revalla, S. Shruthi, V. S. Pon Ananth and N. Mohankumar**

**Abstract** The security of an entire system can be breached owing to a Hardware Trojan attack on the chip. Though many techniques are available to detect the presence of a Trojan, most of them need a reference circuit or require sophisticated Electronic Design Automation tools. Moreover, golden reference circuits are not always available. However, with the usage of average-weighted voting algorithm, the use of reference circuits can be avoided to identify the infected circuits. The automation of the detection test can be achieved by the use of virtual instrumentation. The proposed method ensures the functionality of the circuit. The results obtained assert that this detection system is modular, flexible, and also supports the integrations to accommodate any VLSI circuit off the shelf. This eliminates the use of any complex systems and can act as a standalone Trojan detection system.

**Keywords** Virtual instrumentation · Hardware Trojan · Hardware security Malicious attacks · Weighted average voting algorithm

G. Aishwarya · H. Revalla · S. Shruthi · V. S. P. Ananth · N. Mohankumar (✉)
Department of Electronics and Communication Engineering, Amrita School of Engineering,
Amrita Vishwa Vidyapeetham, Amrita University, Coimbatore 641112, India
e-mail: n_mohankumar@cb.amrita.edu

G. Aishwarya
e-mail: aishwaryaganesan95@gmail.com

H. Revalla
e-mail: hitharavella@gmail.com

S. Shruthi
e-mail: shruthi9598@gmail.com

V. S. P. Ananth
e-mail: ponananthvs@gmail.com

© Springer Nature Singapore Pte Ltd. 2018
J. Anguera et al. (eds.), *Microelectronics, Electromagnetics and Telecommunications*, Lecture Notes in Electrical Engineering 471,
https://doi.org/10.1007/978-981-10-7329-8_43

# 1 Introduction

In the decades before, the security of software, network, and information was of major concern and their protection against malicious attacks was of prime focus. But now, it is evident that a small change in the hardware of a system could manifest an immense security threat. A Hardware Trojan is an example. Hardware Trojan is a malevolent alteration done to the circuitry of the system. It has the capability of generating effects that can be catastrophic to the system. Hence Trojan detection and diagnosis is a must before the chip reaches the market and is a major concern nowadays.

Most of the primitive Trojan detection methods are done by comparing a "Golden IC" to the IC Under Authentication (IUA). Practically, testing with a Golden IC is not feasible due to the reason that the Golden ICs are produced by third-party vendors and are mostly unavailable. The other alternative to a Golden IC is performing reverse engineering on the available chips. This is not easy nor is it cost inefficient. Therefore, a method forgoing the use of any Golden IC is needed. The use of voting algorithm is one method that helps to avoid any references. Voting algorithms are used primarily for error-masking in real-time systems. It can be used for the Trojan detection, by using the weights that have been assigned to the modules before the voter output is obtained.

The power-based analysis [1] requires a set of random patterns to be given to the circuits under test. A Golden IC is reverse engineered and power measurements of IUA are compared to the power measurements of the Golden IC to detect if the IUA is embedded with a Trojan. Multi-supply transient current integration method [1] is another alternative. The Trojan consumes only a negligible amount of current, therefore detecting a Trojan by calculating the entire system's current is not the right method. Hence, the current is measured from multiple power pods that are partitions of the whole circuit. The reference circuit here too shall be obtained after reverse engineering a random set of ICs. The same process of comparison is followed. Region-based power analysis [1] is similar to the above method, where the circuit is partitioned into many regions and power measurements are taken. These measurements are then compared, the results from this reduce the process variations and the leakage current. Even though the method is a good approach in detecting the Trojan, process variations cannot be handled completely.

To effectively categorize the circuits, Principle Component Analysis (PCA) is used in [2]. A reference-free model based on the side channel leakage power analysis by Gate-Level Characterization (GLC) is presented in [3]. A PIC microcontroller-based portable standalone system to identify any malicious activity present in a circuit was put forward in [4]. The system executes the weighed voting for detecting any Trojan in the FPGA implementation of the Circuit under Test. The multiple module redundancy model proposed in [4] is adopted in this work.

A delay-based virtual intelligence model to detect the malicious activity is proposed in [5]. The system was modeled using LabVIEW, which elaborates a delay-based detection scheme. The concept of using virtual instruments for Trojan detection is abstracted from this model. A two-phased voting algorithm which provides double-check using delay arcs and a combination of weighted voting algorithm and hybrid voting algorithm is proposed in [6].

## 2 Proposed Detection Technique

VLSI circuits have a large number of gates that define the functionality and an IP core. If a Trojan is inserted it can make modification and steal information to the same. In this model, a Hardware Trojan which alters the functionality of the circuit depending on the input to the Trojan is considered. This results in the fatal failures when the chip is used in critical systems.

Virtual instrumentation provides the modularity, high flexibility, and cost effectiveness to design the test, measurement, and control systems for any application. Use of virtual instrumentation in this detection system thus greatly reduces the effort needed, is cost friendly and is easily scalable to adapt to any kind of circuit that is to be tested.

The detection system uses the History-based Weighted Average Voting algorithm [7] that is modeled using Virtual Instrumentation. The system block diagram is shown in Fig. 1. The myRIO System Design architecture is exploited by using both the real-time processor and the embedded FPGA on board. ISCAS 85 benchmark circuits are considered for testing the model. The three different modules to be tested are implemented on the FPGA of myRIO. This is done by use of the available circuit in VHDL format and integrated with the help of the IP Integration Module into the LabVIEW code. Thus, eliminating the need for regenerating the code in graphical programming language saving time and reducing complexity. This brings the algorithm and the circuit to the same platform removing the need for any additional requirements. The algorithm is implemented in the processor of the myRIO using LabVIEW.

### 2.1 History-Based Average-Weighted Voting Algorithm

History-based average-weighted voting algorithm [7, 8] is adopted in this work for developing the Trojan detection system. In this algorithm, the same set of random inputs is given to each module in every cycle to generate the module outputs. The algorithm then gives a level of consensus or agreement to each module with respect

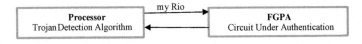

**Fig. 1** Block diagram of the system

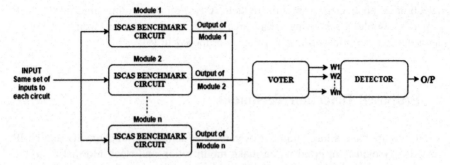

**Fig. 2** Overview of the proposed detection technique

to the other modules, based on their outputs. A history record of the module is obtained by storing these consensus levels. This record is calculated for consecutive cycles in a way so that the complexity is reduced. Based on this history record, the weights are assigned in every cycle. From the final weights obtained, the Trojan-infected module is found. The overview of the proposed detection technique shown in Fig. 2 uses the following steps to arrive at the final weights of the modules.

The following sequence of steps describes the algorithm given in [7].

Step 1: An odd number '$N$' of modules with similar functionality are taken (here $N = 3$). A set of $m$-bit inputs is given to each module and the $n$-bit outputs from them are taken as the voter inputs.

Step 2: The hamming distance, $hd_{ij}$ gives the space between the module outputs.

$$hd_{ij} = |x_i - x_j| \tag{1}$$

Step 3: The number of inputs applied decides the dynamic threshold $v_t$

$$v_t \propto x, \text{ therefore } v_t = x * p \tag{2}$$

where $p$ is the proportionality constant.

Step 4: The degree of nearness between the two modules $i$ and $j$ is calculated as follows:

$$S_{ij} = \begin{cases} 1 & if \quad hd_{ij} \leq v_t \\ \left(\frac{k}{k-1}\right)\left(1 - \frac{hd_{ij}}{k*v_t}\right) & if \quad v_t < hd_{ij} \leq k*v_t \\ 0 & if \quad hd_{ij} \geq k*v_t \end{cases} \tag{3}$$

Step 5:  For the $i$th module, the level of consensus or agreement $S_i$ is computed as

$$S_i = \frac{\sum_{j=1, j \neq i}^{N} S_{ij}}{N-1} \tag{4}$$

Where $i, j = 1, 2, 3 \dots N$, $N$ = no. of modules.

Step 6:  By accumulating the level of agreement in each voting cycle, the History record $H_i$ for $i$th module in $n$th voting cycle is got as below

$$H_i(n) = \sum_{l=1}^{n} S_i(l) \tag{5}$$

Step 7:  Computation of the State indicator, $P_i$ of $i$th module in $n$th voting cycle is done

$$P_i(n) = \frac{H_i(n)}{n} \tag{6}$$

Step 8:  Weight can be computed as based on the level of agreement or consensus

$$w_i = \begin{cases} 2*P_i(n) & \text{if } 0.5 \leq S_i \leq 1 \\ (P_i(n))^2 & \text{if } < 0\, S_i < 0.5 \\ 0 & \text{if } S_i = 0 \end{cases} \tag{7}$$

## 2.2  Flow of Algorithm

The algorithm is started by initializing the values for the variables $p, k, m, n, N, M$; where m is the no. of bits in the input sequence, n is the present voting cycle, $N$ is the no. of modules, and $M$ is the total no. of voting cycles. These variables are fed to generate random inputs. From the outputs generated, hamming distance is calculated. A dynamic threshold is set and the degree of nearness is calculated with respect to this threshold. The history of the record is computed after the level of consensus is calculated. This gives the state indicator. Depending on the level of agreement or consensus, weights are assigned to the modules are calculated from this state indicator. This process is repeated for $M$ no. of voting cycles. The module with the least weight is declared as Trojan infected.

## 2.3   Pseudocode

Declare and set the values of p, m, n, k, M, N
For n=1 to M
  Generate random values for inputs with m bits
  Generate outputs and assign to an array x
  For i=1 to N
    For j=1 to N
      Compute the hamming distance between x (i) and x (j) and store it in h (i) (j)
    End-For
  End-For
  Compute dynamic threshold $v_t$ as a product of p and no. of inputs x
  For i=1 to N
    For j=1 to N
      Assign value to level of nearness S (i) (j) based on the conditions
    End-For
  End-For
  For i=1 to N
    For j=1 to N
      Compute level of consensus $S_1$ (i) using S (i) (j)
    End-For
  End-For
  For i=1 to N
    Compute the value of history record H (n) (i) using $S_1$ (i)
    Compute the value of state indicator P (n) (i) as H (n) (i)/n
    Assign value to weight w (i) based on the conditions
  End-For
  Compute the index of minimum value in w and display it as the Trojan infected
End-For

## 3   Results

Figure 3 shows the LabVIEW model for Detection System. The circuit is implemented on the FPGA of the myRIO and this code is run on the myRIO processor to
find the Trojan infection. The combinational circuits of the ISCAS 85 Benchmark
were used for testing the model. Two types of Trojans were implemented on the
circuits to check the model capabilities, one a gate and another a small circuit. For
verifying the effectiveness of the proposed model, the three ISCAS 85 circuits (c17,
c432, c880 refer Table 1) were introduced with a Trojan at three randomly selected
locations. All these make 18 different types of Trojan infection for testing.

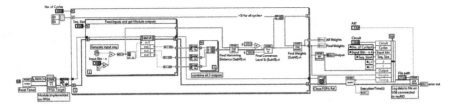

**Fig. 3** Block diagram code of the LabVIEW model

**Table 1** Description of ISCAS 85 circuits used

| Circuit name | Circuits | | | FPGA | |
|---|---|---|---|---|---|
| | Input | Output | Gate count | Slice registers | Bonded IOB |
| **c17** | 5 | 2 | 6 | 6 | 7 |
| **c432** | 36 | 7 | 160 | 160 | 43 |
| **c880** | 60 | 26 | 383 | 357 | 86 |

From Table 1, it can be observed that on realizing the same circuit on the myRIO FPGA, the gate count matches up to the number of slice registers used and the sum of inputs and outputs to the bonded IOBs. This confirms that the circuits have been correctly implemented.

From the tests of the circuits mentioned in Table 1, the optimal input parameters to the algorithm are found. The main inputs to the algorithm are tuning parameter $k$, input proportionality constant $p$ as these depend majorly on the circuit specifications and their level of Trojan infection or functionality error. Another input to the algorithm was the *sequence size* of the random input pattern which when set to a low value of 4 or 5 (all tests were of *sequence size* = 4) for any circuits ensures that the time taken for execution of the test is kept to minimal while yielding proper weight. Low value is taken because as the sequence size increases the time taken for the test run also increases as shown in Fig. 4. The execution time also depends

**Fig. 4** Sequence size versus execution time in seconds for C432 circuit

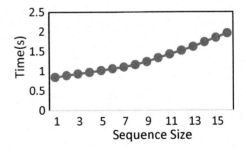

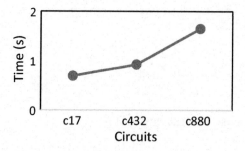

**Fig. 5** Execution time taken in seconds for ISCAS 85 circuits

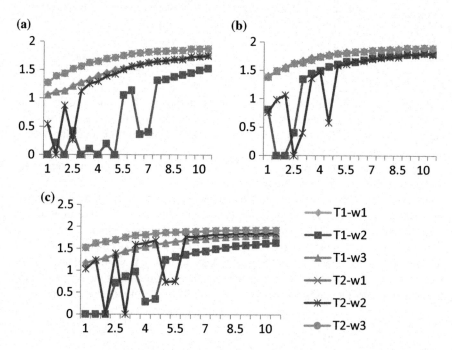

**Fig. 6** K varying graphs of circuit c17 with both Trojan types with $p = 0.1$, number of cycles as 1000, sequence size as 4 at **a** Location 1, **b** Location 2, and **c** Location 3

linearly on the circuit size as shown in Fig. 5. Both types of Trojans used for testing produces negligible differences in execution time.

The parameters $p$ is the input proportionality constant and as this determines the threshold $v_t$, it is kept as low as possible within the range of 0 to 0.1. Based on the results of the tests it can be concluded that lower the $p$ value for larger circuits better the detection rates. The tuning parameter $k$ helps us properly detect the Trojan as it determines the level of difference in the weights. As $k$ increases the difference in weights decreases while the weights as a whole increase eventually reaching their saturation level. This can be seen from the Fig. 6a. Thus by setting the $k$ value, the amount of difference in weights to judge the circuit that is truly Trojan infected can be determined. This is to prevent detection based on minute differences in weights that can occur due to delays in the circuits. The most optimal value of $k$ is kept a 3 while it can be increased for large size circuits. The results from one of the tests are shown in the following figure.

## 4　Conclusion

The history-based average voting algorithm was incorporated to identify the malicious Trojans in the circuit without the use of any references in this work. The application of virtual instrumentation for the Trojan detection replaces the previously used complex system for a simpler one with the help of LabVIEW. Exploitation of the myRIO for the implementation brings all the hardware into a single platform. Thus, the proposed model is flexible, modular, easy to incorporate, and is highly efficient in terms of both cost and time. The proposed detection system was successfully implemented and verified.

## References

1. M. Tehranipoor and F. Koushanfar, "A Survey of Hardware Trojan Taxonomy and Detection", *IEEE Design & Test of Computers*, vol. 27, no. 1, pp. 10–25, 2010.
2. N. Mohankumar and M. Nirmala Devi, "Improving the Classification Accuracy in Detecting Hardware Trojan in ALU Using PCA", *Indian Journal of Science and Technology*, vol. 9, no. 1, 2016.
3. D. K. Karunakaran et al., "Malicious combinational Hardware Trojan detection by Gate Level Characterization in 90 nm technology", in *Fifth International Conference on Computing, Communications and Networking Technologies (ICCCNT)*, China, 2014, pp. 1–7.
4. R. Bharath et al., "Malicious Circuit Detection for Improved Hardware Security, Security in Computing and Communications," in *Communications in Computer and Information Science, Proceedings of Third International Symposium*, SSCC 2015, Kochi, India, vol. 536,pp. 464–472, 2015.
5. Kamala Nandhini et al., "Delay-based Reference Free Hardware Trojan Detection using Virtual Intelligence", in *4th International Conference on Information System Design and Intelligent Applications*, (India–2017) 2017, Vietnam.
6. Venkata Raja et al., " HAPMAD: Hardware Based Authentication Platform for Malicious Activity Detection in Digital Circuits," in *4th International Conference on Information System Design and Intelligent Applications*, (India–2017) 2017, Vietnam.
7. Manasi Das, Samar Bhattacharya, "A Modified History Based Weighted Average Voting with Soft- dynamic Threshold," *International Conference on Advances in Computer Engineering*, 2010
8. Latif-Shabgahi, G., Julian M. Bass, and Stuart Bennett. "History-based weighted average voter: a novel software voting algorithm for fault-tolerant computer systems." *Parallel and Distributed Processing, 2001. Proceedings. Ninth Euromicro Workshop on. IEEE*, pp. 402–409, 2001.

# High-Performance Video Retrieval Based on Spatio-Temporal Features

G. S. N. Kumar, V. S. K. Reddy and S. Srinivas Kumar

**Abstract** Many algorithms have been propounded to retrieve videos from a huge database. Yet, they could not reduce the time consumption and their efficiency could completely not satisfy the users. Unlike the existing systems, the proposed approach integrates spatio-temporal features by exploiting the complete video information and it enhances the efficacy of video retrieval. In this paper, we extract color and motion features to obtain spatio-temporal features. We have employed HSV color histogram method for color feature extraction and motion histogram method for extracting video motion feature. Experimental results have shown better performance of these algorithms compared to the existing algorithms in video retrieval.

**Keywords** Shot boundary detection · Keyframe extraction · HSV color space Color histogram · Motion histogram · CBVR

## 1 Introduction

The invention of the computer has revolutionized the field of communication. The amount of time, labor, and level of difficulty has drastically been reduced in retrieving the text or images or videos by users. It has become possible with the advancement of technology. The size of the video database was insignificant when the technology was not advanced. But nowadays, technology has substantially increased resulting in an immense video database. So, the retrieval of videos has become a challenge [1]. The users use textual commands to retrieve videos from the

G. S. N. Kumar · V. S. K. Reddy (✉)
Department of ECE, Malla Reddy College of Engineering & Technology,
Hyderabad, India
e-mail: vskreddy2003@gmail.com

S. Srinivas Kumar
Department of ECE, Jawaharlal Nehru Technological University, Kakinada, India
e-mail: gsrinivasanaveen@gmail.com

© Springer Nature Singapore Pte Ltd. 2018
J. Anguera et al. (eds.), *Microelectronics, Electromagnetics and Telecommunications*, Lecture Notes in Electrical Engineering 471,
https://doi.org/10.1007/978-981-10-7329-8_44

search engines such as YouTube, Google, Bing, blinkx, etc., videos retrieved from textual commands differ from those retrieved by the actual video content with respect to the needed features of videos [2].

We preferred the content-based video search to the conventional text based video search since the video database has increased substantially. Text is not the content of the video hence the retrieval of videos should be based on actual information of the video. The existing video retrieval systems base on low-level features like color, size, shape, texture, and so on, for the retrieval of videos. But the retrieved videos have less accuracy and efficiency. Moreover, these retrieval systems are not able to judge the semantic content of the video. To do away with these drawbacks, we have to integrate both low-level and high-level features [3]. According to the newly proposed algorithm, we have to extract color and motion features by adopting the methods such as HSV color histogram and motion histogram. The color feature gives the spatial information and the motion feature gives temporal information of videos [4]. The structuring of the remaining sections in this paper is mentioned below: Sect. 2 discusses the proposed system. Section 3 shows the experimental results. Section 4 focuses on the conclusion and the future scope of this paper.

## 2  The Proposed System

A video is a collection of time-sequenced frames. It consists of scenes, shots, and frames. The objective of a video structure analysis is to obtain scene segmentation, shot boundary detection and keyframe extraction [5, 6]. Shot boundary detection is a process of extracting successive video frames. Shot is a continuous video recording between the starting and the stopping of the camera button. There are two types of shot boundaries, one of which is a cut which is an abrupt change and the other is gradual transition [7]. In this proposed system, shot boundaries are detected with the Scale Invariant Feature Transform (SIFT) algorithm [8]. SIFT is a robust algorithm and remains invariant to the changes in illumination, rotation, and the scaling of the image and the addition of noise.

A shot has repeated frames which result in great redundancy that needs to be eliminated so that the shot content can be reflected through selective frames called key frames. The extracted key frames should contain most of the significant content of the shot and avoid as much redundancy as possible [9]. In this paper, key frames are extracted using the Image Information Entropy method [10]. We performed the video retrieval by giving an input, i.e., query sample video to get similar videos from the video database. The query sample video is either a shot or a scene. To make the process of retrieving similar videos from the database, simpler and precise criteria—feature extraction and similarity matching are employed. We carry out the feature extraction by integrating color and motion features. Color feature extraction yields spatial visual content and motion feature extraction gives temporal content of the video. Similarity between the query sample video feature vector and the database videos feature vectors is obtained by using the Euclidean distance method.

## 2.1 Feature Extraction

Comparison of a query shot or a scene with the video database using pixel-based method is cumbersome and time-consuming. Comparing the selective features of a query shot or a scene with those of database videos is reliable and yields relevant results. Then, we verify the matching between the characteristics of the query shot and those of the database shot by comparing their features.

1. **Color Feature Extraction**: The characteristic of color is the mostly employed feature for the extraction of spatial information of the video since it has a large amount of visual content of the video [11]. Mostly, color features are extracted by using the color histogram method which is a statistical representation of the intensity values of the frame. Color histogram depends on the choice of color space and its quantization. RGB color space is well known for the extraction of color histogram. But, in some cases, two different frames may have the same RGB color histogram; Hence, HSV (Hue, Saturation and Intensity value) color space has been chosen for this purpose. We decide the similarity between the two frames by the propinquity of their HSV values [12].

   a. *RGB to HSV Transformation*: At first, the RGB values are normalized with divided r, g, and b values by 255. Let $V_{max}$ correspond to the biggest value of r, g, and b, and $V_{min}$ correspond to the smallest.

Hue Calculations

$$H = \begin{cases} 0, & Vmax = Vmin \\ 60° \times \left(\frac{g-b}{Vmax - Vmin} \, mod 6\right), & Vmax = r \\ 60° \times \left(\frac{b-r}{Vmax - Vmin} + 2\right), & Vmax = g \\ 60° \times \left(\frac{r-g}{Vmax - Vmin} + 4\right), & Vmax = b \end{cases} \tag{1}$$

Saturation Calculation

$$S = \begin{cases} 0, & Vmax = Vmin \\ 0, & Vmax = 0 \\ \left(1 - \frac{Vmin}{Vmax}\right), & Otherwise \end{cases} \tag{2}$$

Value Calculation

$$V = Vmax \tag{3}$$

In the proposed system, hue, saturation, and intensity value have been quantized with 16 bins, 4 bins, and 4 bins consecutively since the human visual system can perceive hue more vividly than saturation and intensity. This quantization produces $16 \times 4 \times 4 = 256$ number of distinct colors (bins). The bins with nonzero count pixels are considered color objects.

2. **Motion Feature Extraction**: In the proposed feature extraction method, the other feature that is considered is the motion feature [13]. It shows the correlation between the frame sequences within a video shot. It is employed to yield temporal variations of a video. In this method, motion histogram is used to obtain motion feature in the shot. The motion histogram is quantized with 60 bins toward positive direction, with the same number of bins toward negative direction and with one bin for no change eventually resulting in 121 bins for horizontal direction and the same number of bins for vertical direction as well. Hence, the number of distinct bins for the motion histogram is $121 \times 121$ [14]. The existing approaches to find the motion vectors between two successive frames are feature matching and optical flow computation, but these approaches make the computation more complex [15]. Therefore, motion vector extracted from MPEG video format is used as a substitute. MPEG format has three types of frames: They are I frame, P frame, and B frame. I frame is intra-pictured and it does not have motion data; the P frame is a predicted picture; it allows the forward motion prediction and B frame is a bidirectional prediction picture; it has both forward and backward direction [16]. In this approach, only the forward direction prediction is considered. So, the p frame has been extracted to compile the motion histogram. If the average of the motion histograms of all the frames in the shot is taken, the final normalized histogram is obtained; this is the motion feature of the shot.

## 2.2   Step-by-Step Implementation of CBVR System

Step-by-step implementation of CBVR system is classified as offline process and online process

Offline Process

1. Input video is taken from video database

2. All the frames are extracted from the video

3. Shot boundaries are detected by using SIFT feature extraction

4. Image information entropy is used to extract key frames

**Fig. 1** Block diagram of the
proposed CBVR system

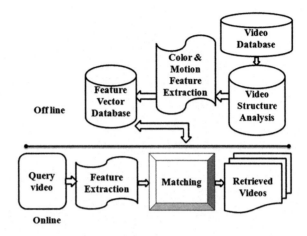

5. HSV color histogram is used to extract color features and motion features are extracted by using motion histogram; both the features are combined to generate integrated feature vector.

6. Repeat Step 1 to 5 for all the videos in the video database and generate feature vector database.

Online

7. Query shot/scene is taken

8. Steps 2, 3, 4, and 5 are repeated to generate query feature vector.

9. The similarity between query feature vector and database feature vector is matched by using Euclidean distance.

10. Videos are retrieved if the distance is less than the threshold value.

The step-by-step implementation of the proposed algorithm is shown in Fig. 1.

# 3  Experimental Results

In this experiment, we have taken 300 videos of diverse categories such as movies, sports, news, and cartoons. First, we extracted color features for each keyframe by using HSV color histogram and motion features have been extracted by using motion histogram. Next, we have integrated color and motion feature vectors and stored them in the feature vector database. We followed the suit for the offline video database and online query for which integrated feature vector was generated. Later, the similarity matching between the query feature vector and the feature vector

**Fig. 2** Query video

database has been calculated by using Euclidean distance. We made use of the threshold value to establish similarity or dissimilarity between query and database feature vector. If the Euclidean distance is less than the threshold value, the two feature vectors are similar otherwise they are dissimilar [17].

The videos that are indexed with similar feature vectors will be retrieved from the video database. The query video is shown in Fig. 2 and the retrieved videos are shown in Fig. 3.

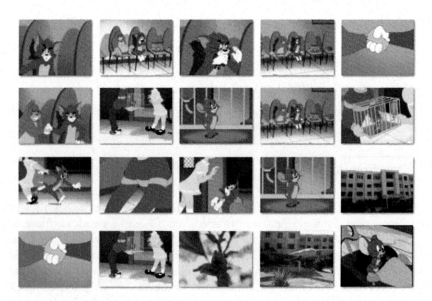

**Fig. 3** Retrieved videos using integrated feature extraction CBVR system

**Table 1** The precision and recall values for different categories of videos

| Type of query video | Color histogram | | Motion histogram | | Proposed method | |
|---|---|---|---|---|---|---|
| | Precision (%) | Recall (%) | Precision (%) | Recall (%) | Precision (%) | Recall (%) |
| News | 73.7 | 75.4 | 80.3 | 79.6 | 93.7 | 91.5 |
| Flowers | 74.4 | 73.9 | 79.6 | 77.5 | 91.6 | 90.1 |
| Cartoon | 72.6 | 71.4 | 76.9 | 75.8 | 91.9 | 94.6 |
| Sports | 75.3 | 72.6 | 77.9 | 74.9 | 93.5 | 86.9 |
| Movies | 73.5 | 75.7 | 78.6 | 76.3 | 92.4 | 92.3 |

The performance measurement of the proposed video retrieval system has been evaluated by using precision and recall rate. The precision and recall are defined as follows:

Precision: It is the ratio of the relevant retrieved videos and the total number of retrieved videos.

$$Precission = \frac{Number\ of\ relevant\ videos\ retrieved}{Total\ number\ of\ videos\ retrieved} \tag{4}$$

Recall rate: It is the ratio of the relevant retrieved videos and the total number of relevant videos in the database.

$$Recall = \frac{Number\ of\ relevant\ videos\ retrieved}{Total\ number\ of\ relevant\ videos} \tag{5}$$

The precision and recall rate of various categories of videos compared to existing algorithms are shown in Table 1.

The above table shows that the proposed algorithm has high precision and high recall rate. Therefore, the proposed algorithm is superior to the existing algorithms. The efficiency or performance of CBVR system depends on the selection of feature extraction. In this paper, an integrated form of both spatial feature and temporal feature has been proposed. Therefore, complete video content can be extracted. Color feature, i.e., spatial feature has been extracted by using HSV color histogram and more visual data has been obtained from this. Hence, relevant videos have been retrieved in this algorithm.

# 4 Conclusion

A new algorithm of color and motion features based on video retrieval has been proposed. The proposed approach exploits the complete video information by integrating spatio-temporal features in contrast to the existing algorithms. The proposed system has been tested with a huge database which contains 300 videos of different categories. We evaluated the system with the precision and recall rate and achieved high precision and high recall rate. The results that have been obtained from the experiments show that the proposed algorithm is being used efficiently with high performance.

# References

1. Hu, Weiming, et al. "A survey on visual content-based video indexing and retrieval." IEEE Transactions on Systems, Man, and Cybernetics, Part C (Applications and Reviews) 41.6 (2011): 797–819.
2. Patel, B. V., and B. B. Meshram. "Content based video retrieval systems." arXiv:1205.1641 (2012).
3. Megrhi, Sameh, Wided Souidene, and Azeddine Beghdadi. "Spatio-temporal salient feature extraction for perceptual content based video retrieval." Colour and Visual Computing Symposium (CVCS), 2013. IEEE, 2013.
4. Gao, Han-ping, and Zu-qiao Yang. "Content based video retrieval using spatiotemporal salient objects." Intelligence Information Processing and Trusted Computing (IPTC), 2010 International Symposium on. IEEE, 2010.
5. Zhao Guang-sheng, A Novel Approach for Shot Boundary Detection and Key Frames Extraction, 2008 International Conference on Multimedia and Information Technology, IEEE
6. Hannane, Rachida, et al. "An efficient method for video shot boundary detection and key frame extraction using SIFT-point distribution histogram." International Journal of Multimedia Information Retrieval 5.2 (2016): 89–104.
7. Wu, Zhonglan, and Pin Xu. "Shot boundary detection in video retrieval." Electronics Information and Emergency Communication (ICEIEC), 2013 IEEE 4th International Conference on. IEEE, 2013.
8. D. G. Lowe, "Distinctive image features from scale-invariant keypoints," International Journal of Computer Vision, vol. 60, pp. 91–110, 2004.
9. Ren, Liping, et al. "Key frame extraction based on information entropy and edge matching rate." Future Computer and Communication (ICFCC), 2010 2nd International Conference on. Vol. 3. IEEE, 2010.
10. Lina Sun and Yihua Zhou, "A key frame extraction method based on mutual information and image entropy," 2011 International Conference on Multimedia Technology, Hangzhou, 2011, pp. 35–38.
11. Daga, Brijmohan. "Content based video retrieval using color feature: an integration approach." In Advances in Computing, Communication, and Control, pp. 609–625. Springer, Berlin, Heidelberg, 2013.
12. Ma, Ji-quan. "Content-based image retrieval with HSV color space and texture features." Web Information Systems and Mining, 2009. WISM 2009. International Conference on. IEEE, 2009.
13. Tahayna, Bashar, Mohammed Belkhatir, and Saadat Alhashmi. "Motion information for video retrieval." Multimedia and Expo, 2009. ICME 2009. IEEE International Conference on. IEEE, 2009.
14. Yi, Haoran, Deepu Rajan, and Liang-Tien Chia. "A new motion histogram to index motion content in video segments." Pattern Recognition Letters 26.9 (2005): 1221–1231.
15. Chun, Young Deok, Nam Chul Kim, and Ick Hoon Jang. "Content-based image retrieval using multiresolution color and texture features." IEEE Transactions on Multimedia 10, no. 6 (2008): 1073–1084.

16. Hu, Rui, Stuart James, and John Collomosse. "Annotated free-hand sketches for video retrieval using object semantics and motion." Advances in Multimedia Modeling (2012), Springer: 473–484.
17. Malik, Fazal, and Baharum Baharudin. "Analysis of distance metrics in content-based image retrieval using statistical quantized histogram texture features in the DCT domain." Journal of king saud university-computer and information sciences 25.2 (2013): 207–218.

# Performance Comparison of Commercially Available RF Analog and Mixed Signal Simulation Tools Using Benchmark Circuits

Vaibhav Ruparelia, Mayank Chakraverty, Sunita S. Desai
and P. S. Harisankar

**Abstract** There are various commercially available Analog/RF simulator tools currently in the market, which have their distinctive applications and advantages. In this paper, some of the most widely used Analog/RF simulators (Cadence Spectre/APS, Keysight ADS and GoldenGate, and Mentor Graphics AFS) have been reviewed with respect to their performance and unique features. An LC-VCO and a CMOS Ring Oscillator are designed using GLOBALFOUNDRIES 45 nm RFSOI technology PDK. They are simulated using all the four listed simulators and their results have been analyzed with respect to performance and circuit design aspects.

**Keywords** Circuit simulator · Radio-Frequency · PDK · Cadence
Spectre · Keysight · ADS · GoldenGate · Mentor graphics · AFS

## 1 Introduction

A wide variety of analyses including DC, AC, noise, transient, etc., that are very useful to baseband circuit designers can be carried out using the SPICE circuit simulator [1]. However, RF designers need more analyses based on a circuit's periodic steady state solution in order to design and analyze RF circuits. In recent years, the RF versions of each of the SPICE-supported analyses and their extensions for RF circuits have been developed, which are built on two distinct core

V. Ruparelia (✉) · M. Chakraverty · S. S. Desai · P. S. Harisankar
RF Analog & Mixed Signal PDK Design Enablement Group, GLOBALFOUNDRIES,
Manyata Embassy Business Park, Nagavara, Bangalore 560045, Karnataka, India
e-mail: vaibhavaruparelia@gmail.com

M. Chakraverty
e-mail: nanomayank@yahoo.com

S. S. Desai
e-mail: desai1988@gmail.com

P. S. Harisankar
e-mail: harisps09@gmail.com

© Springer Nature Singapore Pte Ltd. 2018                                    443
J. Anguera et al. (eds.), *Microelectronics, Electromagnetics*
*and Telecommunications*, Lecture Notes in Electrical Engineering 471,
https://doi.org/10.1007/978-981-10-7329-8_45

algorithms. The first algorithm is harmonic balance and the second is shooting method. A circuit's periodic steady state solution was initially being computed using both harmonic balance and shooting methods. A generalization of these methods with time helped in catering to a wide range of functionalities as required by the RF designer. These methods, in their original form, had been restricted to fairly smaller circuits. For being applied to much larger circuits, both harmonic balance and shooting methods obtained an acceleration recently with the application of Krylov subspace methods [1].

There are various commercial analog/RF simulator tools available currently in the market, in addition to the open-source ones. These proprietary simulator engines have their distinctive applications and advantages. In addition to the good quality device models, that provide best hardware-model correlation, the right circuit simulator used is also of importance to deliver first-time right designs that are matching the hardware. The selection of a right simulator is also important to reduce simulation time for complex post-extracted circuit blocks, and when exhaustive testing (fixed-corner simulations, Monte-Carlo simulation, etc.) is required.

In this paper, some of the most widely used analog/RF simulators will be reviewed with respect to their performance and unique features. The simulators reviewed in this paper are Cadence Spectre/APS, Keysight ADS and GoldenGate, and Mentor Graphics (formerly Berkeley Design Automation) AFS. A case study of 2 circuits (a cross-coupled LC-VCO and a CMOS Ring Oscillator) is shown, where-in the various simulators are compared in some details. GLOBALFOUNDRIES 45 nm RFSOI PDK is used for designing and simulating these 2 circuits and evaluating the above mentioned analog/RF simulators. Note that the optimization of specifications of these circuits is not the goal of this paper. But the focus is to compare various simulators, used for simulating these circuits.

## 2 Basics of Reviewed Simulators

### 2.1 Cadence Spectre

Cadence Spectre represents a family of one of the most innovative and cutting-edge circuit simulators. With the ability to carry out simulations at the differential equation level, Cadence Spectre is ideally suited to simulating both digital and analog circuits. Compared to SPICE, Spectre offers much better simulation speed and convergence characteristics, owing to all the enhancements done to the algorithm that Spectre uses. Spectre RF is also offered by Cadence, to cater specifically to simulation of RF circuit designs, which include analyses related to different types of calculations involving efficient transfer function computation, estimation of the operating point, noise calculation and distortion analysis of common RF and communication circuits that include oscillators, mixers, switch-capacitor filters and so on [2].

Both pre and post layout SPICE simulations of RF analog and mixed signal designs are carried out with high accuracy using the base Spectre simulator. Spectre is the most widely used format for writing models in most of the foundries and hence the Spectre simulator has been accepted as the standard sign-off simulator. Released in around 2010, the Spectre Accelerated Parallel Simulator (Spectre APS) was released by Cadence to replace Spectre Turbo in response to the heavy competition from other EDA companies. For the same accuracy and convergence, a performance improvement of 10–100× over base Spectre can be achieved using APS, as claimed by Cadence. A further boost in performance of APS can be obtained by utilizing multi-core processing capabilities and "++aps" mode (fast APS mode). Taking error tolerances and constraints into account, the performance is improved in fast APS mode by using a transformed time-step control algorithm which is not used in the regular APS mode [3].

## 2.2  Keysight Advanced Design System (ADS)

ADS is an EDA tool, offered by Keysight Technologies, that caters to RF, microwave and high speed digital applications. In addition to Harmonic Balance based RF simulations, ADS can also support an integrated 3D planar EM simulation capability using Momentum tool. The seamless integration of ADS/ Momentum tools enable users to create physical parts (like inductors, transmission lines, etc.) and add them directly into a circuit simulation. Therefore, RF design engineers consider ADS as one of the most preferred design environments in the RF design space. Keysight ADS supports the complete RF design flow, which includes schematic entry, frequency and time-domain simulations, EM simulation and layout [3]. ADS also supports X-parameters, which are mathematically correct extensions of S-parameters, used to characterize the relative phase and amplitudes of harmonics generated by non-linear components under large input power levels at all ports.

While using ADS for RF/analog simulations, many circuit designers prefer using Cadence Virtuoso environment, to utilize the best of both simulation as well as layout suites [3]. Therefore Keysight had released the Dynamic Link flow to allow the designers to use a Cadence schematic in an ADS simulation environment. This requires using Cadence p-cells to create a schematic in its environment and then simulating the block in ADS environment by instantiating its symbol (created in Cadence). While simulating, ADS reads the Spectre netlist and creates a wrapper netlist around it as per the ADS simulation test bench. ADS then runs the simulation and displays the results within its own environment. ADS can also be integrated into Cadence environment using an interoperable PDK approach, wherein ADS p-cells and callbacks are incorporated, to be directly used in Cadence Virtuoso environment.

## 2.3 Keysight GoldenGate

For RF analog and mixed signal IC designs that are integrated fully into the Cadence Analog Design Environment (ADE), Keysight's GoldenGate simulator enables a solution for carrying out advanced simulation and analysis. After the acquisition of Xpedion in 2009, GoldenGate was made available by Keysight (then Agilent). Like ADS, it also supports RF simulations, Momentum etc. Simulation of the complete characterization of the entire RF transceiver inclusive of the parasitic components is a unique capability of GoldenGate. In addition to the ADE functionality, GoldenGate supports the ADS Data Display and X-parameters. The GoldenGate also uses Harmonic Balance simulation methods for the various RF analyses offered.

## 2.4 Analog FastSPICE (AFS)

As part of the acquisition of Berkeley Design Automation, it was in the year 2014 that Mentor Graphics went on to acquire Analog FastSPICE, abbreviated as AFS. Compared to the traditional SPICE simulator, a performance enhancement of 5 to 10 times can be delivered, as claimed by AFS with single-core. Also, an improvement in performance by 2 to 6 times is claimed over parallel SPICE simulators [4]. There is a perfect integration of AFS within the Cadence Virtuoso ADE and the Cadence ViVA waveform tool is ideal at displaying the output results directly without any post processing. Simulation of circuits with more than 100K circuit elements for the full-spectrum PSS and PNOISE analyses is one of the important capabilities that AFS offers. AFS produces outputs in industry standard formats by providing support for standard SPICE netlist formats and standard foundry device models [4].

## 3 Circuits Used for Performance Comparison

### 3.1 Cross-Coupled LC VCO

To begin with, a cross-coupled LC-VCO has been designed and simulated, to compare the simulation time and other aspects of the 4 mentioned simulators. It has been chosen as a starting point for the analyses done in this paper, since it requires some specifications (like frequency of oscillations, output voltage amplitude, power consumption, etc.) to be calculated/measured from time-domain analyses (transient) and some specifications (like phase noise, output power, etc.) to be measured from frequency domain analyses (PSS/PNOISE/etc.). The VCO circuit also requires parametric sweep (to find the tuning range) and initial condition option. Thus, as

**Fig. 1** Cross-coupled
LC-VCO

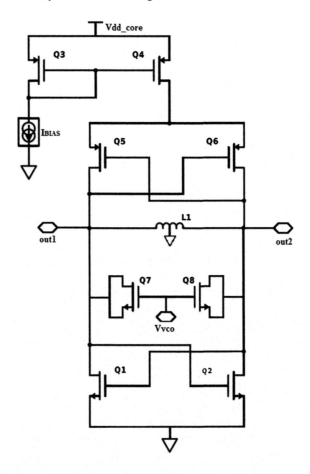

depicted in Fig. 1, the cross-coupled LC-VCO seems to be a good candidate for comparing these different simulators in some detail.

VCO is also an important part of RF transceiver system and PLLs. In an RF Front-End-Module (FEM), it is used as a Local Oscillator (LO) to provide the inputs signal to a mixer, which down-converts the received data signal. Any noise generated or propagated by LO will strongly affect the noise performance of the mixer. Phase noise is an important specification of an oscillator and it corresponds to the variations in the period/frequency of oscillations, termed as Jitter.

In the current LC-VCO design, RF FETs provided in the GLOBALFOUN-DRIES 45 nm RFSOI PDK have been used, which exhibits lesser flicker noise, that will directly correlate with the VCO phase-noise. The complementary structure (PMOS-NMOS) is helpful in maintaining single-ended symmetry for each half circuit to mitigate the up-conversion of flicker noise. A CMOS pair is used in the current design to take the advantages of both low flicker noise in PMOS (~1/10th of NMOS) and higher gain (gm) in NMOS, which results in higher overall

**Table 1** Simulation results—Spectre, ADS, GoldenGate & AFS for cross-coupled LC-VCO

| Specifications | Details | Spectre | AFS | ADS | GoldenGate |
|---|---|---|---|---|---|
| Simulation time | DC-OP | ~1 ms | <1 s | ~10 ms | ~10 ms |
| | TR (1 μs) | ~746 s | ~271 s | ~1217 s | ~486 s |
| | PSS/ PNOISE | ~50.95 s | ~35.33 s | ~48.1 s | ~8.4 s |
| Fosc (GHz) | (from TR) | 31.97 | 31.91 | 32.15 | 32.02 |
| Ivdd_avg (mA) | | 1.2 | 1.29 | 1.16 | 1.26 |
| Vout_pp (V) | | 1.68 | 1.38 | 1.68 | 1.53 |
| Phase noise (dBc/Hz) | @ 10 kHz | −26.61 | −54.66 | −22.59 | −22.66 |
| | @ 1 MHz | −87.37 | −87.79 | −87.69 | −87.65 |
| Peak mem used (MB) | (pnoise/pss) | 75.1 | 75.5 | 75.6 | 54.9 |

Figure-of-Merit (FOM). Both NMOS and PMOS pairs generate negative resistance to the LC-tank. A major demerit of CMOS cross-coupled pair is the requirement of double headroom to keep NMOS & PMOS in saturation, which in turn limits the minimum VDD value. The simulation results of various simulators, from a first-cut design, are as shown in Table 1.

The results shown above are based on multiple single-core simulation runs, to average out the differences due to changing CPU load conditions. The CPU/run time is noted from the simulation log files of the corresponding simulators and averaged out from multiple runs. Transient simulations have been run for 1us simulation time, so as to get notable differences in CPU/run time. The difference in tolerance settings of these simulators can lead to significant differences in run time as well as cause convergence issues. Different integration algorithm options (default, trap, gear, trapgear2, etc.) also change the simulation time and results to some extent. Therefore, the tolerance settings like reltol, vasbtol, iabstol, pivrel, gmin, maximum time step, etc. have been made the same for all the reviewed simulators. Latest available versions have been used for all four reviewed simulators.

It should be noted that there are no PSS/PNOISE analyses in GoldenGate and ADS. In GoldenGate, a CR (Carrier Analysis) is used instead. Carrier Analysis provides the means to compute the steady-state response (voltage, current, power, and oscillation frequency) of a nonlinear circuit. It also provides a very accurate noise analysis functionality for both forced circuits and oscillators under large single tone or multi-tone excitation. Oscillator analysis is also run on top of Carrier Analysis, which performs the computation of the oscillation frequency of any autonomous circuit that produces periodic or quasi-periodic signals [5]. Similarly, for ADS, Harmonic Balance Controller has been used for Phase-noise simulations [6]. Harmonic Balance is a frequency domain method for calculating the steady-state response of nonlinear electrical circuits. It starts with KCL written in frequency domain and a chosen number of harmonics. When a sinusoidal signal is applied to a nonlinear device, harmonics of the fundamental frequency are generated. This method then assumes that the solution can be represented by a linear combination of sinusoids, and then balances current and voltage sinusoids to satisfy Kirchhoff's law [7–9].

From the simulation results, it can see that the specifications from transient analysis are almost matching across all the simulators. The CPU/run time for transient simulation is least for AFS and GoldenGate, and highest for ADS. The Transient simulation run time is much smaller (~365.75 s) using Spectre APS+ + option, with approximately same result values. In addition to Shooting method, Cadence Spectre also started offering Harmonic Balance simulation technique for PSS analysis from MMSIM6 onwards. The Harmonic Balance PSS results for Spectre have been found to be almost similar to the Shooting method results, for this circuit. The PSS/PNOISE run time is least for GoldenGate and largest for Spectre. It should be noted that the AFS Phase noise value @ 10 kHz is significantly off from other 3 simulators. AFS uses Shooting method for PSS/PNOISE simulations, and currently single-tone Harmonic Balance analysis and Harmonic Balance for oscillators is not supported by AFS.

## 3.2 CMOS Ring Oscillator

It has been observed that for circuits with a few frequency components like LNA and mixer, the frequency domain simulation technique is more efficient. And the time domain simulation technique is more efficient for circuits with abrupt edges [3]. Therefore, a CMOS Ring Oscillator was also designed and simulated using various simulators. It was chosen for this work, as its output waveform has abrupt edges (typically square wave, assuming enough number of inverter stages), which contradicts LC-VCO with relatively smoother edges (sinusoidal wave). A 9-stage CMOS Ring Oscillator using minimum NMOS width has been designed using GLOBALFOUNDRIES 45 nm RFSOI technology PDK (refer Fig. 2).

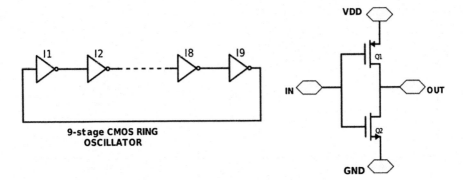

**Fig. 2** 9-stage CMOS ring oscillator and minimum sized CMOS inverter

**Table 2** Simulation results with 9-stage ring oscillator

| Specifications | Details | Spectre | AFS | ADS | GoldenGate |
|---|---|---|---|---|---|
| Simulation time | DC-OP | ~6 ms | ~0 s | ~10 ms | ~20 ms |
| | TR (1us) | ~801 s | ~456 s | ~810 s | ~1198 s |
| Fosc (GHz) | | 17.37 | 18.52 | 18.1 | 17.86 |
| Vout_pp (V) | | 0.986 | 0.964 | 0.984 | 0.957 |
| Ivdd_avg (uA) | | 101.1 | 116.4 | 99.85 | 112.1 |
| Peak mem used (MB) | | 103 | 60 | 102 | 61 |

To ensure symmetrical propagation delays, widths of P-MOSFETs have been sized 1.25 times that of the NMOS FETs. The simulation results for various simulators are as shown in Table 2. Again, the tolerance/accuracy settings are made same for all the simulators, and the CPU/run time is averaged out from multiple simulations. The time taken for netlist expansion and checking out the license is not included in the values mentioned in the results table, as all the simulators do not reveal this detailed breakup of simulation time. The specification values obtained from these four simulators are in close vicinity. The CPU/run time is the least for AFS/Spectre, and largest for GoldenGate/ADS. This shows to some extent that the Harmonic Balance technique is faster for frequency domain analyses and for circuits with limited number of harmonics present. Figure 3 below summarizes the general recommendation for using a particular simulator based on the circuit applications and analyses of interest.

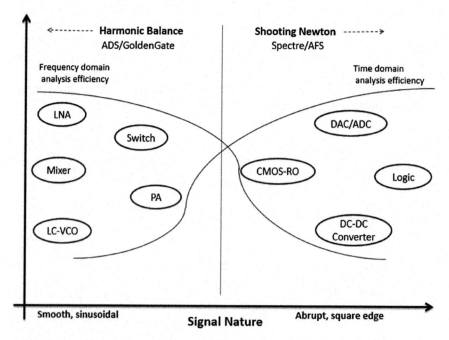

**Fig. 3** Ideal simulator based on circuit/application [3]

# 4 Conclusion

In this work, four most widely used Analog/RF simulators have been reviewed and study of their unique aspects and dissimilarities has been done in some detail. For most of the foundry models, the Spectre simulator is considered the golden simulator, because of its advantages like availability of a complete Virtuoso design suite and seamless integration with other Cadence products. For RF systems, one of the best harmonic balance simulation engines is offered by ADS and GoldenGate. Due to the availability of specific RF analyses, a majority of circuit designers in the RF design space prefer to use ADS or GoldenGate [3]. However, for analog and custom digital circuits, where transient simulations are needed, Spectre APS/APS++ or Mentor Graphics AFS would provide faster results. The specifications, simulation time and peak memory used have been presented, by means of simulating an LC-VCO and a CMOS Ring oscillator. The specification results from the various reviewed simulators match reasonably well, except for AFS in LC-VCO Phase Noise. The circuits with abrupt wave edges will have a very large number of harmonics and therefore might be difficult to converge using Harmonic Balance technique. For such circuits, Shooting Newton and similar techniques might converge better and faster. However, Harmonic Balance algorithm is much faster for mildly non-linear blocks, wherein many linear models are best represented in frequency domain, since simulating such components in time domain could result in more time and/or convergence issues.

# References

1. Kenneth S. Kundert: Introduction to RF Simulation and Its Application. IEEE Journal of Solid State Circuits, Vol. 34, pp. 1298–1319, September 1999.
2. "Virtuoso Spectre Circuit Simulator Reference", Cadence.
3. Wan, Bin, and Xingang Wang: Overview of commercially-available analog/RF simulation engines and design environment. 12th IEEE International Conference on Solid-State and Integrated Circuit Technology (ICSICT), 2014.
4. "Analog FastSPICE Platform–Analog/Mixed-Signal Verification Datasheet", www.mentor.com.
5. "Simulation Types–GoldenGate 2015.01", Keysight Technologies.
6. "Harmonic Balance Simulations–ADS2016.01", Keysight Technologies.
7. Deuflhard, Peter (2006). "Newton Problems for Nonlinear Problems." Berlin: Springer-Verlag. Section 7.3.3.: Fourier collocation method.
8. Nakhla, Michel S.; Vlach, Jiri (February 1976). "A piecewise harmonic balance technique for determination of periodic response of nonlinear systems". IEEE Transactions on Circuits and Systems. CAS-23: 85–91. ISSN 0098–4094.
9. Gilmore, R. J.; Steer, M. B. (1991). "Nonlinear circuit analysis using the method of harmonic balance—A review of the art. Part I. Introductory concepts". Int. J. Microw. Mill.-Wave Comput. Aided Eng. 1: 22–37. https://doi.org/10.1002/mmce.4570010104.

# DWT-PCA Image Fusion Technique to Improve Segmentation Accuracy in Brain Tumor Analysis

V. Rajinikanth, Suresh Chandra Satapathy, Nilanjan Dey and R. Vijayarajan

**Abstract** Because of its high clinical significance and varied modalities; magnetic resonance (MR) imaging procedures are widely adopted in medical discipline to record the abnormalities arising in a variety of internal organs of human body. Each modality of the MRI, such as T1, T2, T2C, Flair, and DW has its own merit and demerits. Hence, in the proposed work, a unique computer-assisted technique (CAT) is proposed to evaluate the abnormalities in MR images, irrespective of its modalities. Proposed CAT has the following stages: (i) Discrete Wavelet Transform Based Principal Component Averaging (DWT-PCA) image fusion, (ii) Tri-level thresholding based on Social Group Optimization and Shannon's entropy, and (iii) Watershed segmentation. This approach is experimentally assessed with MICCAI brain cancer segmentation (BRATS 2013) challenge database. Experimental results confirms that the proposed approach is efficient in offering better values of Jaccard (84.33%), Dice (90.86%), sensitivity (99.93%), specificity (90.67%), and accuracy (95.74%) compared with the single modality registered brain MR images. Hence, the proposed work is extremely significant for the segmentation of abnormal region from the brain MR images registered using Flair, T1C, and T2 modalities.

V. Rajinikanth
Department of Electronics and Instrumentation, St. Joseph's College of Engineering,
Chennai, 600119 Tamil Nadu, India

S. C. Satapathy (✉)
Department of Computer Science and Engineering, P.V.P. Siddhartha Institute
of Technology, Vijayawada 520 007, AP, India
e-mail: sureshsatapathy@gmail.com

N. Dey
Department of Information Technology, Techno India College of Technology,
Kolkata 700156 West Bengal, India

R. Vijayarajan
Department of Electronics & Communication Engineering, GMR Institute of Technology,
Rajam 532127, AP, India

© Springer Nature Singapore Pte Ltd. 2018
J. Anguera et al. (eds.), *Microelectronics, Electromagnetics*
*and Telecommunications*, Lecture Notes in Electrical Engineering 471,
https://doi.org/10.1007/978-981-10-7329-8_46

453

**Keywords** Brain MR image · Image fusion · Social group optimization
Watershed segmentation · Image similarity evaluation

# 1 Introduction

Image processing techniques are extensively adopted in medical discipline to categorize, examine, and treatment scheduling for a range of diseases. Cancer is a life intimidating sickness that frequently begins in the vital internal human parts, such as blood cells, lungs, bone marrow, breast, brain, and external organ like the skin. Cancer revealing in interior organs is relatively complicated than the cancer in external organs. The cancer cells growing within the internal organs will be treated accurately in order to reduce the morbidity and mortality rates.

In recent years, significant amounts of imaging techniques are used to record the infected regions of the internal organs. MRI is the universal imaging practice extensively adopted in imaging clinics due to its assorted modality. Modalities like T1, T2, T1C, Flair, and diffusion-weighted imaging (DWI) are normally considered to record the abnormalities in human brain [1–5]. Due to its clinical significance, brain MR image processing techniques are widely adopted by the researchers based on a chosen modality [6, 7]. The processing system, which works well on a chosen modality, may fail to offer satisfactory result on other modalities due to the reasons like noise, poor contrast, artifacts, and separation of infected section from normal brain section [8]. Hence, it is always necessary to develop a unique tool to analyze the brain MRI irrespective of its modality. The earlier works also confirm that image fusion technique can be considered to enhance the outcome during image examination [9, 10]. In recent years, image fusion procedures are implemented in key areas, like computer vision, remote sensing, and medical domain [11–13].

The aim of the proposed work is to suggest a hybrid procedure to improve segmentation accuracy during brain tumor MR image analysis. The performance of proposed approach is confirmed using the benchmark MR images of BRATS 2013 database [1, 14]. This dataset consists of the MR images captured using T1, T2, T1C, Flair, and also the Ground Truth (GT) images offered by a radiologist.

In this CAT, a three-step procedure is adopted to improve the segmentation accuracy in brain MR images. The preprocessing approach consists of: (i) Image fusion based on the DWT-PCA and (ii) Social Group Optimization (SGO) assisted Shannon's entropy supported tri-level thresholding. During the postprocessing work, the marker-controlled watershed algorithm is adopted to mine the tumor region from the preprocessed brain images. The efficiency of proposed CAT is confirmed by computing the image similarity measures, such as Jaccard, Dice, sensitivity, specificity, and accuracy with the help of relative examination between the segmented tumor section and the GT [4]. The experimental results confirms that the proposed approach is efficient in obtaining the better image similarity measures compared with the results obtained with single MRI modality images.

## 2  Methodology

This section presents the particulars of methodology proposed in this paper to analyze the benchmark brain tumor database. Figure 1 shows various stages involved in the proposed CAT. Test images for the experimental work are chosen from the BRATS 2013 challenge database. In this study, 35 slices of images from each modality (slice number from 95 to 130 of pseudo name 0004) are considered for the investigation. Figure 1 depicts the stages involved in the developed CAT. In which, the amalgamation of fusion and thresholding function act as the preprocessing stage and the segmentation technique acts as the postprocessing stage. Finally, the performance of the CAT is assessed by comparing the segmented tumor region with the GT existing in the chosen database.

### 2.1  Image Fusion

Image fusion is a widely considered practice to employ pixel-level grouping of images in order to improve its key features. In the proposed work, recent technique known as DWT-PCA combination is adopted to enhance the brain MRI information. This technique was developed by Vijayarajan and Muttan [15] to unite the brain image slices recorded with the computed tomography (CT) and MRI techniques.

This procedure can be mathematically expressed as:

$$Fuse = M_1(avg) * image1 + M_2(avg) * image2, \tag{1}$$

where image1 and image2 are test images of size $M \times N$, $M_1$ is the normalized principal components of wavelet coefficients of *image1*, $M_2$ is the normalized principal component of wavelet coefficients of *image2*, and *avg* is the average of normalized principal components.

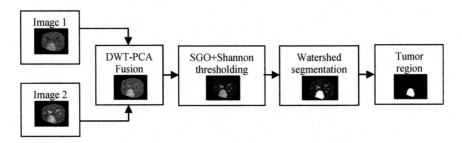

**Fig. 1** Execution of the proposed brain MRI evaluation procedure

## 2.2 Social Group Optimization Assisted Shannon's Thresholding

SGO is a recently invented heuristic approach by Satapathy and Naik [16]. This algorithm is formed by mimicking the behavior in human group. SGO comprises two main phases, namely: (i) improving phase to synchronize the positions of people based on the objective function and (ii) acquiring phase that allows the agents to discover the best potential solution based on the chosen problem.

The mathematical model of SGO can be defined as:

$$Gbest_j = max\{f(X_i) for\ i = 1, 2, \ldots N\}, \tag{2}$$

where $X_i$ is the initial knowledge of people in a group, $i = 1, 2, 3, \ldots, N$ indicate whole people in a group, and $f_j$ is the fitness value.

The improving phase is used to renew the location of every person as in Eq. (3):

$$Xnew_{i,j} = c*Xold_{i,j} + R*(Gbest_j - Xold_{i,j}), \tag{3}$$

where $Xnew$ is the updated position, $Xold$ is the early position, $Gbest$ denotes universal top location, $R$ is an arbitrary numeral [0, 1], and $c$ represents the self-introspection parameter [0, 1]. The $c$ is selected as 0.2 [16, 17].

During the acquiring phase, the agents will find the global solution based on knowledge updating procedure as depicted in Eq. (4).

$$Xnew_{i,j} = Xold_{i,j} + R_a*(X_{i,j} - X_{r,j}) + R_b*(Gbest_j - X_{i,j}), \tag{4}$$

where $R_a$ and $R_b$ are random numbers [0, 1] and $X_{r,j}$ is the arbitrarily selected location of a person in group. In this paper, SGO algorithm is considered to choose the optimal threshold based on the Shannon's entropy [18, 19].

Thresholding procedure is normally considered to group identical image pixels based on user's requirement [20, 21]. Recently, a tri-level thresholding scheme is considered to cluster the digital brain MR images into the background, normal brain region, and the tumor section [22]. The tri-level thresholding will enhance the tumor section, which can be mined by choosing a suitable segmentation procedure. In this paper, Shannon's entropy discussed in [19] is considered to threshold the brain MRI.

The procedure is as follows:

Let us choose an image with size M * N. Then, the gray-level pixel coordinates $(x, y)$ can be expressed as $G(x, y)$, for $x \in \{1, 2, \ldots, M\}$ and $y \in \{1, 2, \ldots, N\}$. Let $L$ be the number of gray levels of the test image and the set of all gray values $\{0, 1, 2, \ldots, L - 1\}$ can be symbolized as $Z$, in such a way that

$$G(x, y) \in Z \,\forall (x, y) \in image \qquad (5)$$

Then, the normalized histogram will be $H = \{h_0, h_1, ..., h_{L-1}\}$.
For tri-level thresholding problem, the above equation can be written as

$$H(T) = h_0(t_1) + h_1(t_2) + h_2(t_3) \qquad (6)$$

$$T^* = \max_T \{H(T)\}, \qquad (7)$$

where $T = \{t_1, t_2, ..., t_L\}$ is the threshold value, $H = \{h_0, h_1, ..., h_{L-1}\}$ is the normalized histogram, and $T^*$ is the optimal threshold.


## 2.3 Watershed Segmentation

In this work, marker-controlled watershed algorithm discussed in [23] is considered to mine the enhanced tumor volume. This approach is the combination of the sobel edge detection algorithm, marker-controlled morphological operation and extraction. A detailed mathematical description of the watershed transform considered in this paper can be found in [24].


## 2.4 Performance Evaluation

A relative analysis between the mined tumor and ground truth is carried out to evaluate the superiority of proposed CAT. The similarity values, like Jaccard, Dice, sensitivity, specificity, and accuracy [1, 4, 5] are computed.
These performance measures are mathematically expressed as

$$\text{Jaccard}(I_{GT}, I_S) = I_{GT} \cap I_S / I_{GT} \cup I_S \qquad (8)$$

$$Dice(I_{GT}, I_S) = 2(I_{GT} \cap I_S)/|I_{GT}| \cup |I_S| \qquad (9)$$

$$Sensitvity = T_P/(T_P + F_N) \qquad (10)$$

$$Specificity = T_N/(T_N + F_P) \qquad (11)$$

$$Accuracy = (T_P + T_N)/(T_P + T_N + F_P + F_N), \qquad (12)$$

where $I_{GT}$ is ground truth image, $I_S$ represents segmented image, $T_P$, $T_N$, $F_P$, and $F_N$ denote true positive, true negative, false positive, and false negative, respectively.

## 3 Result and Discussion

The experimental analysis of the CAT is discussed in this section. This study considers 216 × 160 sized brain MR images registered with a class of modalities. Initially, the SGO and Shannon based tri-level thresholding and watershed segmentation approach is tested on the single modality registered images presented in Fig. 2. This approach presents better result on Flair, T1C, and T2 based MR images and poor result on the T1 modality MR images. Hence, in this study, the images based on Flair, T1C, and T2 are considered for the analysis. Later, the DWT-PCA approach is considered to fuse the test images in order to enhance the tumor section as depicted in Fig. 3. The proposed thresholding and segmentation approach is also tested on the test images depicted in Fig. 3 and the corresponding results are depicted in Fig. 4. From this figure, it can be noted that the proposed segmentation approach is very efficient in extracting the tumor section from Flair, T2, Flair + T2 and Flair + T2 + T1C images compared to T1C and Flair + T1C.

The performance of proposed CAT system is evaluated by comparing the extracted tumor region with the GT. Table 1 presents the image similarity measures computed for the slice_100 image. From this result, it can be noted that the Flair

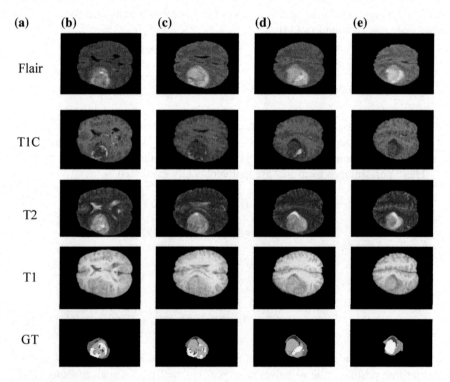

**Fig. 2** Image slices chosen from the BRATS 2013 dataset. **a** Image modality, **b** slice_100, **c** slice_110, **d** slice_120, **e** slice_130

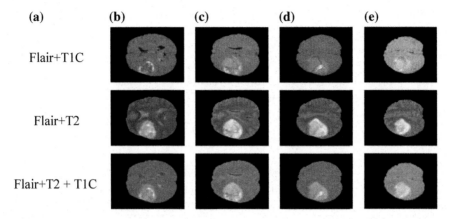

**Fig. 3** Fused image slices of chosen slices from BRATS 2013 dataset. **a** Image modality, **b** slice_100, **c** slice_110, **d** slice_120, **e** slice_130

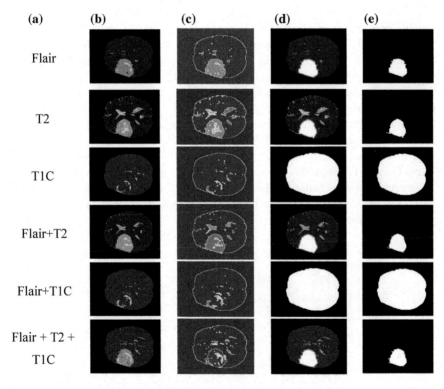

**Fig. 4** Results obtained with the proposed segmentation approach. **a** Image modality, **b** thresholded image, **c** label of detected edges, **d** segmented image, **e** extracted tumor

**Table 1** Image similarity index obtained for slice_100

| Modality | Jaccard | Dice | Sensitivity | Specificity | Accuracy |
|---|---|---|---|---|---|
| Flair | 0.7977 | 0.8874 | 0.9974 | 0.8857 | 0.9399 |
| T2 | 0.7558 | 0.8609 | 0.9989 | 0.8272 | 0.9090 |
| T1C | 0.1544 | 0.2675 | 0.6078 | 1.0000 | 0.7796 |
| Flair + T2 | 0.7818 | 0.8776 | 0.9989 | 0.8542 | 0.9237 |
| Flair + T1C | 0.1544 | 0.2675 | 0.6078 | 1.0000 | 0.7796 |
| Flair + T2 + T1C | 0.7982 | 0.8892 | 0.9994 | 0.8868 | 0.9426 |

**Table 2** Average similarity index obtained with the considered image dataset

| Modality | Jaccard | Dice | Sensitivity | Specificity | Accuracy |
|---|---|---|---|---|---|
| Flair | 0.8417 | 0.9073 | 0.9984 | 0.9012 | 0.9551 |
| T2 | 0.8183 | 0.8974 | 0.9936 | 0.8985 | 0.9272 |
| Flair + T2 | 0.8209 | 0.9003 | 0.9940 | 0.8987 | 0.9295 |
| Flair + T2 + T1C | 0.8433 | 0.9086 | 0.9993 | 0.9067 | 0.9574 |

modality registered test image offers better similarity values compared to T2, T1C, Flair + T2, and Flair + T1C images. But the results offered by the DWT-PCA fused; Flair + T2 + T1C modality image offers superior performance measures compared with other modalities. From this result, it is confirmed that Flair + T2 + T1C modality fused image offers better segmentation accuracy. This approach is further tested on 105 slices (Flair = 35, T1C = 35 and T2 = 35) of BRATS 2013 dataset with the pseudo name 0004 and the average similarity index values obtained are depicted in Table 2 and Fig. 5. From these results, it can be noted that overall performance of Flair + T2 + T1C image is superior than Flair, T2 and Flair + T2. From this work, it is confirmed that fusion based on DWT-PCA is efficient in offering better similarity index values. In future, the performance of DWT-PCA fusion approach can be validated against other fusion approaches existing in the literature.

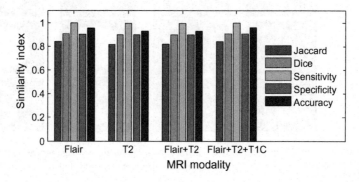

**Fig. 5** Graphical representation of the average similarity index between the ground truth and extracted tumor section

# 4 Conclusion

This work implements a CAT by combining the fusion, thresholding, and segmentation procedures to improve the accuracy during the MRI brain tumor analysis. In this study, benchmark BRATS 2013 MRI dataset is considered to test the performance of the developed CAT. The brain tumor images recorded with the Flair, T1C, and T2 are initially analyzed and the results confirm that flair modality images offered better image similarity measures. In order to improve the outcome of the proposed CAT, a recent image fusion technique, known as DWT-PCA, is considered to generate Flair + T2, Flair + T1C and Flair + T2 + T1C images. Similar experimental work is repeated once again using the fused brain MR images and the average result of Flair + T2 + T1C image offers better values of Jaccard (84.33%), Dice (90.86%), sensitivity (99.93%), specificity (90.67%), and accuracy (95.74%) compared with the single modality registered brain MR images. These results confirm that DWT-PCA image fusion procedure can be used to enhance the outcome during the brain MR image analysis.

# References

1. Menze, B., Reyes, M., Leemput, K.V. et al.: The multimodal brain tumor image segmentation benchmark (BRATS). IEEE Transactions on Medical Imaging, Vol. 34, No. 10, pp. 1993–2024 (2015).
2. Havaei, M., Davy, A., Warde-Farley, D. et al.: Brain tumor segmentation with deep neural networks, Med. Image Anal., Vol. 35, pp. 18–31 (2017). https://doi.org/10.1016/j.media.2016.05.004.
3. Schmidt, M.A. and Payne, G.S: Radiotherapy planning using MRI, Phys. Med. Biol., Vol. 60, pp. R323–361 (2015). https://doi.org/10.1088/0031-9155/60/22/R323.
4. Chaddad, A. and Tanougast, C: Quantitative evaluation of robust skull stripping and tumor detection applied to axial MR images, Brain Informatics, Vol. 3, No. 1, pp. 53–61 (2016). https://doi.org/10.1007/s40708-016-0033-7.
5. Rajinikanth, V., Satapathy, S.C., Fernandes, S.L. and Nachiappan, S: Entropy based Segmentation of Tumor from Brain MR Images—A study with Teaching Learning Based Optimization, Pattern Recognition Letters, Vol. 94, pp. 87–94 (2016). https://doi.org/10.1016/j.patrec.2017.05.028.
6. Rani, J., Kumar, R., Talukdar, F.A. and Dey, N: The Brain tumor segmentation using Fuzzy C-Means technique: A study, Recent Advances in Applied Thermal Imaging for Industrial Applications, IGI global, pp. 40–61 (2017).
7. Moraru, L., Moldovanu, S., Dimitrievici, L.T., Ashour, A.S. and Dey, N: Texture anisotropy technique in brain degenerative diseases, Neural Computing and Applications (2016). https://doi.org/10.1007/s00521-016-2777-7.
8. Vijayarajan, R. and Muttan, S: Local principal component averaging image fusion, International Journal of Imaging & Robotics, Vol. 13, No. 2, pp. 94–103 (2014).
9. Tharwat, A., Gaber,T., Awad, Y.M., Dey, N. and Hassanien, A.E: Plants identification using feature fusion technique and bagging classifier, Advances in Intelligent Systems and Computing, Vol. 407, pp. 461–471 (2016).

10. Dey, N., Das, S. and Rakshit, P: A novel approach of obtaining features using wavelet based image fusion and harris corner detection, International Journal of Modern Engineering Research, Vol. 1, No. 2, pp. 396–399 (2011).
11. Wang, D., Li, Z., Cao, L., Balas, V., Dey, N., Ashour, A. McCauley, P., Dimitra, S. and Shi, F: Image fusion incorporating parameter estimation optimized Gaussian mixture model and fuzzy weighted evaluation system: a case study in time-series plantar pressure dataset, IEEE Sensors Journal, Vol. 17, No. 5, pp. 1407–1420 (2017).
12. Vijayarajan, R. and Muttan, S: Iterative block level principal component averaging medical image fusion, Optik-International Journal for Light and Electron Optics, Vol. 125, No. 17, pp. 4751–4757 (2014).
13. Vijayarajan, R. and Muttan, S: Fuzzy C-means clustering based principal component averaging fusion, International journal of fuzzy systems, Vol. 16, No. 2, pp. 153–159 (2014).
14. Farahani, K., Menze, B., Reyes, M., 2013. Multimodal Brain Tumor Segmentation (BRATS 2013). URL: http://martinos.org/qtim/miccai2013/.
15. Vijayarajan, R. and Muttan, S: Discrete wavelet transform based principal component averaging fusion for medical images, AEU—International Journal of Electronics and Communications, Vol. 69, No. 6, pp. 896–902 (2015).
16. Satapathy, S. and Naik, A: Social group optimization (SGO): a new population evolutionary optimization technique, Complex & Intelligent Systems, Vol. 2, No. 3, pp. 173–203 (2016).
17. Naik, A., Satapathy, S.C., Ashour, A.S. and Dey, N: Social group optimization for global optimization of multimodal functions and data clustering problems, Neural Computing and Applications (2016). https://doi.org/10.1007/s00521-016-2686-9.
18. Kannappan, P.L: On Shannon's entropy, directed divergence and inaccuracy, Probab. Theory Rel. Fields. Vol. 22, pp. 95–100 (1972). https://doi.org/10.1016/S0019-9958(73)90246-5.
19. Paul, S. and Bandyopadhyay, B: A novel approach for image compression based on multi-level image thresholding using Shannon entropy and differential evolution, Students' Technology Symposium (TechSym), IEEE, pp. 56–61 (2014). https://doi.org/10.1109/TechSym.2014.6807914.
20. Rajinikanth, V. and Couceiro, M.S: RGB histogram based color image segmentation using firefly algorithm, Procedia Computer Science, Vol. 46, pp. 1449–1457 (2015).
21. Satapathy, S.C., Raja, N.S.M., Rajinikanth, V., Ashour, A.S. Dey, N: Multi-level image thresholding using Otsu and chaotic bat algorithm, Neural Computing and Applications, (2016). https://doi.org/10.1007/s00521-016-2645-5.
22. Palani, T.K., Parvathavarthini, B. and Chitra, K: Segmentation of brain regions by integrating meta heuristic multilevel threshold with markov random field, Current Medical Imaging Reviews, Vol. 12, No. 1, pp. 4–12 (2016).
23. Kaleem, M., Sanaullah, M., Hussain, M.A., Jaffar, M.A. and Choi, T-S: Segmentation of brain tumor tissue using marker controlled watershed transform method, Communications in Computer and Information Science, Vol. 281, pp. 222–227 (2012).
24. Deng, G. and Li, Z: An improved marker-controlled watershed crown segmentation algorithm based on high spatial resolution remote sensing imagery, Lecture Notes in Electrical Engineering, Vol. 128, pp 567–572 (2012).

# Hybridization of PSO and Anisotropic Diffusion in Denoising the Images

**Azra Jeelani and M. B. Veena**

**Abstract** In the current digital world, image processing plays a great role in various applications. Due to the environmental constraints, noise in image is very common and obvious. Anisotropic diffusion is partial differentiation based mathematical process which has been applied for different types of processing operation in the field of image processing. In this work, challenge of getting the optimal gradient threshold in conduction function for anisotropic diffusion is taken care. A global estimation of threshold value is applied instead of local approach. To achieve this global value, the concept of swarm intelligence is taken. Proposed solution is applied to different types of conduction functions and their relative benefits are analyzed. Hence, particle swarm optimization and anisotropic diffusion are used not only to denoise the images but also sharpen the edges.

**Keywords** Anisotropic diffusion · Conduction function · Swarm intelligence

## 1 Introduction

Image processing is a technique of converting images into digital form to perform various operations either to enhance it or extract some useful information from it. It is one of the rapidly growing technologies, more over a core research area among many other engineering. This technology is used in huge number of applications like robotics, medical diagnosis, weather and climate, photography, television, and lot more. In our research, we are addressing underwater image processing due to the reason being, such images are usually blurred and contain more noise compared to other images. This is mainly due to wrong focus of cameras in moving water. Various types of noise in the images captured under water are salt-and-pepper

A. Jeelani (✉) · M. B. Veena
Department of ECE, BMSCE, Bangalore, India
e-mail: azrajeelani@gmail.com

M. B. Veena
e-mail: veenamb@bmsce.ac.in

© Springer Nature Singapore Pte Ltd. 2018
J. Anguera et al. (eds.), *Microelectronics, Electromagnetics
and Telecommunications*, Lecture Notes in Electrical Engineering 471,
https://doi.org/10.1007/978-981-10-7329-8_47

noise, Gaussian noise and speckle noise. Restoration is a technique of removing the noise. Restoration uses different types of filters for processing images and removing different types of noises. Restoration filters are either linear or nonlinear filters. Linear filters are mean filters whose output is the linear function of its input whereas nonlinear filters are those whose output is not the linear function of its input. Perona-Malik [1] proposed a diffusion coefficient which addresses intra-region smoothing rather than inter-region smoothing. J. Kennedy, R. C. Eberhart et al. [2] described the local and global neighborhood for obtaining the best solution in particle swarm optimization. Y. Shi and R. Eberhart [3] gave the new parameter inertia weight which improves the performance of the particle swarm optimizer. Padmavathi et al. [4] gave the algorithm for modified median filters for denoising in underwater images. Serge Karabchevsky [5] compared all the adaptive filters like Lee, Kuan, Frost, Nonlinear filters, and proved frost filter to be the best and gave the FPGA implementation of the frost filter. Shibin et al. [6] gave a comparative study of different filters for effectiveness of denoising and preserving the edges. In the previous paper [5], different filters were compared and frost filter was the best filter among the classical adaptive filters. But the edges were not preserved. In papers [7–9], work on PSO parameters like inertia weight and threshold function was done. In this paper, with the help of PSO, the optimal solution for K is obtained and it is applied to variety of conduction functions.

## 2 Anisotropic Diffusion

Proposed technique would not just jelly edges yet additionally upgrade edges by repressing dissemination crosswise over edges and permitting dispersion on either side of the edge. This will be versatile and does not use hard edges to change execution in homogeneous areas close edges and little highlights. The new dispersion method depends on a similar least mean square mistake (MMSE) way to deal with sifting as the Lee (Kuan) and frost filter. The benefits of anisotropic dispersion incorporate intra-district smoothing and edge conservation. The accomplishment of the discrete usage of anisotropic dispersion system relies on the precision of the chose estimations of the parameters. The decision of dissemination work is critical in controlling smoothing and even upgrade of edges. The decision of dissemination work and associated parameters inclination edge assume the focal part in exhibitions. The programmed determination of this parameter has been done in this exploration [1–6, 10–14]. The slope limit parameter of the anisotropic dissemination sifting strategy likewise needs adjustment to the denoising needs of the separated picture. The estimation of the parameter chose ought to be to such an extent that each one of the edges is saved over a diminishing edge. The essential thought behind the Perona–Malik anisotropic dispersion is to advance from a unique picture I(x, y), a group of progressively smoothed pictures I(x, y, t) in view of the accompanying fractional differential condition.

$$dI(x, y, t)/dt = div[g(|| \nabla I(x, y, t)||)\nabla I(x, y, t)], \qquad (1)$$

where t is the time parameter, $I(x, y, 0)$ is the original image, $\nabla I(x, y, t)$ is the gradient of the image at time t, and $g(.)$ is conductance function or diffusivity function, minimum value $g(.) = 0$, i.e., no diffusion across edges and its maximum value $g(.) = 1$ results to maximum diffusion within uniform regions. This function is chosen to satisfy lim x tends to infinity $g(x) = 0$, so that the diffusion is stopped across edges. Two diffusivity functions given by Perona–Malik are

$$g_1(x) = \exp[(-\frac{x}{k})^2] \qquad (2)$$

$$g_2(x) = [\frac{1}{1 + (-\frac{x}{k})2}]^{-1} \qquad (3)$$

## 3  Adaptive Particle Swarm Optimization

Swarm optimization and particle swarm optimization [2, 3] are evolutionary computation techniques. In this, a set of potential solutions to the problem at hand is taken as the initial population and is used further to find solutions in the search space. Each candidate solution in the PSO has an adaptive velocity in which its position changes in the solution space. As the individual moves in search space, it carries its best position among all its previous search positions in its memory. Thus, every individual has the aggregated acceleration between its previous best position and toward its topological neighbor. In a D-dimensional search space, the ith particle of the swarm is represented by a D-dimensional vector,

$$X_i = [xi_1, xi_2, \ldots xi_D] \qquad (4)$$

Consider the new velocity with position change of this particle to be represented by D-dimensional vector,

$$V_i = [vi_1, vi_2 \ldots vi_D] \qquad (5)$$

The best previously visited position of the ith particle is represented by

$$P_i = [pi_1, pi_2 \ldots .pi_D] \qquad (6)$$

Let the index of the best particle in the swarm be g, the gth particle in the swarm is the best, "n" is best by that particular particle and let the superscripts denote the iteration number and then the swarm is manipulated according to Eqs. 7 and 8 (Fig. 1).

**Fig. 1** Depiction of the velocity and position updates in particle swarm optimization

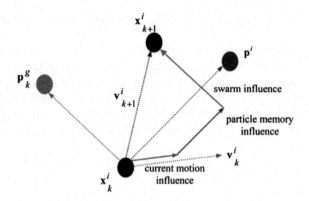

$$v_{k+1}^i = w v_k^i + c_1\, rand\, \frac{(p^i - x_k^i)}{\Delta t} + c_2\, rand\, \frac{(p_k^g - x_k^i)}{\Delta t} \tag{7}$$

$$x_{k+1}^i = x_k^i + v_{k+1}^i \Delta t \tag{8}$$

## 3.1  Selection of Parameters for PSO

Performance of PSO is heavily dependent upon the various parameters available in process. The main parameters are the inertia weights (w), social and cognition constant c1 and c2, maximum value of velocity Vmax, and the size of population S. Generally, these parameters are pre-fixed in most of the applications. But there is no guarantee of optimal convergence in all cases hence there may be a need for some strategy to decide the most optimal value with a particular application.

The inertia weights w play the central role in PSO performance. This parameter is associated with velocity to control the change in the velocity. If the value of w is equal to 0 that is corresponding to change in velocity without using the knowledge of previous velocity. If $w \ll 1$, only a small use of previously available velocity appeared and there is lesser exploration of new solution. It may take long time to converge or may not converge at all to optimal solution. If $w > 1$, it takes a large jump with use of previous velocity information and can be helpful for exploration point of view. But as solution move from one to another, there occurs large change in every iteration and it results in solution to diverge. Hence, it is necessary to have some precaution while placing the value of $w > 1$. It is practically seen that value of w near to 1 can deliver the global exploration while value in the range of [0.2–0.5] can help to explore the local region.

It is advisable to have decreasing tendency of w value with the iteration so that in the beginning, larger value of w could help in exploration while as solution move toward optimality with iteration, decreasing value of w could help to explore the local region. Such approach is generally called as PSO-TVIW (PSO with time-varying inertia weight). There are various other approaches proposed in the past, where Eberhart and Shi have applied the fuzzy controller to control the w with time in adaptive manner.

Vmax has been applied to define the maximum velocity change a solution can have to avoid the divergence. Generally, it is the maximum value of search domain. The use of other controlling parameters like w and $\chi$, has reduced the burden of Vmax value.

PSO can be performed better with use of new parameter called as constriction factor. With the inclusion of $\chi$, there is no necessity to include the other controlling parameter, w and Vmax. Constriction constant helps in faster convergence by decreasing the amplitude of particle oscillation because of more attention toward local and neighborhood best position obtained previously. Depending upon the distance between previous best position and neighborhood best position, focus over local search or global search decide. If the distance is less, local search is performed while global search is performed if the distance is more. With the time, best of neighborhood and best of previous seen are kept changing in result there is shift between local and global search in dynamic manner. Hence, constriction factor can be considered as the balancing parameter between local and global search. Size of population helps in exploring the new solution. If it is too less, there will be no exploration, while too large population will give computational burden and also not much benefit in quality of solution. Generally, it is observed that moderate population size [20–100] can be sufficient in most of the applications.

# 4 Proposed Solution

The quality of anisotropic diffusion heavily depends upon the value of gradient threshold parameter K. The proposed method provides a pre-estimated global value of K instead of obtaining the local K value. With this approach, a single value is estimated earlier from noisy image, which is considered as reference image then the obtained value of K is applied to other images. This approach will save the computational cost. Hence, PSO is applied to estimate the K value from reference image. All eight neighbors are considered in anisotropic diffusion process. For each neighbor, same value of K is considered. In the estimation of K with PSO, exploration of optimal value of K is done so that signal-to-noise ratio could be the maximum. In test phase, the number of applied iterations to denoise the image has resulted in more signal-to-noise ratio. In third phase, the iteration number which has delivered the best value of signal-to-noise ratio is considered for final result. The process of applied PSO and anisotropic diffusion is shown (Figs. 2 and 3).

**Fig. 2** Algorithm for the PSO and anisotropic diffusion

**//DWPSO (Dynamic Weight Particle Swarm Optimization)//**
Pop ← Population of size [m×1] randomly defined
Vpop←velocity population of size [m×1];
Spop←self best population of size [m×1];
Mxit ← Maximum no. of iteration.
[c1,c2,w,χ]←Assign PSO parameters
Fit=Objfun(Pop);
Pg ← Leader selection (Fit)$_{best;}$
*While ( i<Mxit )*
w← Mxw- (i-1)/wd;
        *For j=1:size(Pop,1)*
Rn←[2×1] random value from U[0 1];
Gd← Pg - pop(j);
Sd←Sb – Pop(j);
NVpop(j)←$f$(χ,w,Vpop(j),c1,Gd,c2,Sd,Rn);
*End*
Npop=Pop+Nvpop;
        Fit2←Objfun(Npop);
pg←$f_{best}$(Fit2);
Vpop←Nvpop;
Spop←(pop,Npop);
        Pop=Npop;
  *End*
FS←pg  ;
**// ANISOTROPIC DIFFUSSION**
Img←Read the Image;
// Mask definition for all 8 neighbors
*For  i= 1:8*
M(i)←2D  Convolution Mask
*End*
*For   t = 1: itr*
    *For j=1:8*
GRD(j)←$f_{conv}$ (Img , M(j) );
        *End*
*For  f=1:8*
CF(f)  ← exp(-(Grd(f)/k)^2);
        *End*
Dffimg← Dffimg+ $\lambda$ [$\sum_{m=1}^{8} CF \times GRD$ ]
        PRF=(Dffimg)$_{PSNR}$
*End*

**Fig. 3** Flow chart of the algorithm

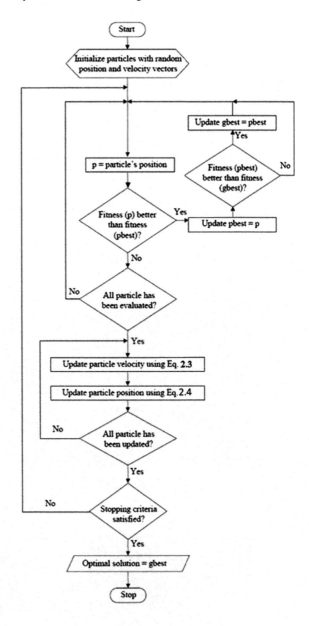

# 5   Experimental Results

The complete environment is implemented in MATLAB. The different parameters under PSO have been defined as

Population size—50, maximum inertia weight—1.2, minimum inertia weight—0.1, $C_1 = C_2 = 0.5$, and constriction factor—0.72.

In the training phase, 5 iterations are applied to estimate the optimal K. while in the test phase, 15 iterations are given (to see the effect of over scanning). Among 15 iterations, the best value of PSNR is taken as final value of denoised image (Figs. 4 and 5).

Test case 1: Training phase (for g1 conduction function)
FINAL gradient threshold value: 49.1101
PSNR: 20.1455
Test phase: ans = 1.0000 Inf 32.9197

**Fig. 4**  Lena image and noisy Lena image (speckle noise)

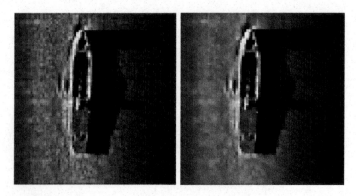

**Fig. 5**  Noisy image and denoised image (using proposed algorithm)

**PSNR = 32.919 (Using g1 conduction function)**
Training phase: (for g2 conduction function)
FINAL gradient threshold value: 49.8867
PSNR: 21.5959
ans = 1.0000 Inf 31.5145
PSNR = 31.514(Using g2 conduction function)

From the above test case, we can observe that the psnr = 32 is improved and the edges are sharpened as compared to the frost filter (psnr = 25) and edges are smoothened from the given figure (Figs. 6 and 7).

Test case 2: Training phase: (For g1 conduction function)
FINAL gradient threshold value: 49.0678
PSNR: 20.1313
Test Phase: ans = 1.0000 Inf 35.1593
PSNR = 35.1593
Training phase: (For g2 conduction function)

original gray image        frost filtered output

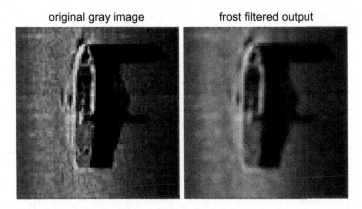

**Fig. 6** Noisy image and denoised image (using frost filter)

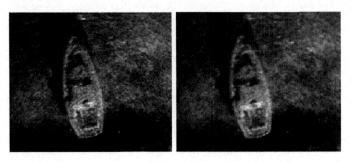

**Fig. 7** Noisy image and denoised image (using proposed algorithm)

FINAL gradient threshold value: 49.8994
PSNR = 21.6180
Test phase: ans = 1.0000 Inf 34.7712
PSNR = 34.7712

# 6 Conclusion

In anisotropic diffusion, the role of gradient threshold is very important and the value of this threshold decides the level of diffusion process. Particle swarm optimization is applied to find the solution to find the optimal threshold value. The value of threshold is obtained under a training phase process in which the noisy reference image threshold value optimizes to increase the PSNR value. This value is applied in anisotropic diffusion process to denoise the noisy test images. Experimentally, it is observed that proposed method is not only efficient in noise reduction but also very much computationally efficient, as there is no need of local estimation of threshold under every scan and for every image. It is also observed that if there is optimal threshold value, performance of simple conduction function is better in comparison with the complex one.

# References

1. Perona, P., and Malik, J.: "Scale-space and edge detection using anisotropic diffusion", IEEE Trans. Pattern Anal. Mach. Intell., 1990, 12, (7), pp. 629–639.
2. J. Kennedy, R.C. Eberhart, et al., "Particle swarm optimization", In Proceedings of IEEE international conference on neural networks, volume 4, pages 1942–1948. Perth, Australia, 1995.
3. Y. Shi and R. Eberhart., "A modified particle swarm optimizer", In Evolutionary Computation Proceedings, 1998. IEEE World Congress on Computational Intelligence., The 1998 IEEE International Conference on, pages 69–73. IEEE, 2002.
4. Dr. G. Padmavathi, Dr. P. Subashini, Mr. M. Muthu Kumar and Suresh Kumar Thakur, "Performance analysis of Non Linear Filtering Algorithms for underwater images", (IJCSIS) International Journal of Computer Science and Information Security, Vol. 6, No. 2, 2009.
5. Serge Karabchevsky, David Kahana, Ortal Ben-Harush, and Hugo Guterman, "FPGA-Based Adaptive Speckle Suppression Filter for Underwater Imaging Sonar", IEEE JOURNAL OF OCEANIC ENGINEERING, VOL. 36, NO. 4, OCTOBER 2011.
6. Shibin Wu, Qingsong Zhu, Yaoqin Xie, "Evaluation of Various Speckle Reduction Filters on Medical Ultrasound Images," in Annual International Conference of the IEEE EMBS Osaka, 2013.
7. Nirmal Singh, Maninder kaur, K.V.P Singh, "Parameter Optimization In Image Enhancement Using PSO", American Journal of Engineering Research (AJER), vol.2, issue 5, 2013.
8. Reena Singh, V.K. Srivastava, "Optimization of Gradient Threshold Parameter in Feature Preserving Anisotropic Diffusion forImage Denoising", International Journal of Innovative Research in Science, Engineering and Technology, vol. 3, issue 2, February 2014.

9. Bodh Raj, Arun Sharma, Kapil Kapoor, Divya Jyoti, "A Novel approach for the Reduction of Noise", International Journal of advance, research, Ideas and Innovations in Technology, vol. 2, issue 3, 2016.

10. Catte, F., Lions, P.L., Morel, J.M., and Coll, T.: "Image selective smoothing and edge detection by nonlinear diffusion", SIAM J. Numer. Anal., 1992, 29, (1), pp. 182–193.

11. James C. Church, Yixin Chen, and Stephen V. Rice Department of Computer and Information Science, University of Mississippi, "A Spatial Median Filter for Noise Removal in Digital Images", IEEE, page(s): 618– 623, 2008.

12. Patidar, Pawan, et al. "Image De-noising by Various Filters for Different Noise", International Journal of Computer Applications 9.4 (2010): 45–50.

13. H.C. Li, P.Z. Fan and M.K. Khan, "Context-adaptive anisotropic diffusion for image denoising", ELECTRONICS LETTRES 5th July 2012 Vol. 48 No. 14.

14. Rohit Verma, Jahid Ali, "A Comparative Study of Various types of Image Noise and Efficient Noise Removal Techniques," vol. 3, no 10, pp. 617–622, 2013.

# Impact of Deep Learning in Image Processing and Computer Vision

**Tilottama Goswami**

**Abstract** With deep learning techniques, a revolution has taken place in the field of image processing and computer vision. The survey paper emphasizes the importance of representation learning methods for machine learning tasks. Deep learning, the modern machine learning is commonly used in the vision tasks—semantic segmentation, image captioning, object detection, recognition, and image classification. The paper focuses on the recent developments in the domain of remote sensing, retinal image understanding, and scene understanding based on newly proposed deep architectures. The author finds it quite intriguing of the classical building blocks of image segmentation (Gabor, K-means), shifting gear, and contributing to image recognition tasks based on deep learning (Gabor convolutional network, K-means dictionary learning). The survey makes an attempt to serve as a concise guide in providing latest works in computer vision applications based on deep learning and giving futuristic insights.

**Keywords** Deep learning · Representation learning · Image processing Computer vision

## 1 Introduction

Computer vision (CV) has been one of the most popular topics for research activity in the past as well as present decade. The computer vision is an interdisciplinary subject which aims to transform data using models to decouple and transform the input data to capture maximum information with the help of mathematics, statistics, physics, biology, and learning theory. Computer vision has a wide variety of applications such as robotic navigation, industrial inspection, process and quality

T. Goswami (✉)
Department of Computer Science and Engineering,
BVRIT Hyderabad College of Engineering for Women, Telangana 500 090, India
e-mail: tilottama.g@bvrithyderabad.edu.in
URL: http://bvrithyderabad.edu.in/

© Springer Nature Singapore Pte Ltd. 2018                                                475
J. Anguera et al. (eds.), *Microelectronics, Electromagnetics
and Telecommunications*, Lecture Notes in Electrical Engineering 471,
https://doi.org/10.1007/978-981-10-7329-8_48

control, military purpose, medical diagnosis, remote sensing, image retrieval, object recognition, building image mosaics, automatic surveillance, and understanding human activity, neuroimaging. Traditional approaches are subjective in nature as they use human expertise for analyzing the best way of feature engineering in extracting the best features according to visual appearance of a given image. This is quite tedious, difficult to cater to large volumes of tasks, and may be prone to human error. The technology developments in the quest of artificial intelligence demand more powerful learning paradigms to mimic the Human Visual System (HVS) to achieve better efficiency and reliability [1]. Most of these applications tackle with similar underlying problem: automatic segmentation and automatic characterization of the visual appearance of the objects irrespective of different application domains. The suitable learning aspects either human engineered learning or automated machine learning bind the success of computer vision tasks. Learning which involves feature engineering is dependent on appropriate hand-crafted features specific for that domain and is application-specific [2].

The survey is organized as follows. Section 2 emphasizes the importance of representation learning and feature learning methods for machine learning tasks. Deep learning is very commonly used in the vision tasks—semantic segmentation, image captioning, object recognition, and image classification. In Sect. 3, the paper focuses on the recent developments in solving the abovementioned vision tasks based on deep learning, in the domain of remote sensing, retinal image under-standing and natural images. In Sect. 4, the author gives a view of the inherent shifting gear of classical approaches (e.g., Gabor, K-means) used in image seg-mentation task, to contributing in a slight different manner for image recognition tasks based on deep learning (Gabor convolution network, K-mean dictionary feature learning method). The authors give a future insight of using orthogonal polynomial operators, in similar terms as Gabor filters are used with CNN for learning texture information; the option has been justified on the basis of seg-mentation results. The paper finally concludes in Sect. 5 providing future directions.

## 2  Representation Learning

Feature learning is representation learning used for machine learning tasks such as recognition, classification, etc. This is an alternative to manual feature engineering or feature extraction, where handcrafted features are designed by knowledge experts and programmers as manual priors; for further processing to achieve the high-level tasks. On the contrary, feature learning discovers discriminative features and builds appropriate representations automatically from the raw data in multiple layers of abstractions using training algorithms. Feature learning or deep learning is modern machine learning whose basis is neural network.

**Objectives of Representation Learning**—General priors prove to be more useful in solving highly complex AI tasks. The learning machine will deal with

**Fig. 1** Deep learning architecture (Bengio, "On the expressive power of deep architectures", Talk at ALT, 2011)

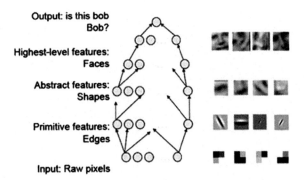

nonspecific tasks to specific tasks in layers of hierarchical organization. The machine learns filters, initially simpler ones and gradually builds deep layers toward more complex filters. The objective of representation learning is learning the invariant features that lead to effective pattern recognition. The deep learning architecture for face recognition is shown in Fig. 1.

**How feature learning takes place?** Representation learning is all about, the way the algorithm learns linear and nonlinear manifolds, based on either parametric or nonparametric methods, depending on either supervised or unsupervised approaches. The machine learning is one of the approaches to achieve high-precision accuracy based on unsupervised learning or supervised learning. Unsupervised learning is bottom-up approach and data driven. Such type of learning is dependent on low level features embedded in the image itself, say X as shown in Fig. 2a. The code Z is intermediate representation of input which is low dimensional and may be sparse. In contrast to unsupervised learning, supervised learning learns by means knowledge extracted from training set of ground-truth segmented images, say Y as shown in Fig. 2b. This is also called top-down approach, model driven, which starts with known-priors about the target object (Y). The central theme of unsupervised feature learning is to learn hierarchy of features,

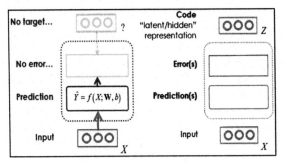

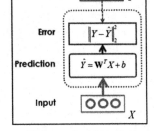

(a) Unsupervised Learning                    (b) Supervised Learning

**Fig. 2** Feature learning methods (Piotr Mirowski, Microsoft Bing London Computational Intelligence Conference, 2014)

one level at a time that can be combined with previously learned transformations. Each level adds a layer of weights to develop a deep supervised predictor. Learning is usually considered as estimating a set of model parameters which maximize regularized likelihood of training data.

The author provides a gist of all the combinations of learning options mentioned in [2] to give the reader a continuity of the topic. Feature extraction algorithms illustrate probabilistic, auto-encoder, and manifold views of representation learning for the unsupervised single layer. Active research is going on in establishing connections between these three views.

*Probabilistic View*—Learned representation with probabilistic model is always associated with latent variables. Sparse coding and its variants relate to latent representation to the data using a linear mapping called dictionary. The undirected graph model (Markov Random Field), Boltzmann distribution, and Restricted Boltzmann Machines (RBM) parameterize joint probability through product of clique potentials describing interactions between visible, between hidden and between visible and hidden elements. Boltzmann distribution and its variants characterize these interactions in context with unsupervised feature learning and applied for modeling natural textures, object classification tasks. Visible elements represent data, latent or hidden elements act as go between the visible elements by interactions indicated by energy function. The major disadvantage with this probabilistic model arises when model has more than two interconnected layers; as the posterior distribution over latent variable becomes intractable.

*Autocoders View*—Autocoders and their many variants such as sparse auto-encoder, regularized auto-encoder, and parameterized representation functions are alternative for non-probabilistic feature learning paradigm, as they have tractable optimization objective. In this, the model learns encoding a parametric map from data to its representation, may also involve decoding function mapping back from representation to input data space. The parameters are learned while performing the task of minimizing the reconstruction error of the original input. Error minimization can be done by stochastic-gradient descent in multilayer perceptrons (MLP).

*Geometric View*—Geometric notion (tangent and orthogonal directions) of manifold is another perspective on representation learning, where real-world data of high-dimensional space is focused on manifold of much lower dimension, capturing the variations in input space. Data-supporting manifold is unsupervised learning task. PCA is a low-dimensional representation learning model which models closest linear manifold to the group of data points. Real-world domains are strongly nonlinear, which have been dealt built on geometric perspective of nonparametric approach relying on training set nearest neighbor graph. The concept of representation learning is summarized here without mathematical details due to page constraints. The references for the works pertaining to these views may be referred from [2].

**How to use the learned features?** A predictor is fed with learned features in order to perform particular high-level tasks such as classification and recognition. A classifier is trained at the top of the learned features. Fine tuning the classifier may take more training iterations that may have computational overhead. There are two options to form a deep architecture—(1) Single layer models combined to form a deep model and (2) Joint training of all levels. Training deep architecture has a general principle of training the intermediate representations layerwise, thus initially learning simpler concepts and building the higher levels. Convolution and pooling operation learns to be invariant to variations and is heart of convolutional neural network, which has been applied to object recognition and image segmentation [3]. Patch-based training is quite efficient as the patches are low dimensional as compared to the whole image. Patches are sampled at random positions of the image. The sampling criteria are crucial in getting the input data points. The author in [4] proposes *progressive sampling* which is flexible and iterative in nature. The positional sampling (patch of pixels) has been done along the object boundary region (grayscale and color variants) and is moved gradually from peripheral to the center of the region. Experiments can be done in near future to see the impact of progressive sampling for selection of patches.

# 3 Recent Deep Learning Application Areas

Deep learning, the modern machine learning is very commonly used in the vision tasks—semantic segmentation, image captioning, object recognition, and image classification that are discussed briefly in this section. Natural images, satellite images, and medical images do not share same structure, still deep network—CNN based classifiers are able to successfully extract the semantics. A recent survey on deep architecture and its variations may be referred [5] for knowing the basic building blocks of convolutional neural networks (CNNs).

## 3.1 Vehicle Localization from High-Resolution Remote Sensing Images

*CV Tasks—Semantic segmentation, object detection, and image classification for vehicle*

Deep learning has been used for all three tasks—segmentation, detection, and classification for aerial images in [6]. The Fully Convolutional Networks (FCN) architecture has been used to learn semantic maps, vehicle segmentation, and vehicle classification. The deep architecture SegNet is used, which is encoder (maxpooling)—decoder (unpooling) architecture, made of convolutional (conv) blocks, batch normalization (BN), and rectified linear units (ReLU) as shown in Fig. 3. Aerial image datasets used are VEDAI, ISPRS Potsdam, NZAM/ONERA Christchurch. Comparisons are done with three CNN architectures LeNet, AlexNet, and VGG-16.

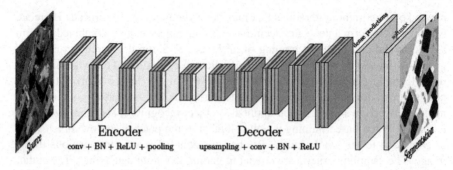

**Fig. 3** SegNet architecture for semantic segmentation in remote sensing

## 3.2   Retinal Image Understanding

*CV Tasks—Semantic segmentation for retinal vessel*

In the field of medical image analysis, retinal image understanding helps in early diagnosis of diseases such as glaucoma and retinopathy. Related to this, segmentation of retinal blood vessels is an open research problem. The works involve classic methods and recently powerful machine learning approaches have also emerged. Classic methods are based on handcrafted filters, morphological operators, and enhancement techniques for vessel extraction. One such classic traditional approach [7] used thin, oriented line operators such as feature vector and Support Vector Machine (SVM) for classifying pixels as vessel or no-vessel. Recently yet, another recent classic automated method is proposed in [8], where curvelet-based edge image is superimposed on optimally designed band pass filtered image that achieved higher average detection accuracy. The datasets provide open access to digital libraries for healthy and pathological retinal images. The paper also classifies vessels as thick, medium, and thin vessels. Detection of different types of lesions is considered as future work. Recently, a novel CNN architecture called Deep Retinal Image Understanding (DRIU) has been proposed [9], for both retinal vessel and optic disc segmentation as shown in Fig. 4. The architecture runs on GPU and the average execution time is in ~100 ms, which has orders of magnitude faster than the classic state-of-the-art methods. The results surpass the performance of human specialists for four pathological retinal benchmark datasets—DRIVE, STARE, DRIONS-DB, RIM-ONE.

## 3.3   Dense Image Captioning for Scene Understanding

*CV Tasks—Object detection, image captioning for natural scenic images*

Image captioning is the process where whole image is annotated. Dense image captioning, on the other hand, produces annotated regions. The regions of interests

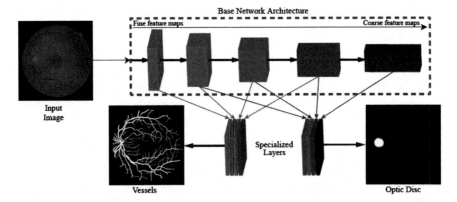

**Fig. 4** DRIU architecture for blood vessel and optic disc segmentation

are captured from the image and translated to sentences describing the vision objects using natural language processing. Various layered architectures like Convolutional Neural Network (CNN), Deep Belief Network, Long Short Term Memory (LSTM), and Recurrent Neural Networks (RNN) are used in the literature for computer vision, image captioning problems. The author in [10] proposes a Multimodal Recurrent Neural Networks model based on combination of CNN over regions, Bidirectional Recurrent Neural Network (BRNN) over sentences, and multimodal embedding for aligning these two. Figure 5 shows BRNN model for evaluating image-sentence score. The authors evaluate the model on Flickr8 K, Flickr30 K and MSCOCO dataset containing 8,000, 31,000 and 123,000 images, respectively.

**Fig. 5** BRNN architecture
for evaluating image-sentence
score

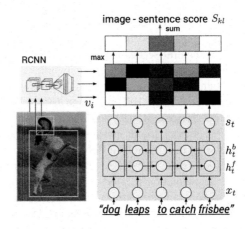

# 4 Shifting Trends—Classical Approaches to Deep Learning

The author finds it quite intriguing of the inherent shift of classical approaches (Gabor, K-means) used in image segmentation task, shifting gear, and contributing in a slight different manner for image recognition tasks based on deep learning (GCN, K-mean dictionary feature learning method). The paper portrays two classical concepts which have been taken into different role play in deep learning techniques. First, Gabor filters used for textural feature extraction have been used differently to form Gabor convolutional network (GCN). Second, K-means clustering used for segmentation are adapted as K-means dictionary feature learning for recognition in deep architecture, which is discussed briefly in the following subsections.

## 4.1 Handcrafted Gabor Filters to Learning Feature Representation

*CV Task—Natural image classification*

Natural images have a large variation in color and texture. The role of color texture as fundamental descriptors proves to be quite effective in the field of color image segmentation and texture characterization. Steerable or adaptive filters are designed from linear combination of basis filters, and are usually tuned to parameters for any arbitrary orientation and phase in order to compute the filter responses. Gabor filter bank, steerable wavelet, and 3-D steerable filters use this concept, which proved to be useful in image processing for edge detection, orientation analysis, and shape analysis. Recent research paves the way to show how the proven traditional handcrafted feature extraction techniques such as Gabor Orientation Filters (GOF) can be used in learning deep feature representations for object recognition [11] in natural scenes using Gabor-CNN, and in natural image classification [12] using Gabor convolutional network (GCN) as seen in Fig. 6, respectively. Recognition tests are done on ImageNet10 dataset. The classification tasks are carried on CIFAR-100 tiny image dataset (60000 32 × 32-sized color images in 10 or 100 classes), MNIST and SVHN as well.

The paper [13] proposes a novel segmentation technique based on K-means clustering on Orthogonal Polynomial (of order 3, 5) and Hybrid Color Space, i.e., OP3-HCS and OP5-HCS. The authors compare various state-of-the-art color texture segmentation methods. The Gabor filter bank generated using scale and orientation parameter settings as mentioned in [14] has been compared with OP-HCS variants on BSD300 image dataset. The results are found to be encouraging as shown in Table 1. In segmentation task, OP3 and OP5 have performed

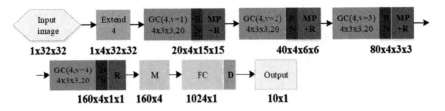

**Fig. 6** GCN architecture for natural image classification

**Table 1** Statistics PRI, BDE, VOI and GCE for segmentation of BSD300 dataset, comparing Gabor filter bank with OP based filters. Higher PRI and lower values of BDE, VOI, and GCE represent better segmentation. Bold marked values are the best

| Filter bank | Color space | PRI | BDE | VOI | GCE | Time (s) |
|---|---|---|---|---|---|---|
| OP3 | HCS | **0.7457** | 10.0457 | 2.4113 | 0.2857 | 38 |
| OP5 | HCS | 0.7431 | 9.6806 | 2.8832 | 0.2496 | 48 |
| OP5 | RGB | 0.7377 | **9.3723** | 3.0320 | 0.2894 | 43 |
| Gabor | HCS | 0.6291 | 13.9447 | 2.3209 | 0.2485 | 38 |
| Gabor | RGB | 0.6340 | 14.6331 | **2.2793** | **0.2167** | 37 |

considerably better than Gabor filters, for RGB and HCS color spaces in terms of PRI and BDE, and results of VOI and GCE are found to be comparable.

Segmentation results exhibit the effectiveness by achieving on an average 74.5 percent on PRI which is quite encouraging without involving any preprocessing, training or smoothing processes. OP3-HCS is better for coarser level image segmentation and can be applied for high-level recognition tasks. OP5-HCS is recommended for identifying heterogeneous highly textured image types as it favors over-segmentation. In future, convolutional neural network can be combined with OP3 or OP5 bank of filters for strengthening the learning of texture information.

## 4.2 K-Means Clustering to K-Means Dictionary Feature Learning

*CV Task—Image recognition*

Many encoding schemes can be found in the literature as mentioned in Sect. 2. The paper [15] proposes K-means trained dictionaries in conjunction with many of them. Sparse coding works better for higher dimensional images, though applying K-means to sub-patches is a realistic one. Spherical K-means also called whitening process is used to make K-means an effective unsupervised dictionary learning method for image recognition as shown in Fig. 7. Experiments are conducted on CIFAR-10, STL-10.

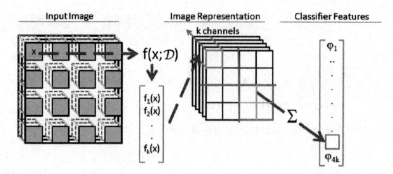

**Fig. 7** K-means dictionary learning for image recognition

## 5 Conclusions

The future computing paradigm paves its way toward achieving application independence in context with image processing and computer vision related applications. Deep learning plays a tremendous role in achieving this objective by means of representation learning. In future, the author perceives the role play of the classical proven feature extractors and segmentation techniques; cater to learned features in representation learning for building precise and effective computer vision applications. GPU plays a pivotal role in parallelization of the deep learning algorithms.

## References

1. Trémeau, A., Tominaga S., Plataniotis, K. N.: Color in image and video processing: Most recent trends and future research directions. Hindawi Publishing Corporation EURASIP Journal on Image and Video Processing (2008).
2. Bengio, Y., Courville, A., Vincent P.: Representation Learning: A Review and New Perspectives: IEEE Transactions on Pattern Analysis and Machine Intelligence, Vol. 35 Issue 8, (2013), 1798–1828, https://doi.org/10.1109/TPAMI.2013.50.
3. Turaga, S. C., Murray, J. F., Jain, V., Roth, F., Helmstaedter, M., Briggman, K., Denk, W., and Seung, H. S. Convolutional networks can learn to generate affinity graphs for image segmentation. Neural Computation, 22, (2010). 511–538.
4. Goswami Tilottama, Agarwal Arun, Rao C.R.: Statistical Learning for Texture Characterization. ICVGIP (2014): 11:1–11:8.
5. Srinivas S, Sarvadevabhatla RK, Mopuri KR, Prabhu N, Kruthiventi SSS and Babu RV A Taxonomy of Deep Convolutional Neural Nets for Computer Vision. Front. Robot. AI 2:36. (2016), https://doi.org/10.3389/frobt.2015.00036A.
6. Nicolas A., Le Saux Bertrand; Lefèvre, Sébastien: Segment-before-Detect: Vehicle Detection and Classification through Semantic Segmentation of Aerial Images: Remote Sensing, Vol. 9 Issue 4, (2017), p 1–18.
7. Ricci, E., Perfetti, R.: Retinal blood vessel segmentation using line operators and support vector classification. IEEE T-MI 26(10), (2007), 1357–1365.

8. Kar Sudeshna S., Maity Santi P., Retinal blood vessel extraction using tunable bandpass filter and fuzzy conditional entropy, Computer Methods and Programs in Biomedicine, Volume 133 Issue C, (2016), 111–132.
9. Kevis-Kokitsi Maninis, Jordi Pont-Tuset, Pablo Andrés Arbeláez, Luc Van Gool: Deep Retinal Image Understanding, MICCAI (2016), https://doi.org/10.1007/978-3-319-46723-8_17.
10. Karpathy A. and Fei-Fei L,: Deep Visual-Semantic Alignments for Generating Image Descriptions, IEEE Trans. Pattern Anal. Mach. Intell 39(4), (2015), 664–676.
11. Hu Yao, Li Chuyi, Hu Dan, Yu Weiyu,: Gabor Feature Based Convolutional Neural Network for Object Recognition in Natural Scene, Information Science and Control Engineering ICISCE, (2016), https://doi.org/10.1109/ICISCE.2016.91.
12. Shangzhen Luan, Baochang Zhang, Chen Chen, Xianbin Cao, Jungong Han, Jianzhuang Liu: Gabor Convolutional Networks. CoRR abs/1705.01450 (2017).
13. Goswami Tilottama, Agarwal Arun, Rao C.R.: Hybrid Region and Edge Based Unsupervised Color-Texture Segmentation for Natural Images, International Journal of Information Processing, 9(1), (2015), 77–92, ISSN:0973-8215.
14. Jain, A. K. and F. Farrokhnia: Unsupervised texture segmentation using gabor filters. Pattern Recognition, 24(12), (1991) 1167–1186.
15. Adam Coates and Andrew Y. Ng, Learning Feature Representations with K-means, G. Montavon, G. B. Orr, K.-R. Muller (Eds.), Neural Networks: Tricks of the Trade, 2nd edn, Springer LNCS 7700 (2012).

# Study of Very Fast Front Surges in Gas Insulated Substation Due to Switching Operation

V. HimaSaila, M. Nagajyothi and T. Nireekshana

**Abstract** Power system is mainly affected by sudden changes in steady-state values of voltage and current, referred to as the transient phenomena, which are the result of lightning stroke or switching operation. The very fast transient overvoltages are generated due to switching operation. This fast front transient seen in gas insulated substation is one of the major concerns for its insulation coordination. In this paper, a 420 kV prototype model is developed and simulated using EMTP software. By operating a disconnector switch, VFTOs at various locations are analyzed. The severity of VFTOs is also analyzed by considering the trapped charge.

**Keywords** Gas insulated substation · Very fast transient overvoltages
Disconnector switch · Switching operation

## 1 Introduction

Compared to the common substations, GIS units are in very much use in India because of its advantages such as less maintenance and less fault probability. As it is enclosed, it is free from environmental problems and less space is required for

V. HimaSaila (✉) · M. Nagajyothi · T. Nireekshana
VNR Vignana Jyothi Institute of Engineering and Technology,
Bachupally, Hyderabad, India
e-mail: v.himasaila@gmail.com

M. Nagajyothi
e-mail: nagajyothi_m@vnrvjiet.in

T. Nireekshana
e-mail: nireekshana_t@vnrvjiet.in

© Springer Nature Singapore Pte Ltd. 2018
J. Anguera et al. (eds.), *Microelectronics, Electromagnetics
and Telecommunications*, Lecture Notes in Electrical Engineering 471,
https://doi.org/10.1007/978-981-10-7329-8_49

installation. The lightning arresters required in GIS are less in number because of its compactness. Despite the advantages, GIS has many issues of concern. The major issue considered in GIS is very fast front transients, which are generated because of switching operation. The VFTOs harm the insulation of internal, external equipment, and also emit the electromagnetic fields [1, 2]. So, a 420 kV GIS is modeled and VFTOs occurring due to opening and closing [3] of disconnector switch are simulated using EMTP software.

A very detail explanation of VFTOs is done in [4, 5], the VFTOs in GIS are mainly considered in systems with higher voltages, where the ratio of the LIWV to the system voltage is less, which is better discussed in [6].

The modeling of GIS is studied from [7–9]. The insulation coordination for the substation in [10–12] involves three major steps, which are discussed above.

## 2   Very Fast Transient Overvoltages (Vfto) in GIS

Switching operation, within nanoseconds, the voltage across the contacts collapses. Therefore, the surge which is initiated between the contacts travels in either direction of bus duct and when it experiences an impedance change, it gets reflected and refracted, i.e., near terminations. So because of these reflections, there develops overvoltages. These overvoltages are known as very fast transient overvoltages [4, 5]. For every restrike, a VFTO is developed, but with different magnitudes for each, having the maximum magnitude 2.5 pu with very short rise time of 4–100 ns. The rise time of the surge is given by

$$t_r = 13.3 \cdot \frac{k_T}{\left(\frac{E}{P}\right)_o \cdot \Box . h}$$

$k_t$     50 kV/ns cm, by Toepler's spark constant
$\left(\frac{E}{P}\right)_o$   860 kV/cm,

Gas pressure of GIS is up to 0.5 Mp.

From the above, we can see that the rise time is in nanoseconds, i.e., because of higher field strength and gas pressure.

## 3   Modeling of GIS Components for VFTO Simulations in EMTP Software

A generalized single line diagram of 420 kV GIS model is as shown in Fig. 1. Feeders are connected to line end, reactor. Different components of the GIS can be modeled into lumped elements due to the traveling nature of the transients [8, 9].

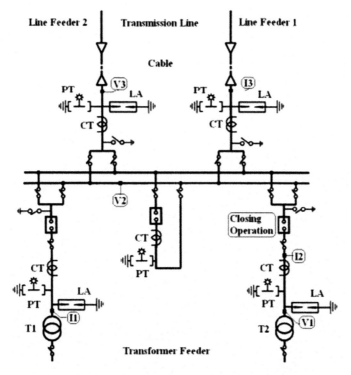

**Fig. 1** Single-line diagram of 420 kV substation

These lumped elements are defined by surge impedances, GIS sections, and wave velocity.

## 3.1 GIS Bus Section

It is coaxial in shape, modeled by distributed parameter transmission line, whose surge impedance is given by

$$Z = 60\ln\left[\frac{b}{a}\right]$$

Z   Characteristic impedance
a   Diameter of inner conductor
b   Inner diameter of outer enclosure

**Fig. 2** Cross section of bus
section

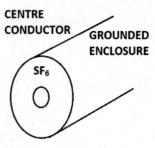

## 3.2 Disconnector

It has three modes of operation: open, close, and arcing mode. Arcing includes dynamic arc resistance also. Open and closing modes are modeled with lumped capacitance (Fig. 2).

## 3.3 Spark in Disconnector

$$R = r + R_0 e^{\frac{-t}{T}}$$

$R_0$   is a high initial resistance
r      is the residual series resistance

## 3.4 Circuit Breaker

The two modes of operation are open and closing operation. Open and closing modes are modeled with lumped capacitance of 15 and 30 pF.

Voltage transformer is also modeled with grounded capacitance of 80 pF.

## 4 Simulation Results

When the disconnector in the substation is operated, VFTOs near the disconnector, reactor, transformer, and at line end are observed as shown below.

Therefore, if we observe from above simulation results Figs. 3a, 4a, 5a and 6a, the lowest peak is observed at reactor terminal and the highest peak of VFTO is at VT1. VFTOs have oscillation frequency form one MHz to several hundred MHz. The frequency of oscillation near the disconnector will be in the range of 1–40 MHz and near the transformer the range will be in between 20 and 100 MHz.

**Fig. 3** VFTO waveform near the disconnector point for closing operation. **a** Time domain. **b** Frequency domain

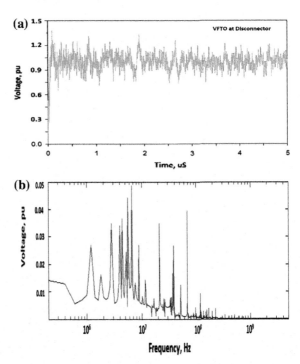

**Fig. 4** VFTO waveforms near voltage transformer for closing operation of disconnector. **a** Time domain. **b** Frequency domain

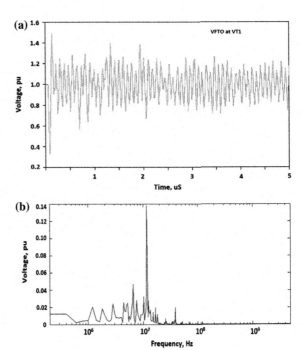

**Fig. 5** VFTO waveform near the reactor for closing operation of disconnector. **a** Time domain. **b** Frequency domain

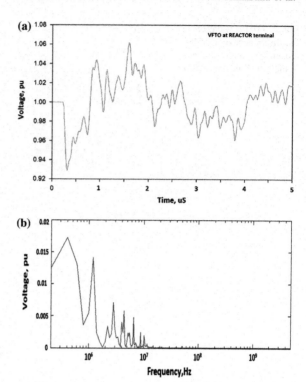

So from Figs. 3b, 4b, 5b, and 6b, we can observe that the range of oscillation is within range.

After complete opening of disconnector switch is done, a trapped charge will be there on the load side. So, because of this trapped charge, the peak of VFTO changes as shown in Table 1.

# 5 Conclusion

A 420 kV GIS is modeled using EMTP software. VFTOs are analyzed with and without considering the trapped charge. If we consider the worst case of trapped charge, i.e., −1 pu, the VFTO peak near the disconnector changes from 1.37 to 1.74.

Highest peak is observed at VT1, compared to the remaining locations. In the most cases, the magnitude of VFTOs will be within 1.5–2.5 pu and from the simulation results, we can observe that magnitude is below 2.5 pu. The oscillation of frequency is also in the range as discussed above.

**Fig. 6** VFTO waveform at line end for the disconnector closing operation. **a** Time domain. **b** Frequency domain

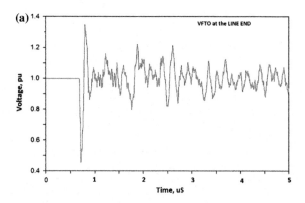

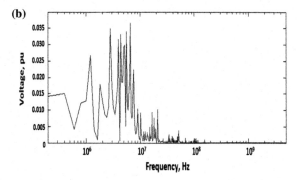

**Table 1** Peak of VFTOs

| Location | Trapped charge (−1 pu) |
|---|---|
| Disconnector | 1.74 |
| VT1 | 1.99 |
| Reactor | 1.11 |
| Line end | 1.69 |
| Circuit breaker | 1.7 |

# References

1. DE Thomas, Wiggins, Nickel, C.D ko & Wright, "Prediction of electromagnetic field and current transients in power transmission and distribution systems" IEEE Trans. on power deliv, Jan 1989, vol. 4, no. 1, pp 744–755.
2. Wiggins, DE Thomas, Nickel & Wright "Transient electromagnetic interference in substations" IEEE Trans. on power deliv, Oct 1994, vol. 9, no. 4, pp 1869–1884.
3. J.A. Martinez (Chairman), P. Chowdhuri, R. Iravani, A. Keri, D. Povh "Modeling guidelines for Very Fast Transients in Gas Insulated Substations", Report Prepared by the Very Fast Transients Task Force of the IEEE Working Group on Modeling and Analysis of System Transients.
4. Working group 33/13-09, "Very Fast Transient Phenomena Associated with GIS", CIGRE, 1988, paper 33-13, pages 1–20.

5. CIGRÉ Working Group 33/13-09: "Monograph on GIS Very Fast Transients", Brochure 35, July 1989.
6. CIGRÉ Brochure 456 Working Group A3. 22: "Background of Technical Specifications for Substation Equipment Exceeding 800 kV AC", April 2011.
7. S. Nishiwaki, Y. Kanno, S. Sato, E. Haginomori, S. Yamashita, and S. Yanabu. "Ground Fault by Restriking Surge of SF6 Gas-insulated Dis- connecting Switch and Its Synthetic Tests," IEEE Trans. on PAS, Vol. PAS-102, No. 1, January 1983, pp. 219–227.
8. D. Povh, H. Schmitt, O. Valcker, and R. Wutzmann, "Modelling and analysis guidelines for very fast transients," IEEE Trans. Power Del., vol. 11, no. 4, pp. 2028–2035, Oct. 1996.
9. Rama Rao, J. Amarnath& S. Kamakshaiah, "Accurate Modeling of Very Fast Transients Overvoltages in A 245kVGIS and Research on Protection Measures", J.N.T.U College of Engineering, Hyderabad, India, IEEE, 2010.
10. Riechert, U.; Neumann, C.; Hama, H.; Okabe, S.; Schichler, U., on behalf of CIGRÉ WG D1.36 and AG D1.03: "Basic Information and Possible Counter Measures Concerning Very Fast Transients in Gas-Insulated UHV Substations as Basis for the Insulation Co-ordination", CIGRÉ SC A2 & D1 Joint Colloquium 2011, Kyoto, Japan, PS3-O-5.
11. IEC 60071-1, Insulation Co-ordination—Part 1: Definition, principles and rules, edition 8, 2006-01.
12. Zaima, E.; Neumann, C.: "Insulation Coordination for UHV AC Systems based on Surge Arrester Application (CIGRÉ C4.306)", The second IEC-CIGRÉ International Symposium on International Standards for UHV Transmission, 29–30 January 2009, New Delhi, India, proceedings pp. 108–118.

# Fuzzy Logic Based Speech Recognition and Gender Classification

Sanjay Dubey, H. Ajay Kumar, R. Abhilash and M. C. Chinnaiah

**Abstract** An approach of recognizing a person based on the individual information present in speech signals is named as speaker recognition. Nowadays, gender classification is a challenging factor in the speaker recognition. Different genders have dissimilar frequency ranges and respective pitch values. Perceptually and biologically, pitch is proved as a good discriminator between male and female voice. More formally, gender classification is done based on the relevant parameters. In this paper, our works aim to classify the gender of the speaker by using the MATLAB Fuzzy Toolbox. Mamdani fuzzy interface system is able to represent the gender classification based on the input variables: frequency and pitch. By the behavior of the input variables on the fuzzy rule based expert system, the output is predicted as male, female, and children. The work also extends to make the fuzzy controller adaptive. The test results show the reliability of performance. The proposed method is build to improve the robustness of the gender classification. Simulation results for male, female, and child accuracy of COA are nearly equal to $0.15 \pm 0.001$, 0.452, and $0.751 \pm 0.001$, respectively.

**Keywords** Gender classification · MATLAB fuzzy toolbox · Mamdani fuzzy interface system

S. Dubey (✉) · H. A. Kumar · R. Abhilash · M. C. Chinnaiah
Department of ECE, B V Raju Institute of Technology, Narsapur, Medak (Dt.)
502313, Telangana, India
e-mail: sanjay.dubey@bvrit.ac.in

H. A. Kumar
e-mail: hanmanthuajay463@gmail.com

R. Abhilash
e-mail: abhilash.r@bvrit.ac.in

M. C. Chinnaiah
e-mail: chinnaaiah.mc@bvrit.ac.in

© Springer Nature Singapore Pte Ltd. 2018
J. Anguera et al. (eds.), *Microelectronics, Electromagnetics and Telecommunications*, Lecture Notes in Electrical Engineering 471,
https://doi.org/10.1007/978-981-10-7329-8_50

# 1 Introduction

Automatic speaker recognition is the advanced technology developed to replace the touch-tone process. Automatic speaker recognition is the process of recognizing the speaker based on the trained voice loaded in the database [1]. The main goal of the speaker recognition is to extract, characterize, and analyze the voice of the speaker. Speaker recognition is classified into two phases namely: speaker identification and speaker verification [2]. Speaker identification is the task of identifying the speaker voice from the set of known voice included in the database. When it comes to speaker verification, it is the process of accepting or rejecting the speaker recognized to be the actual one. In speaker identification, the required voice is compared with the N number of templates, whereas in speaker verification, one-to-one matching is made. Speaker verification is the fastest process when compared with the identification [2]. Speaker identification is further divided into two phases namely: Text-dependent and Text-independent [3].

In the text-dependent process, it has the prior knowledge of what the speaker is going to speak, whereas in the text-independent, it does not have any prior knowledge about what the speaker is able to speak. The classification of the gender is an important factor to be involved in the speaker recognition. As the gender classification is stressed mainly and can be solved by using the fuzzy logic based on the relevant membership functions.

Ramin Halavati et al. proposed a novel approach to speech recognition and decision-making based on fuzzy modeling instead of noise detection and removal it ignores noise [4]. Kunjithapatham Meena et al. proposed the gender classification based on the three main features namely: Zero Crossing Rate (ZCR), Energy Entropy, and Short Time Energy (STE) [5]. In speech processing, classification of the gender plays a major role and can be implemented by using LabView [6]. Classification of the gender mainly focuses on the pitch and frequency values [7]. Three features Zero Crossing Rate (ZCR), Energy Entropy, and Short Time Energy (STE) are given as inputs to the fuzzy and the integrated output is calculated. Parul and R. B. Dubey focused on the improvement of speaker recognition. MFCC is used for feature extraction and modeling of the speaker is done by vector quantization (VQ) [3]. M. Gomathy et al. designed a gender clustering and classification algorithm based on the feature extraction method. Gender clustering is performed by the significant features namely: Euclidean distance, Mahalanobis distance, Manhattan distance, and Bhattacharyya distance methods [8]. Fuzzy local binary patterns are used to build gait-based gender recognition presented by El-Sayed M. El-Alfy et al. Spatiotemporal characteristics of capture, during walking cycle they adopted the gaited energy imager descriptor of gait video [9].

## 2 Proposed Fuzzy Model

The classification of the gender plays an important role in the speaker recognition, as there are many parameters used to classify the genders. Here, the two-input one-output design is designed for the gender classification as shown in Fig. 1. Mamdani fuzzy logic is used here because of its widespread acceptance and it is mainly well suited for human inputs. More formally, the membership functions used here are: one is the frequency membership function and the other is the pitch membership function.

Triangular membership functions are used for the fuzzification process. The three main steps involved in the fuzzy logic are fuzzification, generation of fuzzy rules, and defuzzification. The next step involved here is the generation of the fuzzy rules and is examined by the rules editor followed.

Here, we have used the MATLAB R2012a version. As the MATLAB is the most effective tools in solving the scientific and engineering problems. The MATLAB Toolbox provides the fuzzy logic and all the rules to be developed by the user. All the rules provided by the user will be implemented by the fuzzy interface system. The structure of the model can be gained by using Adaptive Neuro Fuzzy Interface System. Fuzzy logic provides various options in solving many problems. Many parameters are designed to produce the results in easy way. Fuzzy the name itself reveals not clear. The fuzzy logic deals with the partial truth which lies between the complete truth and complete false. All the values of the fuzzy logic lie between 0 (completely false) and 1 (completely true). As the fuzzy logic deals with the membership functions, the uncertainty in membership functions is reduced and designed according to the requirements of the user. The allowance of the multiple membership functions is done by the fuzzy logic. Membership functions behave as the input to the fuzzy logic in retrieving the output. As the fuzzy logic is available in two different shades namely: Sugeno and Mamdani. Mamdani fuzzy logic aims to point out mainly on gender classification to a greater extent. Fuzzy logic is a rule-based expert system that allows us to develop the rules. All the inputs values are fuzzified into the membership functions. The developed rules are executed based on the fuzzy logic and tend to compute the fuzzy output. Defuzzification is done in order to retrieve the crisp output.

**Fig. 1** Proposed fuzzy model

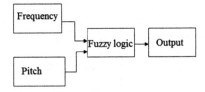

**Table 1** Fuzzy rules

|        | FL       | FM       | FH       |
|--------|----------|----------|----------|
| PL     | Male     | Male     | Male     |
| PM     | Female   | Female   | Female   |
| PH     | Children | Children | Children |

# 3 Generation of Fuzzy Rules

Specifically, fuzzy logic can handle multiple inputs. The two membership functions used here are frequency and pitch and are given as inputs to the fuzzy logic. The variables of the inputs are fuzzified in three different sets namely: low, medium, and high similarly, the output variables are split into male, female, and children, respectively. If-Then rules are designed initially and the next step is to train the fuzzy with the rules generated (as shown in the Table 1). Here, the pitch is used as the main feature, all the rules are made with reference to the pitch. Whenever the pitch is low for frequency values, the gender is turned as male. Similarly, whenever the pitch is medium or high for the values of the frequency, the gender is turned as female or child, respectively.

The rules are designed with the particular frequency and pitch ranges of the male, female, and child as low, medium, and high.

According to the IF-THEN rules, when both the frequency and the pitch show low ranges, gender is shown as male at the output. Similarly, when it shows medium range, output is female and when it shows high range, output is child.

# 4 Calculation of Output

Lastly, with the designed rules of fuzzy based on the membership functions, the output can be calculated by using the Centroid Area.

$$centroidArea\,(COA) = \frac{\int_{x_{min}}^{x_{max}} f(x) * xdx}{\int_{x_{min}}^{x_{max}} f(x)dx} \tag{1}$$

The Centroid Area of the output is calculated by using the formula mentioned above. The output of the fuzzy can be examined by the rules viewer and it gives the output as the variables of male, female, and child and can be compared against the centroid Area result. Figure 2 shows the calculation of specific values by using rule viewer.

The desired output can be classified based on the gender according to the rules specified above. The classified output Fig. 3 is shown.

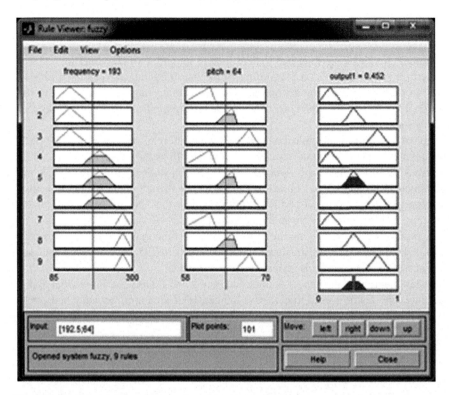

**Fig. 2** Rule viewer

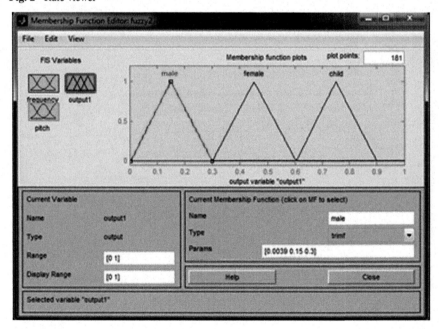

**Fig. 3** FIS output

## 5 Adaptive Neuro Fuzzy Interface System (ANFIS)

Adaptive Neuro Fuzzy Interface system is able to model a fuzzy interface system based on the data provided. According to the provided data ANFIS itself able to design the rules for the model and all the membership functions are tuned to the given data. It is having the capability of integrating both the neural networks and fuzzy logic principles [10].

Figure 4 describes the sample data collected with different frequencies and the pitch values and respective output is loaded into the ANFIS as the trained data and can be tested accordingly. With the given dataset, IF-THEN rules are built by ANFIS and the integrated output is calculated and the respective structure of the model is also exposed. These are the rules which are designed by the ANFIS itself based on the training data in order to perform the operation.

The data collected by varying the two membership functions and respective output is noted. All the values are tabulated and loaded in to ANFIS as the training data and the structure of the model is gained as shown in Fig. 5.

Classification of the gender is done accurately and it has several potential applications. The features used to calculate the center of area are frequency and pitch.

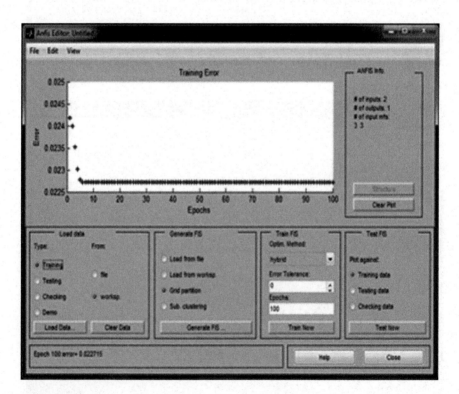

**Fig. 4** ANFIS editor

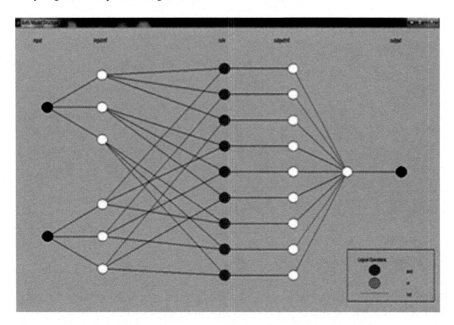

**Fig. 5** Structure of ANFIS model

**Table 2** Calculation values of COA

| Frequency | Pitch | Approx.COA | COA | Gender |
|-----------|-------|------------|-------|--------|
| 86.2 | 58.1 | 0.139 | 0.15 | Male |
| 94.9 | 60.3 | 0.14 | 0.151 | Male |
| 96.5 | 60.6 | 0.147 | 0.152 | Male |
| 108 | 61.7 | 0.150 | 0.151 | Male |
| 120 | 63 | 0.44 | 0.452 | Female |
| 139 | 64 | 0.449 | 0.452 | Female |
| 219 | 66 | 0.67 | 0.752 | Child |
| 250 | 67.4 | 0.74 | 0.751 | Child |
| 280 | 68.9 | 0.743 | 0.75 | Child |

By varying the frequency and pitch input values, output values are noted and are compared against the approximated values of the COA and are tabulated. Simulated graph is plotted with the tabulated values (Table 2).

**Fig. 6** Simulated graph

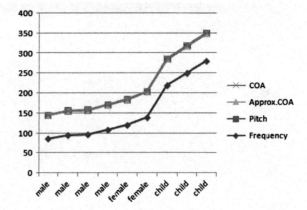

## 6  Conclusion

Current paper investigates the fuzzy logic based gender speech recognition and classification to a greater extent by using Mamdani fuzzy interface system from the fuzzy toolbox. However, based on the relevant parameters, the gender classification is done by using fuzzy rule based expert system. More frequently for error detection, we have used ANFIS by taking two membership functions as frequency and pitch. Calculated centroid area values as shown in Table 2 are compared with the system generated values and respective graph is plotted as shown in Fig. 6.

## References

1. Daniel J. Mashao and John Greene, "Evaluation Of Speaker Recognition Feature-Sets Using The SVM Classifier", STAR Research Group, Department of Electrical Engineering, UCT, Rondebosch, 7701, South Africa, 2003.
2. Miss. Sarika S. Admuthe1, Mrs. Shubhada Ghugardare, Survey Paper on "Automatic Speaker Recognition Systems", International Journal Of Engineering And Computer Science, Vol. 4, Page No. 10895–10898, March 2015.
3. Parul, R. B. Dubey, Automatic Speaker Recognition System , International Journal of Advanced Computer Research, Volume-2, December-2012.
4. Ramin Halavati, Saeed Bagheri Shouraki, Saman Harati Zadeh, "Recognition of human speech phonemes using a novel fuzzy approach", February 2006.
5. Kunjithapatham Meena, Kulumani Subramaniam, Gender Classification in Speech Recognition using Fuzzy Logic and Neural Network, The International Arab Journal of Information Technology, Vol. 10, No. 5, September 2013.
6. Ankush Sharma, Srinivas Perala, Priya Darshni, "Objects Control through Speech Recognition Using LabVIEW" Available Online at http://www.ijecse. International Journal of Electronics and Computer Science Engineering. 102 Available.
7. Benesty, J., Sondhi, M.M., Huang, Y. (eds.): Springer Handbook of Speech Processing, 1176 pp. Springer Science Business Media, Berlin, 2008.

8. M. Gomathy and K. Meena, "Gender Clustering and Classification Algorithms in Speech Processing": A Comprehensive Performance Analysis, International Journal of Computer Applications, Vol. 51 No. 20, August 2012.
9. El-Sayed M. El-Alfy, Amer G. Binsaadoon, "Silhouette-Based Gender Recognition in Smart Environments Using Fuzzy Local Binary Patterns and Support Vector Machines", The 8th International Conference on Ambient Systems, Networks and Technologies, ANT 2017.
10. Mostafa Rahimi Azghadi S., Reza Bonyadi M., Shahhosseini H. (2007) "Gender Classification Based on FeedForward Backpropagation Neural Network", In: Boukis C., Pnevmatikakis A., Polymenakos L. (eds) Artificial Intelligence and Innovations 2007: from Theory to Applications. AIAI 2007. IFIP The International Federation for Information Processing, vol 247. Springer, Boston, MA.

# Slot Positioning on Microstrip Antenna Using Parametric Analysis

**Parsha Manivara Kumar, Dasari Kiran Kumar and Nalam Ramesh Babu**

**Abstract** This paper clearly explains a new methodology with a step-by-step procedure of slot loading technique using parametric analysis. Optimum dimension for the rectangular slot is chosen and then parametric analysis is applied to position the three rectangular slots. End results of this paper show that the slot-loaded microstrip radiation characteristics are better than the unloaded antenna. The radiation characteristics are examined for different rectangular slot positions, and a better design which is suitable for C-band and X-band is proposed. The antenna resonates at 7.1643 and 9.5291 GHz and the total gain of this antenna is 6.20 dB. Variation in parameters like −10dB bandwidth, return loss, VSWR, and gain is presented.

**Keywords** Slot loading · Microstrip antenna · Dual band · Return loss
Parametric analysis

## 1 Introduction

Parametric analysis is being used for determining the size of different shapes of slots [1]. Studies on this topic are carried from past decade from the inception of simulation tools [2]. Operating frequency of the slotted microstrip depends on the size and shape of the slots. Slotted antennas operate at multiple frequencies [3]. We can obtain an increase in the bandwidth using slots of different shapes [4]. This

P. Manivara Kumar (✉)
Department of ECE, Chalapathi Institute of Engineering and Technology,
Lam, Guntur 522034, India
e-mail: manijw80@gmail.com

D. Kiran Kumar · N. Ramesh Babu
Department of ECE, DRK College of Engineering & Technology, Hyderabad
500043, India
e-mail: kirannrcm@gmail.com

N. Ramesh Babu
e-mail: nalam.rameshbabu@gmail.com

© Springer Nature Singapore Pte Ltd. 2018
J. Anguera et al. (eds.), *Microelectronics, Electromagnetics
and Telecommunications*, Lecture Notes in Electrical Engineering 471,
https://doi.org/10.1007/978-981-10-7329-8_51

technique is often used to control the resonance frequency and enhance the bandwidth [5]. Studies are carried out on slots of different shapes and sizes.

In this paper, a new approach of positioning rectangular slots over the patch is presented. The antenna geometry is chosen based on certain standard equations. Initially, a basic line feed microstrip antenna is designed and simulated using High-Frequency Structure Simulator (HFSS). Rectangular shaped slots are kept over the patch diligently by using parametric analysis. Slots are kept with a specific reason to obtain some radiation characteristics. All the three slots are not loaded at once, but a three-stage methodology of loading the rectangular slots one by one is presented in this paper. Parameters like return loss, resonant frequency, and bandwidth are tabulated. At the end, we show the results of a compact dual-band microstrip antenna using three rectangular slots.

## 2 Basic Microstrip Design

For our basic design, we have chosen the operating frequency 7.5 GHz. Rogers RT/Duroid 5880 is suitable for the frequencies from 5 to 10 GHz band, hence it is chosen as substrate. Its dielectric constant is $\in_r = 2.20$.

Width of the microstrip antenna can be determined by using

$$W = \frac{1}{2f_r\sqrt{\mu_0\mu\in_0}}\sqrt{\frac{2}{\in_r+1}} \tag{1}$$

If $f_r = 7.5$ GHz, we obtain the W = 12.45 mm
Length can be calculated from

$$L = \frac{1}{2f_r\sqrt{\in_{reff}}\sqrt{\mu_0\in_0}} - 2\Delta L \tag{2}$$

where

$$\Delta L = 0.412h \times \frac{(\in_{reff}+0.3)\left[\frac{W}{h}+0.264\right]}{(\in_{reff}-0.258)\left[\frac{W}{h}+0.8\right]} \tag{3}$$

where h is the thickness of the substrate

$$\in_{reff} = \frac{\in_r+1}{2} + \frac{\in_r-1}{2}\left[1+\frac{12h}{W}\right]^{\frac{1}{2}} \tag{4}$$

By substituting $\in_r = 2.20$ in (4), we obtain $\in_{reff}$.

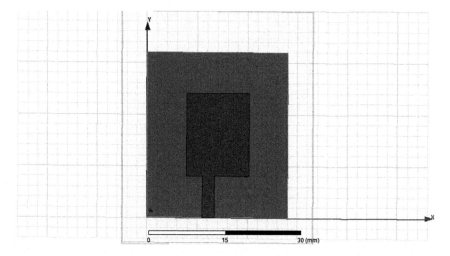

**Fig. 1** Microstrip without slot

Then, we substitute (4) in (3) and (3) in (2) to obtain the length of the patch

$$L = 16\,mm.$$

Ground plane and substrate dimensions are chosen as 28.1 and 32 mm and height of substrate is 0.794 mm. Microstrip antenna geometry with these dimension is shown in Fig. 1.

We used perfectly electric conductor for the patch area, feed line, and ground plane. Substrate is Rogers RT/Duroid 5880, which is suitable for 7.5 GHz. The microstrip antenna geometry with these dimensions is designed in HFSS and it is shown in Fig. 1. Simulation results of S11 for this basic design are shown in Fig. 2.

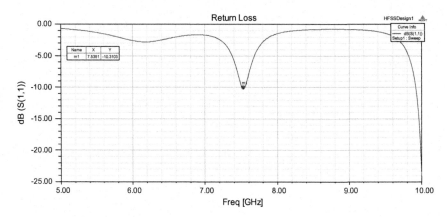

**Fig. 2** Return loss versus frequency graph of the designed microstrip

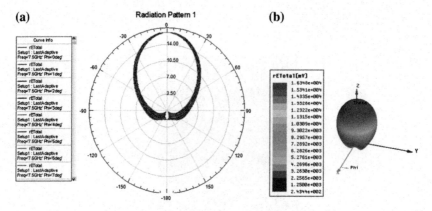

**Fig. 3** **a** Radiation pattern and **b** 3D polar plot of the microstrip antenna

Radiation pattern of this antenna is shown in Fig. 3a, b which shows the 3D polar plot of the basic microstrip. Parametric analysis can also be used to choose feed line dimensions and the position of the feed line on the patch edge.

## 3 Positioning the Slots by Using Parametric Analysis

The previous section discussed about operating frequency and basic geometry of the microstrip antenna. This section focuses on slot loading technique with a three-stage new approach.

### 3.1 Positioning the First Slot

The size of the rectangular slot has some limitation. Simulation results of rectangular slot with different dimensions are examined and the slot size is chosen to be 2.282 mm × 6 mm. Initially, this slot is kept over the patch at some x-position as shown in Fig. 4 and proper y-position needs to be determined. It can be determined by using parametric analysis. If we fix the X-position at 16.825 mm, we need to select y-position where the S11 results for both frequencies are same. If S11 in dB is same at the two frequencies, then practical antenna power requirements for satellite link in X-band and C-band will not differ. S11 results for different y-positions from 10 mm to 17 mm are shown in Fig. 5. By considering the parametric analysis, we have chosen a particular position where the two S11 in dB is same. This y-position for the Slot-1 is 15 mm and its S11 result is shown in Fig. 6.

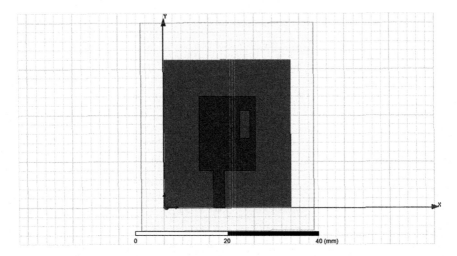

**Fig. 4** Microstrip with slot-1

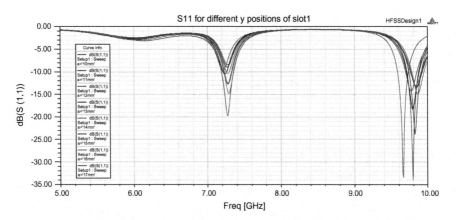

**Fig. 5** S11 of microstrip antenna loaded with rectangular slot for a fixed x-position at 16.825 mm, z-position = 0 mm and different y-positions

## 3.2 Positioning the Second Slot

The next objective is to locate another slot which will give the better radiation. The second rectangular slot is kept horizontally as shown in Fig. 7. Its y-position is again approximated to get better S11 results. In this case, parametric setup consists of different values for y-position of the second slot ranging from 9 to 12 mm in steps of 0.5 mm. Hence, the total number of y-positions simulated will be seven. After the analysis, y-position is chosen to be 9 mm and in the next stage, simulation is carried out with third slot to obtain a better S11. Two-slot geometry is shown in

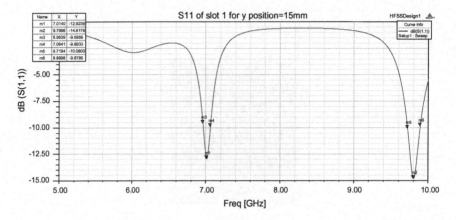

**Fig. 6** S11 of microstrip loaded with rectangular slot for y position = 15 mm

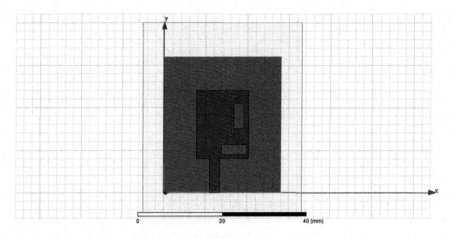

**Fig. 7** Microstrip loaded with rectangular slot-2

Fig. 7. All seven approximations are shown in Fig. 8. Return loss at these two frequencies is not the same. S11 at 7 GHz is different and S11 at approximately 9.7 GHz is different. Hence, we use the third slot to get some change in this behavior.

## 3.3 Positioning the Third Slot

The same procedure is repeated for the third slot. X-position of the slot is kept at 8.825 mm, and parametric setup for different y-positions is taken from 9 to 15 mm in steps of 1 mm. Hence, we obtain a total number of seven approximations again.

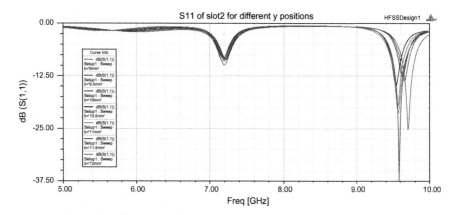

**Fig. 8** S11 of slot-2 for different y-positions

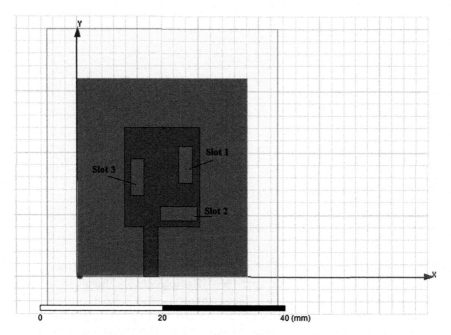

**Fig. 9** Microstrip loaded with slot-1, slot-2, and slot-3 (proposed antenna)

Simulations are carried out and the S11 result can be observed in Fig. 10. Best position for the slot-3 is chosen to be 13 mm and the slot kept at this position is shown in Fig. 9. Now, the positioning of the three slots is completed and the S11 result for this antenna is shown in Fig. 11.

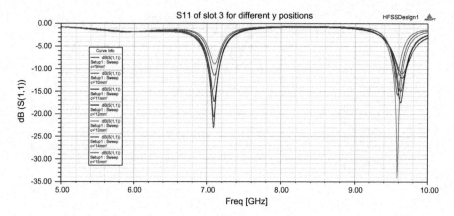

**Fig. 10** S11 of slot-1, 2, and 3 for different y-positions of slot-3

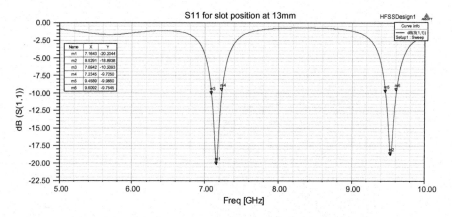

**Fig. 11** S11 of slot-1, 2, and 3 for different y-positions = 13 mm

This antenna resonates at two frequencies 7.1643 GHz and 9.5291 GHz having a −10dB bandwidth of 140 MHz and 150 MHz, respectively. Bandwidth, return loss, and the corresponding VSWR values are tabulated in Table 1.

Radiation characteristics of this rectangular slot loaded microstrip antenna are better when compared to the basic microstrip antenna without any slot as shown in Fig. 1. With three-slotted microstrip, we obtain dual-band characteristics. The return loss at the two resonant frequencies is approximately the same as shown in Table 1. This is a good advantage to design a microstrip antenna suitable for satellite communication at C and X bands and if it requires the same radiation and same return loss at the two bands. Bandwidths at the two frequencies are wide enough and maintain the total gain as shown in Fig. 14.

Figure 13 shows the 3D polar plot of the proposed antenna, which remains same as the unloaded antenna plot shown in Fig. 3b. Figure 12 shows the VSWR values

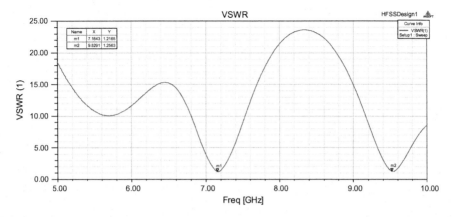

**Fig. 12** S11 of slot-1, 2, and 3 for different y-positions = 13 mm

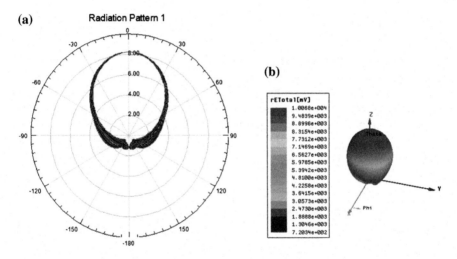

**Fig. 13 a** Radiation pattern **b** 3D polar plot of the proposed antenna

**Table 1** Results of the microstrip loaded with three slots

| Freq (in GHz) | Return loss in dB | −10 dB Bandwidth (MHz) | Shape of radiation pattern | VSWR |
|---|---|---|---|---|
| 7.1643 | 20.2044 | 140 | No change | 1.2165 |
| 9.5291 | 18.8938 | 150 | No change | 1.2563 |

of the proposed antenna that are tabulated in Table 1. Total gain plot is shown in Fig. 13 which gives a gain total of 6.0252 dB (Fig. 14).

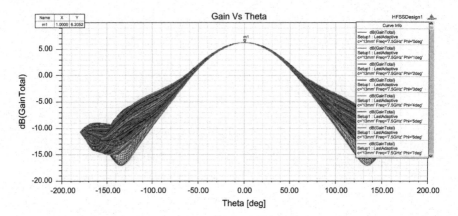

**Fig. 14** Total gain versus theta of proposed antenna

## 4    Conclusion

We have clearly explained the new approach of positioning the slots over the microstrip antenna. Step-by-step procedure by using the parametric analysis is clearly explained. Three rectangular slots are kept one by one over the patch and the simulation results were shown. These results are compared with the unloaded antenna. Dual-band operation at C-Band and X-Band is achieved. Radiation characteristics are quite suitable for these two bands. Variation of different parameters like −10dB bandwidth, return loss, VSWR, and gain is clearly shown in the paper.

## References

1. Hardeep Saini, Amanpreet Kaur, Abhishek Thakur, Rajesh Kumar, Naveen Kumar: A parametric analysis of ground slotted patch antenna for X-band applications, 2016 3rd International Conference on Signal Processing and Integrated Networks, IEEE Xplore Digital library, September 2016, pp 549–552.
2. Haruichi Kanaya: Multi-band miniaturized slot antenna with multi-band impedance matching circuit, 3rd Asia Pacific Conference on Antennas and Propagation, IEEE Xplore Digital library, July 2014 pp 511–554.
3. Prerna Gupta, S.N. Vijay: Design and parametric study of rectangular micro-strip patch antenna for C-Band Satellite Communication, 2016 International Conference on Recent Advances and Innovations in Engineering (ICRAIE), IEEE Xplore Digital library, June 2017, pp 1–5.
4. Jui-Han Lu: Bandwidth enhancement design of single-layer slotted circular microstrip antennas, IEEE Transactions on Antennas and Propagation, Vol. 51, Issue 5, May 2003 pp 1126–1129.
5. Sara Mahmoud, W. Swelam, Mohamed Hassan: Parametric Study of Slotted Ground Microstrip patch antenna, IOSR Journal of Electronics and Communication Engineering Vol 11 Issue 1 ver III (Jan–Feb 2016) pp 1–8.

# Object Removal Using Median Filter in Wavelet Domain

Rajkumar L. Biradar

**Abstract** The technique of repairing and modifying the images with the aid of digital technology in an undetectable form is digital image inpainting. In this paper, an image inpainting is object removal and is carried out by diffusing surrounding information. The challenge is to fill in the "target region" hole left behind by the removal of the object and create the homogeneous background in its place. Object removal may be treated as one of the applications of image inpainting. The median filter is used to diffuse the neighboring pixels into target region. The median filtering process is carried in wavelet domain. If more than one copy of the image is available, then the median diffusion can be estimated more accurately. Hence, we use discrete wavelet transform (DWT) to decompose the damaged image into four wavelet coefficient images. The target region of each coefficient image is filled by diffusing the neighboring information using median filter. After filling in the target region in the wavelet domain, inverse wavelet transform is used to obtain inpainted image in spatial domain. This technique produces better results for removal of medium and large size objects.

**Keywords** Inpainting · Wavelet transform · Object removal · 2D median filter

## 1 Introduction

During ancient period, the reconstruction of missing or damaged portions of images is used extensively in artwork restoration. Restoration can be obtained by image interpolation where a marked region is filled by its surrounding pixels and is called image inpainting. Image inpainting has various applications. It is used in data communication for reconstruction of missing data. In image processing, it is used for removal of natural cracks and scratches which arise due to age and folding of

R. L. Biradar (✉)
Electronics & Telematics Department, G. Narayanamma Institute
 of Technology & Science(for Women), Hyderabad 500104, India
e-mail: rajkumar_lb@yahoo.com

© Springer Nature Singapore Pte Ltd. 2018
J. Anguera et al. (eds.), *Microelectronics, Electromagnetics
and Telecommunications*, Lecture Notes in Electrical Engineering 471,
https://doi.org/10.1007/978-981-10-7329-8_52

image, dust or strain marks in the scanned image due to improper scanning glass, stamped date on letters, red-eye from photographs, the removal of political enemies from a still photograph, etc. To obtain inpainted image, the user selects a region for inpainting and algorithm fills in the region with pixels exterior to it without loss of any perceptual quality. Bertalmio was the first person to introduce the concept of image inpainting [1].

The object removal is one of the applications of image inpainting. In the proposed technique, the target region is filled by diffusing median value of pixels which are surrounding to target region into the target region. The median diffusion is carried out in wavelet domain for accurate diffusion. After filling in of target region in four-wavelet coefficient image, the inverse wavelet transform is found to obtain the image in spatial domain.

## 2 Literature Survey

Bertalmio et al. [1] established the first digital inpainting algorithm using partial differential equation and it is a modified version of level lines using disocclusion developed by S. Masnou and J. M. Morel [2].

The concept of computation fluid mechanics is used to propagate pixels from outside the target region to target [3]. In this technique, computation fluid mechanics uses Navier–Stokes partial differential equations.

Inpainting using anisotropic [4] and Euler–Lagrange modeling was proposed by Chan and Shen [5, 6]. In this technique, the broken edges are not connected, this is a major drawback. Curvature-Driven Diffusion (CCD) model [7, 8] is an extended version [5, 6], in which broken edges are connected using geometric information of isophotes while defining strength of diffusion process. In [6, 9], the filling of target region is achieved by the phase transition in superconductor and Ginzburg–Landau equation. In [10], target region is filled by normal and tangential vectors.

Fast marching method (FMM) algorithm was proposed by A. Telea [11]. This method computes flatness of image from a known image surrounding the pixel as a biased mean to inpaint. The constraint of this technique is producing haze when the region to be inpainted is larger than ten pixels.

All PDE-based inpainting techniques are more complex as they require more number of computations for implementing the techniques such as anisotropic diffusion and multiresolution schemes. Few steps are mathematically non-stable and inpainting process is very slow. For large target regions, blocky effect is observed.

## 3  Median Filter

For smoothing of discrete data, a filter was introduced by Tukey in 1970 is called median filter [12]. It is a nonlinear smoothening filter which removes the impulsive noise effectively while preserving the edges. Preservation of edge is crucial in an image processing due to the nature of visual observation.

Bertalmio diffused exterior pixels of damaged region into it to construct. But the diffusion process, in general, diminishes the edge information. Median filter preserves edges [13, 14]. Hence, it is proposed to use a median filter which retains edges and constructs damaged region by treating it as impulsive noise region.

Consider an observation samples sequence $\mathbf{x} = [x_1, x_2, x_3, \ldots, x_n]$, set of random variable which are independent identically distributed (iid) $x_i$, $i \in \{1, 2, \ldots, n\}$.. $\mathbf{x}$ can be obtained using a window of length $n$ over an arbitrary sequence. Arranging the random variables $x_i$, $i \in \{1, 2, \ldots, n\}$. in an descending order of magnitude such that $x_{(1)} > x_{(2)} > x_{(3)} > \cdots > x_{(n)}$ and then $x_{(i)}$ is called $i$th order statistic. The maximum and minimum of $x_i$, $i \in \{1, 2, \ldots, n\}$. are denoted by $x_{(n)}$ and $x_{(1)}$ respectively. Then, the median of $\mathbf{x}$ is given by

$$y = MEDIAN(\mathbf{x}) = \begin{cases} x_{(D+1)} & for\ odd\ n, D = (n-1)/2 \\ (x_{(D)} + x_{(D+1)})/2 & for\ even\ n, D = n/2 \end{cases} \quad (1)$$

The 2D median filter is given by

$$y(u,v) = MEDIAN \begin{pmatrix} y(u-D, v-D), \ldots, y(u-D, v), \ldots, y(u-D, v+D) \\ \cdot \\ \cdot \\ y(u, v-D), \ldots, y(u, v-1,), f(u,v), \ldots, f(u, v+D) \\ \cdot \\ \cdot \\ f(u+D, v-D), \ldots, f(u+D, v), \ldots, (u+D, v+D) \end{pmatrix} \quad (2)$$

For an image we use, $n \times n$ square window, where $n = 2D + 1$.

## 4  Two-Dimensional Wavelet Transform

Wavelet transforms are constructed using the small waves, called *wavelets,* of varying frequency and limited time duration. This important property of wavelet transform made us to analyze both time and frequency domain information of a function simultaneously [15].

In two dimension, a two-dimensional scaling function, $\phi(u, v)$ and three two-dimensional wavelet functions, $\psi^H(u, v)$, $\psi^V(u, v)$, and $\psi^D(u, v)$ are used. Each is the product of a one-dimensional scaling function $\phi(\cdot)$ and corresponding wavelet function $\psi(\cdot)$. These wavelets measure gray level variation of an image along the different directions: $\psi^H(\cdot)$, $\psi^V(\cdot)$, and d $\psi^D(\cdot)$ measure variations along horizontal edges, vertical edges, and along diagonals, respectively. Hence, the scaled and translated basis functions for two-dimensional are defined as

$$\phi_{j,m,n}(u,v) = 2^{j/2}\phi(2^j u - m, 2^j v - n) \tag{3}$$

$$\psi^i_{j,m,n}(u,v) = 2^{j/2}\psi^i(2^j u - m, 2^j v - n), \tag{4}$$

where $i = \{H, V, D\}$ and identifies the directional wavelets in horizontal, vertical, and diagonal direction.

The two-dimensional DWT of an image $f(u,v)$ of size $N \times N$ is

$$W_\phi(j_0, m, n) = \frac{1}{N} \sum_{u=1}^{N} \sum_{v=1}^{N} f(u,v)\phi_{j_0,m,n}(u,v) \tag{5}$$

$$W^i_\psi(j, m, n) = \frac{1}{N} \sum_{u=1}^{N} \sum_{v=1}^{N} f(u,v)\psi^i_{j,m,n}(u,v) \quad i = \{H, V, D\} \tag{6}$$

The $W_\phi(j_0, m, n)$ coefficient define approximation of $f(u,v)$ at scale $j_0$. The $W^i_\psi(j, m, n)$ coefficients horizontal, vertical, and diagonal details for scales $j \geq j_0$. Normally $j_0 = 0$ and selecting $N = 2^J$ so that $j = 0, 1, 2, \ldots, J-1$ and $m, n = 0, 1, 2, \ldots, 2^j - 1$. The inverse two-dimensional DWT to obtain image $f(u,v)$ from wavelet coefficients is given by

$$f(u,v) = \frac{1}{N}\sum_m \sum_n W_\phi(j_0, m, n)\phi_{j_0,m,n}(u,v) + \frac{1}{N}\sum_{i=H,V,D}\sum_{j=j_0}^{\infty}\sum_m \sum_n W^i_\psi(j, m, n)\psi^i_{j,m,n}(u,v)$$

$$\tag{7}$$

The original image of size $N \times N$ is decomposed into four wavelet coefficient images using two-dimensional wavelet transform—approximated coefficient image, horizontal edge coefficient image, vertical edge coefficient image, and diagonal edge efficient image, each of size $N/2 \times N/2$.

Instead of structure information from single image, we have structured information from four wavelet coefficient images. This allows computing structure information more accurately for filling in the target region, producing better inpainting results. Thus, target region in each wavelet coefficient image is filled by diffusing information available in that particular coefficient image using median filter. After filling in target region in each wave coefficient image in wavelet domain, inverse two-dimensional DWT is found to obtain inpainted image in spatial domain. This process gives a better quality of inpainting results.

## 5 Algorithm

Let an original image $I(i,j)$ be of size $N \times N$ and will be an input to the algorithm. Consider $W_\phi(j, m, n)$ and $W_\psi^i(j, m, n)$ be four wavelet coefficient images of size $(N/2, N/2)$. The inpainted image is substituted by input image at the end of algorithm.

**Start**

Depending on the width of target region, choose the parameter $D$ which is equal to half the width of target region in wavelet coefficient image.

*Range Selection*:

For any pixel $(m, n)$ in the target region $\Omega$, choose the range $R$ $R = ((m - D/2: m + D/2), (n - D/2: n + D/2))$

*Median filtering*:

Take median of all wavelet coefficient images independently over range $R$

$$W_\phi(j, m, n) = median(W_\phi(j, m, n)) \text{ for } \forall (m, n) \in R$$

$$W_\psi^i(j, m, n) = median(W_\psi^i(j, m, n)) \text{ for } \forall (m, n) \in R$$

Repeat above procedure for all pixels $(m, n) \in$ to the target region $\Omega$ in all wavelet coefficient images.

*Obtaining image in spatial domain*:

Find the inverse DWT to obtain the inpainted image in spatial domain.
**End**.

## 6 Results and Discussion

To test performance of the proposed technique, clean undamaged images are artificially degraded and then inpainted. Although, the evaluation of inpainting technique/s is to be assessed by visual appearance and quality which is subjective in nature, a quantitative assessment would be more objective. There is no established quantitative measure for this evaluation. Hence in this work, the quantitative evaluation is performed by peak signal-to-noise ratio (PSNR). The PSNR is defined as

$$PSNR = 20 \log_{10} \left( \frac{255}{RMSE\left(f_{clean}, f_{inp}\right)} \right), \tag{8}$$

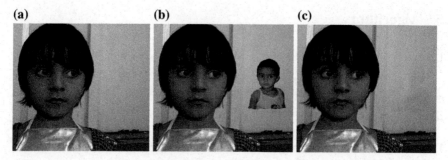

**Fig. 1** Removal of child's face placed artificially on the left side of child's image (**a**). **a** Clean undamaged child's image. **b** Subimage is placed on clean image. **c** Proposed technique with PSNR = 45.6046 dB

where $RMSE\left(f_{clean}, f_{inp}\right) = \sqrt{\dfrac{\sum\limits_{\forall i, j}\left(f_{clean}(u, v) - f_{inp}(u, v)\right)^2}{MN}}$

where $f_{clean}(u, v)$ is clean undamaged image, $f_{inp}(u, v)$ is inpainted image and $M \times N$ is size of $f_{clean}(u, v)$ and $f_{inp}(u, v)$. Please note that this measure is applicable only when the undamaged original image is available.

To test the efficacy of the proposed technique, the natural images (Figs. 1a and 2a) are used. These images are pasted with objects like child's face and Lena's face on original image to obtain synthetic images (Figs. 1b and 2b). These pasted objects are removed using proposed technique and the results are shown in Figs. 1c and 2c with PSNR = 45.6046 dB and PSNR = 44.5434 dB, respectively.

The visual quality and PSNR show a remarkable improvement of the proposed technique. The PSNR measure is well suited for the testing purpose, as we paste objects on original image and then object is removed to calculate PSNR between original and result.

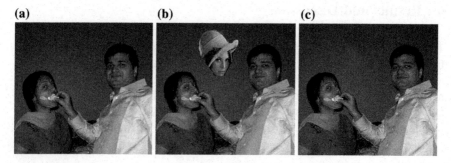

**Fig. 2** Removal of Lena's face placed on pair image. **a** Clean undamaged pair image. **b** A subimage (Lena) is placed on clean image. **c** Proposed technique with PSNR = 44.5434 dB

**(a)**                                                    **(b)**

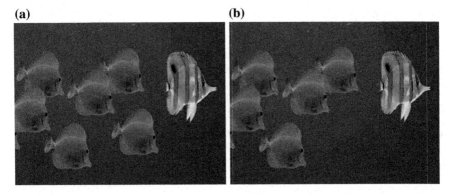

**Fig. 3** Removal of a fish. **a** Original fish image with seven fishes. **b** One fish is removed

As seen, the PSNR measure and qualitative assessments give identical interpretations. Hence, the technique can be extended to all the practical images envisaging similar performance. The unwanted object removal can be done by inpainting the object treating object as the damaged region and constructing suitable background in its place. The object removal may find applications in the areas like removal of logo, to black out some portions of TV frames, airbrushing of enemies, creating magical advertisements, etc. Hence, the algorithm is extended to object/s removal application expecting similar performance as shown in Figs. 3 and 4.

**(a)**                                                    **(b)**

**Fig. 4** Removal of bat from Ganguly still image. **a** Ganguly with bat. **b** Bat is removed

# 7 Conclusion

In this paper, a novel algorithm is introduced, the object is removed in wavelet domain using median filter. The results are better in the context of PSNR and visual appearance.

**Consent regarding Figs. 1 and 2:**

The pictures used in Figs. 1 and 2 belong to the author's family members. Author expresses consent in publishing these and mentions that there are no copyright issues in this regard.

# References

1. M. Bertalmio, G. Sapiro, V. Caselles, and C. Ballester. *Image Inpainting*. Proceedings of SIGGRAPH. Computer Graphics Processing, 2000, pp 417–424.
2. S. Masnou and J.M. Morel. *Level Lines Based Disocclusion*. 5th IEEE International Conference on Image Processing (ICIP), Chicago, IL. Oct 4–7, vol. 3, 1998, pp 259–263.
3. M. Bertalmio, A. L. Bertozzi and G. Sapiro. *Navier-stokes, Fluid dynamics and Image and Video inpainting*. IEEE CVPR, 2001.
4. P. Perona and J. Malik. *Scale-Space Edge Detection Using Anisotropic Diffusion*. IEEE transaction on pattern analysis vol. 12, No. 7, July 1990.
5. F. Chan and J. Shen. *Non-Textured Inpainting by Curvature Driven Diffusion*. UCLA Computational and Applied Mathematics Reports 00-11, March 2000.
6. H. Grossauer. *Digital Inpainting using the complex Ginzburg-Landau equation*. Scale Space method in computer vision, lecturer notes 2695, 2003.
7. F. Chan and J. Shen. *Non-Textured Inpainting by Curvature Driven Diffusion*. UCLA Computational and Applied Mathematics Reports 00-11, March 2000.
8. F.Chan, S.H. Kang, J. Shen. *Euler's Elastica and Curvature-Driven Diffusion*. SIAM J. Appl, Math, vol 63, no 2, pp 564–592, 2002.
9. H. Grossauer and O.Scherzer. *Using complex Ginzburg-Landau equation for image inpainting*. Scale Space method in computer vision, lecturer notes 2695, 2003.
10. X. C. Tai, S. Osher, R. Holm. *Image Inpainting using a TV-Stokes Equation*. IEEE Trans. Image processing, 2005.
11. A. Telea. *An Image Inpainting Technique Based on the Fast Marching Method*. Journal of graphics tools, vol. 9, No. 1, ACM press, 2004.
12. J. W. Tukey. *Nonlinear Methods for Smoothing Data*. In Congr. Rec. EASCON, 1974.
13. G. R. Arce and R. E. Foster. *Detail Preserving Ranked-Order Based Filters for Image Processing*. IEEE Trans. Acoustic., Speech, Signal Processing, volume 37, pp. 83–98, Jan. 1989.
14. B. I. Justusson. *Median Filtering: Statistical Properties in Two Dimensional Digital Signal Processing*. T. S. Huang, Ed. New York, Springer Verlag, 1981.
15. R. C. Gonzalez, R. E. Woods and S.L. Eddins. *Digital Image Processing using Matlab*. Pearson Education, 2004.

# An FPGA-Based Classical Implementation of Branch and Remove Algorithm

Kishore Vennela, M. C. Chinnaaiah, Sanjay Dubey and Satya Savithri

**Abstract** Many solutions came into limelight to explore the mobile robot navigation methods in a known environment with some constraints. Each and every such solution has its own merits and demerits in their simulation as well as implementation. By considering the issues in traveling salesman problem (TSP) which can be adopted in every differential field, we have been through classic implementation of branch and remove algorithm that considers every connected component in the known environment for implementation. With this paper, we came up with simulated results of branch and remove algorithm that exhibits minimum path journey by considering all edges connected among all the nodes in a complete graph. For the simulation in the Verilog HDL environment, the edge length is considered as distance metric or cost of travel between nodes as well as selection of initial node is considered to start the solution of minimum path finding. The results in this paper show that the initial point selection could be random and the implementation works well for any number of nodes and any edge length connecting them.

**Keywords** Branch and remove · Graph-based problem · Verilog HDL
Traveling salesman problem

## 1 Introduction

The mobile robot navigation is the most challenging task for implementing practical autonomous vehicle for human transport. We have to ensure the best solution that gives decay in wastage of available resources. To address this issue, we have

K. Vennela (✉) · M. C. Chinnaaiah · S. Dubey
Department of Electronics and Communication Engineering,
B.V. Raju Institute of Technology Narsapur, Medak, Telangana, India
e-mail: kishore.vennela@bvrit.ac.in

S. Savithri
Department of Electronics and Communication Engineering, JNTU, Hyderabad, India

© Springer Nature Singapore Pte Ltd. 2018
J. Anguera et al. (eds.), *Microelectronics, Electromagnetics
and Telecommunications*, Lecture Notes in Electrical Engineering 471,
https://doi.org/10.1007/978-981-10-7329-8_53

developed an approach that refers traveling salesman algorithms by which we could be able to evaluate the shortest path to the desired location by trimming the unnecessary path in the path evaluation process.

If a robot has to move from one node to another node and wraps each and every node only once and came back to the initial node, then analysis should start with consideration of initial node in a graph or tree structures. So node is the rudiment of a TSP if any differential field is going to take it as reference for implementation. By considering the basic graph structure that represents a network of five cities (nodes in the context of graph theory) and each edge represents distance metric or cost of traveling from one city to other. In the graph nodes, the robot has to visit all the nodes/cities with minimum distance/minimum cost and return to the node from where it has started journey. In the work, we assumed a list of nodes and distances among each one of them and concentrated on evaluating all possible paths as well minimum path without violating the rudiment of visiting all nodes absolutely once and ending at the initial node.

On the other side of this coin, i.e., theoretical aspect of TSP, implementation of these algorithms over FPGA has thrown challenges to the researches. In this context, our work laser focused on the approach of simulation for FPGA-based environment where process has considered distance acquisition through set of sensing elements, comparator module for low-distance/low-cost evaluation, and robot navigation unit for the movement of robot in the designed graph.

## 2   Related Work

Being trending, TSP has attracted many field of research to give solutions, some at simulation level and some other at hardware level. There has been tiny work done on how these algorithms could be transformed into hardware. A couple of hardware implementations of genetic algorithms for the TSP can be seen in the references [1, 2]. Abstraction of TSP theory into the mobile robot navigation has opened new window for researchers to make it keen on the navigation hypothesis. In [3], the authors propose to find first a near optimal sequence of path estimation in three steps: (1) cluster the representative placements in the workspace, (2) solve the resulting TSP in each cluster, and (3) concatenate the resulting paths. Clustering the workspace and solving individual issues and concreting the results rise issues in the view of time-limited executions and loss of data in the concreting process. Involvement of multiple processing modules limits the efficiency of any such algorithm. Sophistication of FPGAs and associated tools could solve such issues. Designing of the pipelined structure for Genetic Algorithms was implemented in [4] for facilitating hardware realization, and other parallel mechanisms were also added to it, thus greatly enhancing the speed. On other side, the neural networks based solutions were implemented over FPGAs to overcome speed and time limitations.

As in [5], the authors proposed a network that is entirely digital and developed over FPGA. The execution is encouraging and the FPGA took not more than 20 μm to solve the task for chosen examples. Authors in [6] proposed the graph-based algorithm for optimization of the path planning which considers the ambiguities in Generalized Lifelong Planning A* algorithms used for path planning. According to authors' proclamation in their observation, the implementation of the graph-based algorithm has some advancement when compared with genetic algorithm.

## 2.1 Dynamic Node Position Estimation

Making the environment under single observation is so tedious in the view of location estimation and processing further. So position estimation took multiple diversion in order to give effective solution and can be broadly comes under supervisory control or self-observatory methods. The position estimation could be carried by broadcasting of the current location by adopting GPS so that path planning could be adoptive. But it is not always flexible to use GPS especially in a closed indoor environment where there will be issues of accuracy. There has been more concentration put on this position estimation in indoor environment. The evaluation of this position through tracking method like Received Signal Strength Indicator (RSSI) from individual mobile elements gives better solution for navigation and low-cost path estimation. Most of the research has taken the ideology of triangulation method for calculating location in mobile robot environment. As in [7], estimation of the location carried through triangulation and averaging model that depend on the RSSI to the observation point. This article elevates the deficiency in using RSSI exclusively in the location estimation and also defines an adaptive algorithm that considers some reference values apart from dynamically changing RSSI from the sender. The signal strength at the observation point is calculated by considering a threshold RSSI from fixed distance 1 m in this case and channel attenuation index along with Gaussian Random variable "$\zeta$". On the other phase of this position calculation, simple mathematical modeling of Angle of Arrival [8] and/or Time of Arrival also give appropriate measurement of mobile node in our case mobile robot in any indoor environment. The same could be achieved through applying sound-based position identification system in the mobile robot navigation and implementing minimum path estimation for dynamically mobile robots environment. Huakang Li and coauthors in [9] constructed wireless network that uses two microphones with every mobile robot for self-positioning. All these stated are completely self-position identification and broadcasting the same through which there could be adoptive solution for continuous position changing inter-robot navigation without compromising in minimum tour length.

# 3  Implementation Strategy

The evaluation of shortest path estimation in the connected node environment has to have initial node selection as base parameter. As mentioned in the following discussions, the path evaluated will be varying based on this step and it is illustrated with examples. The proposed algorithm developed such that it has to work for any number of nodes and any cost of traveling parameters. *Pre-assumptions*: (1) There exists path from every node to other nodes in the graph. (2) Node positions are dynamic and considered crossover's existing in the node(s) connecting process.

## 3.1  Algorithm Description

In the algorithm analysis, it has been observed that when initial or starting node changes then total path changes which influence the total distance/cost metric of the journey. So we concentrated on development of generalized algorithm that should work for any initial node as well any distance/cost metric between two nodes, i.e., edge dimension in a graph. To validate the influence of initial node which is explicitly user specified, assume the initial node or starting node as S3 and process of execution of the BARE algorithm results in the minimum path as S3-S2-S4-S0-S1-S3.

Once initial node is selected (in our case S3), the robot moves from ideal state to initial node. While robot is moving from one node to another node, we need to make a mark of the current (leaving) node as visited. As the movement of the robot is explicitly through distance metric estimation from the current node, the connected edge lengths, i.e., distance/cost metrics which are connected to respective current node are processed for next node estimation. As shown in Fig. 1, comparison among the uncovered edges indirectly the nodes is processed to estimate the next to visit, i.e., node S2. Before leaving the node S3, mark that the node has visited not only for the future reference but also to eliminate the same in next node estimation. After evaluation, the node to be visited in S2 and is shown in Fig. 1b. From the node S2 as shown in Fig. 1c, the same process is repeated to get the next node with minimum distance/cost function. But at node S2, the covered edges are available, i.e., S2 to S3, which we need to eliminate to estimate minimum distance/cost metric. So compare the distances metrics from unvisited nodes as shown in Fig. 1d that selects the edge D9 as it has minimum distance/cost metric and added to our tour plan. The process is repeated at each and every node. This method of estimation gives minimal solution so that all nodes are covered. As shown in Fig. 1f at S0 node, we do not require comparison of distances because we have only one node which is not visited. At this stage we do not care about the distance of corresponding path as they have higher priority to visit.

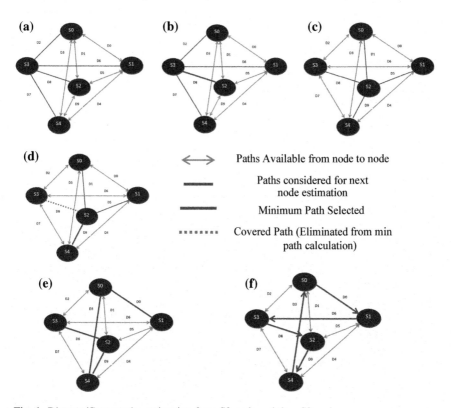

**Fig. 1** Distance/Cost metrics estimation from S3 node and then S2 node

## 3.2 FPGA Implementation

Rising computational constraints for complex problem solutions at hardware level made the essential use of FPGA in the field of robotics apart from DSP processors. In the work, we simulated the algorithm in Vivado14.4 tool and synthesized for Zybo FPGA. The distance/cost metrics could be acquired from the external world using sensing elements and processed further in the algorithm. For simulation environment purpose, we assumed random distance/cost metrics from node to node and executed in the tool. The sensor units used to estimate the distance are connected to standard port/PMOD ports of the Zybo FPGA and motoring action is controlled through relative values of these sensors, i.e., distance/cost metric based movement of the robot that uses BARE algorithm for the estimation.

## 3.3 Algorithm for Dynamic Path Estimation

- The Set of Vertices (S) in overall Graph with n-nodes $S = \{S_0, ..., S_{n-1}\}$ and define edge length or distance/cost metric between each and every vertices/nodes in a complete graph as "m"
  where

$$m = \frac{n(n-1)}{2}$$

- Define Visit_Node at every node, i.e., for n-nodes. It will be $[n - 1: 0]$ Visit_Node

  1. *Choose the initial vertex/node in the graph and move to that vertex/node from ideal node.*
  2. *Consider the edges that are connected to the current vertex/node.*
     *for all connected nodes*
     *begin*
          *checkpoint: check whether the node is visited or not visited.*
     *if (visited)*
       *begin*
            *eliminate the distance/ cost metric of that respective vertex/node from the minimum path estimation process.*
       *end*
       *else*
     *consider the vertex for the minimum path estimation*
     *end*
  3. *Estimate the minimum edge from current vertex/node and move along the minimum edge to reach the next vertex/node. Before moving mark the current vertex/node as visited, viz., Visit_Node[n-1] = True for (n-1)th node visit.*
  4. *Repeat the "Step 2 and 3" at every vertex and make a move to every node based on the estimation result.*
     *for (i = 0; i <= n-1; i ++)*
     *begin*
         *Step 2 and 3;*
     *end*
  5. *If all vertices/nodes are marked as True, then default movement is toward initial vertex/node from current vertex/node.*

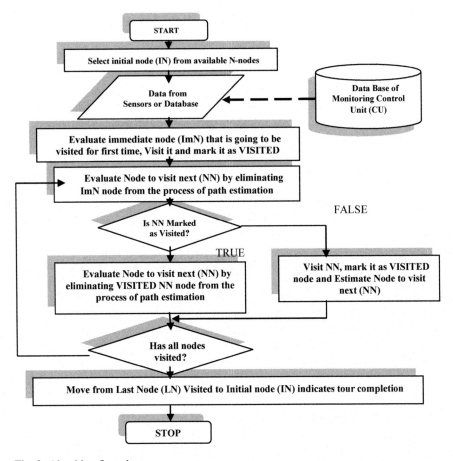

**Fig. 2** Algorithm flow chart

## 4 Dynamic Node Location Estimation

The algorithm so far discussed applied to find minimum path where the connected node is assumed to be static. Apart from their static nature, the proposed algorithm can be applied for any node distance/cost metric without violating the nature of node, i.e., non-mobile node. But practically, for example in geographical monitoring system, where mobile robot(s) are given task of data acquisition for specified area(s) continuously and command robot has to collect the data at regular interval of time, the robots have to move randomly so as to accomplish the assigned task. In such scenario, the mobile node to mobile node distance is dynamic and in turn changes the minimum distance/cost metric path the command robot has to follow.

## 4.1  Node's Dynamic Location Estimation Through SVM

Application of the algorithm gives fruitful results if the node(s) in the graph are maintaining unique distance/cost metric values. Results shown in the following section prove that the static node with constant distance location in the graph can exhibit minimum path solution. There will be situation where the node is replaced by moving object(s) like mobile robot, in such cases, the distance/cost metric is nonuniform and it is changing randomly. To complete the tour with minimum path/cost metric, the algorithm has to act such that it should consider this dynamic movement of the node(s) in the graph by estimating location. So, the primary task to accomplish is to locate mobile robot position in environment. One of such methods is support vector machine (SVM) that deals with analysis of data collected from sensor array and classifies to know the location with respect to activated sensor in the array. The simulated results using MATLAB framed in Sect. 4 give the location of mobile robot indicated by respective sensor status. In the classification of data assurance of mobile robot under observation, movement and location can be given by accuracy factor at each and every sensing unit. Relative values of the data collected from sensor and their accuracy factor execute location estimation. The analysis and results are framed in the following section and are simulated in MATLAB environment with an assumption of four sensing elements.

## 5  Simulation Results

### 5.1  FPGA Simulation Results

By making S3 node as the initial vertex/node simulation of the algorithm has been carried as steps and path estimated is S3-S2-S4-S0-S1-S3.

Step-1:  Node S3 as an initial node, the distance between S3 node to S2 node is (D8 = 4) minimum distance/cost metric over all connected nodes. The marked region (with Arrow) in the simulation waveform Fig. 3 shows the node transition from S3 node to S2 node.

Step-2:  Node S2 as an current node, the distance between S2 node to S4 node is (D9 = 13) minimum distance/cost metric over all connected nodes. The marked region (with Arrow) in the simulation waveform Fig. 3 shows the node transition from S3 node to S4 node.

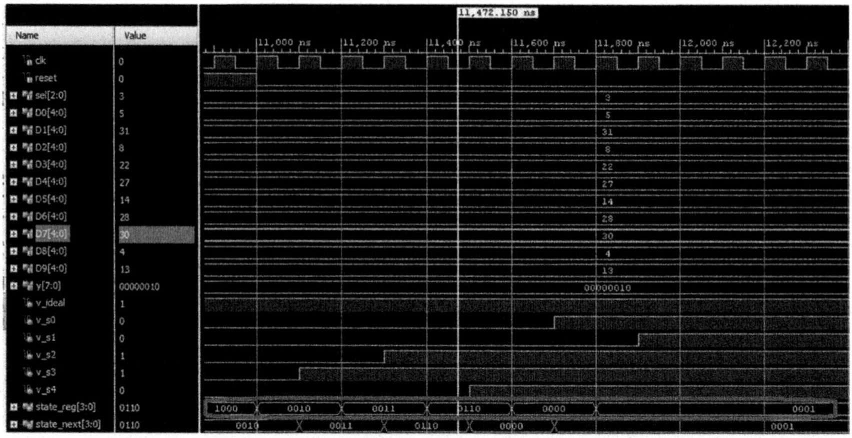

**Fig. 3** State transition by making visit indicator binary value "1" in every visited state

Step-3: Node S4 as an current node, the distance between S4 node to S0 node is (D3 = 22) minimum distance/cost metric over all connected nodes. The marked region (with Arrow) in the simulation waveform Fig. 3 shows the node transition from S4 node to S0 node.

Step-4: Node S0 as an current node, the distance between S0 node to S1 node is (D0 = 5) minimum distance/cost metric over all connected nodes. The marked region (with Arrow) in the simulation waveform Fig. 3 shows the node transition from S0 node to S1 node.

Step-5: Node S1 as a current node, there will not be any minimum distance/cost metric estimation as all other vertices/nodes are marked as visited. So default node to visit after tour completion is initial node. The marked region in the simulation waveform Fig. 3 shows the node transition from S1 node to S3 node along with complete node transitions.

## 5.2 Support Vector Machine Simulation Results

The setup assumed with four-sensor array to estimate the location of the mobile robot. Based on the movement of the robot in the environment, one of the sensors will be active. Even though there is multiple sensors active at a time, the accuracy can be established by application of SVM classifier (Fig. 4).

The simulation result shows the active status of the sensor ID one which indicates the existence of mobile robot in the region of sensor one. Basing on this information from sensor, the mobile path can be changed dynamically such that total tour length is minimum in the view of distance/cost metric. As mentioned

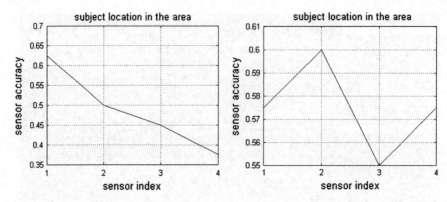

**Fig. 4** Sensor activity

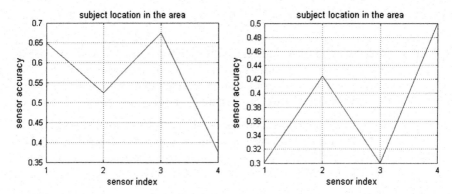

**Fig. 5** Sensor activity

**Table 1** Activity of sensors: active sensors S1 to S4

| If Robot is under S1 | | If Robot is under S2 | | If Robot is under S3 | | If Robot is under S4 | |
|---|---|---|---|---|---|---|---|
| Sensor ID | Accuracy | Sensor ID | Accuracy | Sensor ID | Accuracy | Sensor ID | Accuracy |
| S-1 | **1** | S-1 | 0.600 | S-1 | 0.5750 | S-1 | 0.6250 |
| S-2 | 0.3250 | S-2 | **0.7500** | S-2 | 0.6000 | S-2 | 0.3750 |
| S-3 | 0.3750 | S-3 | 0.4000 | S-3 | **0.6550** | S-3 | 0.6250 |
| S-4 | 0.3750 | S-4 | 0.2750 | S-4 | 0.5750 | S-4 | **0.6500** |

earlier, position estimation is achieved through SVM evolution. Accuracy results at individual sensing elements are tabulated in the following (Fig. 5 and Table 1).

# 6    Conclusion

Work presented in this paper, the branch and remove algorithm worked well for change in number of node/vertices in the graph as well as for any distance/cost metric. Though it has the similarities with neural network based approaches where neuron is treated to be node point and edges as weight function, the approach has its uniqueness in terms of hardware level simulation. Interfacing with the hardware through graph-based method is exhibited in the results section. We have tested this algorithm by changing the mode of execution, viz., static path and dynamic path over a complete graph where every node is connected to other nodes. The results proved that the selection of initial node may change the path to be followed so that minimum cost will occur but work for any distance/cost metric.

# References

1. P. Graham and B. Nelson, "A Hardware Genetic Algorithm for the Traveling Salesman Problem on Splash 2", Brigham Young University.
2. Xiaojun Wang, Brent Nelson, "Tradeoffs of Designing Floating-Point Division and Square Root on Virtex FPGAs", FCCM 2003.
3. L. Gueta, R. Chiba, J. Ota, T. Ueyama, and T. Arai, "Coordinated Motion Control of a Robot Arm and a Positioning Table With Arrangement of Multiple Goals," in IEEE International Conference on Robotics and Automation, 2008 (ICRA 2008), 2008, pp. 2252–2258.
4. Z. Yan-cong, G. Jun-hua, D. Yong-feng and H. Huan-ping, "Implementation of genetic algorithm for TSP based on FPGA," 2011 Chinese Control and Decision Conference (CCDC), Mianyang, 2011, pp. 2226–2231.
5. A. Varma and Jayadeva, "A novel digital neural network for the travelling salesman problem," Neural Information Processing, 2002. ICONIP '02. Proceedings of the 9th International Conference on, 2002, pp. 1320–1324 vol.3.
6. J. Stastný, V. Skorpil and L. Cizek, "Traveling Salesman Problem optimization by means of graph-based algorithm," 2016 39th International Conference on Telecommunications and Signal Processing (TSP), Vienna, 2016, pp. 207–210.
7. H. Fang, M. Liu, F. Li, W. Shen and F. Zhang, "A novel adaptive algorithm for location based on Distance-Loss model in complex environment," 2016 IEEE 20th International Conference on Computer Supported Cooperative Work in Design (CSCWD), Nanchang, 2016, pp. 26–30.
8. Huakang Li, Satoshi Ishikawa, Qunfei Zhao, Michiko Ebana, Hiroyuki Yamamoto and Jie Huang, "Robot navigation and sound based position identification," 2007 IEEE International Conference on Systems, Man and Cybernetics, Montreal, Que., 2007, pp. 2449–2454.
9. J. Zhao, Z. w. Zhou, G. Li, C. l. Li, H. Zhang and W. b. Xu, "The apposite way path planning algorithm based on local message," 2012 IEEE International Conference on Mechatronics and Automation, Chengdu, 2012, pp. 1563–1568.

# A Novel Fractal Stacked Inductor Using Modified Hilbert Space Filling Curve for RFICs

P. Akhendra Kumar and N. Bheema Rao

**Abstract** High quality factor miniaturized inductors are prerequisites of RFIC applications. This paper presents a novel fractal stacked inductors using modified Hilbert space filling curve. The proposed inductor is constructed in series stack fashion according to the process rules to achieve higher inductance values. Using the modified Hilbert structure, lateral flux is eliminated to achieve higher Q values. The results show that more than 90% improvement in L over conventional fractal inductor within same occupying area and more than 10% improvement in Q factor over standard stacked inductor.

**Keywords** High-frequency structural simulator (HFSS) · Inductance value (L) Low-noise amplifier (LNA) · Modified fractal structure · Quality factor (Q) Voltage-controlled oscillator (VCO)

## 1 Introduction

In recent years, RF communication systems demand for high-performance on-chip passive and cost limiting passive components. The inductor is the most critical and crucial element among all the passive components at higher frequencies. Monolithic inductors are the most critical and widely used component in RFIC applications. The noise figure of LNA [1] and phase noise of VCO [2] are improved by the high Q factor inductors.

Spiral inductors suffer from low Q factor [3] due to resistive and capacitive losses [4]. and their inductance values limited by area constraints. Symmetric inductors excited by differential inputs yield improved quality factor due to the

P. Akhendra Kumar (✉) · N. Bheema Rao
Department of ECE, National Institute of Technology, Warangal 506004,
Telangana, India
e-mail: akhendra.p@gmail.com

N. Bheema Rao
e-mail: nbr.rao@gmail.com

© Springer Nature Singapore Pte Ltd. 2018
J. Anguera et al. (eds.), *Microelectronics, Electromagnetics and Telecommunications*, Lecture Notes in Electrical Engineering 471,
https://doi.org/10.1007/978-981-10-7329-8_54

presence of a shorter under pass, lesser number of overlaps, and hallow space in the center of inductor [5, 6, 7].

Series stacked inductors [8] are having an advantage of higher inductance value, but the usage of these inductors is limited to low frequencies due to huge adjacent parasitics. To achieve higher Q factor (Q), usually wider metal width is adopted at the cost of lower self-resonant frequency (SRF).

The Q and SRF of a stacked inductor are improved by adopting up-down series winding [9, 10, 11].

Inductor designed using fractal geometry was highly area efficient [12, 13]. An intuitive study and experimental validation of fractal inductors were carried out in [14–16].

The inductors designed based on fractal space filling curves are well fit into finite area due to its unique iterative property. Ideal fractal inductors cannot be built because of its complexity and technology limitations. In this present study, conventional Hilbert space filling curve is modified to reduce the lateral flux intern improves the Q as well as SRF. Further huge reduction in area is obtained by considering series stack multi-turn construction.

The organization of the paper is as follows. The basic construction of modified Hilbert space filling curve is explained in Sect. 2. Section 3 reports proposed series stack fractal inductor. The details of the simulation results are shown in Sect. 4, followed by conclusion in the summary in Sect. 5.

## 2   Construction of Modified Hilbert Space Filling Curve

In this paper, modified fourth iteration Hilbert space filling curve is derived from conventional Hilbert space Fig. 1 filling curve keeping the first three stages are similar to conventional Hilbert space filling curve. The modified structure shown is in Fig. 2.

The process can be extended up to infinite iterations with the modified fourth-order stage. The modified construction has an advantage of low lateral flux capacitance due to space provided in the construction.

**Fig. 1** Construction of conventional Hilbert curve

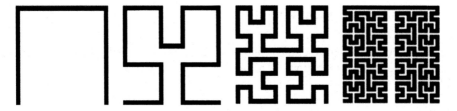

**Fig. 2** Construction of modified Hilbert curve

## 3　Fractal Stacked Inductor Using Modified Hilbert Curve

Figure 3 shows the 3D representation of proposed inductor. It uses four metal layers, adjacent metal layers being connected at the similar ends through a via. With the same chip area, the proposed stacked inductor can achieve huge inductance because of the mutual inductance generated from the metal lines.

In conventional stacked inductor, the parasitic resistance of bottom metal is higher than that of top metal. However, the parasitic resistance and capacitance lead to decrease in Q factor and self-resonance frequency.

The Q factor is the major specification of inductor design, high Q means low loss and high efficiency, the $R_s$ (increased parasitic resistance) is reduced using larger metal width of bottom metal trace which leads to high Q factor.

For a stacked inductor, the interlayer capacitance impacts the resonance frequency four times as much as the bottom layer capacitance. In the proposed structure, metal layers are kept apart from each other to achieve a higher SRF.

The SRF of an inductor is given by Eq. 1

$$F_{SR} = \frac{1}{\sqrt{2\pi L_{eq} c_{eq}}}, \tag{1}$$

where $L_{eq}$ is the equivalent inductance and $C_{eq}$ is the equivalent capacitance of the structure.

**Fig. 3** 3-D representation
series stack modified Hilbert
inductor

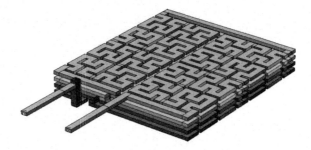

# 4   Results and Discussion

To quantify the advantage of the proposed architecture for series stacked inductors, the top metal and the bottom metal are of 3 μm thick with a 2 μm interlayer dielectric.

$$L = \frac{-1}{2\pi f im\{Y_{11}\}} \tag{2}$$

$$Q - \frac{Im\{Y_{11}\}}{Re\{Y_{11}\}} \tag{3}$$

## 4.1   Impact of Modified Architecture

Five different inductors are considered to validate the proposed inductor structure.

HFSS EM simulations are carried on these inductors to obtain L and Q values using Eqs. 2 and 3.

$L_1$ is a planar inductor. $L_2$ is a stacked inductor. $L_3$ is a single-layered conventional fractal inductor. $L_4$ is the proposed inductor configurations, where all these inductors are designed with an effective conductive width 8 μm and occupying an area of 200 μm × 200 μm.

Figures 4 and 5 show the simulated L and Q characteristics of proposed inductor, while electrical characteristics of four different inductors are summarized in Table 1 based on the EM simulation results. $L_1$ is area limited by area requirements. $L_2$ series stacked inductor which has good inductance value due to

**Fig. 4** Simulated inductance plot for stack modified Hilbert inductor

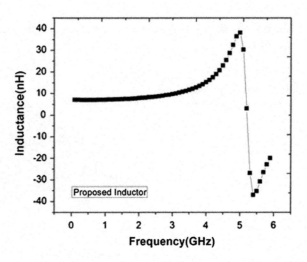

**Fig. 5** Simulated Q factor plot for stack modified Hilbert inductor

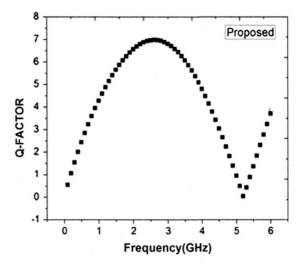

**Table 1** Comparison of standard inductor structures with Series fractal inductor

| Inductor type | Inductance (nH) | Quality factor | SRF |
|---|---|---|---|
| Planar ($L_1$) | 3.2 | 18 | 13 |
| Stacked inductor ($L_2$) | 13 | 4.7 | 2.6 |
| Hilbert fractal ($L_3$) | 3.6 | 19 | 29 |
| Proposed (L4) | 9.65 | 7 | 5.2 |

mutual coupling between the layers but has low self-resonant frequency due to the interlayer capacitance.

Due to self-repetitive nature of fractal inductors, $L_3$ could achieve higher inductance values, but these inductors suffer low Q values as iteration order increases due to lateral flux as well as proximity effects.

The proposed inductor $L_4$ achieves higher inductance value than the previously mentioned inductors and due to novel modified fractal structure, interlayer capacitance as well as proximity effect losses resulting in higher Qmax and SRF values over stacked inductors.

# 5 Conclusions

In this study, a novel fractal stacked inductors using modified Hilbert space filling curve have been proposed. The performance merits such as inductance, Q factor, and SRF have been studied. The results show that proposed structure achieves more than 90% improvement in inductance value and more than 10% improvement in Q factor over standard stacked inductor. The proposed fractal inductor is well suited for RF front end design applications.

# References

1. Baimei, Li., U,C Wang., Minglin, Ma., Shengqiang Guo. An Ultra-Low-Voltage and Ultra-Low-Power 2.4 GHz LNA Design. Radio Engineering (2009), vol. 18, no. 4, 527–531.
2. J Craninckx., Michiel, S J Steyaert. A 1.8-GHz Low-Phase-Noise CMOS VCO using Optimized Hollow Spiral Inductors. IEEE Journal of Solid-State Circuits (1997), vol. 32, no. 5. pp. 736–744.
3. Wang, C., N,Y, Kim. Analytical optimization of high-performance and high-yield spiral inductor in integrated passive device technology. Elsevier Microelectronics Journal (2012), vol. 43, pp. 176–181.
4. H, Darabi., A,A, Abidi. A 4.5-mW 900-MHz CMOS receiver for wireless paging IEEE Journal of Solid-State Circuits(2000), vol. 35, no-8, p. 1085–1096, ISSN: 0018-9200, https://doi.org/10.1109/4.859497.
5. M, Danesh., J,R, Long., R, A, Hadaway., D,L, Harame. A Q factor enhancement technique for MMIC inductors. IEEE MTT-S Int. Microwave Symp. Dig, (1998), vol. 1, pp. 183–186.
6. Marco, Politi., Vito, Minerva., Silvia, Cavalieri D'oro. Multi-Layer Realization of Symmetrical Differential Inductors for RF Silicon IC's. 3rd International Microwave Conference, (2003), ISSN: 1-58053-834-7, https://doi.org/10.1109/EUMA.2003.340914.
7. T,H, Teo., Y B Choi., H, Liao., Y Z Xiong., J,S, Fu "Characterization of symmetrical spiral inductor in 0.35µm CMOS technology for RF application. Solid State Electronics, 2004, vol 48, pp. 1643–1650.
8. A, Zolfaghari., A, Chan., B, Razavi., Stacked inductors and transformers in CMOS technology. IEEE Journal of Solid-State Circuits, (2001), vol. 36, no. 4, p. 620–628, ISSN: 0018-9200, https://doi.org/10.1109/4.913740.
9. Wen, Yan, Yin., Jian, Yong, Xie., Kai, Kang., Jinglin, Shi., Jun Fa, Mao., Xiao Wei, Sun. Vertical Topologies of Miniature Multi spiral Stacked Inductors. IEEE Transactions on Microwave Theory and Techniques, (2008), vol. 56, no. 2, pp. 475-486, ISSN: 0018–9480, https://doi.org/10.1109/TMTT.2007.914624.
10. G, Haobijam., R, Paily. Quality factor enhancement of CMOS inductor with pyramidal inding of metal turns. In Proceedings of IWPSD, (2007), pp. 729–732, ISSN: 978-1-4244-1727-8, https://doi.org/10.1109/IWPSD.2007.4472624.
11. Chih, Chun, Tang., Chia Hsin Wu., Shen Iuan Liu. Miniature 3-D Inductors in Standard CMOS Process. IEEE Journal of Solid-State Circuits, (2002), vol. 37, no. 4, ISSN: 0018-9200, https://doi.org/10.1109/4.991385.
12. Maric, A., Radosavljevic, G., Zivanov, M., Zivanov, L., Stojanovic, G., Mayer, M., Keplinger, F. Modelling and Characterisation of Fractal Based RF Inductors on Silicon Substrate. in International Conference on Advanced Semiconductor Devices and Microsystems, (2008) pp. 191–194.
13. Wang, G., Xu, L., Wang, T. A Novel MEMS Fractal Inductor Based on Hilbert Curve in 4th International Conference on Computational Intelligence and Communication Networks, (2012), pp. 241–244.
14. N, Lazarus., Christopher, D Meyer., Sarah S, Bedai. Fractal Inductors. IEEE Transactions on Magnetics, (2014), vol. 50, no. 4. pp. 1–8.
15. G, Shoute., Christopher., Douglas W, Barlage. Fractal loop Inductors. IEEE Transactions on Magnetics, (2015), vol. 51, no. 6, pp. 1–8.
16. H, Sagan. Space-Filling Curves. New York(USA): Springer-Verlag, (1994). ISBN-10: 0387942653.

# An Approach to Parallel Transformation Technique for High-Efficiency Video Coding

P. Anitha, P. Sudhakara Reddy and M. N. GiriPrasad

**Abstract** Compression plays a vital role in video processing. The reduction or removal of redundant data from raw video stream makes an effective video file transmission and storage with High Efficiency Video Coding (HEVC). HEVC is the standard developed by the Joint Collaborative Team on Video Coding (JCT-VC). HEVC or H.265 includes several modifications compared with its predecessor the H.264. During the HEVC encoding process, HEVC uses transform coding on self-contained and inter-predicted frame residuals, which possess distinct characters compared to residual frame. The residual frame information is performed using transformation, quantization, and entropy coder, and the encoded bit stream is decoded to reconstruct the original video. In this work, by using parallel transformation technique, the residual frame encoding and decoding is implemented for the improvement of HEVC. The results are presented with the use of proposed parallel transformation technique in the framework of the HEVC and performance results of our implementation show better results in video quality metrics. Our proposed implementation on HEVC transform, scaling, and quantization carried out on various video frame formats has shown better results compared to conventional discrete wavelet transform.

**Keywords** HEVC · Residual · Transformation · Discrete wavelet transform (DWT)

P. Anitha (✉) · M. N. GiriPrasad
Department of ECE, Jawaharlal Nehru Technological University,
Ananthapuram, India
e-mail: anithakrishna77p@gmail.com

M. N. GiriPrasad
e-mail: mahendragiri1960@gmail.com

P. Sudhakara Reddy
Department of ECE, Srikalahasteeswara Institute of Technology,
Srikalahasti, India
e-mail: psr_vlsi_dsp@rediffmail.com

© Springer Nature Singapore Pte Ltd. 2018                                        541
J. Anguera et al. (eds.), *Microelectronics, Electromagnetics*
*and Telecommunications*, Lecture Notes in Electrical Engineering 471,
https://doi.org/10.1007/978-981-10-7329-8_55

# 1 Introduction

The development in the video services has become a demand in the present video coding standard, the HEVC. HEVC is also known as H.265, its main feature is in achieving the video quality for high resolutions and reduction in encode video bit stream compared to its predecessor, the H.264/AVC standard, for the same video quality [1–3]. Several improvements were made in HEVC of Motion Estimation (ME) process [1, 4, 5] to achieve better coding efficiency by dividing the video frame in Coding Tree Block (CTB) for Luma, supports the $16 \times 16$, $32 \times 32$, and $64 \times 64$ samples block sizes and Coding Tree Blocks (CTBs) for Chroma samples, grouping these together called as Coding Blocks partition [3]. In HEVC, for the ME process, the Prediction Blocks (PB) are grouped by Coding Blocks (CBs), these Prediction Blocks (PBs) with varying sizes of $8 \times 8$, $16 \times 16$, $32 \times 32$, and $64 \times 64$ samples have within themselves Prediction Units (PUs) with a varying size from $4 \times 4$ to $64 \times 64$ samples. Supporting this, several algorithms were proposed for ME and compensation [6]: block displacement methods [7] fail to capture significant number of pixels, optical flow [8] suffers from aperture problem, feature matching [9] is only calculated for strong features in the image, mesh-based algorithms [10] use only match-based mesh elements in different frames causing unequal motion estimation, and model-based methods [11] fail to calculate the local motion in the video frames.

HEVC, the visual quality [2, 12, 13] is 50% and is reduced compared to H.264; HEVC employs a compression technique to improve compression efficiency with high computation complexity [14]. On the basis of human visual system [15], the visual models have been explored. Supporting this, in [2, 16], the saliency map detection [2] is employed to adjust the quantization parameter of macro-block for H.264. In [17], a model is proposed to reduce the bit rate for the video compression. In [18], a fast intra-prediction mode decision strategy is introduced. In [19], a fast CU size decision algorithm for intra-coding is presented. In [20], a fast CU splitting and pruning method to improve the CU mode decision is proposed.

HEVC [21, 22] is the latest video standard that is developed on top of previous video coding standard H.264/MPEG. In the proposed work, transcoding of the digital video [22] liaises on the latest HEVC coding standard [23]. Compression is the process of reducing the size of video files or audio files. The main purpose of this work is to present a new transformation technique based on frame pixel band extraction and frame sequence matching. The rest of this paper is structured as follows: Sect. 2 shows the Proposed Parallel Transformation Model based HEVC, Sect. 3 describes the implementation of proposed technique, Sect. 4 describes the results and discussions, and finally Sect. 5 summarizes the conclusion.

## 2 Proposed Parallel Transformation Model Based HEVC

HEVC standard was released by ITU-T VCEG to achieve same video subjective performance with half the H.264 data rate. By selecting proper transform, scaling, and quantization techniques, we can improve the process of video information. In the HEVC video encoder, we have chosen to improve the Quantized Transform Coefficients to improve the HEVC.

According to the proposed technique, a new step: Transformation Mode Selection is added in the HEVC. The transform, scaling, and quantization are implemented in four stages as shown in Fig. 1. The four stages are Transformation Mode Selection, Filter Control Analysis, Frame Coding, and Decoding, Predicted Image with Previous Frame.

(a) Transformation Mode Selection: It is the mode selection process of the HEVC encoding. In this process, the input video is converted into number of frames and frame rate. These frames were forwarded to perform a decision, based on frame coefficient band levels. Each frame is analyzed for low-band and high-band coefficients. If any frame is containing high-band coefficients, it will be given a flag "Decision = H", then all these frames will be forwarded to frame coding process. And if any frame is containing low-band coefficients, it will be given a flag "Decision = L", then all these frames will be forwarded to filter control analysis. From Fig. 1, the Transformation Mode Selection is shown in Fig. 1.

(b) Filter Control Analysis: In this block, low-band coefficients are processed to smooth out the artifacts included by the blockwise processing and quantization, by using in-loop deblocking filter (DBF) followed by a sample adaptive offset (SAO). A deblocking filter is similar to the one used in H.264/MPEG AVC is operated in the inter-picture prediction loop.

(c) Frame Coding and Decoding: In this section, we have encoding and decoding process. During encoding process, inter-frame prediction and intra-frame prediction process are performed. Regular I-frame encoding is processed prior the inter- and intra-frames. The group of pictures of the structure specifies how different frame types are ordered by giving successive frames of the video as

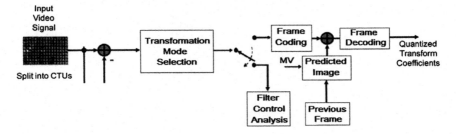

**Fig. 1** The transformation technique implemented in our proposed technique

given. In our proposed technique, we are using the GOP structure as IBBPBBPBB, usually starts with a I-frame, followed by a number of B-frames separated by P-frames. From Fig. 1, the proposed frame coding and decoding is shown.

(d) Predicted Image with Previous Frame: During the encoding, the I-frames with inter- and intra-frames were predicted. If we allow a frame to be predicted from previous frames, we get a Predicted frame (P-frame). To decode a P-frame, the associated motion vectors and the previous frames that are predicted by the P-frame are first decoded. By allowing predictions from both previous and future frames, we get a bidirectional predicted frame (B-frame). Decoding a B-frame thus requires both previous and future frames to be decoded first, together with the associated motion vectors. Usually, only I- and P-frames are used as anchor frames, meaning that only these frame types are used as reference to predict other frames.

The extraction of low- and high-level compressed sub-band coefficients is illustrated in Fig. 2. Here first, the original image frame is transformed into two levels of transformation using Hybrid Wavelets (Both Multi-Wavelet based for frame template matching and Tree Wavelet based for enhancement and smoothing the frame bit level noises). In the proposed technique, we used an arithmetic encoder (refer the references for in-depth knowledge) to encode the high-level compressed sub-band coefficients, and used low-bit level encoder to encode the low-level compressed sub-band coefficients.

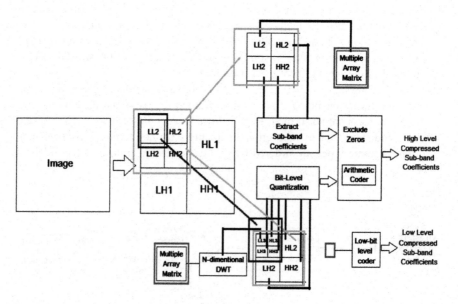

**Fig. 2** Process of L and H band coefficients extraction

**Table 1**  Video sequences considered for analysis of proposed technique

| Sequence | Format (width × height) | Number of frames |
|---|---|---|
| Akiyo (4:2:0): YUV video | QCIF format (176 × 144) | 300 |
| Foreman (4:2:0): YUV video | CIF format (352 × 288) | 300 |
| News (4:2:0): YUV video | QCIF format (176 × 144) | 300 |
| Tulips (4:4:4): YUV video | QCIF format (176 × 144) | 300 |

## 3  Implementation of Proposed Technique

The proposed technique is implemented on MATLAB Tool and tested with several video sequences. To show the efficiency of the proposed technique, we also implemented H.264 in MATLAB and H.265 in HM tool to compare PSNR values. Proposed technique is tested with different videos as shown in Table 1.

In our experimental results, we have taken Movies: AVI Video. For Frame Coding and Decoding process, we have implemented our technique using MATLAB Tool. Our implementation results of the proposed technique are shown in Fig. 3.

## 4  Results and Discussions

For our analysis, we have taken Movies: AVI video file, having 24 frames to extract. The bit rate of proposed technique was implemented. The bytes for each Group of Pictures (GOP) is the bit rate given as input to the encoder, the bytes are available to code sub-band frames of a GOP. The bits available in the LP and HP frames are shown in Table 2.

The Y-PSNR comparative results of proposed technique and HM-5.2 [23] are shown in Table 3.

The quality degradation $Q_d$ of encoded frames results with the comparison with bit rate $B_r$ and visual threshold $T_v$, which is the acceptable perceptual quality of encoded frames. If visual threshold value selected high value, degradation of frame quality increases the low bit rate values. The comparison of bit rate selection, image quality degradation with visual threshold of our proposed technique with H.264 and HEVC is shown in Table 4.

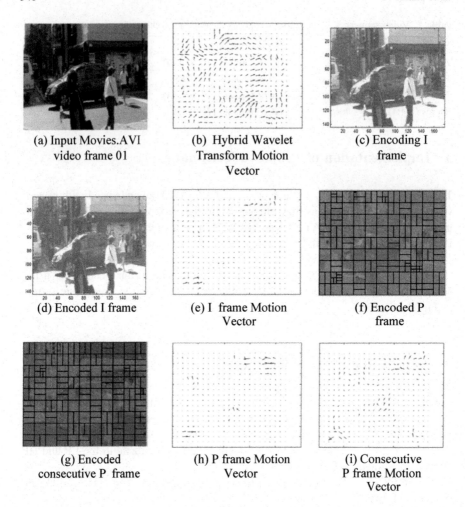

(a) Input Movies.AVI
video frame 01

(b) Hybrid Wavelet
Transform Motion
Vector

(c) Encoding I
frame

(d) Encoded I frame

(e) I frame Motion
Vector

(f) Encoded P
frame

(g) Encoded
consecutive P frame

(h) P frame Motion
Vector

(i) Consecutive
P frame Motion
Vector

**Fig. 3** Proposed HEVC technique implementation of encoding process for 30 Hz frame rate

**Table 2** Percentage of bits in LP and HP frames in GOP

| Sequence | LP (%) | HP (%) |
|---|---|---|
| Akiyo (4:2:0): YUV video | 25 | 75 |
| Foreman (4:2:0): YUV video | 25 | 75 |
| News (4:2:0): YUV video | 25 | 75 |
| Suzie (4:2:0): YUV video | 25 | 75 |
| Football (4:2:2): YUV video | 25 | 75 |
| Tulips (4:4:4): YUV video | 25 | 75 |

**Table 3** Y-PSNR results

| Sequence | | Bit rate (Mbps) | |
|---|---|---|---|
| | | 7 | 8 |
| Akiyo (4:2:0): YUV video | HM-5.2 | 32.69 | 33.27 |
| | Proposed | 41.22 | 41.56 |
| Foreman (4:2:0): YUV video | HM-5.2 | 30.11 | 31.33 |
| | Proposed | 40.01 | 40.26 |
| News (4:2:0): YUV video | HM-5.2 | 38.91 | 39.33 |
| | Proposed | 42.65 | 43.08 |
| Tulips (4:4:4): YUV video | HM-5.2 | 38.90 | 39.35 |
| | Proposed | 40.26 | 40.35 |

**Table 4** Comparative results of rate selection with visual threshold of our proposed technique with H.264 and HEVC

| Sequence | $B_r$ | $T_v$ | H.264 [22] PSNR | HEVC [23] PSNR | Proposed PSNR |
|---|---|---|---|---|---|
| Akiyo | 1 | 5 | 38.34 | 43.01 | 44.21 |
| | | 10 | 38.87 | 42.12 | 43.25 |
| | 3 | 5 | 38.41 | 42.35 | 43.89 |
| | | 10 | 37.89 | 41.15 | 42.69 |
| Foreman | 1 | 5 | 36.26 | 43.35 | 44.56 |
| | | 10 | 38.25 | 42.75 | 44.25 |
| | 3 | 5 | 36.65 | 42.12 | 43.65 |
| | | 10 | 36.89 | 41.25 | 42.36 |
| News | 1 | 5 | 38.25 | 44.25 | 45.26 |
| | | 10 | 37.89 | 43.78 | 44.89 |
| | 3 | 5 | 37.25 | 42.26 | 43.89 |
| | | 10 | 37.35 | 43.36 | 44.98 |
| Tulips | 1 | 5 | 36.21 | 42.75 | 43.98 |
| | | 10 | 36.26 | 41.56 | 42.65 |
| | 3 | 5 | 36.29 | 43.76 | 44.65 |
| | | 10 | 36.22 | 42.36 | 43.98 |

# 5  Conclusions

In this work, a new approach for transform, scaling, and quantization and its applications for video compression is proposed. This parallel transformation technique is based on Hybrid Wavelet Transform Technique and new sets of image sequences. The technique is applied in spatial and wavelet sub-bands for compression. We tested the Hybrid Transformation Technique with several YUV, AVI video formats and sequences are compared the results with the other transformation techniques. An experimental result shows improved PSNR values by using Hybrid Wavelet sub-band technique considerably.

# References

1. Jorge Soto Leon, Carlos Silva Cardenas, Ernesto Villegas Castillo. "A high parallel HEVC Fractional Motion Estimation architecture", 2016 IEEE ANDESCON, 2016.
2. Liyuan Xiong, Wei Zhou, Xin Zhou, Guanwen Zhang, Ai Qing. "Saliency aware fast intracoding algorithm for HEVC", 2016 Asia-Pacific Signal and Information Processing Association Annual Summit and Conference (APSIPA), 2016.
3. Sullivan, G. et al. "High Efficiency Video Coding (HEVC)". Springer, 2014.
4. Maich, H. Afonso, V. Zatt, B. Agostini, L. Porto, M., "HEVC Fractional Motion Estimation complexity reduction for real-time applications," in Circuits and Systems (LASCAS), 2014 IEEE 5th Latin American Symposium, vol., no., pp. 1–4, 25–28 Feb. 2014.
5. Castillo, E.V. Cardenas, C.S. Jara, M.R., "An efficient hardware architecture of the H.264/AVC Half and Quarter-Pixel Motion Estimation for real-time High-Definition Video streams," in Circuits and Systems (LASCAS), 2012 IEEE Third Latin American Symposium on, vol., no., pp. 1–4, Feb. 29 2012-March 2 012.
6. S.J. Choi and J. W. Woods, "Motion-Compensated 3-D Sub band Coding of Video," IEEE Transactions on Image Processing, Vol. 8, No. 2, 1999, pp. 155–167. https://doi.org/10.1109/83.743851.
7. P. Chen and J. W. Woods, "Bidirectional MC-EZBC with Lifting Implementation," IEEE Transactions on Circuits and Systems for Video Technology, Vol. 14, No. 10, 2004, pp. 1183–1194. https://doi.org/10.1109/TCSVT.2004.833165.
8. Y. M. Chi and T. D. Tran and R. Etinne-cummings, "Optical Flow Approximation of Sub-Pixel Accurate Block Matching for Video Coding," Proceedings of IEEE In-ternational Conference on Acoustic, Speech and Signal Processing, Honolulu, 15–20 April 2007, pp. 1017–1020.
9. M. Eckert, D. Ruiz, J. I. Ronda and N. Garcia, "Evaluation of DWT and DCT for Irregular Mesh-Based Motion Compensation in Predictive Video Coding," In: K. N. Ngan, T. Sikora and M.-T. Sun, Eds., Visual Communications and Image Processing, Proceedings of SPIE 4067, 2000, pp. 447–456.
10. B Bross, W Han, J Ohm, et al. WD5: Working Draft 5 of "High-Efficiency Video Coding[C]", ITU-T SG16 WP3 and ISO/IEC JTC1/SC29/WG11 7th Meeting, Geneva, CH, 2011, JCTVC-G1103.
11. Ismail M, Jo H, Sim D. "Fast intra mode decision for HEVC intra coding", [C] Consumer Electronics (ISCE 2014), The 18th IEEE International Symposium on. IEEE, 2014:1–2.
12. Bossen F, Bross B, Suhring K, et al., "HEVC Complexity and Implementation Analysis", [J]. Circuits & Systems for Video Technology IEEE Transactions on, 2012, 22(12):1685 – 1696.
13. Wu, H. R. and Rao, K. R. and Kassim, Ashraf A. "Digital Video Image Quality and Perceptual Coding," Boca Raton, Journal of Electronic Imaging, vol. 16, 2005.
14. Gupta, Rupesh, M. T. Khanna, and S. Chaudhury. "Visual saliency guided video compression algorithm," Signal Processing Image Communication, vol. 28, no. 9, pp. 1006–1022, 2013.
15. Li, Y., Liao, W., Huang, J., He, D., and Chen, Z. "Saliency based perceptual HEVC," Multimedia and Expo Workshops (ICME), 2014 IEEE International Conference on, 2014.
16. Yu Q, Rong Y, He Y. "Fast intra mode decision strategy for HEVC", [C] Signal and Information Processing (ChinaSIP), 2013 IEEE China Summit & International Conference on. IEEE, 2013:500–504.
17. Shen L, Zhang Z, Liu Z. "Effective CU size decision for HEVC intra coding", [J] IEEE Transactions on Image Processing A Publication of the IEEE Signal Processing Society, 2014, 23(10):4232–4241.
18. Seunghyun Cho, Munchurl Kim, "Fast CU Splitting and Pruning for Suboptimal CU Partitioning in HEVC Intra Coding", [J] Circuits and Systems for Video Technology, IEEE Transactions on, 2013, 23(9): 1555–1564.
19. Ke Shen, Gregory W. Cook, Leah H. Jamieson and Edward J. Delp, "An Overview of Parallel Processing Approaches to Image and Video Compression", SPIE, 2000.

20. Soo-Young Lee and K. Aggarwal, "A System Design and Scheduling Strategy for Parallel Image Processing", IEEE Transactions on Pattern Analysis and Machine Intelligence, February 1990 Vol. 12. No.2, pp. 193–204.
21. Christiana Nicoles and Peter Jonker, "A data and task parallel image processing environment", Journal of Parallel computing, Elsevier Science Publishers. 2002, Pages: 945– 965.
22. G. J. Sullivan, J. Ohm, W. J. Han, and T. Wiegand. 2012. "Overview of the High Efficiency Video Coding (HEVC) Standard". IEEE Trans. on Circuits and Systems for Video Technology 22, 12 (2012), 1649–1668.
23. T. Wiegand, G. Sullivan, B. Bjontegaard, and A. Luthra, "Overview of the h.264/avc video coding standard," IEEE Trans. on Circuits and Systems for Video Technology, vol. 13–7, pp. 560–576, July 2003.

# On the Implementation of VLSI Architecture of FM0/Manchester Encoding and Differential Manchester Coding for Short-Range Communications

Siva Jyothirmai Gali and Sudheer Kumar Terlapu

**Abstract** The encryption is the method for converting information into a desired format for a number of information processing needs. It encompasses data transmission storage and compression. In order to have safe data transmission, different types of encoding techniques are developed. Especially for short-range communications, various protocols are used to encode the information. Intelligent transport system needs modernized services for the management of traffic may generally uses Manchester and FM0 encoding. This process improves constancy in signal by dc balancing. The coding Manchester and FM0 methods of code control the VLSI architecture. Also similarly oriented logic simplification (SOLS) technique was proposed for controlling the hardware architecture. By using this technique, the (HUR) hardware utilization rate is increased from 58 to 100%. Also, differential Manchester encoding achieved good dc.

**Keywords** Dedicated short-range communications (DSRC) · FM0
Manchester · Similarly oriented logic simplification (SOLS) · DC balancing

## 1 Introduction

The dedicated short-range communication (DSRC) is a procedure of communication for one way, two way, and medium range [1]. DSRC was divided into two types, they are automobile–automobile and automobile–roadside. In the first type, DSRC activates the message transfer and broadcasts it to the automobiles [2]. The second type of DSRC is mainly targeted on intelligence transportation services, for example, electrical tollgate collection. Basically, the encoding process is done for

S. J. Gali (✉) · S. K. Terlapu
Department of Electronics and Communications, Shri Vishnu Engineering
College for Women (Autonomous), Bhimavaram, India
e-mail: gsivajyothirmai.91@gmail.com

S. K. Terlapu
e-mail: skterlapu@gmail.com

© Springer Nature Singapore Pte Ltd. 2018
J. Anguera et al. (eds.), *Microelectronics, Electromagnetics
and Telecommunications*, Lecture Notes in Electrical Engineering 471,
https://doi.org/10.1007/978-981-10-7329-8_56

551

safe transmission [3] of data. The basic block diagram of DSRC consists of two parts: upper and lower. The top part is for transmission of data and the bottom part is for receiving. Those are transmitter and receiver.

This paper developed Manchester encoding for safe communications [4]. RF-frontend sends and collects the wireless signal using antenna. The signal which is transmitted was supposed to get mean as zero for [5] tough issue. This will be mentioned as dc balancing. The signal which is transmitted contains binary series. Due to this, it is hard to get the dc balance. This [6] problem of dc balancing may be overcome by using FM0 and differential Manchester encoding. Cryptography been popularly used these days in the applications like intelligent transportation and Automated machines [7]. Multipliers of redundant basis were popular due to low complexity [8].

In this paper, first section briefly explains the introduction of the paper. The coding variety for both encodings severely controls the potentiality to draw VLSI structure which is reusable with one another. In this work, the VLSI structure was designed. And second section contains hardware structure of encoder. And its subsections clearly explain dc balance problem and area retime problems and their avoiding methods.

And differential Manchester encoder was explained in third section. Section 4 gives the simulation results of the given encoders with the explanation of operation, also the comparison for conventional and proposed structures.

## 2    Hardware Architecture for Fm0 and Manchester

Here, the two D-flip-flops (DFFs) which are used to store the state code and mux are used to select the logics of A(t) and B(t) by considering clock as select line. And finally, the mux with mode is used to select the type of encoding among FM0 and Manchester here mode is considered as selection line. If mode = 0, then it will select FM0 code and mode = 1 it will give Manchester code. To get the hardware utilization rate, we have to consider both active components and total components. It will be evaluated as HUR

The block diagram mainly contains two portions top and bottom. One is transceiver and receiver given in Fig. 1. The transceiver end consists of three units. They are baseband processing, RF-frontend, and microprocessor. The microprocessor takes the guidelines from media accessing control in order to program the function of baseband processor and RF-frontend. Baseband processing itself is reasonable for correcting errors, modulation, encoding, and clock synchronization.

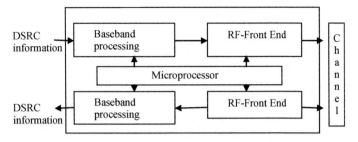

**Fig. 1** Block diagram of DSRC

## 2.1 SOLS Technique

Similarity-oriented logic simplification (sols) is a fully reused structure of FM0 and Manchester. This technique is classified into two parts, they are: (i) Area compact retiming and (ii) Balance logic operation. The functioning of SOLS is given below.

The two encoders are given to mux so that FM0/Manchester output is obtained. The upper portion is FM0 encoding and down portion is Manchester code. The path will be selected as shown in Fig. 2 and the operation is carried out based on the given logic.

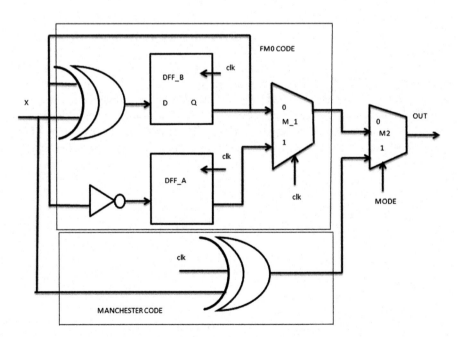

**Fig. 2** Hardware architecture of FM0 and Manchester encoder

## 2.2  Area Compact Retiming

The components which we are used in the architecture are assumed as hardware as they perform operations logically. Here, the active components are nothing but the components which undergo work in the functioning of circuit. And the total components are the components of the entire circuit irrespective of those which involve in functioning. If the number of components which are not involving in operation is present in circuit, some of the hardware will be wasted due to this area and the power will increase. So that in order to make the HUR rate as 100%, the unusable components will be minimized or avoided, so that the count of transistor will also be reduced. For example in FM0 encoding, it consists of two DFFs, those two DFFs are for storing the states of A(t) and B(t). Instead of storing those values, it may be better to store the value which will be selected by mux among A(t) and B (t). So that the value will be stored in single DFF. Due to this, the hardware may reduce and HUR will become 100%. The architecture by considering both DFF is referred as FM0 without area compact retiming as shown in Fig. 3. Finally by reducing hardware that is one DFF is referred as FM0 with area compact retiming as shown below.

## 2.3  Balance Logic Operation Sharing

To avoid the dc balancing problem the architecture is modified by placing inverter after the mux_1. Due to irregular transmission of signal, dc balancing problem may occur which leads to logical errors in code.

Because of dc balancing, unbalancing calculating time may occur between logics and it may result in formation of glitches at mux_1, to avoid this unbalance calculating time, the logic is modified. If more selection of logic modification is done, the XOR operation may lead to inverting of its function at inverter again, to alleviate this problem, XOR is replaced with XNOR results in correct value of XOR at inverter. For selection of logic, mux with mode mux_2 is used so that it acts as

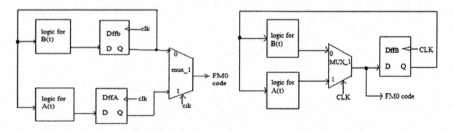

**Fig. 3**  FM0 without area compact retiming and by area compact retiming

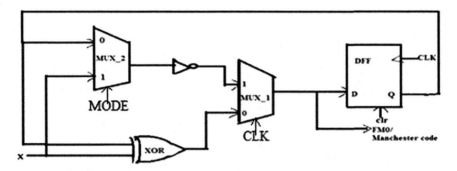

**Fig. 4** Unbalanced computation

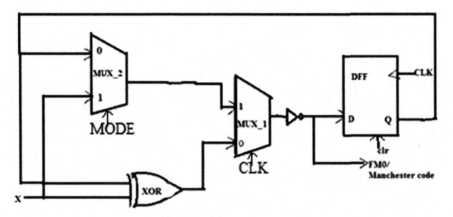

**Fig. 5** VLSI architecture for balance logic

selection line. Finally, the assumption of FM0 or Manchester encoding depends on mode and also the clear (clr) is used in DFF. Hence, every component involved in working of FM0 and Manchester encoding given in Figs. 4 and 5. Further HUR is also increased.

## 3 Differential Manchester Encoding

The Manchester encoding was modified as differential Manchester encoding. Both Manchester and differential Manchester are visibly the same, but for a change in transitions at edge of the bit. Transition in between windows will occur at input "0" and no transition occurs at input "1". The architecture of differential Manchester encoding is as given in below figure. It contains a D-flip-flop which stores current state of A, after that mux in this architecture selects the state which is the next

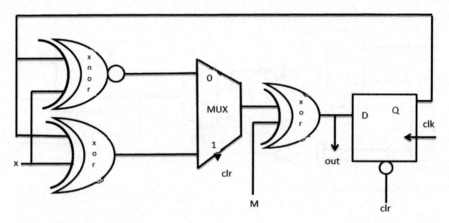

**Fig. 6** Differential Manchester encoding

values of A and B as per the clock. Mode signals, flip-flop clear signal and M are used for selection among Manchester and differential Manchester.

In the mode of Manchester, clear is activated which makes "0" in the place of present value A and M = 1 which turns XOR_2 to play an inverter role. In differential Manchester as shown in Fig. 6, when mode = 0 and clr = 1, the present value of A is transferred as input to next state and XOR_2 behaves like a buffer (Table 1).

**Table 1** Performance comparison table

| Parameters | Existing | Proposed |
|---|---|---|
| Maximum frequency (MHz) | 444.545 | 562.762 |
| Slices | 2 | 1 |
| Flip-flops | 2 | 1 |
| LUTS | 4 | 2 |
| Bonded IOBS | 5 | 6 |
| Registers | 2 | 1 |
| XORS | 2 | 1 |
| Real time to Xst (s) | 1 | 0 |
| Cpu time to Xst (s) | 0.19 | 0.20 |
| Global min fanout | 1000000 | 500 |
| Path delay (ns) | 6.125 | 5.776 |
| Total real time (s) | 5 | 4 |
| Total cpu time (s) | 4.45 | 4.40 |

# 4 Analysis of Experimental Results

The process of simulation is done by considering the Verilog code simulated in Xilinx. Clock (clk), clear (clr), mode, x acts as inputs for this process, whereas the output of DFF will be fed back to XNOR and first mux acts as inputs, output for complete circuit is taken at inverter output given as FM0/Manchester. Here, when the mode is 0 and clr is 1, it will generate the FM0 code, similarly Manchester code is generated for other values of mode and clr. FM0 code is based on clock, i.e., when clock is low it will get internal transitions. When clock is high, it remains constant till the next clock cycle.

Similarly in differential Manchester, encoding is also simulated in Xilinx. Clr and clk act as input in the circuit. Out is taken as output at second XOR. One of the inputs for first XOR and XNOR is fed back from flip-flop. Their simulated results are given in Figs. 7 and 8.

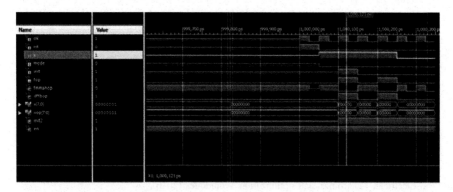

**Fig. 7** Simulation results of FM0/Manchester encoder

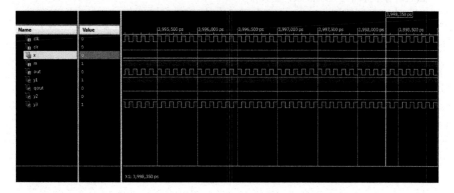

**Fig. 8** Simulation results of differential Manchester encoding

# 5 Conclusion

The work contains completely integrated VLSI architectures of FM0 Manchester and differential Manchester encodings. These encrypting methods are generally used for short-range communications. This work is compared with the existing architecture so that it reduces the delay and hardware usage of circuit is also improved. By assuming the proposed work, the number of transistors is decreased. This work is performed in Xilinx. These architectures are examined with other existed designs. The resource usage estimated values are also reported in a tabular form.

# References

1. F.Ahmed-Zard, F. Bai, S. Bai, C. Basnayake, B.Bellur,B. S. Brovold,etal.: vehicle safety communications Applications (VSC-A) final report. U.S. Dept. Trans., Nat. Highway traffic safety Admin., Washington. DC, USA, REP.DOT HS 810 591. Sep. (2011).
2. J. B. Kenney.: Dedicated short-range communications (DSRC) standards in the United States. Proc. IEEE, vol. 99, no. 7, pp. 1162–1182, Jul. (2011).
3. J. Daniel, V. Taliwal, A. Meier, W. Holfelder, and R. Herrtwich.: Design of 5.9 GHz DSRC-based vehicular safety communication. IEEE Wireless Commun. Mag., vol. 13, no. 5, pp. 36–43, Oct. (2006).
4. M. A. Khan, M. Sharma, and P. R. Brahmanandha.: FSM based Manchester encoder for UHF RFID tag emulator. In Proc. Int. Conf. Comput., Commun. Netw, pp. 1–6., Dec. (2008).
5. M. A. Khan, M. Sharma, and P. R. Brahmanandha.: FSM based FM0 and Miller encoder for UHF RFID tag emulator. In Proc. IEEE Adv. Comput. Conf, pp. 1317–1322., Mar. (2009).
6. Ishwerya P, Nithish Kumar V, Lakshminarayanan G.: An Efficient Digital Baseband Encoder for Short Range Wireless Communication Applications. IEE ICEEOT international Conference on Electrical, Electronics, and optimization Techniques (2016).
7. Leonore Dake, Jyothi, and Kumar Terlapu, Sudheer.: Low Complexity Digit Serial Multiplier for Finite Field using Redundant Basis. Indian Journal of Science and Technology, Volume 9 Number S1, pp. 1–5, December (2016).
8. Jyothi Leonore Dake and Sudheer Kumar Terlapu.: Implementation of high-throughput digit-serial redundant basis multiplier over finite field. IOSR Journal of VLSI and Signal Processing (IOSR-JVSP), vol.6, Issue 4, Ver. I, e-ISSN; 2319-4200, p-ISSN No.: 2319-4197, pp. 35–45, Jul. (2016).

# Dual-Band-Notched CPW-Fed Antennas with WiMAX/WLAN Rejection for UWB Communication

T. Anusha, T. V. Ramakrishna, B. T. P. Madhav and A. N. Meena Kumari

**Abstract** A coplanar waveguide (CPW) fed ultra-wideband (UWB) antenna is designed to operate with dual-band-notched characteristics. In this work, different iterations are proposed with respect to the radiating element structure. The designed antenna models consist of a 50 Ω coplanar waveguide transmission line and different orientations of radiating elements. Single-band and dual-band characteristics of the notch are achieved for WiMAX/WLAN applications with the placement of split-ring resonators on the radiating patch element. The designed models are initially simulated with Ansys HFSS tool, later, validation of prototype model is performed for measuring results.

**Keywords** Dual-band-notched · Coplanar waveguide (CPW)
WiMAX · WLAN · UWB communication

## 1 Introduction

The US Federal Communications Commission (FCC) allotted the ultra-wideband (UWB) spectrum from 3.1 to 10.6 GHz which is extensively used for communication applications. UWB antennas consume less power and are much resistive to multipath interference. These are used in wide range of applications like radars, wireless sensor networks, location tracking, and short-range high data rate wireless communication. The most significant characteristics of UWB antenna design include the matching of impedance, stability in radiation, compactness of antenna, and low manufacturing cost for consumer electronics applications [1–4]. Suppression of unwanted bands can be achieved by using filters in some applications. These UWB antennas with filtering property at different bands are proposed to reduce the potential interference and thus the requirement of an extra band stop

T. Anusha · T. V. Ramakrishna · B. T. P. Madhav (✉) · A. N. Meena Kumari
Department of ECE, K L University, Guntur, AP, India
e-mail: btpmadhav@kluniversity.in

© Springer Nature Singapore Pte Ltd. 2018     559
J. Anguera et al. (eds.), *Microelectronics, Electromagnetics
and Telecommunications*, Lecture Notes in Electrical Engineering 471,
https://doi.org/10.1007/978-981-10-7329-8_57

filter present in the system can be eliminated. Removing of slots with different shapes is the common way to get the band notch characteristics.

UWB transmitters get affected by electromagnetic interference with nearby narrowband communication channels such as wireless local area network (WLAN) applications. Therefore, UWB antennas with notched characteristics in WLAN frequency band can be used to reduce the effect of Interference. Among the different types of UWB antennas, bowtie antenna and the modified structure of bowtie antenna with triangular shape, planar volcano smoke slot antennas have been developed for UWB systems [5–8]. Moreover, UWB antenna desires more than one notch band resulting in the use of noninteracting band notch elements. In recent literature, different structures of multiband notched UWB antenna topologies such as dual, triple, and quadruple have also been reported [9, 10].

In the present work, the compact UWB antenna of area $26 \times 30$ mm$^2$ is proposed first. By placing different shapes of slots, single-band and dual-band-notched characteristics are obtained. The designs of antenna along with simulation details are presented. Measured results are given to provide better understanding of the simulated results.

## 2 Antenna Design and Geometry

The geometry and configuration of an UWB antenna (referred as antenna 1) are shown in Fig. 1. The proposed model is fabricated on FR4 epoxy substrate of height, h = 1.6 mm. The dielectric constant ($\varepsilon_r$) of substrate material is 4.4 and loss tangent (tan $\delta$) is 0.02. Only a single layer of substrate with one-sided metallization is used because both the antenna and the feed method are provided on the same

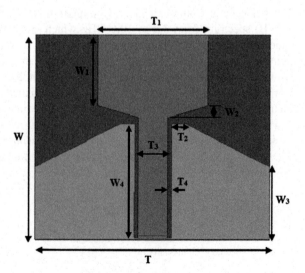

**Fig. 1** Geometry and configuration of antenna 1

**Table 1** UWB antenna geometry

| W | T | $W_1$ | $W_2$ | $W_3$ | $W_4$ | $T_1$ | $T_2$ | $T_3$ | $T_4$ |
|---|---|---|---|---|---|---|---|---|---|
| 26 mm | 30 mm | 9 mm | 1.5 mm | 9.2 mm | 14.7 mm | 14 mm | 1.8 mm | 3.8 mm | 0.5 mm |

plane and thus manufacturing of the antenna is very easy and least expensive. The design parameters of antenna 1 are tabulated as follows in Table 1.

The basic model of the UWB antenna is modified by adding a split-ring resonator (SRR) slot on the radiating element as shown in Fig. 2a. The single split-ring resonator (SRR) contributes for achieving notch band in the UWB design. The notch frequency ($f_{notch}$) depends on the length of slot (L).

The relationship between notch frequency band and slot length is given in Eq. 1. The single SRR is extended with the addition of second SRR around the first SRR

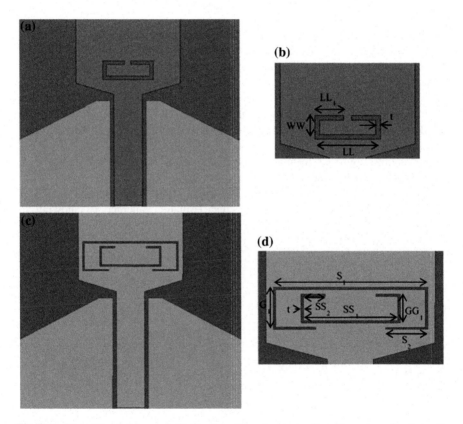

**Fig. 2 a** Geometry of antenna 2. **b** Configuration of antenna 2. **c** Geometry of antenna 3. **d** Configuration of antenna 3

**Table 2** Slot dimensions of antennas 2 and 3

| LL | WW | LL$_1$ | t | S$_1$ | S$_2$ | SS$_1$ | SS$_2$ | GG$_1$ | G$_1$ |
|---|---|---|---|---|---|---|---|---|---|
| 6.8 mm | 2.6 mm | 3 mm | 0.5 mm | 12.5 mm | 3.5 mm | 8.2 mm | 2 mm | 2.6 mm | 3.8 mm |

as shown in Fig. 2c. The dimensional characteristics of the single and double SRR are depicted in Fig. 2b, d and parameters are tabulated in Table 2.

$$f_{notch} = \frac{c}{2L\sqrt{\epsilon_{eff}}} \qquad (1)$$

## 3 Results and Discussion

The basic structure proposed in Fig. 1 is simulated using finite element method (FEM) based HFSS tool and the results are furnished for subsequent iterations in this section. Figure 3 shows the return loss characteristics of basic UWB antenna model, single SRR based UWB model, and double SRR based UWB model with respect to resonant frequency. The reflection coefficient result ensures the operating band of basic UWB antenna in the range of 3.1–10.6 GHz. The single SRR based UWB antenna is behaving like notch band antenna to oppose WLAN band (6–6.8 GHz).

The double split-ring resonator based antenna is blocking two bands, i.e., 3.4–5.5 GHz, which come under WiMAX and WLAN.

The radiation mechanism of single and double SRR based antenna models is presented in Fig. 4 at 8.6 GHz. In the H-plane, antenna is showing a

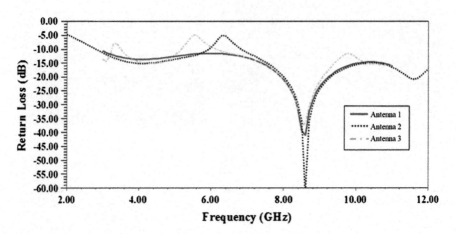

**Fig. 3** Simulated return losses of antennas 1, 2, and 3

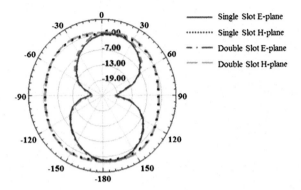

**Fig. 4** Simulated radiation patterns of antenna 2 compared to antenna 3 at 8.6 GHz

quasi-omnidirectional pattern and in E-plane, it is behaving like a dipole antenna with average gain of 3 dB.

Figure 5 shows the modified structure of SRR which is incorporated on the radiating element. The position of the SRR is fixed at same location similar to earlier design in Fig. 2. But, the orientation with respect to the structure has been changed to circular shape. The dimensional characteristics and geometry are presented in Fig. 5b, d. The parametric dimensions in "mm" are tabulated in Table 3.

The performance characteristics of a circular SRR based antenna are analyzed and presented in Fig. 6. It has been clearly observed that single circular split-ring based model is providing single notch band and double circular SRR based antenna is providing dual-notch characteristics in the desired UWB band. The radiation characteristics have been changed in H-plane for the circular split-ring antenna when compared with rectangular split-ring antenna models. The E-plane characteristics of circular double SRR antenna are showing some disturbance when compared with single circular SRR model. The direction of the radiation is changed in other orientation for double SRR model.

The experiment of changing SRR orientation is continued by placing hexagonal shaped slots on the UWB antenna model. The similar kind of characteristics is observed with respect to single and dual SRR structures with respect to the notching characteristics but a change in notch band is observed for hexagonal SRR models.

The geometry and the dimensions are forecasted in Fig. 8 and in Table 4. The main difference that is observed with respect to output parameters is the matching of impedance at desired notch bands. In the earlier designs, the obtained notch bands are not exactly the bands which we will use for WiMAX and WLAN. The results obtained through hexagonal SRR are perfectly matching to 3.6 GHz/5.6 GHz operating application bands. The results furnished in Figs. 9 and 10 are giving the evidence for the proposed notch bands (Fig. 7).

For all the designed models, the gain characteristics are presented in Fig. 11. All the peak gains for antennas are shown using markers which are indicated as marker

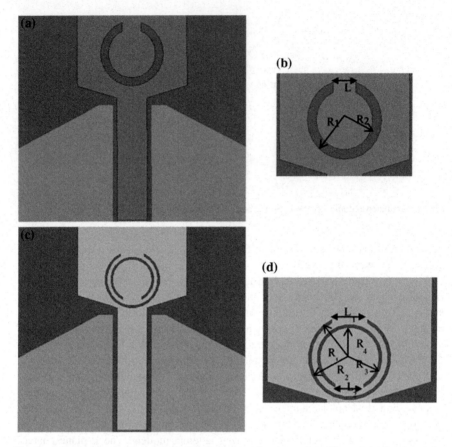

**Fig. 5** **a** Geometry of antenna 4. **b** Configuration of antenna 4. **c** Geometry of antenna 5. **d** Configuration of antenna 5

**Table 3** Slot dimensions of antennas 4 and 5

| $R_1$ | $R_2$ | L | $R_1$ | $R_2$ | $R_3$ | $R_4$ | $L_1$ | $L_2$ |
|---|---|---|---|---|---|---|---|---|
| 4 | 3 | 2.4 | 3.5 | 3.25 | 2.75 | 2.5 | 3 | 2.4 |

1 for antenna 1, marker 2 for antenna 2, and so on. It has been observed that at notch bands, the gain is negative and at the remaining operating bands, the gain is almost stable except for double circular SRR model. A peak realized gain of almost 10 dB is obtained at 8.6 GHz for this model and is indicated using marker 5. Figure 12 is showing the complete analysis related to reflection coefficient for all seven designed models. Notch bands are shown using markers, m1-m2 is the single

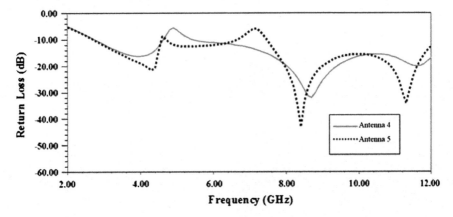

**Fig. 6** Simulated return losses of antennas 4 and 5

**Fig. 7** Simulated radiation
patterns of antenna 4
compared to antenna 5 at
8.6 GHz

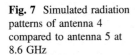

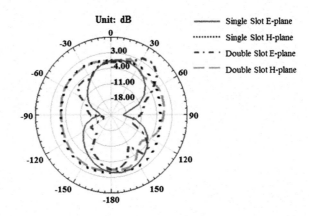

notch region of antenna 2, m3-m4 and m5-m6 are the dual notches of antenna 3, m7-m8 is the single notch region of antenna 4, m9-m10 and m11-m12 are the dual notches of antenna 5, m13-m14 is the single notch region of antenna 6, m15-m16 and m17-m18 are the dual notches of antenna 7.

The surface current distribution provides a clear cut picture of antenna response with respect to its transmission modes. Figure 13 shows the surface current distribution of all the models at 8.6 GHz. The current intensity on the surface of the antenna can be analyzed and the mode of propagation will be predicted from the current results. Table 5 shows the complete list of antenna output parameters with respect to operating and notch bands.

The notch bands can be identified from the Fig. 12 with the aid of markers and the numbers of notch bands are tabulated in Table 5 (Fig. 14).

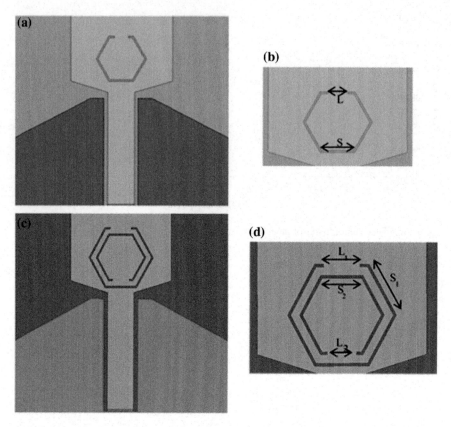

**Fig. 8 a** Geometry of antenna 6. **b** Configuration of antenna 6. **c** Geometry of antenna 7.
**d** Configuration of antenna 7

| **Table 4** Slot dimensions for antennas 6 and 7 | S | L | $S_1$ | $S_2$ | $L_1$ | $L_2$ |
|---|---|---|---|---|---|---|
| | 3.5 | 2 | 4.25 | 3.5 | 3 | 2.4 |

## 4 Conclusion

This paper presents the analysis of different notch band models with respect to SRR
structures. The single split-ring is providing single notch band and double split-ring
is providing the dual-notch band at desired frequency bands. The basic UWB model
is working in the total band with an average gain of 3.3 dB. Antenna model 2 and 3
of rectangular SRR structures are providing gain of 3.2 dB and 2.37 dB, respectively. A peak realized gain of 10 dB is attained for the antenna model with double

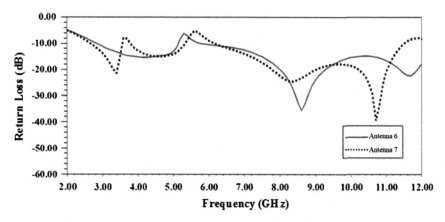

**Fig. 9** Simulated return losses of antennas 6 and 7

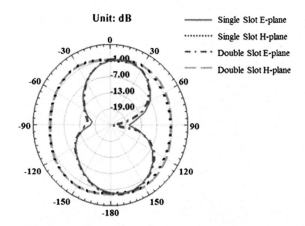

**Fig. 10** Simulated radiation patterns of antenna 6 compared to antenna 7 at 8.6 GHz

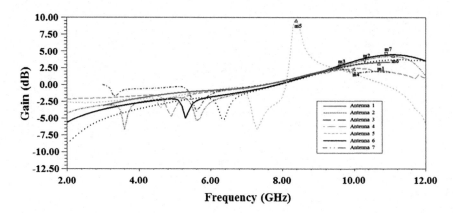

**Fig. 11** Simulated gain of all antennas

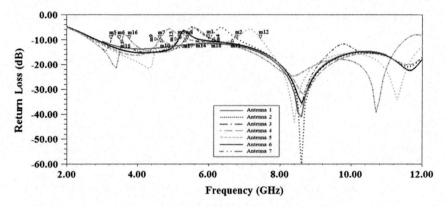

**Fig. 12** Simulated return loss of all antennas

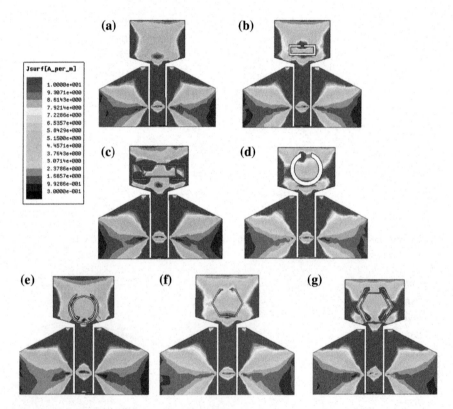

**Fig. 13** Simulated current distribution on radiating patch at 8.6 GHz for all antennas. **a** Antenna 1. **b** Antenna 2. **c** Antenna 3. **d** Antenna 4. **e** Antenna 5. **f** Antenna 6. **g** Antenna 7

**Table 5** Designed antennas overall output parameters

| Antenna | Bandwidth (GHz) | Gain (dB) | Return loss (dB) | No. of notches |
|---------|-----------------|-----------|------------------|----------------|
| Antenna 1 | 3.1–10.8 GHz | 3.3 | −41 | 0 |
| Antenna 2 | 3.1–10.8 GHz with (5.96–6.82) GHz rejection | 3.2 | −60.8 | 1 |
| Antenna 3 | 3.1–10.8 GHz with (3.26–3.5) GHz and (5.2–6.1) GHz rejection | 2.37 | −37.3 | 2 |
| Antenna 4 | 3.1–10.8 GHz with (4.63–5.42) GHz rejection | 2.43 | −31.7 | 1 |
| Antenna 5 | 3.1–10.8 GHz with (4.58–4.7) GHz and (6.67–7.46) GHz rejection | 10 | −42.8 | 2 |
| Antenna 6 | 3.1–10.8 GHz with (5.15–5.71) GHz rejection | 4.45 | −10 to −35 | 1 |
| Antenna 7 | 3.1–10.8 GHz with (3.57–3.77) GHz and (5.35–6.13) GHz rejection | 4.2 | −10 to −38.5 | 2 |

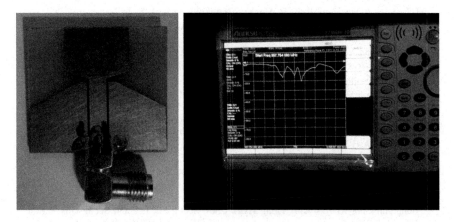

**Fig. 14** Prototyped antenna on FR4 and measured S11 on anritsu MS2037C

circular SRR. The proposed hexagonal shaped SRR based notch band antennas are exactly matching with the impedance and providing better gain characteristics compared to other antenna models. The radiation characteristics and the measured parameters are giving confidence for the applicability of the proposed model in the desired notch band.

**Acknowledgements** Authors express their deep gratitude to department of ECE of K. L. University and DST through grant ECR/2016/000569 and SR/FST/ETI-316/2012.

# References

1. Jae-Min Lee, Ki-Baek Kim, Hong-Kyun Ryu and Jong-Myung Woo, "A Compact Ultrawideband MIMO Antenna With WLAN Band-Rejected Operation for Mobile Devices," IEEE Trans. Antennas propag., vol 11, pp. 990–993, 2012.
2. Debdeep Sarkar, Kumar Vaibhav Srivastava and Kushmanda Saurav, "A Compact Microstrip-Fed Triple Band-Notched UWB Monople Antenna." IEEE Trans. Antennas propag., vol. 13, pp. 396–399, 2014.
3. Wang-Sang Lee, Dong-Zo Kim, Ki-Jin Kim and Jong-Won Yu, "Wideband Planar Monopole Antennas With Dual Band-Notched Characteristics," IEEE Trans. Microwave theory and techniques, vol. 54, pp. 2800–2806, June 2006.
4. Filiberto Bilotti, "Design of Spiral and Multiple Split-Ring Resonators for the Realization of Miniaturized Metamaterial Samples," IEEE Trans. Microwave theory and techniques, vol. 55, No. 8, August 2007.
5. C. C. Lin, P. Jin and R. W. Ziolkowski, "Single, Dual and Tri-Band Notched Ultrawideband (UWB) Antennas Using Capacitively Loaded Loop (CLL) Resonators," IEEE Trans. Antennas propag., vol. 60, No. 1, pp. 102–109, Jan. 2012.
6. K V L Bhavani, Habibulla Khan, D Sreenivasa Rao, B T P Madhav, "Dual Band Notched Planar Printed Antenna with Serrated Defected Ground Structure", Journal of Theoretical and Applied Information Technology, ISSN: 1992-8645, Vol 88, Issue 1, June 2016, pp 28–34.
7. M. L. S. N. S. Lakshmi, B. T. P. Madhav, Habibulla Khan, "Tapered Slot CPW-Fed Notch Band MIMO Antenna", ARPN Journal of Engineering and Applied Sciences, ISSN: 1818-6608, Vol. 11, No. 13, pp 1–7, 2016.
8. D S Ramkiran, et al, "Coplanar Wave Guide Fed Dual Band Notched MIMO Antenna", International Journal of Electrical and Computer Engineering (IJECE), ISSN: 2088-8708, Vol. 6, No. 4, pp. 1732–1741, 2016.
9. P Syam Sundar et al, "Parasitic Strip Loaded Dual Band Notch Circular Monopole Antenna with Defected Ground Structure", International Journal of Electrical and Computer Engineering (IJECE), ISSN: 2088-8708, Vol. 6, No. 4, pp. 1742–1750, 2016.
10. Y. S. V. Raman, B. T. P. Madhav, "Analysis of Circularly Polarized Notch Band Antenna with DGS", ARPN Journal of Engineering and Applied Sciences, ISSN: 1819-6608, Vol. 11, No. 17, September 2016.
11. C Qing-Xin Chu and Ying-Ying Yang, "A Compact Ultrawideband Antenna With 3.4/ 5.5 GHz Dual Band-Notched Characteristics," IEEE Trans. Antennas propag., vol. 56, pp. 3637–3644, December 2008.
12. Young Jun Cho, Ki Hak Kim, Dong Hyuk Choi, Seung Sik Lee and Seong-Ook Park, "A miniature UWB planar monopole antenna with 5-GHz band-rejection filter and time-domain characteristics," IEEE Trans. Antennas propag., vol. 54, pp. 1453–1460, May 2006.
13. Yi-Cheng Lin and Kuan-Jung Hung, "Compact Ultrawideband Rectangular Aperture Antenna and Band-Notched Designs," IEEE Trans. Antennas propag., vol. 54, pp. 3075–3081, November 2006.
14. Peng Gao, Shuang He, Xubo Wei, Ziqiang Xu, Ning Wang and Yi Zhen, "Compact Printed UWB Diversity Slot Antenna With 5.5-GHz Band-Notched Characteristics," IEEE Trans. Antennas propag., vol. 13, pp. 376–379, 2014.

# A Frequency Reconfigurable Spiral F-Shaped Antenna for Multiple Mobile Applications

B. T. P. Madhav, D. Sreenivasa Rao, G. Lalitha,
S. Mohammad Parvez, J. Naveen, D. Mani Deepak
and A. N. Meena Kumari

**Abstract** As the applications of mobile phones are increasing drastically, the need for small physical structures and low volume of antenna with less weight are appreciated. To meet the multiple frequency operations, reconfigurable antennas are suggestible candidates. A reconfigurable antenna is an antenna in which whole volume can be operated at different bands, which results the physical size of the multiband antenna can be reduced. The property of reconfigurability is attained in this paper by combining a Planar Inverted F antenna with a spiral structure appended with diodes. A significant improvement in gain is observed by switching different diodes on the antenna structure and shift in frequency bands is noticed. The simulation study with Ansys HFSS made the task simple to virtualize the antenna performance characteristics before the practical design with PIN diodes. The analysis of designed antenna is clearly analyzed with diode positions and switching operation.

**Keywords** Mobile applications · Reconfigurability · PIN diode

## 1 Introduction

Wireless mobile communication devices play a major role in the modern world where one hardly has a place without these mobile phones [1]. This gives rise to an incredible demand for the smaller wireless or the compact wireless antennas and also a sophisticated structure of antenna that could cover a decent amount of wireless applications [2]. This made the structural or internal cellular antenna designs one of the toughest challenges and also attractive in the history of study. In order to avail all the features of wireless communication services in any country, one needs to design specific antenna to cover major mobile communication bands.

B. T. P. Madhav (✉) · D. Sreenivasa Rao · G. Lalitha · S. M. Parvez · J. Naveen
D. M. Deepak · A. N. M. Kumari
Department of ECE, K L University, Guntur, AP, India
e-mail: btpmadhav@kluniversity.in

© Springer Nature Singapore Pte Ltd. 2018                                  571
J. Anguera et al. (eds.), *Microelectronics, Electromagnetics*
*and Telecommunications*, Lecture Notes in Electrical Engineering 471,
https://doi.org/10.1007/978-981-10-7329-8_58

At the same time, GSM (global systems for mobile) 824–894 MHz, GSM900 (880–960) UMTS (universal mobile telecommunications) 1920–2170 MHz, and LTE (long-term evolution) are the bands which are very likely used in the wireless communications [3–5]. Now, we are interested to make our antenna work in those bands.

In the proposed antenna, the structures included are F-shaped structure and spiral structure. The F-shaped structure is used in planar inverted F antenna popularly known as PIFA. And spiral structure is used in spiral antennas. PIFA is a well-used antenna in mobile phone market. PIFA is also known as short-circuited microstrip antenna. The main advantage of PIFA is that by varying ground plane, bandwidth of antenna can be varied [6, 7]. Idealistically, a reconfigurable antenna has the capability to alter the operating frequency, impedance, polarization, and also the radiation pattern as a result, we can observe the change in operating environment. All proposed operating bands can work on the entire geometry of given compact antenna only [8–10]. So, almost there is no need of an extension of size to create multiband characteristics.

In order to make our antenna tunable or to make it work in multiple operating bands, we need a switching element. Therefore, the switching element we use here is a tuning diode. The type of diode we use here is a PIN diode [11, 12]. This PIN diode gives us the property of reconfigurability. As the effective slot length of the PIN diode is changed, the frequency shift can be observed. The change of frequency band can be observed by connecting and disconnecting the antenna elements. The PIN diode also has some capacitive and resistive values [13, 14]. This paper is divided into the following sections. In Sect. 2, details of the antenna design and geometry are presented. Design analysis is briefly detailed in Sect. 3. Results of the performance of antenna are presented and discussed in Sect. 4.

## 2  Antenna Design and Geometry

The structure of compact reconfigurable antenna which can be embedded in any recent mobile technology is shown below. The 3D substrate of reconfigurable antenna is made of FR4 substrate with a relative permeability of 4.4 and loss tangent of 0.04. The dimensions of FR4 substrate are $120 \times 50$ mm$^2$ with thickness of 1.6 mm. The proposed paper mainly consists of two different types of structures which are together used as a single antenna as shown in Fig. 1. The dimension of antenna is $13.5 \times 35.5$ mm$^2$. The length L1 shown in below figure is of lengths 35.5 mm. The length L2 is fixed to 9.7 mm. Two different structures embedded are F-shaped structure and spiral structure. Antenna also consists of a diode which provides the interconnection between these two structures. The diode used in this antenna is PIN diode (part number HSMP-3860). We also have a RF choke for excitation of spiral antenna when diode is in ON/OFF state (Figs. 1 and 2).

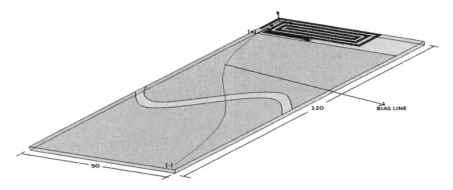

**Fig. 1** Structure of reconfigurable antenna

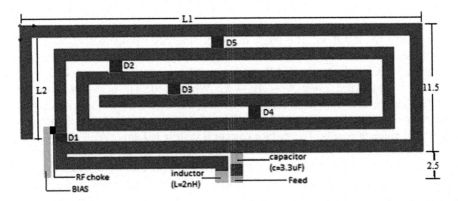

**Fig. 2** Antenna geometry

The detailed discussion on diode ON and OFF state is discussed in theoretical analysis. The resonance frequencies mainly depend on the length of antenna which is proposed to be $13.5 \times 35.5$ mm$^2$. By changing the length L2 or by changing the values of L and C, we can change the peaks of resonance frequencies. Here, we consider the value L as 2 nH and C as 3.3 μF and all line widths of antenna are fixed to 1 mm.

## 3 Design Analysis

In reconfigurable antennas, it is important to understand the current distribution of an antenna. PIN diodes (part number HSMP-3860) act as switches and decide the current distribution in the antenna. Five diodes are placed in the proposed antenna so in 32 combinations, we consider only those cases which are having reasonable bandwidths and more number of resonance frequencies. A positive voltage for PIN

**(a)**                                               **(b)**

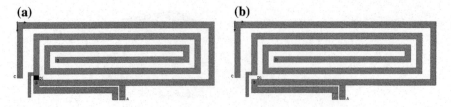

**Fig. 3** **a** Diode D1 in ON state. **b** Diode D1 in OFF state

diode is supplied through bias line and negative voltage is supplied through ground plane. Let the dark spot which is connecting two structures be a PIN diode (D1). The two structures connected are PIFA (planar inverted F shape antenna) and spiral antenna. For PIFA, patch is shorted at far end so that the current at the end is no longer forced to be zero. So that PIFA has same current–voltage distribution as of half-wave patch antenna. In PIFA, the fringing ends are shorted at far end which are responsible for radiation. The patch maintains the basic properties of half-wave patch antenna with size reduction of 50%. When diode D1 is in OFF state, each resonance acts as planar inverted F antenna or PIFA mode. As shown in Fig. 3, the current is distributed along A → B → C. When diode D1 is in ON state, the surface current flows through both PIFA and spiral antennas. So this gives added advantage to reconfigurable structure where two antennas work together and generate the efficient results. The surface current is distributed along A → B → D as shown in Fig. 3.

This diode connection is considered as special because even when diode D1 is in OFF state, the diode D5 gets connection with the inner spiral structure, this makes possibility of increase in diode connections. As stated earlier, the limited cases are considered with more effective results. When diode D1 is in OFF state, we consider two cases which are shown below (Fig. 4).

Whereas in ON state, we have another four diode possibilities as shown in figures below. These diodes can be placed in ON/OFF conditions depending on our bandwidth requirement. The increase in diodes shows that the gain of antenna parameter is increased. So for better performance of antenna and ease of design with good bandwidths, we considered the following cases shown below. We neglect some cases where the presence of diode may not change any gain or bandwidth (Fig. 5).

**(a)**                                               **(b)**

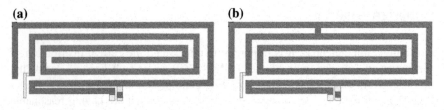

**Fig. 4** **a** Antenna-1 (all OFF). **b** Antenna-2(d5)

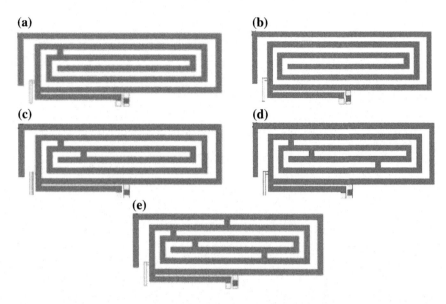

**Fig. 5** **a** Antenna-3(d1). **b** Antenna-4(d1, d2). **c** Antenna-5(d1, d2, d3). **d** Antenna-6 (d1, d2, d3, d4). **e** Antenna-7 (d1, d2, d3, d4, d5)

Table 1 gives the detailed results of all cases considered in the proposed antenna. The results show that as number of diodes increases, the gain increases. In table, each case is specified with a particular name, i.e., antenna 1, antenna 2, and so on.

**Table 1** Antenna parameters for diodes in ON/OFF conditions

| Antenna names | D1 | D2 | D3 | D4 | D5 | Bandwidth (GHz) | Gain |
|---|---|---|---|---|---|---|---|
| Antenna 1 (all diodes in OFF state) | 0 | 0 | 0 | 0 | 0 | 1.00–2.76 | 2.03 |
| | | | | | | 7.53–7.94 | |
| | | | | | | 9.41–9.75 | |
| Antenna 2 (main diode D1 in OFF state and D5 diode in ON state) | 0 | 0 | 0 | 0 | 1 | 1.18–2.69 | 2.4381 |
| | | | | | | 3.84–4.0 | |
| | | | | | | 7.76–8.04 | |
| | | | | | | 8.54–8.68 | |
| | | | | | | 8.73–8.84 | |
| | | | | | | 9.86–9.90 | |
| Antenna 3 (main diode D1 in ON state) | 1 | 0 | 0 | 0 | 0 | 2.95–3.26 | 2.8330 |
| | | | | | | 5.35–5.58 | |
| | | | | | | 7.79–7.92 | |
| | | | | | | 8.49–8.64 | |
| | | | | | | 8.96–9.12 | |
| | | | | | | 9.55–9.99 | |

(continued)

**Table 1** (continued)

| Antenna names | D1 | D2 | D3 | D4 | D5 | Bandwidth (GHz) | Gain |
|---|---|---|---|---|---|---|---|
| Antenna 4 (diode D1 and diode D2 in ON state) | 1 | 1 | 0 | 0 | 0 | 1.44–2.50 | 2.567 |
| | | | | | | 5.39–5.58 | |
| | | | | | | 7.24–7.32 | |
| | | | | | | 9.6–9.96 | |
| Antenna 5 (diode D1, D2, D3 in ON state) | 1 | 1 | 1 | 0 | 0 | 1–3.06 | 3.142 |
| | | | | | | 5.38–5.59 | |
| | | | | | | 5.88–5.90 | |
| | | | | | | 8.45–8.55 | |
| | | | | | | 9.67–9.99 | |
| Antenna 6 (All diodes except D5 in ON state) | 1 | 1 | 1 | 1 | 0 | 1.70–2.56 | 3.0027 |
| | | | | | | 5.37–5.63 | |
| | | | | | | 8.57–8.64 | |
| | | | | | | 9.61–10.04 | |
| | | | | | | 12.24–12.44 | |
| Antenna 7 (All diodes in ON state) | 1 | 1 | 1 | 1 | 1 | 1.48–2.50 | 3.435 |
| | | | | | | 5.8–6.03 | |
| | | | | | | 8.6–9.0 | |
| | | | | | | 9.73–10.16 | |

The table below gives an overview of return loss (s11 parameter) and gain of some particular antenna designs. If diode is in OFF state, it is represented by 0 and if diode is in ON state it is represented by 1.

## 4    Results and Discussion

For the case D5 in ON state, return loss is below $-10$ dB between 1.18 and 2.69 GHz. The impedance bandwidth is as large as 1.5 GHz. The return loss $S_{11}$ is below $-40$ dB (Figs. 6, 7 and 8).

The bandwidth of antenna when all the switches in ON position is almost 1 GHz (1.48–2.5 GHz), 200 MHz (5.8–6 GHz) and the return loss is below $-25$ dB. The bellow formulated results show the VSWR for antenna designs of diodes D1 in ON/OFF cases (Figs. 9, 10 and 11).

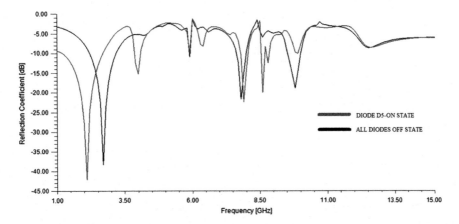

**Fig. 6** Diodes OFF state and D5 in ON state

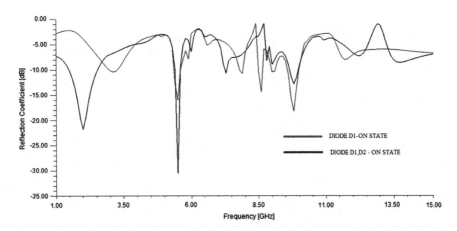

**Fig. 7** Diode D1 and D2 ON state

The figures shown represent the gain total for each antenna shown in Table 1. From above results, the gain is increased as numbers of diodes are increasing. This may be possible even when main diode is in OFF condition, but for ease simulation and better results, we considered limited cases. The maximum gain obtained is 3.4 dB and minimum gain is 2.03 dB obtained when no diode is present. All gain values for particular designs have been tabulated.

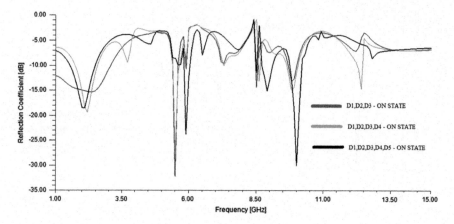

**Fig. 8** Return loss when more than three diodes are in ON state

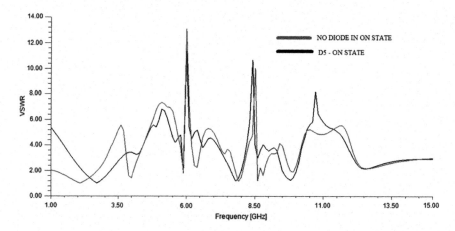

**Fig. 9** VSWR for diode D5 ON state and other diodes OFF state

## 5 Conclusion

In this paper, a frequency reconfigurable antenna is analyzed based on PIN diodes switching. The proposed spiral antenna covers mobile communication bands like PCS (1900), LTE, Wi-Fi, and WLAN (2.4 GHz) with suitable bandwidth 1.5 GHz. Significant change in reflection coefficient and gain is possible by combination of different diodes. Shift in frequency bands is observed while changing the states of diodes. Through the parametric analysis and switching positions, the designed model is verified for its applicability in the mobile communication.

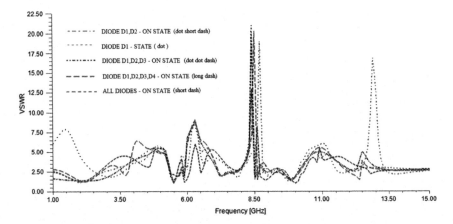

**Fig. 10** VSWR for different diodes ON and OFF states

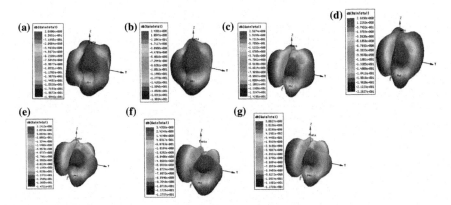

**Fig. 11** Radiation in 3D-Plot for designed antenna iterations with diodes positions

**Acknowledgements** Authors are thankful to Department of ECE of K L University for encouragement and we thank DST for the grant ECR/2016/000569/and SR/FST/ETI-316/2012.

# References

1. S.W. Lee, H.S. Jung and Y.J. Sung, "A Reconfigurable Antenna for LTE/WWAN Mobile Handset Applications," IEEE Antennas and Wireless Propagation letters, VOL. 14, 2015.
2. C.W. Yang, Y.B. Jung, C.W. Jung, "Octal internal antenna for 4G mobile handset," IEEE Antenna wireless propagation letters, VOL. 10, pp. 817–819, 2011.
3. C. Di. Nallo and A. Faraone, "Multi band internal antenna for mobile phones," Electron. Letter, Vol. 41, no. 9, pp. 514–515, 2005.

4. J. Cho, C.W. Jung and K. Kim, "Frequency reconfigurable two port antenna for mobile phone operating over multiple service bands," Electron. Letter. Vol. 45, no. 20, pp. 1009–1011, 2009.
5. D. Sreenivasa Rao, J. Lakshmi Narayana, B. T. P. Madhav, Microstrip Parasitic Strip Loaded Reconfigurable Monopole Antenna, ARPN Journal of Engineering and Applied Sciences, ISSN: 1819-6608 VOL. 11, NO. 19, pp 1–7, 2016.
6. Y. S. Jeong, S. H. Lee, J. H. Yoon, W. Y. Lee, W. Y. Choi, and Y.J. Yoon, "Internal mobile antenna for LTE/GSM 850/GSM 900/PCS1900/WiMAX/WLAN, triple service bands," in Proc. IEEE RWS, pp. 559–562, 2010.
7. B. T. P. Madhav, M. Manjeera, M. S. Navya, D. Sharada Devi, V. Sumanth, "Novel Metamaterial Loaded Multiband Patch Antenna, Indian Journal of Science and Technology, ISSN: 0974-6846, Vol 9, Issue 37, pp 1–9, 2016.
8. C. R. Rowell and R. D. Murch, "A capacitively loaded PIFA for compact mobile telephone handsets," IEEE Trans. Antennas Propag., vol. 45, pp. 837–842, May 1997.
9. N V Seshagiri Rao, Kumari Y, M Ajay Babu, B T P Madhav, "Design and analysis of printed dual band planar inverted folded flat antenna for laptop devices", Far East Journal of Electronics and Communications, ISSN: 0973-7006, Vol 16, No 1, pp 81–88, 2016.
10. G. K. H. Lui and R. D. Murch, "Compact dual frequency PIFA designs using LC resonator," IEEE Trans. Antennas Propag., vol. 49, pp. 1016–1019, Jul. 2001.
11. Angus C. K. Mak, Corbett R. Rowell, Ross D. Murch, and Chi-Lun Mak, "Reconfigurable Multiband Antenna Designs for Wireless Communication Devices", Ieee Transactions On Antennas And Propagation, Vol. 55, No. 7, July 2007.
12. Habibulla Khan, B.T.P. Madhav, K. Mallikarjun, K. Bhaskar, N. Sri Harsha and Muralidhar Nakka, "Uniplanar Wideband/Narrow Band Antenna for Wlan Applications", World Applied Sciences Journal, ISSN 1818-4952, Vol 32, Issue 8, pp 1703–1709, 2014.
13. K. L. Wong, "Planar Antennas for Wireless Communications," Wiley Inter-Science 2003.
14. S.S. Mohan Reddy, P Mallikarjuna Rao, B T P Madhav, "Enhancement of CPW-Fed Inverted L-Shaped UWB Antenna Performance Characteristics using Partial Substrate Removal Technique", 3rd International Conference on Signal Processing and Integrated Networks (SPIN-2016), 978-1-4673-9197-9/16/$31.00 ©2016 IEEE, pp 454–459.

# Octagonal Shaped Frequency Reconfigurable Antenna for Wi-Fi and Wi-MAX Applications

**B. T. P. Madhav, M. Ajay Babu, P. Farhana Banu, G. Harsha Sai Teja, P. Prashanth and K. L. Yamini**

**Abstract** In present day communication systems, reconfigurable antennas have attained much prominence due to the ease of shifting frequency bands without change in the original structure of the antenna. In this paper, a frequency band reconfigurable microstrip octagonal patch antenna has been simulated and analyzed with four switchable states consisting of four narrow bands. By simulating the antenna, four frequency bands are achieved 0.1–1.1, 2.2–3.4, 4.8–5.2, and 7.4–7.8 GHz with return loss $s_{11} < -10$ dB. The designed antenna has plain structure with size of 40 × 40 mm$^2$. To achieve switching, we used PIN diodes which are placed at the slotted structure on the ground plane. The software HFSS is used for simulation. From results, it is evident that there is a change in reconfigurability of the antenna with the change in positions of internal switches. The change in antenna parameters like return loss, directivity, and gain characteristics is evident.

**Keywords** Defected ground · Reconfigurable antenna · Wi-Fi Wi-MAX

## 1 Introduction

In general, reconfigurable antennas are extensively used to meet the requirements of present day's communication system. Reconfigurable antennas are capable of modifying the properties such as frequency and radiation pattern dynamically in a controlled manner. There are several types of reconfiguration techniques such as frequency reconfigurability, polarization reconfigurability, and radiation pattern reconfigurability. In this paper, frequency reconfiguration is adapted by the change in states of PIN diodes which act as switches.

B. T. P. Madhav (✉) · M. A. Babu · P. F. Banu · G. Harsha Sai Teja · P. Prashanth · K. L. Yamini
Department of ECE, K L University, Guntur, AP, India
e-mail: btpmadhav@kluniversity.in

© Springer Nature Singapore Pte Ltd. 2018
J. Anguera et al. (eds.), *Microelectronics, Electromagnetics and Telecommunications*, Lecture Notes in Electrical Engineering 471,
https://doi.org/10.1007/978-981-10-7329-8_59

With swift evolution in the wireless communication technologies, the need for reconfigurable antennas and multiple band antennas is increasing day by day. Reconfigurable antennas are mainly used in cognitive radios [1, 2]. We can use the available spectrum of frequencies available which can be efficiently used reconfigurable antennas [3–6]. By using electromechanical system (MEMS), a switchable quad-band antenna can be designed [7–10] for various applications by switching four different frequency bands. MEMS switches are highly resistant and are easy to modify when compared to others. By using varactor diodes, a reconfigurable mono polar patch antenna has been designed [11] which improves the frequency band characteristics. Varactor diodes are highly reliable when compared to PIN diodes. By using photoconductive diodes a frequency and beam reconfigurable antenna has been designed [12], where return loss of the antenna is significant by using photoconductive switches.

Reconfigurable antennas can be classified according to antenna parameters like operating frequency, radiation pattern, and polarization. In this paper, we have designed a frequency band reconfigurable antenna to operate the antenna in the range of frequency which is used for Wi-Fi and Wi-MAX [13–15].

In this paper, a design of an octagonal patch is configured to obtain four narrow bands by using pin diodes as switches. The four frequency bands include 800–900 MHz which is used for GSM, Wi-Fi in the range of 2.2–3.4 GHz, Bluetooth, and Wi-MAX in the range 4.8–5.2 GHz and 7.4–7.8 GHz, respectively which are used for many applications such as military battle field surveillance, missile control, and ground surveillance radar sets with short or medium range. In this paper, different operating conditions of the diodes are considered as different states and the results are analyzed as per different states.

## 2   Antenna Geometry

The construction of the reconfigured antenna is shown in Fig. 1. It is designed as an octagonal shaped patch antenna on a FR4 substrate which is used as an electrical insulator having huge mechanical strength with relative permittivity 4.4 and thickness 1.33 mm having dimensions of $40 \times 40$ mm$^2$. Below the substrate, ground plane is embedded with dimensions of $20 \times 40$ mm$^2$ and height 0.01 mm, below the microstrip feed line. This microstrip line feeding is simple to match by controlling the inset position.

A slot is introduced in the ground plane with half of its length $L_1$ as 19 mm, width is denoted by $L_2 = 8$ mm and other dimensions $L_3 = 5$ mm and $L_4 = 3.5$ mm. The width of the slots is $w_1 = 1$ mm. By using slotted structure located on the ground plane, the antenna is reconfigured to various frequency bands. Here, the slotted structure acts as a channel, which is arranged to control the frequency bands for various applications. This fixed slot under feed line keeps a pass band and causes stop band in the UWB spectrum. This can be controlled by changing the dimensions of main slot.

**Fig. 1** Octagonal patch antenna

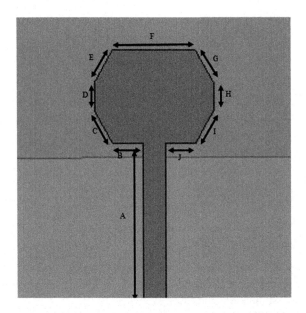

By utilizing the channel structure and including three PIN diode switches inside it as shown in Fig. 2, reconfigurable antenna is introduced. The required structures are created by using 3 PIN diode switches which results in different frequency bands. In this, HPND-4005 beam lead PIN diode is used in which low capacitance of 0.017 pF in ON state and resistance of 4.6 Ω in OFF state is simulated. By

**Fig. 2** Switchable filter on the ground plane

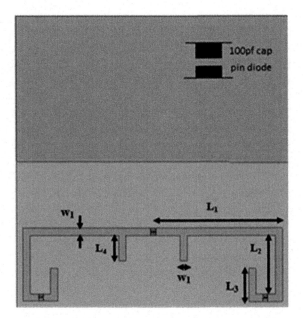

**Table 1** Antenna parameters

| A | B | C | D | E | F | G | H | I | J |
|---|---|---|---|---|---|---|---|---|---|
| 22 mm | 4.4 mm | 5.2 mm | 4 mm | 5.2 mm | 12 mm | 5.2 mm | 4 mm | 5.2 mm | 4.4 mm |

altering the states of PIN diodes, the resultant frequency bands can be obtained. This is designed and simulated using antenna design tool HFSS. The dimensions of the hexagonal patch are tabulated in Table 1.

# 3   Methodology

The process of designing the antenna to analyzing the results is presented in this section with ear marked points.

- Construction of the reconfigurable antenna using electromagnetic tool HFSS.
- Placing PIN diodes in the slots for switching the frequency bands through simulation studies.
- Optimization of the antenna model with parametric analysis and analyzing results to prototype the model on FR4 substrate.

The dimensions of the antenna are given in the following table which is extracted from the different models analyzed with EM simulation tool and optimized values are considered.

# 4   Results and Analysis

By turning all the three PIN diodes OFF, the antenna will operate in triple band where the input reflection coefficient S11 is less than −10 dB and covering frequency bands from 2.4–3.4, 4.8–5.1, and 7.4–7.8 GHz which is depicted in Table 1, when diode D3 is ON and rest are OFF, the antenna operates in dual band covering 0.1–1.1 and 2.2–3.3 GHz frequency bands. The overall band characteristics and gains of different states are presented in the table (Table 2).

The operating bands in state 3 give bandwidth of 1 GHz for 0.1–1.1 and 1.1 GHz for other band operating from 2.2 to 3.3 GHz, in which diode D3 is ON and remaining are OFF. The antenna's operating frequency ranges between 2.5 and 3.5 GHz for state 4 in which all the three diodes are ON. Thus, changing the states of diodes results in changing the frequency band characteristics of the antenna. This switching action can be controlled by operating the different PIN diodes presented on the slotted structure of ground plane. The gain of the antenna is more significant in state 2 and state 3 compared with different states.

**Table 2** Details of combinations of pin diodes and simulated frequency bands in each state

| State | Diode D1 | Diode D2 | Diode D3 | Frequency bands (GHz) | Gain (dB) |
|-------|----------|----------|----------|-----------------------|-----------|
| 1 | OFF | OFF | OFF | 2.3–3.4 <br> 4.8–5.1 <br> 7.4–7.8 | 2.92 |
| 2 | ON | ON | OFF | 0.1–1.1 <br> 2.1–3.4 | 3.04 |
| 3 | OFF | OFF | ON | 0.1–1.1 <br> 2.2–3.3 | 3.02 |
| 4 | ON | ON | ON | 2.5–3.5 | 2.82 |

It can be observed from Fig. 3 that, in state 1, antenna is operated in three narrow bands ranging from 2.3–3.4, 4.8–5.1, and 7.4–7.8 GHz, whereas in state 3, antenna is operated in a narrow band ranging from 2.1 to 3.4 GHZ. When all diodes are OFF (state 1), the antenna operates at its triple band covering S band and C band, but when single diode D3 is OFF, then the antenna results in single narrow band which is suitable for applications in S band.

From Fig. 4, comparison between states 3 and 4 is done to get better understanding about the property of frequency reconfigurability. In state 3, three narrow bands ranging from 0.1–1.1and 2.2–3.3 GHz and a very narrow band of 10 MHz are obtained, whereas in state 4, antenna is operated in a narrow band ranging from 2.5 to 3.5 GHz. When a single diode is ON (state 3), antenna is operating in dual band covering L band and S band suitable for different applications.

In state 3 and 4, antenna is operated in dual operating band with 1 GHz bandwidth each and improvement in the gain of antenna can be observed compared with remaining states. Radiation pattern is defined as the variation of the power radiated by an antenna in terms of direction away from the antenna. The lobe which contains maximum power that exhibits the greatest field strength is the main lobe. Figure 5 shows the radiation patterns of different states. For state 1 and state 2, radiation

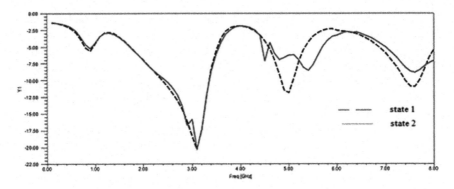

**Fig. 3** Simulated reflection coefficient S11 for the proposed antenna in state 1 and 2

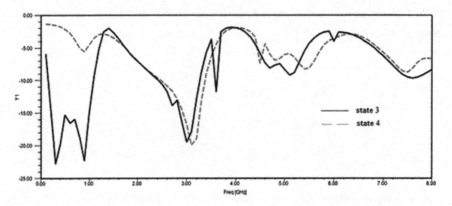

**Fig..4.** Stimulated reflection coefficient S11 for the proposed antenna in state 3 and 4

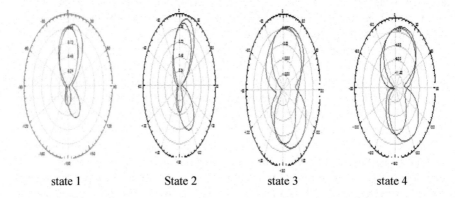

     state 1            State 2             state 3            state 4

**Fig. 5** Radiation patterns for different states

pattern is not bidirectional while bidirectional radiation pattern can be observed for state 3 and state 4.

Antenna's gain gives us information about antenna's directivity and antenna's efficiency. It is clear that from Fig. 6, different gain plots are depicted so as to compare the It can be observed that when the antenna overall gain of the antenna in all the four operating states. A gain of 2.92 dB is achieved for state 1 and almost the same gain is observed in state 4. For state 2 and state 3, the overall gain is about 3.04 GHz and 3.02 GHz, respectively, and thus the radiation pattern and current distributions of these states are very significant.

The current distributions for different states are presented. The intensity of current can be analyzed from the scale which is shown in the figure. It is evident that red color from the plot resembles the intensity of current through the antenna. For state 1, current is distributed along the length of feed line and for state 2 current

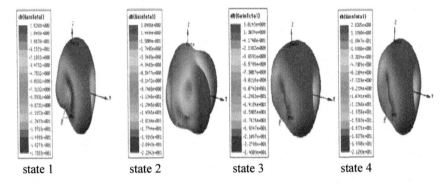

state 1     state 2     state 3     state 4

**Fig. 6** Gain plots for different states

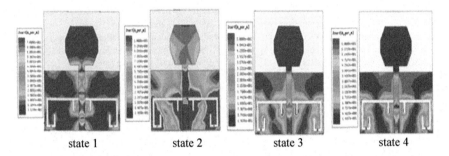

state 1     state 2     state 3     state 4

**Fig. 7** Current distributions for various states of diodes at different operating bands

is distributed along the feed line and on the ground plane. It can be viewed from Fig. 7 that the current is distributed for state 3 and state 4 along the slots placed in the ground structure of the antenna.

# 5 Conclusion

A band reconfigurable octagonal shaped microstrip patch antenna with slotted filter structure placed on defective ground is presented in this paper. The designed model operates at different frequency bands for different states of diodes and the complete analysis is presented in this work. Thus, frequency reconfigurability for different bands in the communication applications is experimented successfully with diodes ON and OFF positions and designed model is switching in the applications of Wi-Fi and Wi-MAX with good impedance bandwidth and radiation characteristics.

**Acknowledgements** Authors are thankful to Department of ECE of K L University for encouragement and also like to thank DST for the grant ECR/2016/000569 and SR/FST/ETI-316/ 2012. We would like to acknowledge Notachi Electronics for technical support in this work.

# References

1. C. Zhang, S. Yang, H.K. Pan, V.K. Nair and A E Fathy, "Frequency reconfigurable antennas for multi radio wireless platforms" IEEE Microwave magazine", vol 10, no 1, pp 66–83, Feb 2009.
2. D. Sreenivasa Rao, et al, Microstrip Parasitic Strip Loaded Reconfigurable Monopole Antenna, ARPN Journal of Engineering and Applied Sciences, ISSN: 1819-6608 VOL. 11, NO. 19, pp 1–7, 2016.
3. AsaJyothi, B. T. P. Madhav, M. Deepthi, C. Sindhoora, V. Jayanth, "A Novel Compact CPW—Fed Polarization Diversity Dual Band Antenna using H-Shaped Slots", Indian Journal of Science and Technology, ISSN: 0974-6846, Vol 9, Issue 37, pp 1–8, 2016.
4. T Y Han and C T Hung, Reconfigurable monopolar patch antenna," Electronics letter", Vol 46, no 3, pp 199–200, Feb 2010.
5. P. SyamSundar, et al, "Radial Stub Loaded Antenna with Tapered Defected Ground Structure", ARPN Journal of Engineering and Applied Sciences, ISSN: 1819-6608 Vol. 11, No. 19, pp 1–7, 2016.
6. P. Gardener, Mr Hamid, P.S. Hall and F. Gahanem, "Switched band Vivaldi antenna", IEEE transaction antenna propagation, vol 59, no 5, pp 1472–1480, 2011.
7. B. T. P. Madhav, A. V. Chaitanya, R. Jayaprada, M. Pavani, "Circular Monopole Slotted Antenna with FSS for High Gain Applications", ARPN Journal of Engineering and Applied Sciences, ISSN 1819-6608, Vol. 11, No. 15, pp 1–7, 2016.
8. Chintana J Panagamuwa, Alford Chauraya, "Frequency and band reconfigurable antenna using photo conductive switches", IEEE transactions on antennas and propagation, vol 54, no 2, Feb 2006.
9. K V L Bhavani, Habibulla Khan, D Sreenivasa Rao, B T P Madhav, "Dual Band Notched Planar Printed Antenna with Serrated Defected Ground Structure", Journal of Theoretical and Applied Information Technology, ISSN: 1992-8645, Vol 88, Issue 1, pp 28–34, 2016.
10. Hamid Baoudaghi, Mohammadnaghiazarmanesh and Mehdi Meharanpour, "A frequency reconfigurable monopole antenna using switchable slotted ground structure", IEEE antenna and wireless propagation letters, vol 12, 2012.
11. N V Seshagiri Rao, Kumari Y, et al, "Design and analysis of printed dual band planar inverted folded flat antenna for laptop devices", Far East Journal of Electronics and Communications, ISSN: 0973-7006, Vol 16, No 1, pp 81–88, 2016.
12. Chirag Gupta, DivyanshuMaheswari, Risteshkumarsraswat and Mithileskumar, A UWB frequency band reconfigurable antenna using switchable ground structure, 2014 Fourth international conference on communication systems and network technologies.
13. S S Mohan Reddy, et al, "Asymmetric Defected Ground Structured Monopole Antenna for Wideband Communication Systems", International Journal of Communications Antenna and Propagation, ISSN: 2039-5086, Vol 5, Issue 5, pp 256–262, 2015.
14. H. Ghafouri-Shiraz and A. Tariq, "Frequency reconfigurable monopole antennas", IEEE transaction antennas propagation, vol 60, no 1, pp 44–50, Jan 2012.
15. B.T.P. Madhav, Harish Kaza, "Design and Analysis of Compact Coplanar Wave Guide Fed Asymmetric Monopole Antennas", Research Journal of Applied Sciences, Engineering and Technology, 10, 3, pp 247–252, Jun-2015.

# Cylindrical Structured Multiple-Input Multiple-Output Dielectric Resonator Antenna

**B. T. P. Madhav, M. Ajay Babu, P. V. S. Praneeth Kumar, M. Venkateswara Rao and D. Padma Srikar**

**Abstract** This paper presents the design and analysis of cylindrical shaped dielectric resonator MIMO antenna. The designed antenna was excited using coplanar waveguide feeding with rectangular L-shaped slots placed on the ground plane. The antenna was characterized using ANSYS Electromagnetic Desktop 17.0 (HFSS). Compared to the conventional microstrip antennas, the proposed dielectric resonator antenna has several advantages such as smaller dimension, high gain, and good radiation characteristics which are more suitable for millimeter wave range applications. The proposed antenna has better radiation characteristics with a peak gain of 9.08 dB, efficiency toward radiation is 97%, return loss bandwidth is 2.3 GHz, and it also offers front-to-back ratio of 16.3 dB at resonating frequency 60 GHz.

**Keywords** Millimeter wave applications · Cylindrical shape · Dielectric resonator antenna (DRA) · Multiple-input multiple-output (MIMO) Coplanar waveguide feed

## 1 Introduction

Dielectric Resonator Antennas (DRAs) have ubiquitous attention because of their advantages such as high radiation efficiency, low dissipation loss, small size, low cost, less weight, and ease of design flexibility. Generally, DRAs are used for millimeter wave applications. Usually, the MIMO antennas are used in order to enhance the data rate. In case of multiple-input multiple-output antennas, different transmitters and receivers are used such that multiple radio signals transmit over a single radio channel. Usually, at higher frequencies, the microstrip patch antennas

B. T. P. Madhav (✉) · M. A. Babu · P. V. S. P. Kumar · M. V. Rao (✉) · D. P. Srikar
Department of ECE, K L University, Vaddeswaram, Guntur, India
e-mail: btpmadhav@kluniversity.in

M. V. Rao
e-mail: mvenkatesh8692@gmail.com

© Springer Nature Singapore Pte Ltd. 2018
J. Anguera et al. (eds.), *Microelectronics, Electromagnetics and Telecommunications*, Lecture Notes in Electrical Engineering 471,
https://doi.org/10.1007/978-981-10-7329-8_60

589

suffer from conductor and substrate losses due to that efficiency and radiation characteristics alleviate. In order to vanquish such losses, the dielectric resonators are used. DR is made of the high permittivity material which is of low cost such that its potential carried out over higher frequencies in order to obtain high radiation efficiency. Over the decade, the research is focusing on bandwidth enhancement using DRA. The microstrip antenna radiates through two narrow edges of the patch whereas the DRA radiates through its entire surface except the grounded portion. These dielectric resonators are widely used in microwave circuits like filters and combiner applications to prevent energy loss in radiation. In order to work at such frequencies ranges, we use different radiating structures. To obtain the frequencies, the factors like dielectric materials, shape of cylinders, and field distribution play a major role in designing such radiating structures. Radiation pattern and resonant frequencies are measured to test the ability of different structures [1]. In order to work in microwave frequency range, many antennas are used some of them are conventional antennas. The dielectric resonators play a vital role in such microwave frequencies. The mode of propagation (like $TE_{111}$ and $TE_{101}$) and input impedance are some of the factors related to properties of the antenna [2]. In order to have the applications of antennas as radiator at higher frequencies, they must possess minimum ohmic losses and wider bandwidth ranges. DRAs with aperture coupling are capable of holding in microwave circuits [3]. DR antennas are the basic building element of many filters and resonator working at microwave frequency range. These have special features of higher radiation efficiency, easy integration, and compact design. The main disadvantage of this antenna is bandwidth, usually for microwave applications wider bandwidth is required. Methods to enhance the bandwidth for DR antennas are presented in [4]. In [5], hybrid resonator using coplanar wave guide feeding in order to improves the gain by introducing the superstrate. It covers the entire industrial scientific and medical radio band (ISM band). The resonator antennas are used in order to obtain high gain, broadband as well as higher efficiency. The stacked DR with slot located beneath the substrate is presented in [6]. Generally, we design small DRAs in order to make the design very precise. To avoid that problem, larger resonators are designed with higher order modes such as $TE_{pqr}$ by using slot coupling [7–9]. Ring-shaped DR with circular microstrip patch is designed using hybrid electric modes (*HE* mode) and *EH* modes as a transverse magnetic mode in order to obtain higher gain [10–12]. Split cylindrical dielectric resonator antenna with different coupling such as microstrip, probe, and aperture [13]. For a multiple-input multiple-output antenna, transverse electric mode $(TE_{011+\delta})$ and hybrid electric mode $(HE_{11\delta})$ must resonate at the identical frequencies. The construction used in [14] is in such way that the DRs are placed one upon another and same dielectric materials are used for both. The design consists of dual feeding such that the bottom resonator is feed by using aperture coupling as well as the top element is fed by using the bottom element.

## 2 Antenna Geometry and Design

The proposed cylindrical MIMO DRA consists of three elements namely conducting surface (ground plane), substrate, and radiating element shown in above Fig. 2, respectively. The high resistivity silicon material with dielectric loss tangent value of $\tan \delta = 0.003$ and relative permittivity $\varepsilon_r$ of 11.9 is used as the substrate material. The dimensions such as width "$W_s$", length "$L_s$" and thickness "$T$" of the substrate used in design are shown in Table 1. The dielectric resonator with radius "$R$", and height "$H$" is made of silicon which is having resistivity of 2000 Ω-cm placed on the top surface of the substrate. The dipole ends DRA whose length (L) and width "$W$" are shown in Table 1. The cylindrical DRA was excited using coplanar wave guide feed line of 50 Ω. The slot width "$S_w$", slot length "$S_L$" are obtained in such a way that they are refrained from the conducting surface (ground plane) and these slots are placed on the underside of the substrate. In case of CPW feeding, the ground as well as the feed is in the same plane itself. The ground plane is excited with perfect electric field which has infinite conductivity and zero resistivity. Figure 1 shows the DRA, which is excited using single port due to that the currents flow through the two slots in the same direction, the electric fields as well as the magnetic fields is generated normally to each other and electromagnetic energy gets radiated through the whole surface of cylindrical DRA. Figure 2 represents proposed multiple-input multiple-output cylindrical dielectric resonator antenna consists of dual port with four "L"-shaped slots placed on the underside of the substrate. The antenna is excited using dual feeding such that the current flows through the four slots in an identical direction and get radiated through dielectric resonator. The radiation boundary of the antenna is placed at a distance of $(\lambda/4)$ from the strong radiating element (dielectric resonator). The resonating frequency of the dielectric resonator antenna can be calculated by using Eq. (1) which is valid for $(0.4 \leq R/H \leq 6)$. From [10], the field and current distributions of the proposed modal are shown in Fig. 4 such that the electric field and magnetic field are normal to each other.

$$F = \frac{c \times 6.324 \times \left[0.27 + 0.36\left(\frac{\varepsilon_r}{2H}\right) + 0.02\left(\frac{R}{2H}\right)^2\right]}{\sqrt{\varepsilon_r + 2}}, \tag{1}$$

**Table 1** Dimensions of the proposed antenna

| Parameter | Dimensions in mm | Parameter | Dimensions in mm |
|---|---|---|---|
| $L_s$ | 6.418 | $L$ | 1.068 |
| $W_s$ | 6.282 | $W$ | 0.265 |
| $T$ | 0.275 | $Ss$ | 0.068 |
| $H$ | 0.4 | $Sw$ | 0.04 |
| $R$ | 1.18 | $S_{L1}$ | 1.961 |

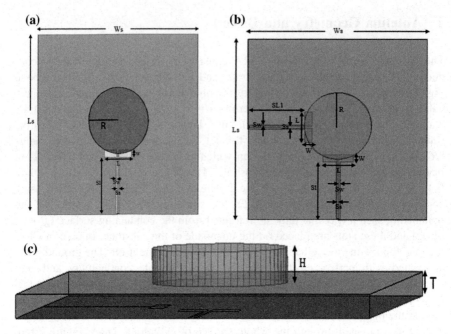

**Fig. 1** **a** CPW-fed DRA **b** CPW fed DRA with MIMO **c** side view of MIMO DRA antenna

where, "*c*" refers to velocity of light, "*R*" is radius of the cylinder, "*H*" is height of DRA, and $\varepsilon_r$ is relative permittivity.

## 3 Simulated Results and Discussion

The characteristics of the antenna such as return loss ($S_{11}$, dB), VSWR, radiation pattern, and gain (dB) are simulated using ANSYS Electromagnetic desktop (HFSS 17) and discussed in this section.

In order to verify the working condition of the antenna, the return loss provides the solution. From Fig. 5 the return loss ($S_{11}$) at port1 is around −16 dB, with return loss $S_{22}$ at port 2 as −42.2 dB and the transmission coefficients $S_{12}$ and $S_{21}$ as −29.9 dB. The return loss bandwidth is around 2.3 GHz at operating frequency 60 GHz. In order to verify whether the antenna is radiating efficiently, the VSWR provides the solution. The value of standing wave ratio is 1.4 dB at 60 GHz (port1) and 1.09 dB at 59.6 GHz (port2). From the obtained results, we calculated the impedance bandwidth and it is around 86% at port 1 and 82% at port 2.

The reflection phase comparison of one-arm SAMC and four-arm SAMC designs is shown in Fig. 3. Its value varies from +180° to −180°, satisfying the basic property of artificial magnetic conductors. From the graph, it is observed that

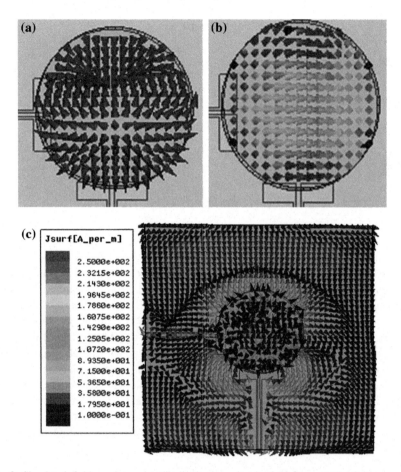

**Fig. 2** Simulated fields and current distributions **a** electric field **b** magnetic field **c** current distribution

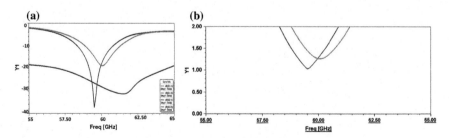

**Fig. 3** **a** MIMO DR antenna simulated S-parameter **b** simulated VSWR of the DR antenna using MIMO

**(a)**                                                        **(b)**

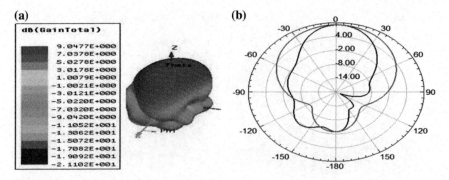

**Fig. 4** Simulated **a** 3D gain **b** radiation pattern

all the four models cross the 0° phase line at four frequencies (1.92, 7.30, 17.8, 25.07 GHz).

In order to measure the ability of the radio energy of the antenna in a particular direction, the gain of the antenna provides the solution. From Fig. 4a, it is clear the peak gain of the proposed antenna is around 9.08 dB and maximum radiation is along the Z direction with radiation efficiency of 97%. Figure 4b presents the normalized antenna pattern of the proposed antenna at 60 GHz. The 3-dB beam widths measured in E-plane, H-plane are 88.02 dB and 44.97 dB, respectively. Peak magnitude of E-plane is 9.6 dB and peak magnitude of H-plane is 9.4 dB.

Figure 5 shows the gain of the proposed antenna with respect to operating frequency (GHz). The realized peak gain of the antenna is 9.08 dB at 60 GHz.

Figure 6 presents the cross-polarization and co-polarization of the proposed antenna at 60 GHz and the cross-polarization plot says that the antenna nearly radiates in omnidirection with respect to H-plane Phi = 90°.

**Fig. 5** Simulated gain versus frequency of the proposed antenna

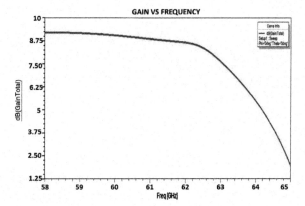

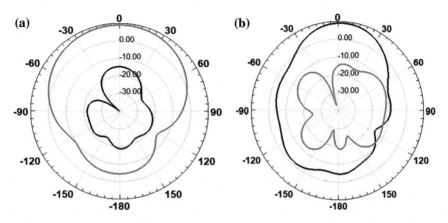

**Fig. 6** Simulated **a** cross-polarization **b** co-polarization patterns at 60 GHz frequency

## 4 Parametric Analysis

The designed antenna by varying the geometrical parameter such as radius (R) height (H) of the cylinder, length (L) and width (W) of the dipole are not varied because by varying these values, the radiation parameters of the antenna may change abruptly such that the efficiency of the antenna will be reduced.

Since the DRA is placed on the top surface of the substrate, by increasing the height of the cylindrical resonator, the substrate thickness will be reduced in order to change the geometry of the antenna and therefore the height as well as parameters related to dipole remains unchanged. The radius of the dielectric resonator is varied by choosing five consecutive radii spacing 0.05 mm. In the first iteration, the radius is chosen as 1.08 mm, the obtained bandwidth is about 2.7 dB, similarly for 1.13 mm BW is of 2.6 dB, for 1.18 mm bandwidth is 2.3 dB, for 1.23 mm the BW is 2.2 dB, for 1.28 mm the BW is 2.1 dB. The parametric variation is done for the proposed antenna in order to get the optimum design values. Since the desired frequency is 60 GHz by choosing the radius value as 1.18 mm, we are able to get the return loss less than (−10 dB). The reflection coefficients at port 1 ($S_{11}$) are around (−16 dB) and the return loss at port 2 ($S_{22}$) is around (−42.5 dB). The transmission coefficients $S_{12}$ and $S_{21}$ values are around (−29.9 dB). So, the optimum values such as R = 1.18 mm, and H = 0.4 mm are used in prototyping the antenna. The reflection coefficient BW can be calculated by using Eq. (2) (Fig. 7).

$$\% \text{ Bandwidth (BW)} = (f_h - f_l) \times 100/f_c. \tag{2}$$

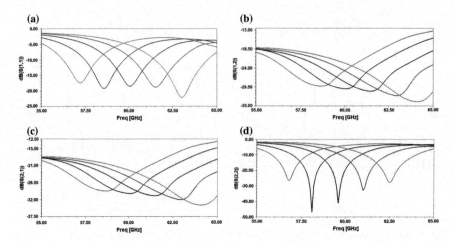

**Fig. 7** **a** Simulated reflection coefficient ($S_{11}$ dB) with varied radius R **b** simulated transmission coefficient ($S_{12}$ dB) with varied radius R **c** simulated transmission coefficient (S21 dB) with varied radius R **d** simulation reflection coefficient ($S_{22}$ dB) with varied radius R

## 5 Conclusion

The cylindrical dielectric resonator MIMO antenna at 60 GHz is analyzed in this paper. The parametric analysis is also done in order to get the optimum design. The proposed model has better radiation characteristics such as high gain, high radiation efficiency, and high front-to-back ratio. This dielectric resonator antenna overcomes the substrate and conductor losses created by microstrip antenna usually at higher frequencies. Due to such appealing characteristics, the antenna is used for millimeter wave range applications.

**Acknowledgements** The authors deeply express their gratitude to ALRC Research Centre, Department of ECE, K L University for their encouragement during this work. Further, Madhav would like to express his gratitude to DST through grant ECR/2016/000569 and FIST grant SR/FST/ETI-316/2012.

## References

1. Long, S., Mark McAllister, Liang Shen: The resonant cylindrical dielectric cavity antenna. In: IEEE Trans. Antennas and Propag. 31, 406–412 (1983)
2. Luk, K. M., et al.: Analysis of hemispherical dielectric resonator antenna. In: Radio science. 28 1211–1218 (1993)
3. Martin, JTH St, et al: Dielectric resonator antenna using aperture coupling. In: Electronics Letters 26 2015–2016 (1990)

4. Coulibaly, Y., et al.: Design of a single circular soft surface applied to an aperture fed Dielectric Resonator antenna for gain and bandwidth improvement. In: Antennas and Propagation (APSURSI), 2011 IEEE International Symposium on. IEEE, (2011)
5. Gong, Ke, XueHui Hu: Low-profile substrate integrated dielectric resonator antenna implemented with PCB process. In: IEEE Antennas and Wireless Propag. Letters. 13 1023–1026 (2014)
6. Adnam Kaya, Irfan Kaya, "U-shape slot antenna design with high-strength $N_{i54}$ $T_{i46}$ alloy," Arabian Journal for Science and Engineering, vol. 41, pp. 3297–3307, (2016)
7. K V L Bhavani, Habibulla Khan, et al.: Wideband CPW Fed Monopole Fractal Antenna with Defected Ground Structure. In: Journal of Engineering and Applied Sciences. ISSN: 1816-949X. 11(11) 2446–2454 (2016)
8. Habibulla Khan, Surendra, L., et al.: Differential Fed MIMO Antenna for Wide Band Applications", ARPN Journal of Engineering and Applied Sciences. ISSN: 1819-6608. 11(21) 1–5 (2016)
9. Lakshmi, M. L. S. N. S., et al.: Tapered Slot CPW-Fed Notch Band MIMO Antenna. ARPN Journal of Engineering and Applied Sciences, ISSN: 1818-6608, 11(13) 1–7 (2016)
10. Pan, Yong-Mei, Kwok Wa Leung, Kwai-Man Luk. Design of the millimeter-wave rectangular dielectric resonator antenna using a higher-order mode. In: IEEE Trans. Antennas and Propag. 59(8) 2780–2788(2011)
11. Agouzoul, M. Nedil, Y. Coulibaly, T. A. Denidni, I. Ben Mabrouk, Talbi, L.: Design of a high gain hybrid dielectric resonator antenna for millimeter-waves underground application. In: IEEE Antennas Propag. Soc. 1688–1691 July (2011)
12. J. Chandrasekhar Rao, N. Venkateswara Rao et al.: Compact UWB MIMO Slot Antenna with Defected Ground Structure. In: ARPN Journal of Engineering and Applied Sciences. ISSN: 1819–6608. 11(17) (2016)
13. A N Obadiah, M R Hamid, "Reconfigurable Bandwidth Antenna for LTE Applications," Arabian Journal for Science and Engineering, vol. 41, pp. 3655–3661, (2016)
14. Ramkiran, D. S., et al.: Coplanar Wave Guide Fed Dual Band Notched MIMO Antenna. In: Int. J. of Electrical and Computer Engg (IJECE). ISSN: 2088-8708 6(4) 1732–1741 (2016)

# Multiband Semicircular Planar Monopole Antenna with Spiral Artificial Magnetic Conductor

B. T. P. Madhav, T. V. Rama Krishna, K. Datta Sri Lekha,
D. Bhavya, V. S. Dharma Teja,
T. Mahender Reddy and T. Anilkumar

**Abstract** An umbrella-shaped planar microstrip patch antenna with defective ground is designed in this paper. The conventional ground of the antenna is replaced by the suitable artificial magnetic conductor (AMC) structures, which is in a defective ground. The AMC structures are the metamaterials which help in reduction of surface waves. The antenna structures are implemented to achieve multiband properties. A spiral artificial magnetic conductor (SAMC) with one arm and four arms is used as ground plane to improve antenna performance. Characteristics like operating bandwidth, antenna gain, and efficiency were analyzed for the proposed antennas. All the proposed antennas operate at frequencies 1.9, 7.3, 17.8, and 25 GHz. The average gain and radiation efficiency are improved by adding rectangular patches in the spiral SAMC ground. The detailed design of the proposed antennas and simulated results obtained using ANSYS HFSS are presented.

**Keywords** Spiral artificial magnetic conductor (SAMC) · Artificial magnetic conductor · Semicircular planar monopole antenna · Multiband

## 1 Introduction

The requirement of multiband antennas has been on constant rise to satisfy the applications of wireless communications. Besides, the designed antenna must have low profile, light weight, and adaptable feeding techniques so that it can be easily integrated with other circuits [1]. To meet these criteria, microstrip patch antennas are used [3]. One way to realize the low-profile antenna design is to use artificial

B. T. P. Madhav (✉) · T. V. R. Krishna · K. D. S. Lekha · D. Bhavya · V. S. D. Teja ·
T. M. Reddy · T. Anilkumar
Department of ECE, K L University, Vaddeswaram, Guntur, India
e-mail: btpmadhav@kluniversity.in

T. Anilkumar
e-mail: t.anilkumar@kluniversity.in

© Springer Nature Singapore Pte Ltd. 2018
J. Anguera et al. (eds.), *Microelectronics, Electromagnetics and Telecommunications*, Lecture Notes in Electrical Engineering 471,
https://doi.org/10.1007/978-981-10-7329-8_61

magnetic conductors (AMC), where conventional metallic ground plane is replaced with a high impedance ground plane.

AMCs are extensively used in modern communications to enhance the bandwidth requirements. It is a special member of high impedance structures (HIS) family, which often serves as a ground plane for low-profile antennas [4]. AMC structures depend on periodic dielectric substrates and various metallization patterns. Also, it does not allow surface wave propagation and reflects the wave's in phase but not out of phase [5]. Using these structures, we can achieve less back-radiation than conventional PECs [6]. Typically, an AMC can be specified as an electromagnetic band gap (EBG) material having a surface as magnetic conductor to operate at a desired frequency band and assist in the reduction of surface waves in certain band gap [7]. Electronic band gap structures are also known as photonic band gap (PBG) structures and are identical to perfect magnetic conductors (PMC) [8]. These EBG structures are efficiently used to overcome the ailments of microstrip patch antennas such as low gain and narrow bandwidth. Other metamaterials such as frequency selective surfaces (FSS) also serve as an alternative to AMCs [9, 10]. AMCs are useful to solve the variety of electromagnetic problems as it has specific properties. These are extensively used as wave guide filters or as planar ground when operated in forbidden frequency band in which it obstructs surface waves to propagate. In-phase reflection of AMC surface indicates that its reflection coefficient magnitude is +1. Reflection phase varies from +180° to −180° with respective frequency [11]. There are several versions of AMCs such as mushroom type, peano-curve type, Jerusalem cross type, Hilbert curve type, spiral type, and uniplanar compact type.

The concept of spiral AMCs belongs to broadband antennas and its performance largely depends on the reflection coefficient. The working of spiral AMC structures is best described by an equivalent LC circuitry. In [13], the spiral defective ground with different feeding techniques was designed to achieve high gain, multi- or ultra-wide bandwidth from 2 to 12 GHz and further proposed four-arm SAMC to eliminate the cross-polarization effect. As the iteration order decreases, the equivalent inductance decreases leading to a higher resonating frequency and also affects the directivity of an antenna. The spacing between the arms of the spiral affects the characteristics of the antenna. Depending on the arrangement of the arms, the spiral AMC is categorized as one arm, two arms, and four arms.

In order to achieve multiband characteristics, an antenna is designed with umbrella-shaped radiator and spiral AMC as a defective ground. First part of this paper focuses on the design of one-arm spiral AMC as ground. Additionally, to reduce the antenna size and cross-polarization, to overcome frequency discontinuities, and to increase the number of resonating frequencies, four-arm spiral AMC is used as a ground. These antennas are used in wide range of applications like fixed satellite, radio navigation, and electromagnetic absorbers.

## 2 Antenna Geometry and Design

A microstrip monopole antenna with two different Spiral Artificial Magnetic Conductor (SAMC) structures is designed. The antenna has a semicircular radiating patch and spiral artificial magnetic conductor (SAMC) as ground plane. The antenna is printed on FR4 epoxy dielectric substrate with thickness of 3.2 mm that has relative permittivity of 4.4. The dimensions of the substrate are $L_s \times W_s$. The radius of the semicircular umbrella shape patch is "$r$" mm. The semicircular umbrella-shaped radiator is placed at a distance of "$L_{feed}$" from one edge of the substrate. The antenna is excited with microstrip line feed, where the width is "$W_{feed}$".

The radiation box is at a distance of $\lambda/4$ from all sides of the substrate. Four models are proposed in the paper as shown in Fig. 1, keeping umbrella-shaped microstrip patch as radiator and varying the ground. The design started by considering conventional one-arm SAMC as the ground plane, as shown in Fig. 1a. The SAMC is one type of structure that exhibits the artificial magnetic conductor property. This AMC structure is modeled in the spiral shape, in this article the rectangular armed spiral is used. The shape of the AMC is considered as spiral for reducing the antenna size and to decrease the surface wave effects. Then, this model is modified by joining the outer turn of one-arm SAMC with rectangular strip on both sides, Fig. 1b. The dimensions of the patch, ground, feed, and the geometry of the spiral are summarized in Table 1.

To minimize the cross-polarization level and to reduce bandwidth discontinuities, four-arm SAMC has been designed, as shown in Fig. 1c. The geometry is formed by joining two number of two-arm SAMCs orthogonally at the center. Further to improve antenna efficiency and gain, the inner turn of four-arm SAMC is joined with a rectangular patch on either side, as shown in Fig. 1d.

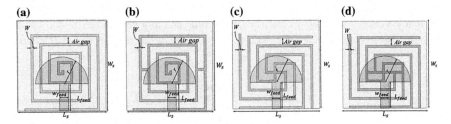

**Fig. 1** Antenna geometries of umbrella-shaped antenna with spiral ground plane (AMC) **a** model 1 with one-arm SAMC, **b** model 2 one-arm SAMC with shorted strips, **c** model 3 with four-arm SAMC, **d** model 4 four-arm SAMC with shorted

**Table 1** Dimensions of the proposed antenna models

| Parameter | $L_s$ | $W_s$ | $L_{feed}$ | $W_{feed}$ | $W$ | Air gap | $R$ |
|---|---|---|---|---|---|---|---|
| Dimensions in mm | 40 | 40 | 12 | 4 | 1 | 3 | 12 |

## 3 Simulated Results and Discussion

The performance of the designed antenna models is measured in terms of return loss, reflection phase, gain, E-plane, and H-plane patterns. The return loss and reflection phase characteristics of four antenna designs are simulated within the frequency sweep range up to 40 GHz and shown graphically in Fig. 2.

It can be observed that all the antennas operate at multiple frequencies, producing multiband characteristics. For antenna model 1 and model 2, i.e., the antennas with single armed SAMC are offered by more return loss beyond 26 GHz frequency. The modified SAMC structure (in Model 3 and Model 4) with the four arms provides a structural symmetry, due to which the reflection power loss is minimized across the 26–40 GHz spectrum. The wideband characteristics can be observed across 12–20 GHz band. It can be observed that Model 4 with the four-arm SAMC has a better performance compared to other antennas. The operating band and resonating frequencies for the four antennas are shown in Table 2.

The reflection phase comparison of one-arm SAMC and four-arm SAMC designs is shown in Fig. 2b. Its value varies from $+180°$ to $-180°$, satisfying the basic property of artificial magnetic conductors. At the resonating frequencies, the reflection phase of the antenna is oscillating between $180°$ and $-180°$. From the graph, it is observed that all the four models cross the $0°$ phase line at four frequencies (1.92, 7.30, 17.8, 25.07 GHz).

Figure 3a depicts the simulation results of peak gain versus frequency of all the simulated one-arm and four-arm SAMCs. The maximum and average gains of the four models are observed and it shows that performance of Model 1 and Model 4 is better and has reliable gain from 6 to 37 GHz. The maximum gain obtained for the antenna with four-arm SAMC as ground is 7.90 dB at 34.8 GHz. The radiation efficiency versus operating frequency of the four antenna geometries is shown in Fig. 3b. The average efficiency across the frequency sweep is observed as 92% for

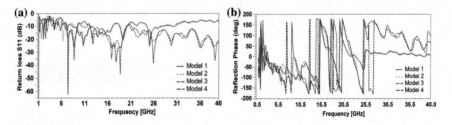

**Fig. 2** Simulation results of **a** return loss and **b** reflection phase versus frequency characteristics of antenna iterations

**Table 2** Operating ranges and resonating frequencies of the four antenna models

| Antenna model iteration | Operating band (GHz) | | Resonant frequencies (GHz) |
|---|---|---|---|
| | Start frequency (GHz) | Stop frequency (GHz) | |
| Antenna model 1 | 1.16 | 1.2690 | 1.19 |
| | 1.8306 | 2.1195 | 1.99 |
| | 5.5347 | 5.8977 | 5.67 |
| | 6.7323 | 7.4160 | 7.14 |
| | 8.4423 | 16.9949 | 9.88, 10.69, 13.74, 15.19, 16.32 |
| | 18.5765 | 20.1553 | 19.2 |
| | 24.4557 | 27.40 | 25 |
| Antenna model 2 | 1.2406 | 1.7170 | 1.52 |
| | 1.9579 | 2.0437 | 2 |
| | 4.4700 | 5.0303 | 4.66 |
| | 6.1029 | 6.7476 | 6.3 |
| | 8.8196 | 12.2270 | 10.67 |
| | 12.5211 | 17.1182 | 14.94 |
| | 18.6607 | 20.1576 | 19.35 |
| | 24.4270 | 27.7139 | 25.07 |
| Antenna model 3 | 2.3125 | 2.5659 | 2.40 |
| | 6.9472 | 9.3649 | 8.66 |
| | 12.1179 | 20.0440 | 12.51, 16.75, 18.60 |
| | 21.1919 | 40 | 25.6, 29.6, 35.81, 39.1 |
| Antenna model 4 | 2.5618 | 2.8890 | 2.7 |
| | 4.2575 | 4.4099 | 4.2 |
| | 7.1401 | 9.4873 | 7.3, 8.5 |
| | 12.07 | 20 | 12.6, 16.8, 17.8 |
| | 20.8 | 40 | 22.9, 25.8, 29.7, 32.9, 36.13, 38.8 |

the antenna in model 4 when compared to earlier iterations. The oscillatory nature of radiation efficiency characteristics can be seen across the 1–6 GHz range is corresponding to the operating bands and notches. It can be observed that the antennas have better efficiency ranging from 1 to 15 GHz and the efficiency is degraded further as the frequency of excitation increases. The vector E-field, current distribution, and gain 3D are shown in Fig. 4.

The E-plane and H-plane radiation patterns of these proposed antennas are simulated at all the four resonating frequencies obtained from reflection phase graph and are shown in Figs. 5 and 6. At lower resonant frequencies, such as 1.92 GHz, the patterns follow the dipole-like radiation patterns and an interchange in E-plane and H-plane patterns is observed for model 1 and model 4 due to incorporation of additional arms of SAMC. At 7.3 GHz resonant frequency, the patterns get the

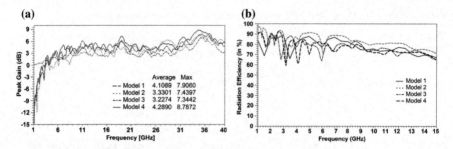

**Fig. 3** Simulation results of **a** peak gain versus frequency and **b** radiation efficiency versus frequency characteristics of antenna iterations

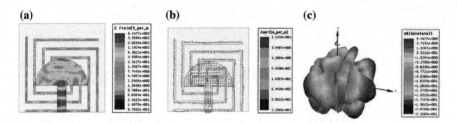

**Fig. 4** Field distributions and radiation patterns of proposed antenna at 20 GHz **a** E-field density (V/m) **b** current distribution **c** 3D polar plot

deformation from omnidirectionality. As the frequency of operation increases, multiple radiating lobes are coming into the scenario. For all of the antenna iterations, the frequency of interest for computation of far-field radiation is based on the applications such as fixed satellite, mobile satellite, and the higher bands such as 17.8, 25 GHz bands are having significance in radio location and radio navigation applications.

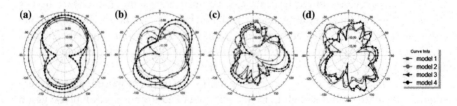

**Fig. 5** Simulated E-plane radiation patterns at different resonating frequencies **a** 1.92 GHz, **b** 7.30 GHz, **c** 17.8 GHz, **d** 25.07 GHz

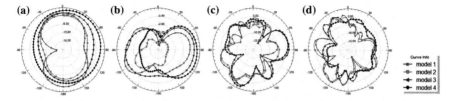

**Fig. 6** Simulated H-plane radiation patterns at different resonating frequencies **a** 1.92 GHz, **b** 7.30 GHz, **c** 17.8 GHz, **d** 25.07 GHz

# 4 Effect of Varying the Position of the Shorting Strips on AMC

The parametric study is performed on the four-arm SAMC structure in the Antenna Model 4. The evolution of antenna models has been incorporated with the shorting of SAMC arms with a metallic strip. The characteristics are studied for the variation in the position of the joining strip. The position of shorting the strip on the four-arm SAMC ground is varied from 0 to 8 mm distance and the analysis results are examined in terms of reflection loss parameter (Fig. 7).

The variation in position "$x_{sh}$" parameter maintains the similar return loss characteristics but the change is observed in the amount of reflection loss offered at the operating bands. The common operating bands can be occurred at the above parametric variations are 2.4, 3.9, 5.9, 7.2–9.7, 12.1–19.8, and 21 GHz to beyond 40 GHz. The parametric study on y-shift parameter of the shorting strip reveals that the variation of position of slot is consistent and makes the antenna geometry operate at 2.33–2.57, 5.87–9, 9.73–20.2, and 22.1–33.3 GHz bands which shows the multiple wideband behavior. Here, the noticeable characteristics are when the horizontal variation is performed, the resonances occurred between 16–21 GHz are shifted when the vertical variation is performed, i.e., the shifting toward the lower bands has been observed.

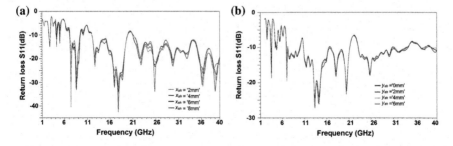

**Fig. 7** Return loss performance of model 4 **a** for parametric study on "$x_{sh}$" (shift in x-direction) parameter **b** for parametric study on "$y_{sh}$" (shift in y-direction) parameter

# 5   Conclusion

The defected ground structure in the form of spiral shape conductor in the ground is investigated with the umbrella-shaped monopole patch element. Experiments are carried on analyzing four types of iterative AMC structures. It is observed that the number of turns incorporated in the spiral shape is the key parameter for occurrence of multi-resonant and multiple operating bands in various regions such as from 2.56–2.88, 4.25–4.4, 7.14–9.48, 12.07–20, and 20.8–40 GHz. These are achieved by controlling the airgap between the spiral elements of AMC. The comparative analysis of all the four models was performed. These models have several applications in wireless communications such as fixed and mobile satellite service, radio location, and navigation applications. The computed antenna gain at different frequencies shows an average value of more than 7 dB. The computed radiation efficiency of the antenna iterations shows gradual decaying characteristics at the higher operating band (Ku, K, Ka bands). The decrease in the radiation efficiency is in synchronous with the return loss characteristics. However, it is needed to be improved for good radiation performance across those bands and it requires further investigations on some other AMC geometries.

**Acknowledgements**  The authors deeply express their gratitude to ALRC Research Centre, Department of ECE, K L University for their encouragement during this work. Further, Madhav would like to express his gratitude to DST through grant ECR/2016/000569 and FIST grant SR/FST/ETI-316/2012.

# References

1. Elsheakh, Dalia Nashaat, et al.: Ultrawide bandwidth umbrella-shaped microstrip monopole antenna using spiral artificial magnetic conductor (SAMC). In: IEEE Antennas Wireless Propag. Lett. 8 1255–1258 (2009).
2. Bhat, Sabzar Ahmad, and R. Madhusudhan Goud.: Analysis of Umbrella Shaped Patch Antenna using Different Ground Shapes for UWB Applications. In: Int. J. of Computer Appl. 114 (2015).
3. Begaud, Xavier, Lepage, A. C.: Wideband low profile antennas and metamaterial. In: International Microwave and Optoelectronics Conference (IMOC 2011), 172, Natal Brazil (2011).
4. P Syam Sundar, Sarat K Kotamraju, T V Ramakrishna, et al.: Novel Miniatured Wide Band Annular Slot Monopole Antenna. In: Far East Journal of Electronics and Communications. 14(2) 149–159 (2015).
5. Lee, S., N. Kim, S. Y. Rhee, Design of novel artificial magnetic conductor as reflector and its SAR analysis. In: Progress In Electromagnetics Research Symposium Proceedings. (2012).
6. Adnam Kaya, Irfan Kaya, "U-shape slot antenna design with high-strength $N_{i54}\,T_{i46}$ alloy," Arabian Journal for Science and Engineering, vol. 41 3297–3307 (2016).
7. Madhav, B. T. P, Harish Kaza, Thanneru Kartheek, Vidyullatha Lakshmi Kaza, Sreeramineni Prasanth, K S Sanjay Chandra Sikakollu, Maneesh Thammishetti, Aluvala Srinivas, K V L Bhavani: Novel Printed Monopole Trapezoidal Notch Antenna with S-Band Rejection. In: Journal of Theoretical and Applied Information Technology. 76(1) 42–49 (2015).

8. M S S S Srinivas, T V Ramakrishna, et al.: Bandwidth enhanced electromagnetic bandgap structure structured closed ground monopole antenna. In: Leonardo Electronic Journal of Practices and Technologies. 28 211–224 (2016).

9. Dalia M. Elsheakh, Esmat A. Abdallah: Compact multiband printed-IFA on electromagnetic band-gap structures ground plane. In: Microwave and Optical Technology Letters. 55(7) 1670–1676 (2013).

10. M Ajay babu, et al.: Flared V-Shape Slotted Monopole Multiband Antenna with Metamaterial Loading. In: International Journal of communications Antenna propagation. 5(2), 93–97 (2015).

11. Van Yem, Vu, Tran The Phuong: Ultra-wide band low-profile spiral antennas using an ebg ground plane. In: International Conference on Advanced Technologies for Communications (ATC). (2015).

12. S S Mohan Reddy et al.: Asymmetric Defected Ground Structured Monopole Antenna for Wideband Communication Systems. In: International Journal of Communications Antenna and Propag. 5(5) 256–262 (2015).

13. Y A N Obadiah, M R Hamid, "Reconfigurable Bandwidth Antenna for LTE Applications," Arabian Journal for Science and Engineering, vol. 41 3655–3661 (2016).

# Implanted Antennas Inside the Human Body: Design, Simulations, and Fabrication

Medikonda Ashok Kumar, Sushanta K. Mandal and G. S. N. Raju

**Abstract** Antennas are used in biomedical applications particularly for EM Radiation energy therapy of different tumors. In this paper, a spiral planar antenna is to operate a MICS (Medical Implanted Communication Service) frequency range of 402–405 MHz. It is designed to implant inside the human body for the treatment of different tumors. The design is carried out using HFSS software. The antenna performance characteristics are analyzed using parameters like i/p impedance, reflection coefficient, return loss, 3D gain, and E-Field distribution. Finally, the prototype and validation are done and theoretical and measured results are compared.

**Keywords** Implanted antenna · Microstrip patch antenna · High-frequency structural simulator (HFSS) · Medical implant communication service (MICS)

## 1 Introduction

Antennas have brought a wide revolution in the wireless applications [1, 2]. The implantable antennas are widely used in medical devices [3, 4]. These antennas have entirely different and specific features when compared with other radiating

M. A. Kumar (✉)
Department of ECE, Centurion University of Technical & Management,
Paralakhemundi, Odissa, India
e-mail: ashok.medikonda@yahoo.com

S. K. Mandal
Department of ECE, Centurion University of Technical & Management,
Bhubaneswar Campus, Rajaseetapuram, Odissa, India
e-mail: skmandal@cutm.co.in

G. S. N. Raju
Department of ECE, Andhra University, Visakhapatnam, AP, India
e-mail: profrajugsn@gmail.com

© Springer Nature Singapore Pte Ltd. 2018
J. Anguera et al. (eds.), *Microelectronics, Electromagnetics and Telecommunications*, Lecture Notes in Electrical Engineering 471,
https://doi.org/10.1007/978-981-10-7329-8_62

systems employed in personal communication system in daily life. The radiating system should be compact and should be capable of operating over multiple bands. They supposed to operate in an environment which is rather complex than free space [3]. In addition, these antennas should be energy efficient and should be consuming minimum power from the integrated battery with the implanted system, thereby leading to enhanced life of the system. Although several antennas have been proposed for implantable medical devices [5], the accurate full human body model has been rarely included in the simulations. In this white paper, an implantable planar is proposed based on the design for communication between implanted medical devices in human body and outside medical equipment. Since the MICS band of 402–405 MHz is a common wireless telemetry for implantable medical devices [6], the proposed PIFA is designed for MICS band in this paper. The main aim of this work is to optimize the proposed implanted antenna inside the skin tissue of human body model and characterize the electromagnetic radiation effects on human body tissues. Simulations have been performed using HFSS [4].

In this paper, a low-profile implantable patch antenna design has been performed. The design is carried out in HFSS EM tool with the design consideration like boundary conditions of human biological system. Reports in terms of S11, current distribution, and SAR are used for analysis. Further, the paper is organized as follows simulation description of the proposed and description of the geometry is presented in Sect. 2. Brief reports presenting to the antenna design along with analysis are presented in Sect. 3. Overall conclusion is mentioned in Sect. 4.

## 2   Design of the Proposed Antenna

The general shape of proposed design for implanted is based on the application. The design of proposed antenna is shown in Fig. 1. The antenna consists of four steps interconnected perpendicular sequentially at the edges while one strip is half the other three and connected to only one strip at one edge. The antenna is excited with a coaxial cable feed system at the edge of the wide strip. The feed partially extends into the substrate area. The dimensions are mentioned in the figure. The substrate has dimensions of 32 mm × 24 mm. the conducting strips facing toward the lengthy substrate side are 24 and 12 mm. Similarly, the conducting strips are 16 mm in length and facing parallel to width of the subtract. The width of the substrate is 4 mm while the material of substrate is Rogers RT Duroid with $\varepsilon r =$ 10.2 and tan $\delta = 0.003$. The other parameters of antenna are considered to be changed within the solution space in order to improve PIFA performance at 402–405 MHz MICS frequency. HFSS Optometric, an integrated tool in HFSS for parametric sweeps and optimizations, is used for tuning and improving the antenna characteristics at the MICS bands inside the ANSYS human body model.

**Fig. 1** Proposed spiral PIF

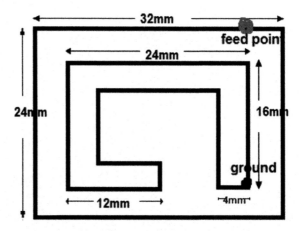

A grounding pin is used at the end of the radiating strip to achieve smaller dimensions. The location of the feed can be optimized along x- and y-axis to match the antenna to 50 Ω over MISC band. The strip lengths are also included in optimizations to provide more degrees of freedom for improving the PIFA performance at the desired frequency band (402–405 MHz).

# 3 Simulation, Results, and Discussion

The simulated model of the antenna is as shown in Fig. 2. For a realistic and more appropriateness, the environment enclosing the antenna is simulated with human skin material like boundary conditions. For simplicity, the case of single-layered skin is considered as the environment encapsulating the antenna. This is corresponding to the case that the antenna is implanted on first layer of the skin.

The current distribution on the surface of the antenna is as shown in Fig. 3. Current toward the short length of the strip from feed point is more than the distribution on the wide length. This is due to the low resistance leading to high current density. SAR is another important parameter which is as shown in Fig. 4.

The resonant characteristics are verified with the corresponding return plots in Fig. 5. The antenna covers the entire MICS band with its max efficiency at 405 MHz.

Validations of the simulated design of the antenna are possible with the fabricated prototype of the same. The simulation strategy is said to be accepted when the simulated results and the corresponding measured results are in good agreement. Accordingly, the design is considered and its corresponding fabrication prototype photograph is shown in Fig. 6. The experimental bench setup is shown in Fig. 7.

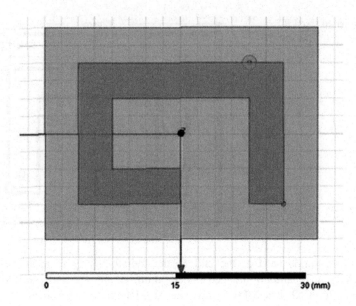

**Fig. 2** Simulated model

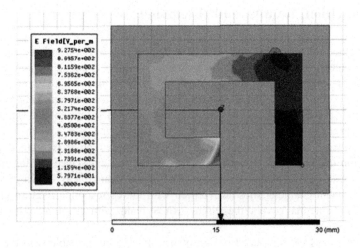

**Fig. 3** Current flow on the surface

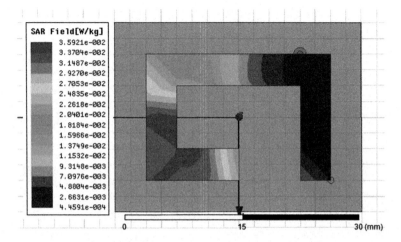

**Fig. 4** SAR distribution of proposed antenna

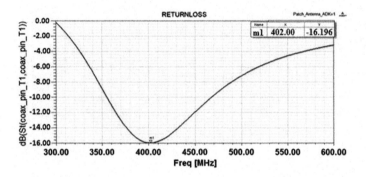

**Fig. 5** Return loss (Frequency = 402 MHz and Total Gain = −16.196 dB)

**Fig. 6** Fabrication of finalized antenna

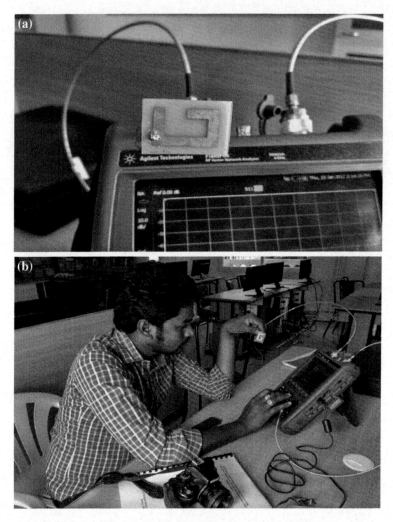

**Fig. 7 a, b** Experimental bench setup of proposed antenna

The measured results pertaining to VSWR and $S_{11}$ are shown in Fig. 8, respectively. The resonant characteristics and the corresponding bandwidth can be read from the $S_{11}$ and VSWR data provided by the network analyzer.

A comparative analysis can be performed using the tabulated data for both simulated and measured reports the designed antenna in Table 1.

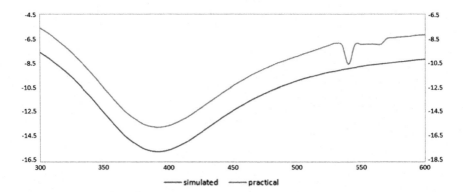

**Fig. 8** Simulated and measured return loss characteristics of fabricated microstrip antenna

| Table 1 Comparison of simulated results with measured result | Case | Frequency (MHz) | $S_{11}$ (dB) |
| --- | --- | --- | --- |
| | Simulated | 402 | −16.196 |
| | Measured | 402 | −14.95 |

## 4 Conclusion

The proposed antenna is designed and validated using the fabricated prototype. The optometric feature of HFSS was employed to optimize the antenna performance at 402–405 MHz. The optimization results demonstrate that the length variation of the radiating strip affects the resonance frequency significantly. The resonance frequency of the implanted antenna was evaluated. For sake of more investigations, an extensive antenna optimization could be performed to obtain miniaturized antenna size and improved antenna performance for implantable medical devices. The designed antennas are very small in size, this favors for miniaturization. The reflection coefficient is as low at −15.2017 dB and −16.196 dB, respectively, for both antennas at 402 MHz. Hence, the geometry is a very good structure for MICS applications.

## References

1. Sudheer Kumar Terlapu, Ch Jaya and GRLVN Srinivasa Raju, "On the Notch Band Characteristics of Koch Fractal Antenna for UWB Applications," International Journal of Control Theory and Applications, Vol. 10, No. 06, pp. 701–707, 2017.
2. P. S. R. Chowdary, A. Mallikarjuna Prasad, P. Mallikarjuna Rao, and Jaume Anguera, "Simulation of Radiation Characteristics of Sierpinski Fractal Geometry for Multiband Applications," International Journal of Information and Electronics Engineering vol. 3, no. 6, pp. 618–621, 2013.

3. W. C. Liu, S. H. Chen, and C. M. Wu, "Implantable broadband circular stacked PIFA antenna for biotelemetry communication," J. of Electromagnetic Waves and Appl., vol. 22, pp. 1791–1800, 2008.

4. Medical implant communications service (MICS) federal register, Rules Reg., vol. 64, no. 240, pp. 69926–69934, Dec. 1999.

5. Changrong Liu, Yong-Xin Guo, and Shaoqiu Xiao "A review of Implanted antennas for wireless Biomedical devices", in FERMAT, Dec. 2012.

6. J. Kim and Y. Rahmat-Samii, "An implantable antenna in the spherical human head: SAR and communication link performance," presented at the IEEE Topical Wireless Communication Technology Conf., Oct. 2003.

# Fusion of Wireless Sensor Images Using Improved Harmony Search Algorithm with Perturbation Strategy and Elite Opposition Based Learning

**H. Rekha and P. Samundiswary**

**Abstract** The idea of image fusion in Wireless Sensor Network (WSN) is to combine the important features of the various images from the multi-focus cameras. Generally, image fusion in WSN consumes more energy and bandwidth to process the images. Hence to reduce the above constraints, it is necessary to reduce the computation time of the image fusion algorithm. In this paper, a histogram-based multi-thresholding with optimization is proposed to fuse the images. Further, an attempt has been made in this paper, by considering the Improved Harmony Search algorithm with Perturbation Strategy (IHSPS) as an optimization technique. In addition to this, the elite opposition based learning is also incorporated with the IHSPS to improve the local search ability. From the simulation results, it is understood that the incorporation of IHSPS with multi-thresholding outperforms the existing multi-thresholding based image fusion algorithms in terms of computation time and image quality.

**Keywords** Image fusion · Entropy · Histogram · Multi-Thresholding
Harmony search algorithm · Opposition based learning · Perturbation strategy

## 1 Introduction

Image fusion has become an important part of image processing applications. It handles the different set of images sensed from various sensors [1–3]. In the last two decades, many researchers have involved in the fusion of multi-focus and multi-sensor images [4–8]. However, as on date, none of the fusion approach is sufficient to fulfill the requirements of the WSN such as energy consumption, processing time

H. Rekha (✉) · P. Samundiswary
Department of Electronics Engineering, School of Engineering and Technology,
Pondicherry University, Pondicherry, India
e-mail: saathvekha16@gmail.com

P. Samundiswary
e-mail: samundiswary_pdy@yahoo.com

© Springer Nature Singapore Pte Ltd. 2018                                          617
J. Anguera et al. (eds.), *Microelectronics, Electromagnetics*
*and Telecommunications*, Lecture Notes in Electrical Engineering 471,
https://doi.org/10.1007/978-981-10-7329-8_63

and bandwidth. Thus, to improve the accuracy of fusion results and to address the above complex problems, some of the existing fusion approaches are integrated with the optimization algorithms. From the literature review [9–11], the image fusion techniques that have utilized the optimization algorithms are successfully reduce the computational complexity. In those studies, wavelet is used as a major part and the optimization is carried out in the approximation band for determining the optimum value of each coefficient. But, in general, the transform-based image fusions are complicated and more time consuming during implementation in WSN. Hence, the researchers are motivated by utilizing the spatial domain based fusion technique for WSN-based applications. For example, Veysel Aslantas and Emre Bendes [12] have appended the Differential Evolution (DE) based optimization algorithm in region-based multisensor image fusion to improve the image quality. Li-Ying Yang [13] has introduced the Particle Swarm Optimization (PSO) in pixel-based image fusion. Another method called bio-geography optimization based multi-focus image fusion is proposed by Ping Zhang et al. [14]. But, the above-mentioned algorithms have not satisfied all the requirements of the WSN because of their global optimization characteristics and low convergence.

To solve the above constraints, this paper proposes a histogram-based multilevel thresholding with optimization for image fusion instead of transform based or spatial based. Here, the histogram of the input images used for fusion is approximated by using the higher number of thresholds. Then, the selection of the optimal thresholds is done by using Shannon's entropy. With an increase in the number of thresholds, the dimension and hence the computational time increases almost exponentially. For this reason, a meta-heuristic optimization algorithm called Improved Harmony Search Algorithm with Perturbation strategy (IHSPS) is used to identify the best thresholds of source images. Elite opposition based learning is also utilized to reduce the computational time by enhancing the local searching ability. The rest of the paper is organized as follows: Sect. 2 deals with the existing image fusion algorithm. Section 3 discusses about the proposed improved harmony search algorithm with perturbation strategy. Section 4 describes the working principle of the proposed image fusion algorithm. Section 5 demonstrates the analysis of the simulation results and comparison of the proposed with the existing techniques and finally the research work is concluded in Sect. 6.

## 2 Discussion About the Image Fusion Algorithm for WSN

The existing multi-thresholding based image fusion algorithm from [15] mainly concentrated on the selection of number of thresholds and the type of optimization algorithm (DE/IHSA). Figure 1 illustrates the general working principle of the existing image fusion algorithm. In this algorithm, a global and objective property of the histogram and Shannon's entropy are used for choosing an optimal threshold from each histogram bin of an input image. By comparing the optimal threshold values of two input images using maximum selection rule, the fusion is performed.

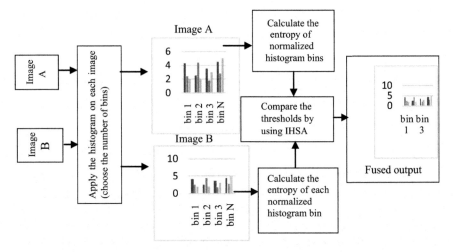

**Fig. 1** Block diagram of the existing multi-thresholding based image fusion algorithm using IHSA/DE optimization

For a perfect fusion, the input images are first partitioned into number of bins using the histogram concept. After the separation of each input image into number of bins, the probability of the pixels from each bin is averaged by using the Shannon's entropy. Although the entropy provides the better result in terms of grouping the pixels, the time required to process the pixels is more and also the average of pixels is not accurate. To improve the probability of the entropy and reduce the computational complexity, the optimization algorithms such as DE and IHSA are tested with the fusion algorithm. However, the computation time by using HMT with IHSA is not enough to save energy consumption for sensor based applications.

## 3 IHSA with Perturbation Strategy and Elite Opposition Based Learning

In order to improve the performance of IHSA, two modifications such as perturbation strategy and elite opposition based learning are included with the IHSA [16, 17]. The detailed explanation of the each modification is given below,

### 3.1 Perturbation Strategy

A perturbation strategy is presented to explore the neighbourhood domain of the current global best harmony. The perturbation strategy is presented here as

$$x_j^{new} = \begin{cases} x_j^{best} + \omega \times rand \times \left(x_j^{r1} - x_j^{best}\right), & if, \ rand_1 < rand_2 \\ x_j^{best} + \sigma \times rand \times \left(x_j^{r2} - x_j^{r3}\right), & otherwise \end{cases}, \qquad (1)$$

where $\omega$ and $\sigma$ are constant number, $rand$, $rand_1$, $rand_2$ belonging to a uniform distribution in the ranges [0, 1], $r_1$, $r_2$, $r_3$ are distinct integers uniformly chosen from the set $\{1, 2,..., size\}$.

## 3.2 Elite Opposition Based Learning

Generally, the opposition-based learning can improve the efficiency of the various optimizations and keep the population diversity. In order to improve the search space, the optimal adaptive value for the IHSA is an elite member which obtained elite opposite solution by opposition-based learning of elite member. Let $B_i$ be the elite members in current populations, then the corresponding of elite opposition based learning is given below

$$\overline{B_{1,j}} = \delta\left(xm_j + yn_j\right) - b_{i,j}, \qquad (2)$$

where $xm_j = \min\left(B_{i,j}\right)$, $yn_j = \max\left(B_{i,j}\right)$

## 4 Proposed Image Fusion Using Multi-thresholding and IHSPS

The working principle of the proposed fusion algorithm is same as above-mentioned existing multi-thresholding based image fusion algorithm. To improve the performance of the existing multi-thresholding based image fusion algorithm, this paper considers IHSPS instead of IHSA. The detailed working process of the proposed image fusion algorithm is explained step by step below,

Step 1: Before implementing the fusion algorithm, the size of the two input images chosen for image fusion has to be checked.

Step 2: The Number of Thresholds (NT) must be assigned before initializing the fusion algorithm.

Step 3: By using the histogram, divide the input images into number of non-overlapping bins according to their pixel values and NT.

Step 4: The threshold value of each bin is calculated by using Shannon's entropy. To maximize the Shannon's entropy the following IHS algorithm with perturbation strategy is incorporated.

Step 5: Initialize the optimization parameters such as Harmony Memory Size (HMS), Harmony Memory Consideration Rate (HMCR), Pitch Adjustment Rate (PAR) and Perturbation parameter (P), stopping criterion (NI).

Step 6: Initialize the Harmony Memory (HM) and elite memory. Consider the individual and elite individual population as $X_j = \{x_{i,1}, x_{i,2} \ldots x_{i,n}\}$ and $X_e = \{x_{e,1}, x_{e,2} \ldots x_{e,n}\}$, respectively,

```
For (i = 0; i < HMS; i ++)
For (j = 0; j < n; j ++)
```

$$X_{i,j}^* = k\left(lb_j + ub_j\right) - X_{e,j},$$

where 'k' is the constant coefficient chosen between [0 1], $lb_j = \min(x_{i,j})$ and $ub_j = \max(x_{i,j})$.

Step 7: Select the fittest individuals from the set of $\{X_{i,j}, X_{i,j}^*\}$ as initial HM.

Step 8: Improvise a new harmony from the HM.

```
For (i=0; i<HMS; i++)
For (j=0; j<n; j++)
If r1≤ HMCR,            %r1,r2,r3 and r ∈(0,1)
```
$$x_j^{new} = x_j^r,$$ %memory consideration
```
If r2≤PAR
```
$$x_j^{new} = x_j^{new} + (2rand - 1) \times (\max(HM_j) - \min(HM_j))$$
```
Else
```
$$x_j^{new} = \max(HM_j) + \min(HM_j) - x_j^r$$ %pitch adjustment
```
End if
Else  
```
$$x_j^{new} = x_{jL} + rand \times (x_{jU} - x_{jL})$$ %random selection
```
End if
End for
 If rand<p % perturbation strategy
```
Select the current best harmony $x^{best}$
```
For j=1 to HMS
```
$$x^{r1}, x^{r2}, x^{r3}$$ are selected from elite memory respectively,
```
If r1<r2
```
$$x_j^{new} = x_j^{best} + \omega \times rand \times (x_j^{r1} - x_j^{best})$$ %ω and σ are constants
```
  Else
```
$$x_j^{new} = x_j^{best} + \sigma \times rand \times (x_j^{r2} - x_j^{r3})$$
```
End if
End for
```

Step 9:   Update the harmony memory. If the fitness of the improvised harmony
          vector is better than that of the worst harmony.
Step 10:  Check whether the stopping criterion (NI) is reached. Otherwise Step 8
          should be repeated.

After the execution of the optimization algorithm, each threshold from the input
images is compared by the maximum selection rule and fused into the resultant
output image. The final output image has more fine details than the individual input
images and also the visual quality of the fused output image is much better than that
of the existing multi-thresholding based image fusion algorithm.

## 5   Simulation Results and Analysis

This section gives the visual and quantitative evolution of the proposed image
fusion method and its comparison with the existing histogram-based
multi-thresholding with IHSA/DE-based fusion method. The proposed method is
tested with pair of multisensor images of size $512 \times 512$ which is shown in Fig. 2.
For better image quality, the number of thresholds taken into consideration is 32.
The parameters used in the simulation of IHSPS are mentioned in the Table 1.

### 5.1   Qualitative Analysis

Generally, the estimation of the performance metrics such as mutual information,
entropy, etc., is not enough to analyse the efficiency of the fused output for some
WSN based applications. Therefore, visual inspections are necessary for the fused
output to judge the amount of recovered information from the input images and
artefacts. From Fig. 3, the visual quality of the existing multi-thresholding image
fusion algorithm using Differential Evolution (DE) and IHSA is poor when compared with the proposed image fusion algorithm with the combination of IHSPS
optimization. Through the qualitative results, the proposed image fusion technique
has produced the better fused output with enhanced image quality than the other

Sensor Image A                                    Sensor Image B

**Fig. 2**  Different set of input sensor images

**Table 1** Parameters of IHSPS

| Parameter | Value |
|-----------|-------|
| HMS | 100 |
| PAR | 0.9 |
| HMCR | 0.95 |
| BW | [0,1] |
| P | 0.25 |
| σ | 0.6 |
| ω | 0.2 |

techniques. Also, the features of the each set of image are perfectly fused with less degradation.

## 5.2 Quantitative Analysis

The visual analysis is not the only performance criterion because sometimes it does not give the clear perception about the image. Hence, some quantitative measurements are also used in this paper to evaluate the performance of the proposed fusion algorithm. They are Mutual Information (MI), Petrovic metric ($Q^{AB/F}$), entropy (H), Fusion Symmetry (FS), Standard Deviation (SD) and computation time [15]. The usage of MI is to calculate how much information of the input images is transferred to the fused image. If the values of the mutual information and the Petrovic metric are higher, it means that the fusion performance of the method is better. Tables 2 and 3 show the various fusion parameter results of the proposed and existing image fusion algorithms for multisensor images. The metrics show that employing the IHSPS optimization with the proposed fusion method leads to the best results in terms of Average Pixel Intensity (API), MI and computation time among all other fusion models. The API and MI are good criteria to justify the quality of the fused output. Further, the proposed method is less expensive and low energy consuming algorithm due to low computation time than that of the existing fusion algorithms.

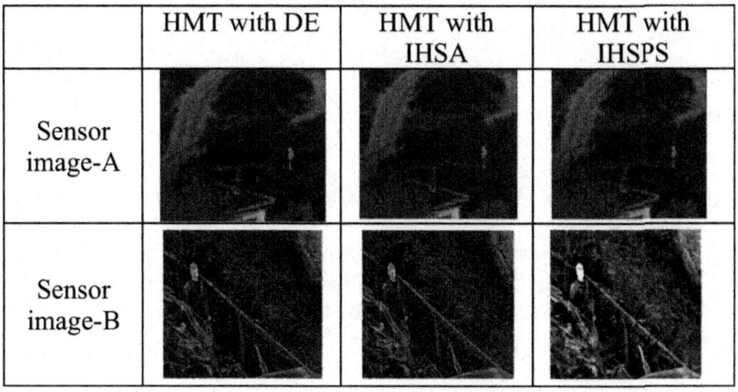

**Fig. 3** Qualitative comparison of the proposed with the existing fusion algorithms

**Table 2** Performance measure of existing and the proposed image fusion algorithm using optimization for sensor image A

| Image Fusion | API | SD | H | MI | FS | CC | $Q^{AB/F}$ | Time (s) |
|---|---|---|---|---|---|---|---|---|
| HMT + DE | 107.26 | 14.52 | 6.572 | 3.091 | 1.895 | 0.652 | 0.652 | 17.33 |
| HMT + IHSA | 107.48 | 15.86 | 6.68 | 3.095 | 1.927 | 0.619 | 0.642 | 2.503 |
| Proposed image fusion algorithm | | | | | | | | |
| HMT + IHSPS | 107.95 | 14.98 | 6.80 | 3.69 | 1.826 | 0.611 | 0.611 | 1.795 |

**Table 3** Performance measure of existing and the proposed image fusion algorithm using optimization for sensor image B

| Image Fusion | API | SD | H | MI | FS | CC | $Q^{AB/F}$ | Time (s) |
|---|---|---|---|---|---|---|---|---|
| HMT + DE | 106.37 | 15.52 | 7.215 | 3.27 | 1.762 | 0.665 | 0.613 | 14.52 |
| HMT + IHSA | 106.43 | 14.91 | 7.027 | 3.96 | 1.854 | 0.675 | 0.621 | 2.061 |
| Proposed image fusion algorithm | | | | | | | | |
| HMT + IHSPS | 107.02 | 14.98 | 7.373 | 3.96 | 1.826 | 0.608 | 0.679 | 1.39 |

## 6 Conclusion

In this paper, an attempt has been made to solve the computational complexity problem of existing multi-thresholding based image fusion algorithm by incorporating the IHSPS optimization. In the IHSPS algorithm, a perturbation strategy and the elite opposition based learning are presented to improve the global and local search capability. The efficiency of this proposed fusion scheme over existing techniques is discussed by conducting the simulation on a different set of sensor images. From the simulation results, it is confirmed that the proposed image fusion algorithm can fuse the sensor images with less computation time and less degradation in image quality than the previously developed image fusion approaches. Hence, the proposed image fusion algorithm is suitable for wireless image sensor-based applications. In the future, the research work can be extended by incorporating the hybrid-based optimization to enhance the fusion performance.

## References

1. M.Abidi and R.Gonzalez, Data Fusion in Robatics and Machine Intelligence. NewYork: Academic, 1992.
2. H. Li, B. S. Manjunath, and S. K. Mitra, "Multisensor image fusion using the wavelet transform," *Graphical Models Image Processing*, vol. 57, no. 3, pp. 235–245, May 1995.
3. G. Piella, "A general framework for multiresolution image fusion: From pixels to regions," *Information Fusion*, vol. 4, pp. 259–280, April 2003.

4. J. J. Lewis, R. J. O'Callaghan, S. G. Nikolov, D. R. Bull, and N. Canagarajah, "Pixel- and region-based image fusion with complex wavelets," *Information Fusion*, vol. 8, no. 2, pp. 119–130, April 2013.

5. Haeberli,P, "A Multi-focus Method for Controlling Depth of Field", *Grafic Obscura*, 1994.

6. Zhi-guo, J., Dong-bing, H., Jin, C., Xiao-kuan, Z, "A Wavelet based Algorithm for Multi-focus Micro-image Fusion", *In Proceedings of International Conference on Image and Graphics (ICIG)*, Hong Kong, China, pp. 176–179, Dec.2004.

7. Ranjith, T., Ramesh, C, " A lifting wavelet transform based algorithm for multi-sensor image fusion",. *CRL Technologies Journal*, vol.3, pp. 19– 22, 2001.

8. Mumtaz, A. & Majid, A. Year, "Genetic Algorithms and Its Application To Image Fusion", *In proceedings of 4th International Conference On Emerging Technologies (ICET 2008)*, 18–19 Oct. 2008.

9. Niu, Y. & Shen, L. "Multi-Resolution Image Fusion Using Amopso-Ii", *Intelligent Computing In Signal Processing and Pattern Recognition*, Springer Berlin/ Heidelberg, 2006.

10. Raghavendra, R., Dorizzi, B., Rao, A. & Hemantha Kumar, G, "Particle Swarm Optimization Based Fusion Of Near Infrared and Visible Images For Improved Face Verification", *Pattern Recognition*, vol.44, pp. 401–411, 2011.

11. X. M. Zhang, L. B. Sun, J. Han, and G. Chen, "An application of swarm intelligence binary particle swarm optimization algorithm to multi-focus image fusion," *Optica Applicata*, vol. 40, no.4, pp. 949–964, 2010.

12. V. Aslantas and R. Kurban, "Fusion of Multi-Focus Images using Differential Evolution Algorithm," *Expert Systems with Applications,* vol.37, no.12, pp. 8861–8870, 2010.

13. Li-Ying Yang, "Pixel level image fusion using Prticle Swarm Optimization with Local Search", *3rd International Workshop on Intelligent Systems and Applications (ISA)*, Wuhan, China, pp. 1–4, May 2011.

14. Ping Zhang, Chun Fei, Zhenming Peng, Jianping Li and Hongyi Fan, "Multi focus image fusion using Biogeography-based optimization," *Mathematical Problems in Engineering,* vol.2015, pp. 1–14, 2015.

15. H.Rekha and P.Samundiswary, "Histogram Driven Fusion of Set of images using Multi-thresholding and Optimization for WSN", *International Journal of Engineering and Technology*, Vol.9, No.2, pp. 548–557, May 2017.

16. Ping Zhang, Haibin Ouyay, Liquan Gao, "Improved Harmony Search Algorithm with Perturbation Strategy", *Proceeding of 27th Chinese Conference on Control and Decision,* Qingdao, China, pp. 5321–5326, May 2015.

17. Luqman Maraaba, Zakariya Al-Hamouz, Hussain Al-Duwaish: Prediction of the Levels of Contamination of HV Insulators Using Image Linear Algebraic Features and Neural Networks, **Arab J Sci Eng, 40 (9) 2609–2617 (2015).**

# A Review of Radio Frequency MEMS Phase Shifters

**G. Srihari and T. Shanmuganantham**

**Abstract** Microelectromechanical systems (MEMS) have been well known in the field of microelectronics and device technology. The future work deals with review of Radio Frequency Microelectromechanical System phase shifter uniqueness and significance in communication structure. A review of common RF Phase shifter topologies using various devices like PIN diodes and Field-Effect Transistors will be addressed. It presents distributed MEMS transmission line phase shifter design with different transmission line architectures and effect of pull-in voltage on RF MEMS phase shifter.

**Keywords** RF MEMS · Phase shifter · FET · CPW · DMTL
ESAs · MMIC · MIM · MAM · BCPW · Actuation voltage
Isolation · Insertion loss

## 1 Introduction

To track, map, and detect the space objects to enable spacecraft to maneuver away from possible collisions, communication links and radar systems are used. Currently, omnidirectional antennas and steerable parabolic reflectors solve these tasks. However, as the new space exploration initiative with prolonged human presence on the Moon and Mars, phased array plays a key role in securing the future of these communication systems. An electronic, mechanical, or automatic switch is employed in Electronically Scanned Arrays (ESAs) to change the phase of every radiating part across an aerial, consequently enabling the radiated beam to guide [1].

G. Srihari (✉) · T. Shanmuganantham
Department of Electronics Engineering, Pondicherry University, Pondicherry, India
e-mail: srihari.nan@gmail.com

T. Shanmuganantham
e-mail: shanmuga.dee@pondiuni.edu.in

© Springer Nature Singapore Pte Ltd. 2018
J. Anguera et al. (eds.), *Microelectronics, Electromagnetics
and Telecommunications*, Lecture Notes in Electrical Engineering 471,
https://doi.org/10.1007/978-981-10-7329-8_64

MEMS technology had wide applications in environmental science, life science, aerospace, and communication for their active characteristics. Mainly, current developments of individual communication devices come with millimeter and microwave frequency range. A network with two ports where the phase differentiation among output terminal and input terminal can be restricted by a dc bias signal [2], known as phase shifter. Present in RF field, MEMS technology has powerfully advantages, since RF MEMS equipments switches, inductors, filters, varactors and describes considerable power decrease for telecommunication, space and radar systems. Ferrite phase shifters have high powers handling capacity, short insertion loss, and high fabrications price as well as complex in nature. Semiconductors such as PIN diode and FET phase shifters are cheap, lesser dimension than ferrites. But due to high insertion loss, their usage was limited. So to overcome the limitations, phase shifters by means of micromechanical system bridges can be used.

Small insertion loss, small drive power, nonstop tenability and small manufacture price, phase shifters are the solution to growth of weightless antennas. Phase shifters design are classified as continuous design and discrete design. Continuous phase shifter gives nonstop changeable phase from 0 to 360° by means of varicaps. Discrete phase shifter gives discrete set of phase delays by means of switches. Two necessities are intended for phase shifters: Stable phase against frequency as well as linear phase against frequency. Radar systems use stable phase designs to process the radar signal and in high accurate automatic systems, and are built with switched networks or else loaded line techniques. True time delay phased arrays uses linear phase designs with switched delay lines [3]. Here, we primarily concentrated on RF phase changing unit cell through episodic assignment of MEMS parallel switches on top of CPW, which is recognized as distributed MEMS t-line [4]. RF control experiences a number of limits, such as restricted process speed, upper operating voltage plus related dependability issues buckling of the ray, and stiction. Miniaturization process was the solution to overcome problem.

## 2 MEMS Phase Shifters

MEMS switches as well as apparatus established abnormal performance on RF as well as UHF frequencies with high isolation, small insertion defeat, and small force power. The MEMS switch produces phase difference by switching signal between two path lengths, shown in Fig. 1. This forms a distributed capacitive switch, which varies useful transmission line capacitance. Bringing MEMS machinery in single design with new dielectric tunable material is able to result in less-weight, low-cost, large-phased arrays through notably reduced production price. The benefit with dielectric tunable material gives continuous alteration in phase. Consequently, signal propagating in two dissimilar path lengths $l°$ and $l° + l$, PIN diode phase shifter produces phase shift. The phase shift corresponding to extra path delay is $\beta l$, with $\beta$ as medium propagation constant [3].

**Fig. 1** Illustration of PIN diode MEMS switched line phase shifter

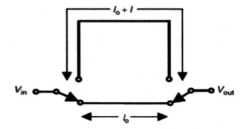

Currently, three types of MMIC phase shifters are available. The switch which is used to pass electrical signals via short or long length of transmission line, is called switched line phase shifters. The later one switches the electrical signal between two circuits that have opposite phase characteristics; typically the two circuits are differentiator and integrator. The last type uses variable capacitors to alter the propagation constant of the transmission line [4]. GaAs MMIC phase shifters have been in development since the early 1980s [5–7], and they are now commercially available as cataloged parts [8–16]. The two, GaAs MMIC phase shifters and MEMS phase shifters have advantages and disadvantages. Because MEMS switches are needy on motion of mechanical arrangement, they are comparatively sluggish toward electronic devices. Electrostatically operated MEMS switch comes with switching rate of 10100 μsec and mechanically operated MEMS switch has switching rate of 1–10 ms. These switching rates are slow in comparison to PIN diode and MESFET switching speeds in the nanoseconds range. Thus, MEMS switches will not be used in ICs such as computer chips, but size of RF IC is dependent on the passive circuits, not the active devices. So, the difference in circuit size between MMIC and MEMS phase shifters is small; X and W-Band phase shifters of each type differ in size by less than a factor of two, as shown in Table 2. Once MEMS technology matures to a commercially viable product, reduction in size will occur.

(1) *Switched delay line phase shifter*: Here, MEMS switches are classified as resistive chain control element (metal–metal) or parallel metal–insulator–metal switch. In high-frequency applications, capacitive parallel switch was regularly used, because it is smaller operating voltage as well as quicker switching rate compared with series switches. Such phase shifters are formed by cascade arrangement of 180, 90, 45, 22.5, and 11.25° phase shifters. Switches are controlled for propagation of RF signals selectively. Distinction between paths lengths gives phase shift.

(2) *Polymer-based MEMS phase shifter*: Because of bridge structure, MEMS phase shifters have a small loss that prevents leakage current. Apart from advantages, lower operating voltage was still a key factor in MEMS structures. Height of MEMS bridge was at least 3 μm to decrease bridge parasitic capacitance and consequences in operating voltage around 100 V through conservative metal bridges. Operating voltage can be decreased through dropping bridge altitude

**Fig. 2** Schematic
representation of switch in
upstate and down state

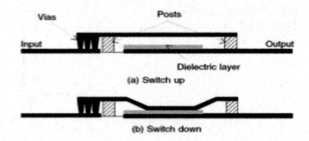

otherwise adopting bridge material comparatively small expandable modulus of polymer.

(3) *Distributed RF MEMS phase shifter*: Here, capacitive parallel switch has slight metal bridge balanced above center electrode of the co-planar waveguide (CPW). DC bias voltage changes bridge height.

A slight dielectric (silicon nitride) coating is placed on base electrode to reduce stiction and gives separation between metal suspension bridge and base electrode. When dc voltage was biased to base electrode of metal bridge, an attractive electrostatic force pulls metal membrane down in direction of base electrode, as shown in Fig. 2.

## 3  Phase Shifter Library

Rebeiz and Scott Barker [17] have given distributed MEMS true time delay transmission line phase shifter using wide band switch for phased array applications (Fig. 3).

A continuous control voltage (10–23 V) applied to middle conductor of co-planar waveguide could alternately achieve up and down states, which causes increasing distributed capacitive loading. The determined results show 118° phase shift at 60 GHz as well as 84° phase shift at 40 GHz.

Barker and Rebeiz [18] presented "Distributed MEMS TTD phase shifter design optimization at 0–60 GHz".

Quartz substrate with varying widths of center conductor was used for phase shifters. The line with 100 μm middle conductor width was exposed to be best for

**Fig. 3** Distributed MEMS
transmission lines on CPW

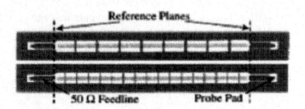

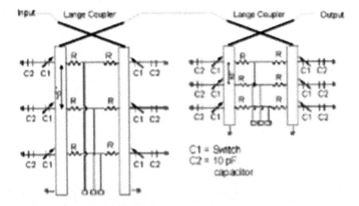

**Fig. 4** Illustration of 4-bit phase shifter with two 2-bit sections

phase shift. The design with Cr = 1.17 produces 360° phase shift at 40 GHz whereas by Cr = 1.3, a phase shift of 1000 is obtained at 40 GHz.

Malczewski et al. [19] presented a paper that gives 4-bit X-band low loss RF MEMS monolithic phase shifter design. These microstrip circuits were created based on reflection methodology via 3-dB couplers on high resistivity silicon. Circuit gives insertion loss 1.4 dB, return loss lower than 11 dB at 8 GHz. The 4-bit phase shifter circuit has two 2-bit reflection sections as shown in Fig. 4. The earlier 2-bit section produces phase shift of 0, 90, 180, and 270° and next 2-bit section gives tiny states: 0, 22.5, 45 as well as 67.5°. Hence, two sections are combined to form a 4-bit phase shifter, switching from 0 to 337.5° in 22.5° steps.

Pillans et al. [20] constructed Ka band 3-bit and 4-bit phase shifter by means of resonant switched t-line microstrip methodology as shown in Fig. 5. Two quarter-wave transformations performed by RF MEMS capacitive switches to allow switching among different delay paths result in phase shifting. 4-bit phase shifter results 0 to 337½° phase shift with 22½° steps and 3-bit phase shifter results 0 to 315° phase shift with 45° steps (Fig. 6).

Hayden et al. [21] draughted the design of 2- and 4-bit wideband distributed microstrip phase shifters on silicon substrate, for DC-18 GHz operation. A microstrip distributed MEMS t-line design, which has periodically stacked series MEMS varactor with predetermined value microstrip radial stub, was presented.

**Fig. 5** Snap of 4-bit MEMS phase shifter

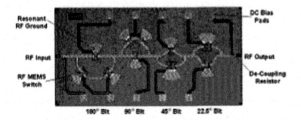

**Fig. 6** Snap of 3-bit MEMS
phase shifter

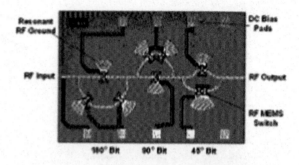

**Fig. 7** Snap of entire 2-bit
phase shifter, 180° part is on
top and 90° part is on bottom

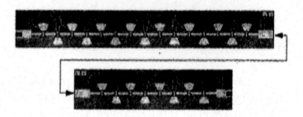

2-bit and 4-bit design results in maximum phase shift of 262° and 333° at 16 GHz
(Fig. 7).

Kim et al. [22] created low loss digital distributed phase shifters by capacitive
parallel switches for V-band applications. To reduce dielectric loss, metal–air–metal
(MAM) capacitors were used in series with MEMS parallel capacitive switches
instead of conventional metal–insulator–metal (MIM) capacitors. By application of
bias through choke spiral inductors to the MEMS shunt switches, operating voltage
of phase shifters was decreased to 15–35 V (Figs. 8 and 9).

Tan et al. [23] gave new RF MEMS series switch low loss phase shifter with
single pole four throw (SP4T) switch design. GaAs substrate was used for fabrication occupies less than 12 mm$^2$ space. 4-bit phase shifter provides near exact
phase shift from 0–18 GHz of 0°, 90.1°, 177.8° and 272° at 10.25 GHz, respectively (Fig. 10).

Tan et al. [24] described low loss X-band true time delay (TTD) MEMS phase
shifter fabricated on GaAs substrate (Fig. 11).

A new approach by means of microstrip t-lines with MIM capacitors as delay
lines was used to decrease circuit dimension and to keep away from high insertion
loss. Here, 2-bit phase shifter attained 90° and 180° phase shifts at 10 GHz,
respectively. It occupies an area of 5 mm$^2$, which was remarkably smaller. Phase
shifter works at 6–14 GHz with improved reflected loss less than-14 dB.

Ramadoss et al. [25] reported an X-band distributed line RF MEMS phase shifter
by means of printed circuit based technology. Hereby application of DC bias
voltage 40 V, the phase shift 31.6° with minimum insertion loss 0.56 dB at 9 GHz
achieved. Such phase shifters are suitable with monolithic integration through large
dimension phased array antenna with low dielectric constant substrates Polyimide

**(a)**                                    **(b)**                                    **(c)**

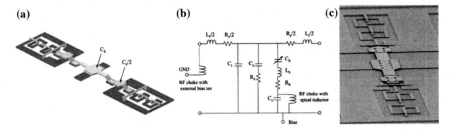

**Fig. 8 a** Diagram **b** correspondent circuit **c** snap of unit cell MEMS bridge and series MAM capacitor phase shifter

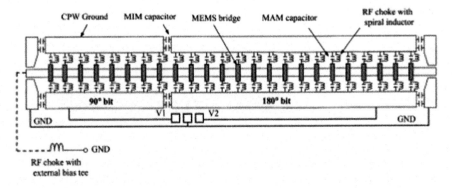

**Fig. 9 a** Diagram of V-band 2-bit MEMS phase shifter

**Fig. 10 a** SP2T and **b** SPST based 2-bit phase shifters

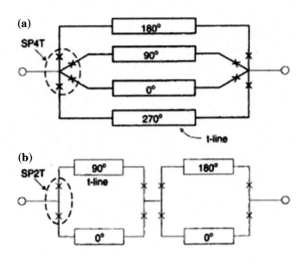

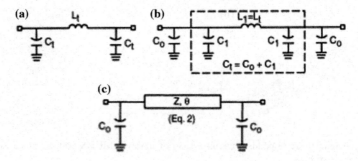

**Fig. 11  a** LC phase shifter network **b** and **c** its semi-lumped networks

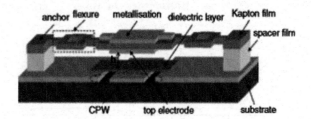

**Fig. 12** Outlook diagram of printed circuit MEMS varactor

or Teflon. Phase shifter units of similar types can be used to give phase shifts of 15°, 30°, 45°, respectively and these are cascaded and could acquire multi-bit phase shifters (Figs. 12 and 13).

Guo et al. [26] suggested a design of MEMS-based millimeter wave phase shifters, consisting a CPW transmission line once in a while stacked with a number of slim metallic membranes. A novel low-resistivity porous silicon wafer (substrate) with low loss micro wave CPW was developed. Oxidized porous silicon (OPS) coated with polyimide-based co-planar waveguide revealed lesser loss than 0–7.5 dB at 0–40 GHz, in contrast with quartz.

Bartolucci et al. [27] gave an experimental consequence of binary DMTL RF MEMS phase shifter using co-planar parallel switches. A new technique based on the image constraint demonstration 2-port network was suggested for modeling device structure. Phase shift of 180° was obtained at $f_0 = 13.7$ GHz. The 180° phase shifter had been realized with six capacitive switches in shunt, which actuates at 50 V (Fig. 14).

Tang et al. [28] gave a new design idea for multi-frequency RF MEMS phase shifter. The suggested method had an insertion loss more than -2 dB and return loss not as much of -10 dB at two working frequency range. Phase shift 180° and 90° was determined for single RF MEMS phase shifter.

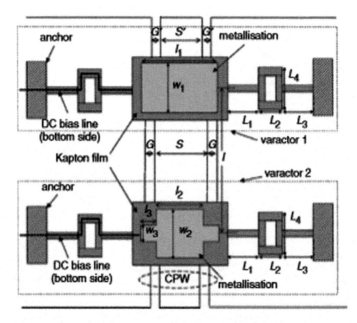

**Fig. 13** Layout of printed circuit based MEMS phase shifter

**Fig. 14** Basic cell circuit
model used for phase shifter

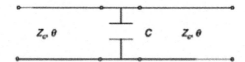

# 4 Conclusion

The review manuscript concludes different MEMS phase shifters. In addition, it also concludes how RF MEMS phase shifters are better than existing phase shifters in dimension, necessary power, and insertion loss. The effect of applied voltage on movement of actuator in different cases was explained. Since MEMS phase shifter comes with small drive power, small insertion loss, and small manufacture cost, electronically scanned array antenna and radar systems are developed with low cost, light weight. Hence, MEMS bridges with different structures act as phase shifter when applied bias voltage is less than pull-down voltage. Among all MEMS phase shifters, DMTL phase shifters are more preferable in all aspects due to reduction of size, required power, insertion loss which show high impact on future research.

# References

1. R. C. Hansen "Phased Array Antennas" John Wiley & Sons Inc., New York, 1998.
2. K. J. Vinoy, K. A. Bose, Vijay K. Vardhan, "RF MEMS and their applications", Pennsylvania State University, USA. K. Elissa, "Title of paper if known," unpublished.
3. Gabriel M. Rebeiz, "RF MEMS Theory, Design, and Technology", John Wiley & Sons Ltd, 2003.
4. Amrita Chakraborty, Sayan Chatterjee "Design of Miniaturized RF MEMS Based Single-Bit Phase Shifter", Comsol conf., Bangalore 2011.
5. Koul and B. Bhat, Microwave and Millimeter Wave Phase Shifters: Volume II Semiconductor and Delay Line Phase Shifters, Artech House, Boston, MA, 1991.
6. S. Ya, C. Jiyi, C. Xiaojian, and L. Jinting, "A compact L band 4-bit MMIC phase shifter," Microwave and Millimeter-Wave Technology Proc. (ICMMT), 1998.
7. M. Rhodes, "Monolithic phase shifter for S-band inter-satellite data relay antenna," IEE Colloquium on Recent Advances in Microwave SubSystems for Space and Sat. Applications, March 18, 1993.
8. A. Lane, "GaAs MMIC phase shifters for phased arrays," IEE Colloquium on Solid State Components on Radar, Feb. 12, 1988.
9. S. Eshelman, A. Malczewski, B. Pillans, J. Ehmke, and C. L. Goldsmith, "X-Band RF MEMS phase shifters for phased array applications," IEEE Microwave and Guided Wave Lett., Dec. 1999.
10. J. Lee, C. Quan, R. Allison, A. Reinehr, B. Pierce, R. Y. Loo, and J. Schaffner, "Array antennas using low loss MEMS phase shifters," 2002 IEEE Ant. and Prop. Society Int. Symp. Dig., June 2002.
11. G. L. Tan, R. E. Mihailovich, J. B. Hacker, J. F. DeNatale, and G. M. Rebeiz,"Low-loss 2 and 4- bit TTD MEMS phase shifters based on SP4T switches" IEEE Trans. Microwave Theory and Tech., Jan. 2003.
12. M. C. Scardelletti, G. E. Ponchak, and N. C. Varaljay, "Ka-Band, MEMS switched line phase shifters implemented in finite ground coplanar waveguide," 32nd European Microwave Conf. Dig., Milan, Italy, Sept. 23–27, 2002.
13. B. Pillans, S. Eshelman, A. Malczewski, J. Ehmke, and C. Goldsmith, "Ka-Band RF MEMS phase shifters," IEEE Microwave and Guided Wave Lett., Dec. 1999.
14. J. B. Hacker, R. E. Mihailovich, M. Kim, and J. F. DeNatale, "A KaBand 3-bit RF MEMS true-time-delay network," IEEE Trans. Microwave Theory Tech., Jan. 2003.
15. L. E. Larson, R. H. Hackett, and R. F. Lohr, "Microactuators for GaAsbased microwave integrated circuits," Int. Conf. on Solid-State Sensors and Actuators Dig. (TRANSDU-CERS'91), June 24–27, 1991.
16. H.T. Kim, J.H. Park, J. Yim, Y.K. Kim, and Y. Kwon, "A compact V-band 2-bit reflection type MEMS phase shifter," IEEE Microwave and Wireless Comp. Lett., Sept. 2002.
17. Gabriel M. Rebeiz, N. Scott Barker, "Distributed MEMS True-Time Delay Phase Shifters and Wide-Band Switches" in IEEE Trans. Microwave Theory and Techniques, November 1998.
18. G.M. Rebeiz and N. S. Barker,"Optimization of Distributed MEMS Phase Shifters" in IEEE MTT-S Dig., 1999.
19. A. Malczewski, S. Eshelman, B. Pillans; J. Ehmke, C. L. Goldsmith "X-band RF MEMS phase shifters for phased array applications", IEEE Microwave and Guided Wave Letters, Dec 1999.
20. "Ka-Band RF MEMS Phase Shifters" S. Eshelman, B. Pillans, A. Malczewski, J. Ehmke, and C. Goldsmith, in Microwave and Guided Letters IEEE, December 1999.
21. J. S. Hayden; A. Malczewski; J. Kleber; C. L. Goldsmith; G. M. Rebeiz "2 and 4-bit DC-18 GHz microstrip MEMS distributed phase shifters" IEEE MTT-S Int. Microwave Sympsoium Dig., Year: 2001.

22. Hong-Teuk Kim, Jae-Hyoung Park, Sanghyo Lee, Seongho Kim, Jung-Mu Kim "V-band 2-bit and 4-bit low-loss and low-voltage distributed MEMS digital phase shifter using metal-air-metal capacitors", IEEE Trans. on Microwave Theory and Techniques Year: 2002.
23. G. L. Tan, R. E. Mihailovich, J. B. Hacker, J. F. DeNatale, G. M. Rebeiz "A very-low-loss 2-bit X-band RF MEMS phase shifter" IEEE MTT-S International Microwave Symposium Dig., (Cat. No.02CH37278) Year: 2002.
24. Guan-Leng Tan, R. E. Mihailovich, J. B. Hacker, J. F. DeNatale, G. M. Rebeiz "A 2-bit miniature X-band MEMS phase shifter", IEEE Microwave and Wireless Components Letters Year: 2003.
25. R. Ramadoss, A. Sundaram, L. M. Feldner "RF MEMS phase shifters based on PCB MEMS technology " Electronics Letters Year: 2005.
26. F. M. Guo, Z. S. Lai, S. Z. Zhu, Z. Q. Zhu, R. J. Zhu, Y. Zheng, G. Q. Yang; A. Z. Li "The study of MEMS millimeter wave phase shifter" Proceedings RAWCON 2002. IEEE Radio and Wireless Conference (Cat. No.02EX573) Year: 2002.
27. G. Bartolucci; S. Catoni; F. Giacomozzi; R. Marcelli; B. Margesin; D. Pochesci "Realisation of distributed RF MEMS phase shifter with very low number of switches" Electronics Letters Year: 2007.
28. Kai Tang, Yu-ming Wu, Qun Wu, Hai-long Wang, Huai-cheng Zhu, Le-Wei Li "A novel dual-frequency RF MEMS phase shifter" Asia-Pacific Symposium on EM Compatibility and 19th Int. Zurich Symposium on Electromagnetic Compatibility Year: 2008.

# Impact Analysis of Blackhole, Flooding, and Grayhole Attacks and Security Enhancements in Mobile Ad Hoc Networks Using SHA3 Algorithm

P. Ramya and T. SairamVamsi

**Abstract** Security is the major concern in mobile ad hoc network (MANET) as it is prone to attacks due to the absence of a centralized authority, dynamic nature, limited resources, scalability, etc. This paper mainly focuses on the effects of various attacks like blackhole, grayhole, and flooding in a MANET. This paper describes the simulation study of various attacks using Ad hoc On-demand Distance Vector (AODV) routing protocol, i.e., how the attacks affect throughput and Packet Delivery Ratio (PDR) using Network Simulator (NS2). The impact of various attacks is also analyzed by increasing the number of attackers. Comparison of various attacks is also presented. The paper also focuses on security enhancements in AODV using SHA3 algorithm.

**Keywords** MANET · AODV · Blackhole · Grayhole · Flooding
Attacks

## 1 Introduction

Mobile ad hoc network (MANET) is a group of mobile nodes [1], they communicate through wireless links. A MANET is a self-configuring, self-healing, and self-organizing network without any centralized authority. However, few nodes may misbehave and therefore can be a significant problem. The nodes may be malicious or selfish in nature. Malicious nodes may drop few or all packets. A selfish node stops forwarding packets in order to conserve its resources like bandwidth, battery life, CPU cycles, etc. Therefore, measures must be taken to isolate these nodes in a MANET [2].

P. Ramya (✉)
GEC, Gudlavalleru, India
e-mail: ramya.sikindhar@gmail.com

T. SairamVamsi
SVECW, Bhimavaram, India
e-mail: vamsi.0438@gmail.com

© Springer Nature Singapore Pte Ltd. 2018
J. Anguera et al. (eds.), *Microelectronics, Electromagnetics
and Telecommunications*, Lecture Notes in Electrical Engineering 471,
https://doi.org/10.1007/978-981-10-7329-8_65

## 2  Security Attacks and Related Work

In this section, we discuss on blackhole, grayhole, and flooding attacks.

**Blackhole attack**: A blackhole intruder [3] initially attacks into the routing path and drops few or all packets instead of forwarding them to its neighbors which causes very small packet delivery ratio. When a source sends a RouteRequest (RREQ) message, blackhole sends a false Route Reply (RREP) message claiming that it has a shortest path to the destination. Whenever the source device receives the message, it sends packets to blackhole node which drops the packets. Figure 1 depicts the blackhole in a MANET. Blackhole attack is implemented by considering few nodes as malicious nodes. The nodes in red are malicious and start dropping packets.

**Flooding attack**: Reactive protocols like AODV depend on RREQ messages. In this type of attack, the attacker either sends lot of RREQ packets or data packets to exhaust the resources like bandwidth which leads to congestion. Figure 2 depicts flooding in MANET. Flooding attack is implemented by sending number of RREQ messages.

**Grayhole attack**: It is a different type of blackhole attack [3]. In this attack, the attacker drops the packets selectively. Initially, the attacker node behaves as a

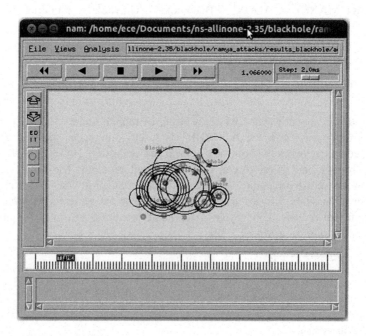

**Fig. 1**  Blackhole in a MANET

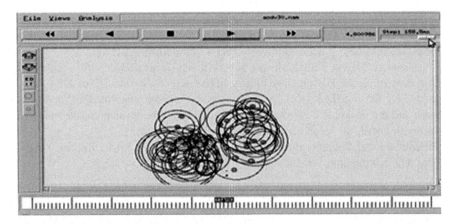

**Fig. 2** Flooding in MANET

genuine node during route discovery and then it may drop all or few incoming UDP packets. As the blackhole node which initially acts as genuine may switch over to malicious node and is difficult to identify it. As stated above AODV is vulnerable to the above attacks.

Many researchers contributed toward the security in MANET. Few of the existing protocols include the following:

SAODV　Secure AODV [4] was proposed to secure AODV messages wherein digital signatures are used to authenticate immutable fields of messages and hash chains to protect information regarding hop count. However, this requires large processing power and often it is slow which are considered as the major drawbacks.

SRP　Secure Routing Protocol [5] incorporates security association between two communicating nodes. The major drawback of this protocol is that it does not consider malicious nodes during communication.

Mane and Gothwal [6] proposed the security improvement for attacks in MANETs using SHA1 algorithm. However, SHA1 is outdated and is prone to vulnerabilities.

## 3　Proposed Algorithm to Improve Security in MANET

In order to enhance security in AODV, SHA3 algorithm is used. SHA3 algorithm was first published by NIST in 2015. SHA3 makes use of Keccak algorithm and also incorporates sponge functions which are a concept of cryptographic hash functions and has arbitrary output length. A Keccak function is represented as Keccak-f[n] where n takes the values 25, 50, 100, 200, 400, 800, and 1600, and uses simple functions like XOR, AND, NOT, and rotate. In the paper, SHA3 is used instead of SHA1 as SHA3 is not vulnerable to few attacks like length extension attacks whereas SHA1 does. Obviously, SHA3 outperforms SHA1.

A Message Digest with hash say K1 as a key is available with source and destination of an AODV message. Then, SHA3 Keccak's hash K2 of K1 is generated, i.e., K2 = SHA3 (K1). AODV message is now encrypted using K2 at source and the packet is now transmitted. Similarly at the destination, the message is decrypted with the same hash, i.e., K2.

Blackhole and flooding attacks can be eliminated using SHA3 Keccak's algorithm. The eliminations are simulated by following the below steps.

- Neighbor nodes are selected for the source and destination using neighbor selection algorithm [7].
- Whenever a node generates a RREQ, RREP, or RERR message, the algorithms and corresponding process are run at sender and receiver which makes use of SHA3 Keccak's algorithm.
- Malicious nodes are injected into the network and are identified as they drop the packets [8]. This is done by using dynamic threshold algorithm. A minimum threshold value is maintained for PDR. If PDR drops below the threshold, it is identified as malicious node.
- All the malicious nodes identified are blacklisted. So a list is created and all the malicious nodes are added to the list. If a route involves a malicious node, the route is discarded and a new route to the destination is identified. Before a route is finalized, there is a need to check the list of malicious nodes and make sure that no malicious node falls in the route. In this way, malicious nodes are isolated and a route which is free of malicious nodes is identified and packets are sent to the destination from the source [9].

## 4   Simulation and Result Analysis

This section deals with the simulation of AODV protocol under various attacks. This paper discusses on the effect of various attacks in MANET through simulation study. The paper demonstrates the impact of various attacks like blackhole, flooding, and grayhole attacks on performance evaluation parameters like throughput, packet delivery ratio (PDR), and end-to-end delay. The simulation is done by varying the network size, i.e., number of nodes and by varying number of attackers also.

A. *Experimental Setup*

The performance analysis is done on Ubuntu operating system using Network Simulator NS2. For performance evaluation, we considered throughput and PDR

**Table 1** Simulation parameters

| Platform | Ubuntu |
|---|---|
| NS2 version | NS2.35 |
| Number of nodes | 10–100 |
| Traffic | CBR (constant bit rate) |
| Packet size | 512 |
| Simulation area | 1200 m × 1200 m |
| Node speed | 30 m/s |
| Mobility model | Two-ray ground |
| Number of attackers | 1–5 |

The analysis is also done by varying number of attackers. Two-ray ground mobility model is considered with node speed of 30 m/s within an area of 1200 × 1200. The packet size considered is 512 bytes and the traffic pattern is constant bit rate. Table 1 shows the all the simulation parameters for network setup.

Simulation Parameters:

B. *Performance Evaluation*

**Throughput**: In general, throughput of the network is the average rate of successful transmission of data packets delivered from source node to destination node. The throughput is as high as possible for any network.

**Packet delivery ratio (PDR)**: PDR of any network is defined as ratio between number of total number of packets received by destination node and total number packets transmitted by source node [10].

**End-to-end delay**: End-to-end delay is stated as the time taken by the packets to reach the destination [11].

The above metrics are considered by changing the size of network and number of attackers.

C. *Simulation Results*

**AODV without attacks**: Figures 3 and 4 depict the variations in throughput and PDR by changing number of nodes. As the nodes are increased, the throughput initially increases and drops when the nodes are further increased. Also the variations in throughput can be observed due to the mobile nature of the nodes.

Figure 5a, b, c depicts the variations in throughput, PDR, and end-to-end delay by varying number of nodes with three blackhole/flooding/grayhole nodes. Figure 6a, b, c shows the throughput, PDR, and end-to-end delay by varying number of attackers. It is observed that there is a significant fall in throughput and PDR when the number of attackers is increased.

Comparison of blackhole, grayhole, and flooding attacks is also done. From the results, it is evident that the attacks have larger impact on MANET. Throughput has

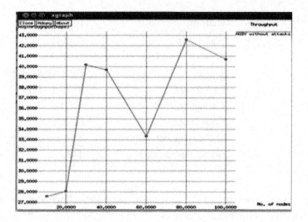

**Fig. 3** Throughput versus number of nodes

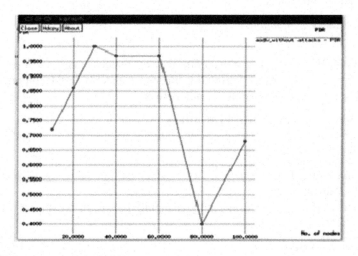

**Fig. 4** PDR versus number of nodes

larger impact in blackhole and end-to-end delay increases in grayhole. As the number of attackers are increased, throughput and PDR drastically fall in both blackhole and grayhole attacks.

**Elimination**: The above attacks are eliminated by using SHA3 Keccak's algorithm as discussed in Sect. 3. As a part of simulation source, destination and neighbors are identified. In the next step, malicious nodes are detected and are added to the blacklist. Then the alarm packets with malicious list are broadcasted so that no malicious node is involved in a route. Finally, a malicious node free route is identified and packets are transmitted from source to destination. Figure 7 shows the complete process of elimination.

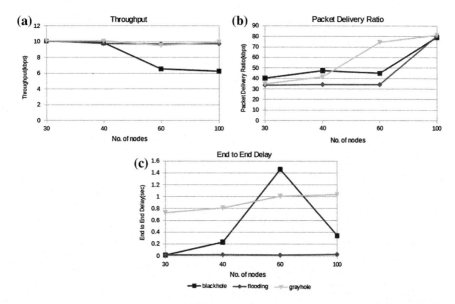

**Fig. 5** **a** Throughput versus number of nodes **b** PDR versus number of nodes **c** End-to-end delay versus number of nodes

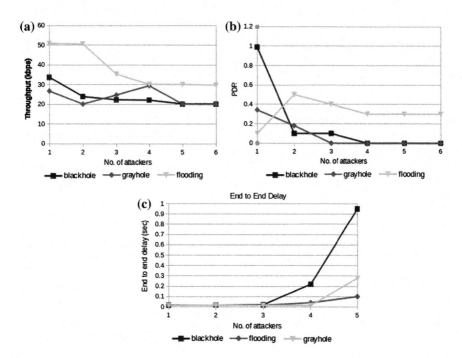

**Fig. 6** **a** Throughput versus number of attackers **b** PDR versus number of attackers **c** End-to-end delay versus number of attackers

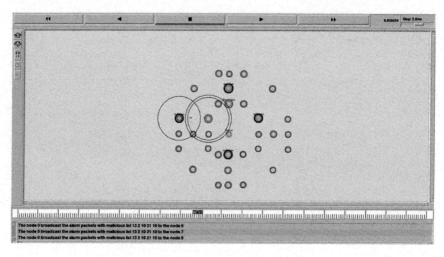

**Fig. 7** Broadcasting alarm packets with malicious list and malicious node free route is selected

## 5 Conclusion

MANET's performance is mainly based on the factors like number of nodes, mobility, number of attackers, bandwidth, propagation model, node positions, etc. The performance degrades when MANETs are prone to vulnerabilities like security attacks. The simulation results conclude that the drastic fall in throughput and PDR is observed as the number of attackers is increased. It is also evident that as the network size increases, it becomes easier for the attackers to intrude and drop the packets. The impact of the attackers also depends on the position of attackers at that instance as they are mobile in nature. We can further improve the work by considering other attacks like rushing and wormhole attacks and also isolate them from the network which is our area of research and the work is in progress.

## References

1. A. Bala, M. Bansal and J. Singh, "Performance Analysis of MANET under Blackhole Attack," *2009 First International Conference on Networks & Communications*, Chennai, pp. 141–145, 2009.
2. Habib, S., Saleem, S., & Saqib, K. M., "Review on MANET routing protocols and challenges", IEEE Student Conference on Research and Development SCOReD, pp. 529–533, 2013.
3. D. Patel and K. Chawda, "Blackhole and grayhole attacks in MANET", *International Conference on Information Communication and Embedded Systems (ICICES2014)*, Chennai, pp. 1–6, 2014.
4. M.G. Zapata, N. Asokan, "Securing ad hoc routing protocols", in Proceedings of ACM Workshop on Wireless Security (WiSe), Atlanta, September 2002.

5. P. Papadimitratos and Z. J. Haas, "Secure routing for mobile ad hoc networks", in Proceedings of SCS Communication Networks and Distributed Systems Modeling and Simulation (CNDS), January 2002.
6. lDadaso Mane and DeepaliGothwal, "Improved Security for Attacks in MANET using AODV", *International Journal of Innovations in Engineering and Technology (IJIET),* vol. 2, Issue 3, pp. 37–44, 2013.
7. Erciyes, K. "Distributed Graph Algorithms for Computer Networks", Computer Communications and Networks, London: Springer, pp. 259– 275, 2013.
8. Mahmoud, Mohamed MEA, and Xuemin Sherman Shen. "Secure routing protocols", Security for Multi-hop Wireless Networks. Springer International Publishing, pp. 63–93, 2014.
9. Ahmed, K. Abu Bakar, M. Channa, K. Haseeb and A. Khan, "A survey on trust based detection and isolation of malicious nodes in adhoc and sensor networks", Frontiers of Computer Science, vol. 9, no. 2, pp. 280–296, 2015.
10. S. Abdel Hamid, H. Hassanein and G. Takahara, "Routing for Wireless Multi-Hop Networks: Unifying Features", SpringerBriefs in Computer Science, pp. 11–23, 2013.
11. Nuha A.S. Alwan, Alaaa S. Mahmood, "Distributed Gradient Descent Localization in Wireless Sensor Networks", Arabian Journal for Science and Engineering (Arab J Sci Eng), Technical note on Electrical Engineering, March 2015, Volume 40, Issue 3, pp 893–899.

# Energy-Efficient and High-Speed Hybrid 1-Bit Full Adder

**Penumatsa Sushma Sri Naga Mowlika and Vemu Srinivasa Rao**

**Abstract** Full adders are one of the best design blocks of the researches to design multiple numbers of applications. This paper presents different types of logic styles of full adders which are been used. The hybrid 1-bit full adder is designed by accepting the different types of logics which are used to get good results. This design is act by applying of cadence virtuoso tools. By using of these logic styles, the capacity and latency of the entire adder are reduced.

**Keywords** Hybrid design · CMOS logic · Transmission gate logic

## 1 Introduction

The full adder is the basic implemented block of any VLSI applications (or) in any normal applications. Adder is an important part of microprocessor. Full adders [1, 2] are mostly used to solve the arithmetic operations like subtraction, multiplication, division, etc. The researchers are also working more on the adder cell so they are keeping a lot of effort on the designs by using the adder cell with different number of transistor count. The logic design styles in general are divided into two categories: (1) Fixed style (2) Unfixed style. The fixed style full adders are dependable, uninvolved with low capacity requirement. The unfixed style full adders are small for on-chip area requirement.

Divergent Logic Styles lean to have the approval when comparing with other Logic Styles. The Significant Logic design styles in the standard dominion are Static CMOS [3], Dynamic CMOS [3], Complementary Pass Transistor Logic (CPL) [1, 4] TGA [5], HPSC adder [6], 4T XNOR [2], Parallel 4 FCA [7], etc.

P. S. S. N. Mowlika (✉) · V. S. Rao
Department of Electronics and Communications, Shri Vishnu Engineering
College for Women, Autonomous, Bhimavaram, India
e-mail: mowlika1102@gmail.com

V. S. Rao
e-mail: vemu1974@gmail.com

© Springer Nature Singapore Pte Ltd. 2018
J. Anguera et al. (eds.), *Microelectronics, Electromagnetics
and Telecommunications*, Lecture Notes in Electrical Engineering 471,
https://doi.org/10.1007/978-981-10-7329-8_66

649

By combining of one (or) more logic styles, it can be referred as hybrid logic design style that may be either of any logic design.

This design can utilize the characteristics of divergent logic styles when compared through execution of the full adder [8]. Coming to the logic styles, the CMOS style adders consist of 28 transistors with ascend and transistor sizing. On the other hand, CPL [5] has the advantages of full swing with 32 transistors but it cannot be applicable for low power applications. This design can be improved with dissimilar specifications as capacity, latency, and number of transistors in this structure. This circuit is implemented by using 180 and 45 nm technology in cadence virtuoso tool. The power dissipation of the full adder is 9.48 μW and delay of the entire adder is 61 ps giving the supply voltage of 1.8 v. This design is also further implemented for 32-bits.

## 2 Details of Adder Structure

This structure is typified by using of three modules. The first and second module represents the XNOR modules which are used to accomplish the sum signal, while the module 3 accomplishes carryout signal which is shown in Fig. 1. In this circuit, every module is formed alone to get the power dissipation, delay and area in optimized way.

### 2.1 Module 1

In this, module 1 is very important for getting of the highest power. Module 1 is also formed because of the reduction of the capacity to the low attainable enhancement for abstaining the reduction of voltage. Figure 2 shows module 1 by using of weak inverters (mp1 and mn1) that means the width of transistors is less.

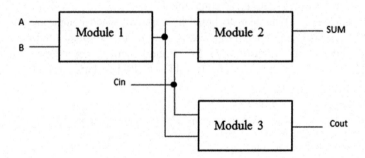

**Fig. 1** Schematic of adder structure

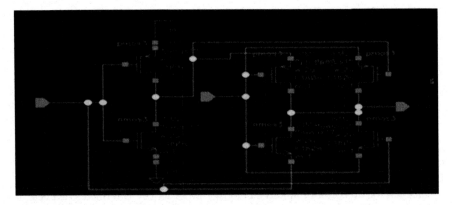

**Fig. 2** Schematic diagram of module 1

By keeping of feeble inverter, the outputs can be occurred to full swing by the transistors of MP3 and MN3. The MP3 and MN3 transistors are connected in an inverter format which can be called as a Level Restoration Circuit.

## 2.2  Module 3

In this full adder circuit, the carry signal can be executed by using the transistor of nmos and pmos which can be shown in Fig. 3. The signal at the input carry promulgates between modules, by scale downing the carry path for the module 3. The resolved use of strong module 3 can be additionally reduction with respect to latency.

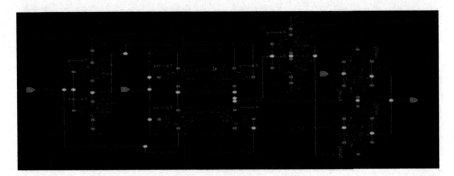

**Fig. 3** Schematic diagram of carry generation module

## 3 Check Pattern Setup for Adder Structure

Figure 4 represents the detailed schematic of the full adder. The results of the adder are sum signal and carryout signal. The sum signal is implemented by XNOR module. The carry output signals are implemented by the module 3. By estimating the logic table of the adder, the carryout signal can be simply written as if both the signals A and B are same then the carry signal cout is equal B else cout will be cin.

The parity inputs A and B can be checked by A XNOR B. The carry-out signal can be simply written as if the signal A and B is same then cout can be occurred as same as B, by using the module 3 by the pull up and pull down nmos and pmos. The carry signal (cin) is mirrored as one of the output of adder by one more module 3 which has the transistor of nmos and pmos. 1-bit adder cell is formed to act best in real-time conditions. This single-bit adder cell is attached in plunge mode; the master cells will not give correct signals to the driven cells. To figure out, the success of the adder cell to its VLSI applications, a simulation test bench setup can be shown in Fig. 7. To provide a practical test bench, the inverters are given at the both of the signals of the check pattern. "The inputs which are given to the cell can be passed between the inverters absorb the effect of signal at the results are passed

**Fig. 4** Schematic diagram of hybrid 1-bit full adder circuit

**Fig. 5** Waveforms of hybrid 1-bit full adder circuit

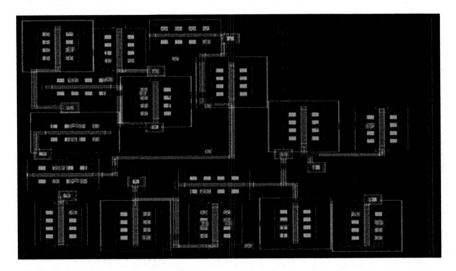

**Fig. 6** Layout of hybrid 1-bit full adder circuit

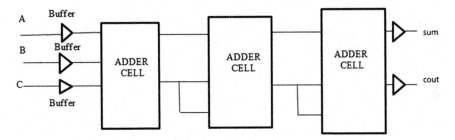

**Fig. 7** Simulation test bench setup

between the inverters which can protect proper loading conditions". The simulation test bench is having three inverters at the side of the input and two buffers at the output side (Figs. 5 and 6).

# 4 Execution Analysis of the Adder

The adder is simulated with respect to using 180 and 45 nm technology and also contrasts with different logic styles which are based on hybrid design. By focusing on the reduction of power, latency of the circuit and PDF, the power consumption of the entire adder cell can be reduced by sizing of the transistor in the inverter circuits. At the same time, the carryout signal can also enhance the sizing of the transistors of transmission gates in the circuit. The width-to-length ratios of the

**Table 1** Results of full adder in 180 nm technology

| Design | Power (uw) | Delay (ps) | Power delay product (PDF) (fj) | Transistor count |
|---|---|---|---|---|
| C-CMOS | 0.349 | 114.1 | 0.003982 | 28 |
| CPL | 24.82 | 257.7 | 6.63961 | 32 |
| TFA | 9.266 | 114.9 | 1.10646 | 16 |
| TGA | 25.3 | 58.55 | 1.48657 | 20 |
| 14T | 11.77 | 89.98 | 1.05906 | 14 |
| 10T | 2.778 | 172.1 | 0.47781 | 10 |
| HPSC | 47.37 | 146.9 | 0.69586 | 22 |
| 24T | 7.113 | 133.6 | 0.09498 | 24 |
| FA_HYBRID | 16.29 | 192.7 | 0.31390 | 24 |
| SERF | 14.18 | 25.19 | 0.03579 | 10 |
| CLRCL | 9.726 | 100.3 | 0.09755 | 10 |

**Table 2** Results of full adder in 45 nm technology

| Design | Power (nw) | Delay (ns) | Power delay product (PDF) (fj) | Transistor count |
|---|---|---|---|---|
| C-CMOS | 23.18 | 8.561 | 0.1984 | 28 |
| CPL | 58.5 | 43.08 | 2.5206 | 32 |
| TFA | 24.77 | 9.107 | 2.255 | 16 |
| TGA | 34.30 | 18.09 | 6.204 | 20 |
| 14T | 20.21 | 8.837 | 0.1785 | 14 |
| 10T | 2.642 | 15.44 | 0.0407 | 10 |
| HPSC | 22.98 | 7.668 | 0.1762 | 22 |
| 24T | 24.17 | 33.81 | 0.81718 | 24 |
| FA_HYBRID | 21.78 | 15.34 | 0.33410 | 24 |
| SERF[2] | 4.24 | 5.108 | 0.21657 | 10 |
| CLRCL[2] | 16.53 | 10.71 | 0.17703 | 10 |

adder circuit are different from one other and also the full adder is also compared with other existed adder are given in Tables 1 and 2. The execution results of the adder cell can be varied from 0.8 to 2.5 v (0.6–1.2 v) in 180 and 45 nm technology.

## 4.1 Power Consumption

Power consumption can be divided mainly into two categories: (1) Static power (2) Dynamic power, and (3) Short circuit power. The static power is formed due to the leakage current. The short circuit power appears from high to low (or) high transition. To reduce the static power, the sizing of the transistor in weak inverter

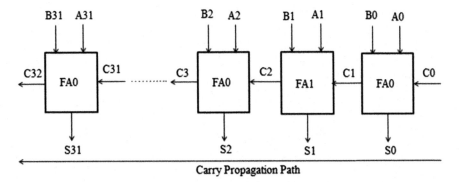

**Fig. 8** 32-bit adder using 1-bit full adder

can be minimized. The static power practically can be occurred when all the supply voltages at the input have given same as that of the dc voltage. When comparing the static and the dynamic of the adder cell, the static power of the adder cell is less.

## 4.2 Delay

The structure of the adder is the basic crew with many of the structures, because if the delays of the carry signal increase, overall speed of the system will be decreased. The input signal is mainly developed because of the transference of the carry and any of the signals A and B. If the signal at the input carry is passing between the module 3, then the track is reduced to decrease the delay.

## 5 Implementation of 32-Bit Adder Structure

The 1-bit full adder is further implemented for 32-bit adder by using the full adder cell circuit in Fig. 8. It looks like a carry ahead structure but, it is does not look like that structure. Power and delay of the 32-bit adder is improved from 1 bit to 32 bit.

## 6 Conclusion

Hybrid 1-bit full adder has very less power and delay when comparing with other logic styles. By using these types of hybrid full adder, the overall power dissipation, area, and delay of entire system will be decreased. In the hybrid full adder, the power consumption is reduced mostly by the XNOR module.

# References

1. C.-K. Tung, Y.-C. Hung, S.-H. Shieh, and G.-S. Huang: A low-power high-speed hybrid CMOS full adder for embedded system, in Proc. IEEE Conf. Design Diagnostics Electron. Circuits Syst., vol. 13, pp. 1–4, Apr. 2007.
2. S. Wairya, G. Singh, R. K. Nagaria, and S. Tiwari: Design analysis of XOR (4T) based low voltage CMOS full adder circuit, in Proc. IEEE Nirma Univ. Int. Conf. Eng. pp. 1–7, Dec. 2011.
3. C. H. Chang, J. M. Gu, and M. Zhang: A review of 0.18-μm full adder performances for tree structured arithmetic circuits, IEEE Trans. Very Large Scale Integr. (VLSI) Syst., vol. 13, no. 6, pp. 686–695, Jun. 2005.
4. R. Zimmermann and W. Fichtner, Low-power logic styles: CMOS versus pass-transistor logic, IEEE J. Solid-State Circuits, vol. 32, no. 7, pp. 1079–1090, Jul. 1997.
5. R. Zimmermann and W. Fichtner: Low-power logic styles: CMOS versus pass-transistor logic, IEEE J. Solid-State Circuits, vol. 32, no. 7, pp. 1079–1090, Jul. 1997.
6. M. J. Zavarei, M. R. Baghbanmanesh, E. Kargaran, H. Nabovati, and A. Golmakani: Design of new full adder cell using hybrid-CMOS logic style, in Proc. 18th IEEE Int. Conf. Electron., Circuits Syst. (ICECS), pp. 451–454, Dec. 2011.
7. P. Prashanth and P. Swamy: Architecture of adders based on speed, area and power dissipation, in Proc. World Conf. Inf. Commun. Technol. (WICT), pp. 240–244, Dec. 2011.
8. K. Navi, M. Maeen, V. Foroutan, S. Timarchi, and O. Kavehei: A novel low-power full-adder cell for low voltage, VLSI J. Integr., vol. 42, no. 4, pp. 457–467, Sep. 2009.

# A Nonvolatile LUT Based on RRAM

Kodamanchili Pavani and Prem Kumar Medapati

**Abstract** The memories which are nonvolatile have been extensively inspected to reinstate SRAM as one of the configuration bits in (FPGA) for long redemption and immediate power on. But, the inequality characteristics in the nonvolatile memories and the extreme logic process convey the dependability problem. In order to overcome the dependability problem NVLUT based on the RRAM (Resistive random access memory) slice is introduced.

**Keywords** Logic in memory · Less power · Nonvolatile LUT
RRAM

## 1 Introduction

The basic nonvolatile memories, such as MRAM (Magnetic Random Access Memory), PRAM (Programmable Random Access Memory), and RRAM (Resistive Random Access Memory), have been examined with best W/L ratios and functional compatibility. Based on the operation in memory criteria, lookup the table, which is the main important block in the FPGA (Field Programmable Gate Array) has been introduced with the nonvolatility. SRAM (Static Random Access Memory) has been used many times but its volatility has only a few applications for long redemption and high power on. Hence, LUT (lookup table) with the nonvolatile memories has been proposed. Different nonvolatile structures with MRAM, RRAM have been considered to straightly restore SRAM in the LUT to gain nonvolatility. But the NV (nonvolatile) SRAM'S Size is greater than SRAM and

K. Pavani (✉) · P. K. Medapati
Department of Electronics and Communications, Shri Vishnu Engineering College
for Women, Autonomous, Bhimavaram, India
e-mail: kodamanchilipavani@gmail.com

P. K. Medapati
e-mail: medapatipremkumar@gmail.com

© Springer Nature Singapore Pte Ltd. 2018
J. Anguera et al. (eds.), *Microelectronics, Electromagnetics
and Telecommunications*, Lecture Notes in Electrical Engineering 471,
https://doi.org/10.1007/978-981-10-7329-8_67

the interruptions for write operations are also crucial to abstain for half selected RRAM cells.

A two-input nonvolatile LUT depended on MRAM in the present mode operation for less power, was suggested by Suziki [1]. Next, he considered a six input nonvolatile LUT by using the serial or parallel magnetic junctions in order to get sufficient margin sensing [2].

Another type of MRAM depended on nonvolatile LUT for runtime reconfigurations was suggested by Zhao [3].

The third type of NVLUT which was defined as hybrid LUT2 was suggested by Ren [4], but in this type, the Roff/Ron of the MRAM is lesser in compared to the PRAM and RRAM, whose output gives the smaller sense margin or more area because of the serial and parallel magnetic junctions. Furthermore the former three types of MRAM-based NVLUT has a disturbance in the parasitic RC, among the preferred line in the MUX (Multiplexer) and the referred path, which may also lead to the failure of nonvolatile LUT. Next, coming to another type that is hybrid LUT2 MRAM cells has the configurations which claim the decoding schematic which is same as before with the functional operation. It contains inputs which are equipped with the different functional blocks and may not be recycled while configuring.

And next when coming to the RRAM, a NVLUT based on the Nano Bridge, was considered by Sakomoto [5], coming to the operation the programming path of the Nano Bridge stakes the single multiplexer with the functional path by accomplishing the W/L of the MOS (Metaloxide semiconductor) transistors in that MUX (Multiplexer), are Appreciable greater than in order to mollify the reset voltage for Ron.

Another type of RRAM-based NVLUT by making the usage of the crossbar array was suggested by Chen [6], but the sneaking currents in the cross-bar array brings greater leakage and very low sensing margin. By considering, the above survey none of them has proved that the capability among the functional variations and memory has less power and low leakage.

## 2 Proposed NVLUT Based on RRAM

In order to rectify the problems mentioned above the NVLUT based on the RRAM is introduced. It consists of SSAVC, TMUX, MRP, RRAM slice, and a Footer transistor (Fig. 1).

### 2.1 Single Stage Sense Amplifier Voltage Clamp (SSAVC)

The single stage sense amplifier is used in diminishing the power it has a sense of basic differential amplifier in compared to the offset voltage, and it is used to

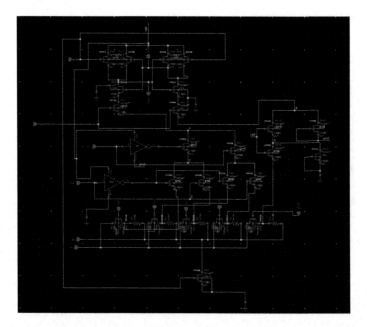

**Fig. 1** Schematic of the NVLUT-based RRAM

convert the state of resistance in RRAM to rail-to-rail operational voltage. The MOS transistors in this type in which two of them are composed of a latch amplifier and the transistors whose input is a CLK in SSAVC are recycled to pre-charge the nodes Q and Q bar to high value, when the value of the CLK, is of low value.

The SSAVC takes very low area, when compared to the two-stage sense amplifier, two transistors are clamped transistors because the same voltage is applied, and the voltage is referred as V bias (Fig. 2).

## 2.2 Tree Multiplexer (TMUX)

Here, a two-input multiplexer is used which is referred as a LUT. It consists of two select lines which are used to select or choose the related RRAM, the operation of the tree multiplexer is same as the NOR operation.

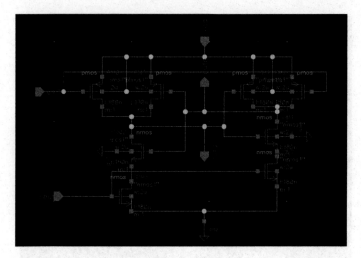

**Fig. 2** Schematic of SSAVC

## 2.3 Matched Reference Path (MRP)

It is used to reduce the parasitic mismatch in RC. There are two reference paths P01 and $P_{reference}$ for better sensing. The parasitic RC's of $P_{reference}$ and P01 must be similar to those transistors in the MRP that take the same size, to mirror the parasitic issues of the off state transistor in the TMUX at the node A.

## 2.4 RRAM Slice (Resistive Random Access Memory)

It consists of five one Transistor and one Resistor combination of Resistive Random Access Memory cells the former four transistors are used for configurations and the remaining transistor is used for reference. It consists of five resistances the first one is considered as Ron and the next three are considered as R off. The Resistance values are in-terms of kilo-ohms and mega-ohms and the last resistance is considered as the R ref, value and it is varied for the values of 10, 20, 50, 150 k, at different voltages of V bias and the power and delay reports are analyzed, the power for different resistance values increases and the delay decreases. The overall product decreases.

## 2.5   Footer Transistor

The Footer transistor is used in-order to start the conversion when the value of the CLK is high. The operation of this transistor is to permit the current to flow while sensing. It is closed in order to prevent the leakage while pre-charging.

## 3   Simulation Results

See Figs. 3, 4, 5, 6 and 7 and Table 1.

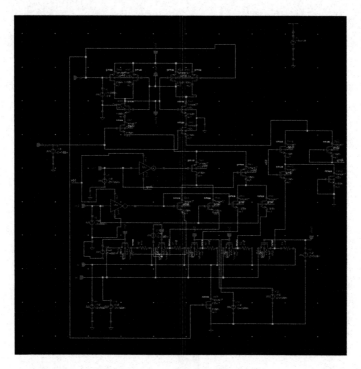

**Fig. 3** Design schematic of NV-LUT-based RRAM when the input voltages are applied

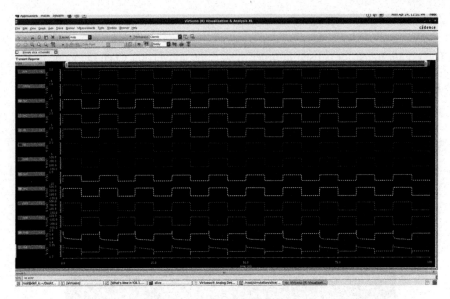

**Fig. 4** Simulation results of the NVLUT-based RRAM when CLK = 1 out b = 0 and Out = 1

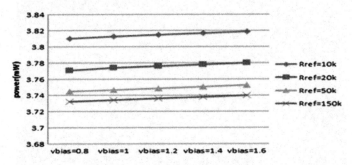

**Fig. 5** Graph for power at different voltages and resistances

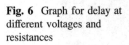

**Fig. 6** Graph for delay at different voltages and resistances

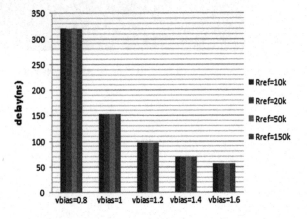

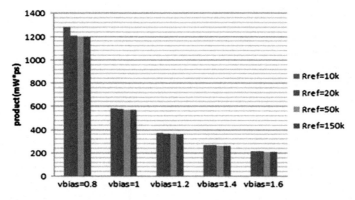

**Fig. 7** Graph for power-delay product of NVLUT

**Table 1** Power and delay reports in 180 (nm) Technology using cadence virtuoso

| V Bias | Rref = 10 k | | Rref = 20 k | | Rref = 50 k | | Rref = 150 k | |
|---|---|---|---|---|---|---|---|---|
| | Power (mW) | Delay (ps) | Power (mW) | Delay (ps) | Power (mW) | Delay (ps) | Power (mW) | Delay (ps) |
| 0.8 | 3.810 | 319.8 | 3.771 | 319.7 | 3.745 | 320.0 | 3.732 | 319.7 |
| 1 | 3.813 | 152.7 | 3.774 | 152.8 | 3.747 | 152.6 | 3.734 | 152.6 |
| 1.2 | 3.815 | 96.7 | 3.776 | 96.54 | 3.749 | 96.81 | 3.736 | 96.81 |
| 1.4 | 3.817 | 70.11 | 3.778 | 70.15 | 3.751 | 70.10 | 3.738 | 70.10 |
| 1.6 | 3.819 | 55.91 | 3.780 | 56.01 | 3.753 | 56.01 | 3.740 | 56.01 |

## 4 Conclusion

The NVLUT based on the RRAM was designed using the cadence virtuoso tool by 0.18 (um) Technology. In 180 nm the power increases for different Rref values and the delay decreases whereas the power-delay Product decreases.

## References

1. D. Suziki.: Fabrication of a non-volatile lookup-table circuit chip using magneto/semiconductor-hybrid structure for an immediate power-up field programmable gate array. in Proc. symp. VLSI Circuits (VLSIC), pp. 80–81, Jun, (2009).
2. D. Suziki, M. Natsuri, T. Endoh, H. Ohno, and T. Hanyu.: Six-input lookup table circuit with 62% fewer transistors using nonvolatile logic-in-memory architecture with series/parallel-connected magnetic tunnel junctions. J. Appl. phys., vol. 111, no. 7, pp. 07E318-1–o7E318-3, (2012).
3. W. Zhao, E. Belhaire, C. Chappert, and P. Mazoyer.: power and area optimization for run-time reconfiguration system on programmable chip based on magnetic random access memory. IEEE Trans. magn., vol. 45, no. 2, pp. 776–780, Feb. (2009).

4. F. Ren.: Energy-performance characterization of CMOS/magnetic tunnel junction (MTJ) hybrid logic circuits. M.S. thesis, Dept. Elect. Eng., Univ. California, Los Angeles, CA, USA, (2011).
5. T. Sakamoto.: A nonvolatile programmable solid electrolyte nanometer switch. in Proc. IEEE Int. Solid State Circuits Conf. (ISSCC), pp. 290–259, Feb. (2004).
6. Y. C. Chen, H. Li, and W. Zhang.: A novel peripheral circuit for RRAM-based LUT, in Proc. IEEE Int. Symp. Circuits Syst. (ISCAS), pp. 1811–1814, May (2012).

# Micro-strip Feed Reconfigurable Antenna for Wideband Applications

M. Dinesh, M. Nandakumar and A. Balachandrareddy

**Abstract** In this paper, a super wideband (SWB) antenna with a bandwidth of 37 GHz (3–40 GHz) with reconfigurable notch band characteristics are presented. The antenna is able to switch from super wide band (SWB), SWB with a notch in the S-band and SWB with a notch in the C-band. This is achieved by incorporating slot and square shaped split ring resonator (SRR) structures which are used to incorporate filtering function to the antenna. The behaviour of the antenna can be reconfigured by activating/deactivating the slot and SRR structure by using photoconductive switches. If both the slot and SRR structure are disconnected from the patch the antenna provides super wideband characteristics with a bandwidth of 3–40 GHz. By activating the slot in the patch structure the antenna produces a notch in the C-band with a bandwidth of 200 MHz in the SWB range. By connecting the SRR structure to the patch, the antenna produces a notch in the S-band with a bandwidth of 700 MHz in the SWB range.

**Keywords** Reconfigurable antenna · Split ring resonator (SRR)
Slot antenna · Super wideband (SWB)

## 1 Introduction

Antennas are an anticipated part of all wireless communication system, which is stand-in as a transducer among transmitter and free space. They are the well-organized radiators of electromagnetic energy into free space. The Recent

M. Dinesh (✉) · M. Nandakumar
Department of Electrical Engineering, Pondicherry University, Pondicherry, India
e-mail: dinesh.isro485@gmail.com

M. Nandakumar
e-mail: nanda.mkumar12@gmail.com

A. Balachandrareddy
Department of ECE, SREC, Tirupati, India
e-mail: balutest@gmail.com

© Springer Nature Singapore Pte Ltd. 2018
J. Anguera et al. (eds.), *Microelectronics, Electromagnetics
and Telecommunications*, Lecture Notes in Electrical Engineering 471,
https://doi.org/10.1007/978-981-10-7329-8_68

wireless communication systems require small profile, low weight, more gain and simple structure antennas to assure reliability, mobility and more efficiency and demand for wireless broadband communication systems with higher data rates have been increasing rapidly. To meet the higher data rates the antenna with broad bandwidth is required [1–4]. The upcoming wireless communication systems require wideband operations, which may cause some interference to the existing systems. To avoid this, broadband micro-strip antennas with band notched characteristics are presented in [5–12]. By considering the space and power requirements of antennas, the antennas with reconfigurable notch band characteristics [13] appropriate for future applications.

The antenna with reconfigurable notch band characteristics is presented in this paper. The antenna consists of 3 modes namely, 'all pass state', 'SWB with a notch in the C-band', and 'SWB with a notch in the S-band'. Depending on the ON-OFF states of the switches the antenna is able to switch from the one mode to others. In this paper, photoconductive switches [14–16] are used because of their better performance compared to other switches. By illuminating the laser light of different power levels the properties of the silicon material changed from semiconductor to conductor. The laser light is coupled through an optical fibre and extends from the ground plane to just underneath the photoconducting element placed on the radiating face of the antenna structure.

This paper structured as follows. Section 2 describes about antenna architecture; Sect. 3 summaries the results and the conclusion can be described in Sect. 4.


## 2 Proposed Structure

The design and analysis of the proposed antenna were performed by using high-frequency structure simulator. The antenna basically consists of rectangular patch antenna with slot and square split ring resonator structure (SRR). The antenna is constructed on FR4 substrate with a dielectric constant of 4.4 and substrate height is 1.6 mm. Figure 1 shows the proposed model of the reconfigurable antenna with dimensions of $35 \times 30 \times 1.6$ mm, on top of substrate patch is constructed with a notch and three switches are placed on the notch to control the current distribution on the patch. The SRR structure is connected to the feed line by using two photoconductive switches connected on either side of the antenna. The parameter description of the proposed antenna is shown in Table 1. The patch, split ring resonator and feed can be calculated based on the standard equations. The feed length can be taken quarter guided wavelength where the resonant frequency is taken a middle value between 3 and 40 GHz.

The photoconductive switches are placed on the slot with dimensions of $1 \times 1 \times 0.28$ mm with a spacing of 2.3 mm each. In this work, photoconductive switches are used for switching purpose. By changing the total electron concentration of the material by illuminating with laser light, the switching action from OFF-state to ON-state is achieved.

**Fig. 1** Proposed structure

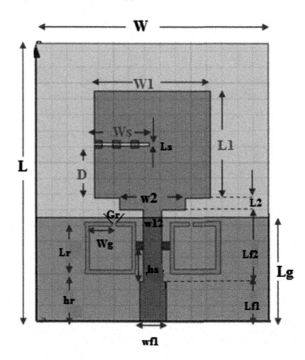

**Table 1** Antenna Parameters

| Parameters | Dimension (mm) | Parameters | Dimension (mm) |
|---|---|---|---|
| L1 | 13.5 | Wf2 | 2.4 |
| W1 | 15 | Lf2 | 9 |
| L2 | 1.5 | Lr | 6 |
| W2 | 8.5 | Gr | 0.5 |
| Ls | 0.4 | Hr | 6.5 |
| Ws | 7.1 | Wg | 1.2 |
| D | 6.6 | W | 30 |
| Wf1 | 3.4 | Lg | 13 |
| Lf1 | 5 | h | 1.6 |

## 3 Results and Analysis

The return loss and radiation pattern measurements are carried out. The return loss characteristics are shown in Fig. 2. The antenna structure consists of 5 switches among which switches 1 and 2 are used to connect the spilt ring resonator structures to the radiating patch and switches 3, 4 and 5 are used to shorten the slot present in the patch. When the switches 1 and 2 are in OFF state and 3, 4 and 5 are in ON state, the SRR structure is not connected to the patch and the slot present in the patch is shorted. In this case, the antenna exhibits 'all pass state' with a band of

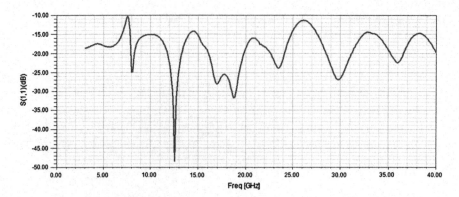

**Fig. 2** Return loss when switches 1 and 2 are in OFF state and 3, 4 and 5 are in ON

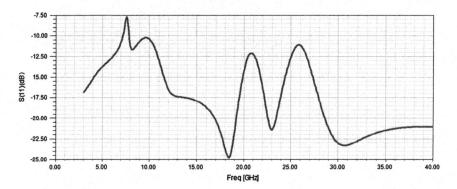

**Fig. 3** Return loss when all switches are in ON state

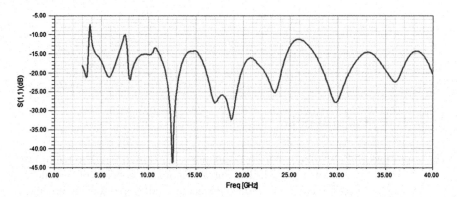

**Fig. 4** Return loss when all switches are in OFF state

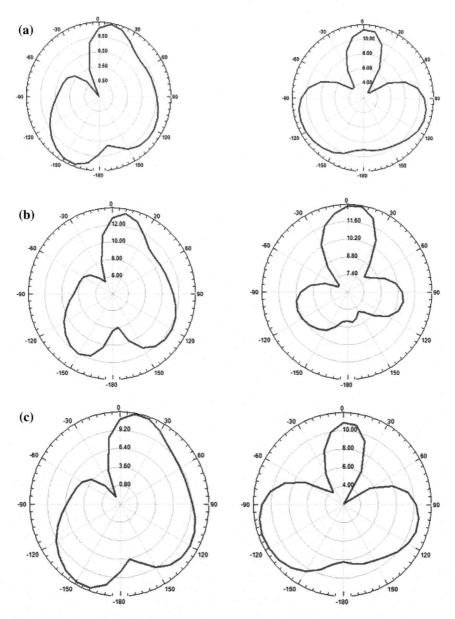

**Fig. 5** Two dimentional E-field (left) and H-field (Right) **a** Switches 1, 2 are in OFF state and 3, 4 and 5 are in ON state **b** All Switches are in ON state **c** All switches are in OFF state

3–40 GHz, i.e. SWB band. When all the switches are in OFF state the SRR structure is disconnected from the patch and slot is present in the patch. In this case, the antenna exhibits a notch band of 200 MHz in the range of C-band in the SWB range and described in Fig. 3. When all the switches are ON state the SRR structure is connected to the patch and slot is not present in the patch. In this case, the antenna exhibits a notch band of 800 MHz in the S-band in the SWB range and described in Fig. 4.

The E-plane and H-plane radiation patterns for three different modes at the resonant frequency of 12.4 GHz is shown in Fig. 5. There is no much difference in the E-plane and H-plane radiation patterns in the first and second modes are shown in Fig. 5a and c but there is a slight difference in the E-plane radiation pattern and noticeable difference in the H-plane radiation pattern in the third mode were shown in Fig. 5b and proposed antenna radiates in bidirectional. In all the three modes moderate gain is obtained.

## 4 Conclusion

In this paper, the antenna with reconfigurable notch band characteristics is presented. The antenna can operate in three modes. The antenna exhibits a huge bandwidth of 3–40 GHz which is used for future UWB applications, such as radar, imaging in medicine and home networking as a wireless personal network (WPAN) in the first mode. In second and third modes, the antenna exhibits different notch bands in the SWB band to avoid the interference caused by the existing systems. The proposed antenna is useful where high bandwidth and large data rates are required.

## References

1. Jianjun Liu, Karu P. Esselle, Stuart G. Hay, and Shunshi Zhong "Achieving Ratio Bandwidth of 25:1 From a Printed Antenna Using a Tapered Semi-Ring Feed" IEEE Antennas and Wireless Propagation Letters, Vol. 10 (2011).
2. Abhik Gorai, Anirban Karmakar, Manimala Pal, and Rowdra Ghatak, "A CPW-Fed Propeller Shaped Monopole Antenna With Super Wideband Characteristics" Progress In Electromagnetics Research C, Vol. 45, pp. 125–135 (2013).
3. Mohamed Nabil Srifi, Symon K. Podilchak, MohamedEssaaidi, and Yahia M. M. Antar, "Compact Disc Monopole Antennas for Current and Future Ultrawideband (UWB) Applications" IEEE Transactions on Antennas and Propagation, Vol. 59, no. 12 (2011).
4. P. S. R. Chowdary, A. Mallikarjuna Prasad, P. Mallikarjuna Rao, and Jaume Anguera, " Simulation of Radiation Characteristics of Sierpinski Fractal Geometry for Multiband Applications," International Journal of Information and Electronics Engineering vol. 3, no. 6, pp. 618–621, 2013.
5. Sudheer Kumar Terlapu, Ch Jaya and GRLVN Srinivasa Raju, "On the Notch Band Characteristics of Koch Fractal Antenna for UWB Applications," International Journal of Control Theory and Applications, Vol. 10, No. 06, pp. 701–707, 2017.
6. J. Liu, K. P. Esselle, S. G. Hay, and S. S. Zhong "Study Of An Extremely Wideband Monopole Antenna With Triple Band-Notched Characteristics" Progress In Electromagnetics Research, Vol. 123, pp. 143–158 (2012).

7. Homayoon orai and nooshinzi valizade shahmirzadi "Frequency and time domain analysis of novel UWB reconfigurable microstrip slot antenna with switchable notch bands" IET Microwaves and Antenna Propagation, Vol. 11 (2017).
8. Anuja V Golliwar, Milind S. Narlawar "Multiple controllable band notch butterfly shaped monopole UWB antenna for cognitive radio" IEEE conference publications, Vol 15 (2016).
9. Yingsong Li, Wenhua Yu "design of a ultrawideband antenna with tunable and reconfigurable band notched characteristics" IEEE 4-th Asia-pacific conference on Antennas and Propagation, pp-142–147 (2016).
10. Naveen Jaglan,, Binod K. Kanaujia, Samir D. Gupta, and Shweta Srivastava "Triple Band Notched UWB Antenna Design Using Electromagnetic Band Gap Structures" Progress In Electromagnetics Research C, Vol. 66, pp. 139–147 (2016).
11. J. T. Bernhard, Reconfigurable Antennas. San Rafael, CA: Morgan & Claypool (2007).
12. D. Peroulis, K. Sarabandi, and L. P. B. Katehi, "Design of reconfigurable slot antennas," IEEE Transaction and Antennas Propagation, Vol. 53, pp. 645–654 (2005).
13. N. Behdad and K. Sarabandi, "Dual-band reconfigurable antenna with a very wide tunability range," IEEE Transaction and Antenna Propagation, vol. 54, pp. 409–416 (2006).
14. Y. Tawk, Alex R. Albrecht, S. Hemmady, Gunny Balakrishnan, and Christos G. Christodoulou "Optically Pumped Frequency Reconfigurable Antenna Design" IEEE Antennas and Wireless Propagation Letters, Vol. 9 (2010).
15. Shou Hui Zheng, Xiongying Liu, manos M Tentzeris "optically controlled reconfigurable band-notched UWB antenna for cognitive radio systems" Electronics Letters, Vol. 50 (2014).
16. Adnan Kaya, Irfan Kaya and Haluk E. Karaca,: U-Shape Slot Antenna Design with High-Strength $Ni_{54}Ti_{46}$ Alloy, Arabian Journal for Science and Engineering, 41(9) 3297–3307 (2016).

# High Performance and Flexible Data Path Architecture for DSP Applications

D. Naga Divya, M. V. Ganeswara Rao, Rajesh K. Panakala
and A. M. Prasad

**Abstract** Digital signal processor (DSP) architecture makes the best use of DSP applications. DSP accelerators are coprocessors to accelerate overall performance of DSP processor by executing certain functions. In the past, several researchers proposed DSP accelerators which make use of data path with flexible computational component (FCC) and flexible pipeline stage (FPS) but, these architectures are not area efficient and low performance. In this paper, a new area efficient, high-speed DSP accelerator is presented. In the proposed architecture, flexible computational unit (FCU) implemented with carry save adder is used to reduce the area and delay. This architecture has implemented on Spartan 3E and results are compared with previous architecture flexible pipelined stage (FPS). The proposed architecture utilized 537 slices out of 4656 slices and critical combinational path delay is 57.47 ns.

**Keywords** Flexible data path · Carry save form · Flexible computational unit (FCU) · Modified booth algorithm · FAMA (Fused add multiply add) unit

D. Naga Divya (✉) · M. V. Ganeswara Rao
Shri Vishnu Engineering College for Women Bhimavaram, Bhimavaram, India
e-mail: devarapallinagadivya@gmail.com

M. V. Ganeswara Rao
e-mail: ganesh.mgr@gmail.com

R. K. Panakala
PVPSIT College, Vijayawada, India
e-mail: rkpanakala@gmail.com

A. M. Prasad
JNTU College for Engineering Kakinada, Kakinada, India
e-mail: a_malli65@yahoo.com

© Springer Nature Singapore Pte Ltd. 2018
J. Anguera et al. (eds.), *Microelectronics, Electromagnetics and Telecommunications*, Lecture Notes in Electrical Engineering 471,
https://doi.org/10.1007/978-981-10-7329-8_69

673

# 1  Introduction

Multimedia and DSP systems gain prominence from past few years. The performance of the computer highly affected by delays produced by the digital signal processing algorithms and digital signal processor performance can be accelerated by using DSP accelerator. In some areas, such as video processing and communication this DSP accelerator used to reduce energy consumption and improve the overall performance. Multipliers are major components in the DSP accelerators due to its high multiplication activity. Multiplier performance is a key to achieve overall system performance, even though the multiplier acts as the slowest element in the system and in DSP the speed of multiplication operation has a huge importance. The reconfigurable architecture is used to execute required operations in DSP applications. The reconfigurable architectures classified into coarse-grained [1] and fine grained architectures. Coarse-grained again categorized as row based [2] and array based. The row-based configuration includes the effective concepts like horizontal parallelism [3], vertical parallelism, and operation chaining [3]. Because of this reason, the row-based reconfigurable architecture chooses for this related work.

In terms of power and performance, the application specific integrated circuits (ASICs) obtain the ideal acceleration, but to accelerate different DSP kernels multiple ASICs are required due to their inflexibility. To avoid this problem, the proposed work focuses on the implementation of flexible accelerator architecture with high performance by using various types of operation templates [4]. A template is called as a group of chained units used to do different operations. So the selection of appropriate template [5] is very crucial to this architecture to execute the large set of operations. This architecture is composed of fused add multiply add (FAMA) unit, where addition and multiplication operations can be done. In this, the multiplier architecture consists of modified booth modules to increase the speed and Wallace tree architecture to reduce the delay. So, by implementing flexible data path area and delay can be reduced when compared to previous data path architecture.

This paper is structured as follows. Section 1 discusses about the DSP accelerator and the types of reconfigurable architecture. In Sect. 2 the proposed flexible data path and its operation are presented. The following section explains the FCU, which is used in the flexible data path to attain the flexibility. In Sect. 4 the experimental results and comparison with existing schemes are given. The final conclusion of this paper is shown in Sect. 5.

## 2  Flexible Data Path Architecture

The flexible data path architecture [6] is shown in Fig. 1. Flexible computational unit (FCU) is the main module in data path architecture used to attain high-performance data path as well as to increase the speed to perform a large set of operations in DSP kernels. Number of FCUs in data path architecture depends upon the limitations of the area that was demanded by the designer.

The architecture of FCU, based on the operation chaining that can be selected from operation templates library, where the operation templates library consists of different templates. Each template comprises of adder or subtractor modules along with multiplier section. For DSP data paths 16-bit length is much more suitable. So, for every template which is selected for the required operations, the input operand bit size is 16-bits. The FCU in data path architecture also operates on 16-bit operands. The arithmetic operations can be done in FCU by consisting of multipliers and adders. Other than FCU, the accelerator architecture consists of Control unit, Register bank, Interconnection Network, and CS to Bin module. Control unit provides the control signals to remaining modules to perform the operations and the register bank in this is used to store the intermediate results obtained from FCUs and the operand sharing values among the FCUs [7]. Interconnection network used to share the values from register bank to FCU modules by using multiplexers in it and to convert carry save operands to binary numbers useful for entire architecture the CS to Bin conversion module is used.

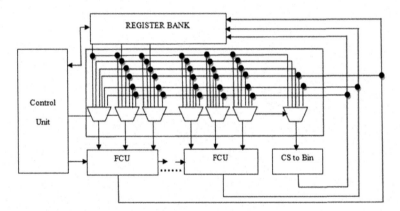

**Fig. 1** Block diagram of the flexible data path. In this whole architecture, operation depends upon the control signals produced by control unit

## 3 Flexible Computational Unit (FCU)

The flexible computational unit is shown in Fig. 2. The FCU operations depend upon the selected operation templates and by placing the FCU in data path architecture it enhances the flexibility of the accelerator.

The FCU comprises adders, multiplier, and multiplexers. In this architecture, the configuration word ($CL_3$, $CL_2$, $CL_1$, and $CL_0$) plays a vital role these bits act as a selection line for multiplexers and carry-in for adders. Data flow model of FCU shown clearly in Fig. 3. Here, MUX0 can be used, it 2's complement the input if

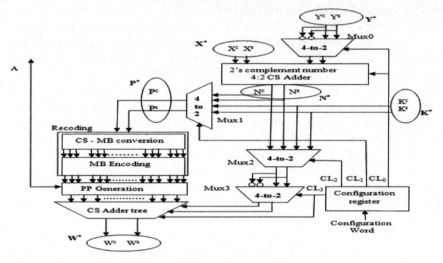

**Fig. 2** Flexible computational unit has been designed in FAMA unit to attain high performance

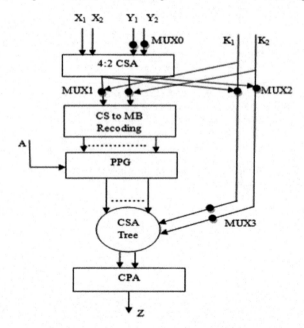

**Fig. 3** Data flow model of FCU

$CL_0 = 0$ or the output is the same as input if $CL_0 = 1$. In this, 4:2 Carry Save Adder (CSA) [8] does the addition operation quickly with implying on minimum carry propagation. This adder enables the addition operation of more than two operands and it executes the two outputs nothing but the sum and carry ($N^* = \{N^C, N^S\}$). $N^* = (X^* + Y^*)$ if $CL_0 = 0$ or $N^* = (X^* - Y^*)$ if $CL_0 = 1$ where ($X^* = \{X^C, X^S\}$, $Y^* = \{Y^C, Y^S\}$). The output of CSA adder and the other operands ($K^* = \{K^C, K^S\}$) are given as input to MUX1 and MUX2. If $CL_1 = 0$, MUX1 output $P^* = N^*$ else $P^* = K^*$ and output of MUX2 is $N^*$ if $CL_2 = 0$ or $K^*$ if $CL_2 = 1$. Then $P^*$ is given as input to recoding technique to generate partial products. For multiplication modified booth multiplier is used to perform high-speed multiplication by using modified booth algorithm [9]. The outputs through CSA are in the form of carry save. By using CS to MB recoding technique operands in carry-save form are converted into MB form and when MB encoding output multiplies with operand A, it produces partial products. From MUX2 the output is given as input to MUX3 and it complements the input if $CL_2 = 1$ and if $CL_2 = 0$ the output occurs without complementing the input. A CSA tree adds MUX3 output and partial products. Finally, the CSA tree output ($W^* = \{W^C, W^S\}$) is added by using Carry Propagate Adder that is a Ripple Carry Adder (RCA). Then, final output Z can be obtained. For different configuration words, different operations can be done as shown in Table 1.

| Table 1 FCU operations for different configuration words | Configuration | FCU operations |
|---|---|---|
| | 0000 | $(X^* + Y^*) * A + (X^* + Y^*)$ |
| | 0001 | $(X^* - Y^*) * A + (X^* - Y^*)$ |
| | 0010 | $K^* * A + (X^* + Y^*)$ |
| | 0011 | $K^* * A + (X^* - Y^*)$ |
| | 0100 | $(X^* + Y^*) * A + K^*$ |
| | 0101 | $(X^* - Y^*) * A + K^*$ |
| | 0110 | $K^* * A + K^*$ |
| | 0111 | $K^* * A + K^*$ |
| | 1000 | $(X^* + Y^*) * A - (X^* + Y^*)$ |
| | 1001 | $(X^* - Y^*) * A - (X^* - Y^*)$ |
| | 1010 | $K^* * A - (X^* + Y^*)$ |
| | 1011 | $K^* * A - (X^* - Y^*)$ |
| | 1100 | $(X^* + Y^*) * A - K^*$ |
| | 1101 | $(X^* - Y^*) * A - K^*$ |
| | 1110 | $K^* * A - K^*$ |
| | 1111 | $K^* * A - K^*$ |

# 4 Experimental Results

XILINX design suite 14.5 ISE is used to implement the architecture, the proposed architecture captured using Verilog HDL and simulated using models in simulator. Figure 4 shows the simulation result for Carry-Save Adder, where it gives two outputs by adding four inputs and simulation results of CS to MB recoding technique shown in Fig. 5. It converts the operands in carry-save to the modified booth form. The simulation results of the FCU shown in Fig. 6. In FCU by using adder four operands can be added and given to multiplier then the output of adder multiplies with multiplicand and finally, added all the partial products by using Wallace tree and that FCU is placed in the Flexible data path. A control unit in the flexible data path gives the control signals to register bank and from the register bank outputs are given to FCU using the interconnection network as shown in Fig. 7.

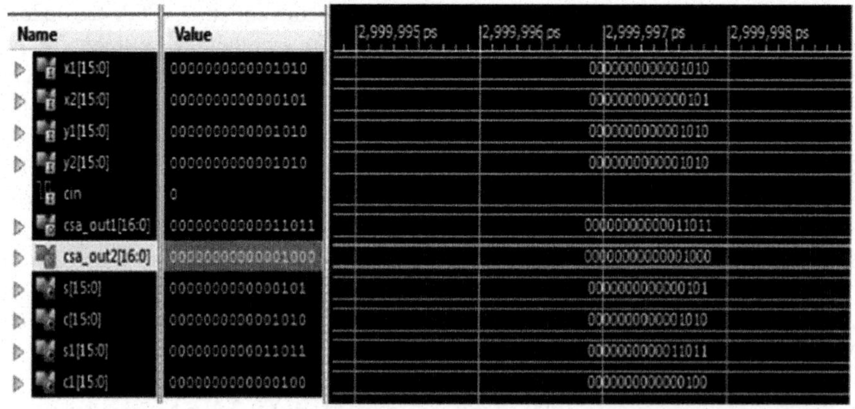

**Fig. 4** Simulation results for carry save adder

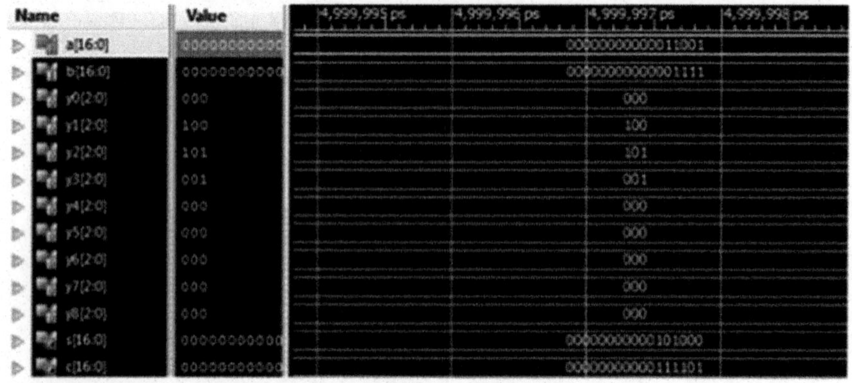

**Fig. 5** Simulation results for CS to MB recoding technique

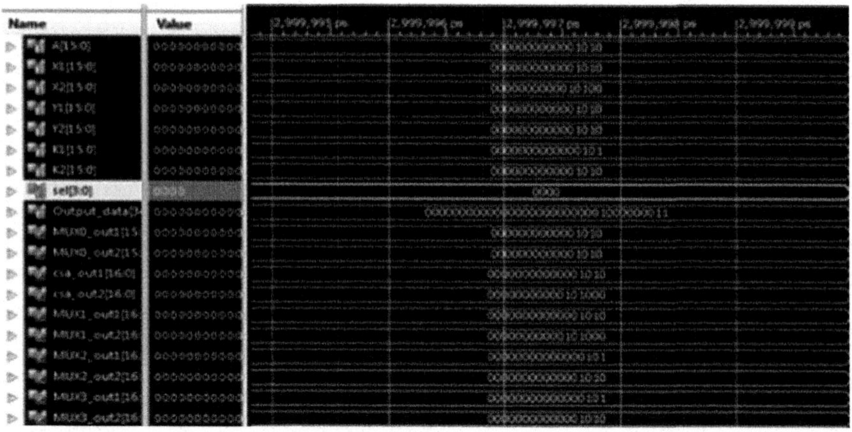

**Fig. 6** Simulation results for FCU

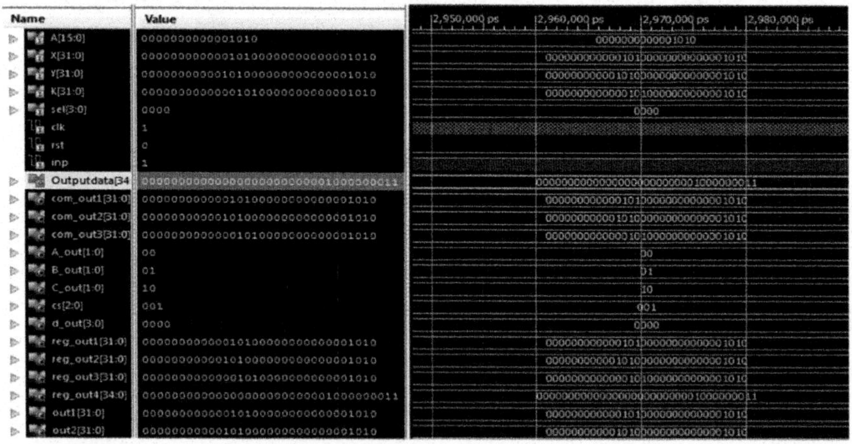

**Fig. 7** Simulation results for flexible data path

The RTL Schematic of FCU shown in Fig. 8 and RTL Schematic of the flexible data path shown in Fig. 9.

The efficiency of proposed FCU drawn from the comparison of the area and delay Table 2, which shows the comparison between existing FPS method implemented using unified cells and proposed FCU implemented using 4:2 carry save adder with respect to area and delay.

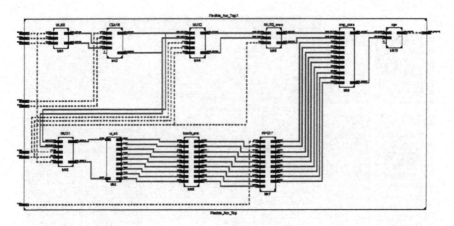

**Fig. 8** RTL schematic of FCU

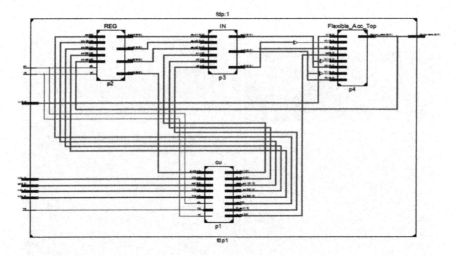

**Fig. 9** RTL schematic of flexible data path

| Parameters | Existing | Proposed |
|---|---|---|
| Area (no of slices) | 699 (100%) | 537 (77%) |
| Delay (ns) | 78.24 (100%) | 57.47 (73.4%) |

**Table 2** Area, Delay of existing and proposed FCU

## 5 Conclusion

In this paper, flexible data path architecture which makes use of the flexible computational unit is implemented to enhance the flexibility of the processor. To make fast additive chaining operations flexible computational unit uses carry save adder and CS to MB recoding technique used to directly recode the input that is in Carry-Save form to modified booth form. For multiplication in this modified booth form is used. It is in radix 4 representation, this representation is better than radix 2 representation. Proposed architecture reduces the area and delay when compared to previous architectures.

# References

1. G. Theodoridis, D. Soudris and S. Vassiliadis.: A Survey of Coarse—Grain Reconfigurable Architectures and Cad Tools. S. Vassiliadis, D. Soudris (Eds.), Springer, Fine and Coarse Grain Reconfigurable Computing, pp. 3–149, (2007).
2. H. Singh, M. H. Lee, Guangming LU, F. J. Kurdahi, N. Bagherzadeh and E. M. Filho: Morphosys: An Integrated Reconfigurable system for Data-parallel and Computation- intensive applications. in computers, IEEE Trans, Vol. 49, no. 5, p. 465, May (2000).
3. Galanis. M, Theodoridis. G, Tragoudas. S and Goutis. C.: A high performance data path for synthesizing DSP kernels. in comp.-Aided Desg. Integ. circuits syst., Vol. 25, no. 6, pp. 1154–1163, June (2006).
4. Lodi. A, Toma. M, Campi. F, Cappelli. A, Canegallo. R and Guerrieri. R.: A VLIW Processor with Reconfigurable Instruction Set for Embedded Applications. In Solid State Circuits, IEEE journal, Vol. 38, no. 11, Nov (2003).
5. Peter A. Milder, Franz Franchetti, james C. Hoe and Puschel. M.: Formal Datapath Representation and Manipulation for implementing DSP Transforms. in DAC, (2008).
6. K. Tsoumanis, S. Xydis, G. Zervakis and K. Pekmestzi: Flexible DSP Accelerator Architecture Exploiting Carry Save Arithmetic. IEEE Trans on Very Scale Integration (VLSI) systems, (2015).
7. Giovanni Ansaloni, Paolo Bonzini and laura pozzi, member: EGRA: A Coarse Grained Reconfigurable architectural Template. IEEE trans on Very Large Scale integer (VLSI) systems, Vol. 19, No. 6, June (2011).
8. Junhyung Um, Taewhan Kim.: An optimal of Carry Save Adders in arithmetic circuits. In computers, IEEE Trans, 50(3):215–233, (2001).
9. K. Tsoumanis, S. Xydis, C. Efstathiou, N. Moschopoulos and K. Pekmestzi.: An Optimized Modified Booth Recoder For Efficient Design of the Add-Multiply Operator. In circuits and systems, IEEE Trans, Vol. 61, no. 4, (2014).

# Ultralow Power 8T Subthreshold SRAM Cell

**Devarapalli Mounika and Akondi Narayana Kiran**

**Abstract** Static Random Access Memory (SRAM) is an important component in these systems therefore ultralow power SRAM has become popular. An Eight-transistor (8T) SRAM cell achieved high data stability in subthreshold operation. The single ended with dynamic feedback control 8T SRAM Cell was implemented with less power consumption verified at all process corners. The standard deviation and mean calculations performed for static noise margins by using Monte Carlo simulation at 300 mV in cadence 45 nm technology.

**Keywords** Single-ended · Ultralow power · Static noise margin (SNM) Static RAM (SRAM)

## 1 Introduction

The embedded memory mainly presents in portable microprocessor control devices, which occupies huge space of the system on chip (SOC). Ultralow power consuming circuits are used in portable systems to maintain batteries for long period of time [1–7]. By using the optimizing architectures and new circuit topologies, we can minimize the power consumption. Mainly, in Static RAMS the circuit operates in subthreshold regime. The SRAM cell has a severe data stability problem in subthreshold regime. It is very difficult to operate the 6T SRAM cell at ultralow voltage power supplies due to read disturb problem. True node decoupling is the effective way to abolish this read disturbs problems in 6T SRAM. The read decoupling approach is put to use by the conventional 8Transistor cell. The read decoupled 8T cell which proposes read static noise margin (RSNM) and hold static

D. Mounika (✉) · A. N. Kiran
Department of Electronics and Communications Engineering,
Shri Vishnu Engineering College for Women, Autonomous, Bhimavaram, India
e-mail: devarapallimounika93@gmail.com

A. N. Kiran
e-mail: narayanakiranakondi@gmail.com

© Springer Nature Singapore Pte Ltd. 2018
J. Anguera et al. (eds.), *Microelectronics, Electromagnetics
and Telecommunications*, Lecture Notes in Electrical Engineering 471,
https://doi.org/10.1007/978-981-10-7329-8_70

noise margin (HSNM). The probability of an increase in read/write failures leads to increase the leakage currents. A single ended 5T SRAM cell was introduced with less area and more stand by the power to reduce the energy exhaustion compared with 6T SRAM cell. In addition, 5T SRAM cell read stability is less when compared with the 6T cell. Various approaches have been introduced to relieve the above issues related to the 5T cell. While trying to optimizing the noise against all process corners at all operations the 5T Cell degrades its read ability compared with 6T SRAM cell. But still, none of the approaches like boosted supply, 7T dual Vth, unsymmetrical 8T, and cross-point data-aware 9T were not satisfied the read and write stability in subthreshold region.

## 2   Subthreshold 8T SRAM Cell Implementation

A new subthreshold 8T SRAM cell was implemented which operates at ultralow voltages to improve the right ability and to avoid read disturb, the 8T cell was executed with by using single ended write with dynamic feedback cutting scheme. The 8T Cell saves more power as compared with other cells. Other than this, Monte Carlo simulations for 8T were plotted to observe the accuracy of the circuit. The circuit simulations are done in CMOS 180 and 45-nm technologies. An 8T SRAM Cell is presented in the Fig. 1.

The 8T consists of one cross-coupled inverter pair, which each inverter may be constructed dependent upon of three cascaded transistors M1–M2–M4 and M8–M6–M5. One nmos transistor M7 was controlled by the write word line (WWL), used to carry the data from write bit line (WBL). The data will be transferred from the cell to output by using the read bit line (RBL). The two extra transistors (M6 and M2)

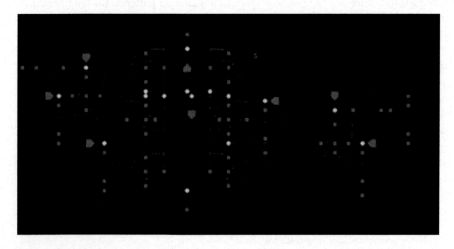

**Fig. 1**  8T SRAM Cell

are connecting in 8T SRAM cell compared with 6T SRAM. Those two transistors are referred as access transistors of the present circuit.

## 2.1 Write Operation

Mainly the write operation will accomplish to write or load the particular information into the memory or system. It is performed by feedback cutting scheme as Write 0 operation and Write 1 operation in this circuit. In this method, At the time of write 1 operation, FCS1 gets low which turns off M6. When the RWL is at low, FCS1 made low and FCS2 made high, then M2 conducts and reciprocal of Q will be connected to VSS. During write 0 operation we should maintain WWL = 1, FCS2 = 0 and makes WBL = 0 by connect to the ground. The low FCS2 makes the QB = 0, then the current from PMOS M1 charges the QB form 0 to 1. The write time mainly depends on power supplies which are given to the circuit. The write time will increase by increasing the power supplies. The write time is high at the process corner of SS. While in case of power consumption, it is high at the process corner of FF. The power consumption is more in write 0 operation compared with write 1 operation (Figs. 2 and 3).

## 2.2 Read Operation

The read operation is executed to observe the information from the circuit. This particular operation is achieved by recharge the RBL and by activates the RWL. During read operation, the FCS1 and FCS2 made high, that turns the QB to share some amount of charge (QB = 1), When WWL is 0 there is no path will be present

**Fig. 2** Write 0 operation simulation results

**Fig. 3** Write 1 operation simulation results

between WBL and Q. The WWL FCS1 and FCS2 were low at a read operation. The read operation mainly deals with the input RWL. If RWL low, due to the positive feedback scheme the respective states are at conditions of (Q = 1 and QB = 0). During read 0 operations, RBL maintains high value then the sense amplifier gives 0 as its output. The read time will be resolute by activating the RWL, The maximum read time is observed at SS corner when compared with all other process corners. Same as write power, the FF corner condition consumes more power when compared with other process corners. The read time is the amount of delay while functioning the circuit. And the read power is the amount of consuming power at particular operation (read operation) (Fig. 4).

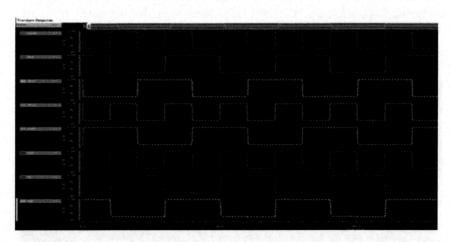

**Fig. 4** Read operation simulation results

## 2.3 Hold Operation

The hold operation is performed to hold the information for a particular time. The functioning is observed by making the wwl, rwl to low, and by making the FCS1, FCS2, WBL, and RBL to high. Although the wbl high, wwl makes the low resistive path by storing 0 at node q. the hold operation is executed by using the inputs (WWL, RWL, FCS1, FCS2, WBL, RBL). By making the (WWL = 0, RWL = 0, FCS1 = 1, FCS2 = 1, WBL = 1, RBL = 1) will operate in hold mode. Depends on these input states the output (Q and QB will get activated and deactivated), i.e., (Q = 1, QB = 0). At hold operation can observe the static noise margin (HSNM) (Fig. 5 and Table 1).

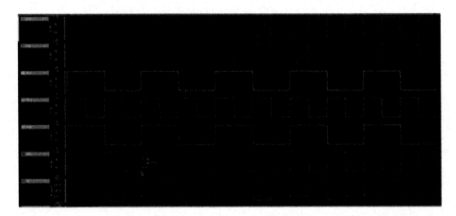

**Fig. 5** Hold operation simulation result

| | | HOLD | READ | WRITE '1' | WRITE '0' |
|---|---|---|---|---|---|
| **Table 1** Observation table of 8T SRAM Cell | WWL | low | high | high | high |
| | RWL | low | high | low | low |
| | FCS1 | high | low | low | high |
| | FCS2 | high | low | high | low |
| | WBL | high | high | High | low |
| | RBL | high | Discharge | High | High |

# 3   Monte Carlo Simulation

Monte Carlo simulation is a process of simulation to simulate the transistors at worst conditions. This is also known as MC Simulation. The Schematic simulation is performed in ADL window and the Monte Carlo is performed in the ADEXL window by choosing the simulation options as Monte Carlo simulation. By using this method, we can perform two simulations at one time (Transient Analysis and DC Analysis). This simulation is used in both Engineering and non-engineering fields and it is also used to measure the accuracy of the circuit. We can also calculate different parameter values at all process corners (FF, SS, SF, FS). While we execute the simulation process, we should mention the No. of points and No. of Bins to operate the schematic as (N = 1000 etc.,). The bins are based on the No. of Points (No. of Bins = No. of Points + starting run number − 1). After the simulation has been done the results will be observed by the histograms.

## 3.1   Write Static Noise Margin (WSNM)

The firmness of SRAM circuit depends on the SNM. A basic SNM is obtained by the inverter characteristics present in the circuit and finding the maximum possible square between them. In a write operation, a battle will happen between access and pull-down transistors. The feedback cutting scheme is used to avoid the disturbance between accesses and pull-down transistors of SRAM. The WSNM will perceive when write operation of the cell will performing. The WSNM is low when compared with RSNM. By using SNM, we can calculate the standard deviation and mean at all process corners (Fig. 6).

**Fig. 6** Monte Carlo simulation results of WSNM

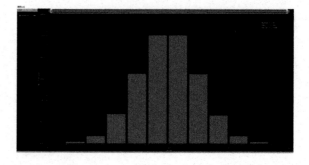

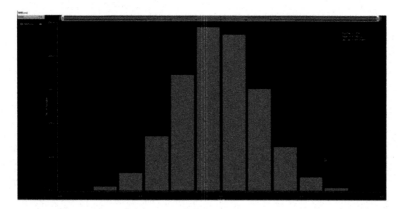

**Fig. 7** Monte Carlo simulation results for RSNM

## 3.2 Read Static Noise Margin (RSNM)

RSNM is to measure the stability of SRAM at a read operation. The RSNM was observed at a read operation of the relevant circuit. This is performed under transistors standard conditions. Simulation results show that the read margin perfectly expresses the SRAM's read stability at process-voltage-temperature (PVT) Conditions. The RSNM is higher than WSNM in 8T SRAM Cell (Fig. 7).

## 3.3 Hold Static Noise Margin (HSNM)

In this mode of operation, the word line is connected to ground. The cross-coupled inverter must be at bi-stable operating points to hold the data properly; this is

**Fig. 8** Monte Carlo simulation results for HSNM

**Table 2** Comparison of SNM at 300 mV and 25 °C

| CADENCE 45 nm | SNM (Monte Carlo simulation) | Mean (MV) | Standard deviation (MV) |
| --- | --- | --- | --- |
| 8T SRAM cell | HSNM | 0.0183 | 1 |
| | RSNM | 83.9 | 0.999 |
| | WSNM | 0.0183 | 1 |

achieved by means of SNM. The SNM equals the noise voltage necessary to flip the state. The HSNM will be Calculated when the circuit is operated at Hold state. The HSNM was quite equal to WSNM (Fig. 8 and Table 2).

## 4 Cell Layout

The layout is designed in the cadence 45 nm technology by using appropriate layers. It is also designed by satisfying the all designing rules provided by the technology file. The layout is the activity of placing the objects in space. And by using this layout, the manufacturing companies may fabricate the circuit. The next step after the layout is called as Circuit Fabrication (Fig. 9).

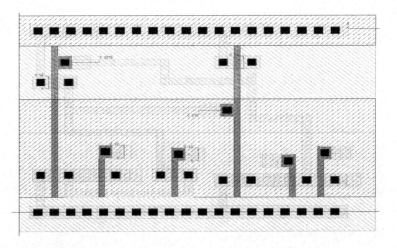

**Fig. 9** Layout representation of 8T SRAM Cell

# 5 Conclusion

The work contains 8T SRAM Cell implementation at ultralow power supplies with improved data stability. This work attained with SNM calculations by using single ended feedback control scheme. The mean and standard deviation are observed by using Monte Carlo Simulation.

# References

1. B.H. Calhoun and A.P. Chandrakasan, A 256-kb 65-nm sub-threshold SRAM design for ultra-low-voltage operation, IEEE J. Solid-State Circuits, vol. 42, no. 3, pp. 680–688, Mar. 2010.
2. J.P. Kulkarni, K. Kim, and K. Roy, A 160 mV robust Schmitt trigger based subthreshold SRAM, IEEE J. Solid-State Circuits, vol. 42, no. 10, pp. 2303–2313, Oct. 2007.
3. C.B. Kushwah, S.K Vishvakarma, A Sub-threshold eight transistor (8T) SRAM cell design for stability improvement, in proc. IEEE Int. Conf. IC Design Technol. (ICICDT), May 2014.
4. M.-F. Chang, S-W. Chang, P-W. Chou, and W.-C Wu, A 130 mV SRAM with expanded write and read margins for subthreshold applications, IEEE J. Solid State Circuits, vol-46, no. 2, pp. 520–529, Feb. 2011.
5. B. Zhai, S. Hanson, D. Blaauw, and D. Sylvester, A Varient-tolerant sub-200 mV 6-T subthreshold SRAM, IEEE J. Solis State Circuits, Vol-43, no. 10, pp. 2338–2348, oct. 2008.
6. J.P Kulakarni, K. Kim, and K. Roy A 160 mV robust Schmitt trigger based subthreshold SRAm, IEEE.J. Solid-State Circuits, vol. 42, no. 10, pp. 2303–2313, oct. 2007.
7. B. Zhai, S. Hanson, D. Blaauw, and D. Sylvester, A variation-tolerant sub-200 mV 6-T subthreshold SRAM, IEEE J. Solid-State Circuits, vol. 43, no. 10, pp. 2338–2348, Oct. 2008.

# Double-Supply Voltage Level Shifter with an Auxiliary Circuit for High-Speed Applications

Jalla Chinnari and Hanumantha Rao Sistla

**Abstract** This paper describes the characteristics of Dual-Supply Voltage Level Shifter. The circuit is fast and power efficient. It converts low input voltage levels into high output voltage level. The effectiveness of the proposed circuit is obtained by the increasing the strength of the NMOS device, it was done when the NMOS transistor is dragging down the output node Q1, Similarly, power of pull-down transistor is also increased by making use of an auxiliary circuit. The results are obtained after the simulation of actual circuit in 0.18-um technology. It determines the overall energy per evolution of 158 fJ, Power utilization in a static mode of operation is 0.4 nW, and propagation delay of 35 ns for an input frequency of 1 MHz, low-supply voltage level of $V_{ddl} = 0.6$ v, and high supply voltage level of $V_{ddh} = 1$ V.

**Keywords** Level converter · Low power · Sub threshold operation
Voltage level shifter

## 1 Introduction

In digital circuits dynamic and short-circuit power consumption is minimized by decreasing the power supply levels. On the contrary propagation delay of the circuit's increases with the reduction of the supply voltage, besides the headroom in analog circuits is limited and consequently, sensitivity to noise of the circuits is increased. Accordingly in balanced–speed composite signal circuits or in digital circuits where various devices start conduction at various speeds because of this reason dual-supply architecture was introduced. It has two supply voltages, one of

J. Chinnari (✉) · H. R. Sistla
Department of Electronics and Communications, Shri Vishnu Engineering
College for Women (Autonomous), Bhimavaram, India
e-mail: chinnari.jalla@gmail.com

H. R. Sistla
e-mail: hanumanth.s@svecw.edu.in

© Springer Nature Singapore Pte Ltd. 2018
J. Anguera et al. (eds.), *Microelectronics, Electromagnetics*
*and Telecommunications*, Lecture Notes in Electrical Engineering 471,
https://doi.org/10.1007/978-981-10-7329-8_71

this is $V_{ddl}$ it is applied to the elements on the noncritical route, where analog and high-speed digital elements are driven with high voltage ($V_{ddh}$). In the architecture with the help of dual-supply voltage level shifting circuits low level of logic values can be converted into higher logic levels for giving the appropriate voltage levels to the succeeding digital blocks for avoiding the decadence of entire efficiency of the circuit. The appropriate level shifter needs to be arranged with less propagation delay, power consumption, and area of the silicon. By using this architecture to preserve the high amount of power in the low-supply sections, the worked level shifter circuit capable of translating the drastically low values of $V_{ddl}$ to voltage lesser than the threshold voltage of the input transistor.

The remaining of this paper is organized as below, in Sect. 2 reviews of the classical architectures are explained those are few of the lately addressed high efficient voltage level shifters. After this in Sects. 3, 4 and 5 explained the high-speed voltage level shifter and shows the effectiveness and experimented report of the designed circuit.

## 2 Review of the Classical Architectures

When coming to the Classical architectures Zhang and Bhide [1] proposed the common level shifting architecture, the operation of this circuit contains conflict at the output nodes. When the applied input voltage is in subthreshold range, In this condition voltage difference between $V_{ddl}$ and $V_{ddh}$ is high because of this affect the circuit is able to convert the voltage levels from low to high only for short period of time but it is not being able to convert for long period of time. In order to overcome these difficulties, many approaches have been described. Among them, one of the attempts is to effort the technology-based approach, by using the semi-static current mirror circuit which consists of strong NMOS utilizing low Vth (threshold voltage) and weak PMOS devices utilizing high, which limits the current and energy of the PMOS devices, proposed by Corsonello and Perri [2]. Still, some constant current flows through the devices, delay, and power consumption is also increased. In order to overcome this Y. Osaki and N. Kuroki proposed the Level shifter with dynamic current mirror [3], subthreshold to above threshold conversion was explained by Lutkemeier and Ruckert [4], of applied input voltage which conducts only within the transition times [5, 6]. This circuit shows the remedy for overcome the constant power utilization, but there is a conflict problem between the devices in the state of high to low transition of the input signal which may lead to the increase in delay and simultaneously overall power utilization.

# 3 High-Speed Voltage Level Shifter

In this, the operation of the circuit has a conflict at the high to low transition to reduce this conflict NM2 transistor is dragged down to the output node. Currents flowing through the transistor NM1 and MN2 are Ip1 and Ip2. At that particular time of action, Ip1 and Ip2 are reduced. This action can be explained in the proposed circuit in Fig. 1. When the applied input signal is from low to high NM1 transistor is turned on and NM4 transistor is turned off at this time of conversion output is correspondingly equal to logic levels of previous output, transition current passes through NM4, NM1, and PM1 (i.e., Ip1), It happens because of overdrive voltage of NM3 is less than that of PM3 (i.e., $V_{ddh}$). The Ip1 current has the mirroring action at PM2 (i.e., Ip2), finally output node is dragged up by PM2, consequently NM3 is dragged down to the gate of NM4, so there will not be any constant current flows in the circuit through NM4, NM1, and PM1. The aspect ratio of PM2 is marked to select larger than that of PM1 by doing this power utilization can be reduced.

When the input signal conversion is from high to low in this state NM1 transistor is on and NM2 transistor is turned off, these attempts to decrease the value at output node by the help of dragging action. Because of this reverse operation, no current passes through PM1. The strength of PM2 is decreased because output node QA is not dragged up to $V_{ddh}$, it reaches the value $V_{ddh}$–Vth, It includes the threshold voltage of PM1. As a result, PM2 have a certain amount of current and

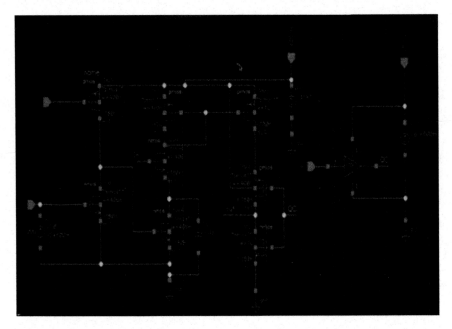

**Fig. 1** Schematic of the voltage level shifter

**Fig. 2** Simulation results of high-speed voltage level shifter when applied input A is high then output voltage is low and Qc is high

simultaneously some conflict was again present. By adding an additional transistor, i.e., PM4 reduces the Ip2. Output node was pulled down by NM2, the voltage across the gate of PM4 is high with the value of $V_{ddl}$ and voltage at the drain-source terminal of PM2 is decreased. It can be observed that the transistors NM2 and PM4 are operated with a voltage higher than that of $V_{ddl}$, this action reduces the currents in the pull-up device(i.e., Ip2) and increase the strength of pull-down device(i.e., NM2). Hence, delay and power consumption of the circuit was reduced (Fig. 2).

## 4 High-Speed Voltage Level Shifter with Auxiliary Circuit

The proposed voltage level shifter has an auxiliary circuit for reducing the power utilization in the circuit. The auxiliary circuit conducts only in high to low transition of the input signal, at this state the node QC drags up to a value higher than the $V_{ddl}$. The conduction of the circuit is described as below. When the applied input signal is formed high-to-low in this condition obtained output is not equivalent to input logic level, in this state the following transistors NM6, NM7, and PM6 are going to enable and NM5 is in the cut-off state. Hence, evolution current passes through the following transistors NM6, NM7, PM6 and same current delivered at PM7 so node QC is being dragged up. When the voltage is higher than $V_{ddl}$ when transistor MP4 is in cut-off state and transistor NM2 is activated, finally output goes to low so NM6 goes to cut-off state and simultaneously stops the current passes through the NM6, NM7, and PM6 it clearly shows the operation of auxiliary circuit which turns on only the high to low transformation (Fig. 3).

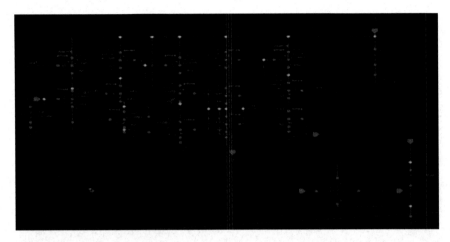

**Fig. 3** High-speed voltage level shifter with Auxiliary Circuit

Because of the design of auxiliary circuit PM7 have very less amount of current flow so conflict can be negligible, with this propagation delay and power consumption also decreased. As a result, the power utilization in the main circuit has also reduced. It shows the power utilization of the overall circuit is 50% of the circuit operated with an auxiliary circuit (Figs. 4 and 5).

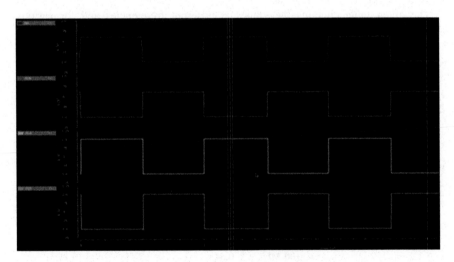

**Fig. 4** Simulation Results of high- speed voltage level shifter with auxiliary circuit when applied input IN is high, INB is low then output voltage is low and Qc is high

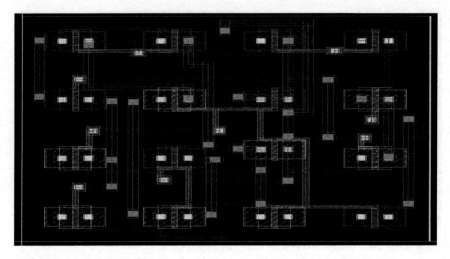

**Fig. 5** Layout of high-speed and power efficient voltage level shifter

## 5 Simulation Results

The proposed voltage level shifter can be verified by using a standard TSMC 0.18 μm CMOS technology. It can be chosen for obtaining the minimum power-delay product (PDP) (Table 1).

The given table contains the values at different frequencies and for various values of $V_{ddl}$, The delay and power are observed, the contention between devices is observed in the subthreshold region. Due to lower temperature, the time required to generate signals will be increased and produce less current.

This graph shows the simulated results at different values of VDDL when supply voltage increases then delay increases and power consumption was reduce. Fig. 6 shows the delay report and Fig. 7 shows the power consumption of the entire circuit.

**Table 1** Power and delay reports in 180 nm technology

| Frequency | VDDL, min(V) | Power (uW) | Delay(ns) |
|-----------|--------------|------------|-----------|
| 5 MHz     | 0.38         | 1.03       | 14        |
| 10 MHz    | 0.41         | 1.56       | 11        |
| 20 MHz    | 0.44         | 2.35       | 7         |
| 50 MHz    | 0.49         | 3.25       | 9         |
| 100 MHz   | 0.54         | 7.45       | 11        |
| 200 MHz   | 0.6          | 19.82      | 12        |
| 500 MHz   | 0.72         | 37.73      | 13        |
| 1 GHz     | 0.9          | 73         | 18        |

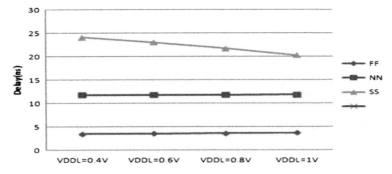

**Fig. 6** Graph for Delay at different Values of VDDL, at various corners

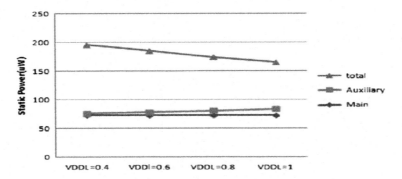

**Fig. 7** Graph for Static power at different values of VDDL

The main circuit consumes less amount of power when compared to auxiliary circuit. If the size of the capacitive load increases it results in an increase of power and delay which is shown in Fig. 7. It also shows the static power consumption of the overall circuit.

# 6   Conclusion

In this paper, the proposed circuit is implemented in cadence virtuoso and observed that the levels of voltages can be shifted from low values to high values with the help of extra element that is an NMOS device. It is named as NM4. It not only develops the ability of pull-down devices but also reduces the capability of the pull-up devices.

# References

1. D. Zhang and A. Bhide.: A 53-nW 9.1-ENOB 1-kS/s SAR ADC in 0.13um cmos for medical implant devices. IEEE j. solid-state circuits, vol. 47, no. 7, pp. 1585–1593, July (2010).
2. M. Lanuzza, P. Corsonello, and S. Perri.: Fast and wide range voltage conversion in multisupply voltage designs. IEEE Trans. Very Large Scale Integr. (VLSI) Syst., vol. 23, no. 2, pp. 388–391, Feb (2015).
3. Y. osaki, T. Hirose, N. Kuroki and M. Numa.: A low power level shifter with logic error correction for extremely low-voltage digital cmos LSIs. IEEE j. solid-state circuits, vol. 47, no. 7, pp. 1776–1783, July (2012).
4. S. Lutkemeier and U. Ruckert.: A sub Threshold to above threshold level shifter comprising a Wilson current mirror hybrid buffer. IEEE trans. Circuits Syst. II Exp. Briefs, vol. 57, no. 9, pp. 721–724., Sep (2010).
5. S.C Luo, C.J Haung and Y.H Chu.: A wide range level shifter using a modified Wilson current mirror hybrid buffer. IEEE Trans. Circuits syst. Ireg. papers, vol. 61, no. 6, pp. 1656–1665, May (2015).
6. S. R. Hosseini, M. Saberi, and R. Lotfi.: A low-power subthreshold to above-threshold voltage level shifter.: for IEEE Trans. Circuits Syst. II, Exp. Brief, vol. 61, no. 10, pp. 753–757, Oct (2014).

# Design and Analysis of Compatible Embedded Antenna for Mini Satellites

M. Kishore Kumar and V. Pradyumna

**Abstract** An embedded antenna is a metallic conductor embedded in a dielectric material whose dielectric constant is greater than 1, Ex: Microstrip patch antenna (MPA). In this paper, minisatellite antenna is presented. MPA antennas offer an ideal solution for satellite communication requirements due to their light weight and low profile. The design includes embedded single layer staircase feed, patch with three shapes square, half-bow tie, and full bow tie with defected ground structures (DGS) with FR4 Epoxy substrate thickness 1.59 mm. The proposed antenna achieved sufficient return loss for LEO satellites. The minisatellite aims to acquire data about high voltage discharge phenomena in LEO. This will enhance the understanding of satellite charging, overall reliability etc. The antennas operate at different frequencies in S-band, C-band, and X-band. The tool employed to design and simulate MPA is HFSS version 13.0.

**Keywords** Embedded MPA · Compatible antenna · Minisatellites
S-band frequency and defected ground · Half-bow tie · Full bow tie
Return loss

## 1 Introduction

The advance low-power miniature electronic components, combined with growing financial pressures for large satellites, gained attention on the use of small satellites to complement large satellite system for many space applications. A minisatellite weighs 700–1500 lb (300–500 kgs) (Figs. 1 and 2).

UoSAT-12 launched by Malaysia weighs 770 lb to take pictures the cameras are mounted on the satellite [1] and Horyu-4 [2] are the examples of minisatellites.

M. K. Kumar (✉) · V. Pradyumna
Department of Electronics and Communication, Sri Vasavi Engineering College,
Tadepalligudem, Andhra Pradesh, India
e-mail: mkishorekumar.hasini@gmail.com; modugu_kishore@yahoo.co.in

© Springer Nature Singapore Pte Ltd. 2018
J. Anguera et al. (eds.), *Microelectronics, Electromagnetics
and Telecommunications*, Lecture Notes in Electrical Engineering 471,
https://doi.org/10.1007/978-981-10-7329-8_72

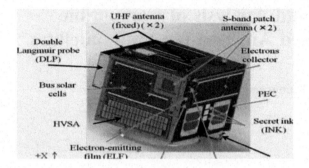

**Fig. 1** Horyu-4 satellite with microstrip patch antenna

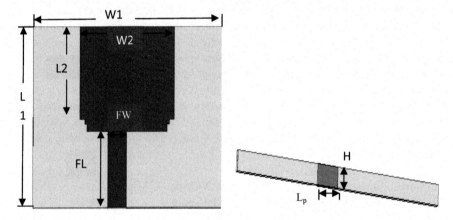

**Fig. 2** Front and Side view of square patch antenna with defected ground

## 2 Literature Survey

Derar Fayez Hawatmeh et al. proposed a design of Embedded 6-GHz 3-D Printed Half-Wave Dipole Antenna it has a return loss of −21dBi at a frequency of 6 GHz. The supporting ABS substrate is manufactured using FDM and provides an inclined surface on which the antenna feed network and balun are deposited using microdispensing of thick-film Ag paste [3]. Aleksandra Markina-Khusid et al. proposed a paper on Design Optimization of a Satellite Communication Terminal developed an integrated methodology for analyzing airborne satellite communication (SATCOM) terminal performance as a function of the antenna [4].

## 3 Designing

The three essential parameters for designing a MPA are Frequency of operation ($f_r$) [5–7] or design frequency which is 4.56 Ghz, the dielectric constant of the substrate is 4.4, the height of the dielectric substrate is 1.59 mm. By substituting these values in design equations [8, 9] dimensions of the antenna are obtained. The dimensions of MPA are 40 × 38 × 1.59 mm the patch shapes of the antennas are obtained in a step-by-step procedure initially the square patch antenna with dimensions 20 × 19.5 mm with staircase feed is designed. The DGS technique [10] is adopted and the full ground is defected as shown in below Fig. 3 the highest return loss value is obtained at the desired S-band frequencies which can be seen clearly in the simulation result section (Figs. 4, 5, and 6).

Then to improve the return loss value the square patch is slotted by using polylines like half-bow tie and full bow tie structures as shown in Figs. 5 and 7. Half-bow tie structure is a simple structure and it is the second step on the part of designing here the slots are made at the bottom corners of the patch and just above the staircase feed. The third part of the design is to make slots at the top corners of the patch which appears similar to full bow tie [11] that is the reason of naming the antenna full bow tie antenna. Bottom slits are similar to each other and top slits are similar to each other (Table 1).

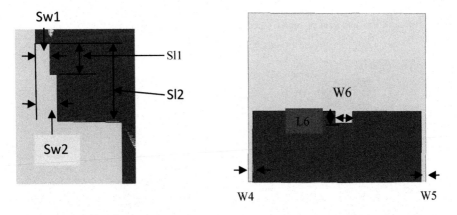

**Fig. 3** Close view of staircase and Back view of square patch antenna with defected ground

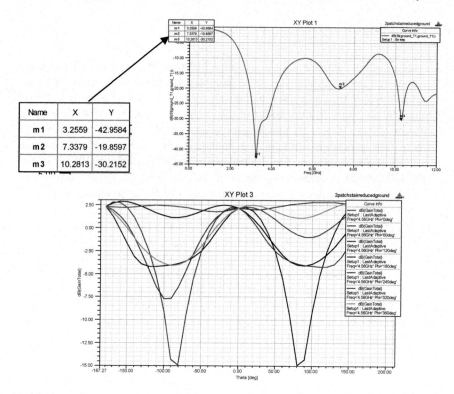

| Name | X | Y |
|------|-------|----------|
| m1 | 3.2559 | -42.9584 |
| m2 | 7.3379 | -19.8597 |
| m3 | 10.2813 | -30.2152 |

**Fig. 4** Return loss and gain of square patch antenna with defected ground

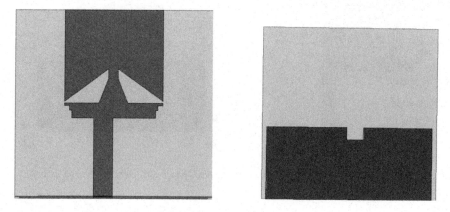

**Fig. 5** Front and back view of HBT antenna with defected ground

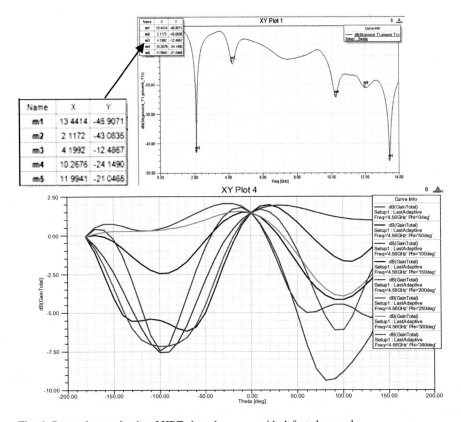

**Fig. 6** Return loss and gain of HBT shaped antenna with defected ground

**Fig. 7** Front and back view of FBT with defected ground

**Table 1** Dimensions of antenna

| Substrate width W1, length L1 and H | 40 mm, 38 mm and 1.59 |
|---|---|
| Substrate width W2 and length L2 | 20 mm, 19.5 mm |
| Feed width FW and length FL | 4 mm, 16 mm |
| Port length $L_p$ | 4 mm |
| Steps: step1 width Sw1 and length Sl1, step 2 width Sw2 and length Sl2 | 1 mm, 1 mm, 1.5 mm, 1.5 mm |
| Ground slots W4, W5, W6, L6 | 1 mm, 1 mm, 3.8 mm, 2.7 mm |

## 4 Simulation Results

The antennas resonated at multiple bands. The highest return loss value obtained at S-band to three antennas designed for analysis, remaining frequencies where the return loss value higher than −10 dB are considered and noted and those bands are C- band and X-band.

### 4.1 Simulation Result of Square Patch Antenna (SPA) with Defected Ground

The SPA with defected ground antenna resonated at three different frequencies in three different bands. The obtained return loss of SPA are −42.9684 dB at 3.2559 GHz, i.e., S-band, −19.8697 dB at 7.3379 GHz, i.e., C-band, and −30.215 dB at 10.2813 GHz, i.e., X-band. The gain of this antenna is 2.499 dB.

### 4.2 Simulation Result of Half Bow Tie (HBT) Antenna with Defected Ground

The patch resonated at four frequencies in S-band, C-band, and X-band, the return loss obtained for HBT are −43.0835 dB at 2.1172 GHz, i.e., S-band, −12.4867 dB at 4.1992 GHz i.e., C-band, −24.1490 dB at 10.2676 GHz X-band and −21.0455 dB at 11.9941 GHz. The gain of FBT is 2.97 dB.

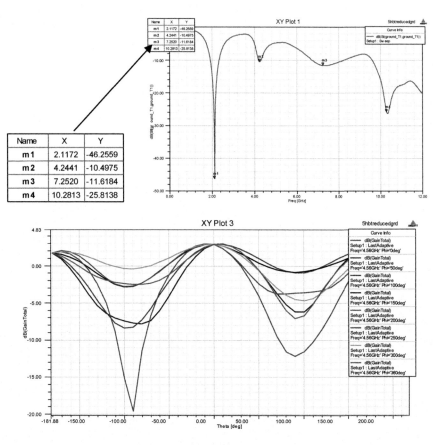

**Fig. 8** Return loss and gain of FBT antenna with defected ground

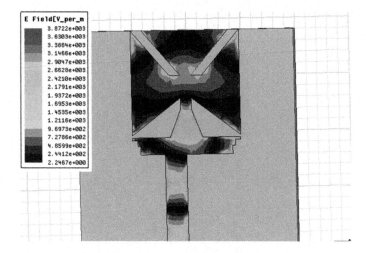

**Fig. 9** Field distribution of FBT with defected ground

**Table 2** Comparision values of various antennas with defected ground

| Parameter | SPA with stair feed defected ground | HBT with stair feed defected ground | FBT with stair feed defected ground |
|---|---|---|---|
| Return Loss | −42.9684 dB at 3.2559 GHz | −43.0835 dB at 2.1172 GHz | −46.2559 dB at 2.1172 GHz |
| Gain | 2.5 dB | 2.97 dB | 2.97 dB |
| Directivity | 3.302 dB | 3.469 dB | 4.281 dB |
| VSWR | 1.0417 | 1.1762 | 1.1458 |

## 4.3 Simulation Result of Full Bow Tie (FBT) with Defected Ground

The FBT antenna with defected ground operated at 4 frequencies. The return loss values obtained are −46.2559 dB at 2.117 GHz, i.e., S-band, −11.6184 dB at 7.2520 GHz, i.e., C-band, and −25.8138 dB at 10.2813 GHz, i.e., X-band. The gain value obtained for FBT antenna is gain 2.976 dB (Figs. 8, 9 and Table 2).

## 5 Conclusion

The proposed antenna consists of three designs, basic square patch, HBT, and FBT antennas. Among all these designs FBT shows better performance in terms of return loss and gain compared to other antenna designs it can be clearly observed from table No. 2. The gain for both HBT and FBT is similar. The gain can be increased further by using other DGS structures.

## References

1. M. N. Sweeting, "Microsatellite and mini satellite pro-grammes" at the University of Surrey for executive technology transfer and training in satellite engineering, 20–22 1995–10.
2. M. T. Islam, Mengu Cho, M. Samsuzzaman, and S. Kibria "Compact Antenna for Small Satellite Applications" IEEE Antennas and Propagation Magazine, Vol. 57, No. 2, page No. 30–36, April 2015.
3. Derar Fayez Hawatmeh, Sam LeBlanc, Paul I. Deffenbaugh, and Thomas Weller "Embedded 6-GHz 3-D Printed Half-Wave Dipole Antenna" IEEE Antennas and Wireless propagation letters, vol. 16, page no. 145–148, 2017.
4. Aleksandra Markina-Khusid, Ke Ning, Bo Yang Yu, and Frank Kolak "Design Optimization of a Satellite Communication Terminal" IEEE Antennas & Propagation Magazine, page no 61–68, June 2017.
5. P. S. R. Chowdary, A. Mallikarjuna Prasad, P. Mallikarjuna Rao, and Jaume Anguera, "Simulation of Radiation Characteristics of Sierpinski Fractal Geometry for Multiband Applications," International Journal of Information and Electronics Engineering vol. 3, no. 6, pp. 618–621, 2013.
6. Sudheer Kumar Terlapu, Ch Jaya and GRLVN Srinivasa Raju, "On the Notch Band Characteristics of Koch Fractal Antenna for UWB Applications," International Journal of Control Theory and Applications, Vol. 10, No. 06, pp. 701–707, 2017.
7. Adnan Kaya, Irfan Kaya and Haluk E. Karaca, "U-Shape Slot Antenna Design with High-Strength $Ni_{54}Ti_{46}$ Alloy", Arabian Journal for Science and Engineering, September 2016, Volume 41, Issue 9, pp 3297–3307.
8. Constantine A. Balanis; Antenna Theory, Analysis and Design, John Wiley & Sons Inc. 2nd edition. 1997.
9. G.S.N.Raju, "Antennas and Wave Propagation" Pearson Education (Singapore) Pvt Ltd., New Delhi, 2005.

10. Rahul Sharma1, A.N. Mishra2; "Design and Analysing of Compact Microstrip Antenna With Defected Ground Structure for UWB Application", International Journal of Research in Engineering and Technology eISSN: 2319–1163 I pISSN: 2321-7308, Volume: 05, Issue: 01, page no. 283–285, January-2016, Available @ http://www.ijret.org.
11. Ankit sharma and Ravi mohan "improvisation of the microstrip single element stepped feed antenna using half and full bow-tie slotting/meandering", ijerstissn 2319–5991vol. 3, no. 1, page No. 133–137, February 2014.

# Power Optimized FFT Architecture to Process Twin Data Streams Using Modified Booth Encoding

## M. Hemalatha and R. Ashok Chaitanya Varma

**Abstract** The main objective of this paper is to design a multipath delay commutator (MDC) using fast Fourier transform (FFT) which has the probability to process twin data streams. The MDC using FFT architecture computes N/2 point decimation in frequency (DIF) and N/2 point decimation in time (DIT) operations simultaneously. The number of registers can be reduced by performing bit reversal operation in the multipath delay commutator. The serial multiplication is performed by using serial multipliers which increases the complexity of the circuit Thus, the complexity of the circuit can be reduced and high throughput can be obtained by using modified booth multiplier used to minimize the power.

**Keywords** Multipath delay commutator (MDC) · Reordering shift registers
Natural order FFT output · Radix-2

## 1 Introduction

In digital signal processing domain, fast Fourier transform (FFT) is one of the most prominent algorithms. The discrete Fourier transform (DFT) is computed effectively by using FFT. At the transmission, end to transmit the data FFT operations are used and at the receiver end, DFT operations are used. To increase the operating speed at the transmission end both DIT and DIF operations are performed simultaneously. The pipelined FFT processor operates as a function of the single path delay feedback, based on the integrating a twiddle factor technique to compute the hardware related radix-2 algorithm [1]. The multimode multipath delay feedback architecture based on the dynamic voltage and frequency scaling (DVFS) is used to

M. Hemalatha (✉) · R. A. C. Varma
Department of Electronics and Communication Engineering, Shri Vishnu Engineering
College for Women (Autonomous), Bhimavaram, India
e-mail: honeyhema.m@gmail.com

R. A. C. Varma
e-mail: r.chaitanyavarma19@gmail.com

© Springer Nature Singapore Pte Ltd. 2018
J. Anguera et al. (eds.), *Microelectronics, Electromagnetics
and Telecommunications*, Lecture Notes in Electrical Engineering 471,
https://doi.org/10.1007/978-981-10-7329-8_73

process the FFT processor for MIMO OFDM applications. As the flexible radix configuration multipath delay feedback (FRCMDF) technique used in the multimode FFT processor which attains high throughput [2, 3]. The architectures of the FFT processor do not have specific bit reversal circuits. Four independent data streams are processed serially. Thus, the outputs of the multiple data streams are not produced in a parallel manner [4, 5]. FFT design based on the dual-path delay feedforward data commutator unit [6] which splits the input stream into two half-word streams. In [7], folding technique and register minimization techniques are used to elaborate the pipelined parallel FFT operations. In [8], the continuous flow parallel bit reversal circuit is proposed in that memory banks are present. Generally, the serial multiplication operation is performed by using half-adders and full-adders. The second stage of the multiplier depends on the output of the first stage. So there is a problem of delay [9–11]. In [12–14] combined SDC-SDF architectures are used to obtain only serial data transmission.

The paper is organized as follows. In Sect. 2 exiting MDC FFT architecture using serial multiplier is discussed. In Sect. 3 proposed radix-8 modified booth algorithm with sign extension is explained. Simulation results are drawn in Sect. 4. In Sect. 5 conclusion, remarks, and references are summarized.

## 2   Distinct Levels of Operation of MDC FFT Architecture

The MDC FFT architecture is shown in Fig. 1 to get the outputs in the natural order. Here, distinct levels of operation are performed, based on the two switches SW1, SW2 modes.

The principle architecture of N-point MDC FFT architecture mainly divides N bit input by a half. Thus, N bit input is divided into two N/2 inputs. The N/2 inputs contain odd and even samples. The 8- point DIF MDC FFT operation is performed

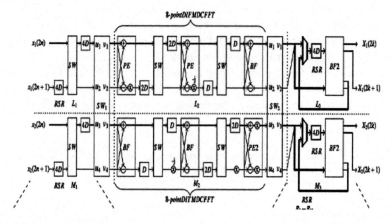

**Fig. 1** MDC FFT architecture

by using even samples in the same way the 8-point DIT MDC FFT operation is performed by using odd samples. Thus, the operation of DIF and DIT by interchanging of switch modes (normal and swap modes). All over the normal mode, the data passes from u1, u2, u3, u4 to v1, v2, v3, v4. All over the swap mode, the data passes from u1, u2, u3, u4 to v3, v4, v1, v2. The reordered bit operation of odd samples is performed before the N/2-point DIT FFT operation. And the reordered bit operation of even sample is performed after the N/2-point DIF FFT operation.

In MDC, FFT architecture performs butterfly operations in serial multiplier, which increases the complexity, increases the number of partial products. The disadvantage of the serial multiplier can be overcome by replacing modified booth multiplier.

## 3 Proposed Architecture RADIX-8 Modified Booth Multiplier

Modified booth algorithm increases the speed because it reduces the partial products by half (Fig. 2).

Modified stand-alone multiplier comprises a modified booth recorder (MBR). MBR is divided into two types booth encoder (BE) and booth selector (BS). The operation of BE is to decode the multiplier signal and BS is to produce the partial product by using the output. Later partial products are added to the Wallace tree adders same as carry-save-adder approach. Table 1 explained about multiplying A by 0, 1, −1, 2, −2, 3, −3, 4, −4, Multiply by zero implies that the product is "0", Multiply by "1" means that the product remains the same as the multiplier, Multiply

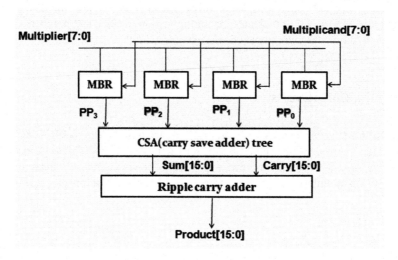

**Fig. 2** Modified Booth module

**Table 1** Booth Recoding
table for radix-8

| Quartet value | Signed-digit value |
|---------------|--------------------|
| 0000 | 0 |
| 0001 | +1 |
| 0010 | +1 |
| 0011 | +2 |
| 0100 | +2 |
| 0101 | +3 |
| 0110 | +3 |
| 0111 | +4 |
| 1000 | −4 |
| 1001 | −3 |
| 1010 | −3 |
| 1011 | −2 |
| 1100 | −2 |
| 1101 | −1 |
| 1110 | −1 |
| 1111 | 0 |

by "−1" means that the product is the complementary form of the number of two, Multiplying With "−2" is to shift left by one bit.

## 3.1 Sign Extension Corrector

The sign extension corrector is designed to increase the booth multiplier capacity by multiplying not only the unsigned number but also the signed number. When symbol unsign s_u = 0, it indicates the multiplication of unsigned numbers. When s_u = 1, it shows the multiplication of the signed number.

## 4 Simulation Results

Here, Xilinx suite design 14.5 version is used to simulate the VERILOG HDL codes. Fig. 3 shows the simulation results for $L_1$ and $M_1$ levels, it gives bit reversed odd bits and passes even bits normally to used bit reversal circuit. Figures 4 and 5 are two types of butterfly operations of FFT, where it gives 8 outputs by performing

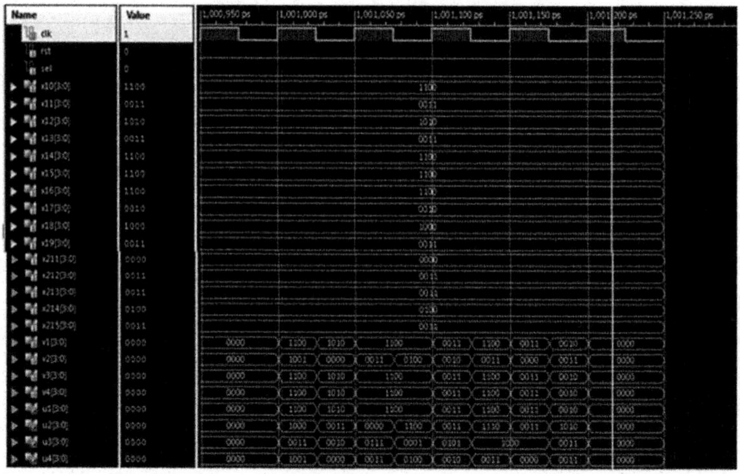

**Fig. 3** Simulation results for $L_1$ and $M_1$ levels

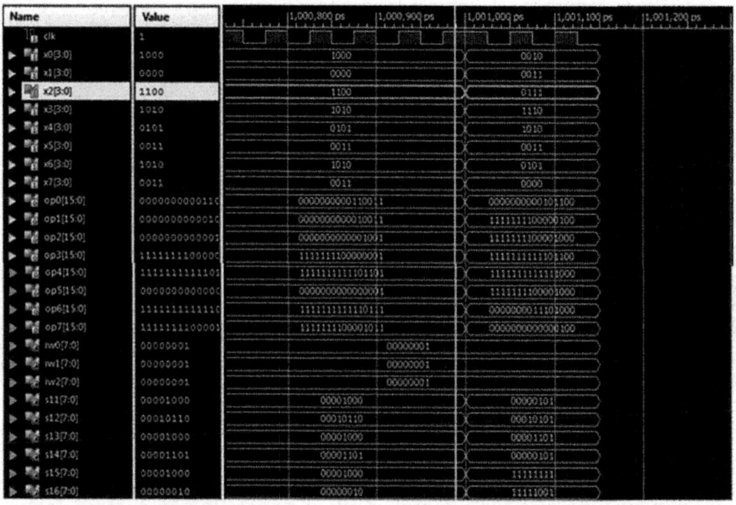

**Fig. 4** Simulation result for Decimation in Frequency (DIF)

butterfly operations with 8 inputs. Figure 6 explains the MDC FFT architecture simulation results for outputs are generated normal order by using input stage, DIF, DIT, and finally, butterfly operations are performed. The RTL schematic of MDC FFT architecture is shown in Fig. 7.

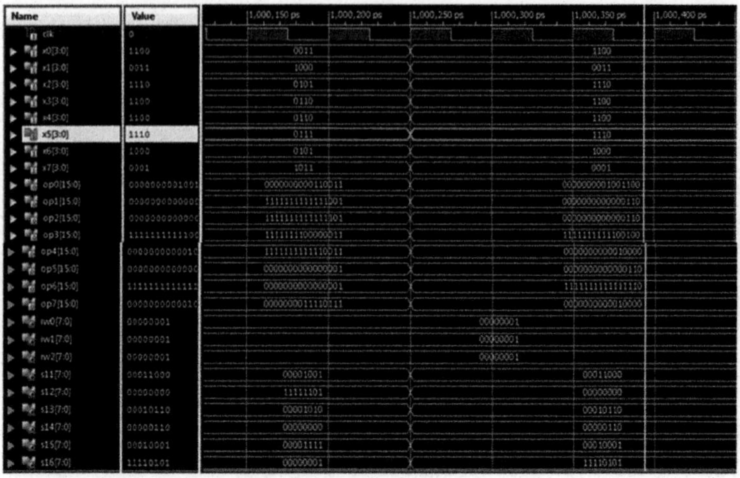

**Fig. 5** Simulation results for Decimation in Time (DIT)

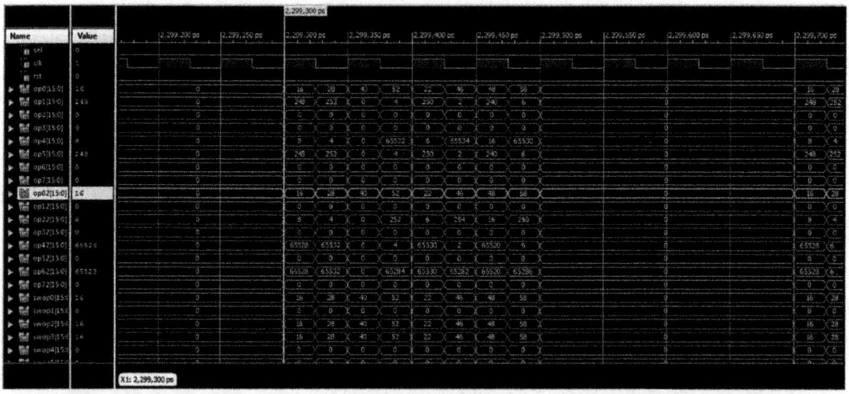

**Fig. 6** Simulation results for MDC FFT architecture

Table 2 explains the comparison power report of existing MDC FFT architecture using the serial multiplier and MDC FFT architecture using modified booth multiplication algorithm to minimize the 23 mw power consumption.

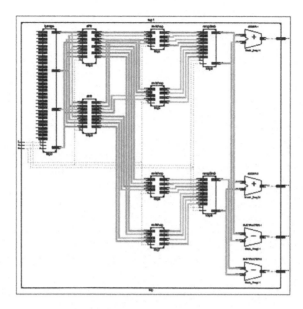

**Fig. 7** RTL Schematic of MDC FFT architecture

**Table 2** Power report comparison of MDC FFT architecture and Modified Booth module MDC FFT architecture

| Name of the system | Power (mw) |
|---|---|
| MDC FFT architecture power report | 228 |
| Modified Booth module MDC FFT architecture power report | 205 |

## 5 Conclusion

In this paper, MDC-based pipelined FFT architecture is implemented by using modified booth multiplier. The number of registers can be minimized and high throughput is obtained by using bit reversal operation in the MDC architecture. Thus, minimal power is obtained by modified booth module.

## References

1. S. He and M. Torkelson.: A new approach to pipeline FFT processor. in Proc. 10th Int. Parallel Process. Symp., 1996, pp. 766–770
2. Y. Chen, Y.-W. Lin, Y.-C. Tsao and C.-Y. Lee.: A 2.4-Gsample/s DVFS FFT processor for MIMO OFDM communication systems. IEEE J. Solid-State Circuits, vol. 43, no. 5, pp. 1260–1273, May 2008

3. S.-N. Tang, C.-H. Liao, and T.-Y. Chang.: An area- and energy- efficient multimode FFT Processor for WPAN/WLAN/WMAN systems. IEEE J. Solid-State Circuits, vol. 47, no. 6, pp. 1419–1435, Jul. 2012

4. K.-J. Yang, S.-H. Tsai, and G. C. H. Chuang.: MDC FFT/IFFT processor with variable length for MIMO-OFDM systems. IEEE Trans. Very Large Scale Integr. (VLSI) Syst., vol. 21, no. 4, pp. 720–731, Apr. 2013

5. M. Garrido, J. Grajal, and O. Gustafsson.: "Optimum circuits for bitreversal", IEEE Trans. Circuits Syst. II, Exp. Briefs, vol. 58, no. 10, pp. 657–661, Oct. 2011

6. Y.-N. Chang.: An efficient VLSI architecture for normal I/O order pipeline FFT design. IEEE Trans. Circuits Syst. II, Exp. Briefs, vol. 55, no. 12, pp. 1234–1238, Dec. 2008

7. M. Ayinala, M. Brown, and K. K. Parhi.: Pipelined parallelFFT architectures via folding Transformation. IEEE Trans. VeryLarge Scale Integr. (VLSI) Syst., vol. 20, no. 6, pp. 1068–1081 Jun. 2012

8. S.-G. Chen, S.-J. Huang, M. Garrido, and S.-J. Jou.: Continuous-flow parallel bit-reversal circuit for MDF and MDC FFT architectures. IEEE Trans. Circuits Syst. I, Reg. Papers, vol. 61, no. 10, pp. 2869–2877, Oct. 2014

9. Jyothi Leonore Dake and Sudheer Kumar Terlapu.: Implementation of high-throughput digit-serial redundant basis multiplier over finite field. IOSR Joural of VLSI and Signal Processing (IOSR-JVSP), vol. 6, Issue 4, Ver. I, e-ISSN; 2319-4200, p-ISSN No.: 2319-4197, pp. 35–45, Jul. 2016

10. Jyothi Leonore Dake, Sudheer Kumar Terlapu.: Low Complexity Digit Serial Multiplier for Finite Field using Redundant Basis. Indian Journal of Science and Technology, Vol. 9, No. S1, pp 1–5 2016

11. Jyothi Leonore Dake and Sudheer Kumar Terlapu.: Implementation of low-complexity redundant multiplier architecture for finite field. International Journal of Cybernetics and informatics (IJCI), Vol. 5, No. 4, pp. 333–340, Aug. 2016

12. Z. Wang, X. Liu, B. He, and F. Yu.: A combined SDC-SDF architecture for normal I/O. pipelined radix-2 FFT. IEEE Trans. Very Large Scale Integr. (VLSI) Syst., vol. 23, no. 5, pp. 973–977, May 2015

13. M. V. G. Rao, P. R. Kumar and A. M. Prasad, Implementation of real time image processing system with FPGA and DSP, 2016 International Conference on Microelectronics, Computing and Communications (MicroCom), Durgapur, 2016, pp. 1–4. https://doi.org/10.1109/MicroCom.2016.7522496

14. Munir A. Al-Absi, Ibrahim A and.As-Sabban, A New Highly Accurate CMOS Current-Mode Four-Quadrant Multiplier, Arabian Journal for Science and Engineering (Arab J Sci Eng)., vol. 40, No.2, pp 551–558, February 2015

# Design of Thinned Rhombic Fractal Array Antenna Using GA and PSO Optimization Techniques for Space and Advanced Wireless Applications

Venkata A. Sankar Ponnapalli, V. Y. Jayasree Pappu
and B. Srinivasulu

**Abstract** Fractal array antennas are repetitive geometry-based structures. These are multi-beam and ultra wideband array antennas having better array factor performance and space filling capability. A big confront in the fractal array antenna design is a large number of antenna elements at larger expansion levels and iterations. This research contribution proposes the design of thinned rhombic fractal array antenna for four different iteration levels with evolutionary optimization methods like genetic algorithm optimization technique and particle swarm optimization technique for space and advanced wireless applications. Owing to the application of evolutionary optimization techniques to the considered fractal array, nearly 25–50% of thinning achieved in various iterations and better array factor properties achieved than the fully populated rhombic fractal array antenna.

**Keywords** Fractal array antennas · Array factor · Thinned arrays
GAT · PSOT

## 1 Introduction

Optimization algorithms are one of the powerful techniques for the synthesis of array antennas. This research contribution introduces thinning of rhombic fractal array antenna for four different iterations with genetic algorithm technique (GAT),

V. A. S. Ponnapalli (✉) · B. Srinivasulu
Department of Electronics and Communication Engineering,
Sreyas Institute of Engineering and Technology, Hyderabad 500068, India
e-mail: vadityasankar3@gmail.com

B. Srinivasulu
e-mail: sreenivasub1979@gmail.com

V. Y. Jayasree Pappu
Department of Electronics and Communication Engineering, GITAM University,
Visakhapatnam 530045, India
e-mail: jayasree.pappu@gitam.edu

© Springer Nature Singapore Pte Ltd. 2018
J. Anguera et al. (eds.), *Microelectronics, Electromagnetics*
*and Telecommunications*, Lecture Notes in Electrical Engineering 471,
https://doi.org/10.1007/978-981-10-7329-8_74

719

and particle swarm optimization technique (PSOT). Thinning of an antenna array is nothing but the switching off some of the antenna elements without corrupting the performance of the fully populated antenna array [1, 2]. This type of thinned array antennas has applications, where narrow beams with less side lobe levels (SLL) and larger side lobe level angles (SA) are required with simple antenna array structures, such as satellite receiving antennas, multi-input, and multi-output radars, and other advanced wireless communication systems. Basically, fractal arrays are thinned arrays, due to their interleaving property [3–5]. To enhance the performance of these arrays further, optimization algorithms are needed. The GAT and PSOT methods are used in the thinning process. The GAT and PSOT is used to establish the finest set of "ON and OFF" antenna elements that offers a radiation pattern with excellent array factor properties [6, 7]. This type of optimization techniques and application of this type of techniques to the array antennas have become more and more accepted in this area as they are applicable to both deterministic and random array antennas [8–10].

Application of optimization techniques for the thinning of fractal arrays is prominently less in the literature. Nature-based linear array antennas of aperiodic nature have proposed and these array antennas are synthesized for thinning, lesser side lobe ration and mutual coupling losses using a hybrid model of neural networks and genetic algorithm approach [11] and other evolutionary computing techniques [12–14]. An iterative Fourier method has presented and implemented in [15], to develop the fractal array element response from the prescribed fractal array factor and to remove some of the antenna elements based on this peculiar property. Thinned Sierpinski and Haferman carpet array antennas have designed using evolutionary optimization techniques in [16]. Owing to this application of optimization techniques a better amount of side lobes reduced with nearly 50% of thinning.

This research article organized as follows. Section 2 introduces the optimization techniques for the thinning of fractal array antennas and thinning procedure. Section 3 explains the geometrical construction, array factor behavior, and analytical equation of rhombic fractal array antenna for four different iterations. This section also deals with the application procedure of optimization techniques to the rhombic fractal array. Section 4 exemplifies the results of the proposed thinned array. Lastly, conclusion draws in Sect. 5 of this research article.

## 2  Optimization Techniques for the Thinning of Proposed Fractal Arrays

Optimization techniques are useful in finding the optimum solution for a particular system or method. Actually, these optimization techniques are the powerful tools to synthesize the size and input variables of antennas and antenna arrays. The input variables may be current excitation, phase, and spacing between the elements. Length and width of patch antennas, input current and phase excitations of antenna arrays, the position of the antenna elements in an array, and some more antenna parameters can be optimized using these optimization techniques. Depending on the behavior and the inspiration of algorithm, optimization algorithms can be classified into different ways. This research work considered genetic algorithm technique (GAT) [2], and particle swarm optimization technique (PSOT) [8] optimization techniques for the thinning of proposed fractal array antennas and these algorithms comes under evolutionary optimization techniques. An evolutionary algorithm optimization technique is a subset of evolutionary computation method. An evolutionary algorithm uses methods motivated by biological evolutional nature, such as mutation, choice, recombination, reproduction.

## 2.1  GA Technique

GAT optimization technique is used to locate a group of parameters that reduces the output of a considered function. The following steps are shown in Fig. 1, depict the details of each step in the optimization technique. A thinned fractal array has discrete parameters. One bit represents the antenna element position as "ON" or "OFF". The fitness function related to this gene is the maximum SLL of its related far field pattern. By thinning of fractal array antennas, a lesser number of antenna elements contributes to the configuration of the radiating beam in comparison to fully populated fractal array antenna. Thinning factor is defined as

$$\text{Thinning factor} = (N - NA)/N \tag{1}$$

$$NIN = N - NA \tag{2}$$

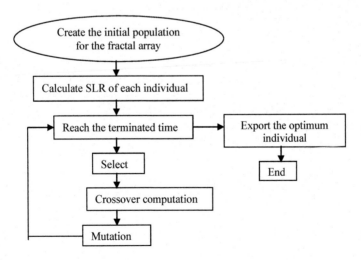

**Fig. 1** Flow chart of GAOT for proposed fractal array

Here N, NA, and NIN are a total number of elements in the array and a total number of active and inactive elements in the thinned array.

## 2.2 PSO Technique

Particle swarm optimization technique (PSOT) algorithm is a population-based random optimization method which relies on the social presentation of the birds. PSOT has been flourishingly applied in various research works of Science and Technology. It is exhibited that PSOT obtains superior results in a quicker, despicable way compared with other optimization techniques and methodologies. Another cause that PSOT is attractive because of its little factors to regulate. PSOT has been used for approaches that will be used across a broad range of applications, as well as for explicit applications focused on an exact form. Each particular solution is a bird in the investigative space and it is known as "particle". Particles of this algorithm have fitness, estimated from the fitness function to be optimized, and particles have velocities which express the flying of the particles. PSOT does not utilize genetic algorithm operation functions like, mutation and crossover functions, but update themselves with the inner velocities. First, originate each particle with random velocity, and position. Compute the cost of each particle. If the current cost is less than the finest value so far, consider this position. Select the particle with the less cost function of all particles. The position of this particle is g-Best. Evaluate, for each particle, the novel velocity and position. Repeat steps 2–4 until maximum iteration or least error criteria is not reached.

## 3  Array Factor Equations of Rhombic Fractal Array Antenna

Rhombic fractal array antennas generated by concentric elliptical ring sub-array geometric design for an expansion level of one has considered for the application of optimization techniques. The generalized design equation of rhombic fractal array antenna for an expansion level of one is represented in Eq. (3) and pictorial representation of rhombic fractal array antenna of an expansion level one for four successive iterations is shown in Fig. 2 [5].

$$A \cdot F_P(\theta, \varphi) = \prod_{P=1}^{P} \left[ \sum_{m=1}^{M} \sum_{n=1}^{N} I_{mn} e^{jk(1)^{P-1} \Psi_{mn}} \right], \tag{3}$$

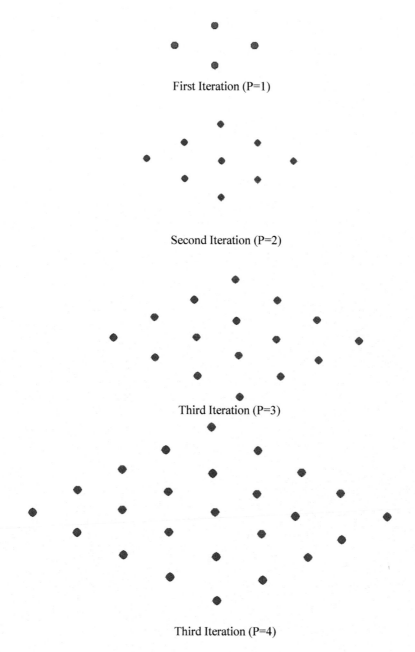

First Iteration (P=1)

Second Iteration (P=2)

Third Iteration (P=3)

Third Iteration (P=4)

**Fig. 2** Rhombic fractal array antenna for an expansion factor of one for different iterations

**Table 1** Array factor properties of thinned rhombus fractal array generated with GA, PSO techniques up to four successive iterations (S = 1)

| Total no. of elements | Thinned rhombic fractal array generated with GA | | | Thinned rhombic fractal array generated with PSO | | |
|---|---|---|---|---|---|---|
| | HPBW (Deg.) | SLL (dB) | Thinned elements | HPBW (Deg.) | SLL (dB) | Thinned elements |
| 4 | 20.6 | −1.3 | 02 | 36 | −15.5 | − |
| 9 | 34 | −6 | 05 | 34 | −20.5 | 05 |
| 16 | 33 | −16.8 | 08 | 33 | −19.3 | 08 |
| 25 | 40 | −11 | 13 | 41 | −12 | 10 |

where $I_{mn}$ is the current amplitudes, k is the wave number, S is the expansion level, here S = 1, P is the iteration number and here four successive iterations have considered, m is the number of concentric rings, and n is the number of elements presented at each iteration level.

# 4 Results and Discussion

Application of GAT and PSOT optimization techniques to the rhombic fractal array antenna of expansion factor one is discussed in this research contribution. The array factor behavior of the thinned rhombic array antenna is compared with its fully populated counterparts, and all these results are consolidated in Table 1. Owing to these optimization techniques nearly 25–50% of thinning achieved in each successive iteration. Figure 3 depicts array factor behavior of a rhombic fractal array of the first iteration. In this case, two and no antenna elements thinned in GAT and

**Fig. 3** Array factor of GA and PSO optimized thinned rhombic fractal array antenna of first iteration with S = 1

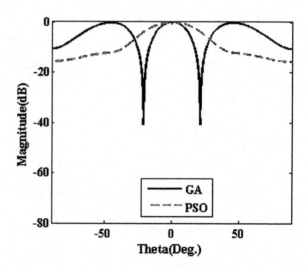

**Fig. 4** Array factor of GA
and PSO optimized thinned
rhombic fractal array antenna
of second iteration with S = 1

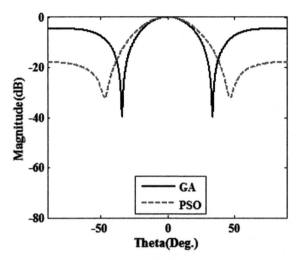

**Fig. 5** Array factor of GA
and PSO optimized thinned
rhombic fractal array antenna
of third iteration with S = 1

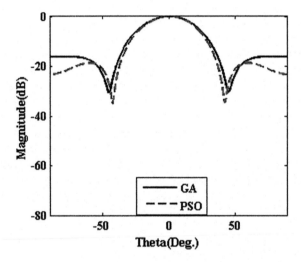

PSOT, respectively. Figure 4 exemplifies thinned array factor behavior of rhombus fractal array of the third iteration. Five antenna elements switched off in both cases of this array. Figure 5 shows array factor behavior for third iteration and eight elements are thinned in both cases. Figure 6 shows last iteration and 13 and 10 antenna elements are thinned in GAT and PSOT cases, respectively.

**Fig. 6** Array factor of GA and PSO optimized thinned rhombic fractal array antenna of fourth iteration with S = 1

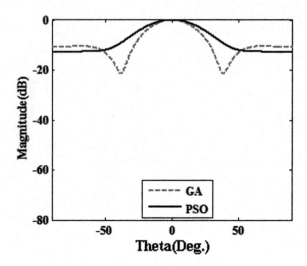

# 5 Conclusions

GAT and PSOT optimization techniques for the thinning of the antenna elements in rhombic fractal array antenna have been introduced and analyzed. In both cases, nearly 25–50% of antenna elements thinned with nominal array factor behavior. Both optimized arrays performed well in all array factor properties except in the case of beam width. The beam widths obtained in these cases are a little bit wider than the fully populated rhombic array antenna. Among the two optimization techniques, GAT has achieved nearly 50% of thinning rate in all iterations with fine array factor behavior. In PSOT case, a fine amount of side lobe levels achieved and these value are highest among the all other side lobe levels in this research work. Proposed arrays are suitable for space and other advanced communication systems owing to the better percentage of thinning at higher expansion levels.

# References

1. R. C. Hansen: Phased Array Antennas. 2nd edn. Second Edition. John Wiley & Sons, New Jersey, USA (2009).
2. Randy L. Haupt: Thinned Arrays Using Genetic Algorithms. IEEE Transactions on Antennas and Propagation. 42 (1994) 993–999.
3. D. H. Werner, R. L. Haupt and P. L. Werner: Fractal Antenna Engineering: The theory and design of antenna arrays. IEEE Antennas and Propagation Magazine. 41 (5) (1999) 37–59.
4. Omar Elizarraras et al: Design of Circular antenna arrays of circular subarrays exploiting rotational symmetry. Journal of Electromagnetic Waves and Applications. 2017 (2017) 1–12.
5. V. A. Sankar Ponnapalli and P. V. Y. Jayasree: Design of multi-beam rhombus fractal array antenna using new geometric design methodology. Progress In Electromagnetics Research C. 64(2016) 151–158.

6.  J. S. Petko and D.H. Werner: The Pareto optimization of ultra wide band polyfractal arrays. IEEE Trans Antennas Propag. 56 (2008) 97–107.
7.  J. S. Petko and D. H. Werner: The evolution of optimal linear polyfractal arrays using genetic algorithms. IEEE Trans Antennas Propag. 53 (2005) 3604–3615.
8.  J. Nanbo and Y. Rahmat-Samii: Advances in particle swarm optimization for antenna designs: Real-number, binary, single-objective and multi objective implementations. IEEE Trans Antennas Propag. 55 (2007) 556–567.
9.  T.H. Ismail and Z.M. Hamici: Array pattern synthesis using digital phase control by quantized particle swarm optimization. IEEE Trans Antennas Propag. 58 (2010) 2142– 2145.
10. V. Murino, A. Trucco, and C.S. Regazzoni: Synthesis of unequally spaced arrays by simulated annealing. IEEE Trans Signal Process. 44 (1996) 119–122.
11. C. S. DeLuccia and D. H. Werner: Nature-Based Design of Aperiodic Linear Arrays with Broadband Elements Using a Combination of Rapid Neural-Network Estimation Techniques and Genetic Algorithms. *IEEE Antennas and Propagation Magazine.* 49 (2007) 13–23.
12. Chakravarthy, V.V.S.S.S., Sarma, S.V.R.A.N, Babu K.N., Chowdary, P.S.R., & Kumar, S.T. (2015). Non-uniform circular array synthesis using teaching learning based optimization. Journal of Electronics and Communication Engineering, doi:10.6084/m9.figshare.1452770.
13. V. V. S. S. S. Chakravarthy and P. M. Rao, "On the convergence characteristics of flower pollination algorithm for circular array synthesis," *2015 2nd International Conference on Electronics and Communication Systems (ICECS)*, Coimbatore, 2015, pp. 485–489. https://doi.org/10.1109/ECS.2015.7124953.
14. V. V. S. S. S. Chakravarthy and P. Mallikarjuna Rao, "Amplitude-only null positioning in circular arrays using genetic algorithm," *2015 IEEE International Conference on Electrical, Computer and Communication Technologies (ICECCT)*, Coimbatore, 2015, pp. 1–5. https://doi.org/10.1109/ICECCT.2015.7226120.
15. Rowdra Ghatak, et al: Evolutionary Optimization of Haferman Carpet Fractal Patterned Antenna Array. International Journal of RF and Microwave Computer—Aided Engineering. 25 (2015) 719–729.
16. Anirban Karmakar, et al: Sierpinski carpet fractal-based planar array optimization based on differential evolution algorithm. Journal of Electromagnetic Waves and Applications. 2015 (2015) 1–16.

# Design and Analysis of Single Precision Floating Point Multiplication with Vedic Mathematics Using Different Techniques

K. V. Gowreesrinivas and P. Samundiswary

**Abstract** In this paper, multiplication for single precision floating point numbers is analyzed using Vedic multiplier with different techniques. In Vedic multiplier, the full adder is designed using modified $2 \times 1$ and $4 \times 1$ multiplexers, 3:2 and 4:2 compressors, and various prefix adders, such as Brent-Kung, Sklansky and Knoules adders for partial products addition. Furthermore, the performance metrics in terms of area and delay comparison is done. From the results, it is concluded that compressor-based Vedic multiplier requires less hardware and prefix adder-based Vedic multiplier is better in terms of delay. The newly introduced changes in Vedic multiplier makes the Vedic multiplier better in performance for the floating point multiplication for single precision numbers using different methods. All modules are coded with Verilog Hardware Description Language and simulated with Xilinx ISE tool.

**Keywords** Single precision · Double precision · Vedic multiplier
Multiplexer · Compressor · Verilog

## 1 Introduction

The advancement in present technology requires high-performance floating point units (FPUs) with greater throughput. In this regard, many high-speed FPUs are developed to improve the speed. Floating point operations are mainly influenced by multiplication and addition operations, Generally, in floating point multiplication, first comparison of the two exponents of given numbers is done, after that swapping and shifting operations are performed if necessary [1]. If any one of the numbers is

K. V. Gowreesrinivas (✉) · P. Samundiswary
Department of Electronics Engineering, Pondicherry University, Puducherry 605014, India
e-mail: srinuu43306@gmail.com

P. Samundiswary
e-mail: samundiswary_pdy@yahoo.com

© Springer Nature Singapore Pte Ltd. 2018
J. Anguera et al. (eds.), *Microelectronics, Electromagnetics and Telecommunications*, Lecture Notes in Electrical Engineering 471,
https://doi.org/10.1007/978-981-10-7329-8_75

signed number then adjusting the sign is compulsory, then mantissas of given numbers are added and that result needs another sign adjustment if it is a negative result. Finally, the adder renormalizes the summation result and adjusts the exponent accordingly, and truncates the resulting mantissa using an appropriate rounding scheme [2].

The floating point number representation is given by [3]:
$$Z = (-1)^S * 2^{(E - bias)} * (1.M)$$

Here, S = Sign value; E = exponents sum, i.e., e1 + e2; M = mantissa multiplication product, i.e., M1*M2

The generalized steps involved in multiplication for two floating point numbers are explained below:

    i. Multiplication of Significands, i.e., $[1.M_1 * 1.M_2]$

   ii. Decimal point Placement in multiplication result

  iii. Addition of exponents, i.e., $[E_1 + E_2 - Bias]$

  iv. Sign calculation using XOR operation of MSB bits

   v. Normalizing the result and rounding the implementation

  vi. Verification of Underflow and Overflow.

## 2 Different Floating Point Multipliers

Multiplication is an important fundamental function in arithmetic operations. Multiplication-based operations, such as multiply and accumulate (MAC) and inner product are among some of the frequently used computation intensive arithmetic functions (CIAF) currently implemented in many digital signal processing (DSP) applications, such as convolution, fast Fourier transform (FFT), filtering, and in microprocessors m its arithmetic and logic unit. Since multiplication dominates the execution tune of most DSP algorithms, so there is a need for high-speed multiplier.

### 2.1 FP Multiplication Using Existing Multipliers

In this section, different existing multipliers used for multiplication are discussed:

- Array multiplier needs N−1 stages to complete the process, where N indicates the number of bits in the multiplier. It needs lesser area occupancy and less speed. In this, speed is depended on the previous stage carry. To eliminate the dependency of the current stage on the previous stage, a Wallace Tree multiplier is introduced [4].

- Wallace tree multiplier needs M + 1 stages to complete the M-bit multiplication process and it uses carry-save technique for better improve in speed with compromising the area [4].
- Dadda multipliers are slightly faster than the corresponding Wallace tree multipliers for each size considered despite the requirement of large no. of carry lookahead adders. For the smallest pair of multipliers, the dadda multiplier requires two levels of carry look ahead logic, while the Wallace multiplier requires only one [5, 2]. Booth multiplication follows 2's compliment method to multiply two signed binary numbers.
- In Vedic, mantissa bits of the two numbers is done using Urdhava Triyakbhayam sutras [3, 4, 6–9]. In this, the generation of partial products is done using AND logic and their addition is done in vertical and crosswise manner. It requires single stage to perform both so which decreases the carry propagation from LSB to MSB [10, 11].

## 2.2 Summarized Existing Work

In this section, existing multiplication techniques for floating numbers are summarized briefly:

a. First, floating point multiplication for single precision numbers is done with a regular full adder and further, the performance metrics are summarized.
b. Next, floating point multiplication single precision are done with different existing 3:2 and 4:2 compressor techniques and further the performance metrics are summarized.
c. Furthermore, single precision floating point multiplication is done with different existing parallel prefix adder techniques, such as sklansky, brent-kung, knoules adders which are used in multiplication to add partial products and further the performance metrics are summarized.
d. Finally, performance metrics are tabulated for single precision floating point multiplication with different existing methods.

## 3 Proposed Work

In this paper, floating point multiplication for both single precision and double precision is analyzed using Vedic multiplier with different modified techniques like 4 × 1 and 2 × 1 multiplexers, 3:2 and 4:2 compressors. Generally, floating point

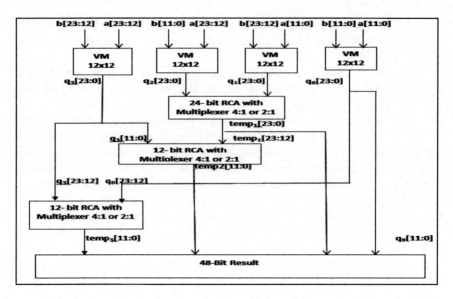

**Fig. 1** 24 × 24 bit Vedic multiplier for SPFPM using a multiplexer

multiplication requires mantissa multiplication and exponent addition to get the final result. In this, mantissa multiplication part is analyzed with existing and modified multiplier techniques. The modified techniques are as follows:

1. First, full adder is replaced with 4 × 1 multiplexer and 2 × 1 multiplexer to improve the area and speed
2. Next, 3:2 and 4:2 compressors are replaced with XOR-XNOR logic and 2 × 1 multiplexer to improve the area and speed
3. Finally, performance metrics are summarized with respect to delay and area.

Block diagrams of single precision floating point multiplication is illustrated in Fig. 1.

### 3.1 Floating Multiplication Using Proposed Techniques

In this section, multiplication using different techniques such as modified 4 × 1 and 2 × 1 multiplexer, using modified 3:2 compressor and 4:2 compressor, prefix adder-based multiplication is explained.

a. **Multiplexer-based Multiplication**

In this method, how full adder is designed using two multiplexers such as 4 × 1 and 2 × 1 to perform addition operation to generate sum and carry signals.

**Fig. 2** Full adder using
4 × 1 MUX

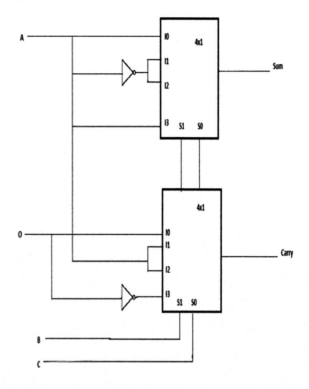

- In this, full adder using 2 × 1 multiplexer is illustrated in Fig. 3. For the first multiplexer, XOR and XNOR of A, B are the inputs to the multiplexer and it gives sum output. Second, multiplexer having AND and OR of A, B are the inputs to the multiplexer and it gives carry output. Here, $C_{in}$ acts as a select line for both multiplexers.

- In this full adder using 4 × 1 multiplexer is illustrated in Fig. 2. In 4 × 1 multiplexer, i0, i3 are connected to input A and i1, i2 are tied together and connected to the inverted output of A input in the first multiplexer. Whereas in the second multiplexer, i0 tied to logic '0' and i3 tied to logic '1', i1, i2 are tied together and connected to input A. In this, input B and C are acted as select lines for both multiplexers.

Finally, comparison analysis is done with respect to area and delay for multiplexer-based single precision multipliers and normal full adder-based single precision floating point multipliers.

**Fig. 3** Full adder using
2 × 1 MUX

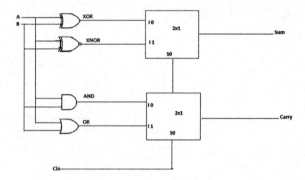

**Fig. 4** FA using modified
3:2 compressor

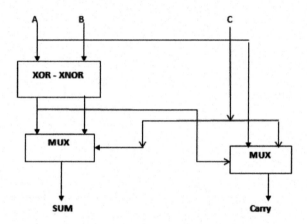

## b. **Compressor-based Multiplication**

The design of full adder using 3:2 and 4:2 compressor is illustrated in Figs. 4 and 5. In DSP applications, compressor circuit plays a major role to improve the performance [12]. In general, multiplication is done in three stages: with logical AND gates partial products generation, accumulation of generated partial products, and finally, the summation of partial products by using various adders. Among these three stages, partial product accumulation stage has a major contribution with respect to, delay and area. In this paper, regular compressor circuit is designed with Logical XOR and XNOR with a multiplexer to improve the performance.

**Fig. 5** FA using modified
4:2 compressor

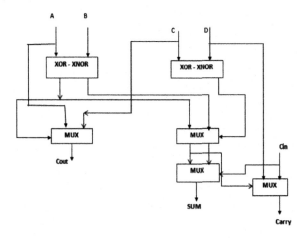

## 4 Simulation Results

The simulation and synthesis of the single precision floating point multiplication are designed using Xilinx ISE and performance parameters are summarized in terms of delay and complexity. In this, array, Wallace, and Vedic multipliers are considered for functional verification and results are tabulated. The comparison analysis of various multiplier techniques are done in terms of area and delay and listed in the below table.

### 4.1 Multiplexer-Based Floating Point Multiplication Using Vedic Multiplier

From the Tables 1 and 2, it is inferred that Vedic multiplier using 2 × 1 multiplexer needs 550 slices and 960 lookup tables and Vedic multiplier using 4:2 compressor with XNOR requires 274 slices and 532 lookup tables. In case of Propagation delay, Vedic multiplier using 4:2 compressor using XNOR-Mux logic requires 14.29 ns and whereas Vedic multiplier using 4:2 compressor using XOR-Mux needs 17.42 ns

From the Table 3, it is inferred that floating point multiplication using Vedic multiplier with 4:2 compressors with XNOR-Mux is better in terms of both area and delay whereas floating point multiplication using Vedic multiplier with 4:2 compressors with XOR-Mux is better in terms of area.

**Table 1** Area and delay of SPFPM using Vedic multiplier with 4 × 1 Mux and 2 × 1 Mux

| Using multiplexers | Device utilization summary and delay | | |
|---|---|---|---|
| | No. of slices | No. of 4-i/p LUTs | Delay (ns) |
| 4 × 1 Mux | 538 | 937 | 36.59 |
| 2 × 1 Mux | 550 | 960 | 36.61 |
| Full adder | 546 | 952 | 36.83 |

**Table 2** Area and delay of SPFPM using Vedic multiplier with 4:2 compressor

| Using 4:2 compressor | Device utilization summary and delay | | |
|---|---|---|---|
| | No. of slices | No. of 4-i/p LUTs | Delay (ns) |
| Vedic multiplier (4:2 compressor) | 307 | 537 | 16.33 |
| Vedic multiplier (using XOR-Mux) | 250 | 438 | 17.42 |
| Vedic multiplier (using XNOR-Mux) | 274 | 532 | 14.29 |

**Table 3** Performance comparison of SPFPMs with different techniques

| Floating point multiplication | Area (No. of slices/4-i/p LUTS) | | | Delay (ns) | | |
|---|---|---|---|---|---|---|
| | 2 × 1 Mux | 4:2 compressor (XNOR-Mux) | Knoules PPA | 2 × 1 Mux | 4:2 compressor (XNOR-Mux) | Knoules PPA |
| SPFPM | 550/ 960 | 274/532 | 529/936 | 36.61 | 14.29 | 58.09 |

# 5   Conclusion

The single precision floating point multipliers are generally used in digital signal processing applications. However, the performance of floating point multipliers in terms of complexity and propagation delay is the main problem. Hence, the whole system performance can be enhanced by minimizing occupied area. In this paper, a single precision floating point multipliers are developed with Vedic multiplier by using different techniques for the multiplication of two mantissa numbers. Further, the performance analysis of floating point multiplication for single precision numbers with existing techniques and proposed techniques are done. From the result, it is observed that Vedic multiplier using 4:2 compressor using XOR-XNOR-Mux logic is better in terms of delay and Vedic multiplier with 4:2 compressors with XOR-Mux is better in terms of area.

**Acknowledgements** Authors are very grateful to DEITY Scheme, New Delhi (Visveswaraya Ph. D. Fellowship) for assisting financial support to do the Research work.

# References

1. Sushma, R., Huddar, Sudhir, R., Kalpana, M., Surabhi, M.: "Novel High Speed Vedic Mathematics Multiplier using Compressors", Proceedings of International Conference on Communication and Signal Processing, pp. 465–469, Apr 2013.
2. Jitendra Babu, N., Rajkumar, S.: "A Novel Low Power and High Speed Multiply-Accumulate Unit Design for Floating-Point Numbers", Proceedings of International Conference on Smart Technologies and Management for Computing, Communication, Controls, Energy and Materials, Chennai, pp. 411–417, May 2015.
3. Gowreesrinivas, K.V., Samundiswary, P.: " High Speed Single Precision Floating Point Multiplier Using Vedic Multiplier and Knowles Adder" Proceeding of International Conference on Advance Computing and Communication Systems, Coimbatore, pp. 1519–1524, Jan. 2017.
4. Gowreesrinivas, K.V., Samundiswary, P.: "Comparative Study Analysis of Single Precision Floating Point Multipliers, Proceeding of International Conference on Electronics and Communication Systems, Coimbatore, Feb. 2017.
5. Vinod, B., Prasanna, P., Prachi, P.: "Design and Verification of Dadda Algorithm Based Binary Floating Point Multiplier", Proceedings of International Conference on Communication and Signal Processing, Chennai, Apr 2014.
6. Gowreesrinivas, K.V., Samundiswary, P.: "Comparative Study of Performance of Single Precision Floating Point Multiplier Using Vedic Multiplier and different types of Adders" Proceedings of International Conference on Control, Instrumentation, Communication and Computational Technologies, Kanyakumari, pp. 558–563, Dec 2016.
7. Sivanandam, K., Kumar, P.: "Run Time Reconfigurable Modified Vedic Multiplier for High Speed Applications", Proceedings of International Conference on Computing, for Sustainable Global Development, Mar 2015.
8. Ragini, P., Jitendra, J.: "Analysis of Effects of using Exponent Adders in IEEE-754 Multiplier by VHDL", Proceedings of International Conference on Circuit, Power and Computing Technologies, Mar 2015.
9. Arish, S., Sharma, R.K.: "An Efficient Floating Point Design for high Speed Applications using Karatsuba Algorithm and Urdhva Tiryagbhyam algorithm", Proceedings of International Conference on Signal Processing and Communication", Vijayawada, pp. 99–104, Mar 2015.
10. Mr. Mohanasundaram, S.S., Nirmal kumar, A., Arul parka., T.: "Design of Floating Point Multiplier Using Vedic Mathematics" International Journal of Innovative Science, Engineering & Technology, Vol. 2 Issue 1, January 2015.
11. Swapnil Suresh, M., Sanket Sanjay, N., Madhav Makarand B., Mrs. Rashmi Rahul K..: "32 Bit Floating Point Vedic Multiplier", IOSR Journal of VLSI and Signal Processing, Vol.6, Issue 2, pp 16–20, Apr. 2016.
12. Ramya sri, Y., Aruna, V B K L.: "Implementation of Double Precision Floating Point Multiplier Using Wallace Tree Multiplier", International Journal of Innovative research in science, Engineering Technology, vol.4, issue 7, July 2015.

# Weighted Averaging SWT Technique for Enhanced Image Fusion in X-ray Mammography

**M. Prema Kumar, N. Sowjanya and P. Rajesh Kumar**

**Abstract** X-ray Mammography has been a common technique of breast cancer identification. A single X-ray mammogram will not be able to convey full information about cancer to the radiologist. In this, an image fusion using Weighted Average SWT is proposed and histogram equalization is performed to enhance the quality of the fused X-ray mammogram. The resultant X-ray mammogram is same as conventional X-ray mammogram but with appreciably superior detail and is then reconstructed by using its inverse transform. This fused X-ray mammogram is well-suited for clinical settings and equips the radiologist to use lifetime diagnosis experience in X-ray mammography.

**Keywords** X-ray mammography · SWT · Image fusion

## 1 Introduction

Image fusion integrates information from numerous image sensor data and these fused images are apt for the intention of human visual awareness and further computer vision. The successful image fusion acquired from different modalities is of enormous significance in many applications like medical, remote sensing computer vision, and robotics. The use of various medical imaging systems is rapidly increasing so multi-modality image fusion plays a vital role in the medical field. The blend of the medical images can often lead to added clinical information not noticeable in the detach images [1].

The functional and the anatomical information are combined into a single image. Most of the equipment is not providing such data. The process of image fusion

M. P. Kumar (✉) · N. Sowjanya
Department of ECE, SVECW (A), Bhimavaram, Andhra Pradesh, India
e-mail: medapatiremkumar@gmail.com

P. R. Kumar
Department of ECE, AUCE (A), Andhra University, Visakhapatnam, India

© Springer Nature Singapore Pte Ltd. 2018
J. Anguera et al. (eds.), *Microelectronics, Electromagnetics*
*and Telecommunications*, Lecture Notes in Electrical Engineering 471,
https://doi.org/10.1007/978-981-10-7329-8_76

allows the incorporation of various image sources. The fused image be capable of containing balancing spatial and spectral resolution characteristics [2].

Image fusion in mammography has been rarely used. Generally, one X-ray mammogram has some information embedded for the radiologist to diagnose the problem. If more than one X-ray mammogram taken and fused using the image fusion techniques, it would further enhance the data embedded in the fused image. This will make the radiologist convenient and provide with an enhanced image for painless diagnosis of breast cancer.

In this paper, a new image fusion technique based on the weighted average SWT with intensity enhancement is proposed. This technique shows better experimental results in identifying breast cancer when compared to regular weighted average DWT technique of image fusion [3].

This paper is organized as follows: Sect. 2 presents image fusion using SWT, Sect. 3 presents Implementation, Sect. 4 presents image quality parameters, Sect. 5 presents experimental results, and finally, Sect. 6 reports conclusion.

## 2  Image Fusion Using Stationary Wavelet Transform (SWT)

The discrete wavelet transform is to renovate the transformation invariance to slightly special DWT, named undecimated DWT, to identify the stationary wavelet transform (SWT) which is shift invariant, It holds back the down-sampling process of the decimated algorithm and instead of up-sampling the filters coefficients by padding zeros. In this, the filter is up-sampled, i.e., "Stationary Wavelet Transform (SWT) is same as DWT but the only practice of down-sampling process is concealed [4, 5]".

### 2.1  Stationary Wavelet Transforms

In 1-D SWT the approximation and detail coefficients at the first level are both of size $N$, which is the signal length. In SWT first level, all the decimated DWT for a

**Fig. 1** 2-D SWT

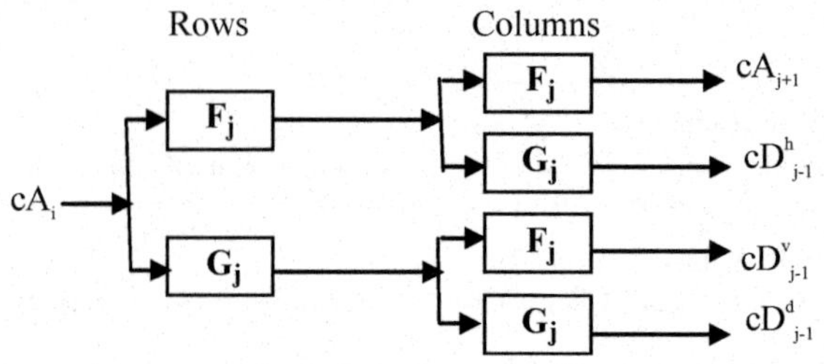

**Fig. 2** Flowchart

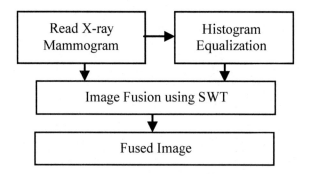

given signal can be found by doing the convolution operation between the signal and the appropriate filters coefficients as in the DWT but without including down-sampling (Figs. 1 and 2).

## 3  Implementation

In the proposed method of fusion two X-ray mammograms from one subject are taken and fused as per the following flowchart:

The fusion of images is performed using simple and weighted averaging method as per the following equations:

*Simple Average SWT-based image fusion.*

$$I(a,b) = \frac{SWT\{I_1(a,b)\} + SWT\{I_2(a,b)\}}{2} \tag{1}$$

*Weighted Average SWT-based image fusion.*

$$I(a,b) = \frac{w_1 * SWT\{I_1(a,b)\} + w_2 * SWT\{I_2(a,b)\}}{w_1 + w_2}, \tag{2}$$

where

$I(a,b) \rightarrow$ Fused image          $I_1(a,b) \rightarrow$ Input image 1

$I_2(a,b) \rightarrow$ Input image 2          $w_1, w_2 \rightarrow$ Weights added.

The end fused image is measured for quality with the Image quality measuring parameters (Fig. 3).

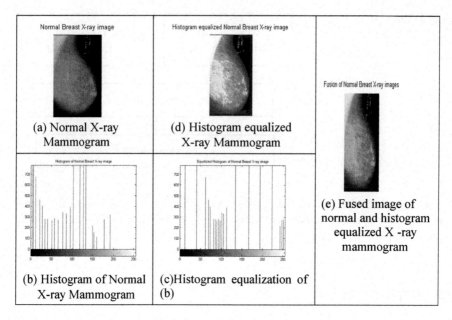

**Fig. 3** Image fusion using simple average SWT method for normal X-ray mammogram

## 4 Image Quality Parameters

The fused image is assessed using image quality parameters like peak signal-to-noise ratio (PSNR), root mean square error (RMSE), normalized cross-correlation (NCC), image quality index (IQI), fusion factor (FF), fusion symmetry (FS), and Mutual Information (MI) [6–8]. Equations are given below and these parameters are tabulated in Table 1.

$$PSNR = 10 * \log 10(255^2/MSE) \tag{3}$$

**Table 1** Image fusion using simple average SWT method parameter analysis

| S. no. | Parameter | Normal X-ray mammogram | Benign X-ray mammogram | X-ray mammogram with microcalcification |
|--------|-----------|------------------------|------------------------|------------------------------------------|
| 1 | RMSE | 22.305 | 17.643 | 13.867 |
| 2 | PSNR | 21.162 | 23.195 | 25.290 |
| 3 | NCC | 0.873 | 0.823 | 0.917 |
| 4 | MI | 1.642 | 1.985 | 1.583 |
| 5 | IQI | 0.818 | 0.884 | 0.843 |
| 6 | FS | 0.031 | 0.085 | 0.0321 |

$$MSE = \frac{1}{MN} \sum_{i=1}^{M} \sum_{j=1}^{N} (I_{ij} - I_{1ij})^2 \tag{4}$$

$$\text{RMSE} = \sqrt{\text{MSE}} \tag{5}$$

$$NCC = \sum_{i=1}^{M} \sum_{j=1}^{N} \frac{I_{ij} * I_{1ij}}{I_{ij}^2} \tag{6}$$

$$IQI = \frac{\sigma_{xy}}{\sigma_x \sigma_y} \frac{2\bar{x}\bar{y}}{\bar{x}^2 + \bar{y}^2} \frac{2\sigma_x \sigma_y}{\sigma_x^2 + \sigma_y^2} \tag{7}$$

$$MI = \sum_{(f,a)} P_{FA}(f,a) \log_2 \frac{P_{FA}(f,a)}{P_F(f)P_A(a)} + \sum_{(f,b)} P_{FB}(f,b) \log_2 \frac{P_{FA}(f,b)}{P_F(f)P_b(b)} \tag{8}$$

$$FS = abs\left(\frac{MI_{FA}(f,a)}{MI_{FA}(f,a) + MI_{FA}(f,b)} - 0.5\right) \tag{9}$$

## 5  Results

The method is tested for three different types of X-ray mammograms namely, Normal, Benign and Microcalcification X-ray mammograms. The fused images for different inputs using the simple average method and weighted average SWT method are shown in Figs. 4, 5, 6, 7, 8, 9 (Fig. 10).

It has been observed that for weighted average SWT method gives better results when compared with simple average SWT method for Normal, Benign, and Microcalcification Breast X-ray mammogram.

The same was also assessed using the image quality parameters and tabulated as in Tables 1 and 2.

## 6  Conclusion

It can be concluded that from the Tables 1 and 2 the weighted average SWT method of image fusion is giving better results when compared to Simple average SWT method of image fusion. This can be prominently observed from the PSNR and RMSE values.

It can be seen that the PSNR value is consistently increasing for Benign and Microcalcification X-ray mammograms. The parameter comparison can also be seen from Figs. 8 and 9.

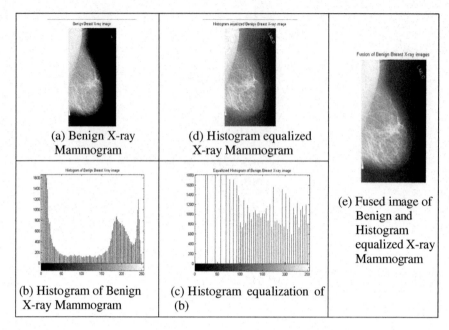

**Fig. 4** Image fusion using simple average SWT method for Benign X-ray mammogram

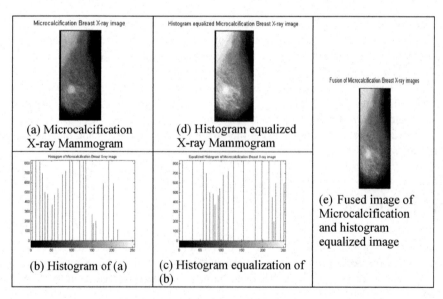

**Fig. 5** Image fusion using simple average SWT method for microcalcification X-ray mammogram

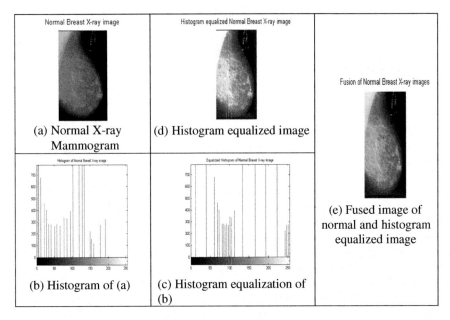

**Fig. 6** Image fusion using weighted average SWT method for normal X-ray mammogram

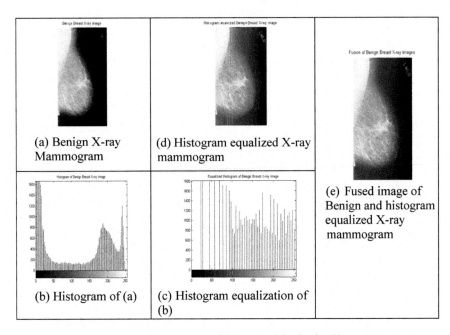

**Fig. 7** Image fusion using weighted average SWT method for Benign X-ray mammogram

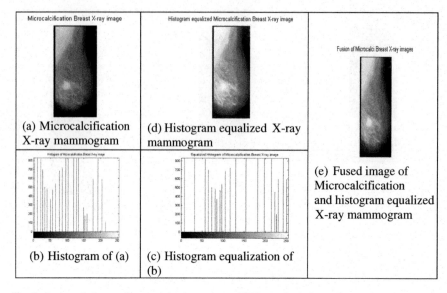

**Fig. 8** Image fusion using weighted average SWT method for microcalcification X-ray mammogram

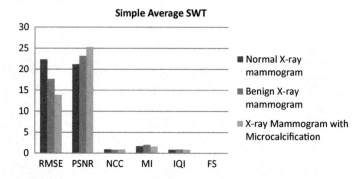

**Fig. 9** Parameter analysis of simple average SWT method

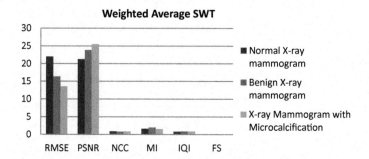

**Fig. 10** Parameter analysis of weighted average SWT method

**Table 2** Image fusion using weighted average SWT method parameter analysis

| S. no. | Parameter | Normal X-ray mammogram | Benign X-ray mammogram | X-ray mammogram with microcalcification |
|--------|-----------|------------------------|------------------------|------------------------------------------|
| 1 | RMSE | 21.984 | 16.367 | 13.532 |
| 2 | PSNR | 21.288 | 23.850 | 25.503 |
| 3 | NCC | 0.894 | 0.831 | 0.919 |
| 4 | MI | 1.649 | 1.989 | 1.593 |
| 5 | IQI | 0.836 | 0.892 | 0.924 |
| 6 | FS | 0.001 | 0.082 | 0.030 |

# References

1. V.P.S. Naidu and J.R. Rao, "Pixel-level Image Fusion using Wavelets and Principal Component Analysis", Defence Science Journal, Vol. 58, No. 3, May 2008, pp. 338–352.
2. A. P. James, B. V. Dasarathy, "Medical Image Fusion: A survey of the state of the art", *Information Fusion*, 2014.
3. Jagalingam P., ArkalVittalHegde, "Pixel Level Image Fusion–A Review on Various Techniques", 3rd World Conference on Applied Sciences, Engineering &amp; Technology 27–29 September 2014, Kathmandu, Nepal.
4. Z. Wang, C.A. Clavijo, E. Roessl, U. van Stevendaal, T. Koehler, N. Hausergy and M. Stampanoni, "Image fusion scheme for differential phase contrast mammography", 7th Medical Applications Of Synchrotron Radiation Workshop (MASR 2012) Shanghai Synchrotron Radiation Facility (SSRF), 17–20 October, 2012, Published By IOP Publishing For Sissa Media lab.
5. Tania Stathaki, 2008. "Image Fusion Algorithms and Applications", Elsevier.
6. V. Jyothi, B. Rajesh Kumar, P.K. Rao, D.V.R.K. Reddy, "Image Fusion using Evolutionary Algorithms (GA)", International Journal of Computer Technologies and Applications, 2(2), 2012.
7. M Prema Kumar and P Rajesh Kumar, "Image Fusion of Mammography Images using Genetic Algorithm (GA)", Australian Journal of Basic and Applied Sciences, 9(33) October 2015, Pages: 45–50.
8. Luqman Maraaba, Zakariya Al-Hamouz, Hussain Al-Duwaish: Prediction of the Levels of Contamination of HV Insulators Using Image Linear Algebraic Features and Neural Networks, **Arab J Sci Eng, 40 (9) 2609–2617 (2015).**

# The Reconfigurable Multi-ASIP
# Architecture for Turbo Decoding

**A. L. Sruthi and P. Ravi Kumar**

**Abstract** The past few years had seen a big evolution in the wireless standards. This evolution aims to improve the parameters, such as flexibility and time delay, which make it reusable for different modes and standards. In order to reach these requirements the multiprocessor architecture ASIP (Application-specific instruction set processor) has been developed in the decoding process. This reconfigurable ASIP was implemented with less power consumption by using Xilinx design suite in 12.2 version.

**Keywords** ASIP · Wireless communication · Turbo decoding
Multiprocessor

## 1 Introduction

In wireless communication systems, the channel coding block is an essential part to improve the channel quality. Over the past few years, wireless communication systems had used turbo codes to improve the performance [1–8]. Generally, it uses dual encoders at the primary session named as encoder1 and encoder2, and uses dual decoders at the ending session. By using divide and conquer rule the output performs the previous error correction codes. The message bits enter the transmitter and send to encoder1 and encoder2. Before entering encoder2 the message bits are scrambled by an interleaver which reorders the data to be transmitted and to reduce the burst errors. Every encoder produces some calculations and sends the data to the receiver. The original message bits and the two strings of parity bits are gathered to a single block and then send to media, where noise can cause some errors in the

A. L. Sruthi (✉) · P. Ravi Kumar
Department of Electronics and Communications, Shri Vishnu Engineering
College for Women Autonomous, Bhimavaram, Andhra Pradesh, India
e-mail: allurilakshmisruthi@gmail.com

P. Ravi Kumar
e-mail: ravikumar_TNK@svecw.edu.in

© Springer Nature Singapore Pte Ltd. 2018
J. Anguera et al. (eds.), *Microelectronics, Electromagnetics
and Telecommunications*, Lecture Notes in Electrical Engineering 471,
https://doi.org/10.1007/978-981-10-7329-8_77

transmitted data. The decoders at the receiver side exchange this data continuously. So that, after decoding the data with some iteration the final output gets back to binary bits.

Later, a single configurable engine (reconfigurable turbo decoder) is used for multiple standards [6–10]. The turbo decoder is the most difficult blocks in any communication chain which requires a large area and low power. The use of an ASIP processor is to reduce the time when the information is changing from one application to the other. In order to reach the targeted applications, the suitable ASIP architecture emerges for multiple turbo decoders. The main architectures had able to be dynamically adapted to face emerging things. Starting from this session, it aims to propose novel contributions to maintain very fast reconfiguration of flexible decoder architecture. This ASIP is named as Decoder ASIP (DecASIP), which supports many standards and it is integrated to a flexibility multiprocessor platform named as Universal channel Decoder (UDec).

## 2 Reconfigurable UDec Architecture

The considered multiprocessor is shown in Fig. 1, where it contains four Reconfigurable Decoder ASIPs in which two ASIPs are connected in one column and the other two ASIPs are connected in another column. These are interconnected through the network on chip interface. Each RDecASIP is affiliated with input memory, program memory, configuration memory, extrinsic memory, and cross-metric memory. Initially, the input memory stores the Log Likelihood Ratio (LLR) values. Program memory contains the commands which are used for the task decoding and instructions to initiate the ASIP will be stored in the configuration memory. Finally, the outcome data will be sent to extrinsic memory.

The entire structure is connected to configuration architecture. It mainly consists of four blocks, i.e., master, slave, interconnect, and selector. The configuration manager works depending on the internal and external instructions. The master interface works when it receives the destination address and base address from configuration memory. The control signal is to send the data to slave interface so the transfer is on state, then slave interface checks whether the received destination address is related to its own address or not. The selector block is used to send the information from slave interface to configuration memories which are connected to RDecASIPs.

## 3 Flexible UDec Platform

In this section, several methods are used to improve the flexibility features of UDec architecture. The techniques which concern different communication networks connecting the ASIP blocks at the upper session. In this considered section, the

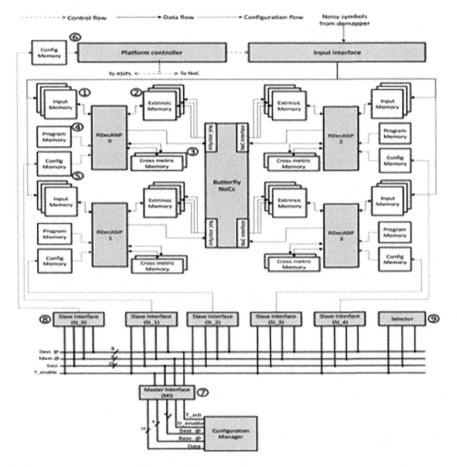

**Fig. 1** Reconfigurable UDec architecture

possibility to adapt at the instant of time the location and number of the enabled RDecASIPs are specified.

## 3.1 Ring Buses Dynamic Adaptation Examples

Figure 2 shows the examples of ring buses. In order to exchange the data and to know the numbers of active RDecASIPs, these ring buses are used. Here, four RDecASIPs are considered in each ring bus. In Fig. 2a the data is exchanged from RDecASIP1 to RDecASIP2 and vice versa. So, in this case, we can say the no of activated RDecASIPs are 2 and the ASIP shift value will be 1. Similarly, in Fig. 2b the data is exchanged from three ASIPs, so the no of activated ASIPs are 3 and the ASIP shift value is 3. In Fig. 2c the selection vector drives multiplexers to find out the ring bus adaptation.

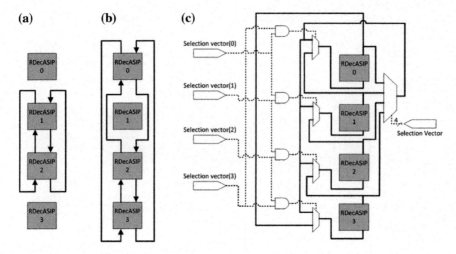

**Fig. 2** Adaptation of ring buses **a** Two ASIPs are selected. **b** Three ASIPs are selected. **c** Flexible architecture is used for one ring bus

## 3.2 Butterfly NoCs Adaptation

Figure 3 shows the principle of routing to the butterfly topology network on chip interface. Here, 8 ASIPs are considered where 4 ASIPs are placed at transmitter section other four ASIPs at receiver section. The butterfly network on chip is a multistage network of an indirect topology: nodes at the last session and routers are in the center. Each RDecASIP is connected to transmission network interface which is a complete routing information generator, from this interface the data will be sent

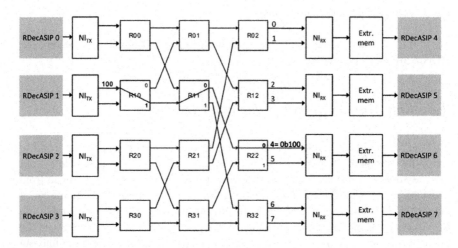

**Fig. 3** Butterfly topology routing principle

to routers and it consists of two inputs and two outputs. The first output bit will be selected when it is '0' and the second output bit will be selected when it is '1'.

The data which is to be routed is composed of 3 bits data. Every bit indicates the output of the router at multiple stages during transmission. In the example of Fig. 3, number 4 is targeted at the output. So the routing information for the first router is 1, for the second one is 0 and for the last one is also 0 corresponding to the binary value of 4. The routing data is passed to receiver network interface. Finally, the information will be sent to the receiver through extrinsic memory. Moreover, the ASIP shift value shows the position of the selected ASIPs, and consequently, it also shows the position of the destination ASIPs.

This topology allows every turbo decoder to connect to different registers so that only all turbo decoders are enabled without any disruptions. This butterfly topology can handle multiple turbo decoders, so it has good performance. The butterfly network structure has the advantage of reducing network latency. This topology is commonly preferred due to asynchronous loads. If heavy loads are given as input then this type of topology is used. The advantages of using this network are it has high no of routes, reduces the network latency and no path diversity.

# 4 Experimental Results

Turbo decoders are implemented using butterfly topology in XILINX design suite which is used for simulating the model in this by using VERILOG HDL code. A reset signal is given first to clear the register and output will be zero for the reset operation. The decoder output will be produced for a subsequent clock cycle.

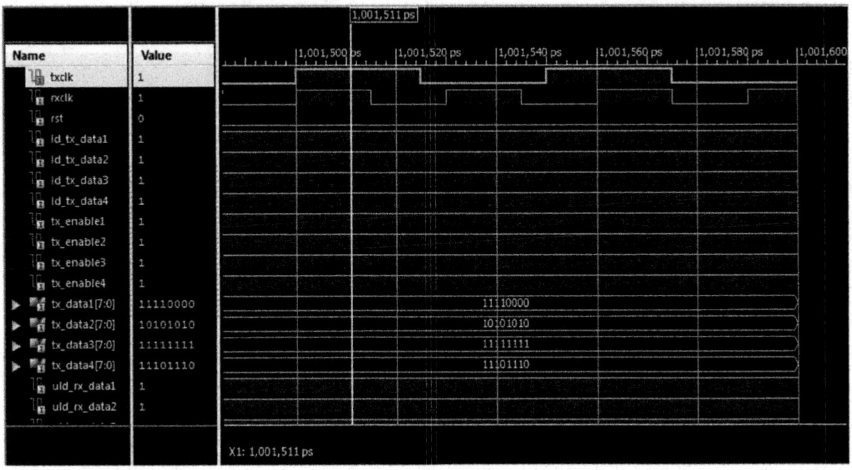

**Fig. 4** Simulation results for butterfly topology at transmitter

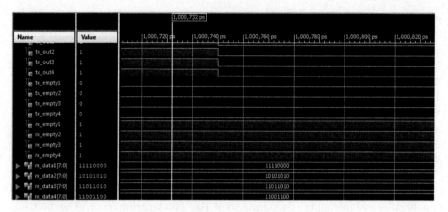

**Fig. 5** Simulation results for butterfly topology at receiver

Figures 4 and 5 show the simulation results for butterfly topology at transmitter and receiver sections of butterfly topology. Four turbo decoders are interconnected through the butterfly topology.

The input data bit is represented by tx_data [7:0]. Then, the data bits are generated for four processors, then during transmission over the channel, the data bits get corrupted and finally, after the decoding process the final output is recovered and it is represented by rx_data [7:0]. Figure 6 shows the power report (0.139 W) of this method. The RTL schematic of using four turbo decoder is shown in Fig. 7. The input signal is represented by tx_data, clk, reset. The output signal of turbo decoder is represented by rx_data1, rx_data2, rx_data3, rx_data4.

Table 1 shows the power results comparison between the existing and proposed methods. An efficient processor is proposed to implement a scalable low-power processor capable of supporting a multi-standard turbo decoder. The proposed method consumes less power compared with existing decoders. It presents a low-power delay architecture which adopts several techniques to reduce power consumption.

## 5 Conclusion

The turbo decoder is implemented for parallel processing to increase the speed of the system. Butterfly topology is developed for asynchronous loads and to reduce the network latency. The existing techniques had developed the flexible parameters to decrease the power, area, and time delay. In this method, the architecture implementing the 128 RDecASIPs can be executed in 10.5 ns by using XILINX design suite. This architecture improves the time delay, power, and overall performance of the processor when compared with previous techniques.

**Fig. 6** RTL schematic of turbo decoder

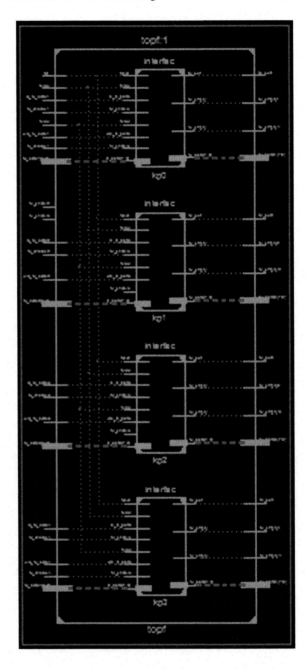

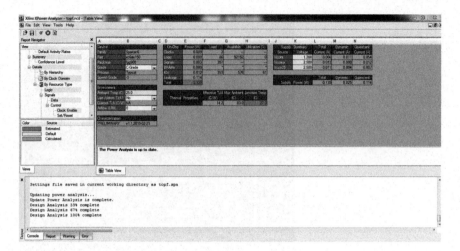

**Fig. 7** Power report

**Table 1** Comparison results of power and delay reports

| Methods | Power (mW) | Delay (ns) |
|---|---|---|
| Existing method | 0.139 | 14.5 |
| Proposed method | 0.118 | 10.5 |

# References

1. C. Berrou, A. Glavieux, and P. Thitimajshima, Nearshannon limit error- correcting coding and decoding: Turbo-codes. 1, in IEEE International Conference on Communications, ICC 93., vol. 2, pp. 1064–1070 vol.2, 1993.
2. P. Murugappa, Towards optimized flexible multi-asip architectures for ldpc/turbo decoding," Ph.D. dissertation, Institut Mines-T´el´ecom-T´el´ecomBretagne-UEB, 2012.
3. V. Lapotre et al., Optimizations for an efficient reconfiguration of an ASIP-based turbo decoder, in Proc. IEEE Int. Symp. Circuits Syst. (ISCAS), May 2013, pp. 493–496.
4. V. Lapotre, P. Murugappa, G. Gogniat, A. Baghdadi, M. Hubner, and J.-P. Diguet, Stopping-free dynamic configuration of a multi-ASIP turbo decoder," in Proc. Euromicro Conf. Digit. Syst. Design (DSD), pp. 155–162. 2013.
5. T. Vogt and N. Wehn, "A reconfigurable asip for convolutional and turbo decoding in an sdr environment," IEEE Transactions on Very Large Scale Integration (VLSI) Systems, vol. 16, no. 10, pp. 1309–1320, 2008.
6. C. Condo, M. Martina, and G. Masera, "VLSI implementation of a multimode turbo/LDPC decoder architecture," *IEEE Trans. Circuits Syst. I, Reg. Papers, vol. 60, no. 6, pp. 1441–1454, Jun. 2013.*
7. C.-H. Lin, C.-Y. Chen, E.-J. Chang, and A.-Y. Wu, "Reconfigurable parallel turbo decoder design for multiple high-mobility 4G systems, J. Signal Process. Syst., vol. 73, no. 2, pp. 109–122, Nov. 2013.
8. C. Condo, M. Martina, and G. Masera, "VLSI Implementation of a Multi-Mode Turbo/LDPC Decoder Architecture," IEEE Transactions on Circuits and Systems I: Regular Papers, Early Access Articles, 2012.

9. D. Lattard, E. Beigne, F. Clermidy, Y. Durand, R. Lemaire, P. Vivet, and F. Berens, "A reconfigurable baseband platform based on an asynchronous network-on-chip," IEEE Journal of Solid-State Circuits, vol. 43, no. 1, pp. 223–235, Jan. 2008.
10. G. Prasad kumar, M. Vijaya Laxmi, Flexible UDec ASIP Processor with Multimode Turbo Decoder, International Journal of Science and Research (IJSR), volume 5 Issue 6, June 2016.

# Adaptive Beam Steering of Smart Linear Array Using LMS and RLS Algorithms

**Bammidi Deepa and B. Roopa**

**Abstract** Smart antenna improves the gain of the main lobe in a direction of arrival and null generation toward the interference. Using this technique, the direction of arrival (DOA) of the antenna array can be improved and array factor can be derived in the desired direction of Angle of arrival. A report on performance evaluation of adaptive beam steering generation using Least Mean Square (LMS) and Recursive Least Mean Square (RLS) algorithms is presented. LMS algorithm is simple in the computation of Beam Forming. By repeated corrections of the weights, in an iterative procedure, the LMS algorithm finds the best weights. RLS algorithm exhibits very fast conjunction though at the cost of high complexity of computation. The effectiveness of these optimization algorithms would be compared with respect to run time. The two algorithms are compared with respect to less run time while maintaining the required specifications of the antenna is discussed. The simulation of all the results would be carried using MATLAB.

**Keywords** LMS · RLS · Direction of arrival (DOA) · Beam forming

## 1 Introduction

A smart antenna is defined as an antenna array, with a signal processor that can adjust its beam pattern for the optimized gain and directivity while reducing the interference [1].

Two basic types of smart antennas are well known. They are switched beam system and beam-formed adaptive systems. Different beam patterns are obtained from switched beam systems and based on the need, one of the beams are accessed

B. Deepa (✉)
Anil Neerukonda Institute of Technology and Sciences, Visakhapatnam, India
e-mail: deepakundala@gmail.com

B. Roopa
Chaitanya Bharathi Institute of Technology, Hyderabad, India
e-mail: roopabammidi@gmail.com

© Springer Nature Singapore Pte Ltd. 2018
J. Anguera et al. (eds.), *Microelectronics, Electromagnetics and Telecommunications*, Lecture Notes in Electrical Engineering 471,
https://doi.org/10.1007/978-981-10-7329-8_78

at a point of time. Whereas, the beam forming type systems enhance the beam to required direction of application and reduces the back lobes and side lobes [2]. In this paper, the beam steering characteristics of the smart antenna array are discussed with two adaptive algorithm techniques like LMS and RLS algorithms. Further, the algorithms are compared in terms of Angle of Arrival (AOA), run time, and beam steering [3].

## 2 LMS and RLS Algorithms

### 2.1 Least Mean Square (LMS) Algorithm

The LMS algorithm assumes small weights and, at every iteration, the mean square error gradient is calculated and the weight vectors are revised. If the gradient of mean square error (MSE) is positive, it is needed to reduce the weights otherwise, the error would keep on rising positively if the same weight is used for next computations. If the gradient is negative, it is not required to increase the weights. The weight revision equation is $W_{n+1} = W_n - \mu \varepsilon[n]$, in which $\varepsilon$ shows the mean square error and $\mu$ represents the coefficient of convergence [1]. The minus symbol shows a change in weights in opposite direction to the slope of the gradient. With the optimal weight vectors, the mean square error is minimized as it is a quadratic function of weights of the filter and has only one extreme. By increasing or decreasing the mean square error for corresponding filter weight curve, the LMS approaches toward the optimal weights [4] (Figs. 1 and 2).

LMS algorithm reduces the value of e(n) by an iterative process.

$$\text{The output } y(n) = W^H x(n); w = [w_1, w_2, \ldots, w_N]^H \tag{1}$$

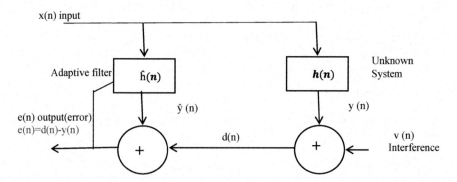

**Fig. 1** Block diagram of LMS algorithm

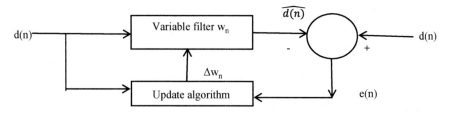

**Fig. 2** Block diagram of RLS algorithm

H represents the Hermitian transpose (complex conjugate) and 'w' shows a complex weight vector, for N number of elements. Received signal of antenna array is

$$X(n) = [x_1(n), x_1(n), \ldots, x_N(n)], \tag{2}$$

Weight vectors are revised according to the equation of LMS algorithm

$$(n+1) = w(n) + \mu x(n) e^*(n), \tag{3}$$

where μ is the step size. If μ is chosen to be large, the gradient estimate determines the change of the weights, and these weights may undergo large value change and thus the negative gradient may become positive [5–8]. If μ is selected to be very small, then the time to conclude the optimal values of weights will be too longer. Therefore, μ requires an upper boundary and is equal to $0 < \mu < \frac{2}{\lambda_{max}}$ [5].

## 2.2 Recursive Least Mean Square (RLS) Algorithm

In RLS, the cost function can be reduced by selecting the filter coefficients appropriately, and then updating the filter matrix. The error depends on the filter coefficients $\hat{d}(n)$. Error coefficients [2]

$$e(n) = d(n) - \hat{d}(n) \tag{4}$$

A gain matrix $\widehat{R^{-1}}(n)$ replaces the gradient step size μ in RLS algorithm, at the nth iteration with the weight vector update equation is

$$w(n) = w(n-1) - \hat{R}^{-1}(n)x(n) \in^* (w(n-1)) \tag{5}$$

$$\delta_o \hat{R}^{-1}(n-1) + x(n)^* x^H(n) \tag{6}$$

and $\delta_o$ representing a real scalar less than but near to 1 [4]. $\delta_o$, the forgetting factor is used for exponential weight vector of past data and the revised equation tends to de-emphasize the old samples. Algorithm memory is given by $\frac{1}{1-\delta_o}$. For example, for $\delta_o = 0.99$ the algorithm memory last to 100 samples. The initialization of the matrix is

$$R^{-1}(0) = \frac{1}{\varepsilon_o} \qquad (7)$$

## 3 Results and Discussions

### 3.1 Results Obtained Using LMS Algorithm

Figure 3 shows the AOA versus Array Factor plot if the desired signal Angle of arrival = 30°, interfering signals are at 0° and 90° number of elements = 8, spacing between the elements = 0.5λ, step size = 0.001, elapsed time = 2.664 s.

Figure 4 shows the AOA versus Array Factor plot if the desired signal Angle of arrival = 45°, interfering signals are assumed at 22.5° and 0° number of elements = 16, spacing between the elements = 0.25λ, step size = 0.01, elapsed size = 1.958 s.

Figure 5 shows the AOA versus AF plot when desired signal Angle of arrival = 22.5°, interfering signals are assumed at 7° and 60° number of elements = 16, spacing between the elements = 0.125λ, Step size = 0.1, elapsed time = 1.750238 s.

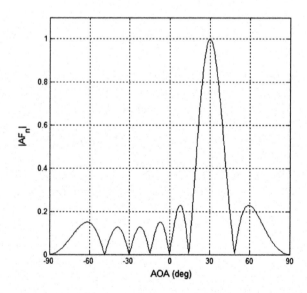

**Fig. 3** AOA versus array factor plot-case 1

**Fig. 4** AOA versus array
factor plot-case 2

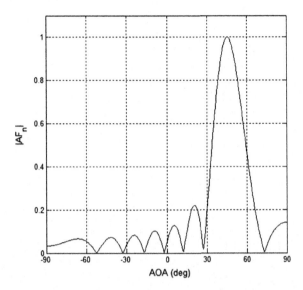

**Fig. 5** AOA versus array
factor plot-case 3

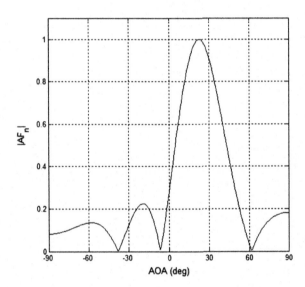

## 3.2 Results Obtained Using RLS Algorithm

Figure 6 shows the AOA versus AF plot when the desired signal Angle of
arrival $= 45°$, interfering signals are assumed at $0°$, and $22.5°$ number of elements $= 12$, spacing between the elements $= 0.5\lambda$, forgetting factor $= 0.95$,
elapsed time $= 0.130281$ s.

**Fig. 6** AOA versus array
factor plot-case 4

**Fig. 6** AOA versus array
factor plot-case 4

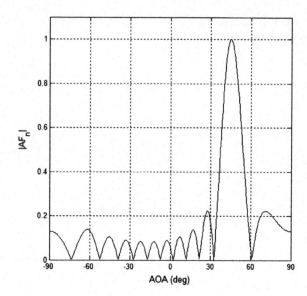

Figure 7 shows the AOA versus AF plot when the desired signal Angle of arrival $= 0°$, Interfering signals are assumed at $30°$ and $45°$ number of elements $= 8$, spacing between the elements $= 0.5\lambda$, Forgetting factor $= 0.95$, Elapsed time $= 0.132754$ s.

Figure 8 shows the AOA versus AF plot if the desired signal Angle of arrival $= 0°$, interfering signals are assumed at $30°$ and $60°$ number of elements $= 10$,

**Fig. 7** AOA versus array
factor plot-case 5

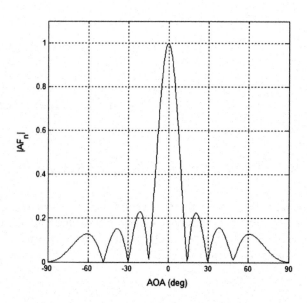

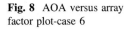

**Fig. 8** AOA versus array factor plot-case 6

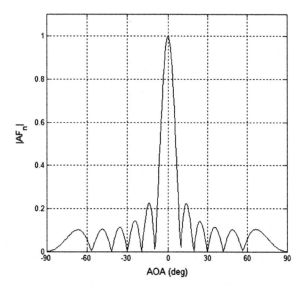

spacing between the elements = $0.6\lambda$, Forgetting factor = 0.95, Elapsed time = 0.126440 s.

Figure 9 shows the Comparison of LMS and RLS algorithms plot if the desired signal AOA = 0°, interfering signals are assumed at 30° and 60° number of elements = 36, spacing between the elements = $0.5\lambda$, step size = 0.01, forgetting factor = 0.95. By using RLS algorithm, the above radiation pattern is achieved in 132 ms and for the same radiation pattern, the run time for LMS algorithm is 1.95 s. Also, it is observed from the results that beam steering and directivity are optimized for some values of AOA.

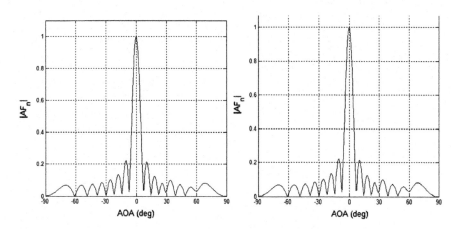

**Fig. 9** Comparison of LMS and RLS algorithm performance

# 4  Conclusion

The array factor is derived in the desired direction of AOA and it is analyzed by using LMS and RLS algorithms, by generating the adaptive beam steering. The performance of these optimization algorithms is compared and analyzed with respect to run time and beam steering. The Simulation results are carried using MATLAB by changing a number of elements and spacing between elements. If the spacing between the elements is less than 0.5, then phase deviation in null formation is observed. For the same specifications of the antenna, the time complexity in case of LMS is observed to be 1–2 s and for RLS it is about less than 200 ms. The behavior of the smart antenna in adaptive beam forming for various AOA is studied, observed, and the plots are generated using MATLAB.

# References

1. Frank B. Gross, "Smart Antennas with Matlab", 2nd edition, Publication, McGraw-Hill Education, section: 8.1–8.4.
2. J.D. Kraus, "Antennas and Wave Propagation", 2nd edition, Publication, McGraw-Hill Education, Page numbers: 8–22.
3. Constantine A. Balanis, Publication: A JOHN WILEY & SONS, Page numbers: 257–288.
4. Subhashree Nibedita Baliarsingh, Anupama Senapati, Arindam Deb, Jibendu Sekhar Roy, "Adaptive Beam Formation for Smart Antenna for Mobile Communication Network Using New Hybrid Algorithms", International conference on communication and signal processing, page numbers: 2146–2151.
5. Ch. Santhi Rani1, P. V. Subbaiah, K. Chenna kesava Reddy and S. Sudha Rani, "LMS and RLS algorithms for smart antennas in a w-cdma mobile communication environment", ARPN journal of Engineering and Applied Sciences, ISSN: 1819–6608, Page numbers: 78–88.
6. Omar Khaldoon Abdulrahman, Md. Mijanur Rahman: Modifying MVDR Beamformer for Reducing Direction-of-Arrival Estimation Mismatch, Arabian Journal for Science and Engineering, 41(9) 3321–3334 (2016).
7. VVSSS Chakravarthy, PSR Chowdary, Ganapati Panda, Jaume Anguera, Aurora Andujar, Babita Majhi: On the Linear Antenna Array Synthesis Techniques for Sum and Difference Patterns Using Flower Pollination Algorithm, Arabian Journal for Science and Engineering, Aug (2017).
8. Adnan Kaya, Irfan Kaya and Haluk E. Karaca,: U-Shape Slot Antenna Design with High-Strength $Ni_{54}Ti_{46}$ Alloy, Arabian Journal for Science and Engineering, 41(9) 3297–3307 (2016).

# A Level Cross-Based Nonuniform Sampling for Mobile Applications

## R. Viswanadham, T. Sudheer Kumar and M. Venkata Subbarao

**Abstract** The main objective of this chapter is to implement a level cross-based nonuniform sampling for cellular mobile systems to reduce computational complexity and bandwidth. All classical mobile systems sample and process the signals based on the **Nyquist** signal processing architectures. These systems do not consider the speech signal variations and they sample the signal at a fixed rate. It causes to process more number of samples without any significant information. As a result, they need more transmission bandwidth to transfer the signal and they take more number of computations to process. As the number of computations increases, the system complexity and power consumption will increase. In this chapter, we consider several realistic signals like speech signals to verify the performance of proposed sampling technique.

**Keywords** Cellular · Uniform sampling · Adaptive filtering
Activity

## 1 Introduction

In the past decade, cellular mobile communication is one of the most active areas of technology development. This development brings services, such as the sharing of videos, images, and data along with basic voice telephony. Transmitter power and channel bandwidth are the basic resources to add capacity to wireless. But these resources are not rising at rates that can bear estimated demands for capacity. The transmission bandwidth of a signal depends on the sampling rate. The power requirement depends on the number of computations. Uniform sampling is suitable for stationary signals. But for time-varying signals like speech, audio, etc., uniform

R. Viswanadham (✉) · T. Sudheer Kumar (✉) · M. Venkata Subbarao
Department of ECE, Shri Vishnu Engineering College for Women, Bhimavaram, India
e-mail: ravuri.viswanadh@gmail.com

T. Sudheer Kumar
e-mail: skterlapu@gmail.com

© Springer Nature Singapore Pte Ltd. 2018
J. Anguera et al. (eds.), *Microelectronics, Electromagnetics
and Telecommunications*, Lecture Notes in Electrical Engineering 471,
https://doi.org/10.1007/978-981-10-7329-8_79

sampling is not suitable because there may be a lot of signal absences in time. So if uniform sampling is considered for voice signals, it leads to larger bandwidth requirement and high power consumption. All classical mobile systems sample and process the signals based on the Nyquist signal processing architectures. These systems do not consider the speech signal variations and they sampled the signal at a fixed rate. It causes to process more number of samples without any significant information. As a result, they need more transmission bandwidth to transfer the signal and they take more number of computations to process. As the number of computations increases, the system complexity and power consumption will increase.

The main motive of this chapter is to reduce the transmission bandwidth of the voice signal and number of computations. As the number of samples decreased the number of computations gets condensed results a great reduction in power consumption. To reduce the system complexity and energy cost, in this chapter, we consider level cross-based nonuniform sampling along with signal processing techniques including filter designs.

This chapter is organized as follows. Section 2 describes the brief summary of existing methods for nonuniform sampling along with literature survey. Section 3 describes the proposed method along with the graphical explanation. Simulation results and computational complexity of nonuniform sampling are presented in Sect. 4. Conclusion and remarks are presented in Sect. 5.

## 2  Literature Survey

Almost all natural signals like speech, seismic, and biomedical are time varying in nature. K.M. Guan proposed adaptive reference levels in a level-crossing analog-to-digital converter [1]. In the nonlinear quantization functions, the number of quantization levels is dynamically assigned depending on the importance of the given amplitude range [2, 3]. A. C. Singer proposed a stable algorithm to perfectly reconstruct signals of finite rate of innovation using level-crossing samples [4]. M. Malmir Chegini suggested level cross ADC is a substitute for traditional schemes. They also suggested an alternative and multi-level adaptive level cross schemes to improve the performance of converters [5]. David G. Nairn provided the current research trends of time-interleaved A/D converters [6].

T. Wang et al. [7] described the conversion of audio signals with good resolution using small number of threshold levels and interpolation. Here to produce uniform samples, the samples are taken at nonuniform in time and then interpolated. Mariya Kurchuk [8] presented a new variable-resolution quantizer and also advantages of variable-resolution ADC were discussed. M. Sun S. Senay et al. [9] proposed a variable LC scheme for the sampling and reconstruction. This process is particularly matched for applications where the signal occurs in bursts [10]. Modris Gretitans et al. [11] proposed the modification of the traditional level-crossing sampling technique, which allows reducing the number of obtained samples, if the levels are placed too densely within the dynamic range of the input signal. The reduction is based on finding and keeping only those level-crossing samples, which are most closely located to signal peaks. All the other samples are discarded. In such a way, the obtained result is similar to peak sampling with samples being taken only at peak points (local extreme) of the signal. The recovery of the continuous signal can be based on piecewise linear interpolation, which provides good results for the speech signal [12]. The proposed sampling technique can be used in data acquisition systems to reduce the amount of data being transferred at peak points (local extreme) of the signal [13].

S. M. Quisar et al. [14] proposed multirate filtering approach based on adaptive rate filtering techniques. In level cross sampling scheme, based on local variations of the time-varying signal sampling rate will be decided. Due to great reduction in unwanted samples, the computational complexity will be greatly reduced in post-signal processing sections. S. Patil and Y. Tsividis et al. [15] proposed a new sampling technique, called Derivative LCS. Instead of direct level-crossing of input, here they consider the derivative of the input, and then result is transmitted.

## 3 Nonuniform Sampling-Level Cross Sampling

The LCSS is one of the techniques for sampling the time-varying signal. In LCSS technique, the samples are considered only when the speech or time-varying signal $x(t)$ crosses the fixed threshold level. In uniform sampling, all the samples are equally spaced but here samples are nonuniformly spaced along the time axis. The samples are considered depend on signal variations. When the slope of the signal is high then samples are very close to one another whereas if the slope of the signal is very low then samples are largely spaced. The functional representation of LCSS is shown in Fig. 1.

### 3.1 LCSS-Based ADC

In LCSS-based ADC process, for reconstruction of the original signal we must know the sampling time periods, whereas the sample magnitudes are quantized based on the number of bits M. The value of M decides the number of quantization levels and it is considered with a large value so that the quantization error should be minimum and to ideal reconstruction of original signal. The detailed reconstruction block diagram is shown in Fig. 2.

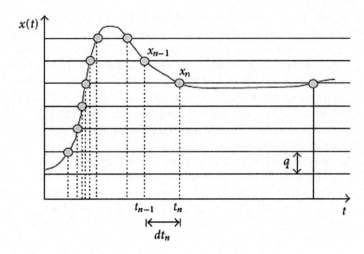

**Fig. 1** Graphical representation of LCSS

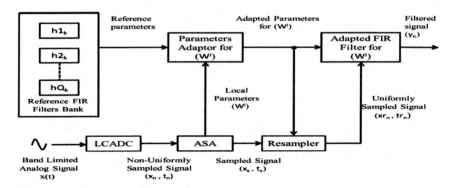

**Fig. 2** Reconstruction of signal from LCS signal

In this method sampling frequency $(f_s)$ is depends on 'M' and signal variations. Due to the reduction of number samples using LCSS, the energy will save because of less number of computations. The threshold levels differ as

$$q = \Delta V / (2^M - 1) \tag{1}$$

where M is a number of bits of ADC, $\Delta V$ is a dynamic range of the input signal $x(t)$. The maximum sampling frequency $(Fs_{max})$ and minimum sampling frequency $(Fs_{min})$ are given as

$$Fs_{max} = 2f_{max}\left(2^M - 1\right) \tag{2}$$

$$Fs_{min} = 2f_{min}\left(2^M - 1\right) \tag{3}$$

## 3.2 Activity Selection Method

The main drawback of LCADCs is based on signal characteristics and some parts of the signal are sampled at higher rates than existing methods. By considering LCS nonuniformly, this limitation is reduced up to some extent. Based on local features of the signal, it selects only appropriate signal parts. Further, each selected portion characteristics are analyzed independently and then they used these characteristics to adapt parameters of the system and this procedure is called Activity Selection Algorithm [16].

## 4 Simulation Results and Discussions

In this chapter, we consider several realistic signals like speech signals to verify the performance of proposed sampling technique. To illustrate the performance proposed method, we consider a nonstationary signal with three active parts. The parameters of the time-varying signal are shown in Table 1. Every active part has different low and high frequencies.

Here, the signals have minimum frequency component $f_{min}$ is 5 Hz and maximum frequency component $f_{max}$ is 1 kHz. Here, we consider numbers of quantization levels are 8. For 3-bit resolution, $Fs_{min}$ and $Fs_{max}$ are 70 Hz and 14 kHz respectively. Here, the dynamic range of the signal $\Delta V$ is 1.8 V and step size is 0.2571 (Fig. 3).

The uniform-sampled signal and the LCADC output is shown on the Figs. 4 and 5. From Fig. 5, it can be observed that the number of samples obtained from uniform sampling is more than the Level-Crossing-Sampled signal.

| Table 1 Active parts information of the signal | Active part | Components | Duration (s) |
|---|---|---|---|
| | I | $(1/2)\sin(2\pi20t) + (2/5)\sin(2\pi1000t)$ | 1 |
| | II | $(2/5)\sin(2\pi10t) + (2/5)\sin(2\pi150t)$ | 1 |
| | III | $(3/5)\sin(2\pi5t) + (3/10)\sin(2\pi100t)$ | 1 |

**Fig. 3** The input simulated
signal with three active parts

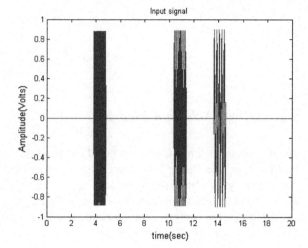

**Fig. 4** Uniformly sampled
signal

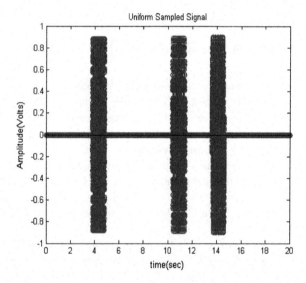

The ASA output is as shown in Fig. 6. Here, only active parts of the signal are extracted. To get high-frequency components in each active part, a bank of 11 LPF FIR filter is developed. The parameters considerations are shown in Table 2.

The ASA delivers three selected windows for the whole $x(t)$ span of 20 s. The windows specifications are displayed in Table 3.

After performing the filtering operation, by considering the filter coefficients obtained by Activity Reduction by Filter Decimation/Interpolation technique, the reconstructed signal is as shown in below Fig. 7.

To prove the superiority of the proposed LCS with ASA, here we consider a speech signal for 2 s duration as shown in Fig. 8. In this case, the signal is given as

**Fig. 5** Level cross sampled signal

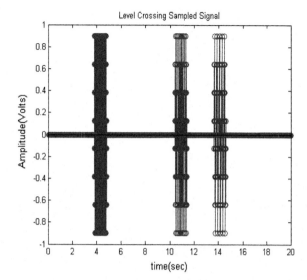

**Fig. 6** Active parts selection with ASA

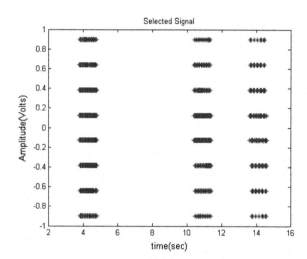

**Table 2** Filter parameters

| Parameter | Transmission B.W. | Fc | Rp (dB) | Rs (dB) | Fref | P |
|-----------|-------------------|-----|---------|---------|------|-----|
| Value | 30:80 | 25 | −80 | −25 | 2500 | 127 |

input to LCADC of resolution 5 bits. Thus, the $Fs_{min}$ and $Fs_{max}$ are 3100 Hz and 248 kHz respectively. The signal is sampled at the ADC resolution of 5-bits. The uniform sampled speech signal is as shown in Fig. 9.

**Table 3** After ASA number of samples in active parts

| Wi | T$^i$ (s) | N$^i$ (Samples) | Fs$^i$ (Hz) |
|---|---|---|---|
| I | 0.9968 | 1080 | 1083 |
| II | 0.9992 | 800 | 801 |
| III | 0.9988 | 470 | 471 |

**Fig. 7** Reconstructed signal

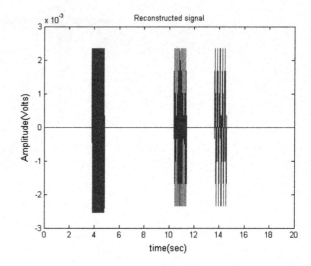

**Fig. 8** Speech signal

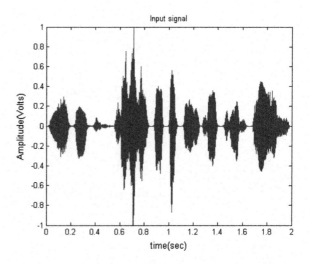

The level cross sampled signal and ASA outputs are shown in Figs. 10 and 11. By observing Figs. 9, 10 and 11 the number samples in LCS-based ASA output is greatly reduced.

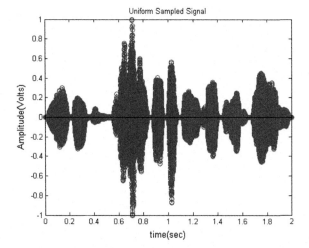

**Fig. 9** Uniformly sampled speech signal

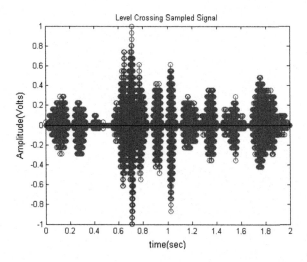

**Fig. 10** Level-crossing-sampled speech signal

Due to this activity-based signal processing, the signal information is not disturbed but numbers of samples will be greatly reduced results saving bandwidth and reduction in power consumption.

After performing the filtering operation, by considering the filter coefficients obtained by ARDI technique, the reconstructed signal is as shown in Fig. 12. From the above two cases of signals, it is clear that LCS-based activity selection algorithm produces very less number of samples after sampling results saving bandwidth and reduction in power consumption.

**Fig. 11** LCS-ASA speech signal

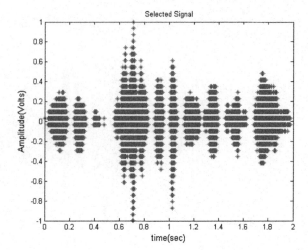

**Fig. 12** Reconstructed signal

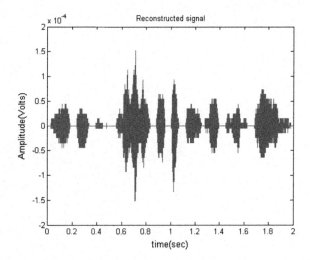

## 5   Conclusion

A new nonuniform sampling technique called level cross-based activity selection algorithm for cellular mobile systems is presented in this chapter. Due to this nonuniform sampling, there is a great reduction in computational complexity and bandwidth. All classical mobile systems do not consider the speech signal variations and they sampled the signal at a fixed rate. It causes to process more number of samples without any significant information. As a result, they need more transmission bandwidth to transfer the signal and they take more number of computations to process. To verify the performance of proposed sampling technique, speech signals are sampled nonuniformly then processed and finally reconstructed with filter banks.

# References

1. K. M. Guan and A. C. Singer, A Level-Crossing Sampling Scheme for both Deterministic and Stochastic Non-Band limited Signals, in Proceedings of Sarn off Symposium, March, 2006.
2. Z. Duan, J. Zhang, C. Zhang, E. Mosca, A simple design method of reduced-order filters and its application to multi-rate filter bank design, Elsevier Journal of Signal Processing Volume 86, Issue 5, Pages 1061–1075, May 2006.
3. E. Kofman and J. H. Braslavsky, Level crossing sampling in feedback stabilization under data rate constraints, Proc. IEEE Conf. Decision and Control, pp. 9423–4428, Dec. 2006.
4. K. M. Guan and A. C. Singer, Opportunistic Sampling by Level-Crossing, ICASSP'07, pp. 1513–1516, April 2007.
5. M. Malmir Chegini, F. Marvasti, Performance improvement of level-crossing A/D converters, IEEE International Conference on Telecommunications and Malaysia International Conference on Communications ICT-MICC, 438–441, May 2007.
6. D. Nairn, Time-interleaved analog-to-digital converters, IEEE 2008 Custom Integrated Circuits Conference (CICC), pp. 289–296, Sept. 2008.
7. T. Wang, D. Wang, P. J. Hurst, B. C. Levy, S. H. Lewis, A level-crossing analog-to-digital converter with triangular dither, IEEE Trans. Circuits Syst. I Reg. Papers, vol. 56, no. 9, pp. 2089–2099, Sep. 2009.
8. M. Kurchuk and Y. Tsividis, Signal-dependent variable-resolution clock less A/D conversion with application to continuous-time digital signal processing, IEEE Transactions on Circuits and Systems, vol. 57, pp. 982–991, May 2010.
9. M. Sun S. Senay, L. F. Chaparro and R. J. Sclabassi. Adaptive level-crossing sampling and reconstruction. *Proc. Of EUSIPCO 2010*, pages 1296–1300, Aug. 2010.
10. M. Trakimas, S. R. Sonkusale, An adaptive resolution asynchronous ADC architecture for data compression in energy constrained sensing applications, IEEE Trans. Circuits Syst. I Reg. Papers, vol. 58, no. 5, pp. 921–934, May 2011.
11. Modris Greitans, Rolands Shavelis1, Laurent Fesquet, TahaBeyrouthy, Combined Peak and Level-Crossing Sampling Scheme,... IEEE Transactions on Communications, May 2011.
12. M. Kafashan, M. Ghorbani, and F. Marvasti, A sigma-delta analog to digital converter based on iterative algorithm, EURASIP Journal on Advances in Signal Processing, vol. 2012, no. 1, p. 149, 2012.
13. C. Weltin-Wu, Y. Tsividis, An event-driven clock less level-crossing ADC with signal-dependent adaptive resolution, IEEE J. Solid-State Circuits, vol. 48, no. 9, pp. 2180–2190, Sep. 2013.
14. S. M. Qaisar, L. Fesquet and M. Renaudin, Adaptive rate filtering a computationally efficient signal processing approach. EURASIP Journal on Advances in Signal Processing. 2014 January; vol. 94:620–630.
15. Martinez-Nuevo. P, Patil. S, Tsividis. Y, Derivative Level-Crossing Sampling, IEEE Transactions on Circuits and Systems II. 2015 January; 62(11):11–15.
16. Saied M. Abd El-atty, Z. M. Gharsseldien, Saied M. Abd El-atty: Mobile Traffic Offloading in Heterogeneous Networks-Based Small Cell Technology, Arab J Sci Eng, 41(2) 555–567 (2016).

# SIW-Based Different Anchor-Shaped Slot Antennas for 60 GHz Applications

M. Nanda Kumar and T. Shanmuganantham

**Abstract** Substrate-integrated waveguide is a good candidate for millimeter communication. In this chapter, SIW-based two different anchor-shaped slot antennas for 60 GHz applications is proposed, and designed by using Rogers dielectric material with a dielectric constant of 2.2 and height of substrate is 0.381 mm. One of the structures will provide 5 GHz impedance bandwidth with respect to $-10$ dB reference line (range is 59.831–64 GHz), resonant frequency is approximately 60 GHz and their results like reflection coefficient, gain, VSWR, radiation efficiency, transmission efficiencies are $-23.5$ dB, 5.5 dBi, 1.165, 91%, and 81%.

**Keywords** Substrate-integrated waveguide (SIW) · Microstip
GIFI · Wireless LAN (WLAN)

## 1 Introduction

Millimeter wave frequency, i.e., 30–300 GHz plays an important role in communications due to increasing improvements in academia and industry applications. The fixed unlicensed frequencies of millimeter wave technologies are 60 GHz (wireless communication networks) [1], 79 GHz (Automotive radar systems) [2], and 94 GHz (millimeter wave imaging). To develop this, an antenna requires broadband and high gain.

57–64 GHz band (7 GHz) is an unlicensed band which is assigned by federal communication commission (FCC) to access unlicensed devices and the resonant frequency is 60 GHz. this frequency band is used for high data rates [3], short-range

M. Nanda Kumar (✉) · T. Shanmuganantham
Department of Electronics Engineering, Pondicherry University, Pondicherry, India
e-mail: nanda.mkumar12@gmail.com

T. Shanmuganantham
e-mail: shanmugananthamster@gmail.com

© Springer Nature Singapore Pte Ltd. 2018
J. Anguera et al. (eds.), *Microelectronics, Electromagnetics and Telecommunications*, Lecture Notes in Electrical Engineering 471,
https://doi.org/10.1007/978-981-10-7329-8_80

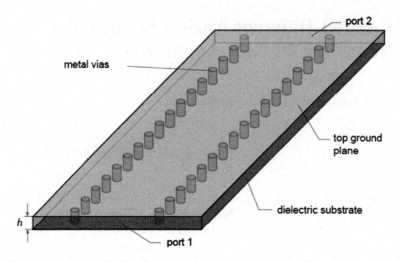

**Fig. 1** Substrate integrated waveguide

applications such as wireless local area network (WLAN), automotive applications, and GIFI.

SIW is mainly invented to implement for high-frequency applications like millimeter and centimeter applications and one type of substrate-integrated circuits (SICs) [4–10]. The shape of SIW is a waveguide, integrated with help of two rows (periodic) of holes or vias interlinked with the bottom and top ground planes of a substrate. Figure 1 describes the simple structure of an SIW. Compared to microstrip lines, the fabrication process of SIW is very simple, low weight, moderate size, and cost effective and compare to a rectangular waveguide, it has high quality factor, more power handling, low interference. The main advantage of SIW is integrated into planar forms, includes passive components, active components, and antennas.

In this chapter, different anchor-based SIW slot antenna combined with a microstrip line feed with the input impedance of 50 Ω for 60 GHz frequency applications is proposed. The structure of chapter is as follows. Section 2 describes about a design of SIW; Sect. 3 describes about the Antenna structure. Finally, Sect. 5 describes about the conclusion.

## 2 SIW Design

SIW is type of transmission line. It is derived form rectangular waveguide and sandwiched between dielectric-filled waveguide (DFW) and waveguide. The standard width of SIW is measured with help of Eq. 1 and mentioned below [11–14].

$$a_R = a_S - \frac{d^2}{0.95s} \qquad (1)$$

Radiation loss plays a very important role in SIW. The standard equations used to minimize the radiation loss is mentioned in Eqs. 2 and 3 [12–14].

$$d \le \frac{\lambda_g}{5} \qquad (2)$$

and

$$s \le 2d \qquad (3)$$

## 3  Proposed Antenna Structure

The representation of the proposed structure is shown in Fig. 2. Figure 2a represents the top view of two different anchor-shaped slot antennas and Fig. 2b represents the bottom view proposed antennas but bottom views of the two antennas are same. In that sky blue color represents the substrate, i.e., Rogers RT/Duriod 5880 with a dielectric constant of 2.2 and chooses the thickness (height) is 0.381 mm, yellow color represents the copper and thickness is 35 μm. The D, S are the diameter of hole and spacing between two holes, their values are 0.1, 0.2 mm.

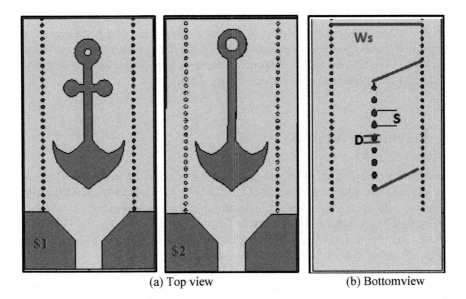

(a) Top view                    (b) Bottomview

**Fig. 2** Proposed structure

**Table 1** Antenna parameters

| Parameter | Value (mm) | Parameter | Value (mm) |
| --- | --- | --- | --- |
| W1 | 0.83 | L1 = L2 | 0.842 |
| W2 | 2.5 | L3 | 5.7 |
| W3 | 0.25 | L5 | 4.2 |
| Ws | 2.8 | L4 | 0.67 |
| R1 | 0.5 | R3 | 0.25 |
| R2 | 0.1 | | |

**Fig. 3** Parameter description of the slot

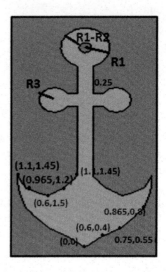

The w1, w2 are the width of microstrip, tapering width of microstrip, which can be derived by basic equations of microstrip. In Table 1 the parameters are used to design the two different anchor-shaped slot antennas. The parameter representation of the proposed antenna is represented in Fig. 2 (Fig. 3).

## 4 Results and Analysis

The reflection coefficient of proposed antennas is mentioned in Fig. 4. The first antenna ($1) will give the bandwidth of 4.12 GHz (in between 59–63.12 GHz) with respect to −10 dB reference line, the resonant frequency is approximately 60 GHz, represented in green color. Next, antenna ($2) will give bandwidth of 5 GHz, i.e., 59–64 GHz, the resonant frequency is approximately 60 GHz and their reflection coefficient value is −23.5 dB. Figure 5 represents VSWR for the proposed structures ($1, $2). All two structures bandwidths are match with impedance bandwidth with respect to VSWR (2:1). Compared to two structures, the second antenna has more bandwidth, i.e., more than 800 GHz. Bandwidth improvement is a major

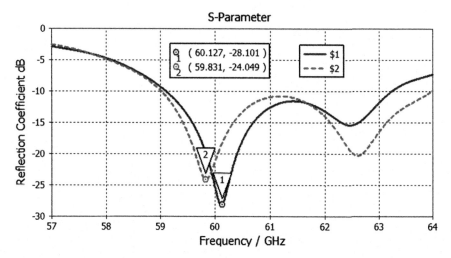

**Fig. 4** Frequency versus reflection coefficient ($S_{11}$)

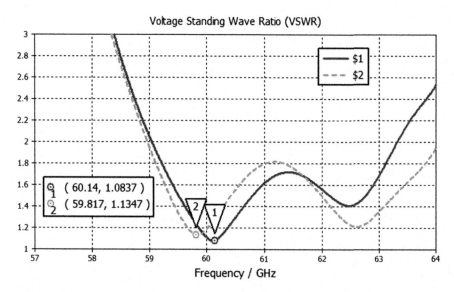

**Fig. 5** Voltage standing wave ratio

parameter in 60 GHz band. So, all parameters like surface current, efficiencies, gain, and radiation patterns are represented for antenna2 ($2) why because it as more bandwidth compare to antenna1 ($1).

The two-dimensional radiation patterns of 60 GHz are represented in Fig. 6. Figure 6a represents E-field, radiates in bidirectional and Fig. 6b represents the H-field, radiates in all direction. The efficiencies of the proposed antenna in between

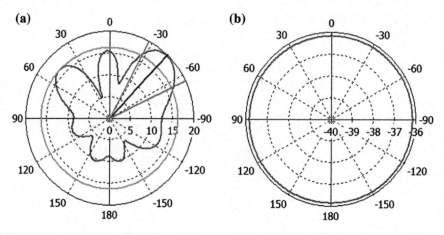

**Fig. 6** 2D radiation patterns for all resonant frequencies **a** E-field, **b** H-field

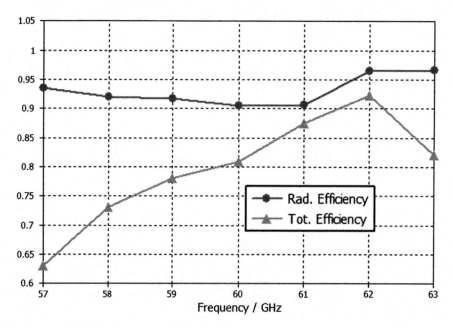

**Fig. 7** Efficiencies in between 57–64 GHz

57 and 64 GHz was represented in Fig. 7. In that red color indicates radiation efficiency and green color indicates for transmission efficiency. This antenna has good radiation as well as transmission efficiency and their values are 91, 81% at 60 GHz.

**Fig. 8** Maximum gain over frequency in between 57 and 64 GHz

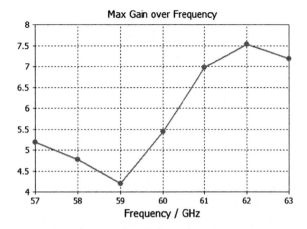

**Fig. 9** Three-dimensional gain pattern at 60 GHz

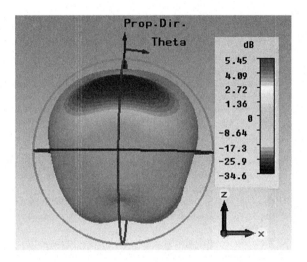

Figure 8 represents the maximum gain over frequency in between 57 and 63 GHz. The maximum gain is 7.5 dBi at 62 GHz but a little bit less at 60 GHz, i.e., 5.45 dBi that is shown in Fig. 9 in that direction of propagation is in the direction of Z (slot is etched in the direction of z).

# 5 Conclusion

In this chapter, two different anchor-shaped SIW slot antennas for millimeter wireless communication (60 GHz application) fed by microstrip was introduced. One of antenna structure produces 5 GHz bandwidth in between 59 and 64 GHz with a resonant frequency of 60 GHz with respect to −10 dB reference line and also match with the VSWR (2:1) and also observes computer simulation results like VSWR, reflection coefficient, gain, efficiencies, and radiation. Due to high bandwidth this can be implemented for WLAN, GIFI, and automotive applications. Furthermore, SIW slot antenna is also used for another two unlicensed bands in millimeter wave frequencies like automotive radar and millimeter wave imaging.

# References

1. Purva Shrivastava and T. Rama Rao "Performance Investigations with ATLSA on 60 GHz Radio Link in a Narrow Hallway Environment" Progress in Electromagnetics Research, vol. 58, pp. 69–77 (2015)
2. Shi Cheng, Hanna Yousef, and Henrik Kratz "79 GHz Slot Antennas Based on SIW in a Flexible Printed Circuit Board" IEEE Transaction on Antennas and Propagation, vol. 57, no. 1 (2009)
3. D. Lockie and D. Peck, "High data rate Millimeter wave Radios," IEEE Microwave Mag., vol. 10, o. 5, pp. 75–83 (2009)
4. S. Ramesh and T. Rama Rao "Planar High Gain Dielectric Loaded Exponentially TSA for Millimeter Wave Wireless Communications" Wireless Press Communication (Springer), pp. 3179–3192 (June 2015)
5. Yujian Li and Kwai-Man Luk "Low Cost High Gain and Broadband Substrate-Integrated-Waveguide-Fed Patch Antenna Array for 60-GHz Band" IEEE Transactions on Antennas and Propagation, vol. 62, no. 11, pp. 5531–5538 (2014)
6. Junfeng Xu, ZhiNingChen, and Xianming Qing "CPW Center-Fed Single-Layer SIW Slot Antenna Array for Automotive Radars" IEEE Transactions on Antennas and Propagation, vol. 62, no. 9, pp. 4528–4536 (2014)
7. Soumava Mukherjee, Animesh Biswas, and Kumar Vaibhav Srivastava "Substrate Integrated Waveguide Cavity-Backed Dumbbell-Shaped Slot Antenna for Dual-Frequency Applications" IEEE Antennas and Wireless Propagation Letters, vol. 14, pp. 1314–1317 (2015)
8. Dongquan Sun, Jinping Xu and Shu Jiang "SIW horn antenna built on thin substrate with improved impedance matching" ELECTRONICS Letters, vol. 51, No. 16 pp. 1233–1235 (2015)
9. Maurizio Bozzi, Luca Perregrini, Ke Wu, Paolo Arcioni "Current and future research trends in substrate integrated waveguide technology" Radio engineering, vol. 18, no. 2, pp. 201–207 (2009)
10. Nanda kumar M and T Shanmuganantham "Substrate Integrated Waveguide Cavity Backed Bowtie Slot Antenna for 60 GHz Applications" IEEE International Conference on Emerging Technology Trends (2016)
11. Nanda Kumar M and T Shanmuganantham "Substrate Integrated Waveguide Cavity Backed with U and V Shaped Slot Antenna for 60 GHz Applications" International Conference on Smart Engineering Materials (2016)
12. Nanda Kumar M and T Shanmugnantham "Current and Future Challenges in Substrate Integrated Waveguide Antennas–An overview" IEEE International Conference on Advanced Computing (February-2016)
13. Nanda Kumar M and T Shanmugnantham "Substrate Integrated Waveguide $\pi$ Shaped Slot Antenna for 57–64 GHz Band Applications" lecturer notes in electrical engineering (Springer) (Accepted for Publications)
14. Nanda Kumar M and T Shanmugnantham "Substrate integrated waveguide tapered slot antenna for 57–64 GHz applications" ICCCSP (Jan-2017)

15. Nanda Kumar M and T Shanmugnantham "Neptune shaped slot antenna with SIW cavity for 60 GHz applications" International Journal of Control Theory and Applications, vol. 10 (2017)
16. Nanda Kumar M and T Shanmugnantham "SIW based Crown shaped slot antenna for 60 GHz Applications" International Journal of Control Theory and Applications, vol. 10 (2017)

# Microstrip Feed Dumbbell-Shaped Patch Antenna for Multiband Applications

K. Yogaprasad and R. Anitha

**Abstract** The patch antenna designs have a good resolution for microwave applications. In this chapter, we introduce a dumbbell-shaped patch ring slot with a triangular patch antenna for multiband applications and the coplanar waveguide is used as a feed. The size of the proposed structure is 14 mm $\times$ 14 mm $\times$ 1.6 mm$^3$, which is intended by using a FR-4 substrate (dielectric constant is 4.4) with the height of 1.6 mm. This antenna produces four resonant frequencies 1.4745, 8.714, 13.726, 15.796 GHz and their reflection coefficient values are $-14.799$, $-26.583$, $-24.597$, and $-17.255$ dB. The simulation results like return loss, VSWR, gain, radiation patterns, efficiencies, and surface current are observed.

**Keywords** CPW $\cdot$ Multiband $\cdot$ Circular patch antenna $\cdot$ Microstrip

## 1 Introduction

In wireless communication, antennas play a very important role, it is stand-in as a transducer among transmitter and free space. They are well-organized radiators of electromagnetic energy into free space.

Recent wireless communication systems require small profile, low weight, more gain, and simple structure antennas to assure reliability, mobility, and more efficiency [1–3]. A microstrip patch antenna is very simple to construct. Microstrip antennas consist of a patch of metallization on a grounded dielectric substrate. They are low profile, lightweight antennas, most suitable for aerospace and mobile applications [4, 5].

K. Yogaprasad (✉)
Department of ECE, SITAMS, Chittoor, India
e-mail: kyogaprasad@gmail.com

R. Anitha
Department of ECE, SVCE, Chittoor, India
e-mail: anithavr@gmail.com

© Springer Nature Singapore Pte Ltd. 2018
J. Anguera et al. (eds.), *Microelectronics, Electromagnetics and Telecommunications*, Lecture Notes in Electrical Engineering 471,
https://doi.org/10.1007/978-981-10-7329-8_81

In the available literature, some researchers have used FR-4 epoxy as the antenna substrate to reduce the antenna cost [6–14] and some have used different substrates due to lossy characteristics of FR-4 at millimeter frequencies.

This chapter is structured as follows. First, the design of antenna is described in Sect. 2 which is followed by results and discussion in Sect. 3 and finally, conclusion can be described in Sect. 4.

## 2 Proposed Antenna Structure

The antenna designs of the proposed structure are shown in Fig. 1. First dumbbell-shaped patch antenna fed by microstrip line [Ant $1] is introduced, which will give three resonant frequencies (1.6373, 9.113, and 13.912 GHz). Next section, introduces ring-shaped patch instead of circular-shapeddumbbell with a width of 0.3[Ant $2], again produces three resonant frequencies (7.9724, 13.662, and 15.675 GHz). Finally, we introduce triangular patch in between ring patch [Ant $3] with a width of 4.6 and produces four resonant frequencies (1.475, 8.714, 13.726, and 15.796 GHz). The proposed structures are intended by using flame retardant (FR) 4 substrate with dielectric constant of 4.4 and dimensions of proposed structures are $19 \times 12 \times 1.6 \text{ mm}^3$. In Fig. 2, yellow represents the substrate and red represents the copper with a thickness of 0.1. The parameters used to design proposed antenna is described in Table 1.

## 3 Results and Analysis

Figure 3 shows the frequency (GHz) versus reflection coefficient (dB) of the proposed antenna, which is observed between 1–18 GHz frequency range, achieves four resonant frequencies 1.4745, 8.714, 13.726, 15.796 GHz, and their reflection

**Fig. 1** Circular patch antenna

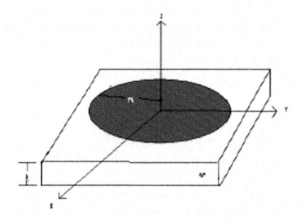

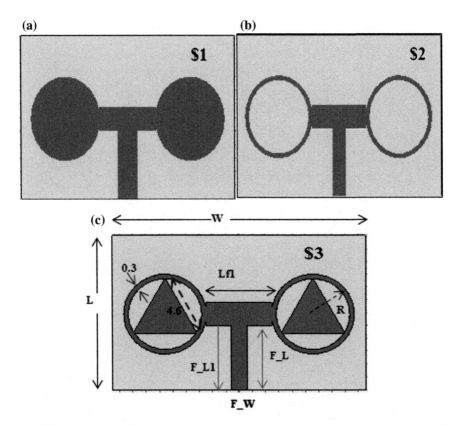

**Fig. 2** Different antenna structures

**Table 1** Antenna parameters

| Parameter | Value (mm) | Parameter | Value (mm) |
|---|---|---|---|
| W | 19 | Lf1 | 5 |
| L | 12 | R | 3.1 |
| F_L | 5 | h | 1.6 |
| F_L1 | 6.8 | t | 0.1 |
| F_W | 1.8 | | |

coefficient values are −14.799, −26.583, −24.597, −17.255 dB. This antenna is useful for L-band (1.4745 GHz), X-band (8.714), and Ku-band (13.726 and 15.796 GHz) Applications.

The reflection coefficient values of the three antennas are demonstrated in Fig. 4. Green indicates the first antenna that is microstrip feed dumbbell-shaped patch antenna ($1), blue indicates the second antenna that is after introducing ring shape patch instead of circular shapes in dumbbell ($2) and third one is the proposed

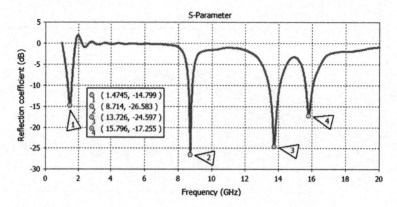

**Fig. 3** Frequency versus reflection coefficient ($S_{11}$) of a proposed antenna

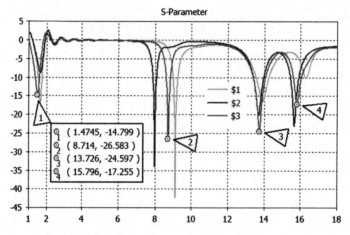

**Fig. 4** Frequency versus reflection coefficient ($S_{11}$) for three antenna structures

antenna that is after introducing triangular patch in circular rings of dumbbell, which is indicated by red color ($3).

The dumbbell-shaped patch antenna (Ant $1) will provide three resonant frequencies are 1.6373, 9.113, 13.912 GHz and their reflection coefficient values are $-17.316$, $-42.217$, $-16.414$ dB next introducing ring patch instead of circular shape in dumbbell (Ant $2), this antenna also provides three resonant frequencies that are 7.9724, 13.662, 15.675 GHz and their reflection coefficient values are $-33.5$, $-19.621$, 15.675 dB after introducing triangular patch in ring shape (Ant $3), finally got four resonant frequencies that are 1.475, 8.714, 13.726, and 15.796 GHz and their reflection coefficient values are $-14.799$, $-26.583$, $-24.597$, and $-17.255$ dB. The proposed antenna is useful for L-band, X-band, and Ku-band applications.

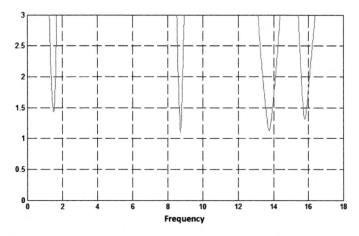

**Fig. 5** Voltage standing wave ratio

Voltage standing wave ratio versus frequency has been presented in Fig. 5. The obtained four resonant frequencies 1.4745 GHz (covers a band of 1.363–1.5944 GHz, bandwidth is 0.2315 GHz (VSWR2:1)), 8.714 GHz (covers a band of 8.6074–8.8363 GHz, bandwidth is 0.2289 GHz (VSWR(2:1)), 13.726 GHz (covers a band of 13.389–14.069 GHz, bandwidth is 0.68 GHz (VSWR(2:1)), and 15.796 GHz (covers a band of 15.608–16.056 GHz, bandwidth of 0.448 GHz (VSWR(2:1)), their VSWR values are 1.4422, 1.1131, 1.1517, and 1.3303.

Figure 6 describes the two-dimensional radiation pattern in that Fig. 6a represents E-field and Fig. 6b represents the h-field for all resonant frequencies and also observed bidirectional propagation in e-field and all direction propagation in h-field.

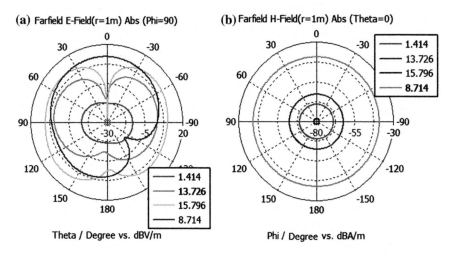

**Fig. 6** 2D radiation patterns for all resonant frequencies **a** E-field, **b** H-field

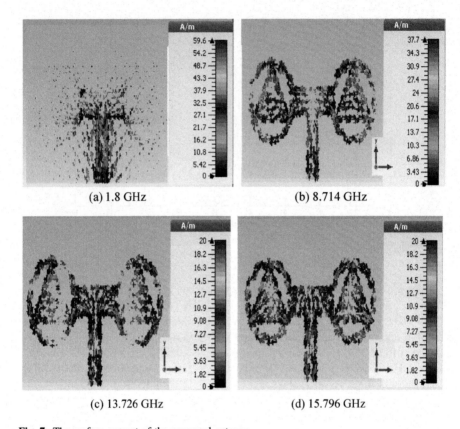

(a) 1.8 GHz                                         (b) 8.714 GHz

(c) 13.726 GHz                                      (d) 15.796 GHz

**Fig. 7** The surface current of the proposed antenna

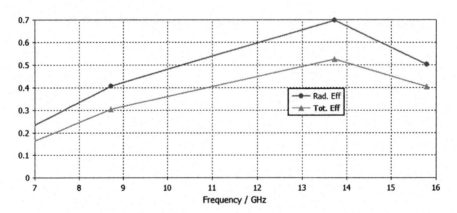

**Fig. 8** Efficiencies of the proposed antenna

**Fig. 9** 3D gain pattern at
13.726 GHz

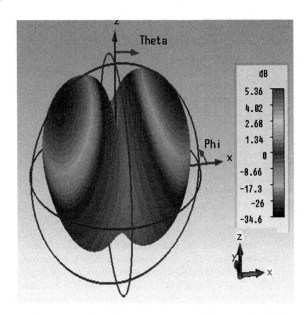

The surface currents of the proposed structure with different resonant frequencies are described in Fig. 7. In Fig. 7a, the current flow is very high at a feed and all other figures current flow is high at entire patch and feed and we also observe that current flow is gradually decreasing while increasing frequencies. The efficiencies of the proposed structure were presented in Fig. 8. Radiation and total efficiencies are 40, 70, 50% and 30, 52, 40% at 8.714, 13.726, 15.796 GHz.

Figure 9 describes the three-dimensional gain pattern at 13.726 GHz and gain value is 5.36 dB and also observed for another three resonant frequencies gain values are 1 dB at 1.4 GHz, 3 dB at 8.714 GHz, and 2.5 dB at 15.976 dB.

# 4  Conclusion

In this chapter, microstrip feed dumbbell-shaped patch antenna is introduced. This antenna will provide four resonant frequencies that are 1.475, 8.714, 13.726, and 15.796 GHz and their reflection coefficient and VSWR values are −14.799, −26.583, −24.597, −17.255 dB, and 1.4422, 1.1131, 1.1517, 1.3303 and will also observes the other parameters like the surface current, radiation patterns, and efficiencies. This antenna is useful for L-band, X-band and Ku-band applications.

# References

1. N. Gunavathi, D. Sriram Kumar "A Simple CPW-Fed Slot Antenna with Octagon Shaped Tuning Stub for HiperLAN/2 and WLAN Applications" ICECS (2014)
2. Mr. T. Shanmuganantham, Dr. S. Raghavan, "Novel Printed CPW-Fed Slot Antenna for Wireless Letters, Wiley interscience (MOTL), USA, Vol. 52 No. 6, pp. 1258–1261 (2010)
3. Bashar B. Qas Elias, "Design of Broadband Circular Patch Microstrip Antenna for KU-Band Satellite Communication Applications", International Journal of Microwave and optical Technology, Vol. 11, No. 5, pp. 362–368 (2016)
4. B. J. Kwaha, O. N Inyang and P. Amalu "The Circular Microstrip Patch Antenna – Design and Implementation" IJRRSS, Vol. 8, No. 8, pp 88–95 (2011)
5. Sarthak Singhal and Amit Kumar Singh "Modified Star-Star Fractal Super-Wideband Antenna" MOTL, Vol. 59, No. 3 (2016)
6. Naveen Jaglan, Binod K. Kanaujia, Samir D. Gupta, and Shweta Srivastava "Triple Band Notched UWB Antenna Design Using Electromagnetic Band Gap Structures" Progress In Electromagnetics Research C, vol. 66, pp. 139–147 (2016)
7. Y. Tawk, Alex R. Albrecht, S. Hemmady, Gunny Balakrishnan, and Christos G. Christodoulou "Optically Pumped Frequency Reconfigurable Antenna Design" IEEE Antennas and Wireless Propagation Letters, Vol. 9 (2010)
8. Rahul Tyagi, SaurabhKohli, "Design and Development of Square Fractal Antenna for Wireless Application", International Journal of Advanced Research in Computer Science and Software Engineering, Vol. 5 (2015)
9. Naima A. Touhami, Yahya Yahyaoui1, Alia Zakriti2, Khadija Bargach1, Mohamed Boussouis1, Mohammed Lamsalli1, and Abdelwahid Tribak3, "A Compact CPW-Fed Planar Pentagon Antenna for UWB Applications" Progress In Electromagnetics Research C, Vol. 46, 153–161 (2014)
10. J. Malik and M. V. Kartikeyan, "Metamaterial Inspired Patch Antenna With L-Shape Slot Loaded Ground Plane For Dual Band (Wimax/Wlan) Applications", Progress In Electromagnetics Research Letters, Vol. 31, 35–43 (2012)
11. Li-Ming Si, Weiren Zhu, and Hou-Jun Sun, "A Compact, Planar, and CPW-Fed Metamaterial-Inspired Dual-Band Antenna," IEEE Antennas and Wireless Propagation Letters, vol 12 (2013)
12. X. L. Quan, R. L. Li, Y. H. Cui, and M. M. Tentzeris, "Analysis and design of a compact dual-band directional antenna," IEEE Antennas and Wireless Propagation Letters, vol. 11, pp. 547–550 (2012)
13. Y.-L. Kuo and K.-L. Wong, "Printed double-T monopole antenna for 2.4/5.2 GHz dual-band WLAN operations," IEEE Transaction and Antennas Propagation, vol 51, no. 9, pp. 2187–2192 (2010)
14. Adnan Kaya, Irfan Kaya and Haluk E. Karaca: U-Shape Slot Antenna Design with High-Strength $Ni_{54}Ti_{46}$ Alloy, Arabian Journal for Science and Engineering, 41(9) 3297–3307 (2016)

# Arc-Shaped Monopole Liquid-Crystal Polymer Antenna for Triple-Band Applications

S. S. Mohan Reddy, B. T. P. Madhav, B. Prudhvi nadh,
K. Aruna Kumari, M. V. S. Praveen, M. Hemachand
and E. Mounika

**Abstract** An arc-shaped compact monopole antenna with defected ground structure is designed for triple-band applications with gain enhancement are observed in this chapter. The antenna structures are implemented to achieve triple-band properties. By using linear array technique, characteristics like operating bandwidth, antenna gain, and efficiency were analyzed and improved for the designed antennas. The designed antenna has better radiation characteristics with a peak gain of 8.49 dB, efficiency toward radiation is 97%, return loss bandwidth is 2.3, 3.4, 5.2 GHz and it also offers front-to-back ratio of 4.1257. The average gain and radiation efficiency is improved by applying linear array to the proposed antenna. The detailed design of the proposed antenna is simulated using ANSYS HFSS 17.

**Keywords** Linear array · Triple band · Monopole · Gain

## 1 Introduction

The Microstrip Patch Antenna (MSA) comprises of a radiating patch on the head side of a dielectric substrate which has a ground plane on the tile side. For improved analysis, the patch has been taken in shapes like square, rectangular, round, triangular, curved, or some other normal shape. For better antenna performance, the antenna must contain a thick dielectric substrate with a low dielectric constant is needed for data transmission and better radiation [1]. In order to design a compact MSA, higher dielectric constants we must use materials having less efficient and result in a narrower bandwidth. The patch antennas are widely used in wireless

S. S. Mohan Reddy (✉)
Department of ECE, SRKR Engineering College, Bhimavaram, AP, India
e-mail: rahulmohan720@gmail.com

B. T. P. Madhav · B. Prudhvi nadh · M. V. S. Praveen · M. Hemachand · E. Mounika
Department of ECE, K L University, Guntur, AP, India

K. Aruna Kumari
Department of CSE, SRKR Engineering College, Bhimavaram, AP, India

© Springer Nature Singapore Pte Ltd. 2018                                        797
J. Anguera et al. (eds.), *Microelectronics, Electromagnetics
and Telecommunications*, Lecture Notes in Electrical Engineering 471,
https://doi.org/10.1007/978-981-10-7329-8_82

applications due to their low-profile structure for that they are extremely compatible with compact antennas in handheld wireless devices such as cellular phones, pagers, etc. The MSAs has the advantages such as lightweight and low volume, low fabrication cost, and feasibility to integrate into Microwave-Integrated Circuits (MICs) and shows of dual and triple-frequency characteristics. With all these advantages, of microstrip patch antenna there are also disadvantages such as narrow bandwidth, low efficiency, low gain, low power handling capacity, and surface wave excitation [1, 2]. The designed antenna has arc-shaped monopole antenna with defected ground structure has triple-band characteristics. The triple-band antenna which is operated at WLAN (5.15–5.35 GHz) and WiMAX (3.4–3.6 GHz) band frequencies is designed to improve the narrow band characteristics of the circular patch antenna to multiband frequency.

Triple band characteristics of the antenna are achieved by arc-shaped circular ring and one rectangular patch inside the circular ring [3]. The radiation performance and gain analysis is observed by experimental data obtained by the simulation of antenna design. The results and design parameters of the proposed antenna are discussed in the following sections along with dimensions. Finally, to improve the gain and radiation characteristics of the antenna, linear array antenna setup is included in the design of the proposed antenna.

## 2 Development of the Proposed Antenna

The below figure states the return loss characteristics of the designed antenna. The frequency band obtained is a triple band. The return loss versus frequency plot is taken in the range of 1–7 GHz. Between that range, three frequency bands are obtained the band-1 is ranging from frequency 2.3 to 2.5 GHz with the bandwidth of 0. 2 GHz. The second band is ranging from frequency 3.4 to 4 GHz with the bandwidth of 0.6 GHz. And the third band is ranging from frequency 5.2 to 5.9 GHz with the bandwidth of 0.7 GHz. The resonant frequencies obtained at the three frequency bands are 2.4 GHz, 3.6 GHz, and 5.5 GHz, respectively, with a notch frequency of 4.6 GHz between 3.6 and 5.5 GHz (Figs. 1, 2, 3, and 4; Table 1).

The VSWR characteristic of the proposed antenna is observed below 2 GHz. The VSWR versus frequency plot is taken in the range of 1–7 GHz. Between that range, three frequency bands are obtained in which the band-1 is ranging from frequency 2.3 to 2.5 GHz with the bandwidth of 0.2 GHz. The second band is ranging from frequency 3.5 to 3.8 GHz with the bandwidth of 0.3 GHz. And the third band is ranging from frequency 5.3 to 5.7 GHz with the bandwidth of 0.4 GHz.

The input impedance given of the designed antenna is 50 ohms. The impedance characteristics of the proposed antenna shown in the Fig. 5 gives the magnitude, real and imaginary impedance.

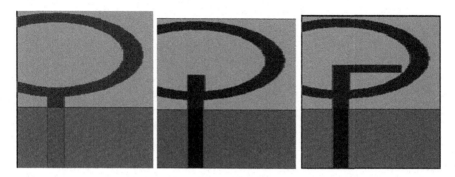

**Fig. 1** Antenna iterations

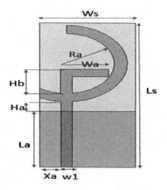

**Fig. 2** Layout of the designed antenna

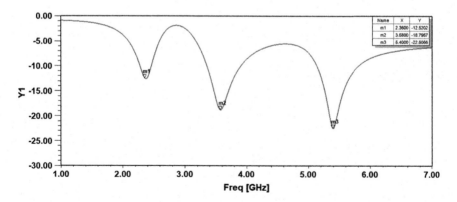

**Fig. 3** $S_{11}$ characteristics of the designed antenna

Gain characteristics of the antenna before applying four element arrays are presented in Fig. 6.

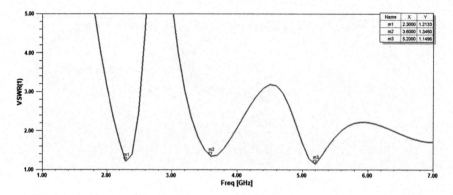

**Fig. 4** VSWR characteristics of designed antenna

**Table 1** Antenna parameters

| Variable | Dimensions (mm) | Variable | Dimensions (mm) |
|----------|-----------------|----------|-----------------|
| Ws       | 16              | Hb       | 6.5             |
| Wa       | 8               | Ls       | 38.5            |
| Wb       | 4.5             | La       | 15              |
| Ra       | 8               | W1       | 1.9             |
| Ha       | 2.5             | Xa       | 3.55            |

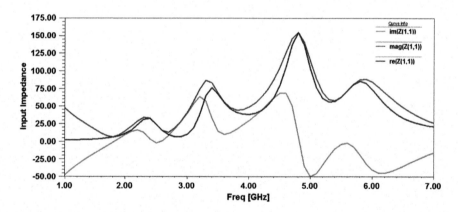

**Fig. 5** Impedance characteristics of designed antenna

   Linear Array Antenna Arrangement is in Fig. 7.
   The below figure shows the linear array arrangement of the designed antenna. Before applying the linear array technique [4], the gain of the antenna is less as shown in Fig. 6 at three frequencies when applying the linear array technique, the gain is enhanced as Fig. 7. The gain has increased in three bands [5–8] gradually, i.e., at 2.4 GHz the gain is 5 dB and at 3.6 GHz gain is 7 db and at 5.5 the gain is

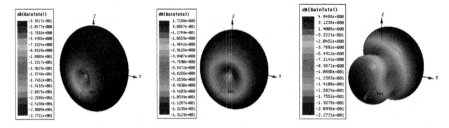

**Fig. 6** **a** 2.4 GHz, **b** 3.6 GHz, and **c** 5.5 GHz

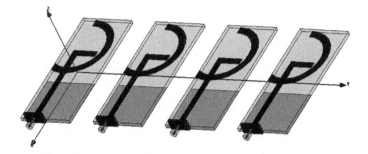

**Fig. 7** Linear array arrangement of antenna

8.4 db. The increase of gain establishes a good communication link between the transmitter and receiver antenna.

Figure 8 shows that the antenna at 2.4 GHz frequency radiated in the bidirectional pattern (Fig. 7a) and at 3.6 GHz frequency antenna radiated at omnidirectional pattern in phi = 0° plane (Fig. 6b). At 5.5 GHz frequency antenna we can see the directional patterns (Fig. 7c). Radiation patterns of the antenna before applying linear array are presented in Fig. 9.

Radiation patterns at three resonant frequencies after using 4 linear array elements direction is presented in Fig. 10.

The parametric analysis-1 and 2 of the designed antenna is observed in Figs. 11 and 12 by varying the width and height of the L-shaped stub. By changing the

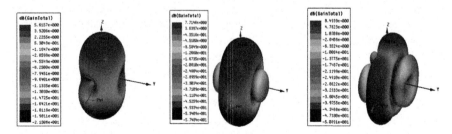

**Fig. 8** 3D pattern **a** 2.4 GHz, **b** 3.6 GHz, and **c** 5.5 GHz

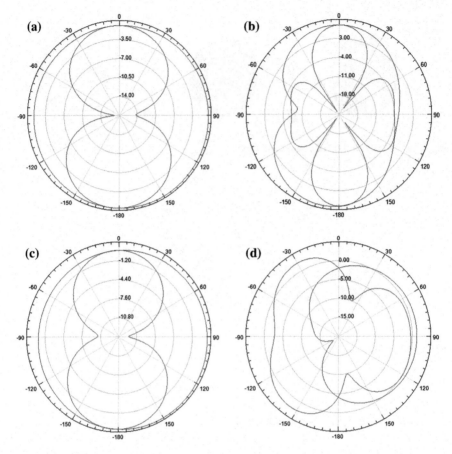

**Fig. 9** Polar plots **a** 2.4 GHz, **b** 4.6 GHz, **c** 3.6 GHz, and **d** 5.5 GHz

width of the feed to 1.5, 2, and 2.5 mm the resonant frequency values obtained at 1.5 mm is 2.4, 3.4, and 5.5 GHz, and on increasing the width to 2 and 2.5 mm the resonant frequency values obtained are 2.4, 3.6, 5.5 and 2.4, 3.7,5.5 GHz, respectively, and the maximum return loss obtained is around −19 db at 2.5 mm. By varying the height of the L-stub to 4.5, 5.5, and 6.5 mm the resonant frequency values obtained at 4.5 mm is 2.5 and 5.2 GHz and the maximum return loss obtained is −32 db at 4.5 mm, and on increasing the height to 5.5 and 6.5 mm the resonant frequency values obtained are 2.5, 4.9 and 2.4, 3.6, and 5.5 GHz respectively. It was observed that for 4.5 and 5.5 mm values only dual-band frequency spectrum is obtained of the arc. By changing the width of the L-shaped stub to 6, 7, and 8 mm the resonant frequency values obtained at 6 mm is around 2.4, 4.2 and 5.6 GHz, and on increasing the width to 7 and 8 mm the resonant frequency values obtained are 2.4, 3.9, 5.6, and 2.4, 3.6, 5.5 GHz, respectively, and the maximum return loss obtained is around −23 db at 6 mm. By varying the radius

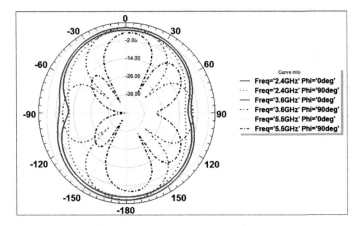

**Fig. 10** E-plane and H-plane radiation patterns at resonating frequencies

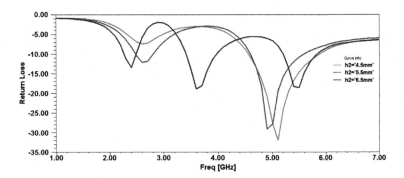

**Fig. 11** Analysis by varying height of the L-shaped stub

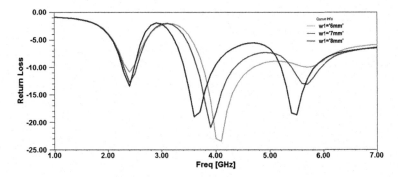

**Fig. 12** Analysis by varying width of the L-shaped stub

of the arc to 6, 7 and 8 mm the resonant frequency value obtained at 6 mm is 3.2, 5.8 GHz, and on increasing the radius to 7 and 8 mm the resonant frequency values obtained are 2.8, 5.3, and 2.4, 3.6, and 5.5 GHz respectively. It was observed that for 6 and 7 mm values only dual-band frequency spectrum is obtained [9, 10] (Figs. 13, 14, 15 and 16).

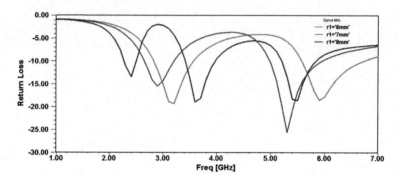

**Fig. 13** Analysis by varying radius of the arc

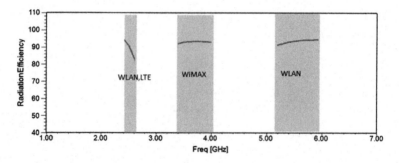

**Fig. 14** Radiation efficiency of the designed antenna

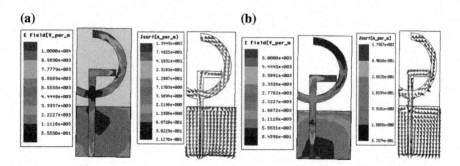

**Fig. 15** E-field and current distribution of the proposed antenna at different frequencies **a** 2.4 GHz, **b** 3.6 GHz

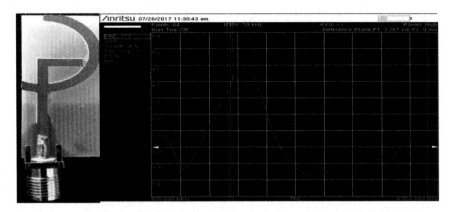

**Fig. 16** Fabricated design and measured VSWR of the proposed antenna

# 3 Conclusion

This antenna is designed to work at three different bands which cover the WLAN and LTE, WiMAX and WLAN applications A combination of arc-shaped patch and DGS ground makes the antenna to work efficiently. A unique application by adding a linear array technique structure is adjoined for which antenna gain is improved to 8 dB, The proposed antenna shows a variation in gain from −2 to 8 dB with omnidirectional radiation pattern in the operating band with the linear array technique. The antenna is small in size and fabricated on the FR4 dielectric substrate.

**Acknowledgements** The authors deeply express their gratitude to Department of ECE, SRKR Engineering College, Research Centre, Department of ECE, K L University for their encouragement during this work. Further, Madhav would like to express his gratitude to DST through grant ECR/2016/000569 and FIST grant SR/FST/ETI-316/2012.

# References

1. Lee, Kai Fong, and Kwai Man Luk. Microstrip patch antennas. World Scientific, 2011.
2. Shi, Ya Wei, Ling Xiong, and Meng Gang Chen. "Compact Triple-Band Monopole Antenna for WLAN/WiMAX-Band USB Dongle Applications." ETRI Journal 37.1 (2015): 21–25.
3. Thomas, K. G., and M. Sreenivasan. "Compact triple band antenna for WLAN/WiMAX applications." Electronics letters 45.16 (2009): 811–813.
4. Khodier, Majid M., and Christos G. Christodoulou. "Linear array geometry synthesis with minimum sidelobe level and null control using particle swarm optimization." IEEE Transactions on Antennas and Propagation 53.8 (2005): 2674–2679.
5. Dang, Lin, et al. "A compact microstrip slot triple-band antenna for WLAN/WiMAX applications." IEEE Antennas and Wireless Propagation Letters 9 (2010): 1178–1181.
6. Alkanhal, Majeed AS. "Composite compact triple-band microstrip antennas." Progress In Electromagnetics Research 93 (2009): 221–236.

7. Pei, Jing, et al. "Miniaturized triple-band antenna with a defected ground plane for WLAN/WiMAX applications." IEEE Antennas and Wireless Propagation Letters 10 (2011): 298–301.
8. Cai, L. Y., G. Zeng, and H. C. Yang. "Compact triple band antenna for Bluetooth/WiMAX/WLAN applications." Signals Systems and Electronics (ISSSE), 2010 International Symposium on. Vol. 2. IEEE, 2010.
9. Qi, Dongsheng, Binhong Li, and Haitao Liu. "Compact triple-band planar inverted-F antenna for mobile handsets." Microwave and optical technology letters 41.6 (2004): 483–486.
10. Zhai, Huiqing, et al. "A compact printed antenna for triple-band WLAN/WiMAX applications." IEEE Antennas and Wireless Propagation Letters 12 (2013): 65–68.

# Design and Characterization of an ASIC Standard Cell Library Industry–Academia Chip Collaborative Project

M. Naga Lavanya and M. Pradeep

**Abstract** Standard cell design approach was important for allowing designers to scale ASICs correspondingly simple single-function ICs (of several thousand gates) to complex multi-million gate devices (SoC). Standard cell libraries are required for any IC design or chip fabrication process. The proposed work involves the design and development of an ASIC standard cell library for the 90 nm technology node using Cadence tool. The work involves identifying the optimal circuit topology for the defined logic functions, Circuit design for the functional, performance, and power dissipation specifications, Formulation of circuit simulation index for circuit characterization, design of a standard cell layout template, Layout Engineering for all the sets (DRC, LVS compliant), Parasitic Extraction and back annotation, Post layout characterization for DC, transient and power dissipation performance, and Scripting for Liberty format (Tool views). Similarly, according to our chip specifications the objective is to design optimal circuit topology like area, timing, and power.

**Keywords** ASIC design standard cell library schematic design
Characterization and layout (DRC, LVS)

## 1 Introduction

Integrated circuit devices play an essential role in today's industry. Integrated circuit technology has passed through a spectacular revolution within the past 20 years. Using Moore's law, the quantity of transistors which will be integrated on one die has been exponentially increasing with time [1]. Integrated circuits have many applications from automotive controls, televisions, computers, microwaves,

M. Naga Lavanya · M. Pradeep (✉)
Department of ECE, Shri Vishnu Engineering College for Women, Bhimavaram, AP, India
e-mail: pradeepm999@gmail.com

M. Naga Lavanya
e-mail: m93lavanya@gmail.com

© Springer Nature Singapore Pte Ltd. 2018
J. Anguera et al. (eds.), *Microelectronics, Electromagnetics
and Telecommunications*, Lecture Notes in Electrical Engineering 471,
https://doi.org/10.1007/978-981-10-7329-8_83

and televisions, play stations, cameras, mobile phones, and aeroplanes. Different integrated circuit implementation approaches are adopted ranging from custom design approach used for microprocessors and memories to the absolutely programmable designs for medium to low-performance applications to meet all the design specifications [2]. A number of different approaches are used to a digital integrated circuit that can be manufactured, but they all are some basic steps to be followed. The steps starts with sizing all the transistors in the schematic design, connecting wires, and all the input and output pins are arranged then the schematic being put down in a layout. First design schematic and layout were designed and finally that layout was fabricated on chip.

**Full Custom Design**: Full custom means that you are donning the work started from zero that is everything doing in the schematic and layout is created manually. Each transistor used in the schematic can be arranged as some desired specifications, decreased area, speed, load capacitor, etc. The designer has total control on layout; each and every wire connection in the layout is placed our own way. The advantage of full custom design layout can be designed manually and carefully to meet our design specifications and fit into the cell. The disadvantage of full custom design is it takes a lot of work and time to design [3].

**Standard Cell Library**: "Standard cell library have a group of logic or gate level components, functional level components that are systemized and consists of cells based on the individual layout." Standard cell libraries are basic building blocks of ASIC design flow because of its basic interface execution and consistent structure [3].

In this chapter, standard cell library development was done by using full custom design approach. Because everything is done manually (schematic and layout design). This chapter's main aim is to design and implement high-speed standard cell library corresponding to the design specifications and to characterize the transient analysis and DC analysis on the executed schematics. Transistors sizing are very small compared to other standard cell libraries. Area of the standard cell template is also minimum.

## 2   Standard Cell Library Design Flow

Standard cell library design flow is shown in Fig. 1. This design flow represents total standard cell library design in a number of steps. Each and every step in the design flow is very important. These design flow steps are followed to design any standard cell library. But in this design, extra design steps are added, according to my library development. Those are liberty generation and Verilog and VHDL Simulation models.

**Fig. 1** Standard cell design
flow

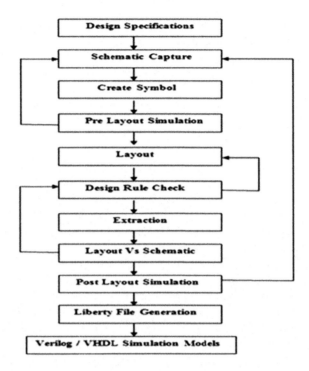

**Fig. 1** Standard cell design
flow

## 3 Specifications of Standard Cell Library

### 3.1 Design Characteristics of Standard Cell

According to our technology node, the amounts of routing layers are chosen for the
design of cells in the library. In few cases, the design of the cells will rely on the
functions of the obtainable metal layers with design rules. The following characteristics are standard to all cell libraries. Standard cell library specifications are
observed in Table 1.

**Design Characteristics of Circuit**: First, each cell in this library is tested and
the functional specifications and the electrical characteristics are described. By

**Table 1** Standard cell library
specifications

| Parameter | Description |
|---|---|
| Technology node | 90 nm |
| Supply voltage | 1 V |
| Rise and fall times | 137.6 ps |
| output frequency | 100 MHz |
| Number of tracks | 15 |
| Cell height | 7.04 μm |
| Temperature range | −40 to 125 °C |

using Drive strength concept sizing all the transistors in schematic. **"Drive strength means it is the capacity of a cell to drive a value to the cell connected to its output."** Each cell is designed with multiple drive strengths using different drive strengths sizies of the transistors easily and layout is designed easily with minimum area.

**Cell Shape Characteristics**: Standard cell layout design is constructing by using a predefined design. It will make sure that all the design requirements are met. The layout design consists of the cell height, the placing of the N-wells/P-wells, N, P-type transistors widths, by using layout guidelines to flip the cell vertically or horizontally without generating any errors. The shape of the cells is in rectangular. Cells for particular columns or chip regions are all at a similar height a library may contain various arrangements of cells. A particular design rule is the width of cell is minimum because placement may be easier and faster [4]. The power supply lines VDD and GND have the same width for the whole standard cell; the width of the supply over the cell length is constantly steady.

**Interface of the Cell Characteristics**: All IN and OUT ports have a predefined sort, layer, position, size, and interface focuses. These attributes are resolved in view of the placer and additionally switch to be utilized to execute the plan. Router always think about how to simplify the design and gives best outcomes. By utilizing layer spacing design rule to outline, all non-shared polygons must be dispersed from the limit of the cell. According to this design rule, abutting cells will be correct by construction.

## 3.2    Standard Cell Layout Guide Lines

**How to Consider Wire Track Spacing**: The format of standard cell layout starts with measuring horizontal and vertical wire tracks. Wire tracks are utilized to control, and place routing devices to perform interconnection between cells. Following all the design rules, for example, width and separating of the initial two leading layers (e.g., metal one and metal two) are utilized to set appropriate wire track dispersing [5]. There are three ways to discover wire track spacings, i.e., **Line-to-Line (d1), Line-to-Via (d2), Via-to-Via (d3)** as shown in Fig. 2.

**Procedure for Standard Cell height Measurement**: The height and width of a standard cell is determined by multiple of the vertical wire tracks and horizontal wire tracks. The standard cell height is fixed in the entire library, but the standard cell width will change according to the width of transistors [5]. The height of the standard cell consider all the parameters are like diffusion spacing between P- and N-type transistors, width of P-Type, N-Type transistors and the Power(VDD) and Ground (VSS) buses Width (Fig. 3).

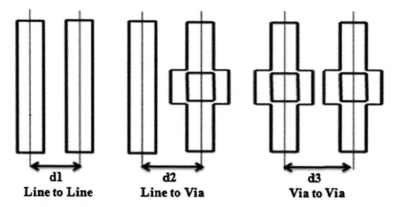

**Fig. 2** Wire track spacing

**Fig. 3** Generalized height of standard cell

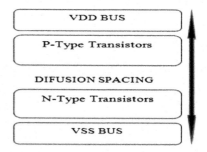

## 4 Standard Cell Characterization

Both combinational and sequential circuits are implemented in this standard cell library design. Combinational circuits are INV, BUFFER, NAND, NOR, AND, OR, MUX, XOR, etc., circuits are designed with maximum Drive strength (1X-200X). Sequential circuits are D Latch, D Flip flop, etc. Here, only one basic CMOS INVERTER with Drivestrength1X is explained using logical effort concept sizing all the transistors in this library.

### 4.1 CMOS Inverter

The CMOS inverter cell gives the inverse of input (IN). The output (out) is mentioned by the functional equation given below. $Y = \bar{A}$. The schematic diagram of INV1X with width of PMOS ($W_p$) and NMOS ($W_n$) as Wp/Wn = 1.25 are shown in Fig. 4. Symbol generation, transient analysis of INV1X are shown in Figs. 5 and 6.

**Fig. 4** Fastest CMOS
INV1X schematic

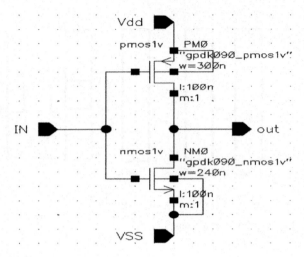

**Fig. 5** INV1X symbol

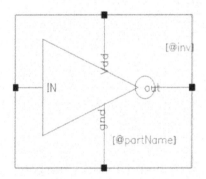

**Schematic Diagram of CMOS Inverter (INV 1X)**
See Fig. 4.
**Symbol Generation**:
See Fig. 5.
**Simulation Results for INV1X**:
See Fig. 6.

## 4.2 Layout Representation

Before going to the Inverter (INV1X) Standard Cell Layout, first we consider the Horizontal tracks, Vertical tracks, and fix the Height of the Standard Cell Layout.

Cell height is chosen as the most complex cell in this library such as MUX4X1. In this standard cell library, number of horizontal tracks are 15 and vertical tracks depend on width of the transistor as shown in Fig. 7. For measuring purpose, the set

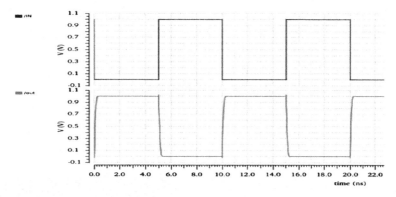

**Fig. 6** Transient analyses of INV1X

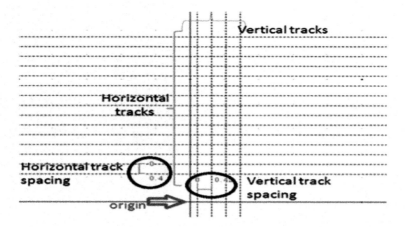

**Fig. 7** Representation of horizontal and vertical tacks

of design rules are required. Here, GPDK 90 nm DRC manual is used for the reference [7]. Physical specifications of this template are shown in Table 2. After generating INV1X Standard Cell Layout measure area also. INV1X standard cell characteristics are shown in Table 3. Once DRC and LVS checks are finished for INV1X, next we are doing RC Extraction (QRC). Once QRC was done particular cell av-extracted layout will be generated (Fig. 8).

**Table 2** Physical specifications of standard cell template

| Physical attribution | Characteristics |
| --- | --- |
| Number of tracks | 15 |
| Cell height | 7.04 µm |
| Horizontal grid spacing | 0.4 µm |
| Vertical grid spacing | 0.42 µm |
| Power/ground rail width | 0.84 µm |

**Table 3** INV1X standard cell characteristics

| INV1X cell characteristics |
| --- |
| Sizes of the PMOS and NMOS—Wp—300 nm, Wn—240 nm |
| Shape ratio ($\gamma$)—1.25 |
| Rise time—137.6 pF and fall time—137.6 pF |
| Propagation delay—78.435 ps |
| Cell height—7.04 µm, cell width—1.72 µm |
| Area of the cell—12.07 µm$^2$ |

**Fig. 8** INV1X standard cell layouts

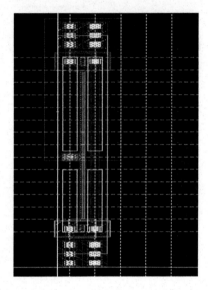

## 5  Conclusion

The main objective of this chapter is to design high-speed standard cell library and it is accomplished by meeting the given design specifications. According to that sizing, the transistors are also very small. The specifications are like Height of the standard cell template, rise time and fall time values, and propagation delay for INV1X. For any chip fabrications process, a standard cell library is required. This project represents standard cell template design, by using this template less number of transistors and has more possibilities for superposition of logic. Here, standard cell template height is measured and is fixed by using a number of tracks and area of the cell also minimum. This venture meets pre-format and post-design comes about according to the design specifications. The standard cell design is done implementation by using Cadence software. In the later work liberty file for all the cells in our library will be generated.

# References

1. Sung-Mo Kang, Leblebici.: CMOS Digital Integrated Circuits, 2nd Edition, Mc Graw Hill.
2. Prof. Poornima H S, Prof. Chethana K S.: Standard Cell Library Design and Characterization using 45 nm technology.
3. Mr. Narhari, R. Kotkar M.E.: Development of High Performance Standard Cell Library in Umc 180 nm Technology.
4. Dan Clein.: CMOS IC Layout: Concepts, Methodologies, and Tools, Third edition December (2006).
5. Khosrow Golshan.: Physical Design Essentials an ASIC Design Implementation Perspective (2006).
6. Neil Weste, Kamran. E.: Principles of CMOS VLSI Design, Addison Wesley Publication.
7. GPDK 90 nm Mixed Signal Process Specifications.
8. Nidhi, Agnihotri.: Ultra Low Power Standard Cell Library Development in 90 nm Technology.

# Design and Characterization of 6T SRAM Cell Industry-Academia Collaborative Chip Design Project

**Hema Thota and G. R. L. V. N. Srinivasa Raju**

**Abstract** Static Random Access Memory (SRAM) being a volatile semiconductor memory, it is used only when power is supplied, to store a bit of binary logic '0' and '1'. Asia Pacific is the largest and fastest growing region in SRAM market. The present key challenges faced by this market are, large cell sizes, the high cost of designing and lower cell stability. This paper focuses on the motive to meet such key challenges of SRAM design. So, it discusses about 6T SRAM design, stability analysis and cell characterization of SRAM cell. Stability analysis signifies the Static Noise Margin (SNM) of the memory cell. SNM, when both write as well as read operations are done in different ways describing the necessity of each method as a model for stability analysis, is discussed in this paper. The simulations are carried out on Cadence—virtuoso, using GPDK (Generic Process Design Kit) 180 nm CMOS technology.

**Keywords** GPDK 180 nm CMOS technology · SRAM · Bitcell
Wordline · Bitline · Marginality analysis · Stability · SNM

## 1 Introduction

As the demand for portable device size and battery operated system are increasing with greater scale, the demand for on-chip memory also increases. According to the research analysis studies on the Static Random Access Memory (SRAM) Market, Industry ARC (Analytics, Research and Consulting) reported that the largest and fastest growing region during the period 2014–2021 in this market will be Asia Pacific. There is high scope for the SRAM Market in various regions across the

H. Thota (✉) · G. R. L. V. N. Srinivasa Raju
Department of Electronics and Communication, Shri Vishnu Engineering College
for Women, Bhimavaram, India
e-mail: hemathota39@gmail.com

G. R. L. V. N. Srinivasa Raju
e-mail: hodece@svecw.edu.in

© Springer Nature Singapore Pte Ltd. 2018                                     817
J. Anguera et al. (eds.), *Microelectronics, Electromagnetics
and Telecommunications*, Lecture Notes in Electrical Engineering 471,
https://doi.org/10.1007/978-981-10-7329-8_84

globe during 2014–2021 as shown in Fig. 1 (from the report given by Industry ARC). The basic idea of choosing memory design as an academic project is, to bring out a complete Memory IP for Indian academic research usage as a part of the industry-academia partnership-based collaborative chip or IP design projects. Figure 2 shows the conventional 6T-SRAM cell [1]. It uses six transistors, out of which four transistors ($M_3$–$M_6$) are used for storing a bit and two other transistors ($M_1$ and $M_2$) are used to access the bit stored in the cell. The four transistors used for storing the data are arranged in the form of two inverters connected back to back (cross-coupled) forming a bi-stable latching circuitry. The read and write operations are explained [2]. The stability [3–9] of the cell is also an important key factor that affects the data stored in the memory. The stability of the cell is determined by the parameter variations in the memory to process variations, namely, sensitivity and the conditions at which the circuit is operated. Both the cell area and stability are interdependent perspectives for a cell design, because improved stability requires a larger area of the cell (larger sized transistors). Over past several years, there has been considerable effort to get through and then create the models for the stability of memory cells. Though, the inverters connected back to back seems to be simplest in appearance, it can be attempted to create the model for the stability of the cell analytically, that has been gone to some limits of success [5].

This paper organization has been carried out in five sections. The 6T SRAM cell is introduced in Sect. 1. Section 2 describes the operation of the cell—read and write, both the bits from and to the cell, respectively. Cell characterization and

**Fig. 1** SRAM market value, reported by industry ARC

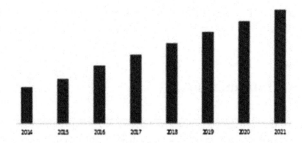

**Fig. 2** Six-transistor SRAM cell

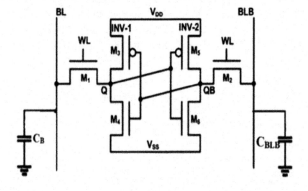

design are discussed in Sect. 3. Section 4 is followed by simulation test setup and results. The paper discussion is then, finally concluded in Sect. 5.

## 2 Operation of 6T SRAM Memory Cell

### 2.1 Read Operation

To read a bitcell, both the bitlines are first precharged to the supply voltage ($V_{DD}$). Then WL is made high and logic '0' is stored at node Q; the bitline at this node is discharged from access and pulldown transistors. In 6T SRAM, shown in Fig. 2, the transistors $M_1$ and $M_4$ are used for discharging the bitline which is precharged. If BLB is at low voltage (or discharged), then logic '1' is held at node Q. To be clear on the states which the cell stores, it depends on the discharge of bitlines. The differentiated signals stay on bitlines to be converted to a single logic output by the sense amplifier. Finally, the wordline is de-asserted back to '0'. Figure 3 shows the read operation of bit '0'. For good read stability,

$$\beta = \frac{\left(\frac{W}{L}\right)_{Pull\,down\,Transistor}}{\left(\frac{W}{L}\right)_{Access\,Transistor}} > 1 \tag{1}$$

### 2.2 Write Operation

The write operation can be clearly understood by assuming the initial voltages at nodes Q and QB as at, logic '0' and logic '1' voltages, respectively. The wordline (WL) is initially set to '0' and BL, BLB are precharged to supply potential. After precharged, the two bitlines are removed from $V_{DD}$. Then, WL is given a high voltage

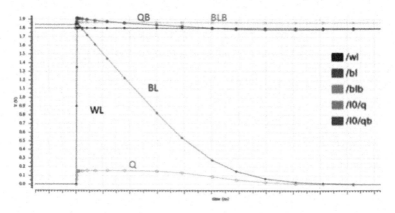

**Fig. 3** Read '0' operation of a 6T SRAM cell

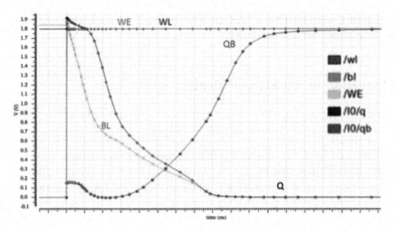

**Fig. 4** Write '0' operation of a 6T SRAM cell

(content is stored in the cell in this process). Now the actual data value is placed on BL which can be done by pulling down the Voltage at that node using the write driver circuitry. BLB which is in connection with node QB through $M_2$ is driven to the very low voltage using a write driver circuit by $M_2$ access transistor, then the bitline at node Q is held high to pull it down through $M_1$ access transistor. As soon as nodes Q and QB invert their states, WL is brought back to logic '0'. For writability (Fig. 4),

$$P = \frac{\left(\frac{W}{L}\right)_{Pull\ up\ Transistor}}{\left(\frac{W}{L}\right)_{Access\ Transistor}} < 1 \tag{2}$$

## 3  Characterization of SRAM Cell

The SRAM cell can be characterized by its stability analysis. The marginality of cell signifies the stability, which is determined by Static Noise Margin (SNM). It can be termed as the amount of DC noise ($V_N$) in maximum that can be tolerated by the inverters connected back to back so that, the cell is retained with its data. All the respective methods mentioned below are done for the design and achieved far better results from the optimized design.

### 3.1  Analytical-Based Approach to Determine SNM of a 6T SRAM Cell

SNM can be found out analytically by using KCL and KVL equations and applying one of the mathematically equivalent of the noise margin criteria [6]. The SNM for an SRAM cell can be determined analytically and is given by Evert Seevinck, the equation given below:

$$SNM_{6T} = V_T - \left(\frac{1}{K+1}\right)\left\{\frac{V_{DD} - \frac{2\beta+1}{\beta+1}V_T}{1 + \frac{r}{k(\beta+1)}} - \frac{V_{DD} - 2V_T}{1 + k\frac{\beta}{P} + \sqrt{\frac{\beta}{P}\left(1 + 2k + \frac{\beta}{P}k^2\right)}}\right\}, \quad (3)$$

where, $\beta$ and P are the bitcell and pullup ratio, respectively, $V_T$ is the threshold voltage.

$$V_s = V_{DD} - V_T, .V_r = V_s - \left(\frac{\beta}{\beta+1}\right)V_T, ..k = \left(\frac{\beta}{\beta+1}\right)\left\{\sqrt{\frac{\beta+1}{\beta+1 - \frac{V_s^2}{V_r^2}}} - 1\right\} \quad (4)$$

With the analytical approach, it becomes difficult to find the expression for writability with appropriate assumptions and approximations, because of the complex equations are involved. The following simulation-based approaches are more efficient; as spice MOS models include second-order effects. The analytical and simulated values are tabulated in Table 2 (Fig. 5).

## 3.2 Simulation-Based Approach to Determine SNM of an SRAM Cell

SNM for an SRAM cell can be determined in many ways. Some of the simulation-based approaches to find SNM are given below. Each of the approaches is followed by its details to determine SNM value and some limitations of that particular approach.

**Fig. 5** Standard 6T SRAM for modelling the SNM

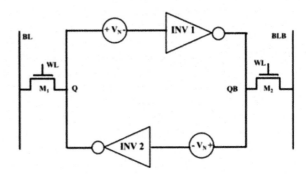

**Approach-I; Butterfly Curves:** SNM, in the conventional SRAM is determined by butterfly curves. The maximum side of the square fitted in the lobes of VTC, determines the SNM as shown in Fig. 7. But, there are certain limitations for determining the SNM using the butterfly curves. The SNM is delimited to a maximum of 0.5 $V_{DD}$, when the Voltage transfer characteristic (VTC) has been obtained from the butterfly curves, There is no ability to measure SNM with an automatic inline tester, Not able to provide the information statistically about the rate of failure in SRAM cells, lack of direct SNM availability, For reading stability and writability measure, separate analyses are required, stability analysis also requires the information regarding the current flow, which is not provided (Fig. 6).

**Approach-II; N-curve Stability Metrics:** The alternative approach to determine the stability margin is to implement N-curve for the Memory cell. N-curve not only gives the information of the current in the circuit besides, the RSNM, WSNM can also be determined from the same curve. The N-curve that is extracted, had three points of intersection named A, B and C. While, the points A and C lie at stable states, B is at meta-stable state. These correspond to the butterfly curves plotted above the N-curve, given by Fig. 7. The difference potential between points A and B represents the Read SNM and that between B and C gives Write SNM. The above curve also determines Static Power Noise Margin (SPNM) which is termed as the maximum allowable DC noise power at internal nodes of the cell before the content gets changed. Writability of a cell can be characterized by the parameter Write Trip Power (WTP), which can be actually achieved by calculating the area above the curve, between points B and C. The amount of power required in minimum to invert the data at storage nodes is termed as WTP. However, a major limitation of the N-curve stability metric is, it gets access with the nodes Q and QB as done to obtain butterfly curves.

**Approach-IV; Bitline measurement stability metrics:** In this way to obtain read stability and writability for a memory cell is characterized by accessing only to supply voltage ($V_{CELL}$), Wordline ($V_{WL}$) and bitlines ($V_{BL}$ and $V_{BLB}$).

*Supply Read Retention Voltage (SRRV):* In this approach, the stability under read operation is achieved by supply power needed in minimum for data retention and it

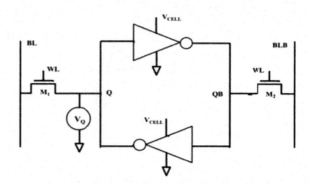

**Fig. 6** Test bench-1 to determine SNM for Simulation-based approaches I, III and IV

**Fig. 7** N-curve with
corresponding particular
butterfly curves of the 6T
SRAM cell

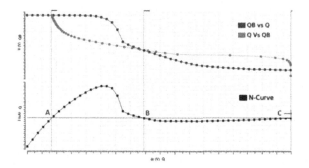

is termed as SRRV. When both bitlines are left floating floating around supply
voltage, WL is brought high and $V_{CELL}$ is brought down to a sufficiently lower
potential. Then, the bitcell loses the stable state and makes nodes Q and QB to
remain in the same state. At this point, $M_1$ overcomes $M_4$ (refer Fig. 2) so that node
Q, actually storing '0' increases above the threshold of INV-2 and inverts the state
of bitcell. It can also be depicted by the sudden fall in the current flowing through
the bitline, $I_{BL}$, as given in Fig. 8. This all of a swift in current is mainly because of
increase in the voltage at node Q holding '0', the time when potential reaches to
meta-stable state or that the bitcell state gets inverted due to the drop in $V_{CELL}$.

*Wordline Write Trip Voltage (WWTV)*: It is termed as the minimum word
potential needed to invert the cell content during a write cycle and it can be used to
find the writability of the cell. In other terms, it represents the maximum allowable
DC slack on WL to write to the cell. As WL potential is increasing towards a high
potential, initially, the current monitored resembles the $I_D$–$V_G$ characteristics of the
access transistor $M_1$ (refer Fig. 2). When WL is sufficiently increased to the highest
potential, it causes the cell content to invert. It is represented by the sudden fall in
the magnitude of $I_{BL,}$ as shown in $I_{BL}$ versus the $V_{WL}$ of Fig. 9. WWTV is termed
as the difference in the voltage of $V_{DD}$ and that of $V_{WL}$, which results in the swift in
the current, $I_{BL}$.

**Fig. 8** SRRV representing
the read stability of the cell

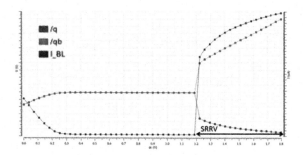

**Fig. 9** WWTV representing
the writability of the cell

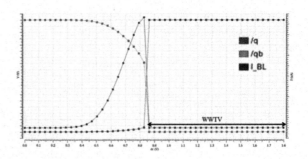

# 4   Simulation Plan and Analysis

All the above simulations are performed with bitcell ratio as 5 and pull-up ratio as
1.25. PVT and Monte Carlo analysis are performed with sizes of transistors being
parametrized by varying the bitcell and pullup ratios at NN, SS, SF, FS and FF
corners at −40, 27 and 125 °C with Voltage of 1.62, 1.8 and 1.98 V (45 Corners).

**Table 1**  Voltage conditions for stability models

| Method | Butterfly curves | | N-curve | Bitline measurement | | | |
|---|---|---|---|---|---|---|---|
| Parameter | RSNM | WSNM | RSNM and WSNM | RSNM | | WSNM | |
| | | | | SRRV | WRRV | WWTV | BWTV |
| $V_{CELL}$ | $V_{DD}$ | $V_{DD}$ | $V_{DD}$ | Swept from 0 to $V_{DD}$ | $V_{DD}$ | $V_{DD}$ | $V_{DD}$ |
| $V_{WL}$ | $V_{DD}$ | $V_{DD}$ | $V_{DD}$ | $V_{DD}$ | Swept from $V_{DD}$ and above | Swept from 0 to $V_{DD}$ | $V_{DD}$ |
| $V_{BL}$ | $V_{DD}$ | $V_{DD}$ | $V_{DD}$ | $V_{DD}$ | $V_{DD}$ | $V_{DD}$ | $V_{DD}$ |
| $V_{BLB}$ | $V_{DD}$ | $V_{SS}$ | $V_{DD}$ | $V_{DD}$ | $V_{DD}$ | $V_{SS}$ | Swept from $V_{DD}$ to 0 |
| $V_Q$ | Swept from 0 to $V_{DD}$ | Swept from 0 to $V_{DD}$ | Swept from 0 to $V_{DD}$ | [a] | [a] | [a] | [a] |
| Measured parameter | Q & QB-Max Square | Q & QB-Min Square | $I_{BL}$ and points of zero crossing | Q, QB and $I_{BL}$ | Q, QB and $I_{BL}$ | Q, QB and $I_{BL}$ | Q, QB and $I_{BL}$ |
| Test bench | 6 | 6 | 6 | 6[a] | 6[a] | 6[a] | 6[a] |

[a]$V_Q$ can be removed from the test bench

**Table 2** Simulated results for SNM from all the methods

| Approach | | RSNM | WSNM |
|---|---|---|---|
| Butterfly curves | | 0.5800 | 0.870 |
| N-curve | | 0.5602 | 1.071 |
| Bitline measurement | SRRV | 0.5760 | a |
| | WWTV | a | 0.898 |
| Analytical | | 0.4902 | a |

[a]The parameter cannot be measured by that approach

## 4.1 Technology Parameters and CAD Tools

*Technology Parameters*: *Technology Specification*: GPDK (Generic Process Design Kit) 0.18 μm CMOS Technology, *Supply Voltage range*: 0–1.8 V, *SPICE MOS Model*: BSIM (Berkeley Short Channel IGFET Model), Version: 3v3.

*CAD Tools*: *Schematic Entry*: Cadence Schematic Entry ADE (Analogue Design Environment), *Simulator*: Cadence Spectre.

All the Simulations performed are shown in the above descriptions as Fig. 3 through Fig. 9. The voltage conditions for the Stability models to all the above-mentioned approaches are tabulated in Table 1. For all the approaches dealt above, the simulated results are listed in Table 2. Approach III, (N-Curve) is also capable of measuring Static power under both write and read. SPNM is measured to be 61.79 μW and WTP is 25.67 μW.

## 5 Conclusion

The cell is characterized by varying the various voltages of the cell. The current $I_{BL}$ is observed by varying all the parameter voltages mentioned above. It signifies the characterization of the cell to determine the read stability and writability. Stability models are presented by listing down all the advantages and disadvantages of different approaches. Stability of the cell is increased for larger sized transistors. The static powers of the cell for both read and write operations are also observed. PVT and Monte Carlo are also performed using N-curve to obtain the simulated results with bitcell ratio as 5 and pullup ratio as 1.25. The minimum RSNM is 0.550 at FS corner, 1.62 V, 125 °C and WSNM is 0.834 at SF corner, 1.62 V, −40 °C. The maximum RSNM is 0.7606 at SF corner, 1.98 V, −40 °C and WSNM is 1.313 at FS corner, 1.98 V, 125 °C. Thus, stability is improved as compared to that presented [3, 4].

# References

1. Jawar Singh, Saraju P. Mohanty, Dhiraj K. Pradhan.: Robust SRAM Designs and Analysis. Springer (2013)
2. Neil H. E. Weste and David Money Harris.: CMOS VLSI Design, a Circuits and Systems Perspective. Pearson Education publishing as Addison-Wesley. 4th Edition (2011)
3. Debasis Mukherjee, Hemanth Kr. Mondal and B.V. R. Reddy.: Static Noise Margin Analysis of SRAM Cell for High Speed Application. International Journal of Computer Science Issues. Volume 7. Issue 5. September 2010
4. S. K. Singh, S. V. Singh, B.K. Kausik, T. Tripathi.: Characterization & Improvement of SNM in Deep Submicron SRAM Design. IEEE Transactions. DOI:978-1-4799-2866-8.2014
5. R. C. Jaeger and R. M. Fox.: Phase plane analysis of the upset characteristics of CMOS RAM cells. Industry Microelectronic. Symposium. pp. 183–187, June 1985
6. Evert Seevinck, Frans J. List and Jan Lohstroh.: Static-Noise Margin Analysis of MOS SRAM Cells. IEEE Journal Of Solid-state Circuits, Vol. Sc-22, No. 5, October 1987
7. Andrei Pavlov, Manoj Sachdev.: CMOS SRAM Circuit Design and Parametric Test in Nano-scaled Technologies, Process Aware SRAM Design and Test. Springer (2008)
8. Mohamed H. Abu – Rahma Mohab Anis.: Nanometer Variation –Tolerant SRAM, Circuits and statistical Design for Yield, Springer (2013)
9. Attuluri R. Vijay Babu, Pantalingal Manoj Kumar, Gorantla Srinivasa Rao: Effect of Design and Operating Parameters on the Performance of Planar and Ducted Cathode Structures of an Air-Breathing PEM Fuel Cell, Arab J Sci Eng, 41(9) 3415–3423 (2016)

# Spectrum Sensing in Cognitive Radio Networks Using Time–Frequency Analysis and Modulation Recognition

M. Venkata Subbarao and P. Samundiswary

**Abstract** Spectrum sensing is the most important step in the cognitive radio. It involves spectral detection, channel estimation, and channel state prediction. Most of the traditional spectrum sensing techniques are used for narrowband sensing. At the same time, these techniques cannot distinguish the available user either as primary or secondary. Under the fading conditions, these conventional methods give a false alarm. This chapter presents a new wideband sensing algorithm using Time–Frequency Analysis. Using this method, it is possible to visualize entire spectrum scenario at any instant of time. Further, the primary user and secondary user are distinguished by using Modulation Recognition-based Spectrum sensing which is also presented in this chapter. Several realistic cases are also considered to verify the superiority of the above mentioned proposed methods of spectrum sensing.

**Keywords** Modulation recognition · Time–frequency transforms
Spectrum sensing · Cognitive radio · Energy detection

## 1 Introduction

Wireless communication is one of the fastest growing areas of communication in the past decade. In recent years, there is a drastic increase in a number of users, which results in increased demand for radio spectrum increase proportionally. Static spectrum allotment is the major hurdle in the existing spectrum scarcity. Due to fixed allocation, it is very hard to drive new users; this shortage of spectrum creates various research challenges to researchers. Cognitive Radio (CR) is one of the

M. Venkata Subbarao (✉) · P. Samundiswary
Department of Electronics Engineering, School of Engineering & Technology,
Pondicherry University, Pondicherry, India
e-mail: mvsubbarao@svecw.edu.in

P. Samundiswary
e-mail: samundiswary_pdy@yahoo.com

© Springer Nature Singapore Pte Ltd. 2018
J. Anguera et al. (eds.), *Microelectronics, Electromagnetics*
*and Telecommunications*, Lecture Notes in Electrical Engineering 471,
https://doi.org/10.1007/978-981-10-7329-8_85

emerging technologies, which gives superior solution for the existing underutilization of spectrum. The centralized cooperative CR network allows Secondary Users (SUs) to access unused spectrum by primary users (PUs) [1–3].

For the implementation of CR network, the foremost step is spectrum sensing. Most spectrum sensing techniques follow different methodologies like channel state prediction and spectral detection [4]. The block diagram representation of a cognitive radio system is shown in Fig. 1. Here fixed carriers/channels are allotted to PUs. Using these fixed carriers, PUs send their information through the common communication channel. In a cooperative network, a centralized spectrum sensing unit scans the entire channel and identifies the spectrum holes. Based on spectrum sensing result, the centralized unit takes care the allocation of available frequencies to SUs.

Cooperative-sensing techniques like matched filter detection, energy detection, and cyclostationary detection schemes are narrowband spectrum sensing techniques. These narrowband sensing algorithms find the availability of a single channel with the prior information of PU [5–8]. For wideband sensing, these techniques are not suitable. Under non-cooperative conditions these techniques more false alarm. If SU is present in a channel then all the existing techniques give more false alarms, because they consider PU data as a reference to take a decision about the channel. The wavelet transform based spectrum analysis techniques are used for wideband spectrum sensing and to reduce complexity in computations [9, 10]. These techniques give poor performance in noisy conditions and also the mother wavelet function significantly affects the performance of sensing.

Based on motivations, a novel narrowband and wideband spectrum sensing methods using Time–Frequency Analysis (TFA) are presented in this chapter. TFA deals with all realistic signals like power, seismic, biomedical and communication signals [11–13]. This chapter also introduces a new approach for narrowband

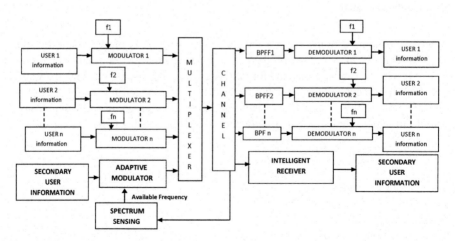

**Fig. 1** Block diagram of a simple cognitive radio system

sensing with modulation detection. For wideband sensing, the sensed composite signal is applied to Modified S-Transform (MST).

The rest of the chapter is organized as follows: Wideband signal analysis using Modified S-Transform is described in Sect. 2. Section 3 describes narrowband sensing using modulation detection. Simulation results of narrowband and wideband spectrum sensing in fading environment with MST is discussed in Sect. 4. Conclusion and remarks are presented in Sect. 5.

## 2 Signal Analysis with Modified S-Transform (MST)

Modified S-Transform is a time–frequency transform which analyzes the signal in time and frequency domain simultaneously. MST converts time domain sequence into time–frequency domain.

For a signal $z(t)$ the S-transform output $s(t, f)$ is given by

$$
\begin{aligned}
s(t, f) &= \int_{-\infty}^{\infty} z(\tau) p(t - \tau, f) e^{-2i\pi f \tau} d\tau \\
&= \int_{-\infty}^{\infty} z(\tau) \frac{1}{\sigma(f)\sqrt{2\pi}} e^{\frac{-(t-\tau)^2}{2\sigma(f)^2}} e^{-2i\pi f \tau} d\tau
\end{aligned}
\tag{1}
$$

where $z(t)$ is the sensed signal from the channel, and $\sigma(f)$ is standard deviation of the Gaussian window $p(t)$ and it is given by

$$
\sigma(f) = 1/|f|
\tag{2}
$$

For wideband spectrum sensing, we have consider the standard deviation of the window is

$$
\sigma(f) = k/(l + m/\sqrt{f})
\tag{3}
$$

where $l$ and $m$ are any positive constants, $f$ is fundamental frequency of the signal $z(t)$, and $k \leq \sqrt{l^2 + m^2}$.

With new standard deviation, the modified Gaussian window can be expressed as

$$
p(t, f) = \frac{l + m\sqrt{|f|}}{k\sqrt{2\pi}} e^{-\frac{(l + m\sqrt{f})^2 t^2}{2k^2}}, k > 0
\tag{4}
$$

Here $t$, $\tau$ are time variables $l$ is a constant and $k$, $m$ are scaling factors that control the number of oscillations in the window.

Short Time Fourier Transform (STFT) can be obtained with MST by setting $m = 0$ and $k = 1$. When $k$ is increased, the window broadens in time domain which results in great improvement of frequency resolution in the frequency domain.

S-Transform with modified Gaussian window is given as

$$S(\tau,f) = \int\limits_{-\infty}^{\infty} Z(\alpha+f)e^{\left(-2\pi^2\alpha^2K^2\right)/\left(l+m\sqrt{|f|}\right)^2}e^{2i\pi\alpha\tau}d\alpha \qquad (5)$$

The discrete version of the MST of a signal is given as

$$S[j,n] = \sum_{a=0}^{N-1} Z[a+b]e^{\left(-2\pi^2a^2K^2/\left(l+m\sqrt{|f|}\right)^2\right)}e^{i\frac{2\pi aj}{N}} \qquad (6)$$

where $Z[a+b]$ can be obtained by shifting the Discrete Fourier Transform (DFT) of $z(k)$ by b.

## 3 Spectrum Sensing Using Modulation Detection

Existing spectrum sensing techniques fails to identify a secondary user when SU occupied any vacant channel. All cooperative-sensing techniques fail to distinguish between PU and SU after they occupied a channel.

To overcome this limitation, here a new approach called spectrum sensing using modulation detection is introduced in this chapter. To distinguish PU with SU, different sets of modulation techniques are allotted to primary users and secondary users. All PUs use a fixed set of modulation schemes and these are not allowed to

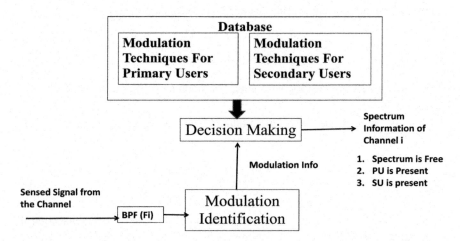

**Fig. 2** Narrowband sensing with modulation detection

use by SUs. With modulation detection scheme, the identified modulation type of channel data is compared with fixed sets of modulation types. Through this, the decision-making device gives three different cases about the spectrum. If modulation type is not matched with any of the set, then it shows that spectrum is vacant. If it matches with PUs set, then it understood that PU is present else it is noted that SU is present. The block diagram representation of narrowband sensing with modulation detection is shown in Fig. 2.

Modulation recognition of unknown signal using TFA is presented in simulation results. Using TFA, it is possible to trace the variations of signal characteristics like amplitude, frequency, and phase. These variations are traced with respect to time and these plots are called time–frequency contours. With these contours, it is possible to identify the modulation type of the signal.

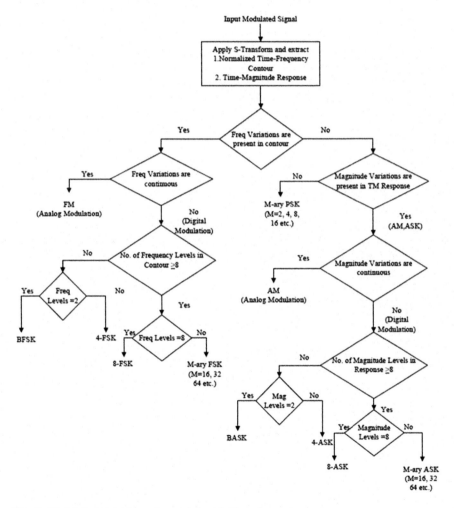

**Fig. 3** Modulation detection using variations in time–frequency contour

The detailed flow chart of modulation detection using MST is shown in Fig. 3. If the signal has only frequency variations in the contour it is treated as m-ary FSK. The value of m is identified based on no of frequency variations preset in the signal. Similarly, if the signal has amplitude variations then it is treated as m-ary ASK. If the signal has only phase variations it is treated as m-ary PSK modulation.

## 4   Simulation Results and Discussions

To illustrate the performance narrowband and wideband spectrum sensing with proposed methods, we consider a wideband signal with four frequencies and five active parts. The parameters of the wideband signal are shown in Tables 1 and 2. Here we consider TV broadcasting frequencies for the formation of wideband signal.

In each active part different set of users are present and some users are absent. The time–frequency contour of wideband signal is shown in Fig. 4.

From the Fig. 4 it is clearly shown that where the users are present where the users are absent. In the first active part, all PUs are present so there is no free channel for SUs. Similarly, in each active part, the vacant channels are represented by black dots. Figures 5 and 6 show performance of sensing in AWGN channel at different SNR conditions.

From Figs. 5 and 6 it clearly shows that even at very high noise conditions MST based wideband sensing gives a better result than existing methods. The performances of all existing methods are very poor under low SNR cases and different fading conditions.

Figure 7 shows the wideband sensing under fading conditions without any doppler frequency and phase shifts. Here we consider 4 paths and the path gains are selected randomly. The path gains are 1.0, 0.6663, 0.8661 and 0.7618 respectively.

**Table 1** Frequency information of users

| S. no. | User (U) | Carrier frequency (MHz) |
|--------|----------|-------------------------|
| 1 | $U_1$ | 150 |
| 2 | $U_2$ | 250 |
| 3 | $U_3$ | 350 |
| 4 | $U_4$ | 450 |

**Table 2** Active parts information

| S. no. | Active part | User information (U) |
|--------|-------------|----------------------|
| 1 | $A_1$ | $U_1, U_2, U_3,$ and $U_4$ |
| 2 | $A_2$ | $U_2, U_4$ |
| 3 | $A_3$ | $U_1, U_3$ |
| 4 | $A_4$ | $U_2, U_3,$ and $U_4$ |
| 5 | $A_5$ | $U_1, U_2,$ and $U_4$ |

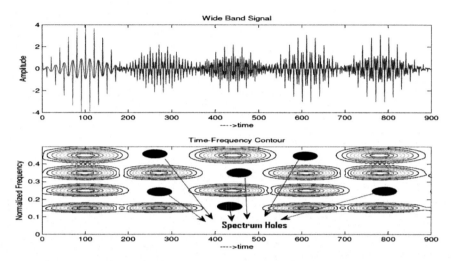

**Fig. 4** Time–frequency contour of wideband signal using MST

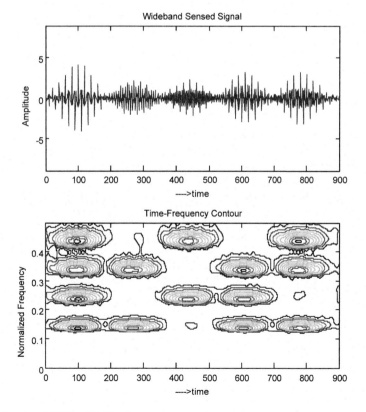

**Fig. 5** Sensing at SNR = 10 dB

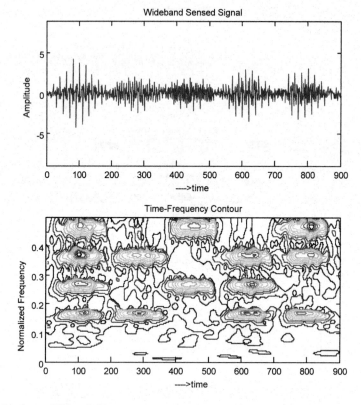

**Fig. 6** Sensing at SNR = 5 dB

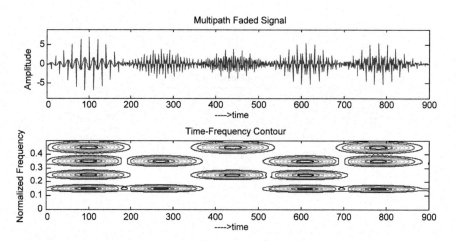

**Fig. 7** Wideband sensing with MST under fading conditions

Figure 8 shows sensing under fading conditions with multipath gains and Doppler phase shifts. The path gains are 1.0, 0.0690, 0.5519, and 0.4038, the phase shifts of the 4 paths are 0, 6π/100, 21π/100, and 54π/100 respectively.

Spectrum sensing under the doppler frequency and phase shift conditions is shown in Fig. 9. The path gains and Doppler shifts are Path Gains: [1.0000 0.6104 0.2480 0.5180]; Phase shifts: [0 54π/50 21π/50 31π/50]; Frequency shifts: [0 242.5662 151.9738 220.9372].

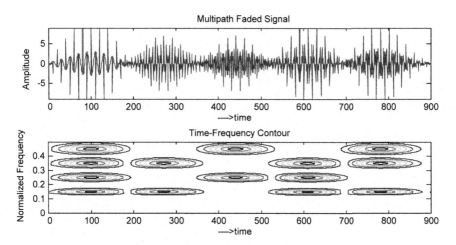

**Fig. 8** Sensing under fading conditions having Doppler phase shifts

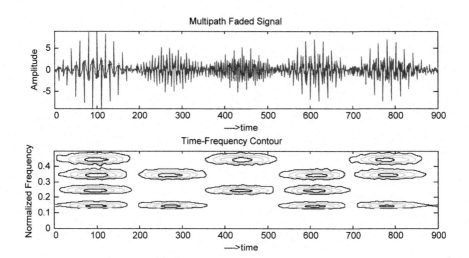

**Fig. 9** Sensing under Doppler conditions

From these simulation results, it can be proved that the proposed wideband sensing algorithm is superior to the existing methods under noisy and noiseless conditions.

# 5 Conclusion

A new wideband sensing approach is presented in this chapter. A new narrowband sensing approach is introduced in this chapter and it also distinguishes the primary user from the secondary user. Most of the cooperative-sensing techniques are narrowband sensing techniques. These techniques consider only one channel information at a time and search whether the channel is vacant or not. For wideband sensing, these techniques need parallel filter banks and that leads to increase in system complexity. With TFA proposed method gives the best solution for wideband spectrum sensing even under fading channel conditions.

# References

1. S. Haykin et.al. "Cognitive radio: Brain-empowered wireless communications", IEEE J. Sel. Areas Communication, vol. 23, pp. 201–220, Feb. 2005.
2. S. Haykin "Fundamental Issues in Cognitive Radio" in Cognitive Wireless Communication Networks, Springer, 2007, pp. 1–43, https://doi.org/10.1007/978-0-387-68832-9_1.
3. S. Haykin, D. J. Thomsom, and J. H. Reed, "Spectrum Sensing for Cognitive Radio" Proceedings of the IEEE, vol. 97, no. 5, pp. 849–877, May, 2009.
4. T. Yucek and H. Arslan, "A Survey of Spectrum Sensing Algorithms for Cognitive Radio Applications," IEEE Communication Survey & Tutorials, vol. 11, no. 1, pp. 116–130, 2009.
5. Budiarjo M, K Lakshmanan and H Nikookar, "Cognitive Radio Dynamic Access Techniques," Wireless Pers. Communication., vol. 45, pp. 293–324, Feb., 2008.
6. Tazeen S. Syed and Ghazanfar A. Safdar, "On The Usage of History for Energy Efficient Spectrum Sensing" IEEE Communications Letters, vol 19, NO 3, pp. 407–410, March 2015.
7. P. Karunakaran, T. Wagner, A. Scherb and W. Gerstacker, "Sensing for spectrum sharing in cognitive LTE-A cellular networks," 2014 IEEE Wireless Communications and Networking Conference (WCNC), Istanbul, 2014, pp. 565–570.
8. Goutam Ghosh, Prasun Das, and Subhajit Chattergee "Cognitive Radio and Dynamic Spectrum Access-A Study" International Journal of Next-Generation Networks (IJNGN) Vol. 6. No. 1, pp. 43–60, March 2014.
9. Z. Tian and G. B. Giannakis, "A Wavelet Approach to Wideband Spectrum Sensing for Cognitive Radios," 2006 1st International Conference on Cognitive Radio Oriented Wireless Networks and Communications, Mykonos Island, 2006, pp. 1–5.
10. M.F. Duarte and R.G. Baraniuk, "Spectral compressive sensing," Applied and Computational Harmonic Analysis, vol. 35, no. 1, pp. 111–129, 2013.
11. M. V. Subbarao and P. Samundiswary, "Time-frequency analysis of non-stationary signals using frequency slice wavelet transform," 2016 10th International Conference on Intelligent Systems and Control (ISCO), Coimbatore, 2016, pp. 1–6. https://doi.org/10.1109/ISCO.2016.7726999.

12. M. Venkata Subbarao, P. Samundiswary, "An Intelligent Cognitive Radio Receiver for Future Trend Wireless Applications", International Journal of Computer Science and Information Security (IJCSIS), Vol. 14, CIC 2016, Track 2, pp. 7–12, Oct 2016.
13. M. Venkata Subbarao, P. Samundiswary, "Wideband Spectrum Sensing for Cognitive Radio Networks using Time-Frequency Analysis", International Journal of Control Theory and Applications (IJCTA), Vol 9, Issue 41, pp. 671–677, Dec-2016.

# Synthesis of Linear Antenna Array Using Cuckoo Search and Accelerated Particle Swarm Algorithms

M. Vamshi Krishna, G. S. N. Raju and S. Mishra

**Abstract** Array pattern synthesis has a lot of importance in most of the communication and radar systems. It increases in defining the appropriate configuration of the array, which produces desired radiation pattern. Low sidelobe narrow beams are very useful for point-to-point communication and high-resolution radars. In this chapter, two evolutionary computing techniques like cuckoo search algorithm and accelerated particle swarm optimization are used. The desired amplitude levels are achieved by the algorithm with element spacing d = 0.40 and 0.45. The main objective is to generate patterns with fixed beam width with acceptable sidelobe level. The results are compared with conventional Taylor method. The array patterns are numerically computed for 100 number of elements.

**Keywords** Sum pattern · Sidelobe · Antenna array · Taylor
Cuckoo search algorithm (CSA) · Accelerated particle swarm optimization
(APSO)

## 1 Introduction

Array antennas have a great importance because its radiators have the ability to exhibit beam scanning with enhanced gain and directivity [1]. For a desired pattern, appropriate weighting vector is used. The major advantages of the use of array are that the mainlobe direction and sidelobe level of radiation pattern are controllable and function of the magnitude and the phase of the excitation current and the position of each array element [2].

Various analytical and numerical techniques have been developed to meet this challenge [3]. Analytical techniques converge to local values rather than global optimum values, which optimize further when compared to mathematical tech-

M. Vamshi Krishna (✉) · G. S. N. Raju · S. Mishra
Department of ECE, Centurion University of Technology and Management,
Visakhapatnam, Andhra Pradesh, India
e-mail: vamshikrishna@cutmap.ac.in

© Springer Nature Singapore Pte Ltd. 2018
J. Anguera et al. (eds.), *Microelectronics, Electromagnetics
and Telecommunications*, Lecture Notes in Electrical Engineering 471,
https://doi.org/10.1007/978-981-10-7329-8_86

niques [4]. Hence, there is need of evolutionary techniques so we can achieve the desired patterns with minimum sidelobe level [5].

The most used optimization techniques in array pattern synthesis are steepest descent algorithms. In this chapter, an effective method based on Cuckoo Search Optimization [6] (CSA) and Accelerated Particle Swarm Optimization [7] (APSO) is proposed for synthesizing of linear antenna array. As an excellent search and optimization algorithm, CSA has gained more and more attention and has very wide applications.

In this chapter, a CSA and APSO are applied for array synthesis, to control the desired pattern, for linear geometrical configuration with various spacing ranging between 0.40 λ and 0.45 λ relative displacement between the elements, the excitation amplitudes of individual elements and with no additional phase are computed.

## 2   Array Formulation and Fitness Calculation

### 2.1   Linear Array

Because of its simple design, most commonly a linear array is synthesized for many communication problems [8]. The representation of such geometry is as shown in Fig. 1. Considering a linear array of N isotropic antennas, where all the antenna elements are identically spaced at a distance d from one another along the x-axis [9].

The free space far-field pattern E (u) is given by

$$E(U) = 2 \sum_{n=0}^{N} A_n \cos[k(n-0.5)d(u-u_o)] \qquad (1)$$

where

$A_n$   excitation of the nth element on either side of the array

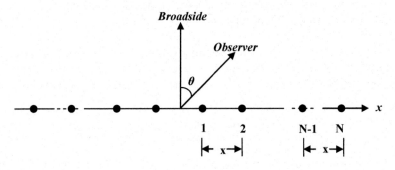

**Fig. 1** Linear array antenna

K    wave number $= 2\pi/\lambda$
$\lambda$    wavelength
$\theta$    angle between the line of observer and broadside
$\theta_0$    Scan angle
d    spacing between the radiating elements
u    $\sin \theta$
$u_0$    $\sin \theta_0$

Normalized far field in dB is given as

$$E(U) = 20\log_{10} \frac{|E(U)|}{|E(U)_{max}|} \qquad (2)$$

The excitation amplitudes are taken as parameters to be optimized with the objective of achieving reduced sidelobe level. Equation (1) is used to find the far-field pattern information of current amplitude excitation $A_n$ for all the elements, with element spacing as d $=$ (0.40 and 0.45) with zero additional phase.

In this optimization process, a design is made to minimize the sidelobe level of the radiation pattern without disturbing the gain of the main beam. The problem of minimizing the maximum SLL in the pattern with prescribed beamwidth by varying wavelength spaced array is solved using the fitness function. An appropriate set of element amplitudes are achieved by reduced sidelobe levels.

Thus, the fitness function is formulated as

$$\text{Fitness} = \text{Obtained Peak SLL} - \text{Desired Peak Side SLL}$$
$$-1 \leq u \leq 1 \ u \neq u_0$$

$$\text{Here Obtained Peak SLL} = \max[20\log_{10} \frac{|E(U)|}{|E(u_0)_{max}|}]$$
$$\text{Desired Peak SLL} = -35 \text{ dB}$$

# 3 Optimization Techniques

## 3.1 Cuckoo Search Algorithm (CSA)

The CSA mimics the natural behavior of cuckoo birds [10]. The principle depends on reproduction strategy of cuckoos. The algorithm has 3 idealized assumptions:

1. Egg laid by a cuckoo in a specific time is reserved for hatching.
2. Depending on the nest, the quality of the egg is defined.
3. Host nests are finite and the probability of identifying eggs lies between (0 and 1).

Random-walk style search is implemented by by Lévy flights [11]. Single parameter in Cuckoo Search Algorithm makes it simpler when comparing the other agent-based metaheuristic algorithms. The new generation of excitation current amplitude are determined by the best nest. The updating procedure is mentioned in the following Eq. (3)

$$X_i^{(t+1)} = X_i^{(t)} + \alpha \oplus \text{Lévy}(\lambda),$$  (3)

The Levi flight equation represents the stochastic equation for random-walk as it depends on the current position and the transition probability (second term in the equation). Where $\alpha$ is the step size, generally $\alpha = 1$. Element wise multiplications is given as:

$$\text{Lévy} \sim u = t^{-\lambda}, \quad (1 < \lambda \leq 3).$$  (4)

Here, the term $t^{-\lambda}$ refers to the fractal dimension of the step size and the probability Pa in this paper is taken as 0.25 [12].

## 3.2 Accelerated Particle Swarm Optimization

The standard PSO uses both the individual personal best and the current global best but APSO uses global best only [13]. This technique interestingly accelerates the search efficiency and iteration time.

It decreases the randomness as the iterations proceed. The APSO starts by initializing a swarm of particles with random positions and velocities. The fitness function of each particle is evaluated and the best g value is calculated [14].

Later, actual position is updated for each and every particle. This process is repeated until the optimum best g value is obtained. Some of the advantages of APSO over other traditional optimization techniques, it has the reliability to modify and find a balance between the global and local exploration of the search space, and it has implicit parallelism.

$$Vel_n(t+1) = w \cdot Vel_n(t) + c_1 \cdot r_1(pbest_n - X_n(t)) + c_2 \cdot r_2(gbest - X_n(t))$$  (5)

$$X_n(t+1) = X_n(t) + Vel_n(t+1)$$  (6)

Here, w is the inertia coefficient of the particle which play a vital role in PSO. $Vel_n(t+1)$ is present particle's velocity, $Vel_n(t)$ is the earlier particle's velocity, $X_n(t)$ is the present particle's position, $X_n(t)$ is the earlier particle's position. $r_1$ and $r_2$ are random in nature and lies in the range [0 and 1] and uniformly distributed [15].

$c_1$ and $c_2$ are the acceleration constants which manage the relative effect of the pbest and gbest particles. $pbest_n$ is the present pbest value, gbest is the present gbest value.

The PSO algorithm is defined in 4 steps which will terminate when the exit criteria are met. The velocity vector is produced by using the below expression.

$$Vel_n(t+1) = w \cdot Vel_n(t) + \alpha \cdot c_n + \beta(gbest - X_n(t)) \tag{7}$$

Here $c_n$ value lies in between $(0, 1)$ and in random nature.
The position vector is modified using the following expression

$$X_n(t+1) = X_n(t) + Vel_n(t+1) \tag{8}$$

Combining the above two equations yield the following expressions.

$$X_n(t+1) = (1 - \beta)X_n(t) + \alpha \cdot c_n + \beta(gbest) \tag{9}$$

The distinctive values of APSO are $\alpha = 0.1–0.4$ and $= \beta$ 0.1–0.7. Here $\alpha$ is 0.2 and $\beta$ is 0.5.

$$\text{while } \alpha = \gamma^t \quad (0 < \gamma < 1) \tag{10}$$

$\gamma$ is referred to a control parameter with magnitude 0.9 coherent in the iteration number.

## 4 Results

Cuckoo search algorithm and Accelerated Particle Swarm Optimization are applied to evaluate amplitude distribution required to maintain sum patterns with Sidelobe level at −35 dB. The patterns are numerically computed for N = 100 arrays of

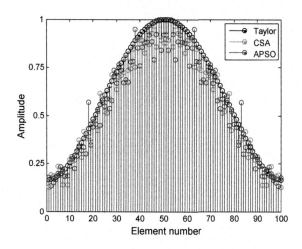

**Fig. 2** Amplitude distribution for N = 100 element array with d = 0.40 using Taylor, CSA, and APSO

**Fig. 3** Sum pattern
optimized for N = 100 array
with d = 0.40 and nbar = 6
using Taylor, CSA, and
APSO

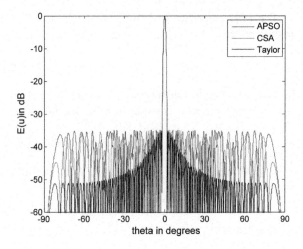

**Fig. 4** Amplitude
distribution for N = 100
element array with d = 0.45
using Taylor, CSA, and
APSO

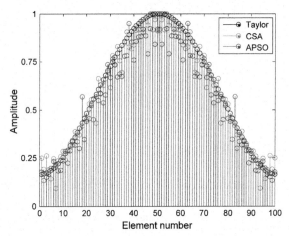

**Fig. 5** Sum pattern
optimized for N = 100 array
with d = 0.45 and nbar = 6
using Taylor, CSA, and
APSO

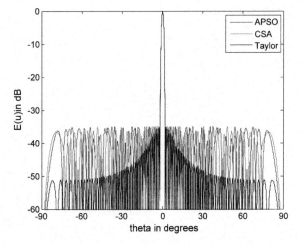

elements by varying the spacing between elements as d = 0.40 and 0.45. As the number of elements are increased in the array, the Null to Null beamwidth is found to decrease (Figs. 2, 3, 4 and 5).

# 5 Conclusion

The synthesis of uniform linear arrays for sidelobe level reduction is considered in the present work. The Algorithms is found to be useful to generate desired radiation pattern. The method is useful to solve multi-objective array problems involving with specified number of constraints. The sidelobe level is decreased to −35 dB. The results are extremely useful in communication and radar systems where the mitigation of EMI is a major concern. The beamwidth remains unaltered even after reducing the sidelobe level. The rise of far away sidelobes is not a problem in the system of present interest.

# References

1. Raju G S N, Antennas and wave propagation, 3rd ed., Pearson Education, 2005
2. Stienberg B D. Principles of Aperture and Array System Design. New York: John Wiley and Sons, 1976
3. Unz H "Linear arrays with arbitrarily distributed elements, IEEE Transactions on Antennas and Propagation, 1960, 1(8), pp. 222–223
4. Harrington R., "Sidelobe reduction by non-uniform element spacing" IRE Transactions on Antennas and propagation, 1961, 9(2), pp. 187–192
5. V. V. S. S. S. Chakravarthy, P. S. R. Chowdary, Ganapati Panda, Jaume Anguera, Aurora Andújar, and Babita Majhi: On the Linear Antenna Array Synthesis Techniques for Sum and Difference Patterns Using Flower Pollination Algorithm, Arabian Journal for Science and Engineering, Aug (2017)
6. Yang, X.-S., and Deb, S, "Engineering Optimization by Cuckoo Search", Int. J. Mathematical Modelling and Numerical Optimization, Vol. 1, No. 4, 330–343, 2010
7. Amir Hossein Gandomi., Gun Jin Yun., Xin-She Yang., Siamak Talatahari.: Chaos − enhanced accelerated particle swarm optimization. J. Commun Nonlinear Sci Numer Simulat. Vol. 18. (2013) 327–340
8. Khairul Najmy Abdul Rani, Mohd. Fareq Abd Malek, Neoh Siew-Chin", Nature-Inspired Cuckoo Search Algorithm For Side Lobe Suppression In A Symmetric Linear Antenna Array, Radioengineering, Vol. 21, No. 3, September 2012
9. Haffane Ahmed, Hasni Abdelhafid, "Cuckoo Search Optimization For Linear Antenna Arrays Synthesis", Serbian Journal Of Electrical Engineering Vol. 10, No. 3, 371–380, October 2013
10. K N Abdul,F Malek, "Symmetric Linear Antenna Array Geometry Synthesis Using Cuckoo Search Metaheuristic Algorithm", 17th Asia-Pacific Conference On Communications (Apcc) 2nd − 5th October 2011
11. Urvinder Singh and Munish Rattan: "Design of Linear and Circular Antenna Arrays Using Cuckoo Optimization Algorithm", Progress in Electromagnetics Research C, Vol. 46, 1–11, 2014

12. M. Khodier And M. Al-Aqeel, "Linear And Circular Array Optimization: A Study Using Particle Swarm Intelligence", Progress In Electromagnetics Research B, Vol. 15, 347–373, 2009
13. Majid M. Khodier, Christos G. Christodoulou, "Linear Array Geometry Synthesis With Minimum Sidelobe Level And Null Control Using Particle Swarm Optimization", IEEE Transactions On Antennas And Propagation, Vol. 53, No. 8, August 2005
14. N. Pathak, G. K. Mahanti, S. K. Singh, and J. K. Mishra A. Chakraborty," Synthesis Of Thinned Planar Circular Array Antennas Using Modified Particle Swarm Optimization", Progress In Electromagnetics Research Letters, Vol. 12, 87–97, 2009
15. Pinar Civicioglu, Erkan Besdok: A conceptual comparison of the Cuckoo-search, particle swarm optimization, differential evolution and artificial bee colony algorithms, Springer Science + Business Media B.V. 2011

# A New VLSI Architecture for Skin Tone Detection in an Uncontrolled Background

M. V. Ganeswara Rao, Rajesh K. Panakala
and A. Mallikarjuna Prasad

**Abstract** Human face detection in image sequence plays a crucial role in the applications such as video surveillance, security monitoring, human computer communication, smart homes, autonomous robots, and medical image analysis. Human recognition is based on identification and locating a human face in images or image sequence in spite of background, size, position, and lighting stipulation. The state-of-the-art face detection algorithms make use of skin tone filter to enhance the performance of human face detection and recognition algorithms. In this paper, a new parallel hardware architecture for skin tone detection has been proposed, where meanCr, meanCb, etc. are computed concurrently. Hence, the proposed architecture achieves high throughput compared to the DSP implementation of the same. The proposed architecture has been implemented and validated using Xilinx Spartan 3E XC3S500E FPGA chip. The implementation also occupies only 40.19% device area. The critical path delay in FPGA implementation is only 11.5 ms.

**Keywords** Face detection · Skin tone · Field-programmable gate array (FPGA)
VHDL

M. V. Ganeswara Rao (✉)
Shri Vishnu Engineering College for Women, Bhimavaram, India
e-mail: mgr_ganesh@svecw.edu.in

R. K. Panakala
PVP Siddhartha Institute of Technology, Vijayawada, India
e-mail: rkpanakala@gmail.com

A. Mallikarjuna Prasad
JNTU College of Engineering, Kakinada, India
e-mail: a_malli65@yahoo.com

© Springer Nature Singapore Pte Ltd. 2018
J. Anguera et al. (eds.), *Microelectronics, Electromagnetics
and Telecommunications*, Lecture Notes in Electrical Engineering 471,
https://doi.org/10.1007/978-981-10-7329-8_87

# 1  Introduction

Human interface is an important factor in extensive collection of applications such as face recognition, video surveillance, and face database management systems. Detections of faces in images play a very vital role. Face detection can be viewed as a binary classification problem (face and non-face). However, some techniques are developed for face detection and face reorganization, for example, template approaches, featured based approaches, and their combination. But these techniques are not satisfactory and usually contain false alarms. Later, Kanade and Schneiderman proposed neural network based approach [1]. However, face detection speed is not good enough to use in real-time applications. In the year 2000, a real-time face detection algorithm is introduced by Viola and Jones [2] and it offers satisfactory results in terms of performance and accuracy for small and mid-size images. However, in the state-of-the-art applications such as video surveillance, the image data to process for face detection becomes huge. In their approach, it is required to inspect entire image in various window sizes, and to examine them these areas could be passed through cascaded classifiers constructed during the training stage. But the speed of response does not satisfy the real-time applications [3, 4]. This problem is more serious when it is implemented using embedded platforms, where resources are limited.

A skin tone filter is proposed to attack speed problem in the above scenario. This filter identifies the skin regions, which is small portion of the whole image. If skin regions are filtered before being applied to Viola and Jones algorithm, the significant amount of computational complexity is reduced and obviously face detection achieves the real-time performance [5–7].

This paper presents a new approach to implement skin tone detection which is used to accelerate the face detection systems. Normally, in order to identify a face (s), it is essential to check each and every pixel in an image. Instead, prior to face detection, we can use skin tone detector to identify skin patches and non-skin patches in the image (see Fig. 1). This approach simplifies face detection systems by reducing computations at detection stage. This skin tone detection module was implemented on Xilinx Spartan 3E board.

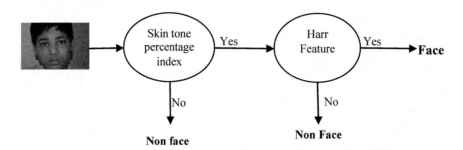

**Fig. 1** Skin detector integrated into face detection

## 2 Theory and Algorithm

### 2.1 Color Space for Face Detection

Most of the images are in RGB color space; based on the application this, color space is converted into other space in order to build the efficient application. In the analysis of face detection, RGB color space is not effective because R, B, and G components are highly correlated with each other. The inefficiency of the RGB color model for face detection is shown by the literatures. Here, we used YCrCb color space to filter skin regions in image. The color space color components are correlated to achieve very effective face detection within a short period of time.

### 2.2 YCrCb Color Model

In this color model, Y represents luminance and Cr and Cb represent chrominance values. The advantage of YCrCb color space over RGB and HSV is luminance and Chroma spaces are separated in YCrCb space. In real time, images are in RGB color space, and these RGB color models are converted into YCrCb color model using the following relation:

$$Y = 0.257\,R + 0.504\,G + 0.098\,B + 16 \tag{1}$$

$$Cb = -0.148\,R - 0.291\,G + 0.439\,B + 128 \tag{2}$$

$$Cr = 0.439\,R - 0.368\,G - 0.071\,B + 128 \tag{3}$$

$$140 < = Cr < = 165 \tag{4}$$

$$140 < = Cb < = 195 \tag{5}$$

$$
\begin{aligned}
[Cr, Cb] &= \text{skintonepixel} \quad && \text{if } Cr1 \leq Cr \leq Cr2 \\
& && \text{and} \\
& && Cb1 \leq Cb \leq Cb2 \\
&= \text{nonskintonepixel} \quad && \text{Otherwise}
\end{aligned}
\tag{6}
$$

### 2.3 Nonlinear Transformation and Skin Model

In YCrCb color space, Chroma components Cb and Cr are function of luma component Y; let the transformed Chroma be TransCr and TransCb and spread of the cluster widthCr and widthCb, the skin color model specified by the centers

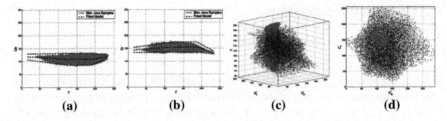

**Fig. 2** **a** YCb subspace, **b** Cr subspace, **c** transformed YCrCb color space, **d** a 2D projection of (c) in the transformed subspace red dots skin cluster and blue dots are non-skin pixels

meanCr and meanCb. Below equations are used to compute transformed Chroma values TransCr and TransCb (see Fig. 2)

$$TransCr = Cr \quad \text{if } 125 < = Y < = 188$$
$$= [Cr - meanCr] * \left[ \frac{38.76}{width\ Cb} \right] + meanCr(188) \tag{7}$$

$$TransCb = Cb \quad \text{if } 125 < = Y < = 188$$
$$= [Cb - meanCb] * \left[ \frac{46.97}{widthCb} \right] + meanCb(188) \tag{8}$$

The elliptical model for skin tone in transformed Chroma components [TransCr, TransCb] space is described by the following equations:

$$\begin{bmatrix} K \\ L \end{bmatrix} = \begin{bmatrix} \cos\theta & \sin\theta \\ -\sin\theta & \cos\theta \end{bmatrix} \begin{bmatrix} transCb - c_x \\ transCr - C_y \end{bmatrix}$$
$$\frac{(K - ec_x)^2}{a^2} + \frac{(L - ec_y)^2}{b^2} = 1$$

## 3   Implementation

The skin tone detection algorithm presented in the previous section is captured using VHDL, simulated using MULTI SIM simulator, and implemented on Xilinx Spartan 3E FPGA [8]. The RGB input image is loaded into BRAM of FPGA and a block RAM controller is implemented to read the BRAM by subsequent modules. This process is shown in Fig. 3.

**Fig. 3** BRAM loading process

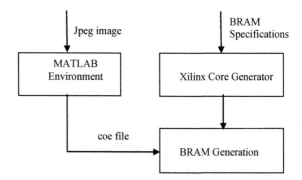

## 3.1 Hardware Architecture of Skin Tone Detector

The RGB image stored in BRAM is read by the RGB to YCrCb converter and luma component Y is fed to meanCr, meanCb, widthCb, and widthCr modules concurrently. The output of these modules is connected to next subsequent modules to produce TransCr and TransCb. These transformed Chroma values are used to find out skin score of each pixel. Figure 4 shows the pipelined architecture of skin tone detector.

## 4 Experimental Results

We have implemented the proposed architecture on Xilinx Spartan 3E FPGA (see Fig. 4 and experimental setup is shown in Fig. 5); chip and evaluated system uses 50 image databases, including family and single face image databases. The face

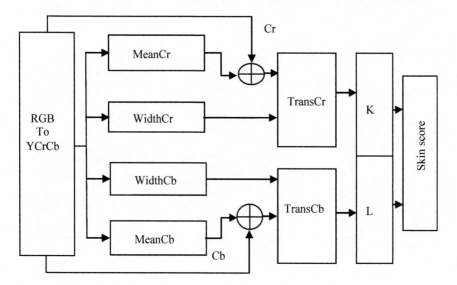

**Fig. 4** Block diagram of skin tone detector

**Fig. 5** Experimental setup

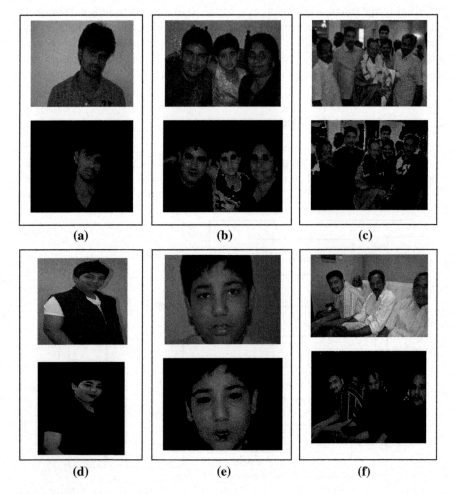

**Fig. 6** Skin detection examples

**Table 1** FPGA design results

| Parameter | Value |
|---|---|
| Execution time at 600 MHz clock | 1.4 ms |
| Slices | 984/2,448 (available) |
| CLBs | 246/612 (available) |
| Core power dissipation | 90.2 mW |

**Table 2** TI DSP 6416DSK design results

| Parameter | Value |
|---|---|
| Execution time at 600 MHz clock | 12.9 ms |
| Hardware utilization | 90 KB |
| Core power dissipation | 255 mW |

databases created for face detection include only grayscale images, which are not suitable for this algorithm. Therefore, we have collected our database for skin tone detection from the world wide web and personal photo collections. These color images are taken under varying illumination conditions and with complex background. Some of the results obtained by the proposed architecture are shown in Fig. 6.

The outputs of the design are compared with DSP implementation to ensure the effectiveness of the proposed architecture. The synthesis report of proposed architecture is shown in Table 1 and results are obtained from the TI DSP processor. The results are given in Table 2.

# 5 Conclusion

In this paper, a novel VLSI architecture for skin tone detection is proposed, which is based on nonlinear transformation and pipeline approach. The proposed architecture first converts the RGB color images into YCrCb images for further processing, and later these pixels data are transformed to elliptical space to detect skin patches in the image. The proposed design is evaluated by applying a large range of image databases, and results show that the FPGA-based implementation of the design is superior in terms of execution time compared to that of the DSP-based implementation.

**Declaration** I declare that the images included for simulation and presented in this paper are mine and have no conflicts of interest with any others. Hence, I have the consent and transfer the same to the publisher.

# References

1. H. Schneiderman and T. Kanade.: Probabilistic modeling of local appearance and spatial relationships for object recognition. In Proceedings of IEEE Computer Society Conference on Computer Vision and Pattern Recognition, 1998
2. R.L. Hsu, M. Abdel-Mottaleb, A.K. Jain, "Face Detection in Color Images", IEEE Transactions on Pattern Analysis & Machine Intelligence, vol. 24, no., pp. 696–706, May 2002, https://doi.org/10.1109/34.1000242
3. Luqman Maraaba, Zakariya Al-Hamouz, Hussain Al-Duwaish: Prediction of the Levels of Contamination of HV Insulators Using Image Linear Algebraic Features and Neural Networks, Arab J Sci Eng, 40 (9) 2609–2617 (2015)
4. B. Dahal, Abeer Alsadoon, P.W.C. Prasad, Amr Elchouemi, "Incorporating skin color for improved face detection and tracking system", 2016 IEEE Southwest Symposium on Image Analysis and Interpretation (SSIAI), vol. 10, no. 05, pp. 173–176, 2016, https://doi.org/10.1109/SSIAI.2016.7459203
5. Matti Pietikäinen, Birgitta Martinkauppi, Maricor Soriano, "Detection of Skin Color under Changing Illumination: A Comparative Study", Image Analysis and Processing, International Conference on, vol. 07, no. 7, pp. 652, 2003, https://doi.org/10.1109/ICIAP.2003.1234124
6. Rao, MV Ganeswara, Rajesh K. Panakala, and A. Mallikarjuna Prasad. "Recent Advances in Face Detection Techniques: A Survey." I J C T A, 8(5), 2015, pp. 2145–2152
7. Sadek, M.M., Khalifa, A.S. & Mostafa, M.G.M. Multimed Tools Appl (2017) 76: 3065. https://doi.org/10.1007/s11042-015-3170-8
8. Kandadai, V., Sridharan, M., Manickavasagam Parvathy, S. et al. Arab J Sci Eng (2016) 41: 3355. https://doi.org/10.1007/s13369-015-1878-4

# Efficiently Secure Data Privacy on Hybrid Cloud Using Novel Image Scrambling and Modified SPIHT

**T. M. Praneeth Naidu and G. Spandana**

**Abstract** Technology has grown to such an extent that anything can be stored in cloud. But the extension of the adoption of the cloud is hindered as there are some major concerns regarding the privacy of data stored and data size in the public cloud. In hybrid cloud computing (HCC), the images are preserved in dedicated cloud. However, this technique avoids the basic feature of the cloud computing as it increases the computation and storage overhead on the cloud. A novel algorithm approach is used to compute the private image data, where 0.001 time of AES algorithm is used and the delay is 3–5% when compared to other traditional public cloud approaches. The data compression algorithm used reduces the data size by 87%, thus reducing the time of uploading by 70%.

**Keywords** HCC · SPIHT · Data privacy

## 1 Introduction

The advent of technology in information and communication has grown to such an extent that a massive amount of data has been produced by the organizations and has become difficult to store and manage this data in a cost-effective way. Cloud computing is one of the cost-effective solutions to store the data in an effective way. Cloud computing is one of the recent trends which has gained a lot of attention in the recent years [1–5]. It uses a provisioning mechanism when on demand and usage-based payment. Security and privacy are major concerns in this model because most of the individuals and organizations use the third party to store their

T. M. Praneeth Naidu (✉)
Incline Inventions Pvt. Ltd., Hyderabad, India
e-mail: praneethtm@gmail.com

G. Spandana
Department of Electronics and Communication, Maturi Venkata Subba
Rao Engineering College, Hyderabad, India
e-mail: spandana.coign@gmail.com

© Springer Nature Singapore Pte Ltd. 2018
J. Anguera et al. (eds.), *Microelectronics, Electromagnetics
and Telecommunications*, Lecture Notes in Electrical Engineering 471,
https://doi.org/10.1007/978-981-10-7329-8_88

855

data such as cloud service provider (CSP), who has the total control over the data, where there is a scope for attacks which may possibly lead to data insecurity [6–10].

To protect the data, the existing solutions are to encrypt/decrypt the data which coordinates the access control in order to establish privacy for the data preserved on the cloud. But this introduces a heavy computation through processes like key distribution, data query, data management, etc.

In this work, a different method is suggested of obtaining typical data privacy by using the hybrid cloud. The HCC comprises both public and private clouds which is generally owned and controlled by its owner. The user data is spread into non-sensitive data and sensitive data which are preserved in private and public clouds, respectively. Data such as medical image can be stored in private cloud, but this might require a lot of storage space on the private cloud, but a user would like to minimize the storage space. The above challenge is addressed and made sure that hybrid cloud is effectively used to store the data in private and public clouds without creating any communication overhead between them. The main intension is to achieve privacy and to reduce the storage and computation in privacy cloud and to reduce communication overhead between private and public clouds. The algorithm used in this work divides the image into blocks, and noise is added. Considering the balance between the complexity of recovering the image and communication overhead, the size of the block is determined. Now, a random shuffle operation is performed on the blocks, making the image hard to recognize. The relationship among tables stored in public cloud is removed using hash functions with different keys; this makes the analysis of the stored data difficult.

## 2  Proposed Technique

In the proposed technique, a novel algorithm for image scrambling is proposed for enhancing the security of the input image. The encrypted data is then compressed via lossless compression technique called set partitioning in hierarchical trees which reduces the size of the data considerably. The proposed technique can be explained in four steps such as reading the input image, later scrambling it in order to secure it which is followed by compressing and then at last uploading it to the cloud.

There are several ways to secure the image data. Some of the incorporated techniques are mentioned here as follows.

### 2.1  Modifying Image

Typically, the image data occupies more volume than simple text data. It is difficult to perform operations based on pixels with any kind of encryption used. In order to overcome this problem, the image is divided into large number of blocks having a size of $N \times N$.

## 2.2 Random Shuffling of the Blocks

In an attempt to make the image secured by making it unrecognizable, all the blocks are shuffled and manipulated. The n-blocks are clustered and are added into the group with randomly chosen strides, where each cluster has the same size m. The steps are given in detail below.

## 2.3 Recovering Images

During the image query, a request is processed through both private and public clouds simultaneously. To recover the image, the information is taken from the private cloud. The shuffle order is obtained from the permit using the algorithm 1 and the blocks are reordered using the shuffled blocks from the public cloud, which gets the modified image. The original image is obtained from the modified image using the random values from the private cloud.

## 3 SPIHT Algorithm

Set Partitioning in Hierarchical Trees (SPIHT) is one of the powerful wavelet-based image compression methods. It has gained a lot of attention worldwide, and many consumers and researchers have tested and used SPIHT. It is considered as one of the best advancements in the field and therefore requires special attention as it provides the several characteristics like efficient image quality and high PSNR, completely embedded coded file, simple and fast quantization algorithm, coding and decoding is very fast, compression without loss, exact bit rate coding, protection from errors, progressive image transmission, and completely adaptive (Fig. 1).

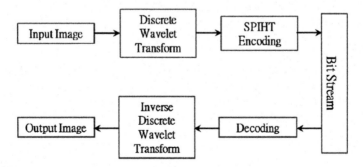

**Fig. 1** SPIHT image compression

```
Initialization
Sorting pass
  For x=1 to end LIP
  From LIP1 it goes to |c(u,v)|>2^k,
    If the condition is "yes" it goes to output 1
    sgn(c(u,v)) and then to LSP2, from LSP2 to the 3^rd
    step.
    If the condition is "no" the output is 0 and then
    goes to LIP2 to LIP1 and the process repeats
    again.
  For x=1 to end LIS
  From LIS1 it goes to Type(i,j) and then checks for
  the condition,  max|D(i,j)|>=2^k
  If the condition is "yes" then the output is 1 it
  goes to the next step where each(k,l) in O(i,j) is
  of type L then goes to LIS1.

    If the condition is "no" the output is 0 and then
    goes to LIS2 and then LIS1 and the process repeats.
  Refinement pass
  For x = 1 to endLSP from LSP1 to output n^th bit c(i,j).
```

**Fig. 2** Algorithm description

At least one of these objectives was tried to be developed by using other compression methods; now, this makes SPHIT as an outstanding technique as it obtains all the above qualities simultaneously. The algorithm flow is demonstrated as follows (Fig. 2).

## 4  Experimental Analysis

The user selects an image from the computer. The algorithm then reads the image and starts the processing. The input image is shown in Fig. 3, while the input key pop-up is shown in Fig. 4. The subtracted image is shown in Fig. 5. The shuffled image is shown in Fig. 6.

At the beginning of the random number generators for noise addition, the user is asked a key between 1000 and 2000. Then, the random number generator creates a map with random values from 0 to 255. These values are then subtracted from the image. Then, the blocks are divided and shuffled.

**Fig. 3** Input image

**Fig. 4** Enter the random input key for shuffling

As discussed in algorithm, the second stage is to compress the image data. A variety of algorithms have been tested for this purpose. The results are tabulated in Tables 1 and 2.

**Fig. 5** Subtracted image

**Fig. 6** Block shuffled image

**Table 1** Input images, shuffled images, image dimensions, and image size

| Input image | Shuffled image | Image dimensions | Image size (Kb) |
|---|---|---|---|
| | | 256 × 256 | 194 |
| | | 256 × 256 | 194 |
| | | 256 × 256 | 194 |

**Table 2** Performance of various techniques

| Compression type | Image ID | Original Size (Kb) | Compressed size | PSNR | MSE | Time of execution | Time to upload |
|---|---|---|---|---|---|---|---|
| Block truncation coding | 1 | 194 | 193 | 44.399 | 767.03 | 1.2307 | 5.494 |
| | 2 | 194 | 193 | 45.0556 | 718.35 | 0.696 | 5.494 |
| | 3 | 194 | 193 | 45.5055 | 686.75 | 0.6518 | 5.494 |
| Discrete cosine transform | 1 | 194 | 192 | 40.829 | 1096.1 | 0.3176 | 5.49 |
| | 2 | 194 | 192 | 45.608 | 679.54 | 0.2984 | 5.494 |
| | 3 | 194 | 192 | 43.107 | 872.86 | 0.2874 | 5.493 |
| Singular value decomposition | 1 | 194 | 190 | 40.5005 | 1132.8 | 0.4909 | 5.487 |
| | 2 | 194 | 189 | 45.8945 | 660.54 | 0.4339 | 5.486 |
| | 3 | 194 | 189 | 42.7461 | 904.98 | 0.4782 | 5.486 |
| Pyramid | 1 | 194 | 184 | 29.2315 | 3496.0 | 3.3096 | 5.48 |
| | 2 | 194 | 184 | 30.2226 | 3166.1 | 3.2769 | 5.48 |
| | 3 | 194 | 184 | 29.6109 | 3365.8 | 3.2972 | 5.48 |
| Proposed algorithm with SPIHT | 1 | 194 | 24 | 52.4359 | 617.93 | 9.7531 | 1.21 |
| | 2 | 194 | 24 | 51.9515 | 514.43 | 9.4271 | 1.21 |
| | 3 | 194 | 24 | 51.8418 | 448.59 | 9.0184 | 1.21 |

# 5 Conclusion

The proposed method uses a hybrid method combining the image security by using encryption and then uploads the data to the cloud after compression. The proposed SPIHT-based technique effectively compresses the data, thus reducing the data storage in the cloud.

# References

1. Huang, X., & Du, X. (2013b). Efficiently secure data privacy on hybrid cloud. In IEEE International Conference on Communications (pp. 1936–1940).
2. P. Mell and T. Grance, "Draft nist working definition of cloud computing," Referenced on June. 3rd, 2009.
3. M. D. Ryan, "Cloud computing privacy concerns on our doorstep," Communications of the ACM, 2011.
4. A. Feldman, W. Zeller, M. Freedman, and E. Felten, "Sporc: Group collaboration using untrusted cloud resources," OSDI, Oct, 2010.
5. C. Wang, Q. Wang, K. Ren, and W. Lou, "Privacy-preserving public auditing for data storage security in cloud computing," in INFOCOM, 2010 Proceedings IEEE, 2010.
6. S. Yu, C. Wang, K. Ren, and W. Lou, "Achieving secure, scalable, and fine-grained data access control in cloud computing," in INFOCOM, 2010 Proceedings IEEE, 2010.

7. V. Goyal, O. Pandey, A. Sahai, and B. Waters, "Attribute-based encryption for fine-grained access control of encrypted data," in Proceedings of the 13th ACM conference on CCS, 2006.
8. E. Demaine and M. Demaine, "Jigsaw puzzles, edge matching, and polyomino packing: Connections and complexity," Graphs and Combinatorics, vol. 23, 2007.
9. T. Cho, S. Avidan, and W. Freeman, "A probabilistic image jigsaw puzzle solver," in Computer Vision and Pattern Recognition (CVPR), 2010 IEEE Conference on, 2010.
10. Kefei Zhang, Fang Yuan, Jiang Guo: A Novel Neural Network Approach to Transformer Fault Diagnosis Based on Momentum-Embedded BP Neural Network Optimized by Genetic Algorithm and Fuzzy c-Means, Arabian Journal for Science and Engineering, 41(9) 3451–3461 (2016).

# Performance Analysis of Reconfigurable Antenna with Notch Band Characteristics

Pavani Kollamudi

**Abstract** Reconfigurable antennas can combine wideband and narrowband characteristics in a single antenna. In this paper, a configurable antenna with six possible configurations is presented with the base structure as a hexagonal patch antenna. The antenna is simulated and analyzed in computer simulation tool (CST). The analysis is based on return loss reports for each configuration. The antenna is designed on a low-cost FR4 substrate with dielectric constant 4.4 with a thickness of 1.6 mm.

**Keywords** Reconfigurable antenna · Wideband · Narroband
Hexagonal patch antenna

## 1 Introduction

Advanced communication systems demand for antennas with multi-functionality that refers to serving for different applications on a single terminal. Antennas for general personal communication are the best example of such applications involving in operating frequency covering WLAN, Wi-Fi, Bluetooth, GSM, GPS, etc. In order to facilitate for such applications, we need to possess certain specific radiation characteristics like wideband and multiband. Reconfigurable antennas have been developed to suit for such applications. These reconfigurable antennas can be designed to comply with many advanced communication system requirements like compactness and conformness. In the recent literature openly available,

P. Kollamudi (✉)
Department of Electronics and Instrumentation Engineering, LBRCE, Mylavaram,
Vijayawada, Andhra Pradesh, India
e-mail: pavani.be@gmail.com

© Springer Nature Singapore Pte Ltd. 2018
J. Anguera et al. (eds.), *Microelectronics, Electromagnetics
and Telecommunications*, Lecture Notes in Electrical Engineering 471,
https://doi.org/10.1007/978-981-10-7329-8_89

several reconfigurable antennas are designed to switch between multiple similar bands [1–4]. Similarly, these antennas are even capable of switching from wideband to narrowband and vice versa through multiband characteristics [5–11].

In this paper, such reconfigurable antennas which sweep both wide and narrow and multiple bands are presented. The simulation of the antenna is carried out in CST. Further, the paper is organized into four sections. Brief description of the proposed antenna geometry is given in Sect. 2 and the simulation details are mentioned in Sect. 3. Discussions on the simulated results are given in Sect. 4. Overall conclusions are mentioned in Sect. 5.

## 2 Geometry of Proposed Antenna

The typical geometry of the proposed reconfigurable is shown in Fig. 1. The geometry consists of a hexagonal patch to the end of the substrate. Along the feed line strip on either sides, three horizontal strips are arranged out of which the strip close to the ring is loaded with hexagonal patch. All these six steps are not in physical contact with the feed line strip but connected through a PIN diode which can be switched. The ground plane is shown in Fig. 1b with a defective ground structure including a small slit.

The typical dimensions of the proposed geometry are empirically determined using the tuning module in the EM tool. These dimensions are shown in Table 1.

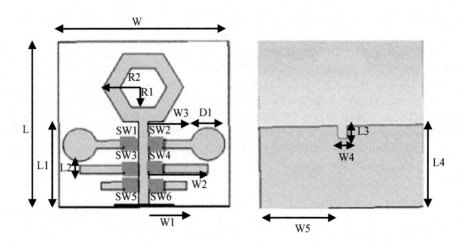

**Fig. 1** Geometry of the proposed reconfigurable antenna

**Table 1** Dimensions of the antenna

| S. no. | Parameter | Optimized values in mm |
|--------|-----------|------------------------|
| 1 | W | 40 |
| 2 | W1 | 10 |
| 3 | W2 | 15 |
| 4 | W3 | 11.25 |
| 5 | W4 | 2.25 |
| 6 | W5 | 20 |
| 7 | L | 40 |
| 8 | L1 | 20 |
| 9 | L2 | 2 |
| 10 | L3 | 3 |
| 11 | L4 | 20 |
| 12 | R1 | 10 |
| 13 | R2 | 5 |
| 14 | D1 | 4 |

# 3 Simulation and Operation

The proposed geometry of the antenna is simulated in CST which is a method of moment-based tool. During the analysis of the simulated antenna, the mesh size is dynamically chosen and taken to be at least 10 cells per wavelength. The antenna operates with different functionalities in six configurations. Each configuration is determined by the associated switch. The typical switching process and concerned configuration are listed in Table 2.

The simulated geometries starting from configuration I to configuration VI are presented in Fig. 2a through Fig. 2f.

**Table 2** Six configurations

| Configuration | Switches (SW1/SW2/SW3/SW4/SW5/SW6) |
|---------------|-------------------------------------|
| I | All OFF |
| II | All ON |
| III | SW1 and SW2 ON and all others OFF |
| IV | SW3 and SW4 ON and all others OFF |
| V | SW5 and SW6 ON and all others OFF |
| VI | SW2, SW4, and SW6 ON and all others OFF |

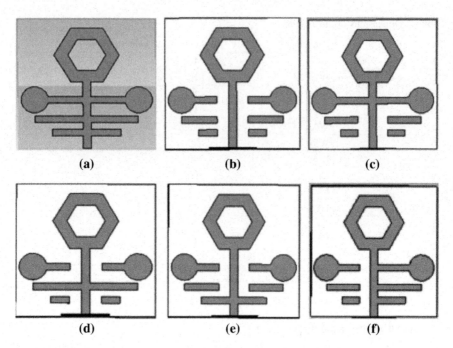

**Fig. 2** Simulated geometries in six configurations

# 4 Results and Discussion

Simulation results pertaining to the proposed reconfigurable antenna are presented in this section. The analysis of the antenna is carried out using the simulated reports known as return loss (S11) for each configuration. In its basic form, the antenna operates as a wideband antenna with all the horizontal strips disconnected as in configuration I. It is possible to conclude the UWB characteristics with wideband covering the entire UWB region which is evident from the S11 plot in Fig. 3a.

Similarly, in configuration II, the antenna exhibited two bands of operation as dual band. This is evident from the S11 plot for configuration II in Fig. 3b. All the six branches on either side of the feed line act as stubs in this configuration. In the configuration III, only the stub immediate to the ring only is connected to the feed line, while all the remaining is disconnected. Under this configuration, the antenna exhibits triple-band characteristics as shown in Fig. 3c.

Similarly, in configuration IV, the center horizontal stub is only connected to feed line and the resultant geometry exhibits dual band with one wideband around the second resonant frequency 11.5 GHz and the same is evident from Fig. 3d. In configuration V, the last stub is active. The radiation characteristics exhibit one

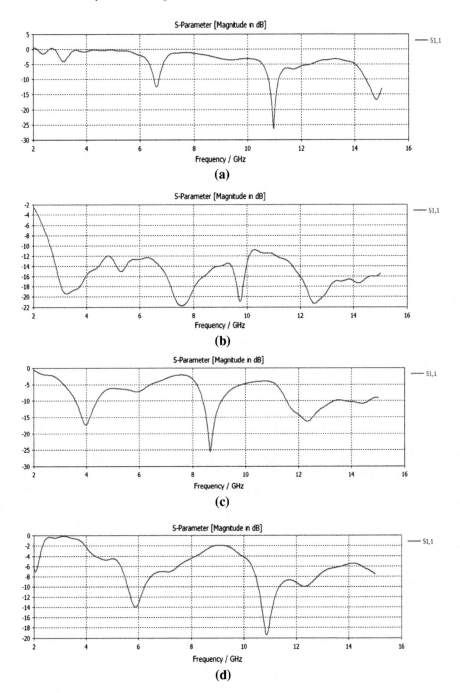

**Fig. 3** **a** S11 of configuration I **b** S11 of configuration II **c** S11 of configuration III **d** S11 of configuration IV **e** S11 of configuration V **f** S11 of configuration VI

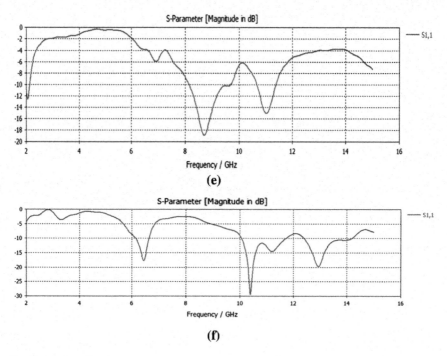

**Fig. 3** (continued)

narrow and one wideband around 2 GHz and 9 GHz, respectively. The third band with poor gain is also visible in the S11 plot as shown in Fig. 3e. In configuration VI, only the stubs located on the right-hand side of the feed line are active resulting in one narrow and two widebands as shown in Fig. 3e at 6.5 GHz, 10.5 GHz, and 13 GHz, respectively.

## 5  Conclusion

The proposed reconfigurable antenna with different switching mechanisms exhibits wideband and narrowband features. Multiple bands with different combinations of wideband and narrowbands continuously shift their location in the frequency response from configuration to configuration. This is due to the switching between the stubs which are actively connected to feed line. Future scope of this project would be proper validation using the fabricated prototypes.

# References

1. Z. Mei, G. Fu, J. Gong, Q. Li, and J. Wang, "Printed monopole UWB antenna with dual band-notched characteristics" in IEEE International Conference on Ultra-Wideband (ICUWB 2010), Vol. 2, pp. 1–3, Sep. 2010.
2. Y. Tawk, and C. G. Christodoulou, "A new reconfigurable antenna design for cognitive radio," in IEEE Antennas and Wireless Propagation Letters, Vol. 8, pp. 1378–1381, Dec. 2009.
3. Y. Tawk, J. Costantine and C. G. Christodoulou, "A rotatable reconfigurable antenna for cognitive radio applications," in IEEE Radio and Wireless Week (RWW), Phoenix, Arizona, Jan. 2011.
4. P. Marshall; Quantitative Analysis of Cognitive Radio and Network Performance. Artech House, 2010.
5. T. Aboufoul, K. Ali, A. Alomainy, and C. Parini, "Combined pattern and frequency reconfiguration of single-element ultra-wideband monopole antenna for Cognitive Radio devices," in 7th European Conference on Antennas and Propagation (EuCAP 2013). pp. 932–936, Gothenburg, Sweeden, Apr. 2013.
6. H. F. AbuTarboush, et al., "A reconfigurable wideband and multiband antenna using dual-patch elements for compact wireless devices," IEEE Transactions on Antennas and Propagation, Vol. 60, No. 1, pp. 36–43. Jan. 2012.
7. T. Aboufoul, and A. Alomainy, "Reconfigurable printed UWB circular disc monopole antenna," in Loughborough Antennas and Propagation Conference (LAPC). pp. 1–4, Loughborough, Nov. 2011.
8. Adnan Kaya, Irfan Kaya and Haluk E. Karaca,: U-Shape Slot Antenna Design with High-Strength $Ni_{54}Ti_{46}$ Alloy, Arabian Journal for Science and Engineering, 41(9) 3297–3307 (2016).
9. Chakravarthy, V.V.S.S.S., Sarma, S.V.R.A.N, Babu K.N., Chowdary, P.S.R., & Kumar, S.T. (2015). Non-uniform circular array synthesis using teaching learning based optimization. Journal of Electronics and Communication Engineering, https://doi.org/10.6084/m9.figshare.1452770.
10. Euardo J.B. Rodrigues, hertz W.C. Lins, Adaildo G.D Assunncao, "Reconfigurable circular Ring Patch Antenna for UWB and Cognitive Radio Applications", 8th European conference on Antennas and Propagation, 2014.
11. V. V. S. S. S. Chakravarthy, P. S. R. Chowdary, Ganapati Panda, Jaume Anguera, Aurora Andújar, and Babita Majhi: On the Linear Antenna Array Synthesis Techniques for Sum and Difference Patterns Using Flower Pollination Algorithm, Arabian Journal for Science and Engineering, Aug (2017).

# Circular Array Synthesis Using Cuckoo Search Algorithm

**Suraya Mubeen**

**Abstract** Circular arrays are very much preferred due to their obvious reasons and capability to control the main beam position inherently. In this paper, synthesis of circular array using amplitude spacing technique is demonstrated. The analysis is carried out using the simulated radiation patterns with sidelobe level suppression. The simulation is carried out in MATLAB.

**Keywords** CSA · Circular array · Array factor

## 1 Introduction

Array antennas are capable of several features like beam scanning, beam shaping, beam steering, sidelobe level control, and null control which are not possible with any single-element antenna because of its obvious reasons. An array antenna is preferred to single antenna. Moreover, complex mechanical activity based system is not required in array to accomplish the above-listed tasks. Typically, there are three types of array geometries like 1D, 2D, and 3D [1, 2]. Linear array belongs to the class of 1D, while planar arrays like circular, square, and rectangular array geometries come under 2D. Similarly cylindrical and cubic geometries belong to 3D.

Synthesis of array involves in determining the steering parameters like current and phase excitation of each element as well as inter-element spacing of the elements in the array. Several numerical techniques like Taylors, Chebyshev, and Schelkunoff are proposed in the synthesis of this array [3]. However, they are computationally complex and often fail to overcome the local minima. Considering these, several evolutionary computing techniques are proposed. Genetic algorithm

S. Mubeen (✉)
CMRTC, Medchal, Secunderabad, Telangana, India
e-mail: suraya418@gmail.com

© Springer Nature Singapore Pte Ltd. 2018      873
J. Anguera et al. (eds.), *Microelectronics, Electromagnetics
and Telecommunications*, Lecture Notes in Electrical Engineering 471,
https://doi.org/10.1007/978-981-10-7329-8_90

[4], particle swarm optimization [5], flower pollination algorithm [6–8], firefly algorithm [9], teaching learning-based optimization [10], and several other nature-improved techniques are applied for array synthesis with different objectives. In this paper, circular array synthesis is performed using cuckoo search algorithm. The circular array is considered over linear array because of its inherent beam steering characteristics.

Further, the paper is organized as follows. Circular array geometry and the array formulation are given in Sect. 2. Description of the algorithm is mentioned in Sect. 3. Simulated results are discussed in Sect. 4. Overall conclusion is given in Sect. 5.

## 2 Formulation of the Design Problem

### 2.1 Array Factor Formulation

The geometry of the circular array considered in this work is shown in Fig. 1. The corresponding problem statement can be defined as to find appropriate set of excitation amplitudes for the elements in the array which can produce desired radiation pattern with suppressed sidelobe levels without any beamwidth constraint. Hence, the design of the array can be considered as a nonuniform circular array.

The corresponding array factor of this geometry is given as

$$AF(\phi) = \sum_{n=1}^{N} I_n \cdot \exp(j \cdot (kr \cdot \cos(\phi - \phi_n) + \beta_n)), \tag{1}$$

**Fig. 1** Geometry of the circular array

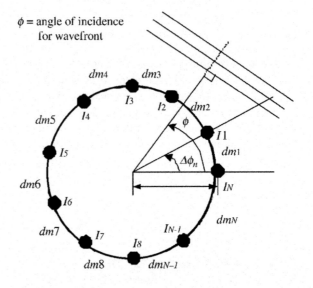

$\phi$ = angle of incidence for wavefront

where $n$ refers to the element number, while $N$ is the total number elements in the array. Similarly, $I_n$ is the nth element current excitation and $\beta_n$ is the corresponding phase excitation, whereas $d_n$ refers to the inter-element spacing function. Similarly, "$kr$" and $\varnothing_n$ are given as

$$kr = \frac{2\pi r}{\lambda} = \sum_{i=1}^{N} d_i \tag{2}$$

$$\phi_n = \frac{2\pi}{kr} \sum_{i=1}^{n} d_i \tag{3}$$

## 2.2 Fitness Formulation

The formulation of the fitness function corresponding to the objective of the work is presented in this section. Magnitude of the SLL is obtained from the radiation pattern plot used in the objective function.

The fitness function is given as

$$f = SLL_{desired} + \max\left(SLL_{\theta = -\pi/2\, to\, \pi/2}\right), \tag{4}$$

where

$SLL_{desired}$ is the positive value of the desired SLL. For example, in this paper, the desired SLL is $-25$ dB, and hence $SLL_{desired} = 25$.

$SLL_{\theta=-\Pi/2\ to\ \Pi/2}$ is the observed SLL between the range of $-90°$ and $90°$ excluding the region covered by the principal beam.

The expression results in a single positive error value. Convergence is said to be achieved if this value minimizes to 0.

## 3 Cuckoo Search Algorithm (CSA)

The CSA is yet another nature-inspired technique which mimics the behavior of cuckoo birds [11]. The structure follows the reproduction mechanism of cuckoo birds. The algorithm typically follows three rules. Accordingly, the eggs laid by the cuckoo birds are safely stored in nests which are randomly chosen. The quality of the nest defines the quality of the egg and the probability of identity of the egg, and the probability to identify the nationality of the eggs is within the range (0, 1).

Over every iteration, every individual is updated with a step size. The updating procedure is explained as follows [11]:

$$I_i^{(t+1)} = I_i^{(t)} + \alpha \oplus L(\lambda)$$

$L(\lambda)$ refers to Levy flight, I is an individual, and t is the iteration number. Step size is given by "$\alpha$" and the value of $\lambda$ lies within (1, 3).

The implementation of the algorithm involves in population initialization, where each individual is considered as a vector of current excitation coefficients. Over the iterations, these coefficients are modified till the termination criterion is achieved. In this work, the termination criterion is the desired convergence or the computation time.

## 4   Results and Discussions

Results pertaining to the above discussion and objectives of the proposed work are presented in this section. Results are in terms of the obtained radiation plots using the current excitation vector given by the algorithm corresponding to the fitness function. In addition for the sake of analysis, the respective convergence plots along with amplitude distribution plots are also given. Two different cases are considered for simulation-based experimentation. In the first case, the main beam is positioned at 0°, while in the second case the beam is steered to 15°.

### 4.1   Unscanned Patterns

In this case, the objective is to suppress the sidelobe level with the main beam positioned at 0°. The corresponding radiation pattern for a 50 element array is as shown in Fig. 2. A very low sidelobe level of −25 dB is reported in the plot. The convergence characteristics can be studied from the convergence plot shown in Fig. 3. It can be inferred that it took more than 7000 generations to reach the convergence low value. The amplitude distribution responsible for the radiation pattern with suppressed sidelobe levels is given as distribution plot shown in Fig. 4.

### 4.2   15° Scanned Patterns

Similar to the previous case, the objective is to suppress the sidelobe level. However, the main beam is positioned in this case at 15° in order to represent the case of

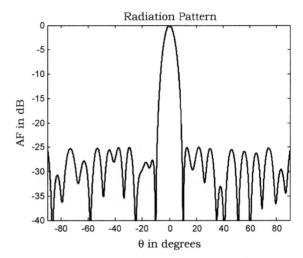

**Fig. 2** Radiation pattern of unscanned 50-element array

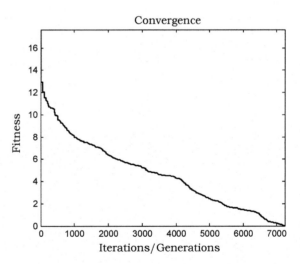

**Fig. 3** Convergence plot of unscanned 50-element array

scanned beams. The corresponding radiation pattern for a 50 element array is shown in Fig. 5. A very low sidelobe level of −25 dB is evident from the plot. The corresponding convergence characteristics plot is shown in Fig. 6. It can be inferred that it took more than 9000 generations to reach the convergence low value. This appears to be more time consuming than the earlier case of unscanned beams. The amplitude distribution responsible for the radiation pattern with suppressed sidelobe levels is given as distribution plot shown in Fig. 7.

**Fig. 4** Amplitude
distribution of unscanned
50-element array

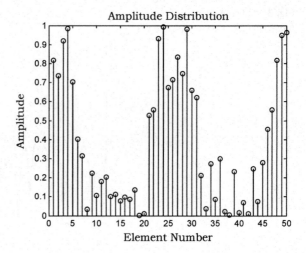

**Fig. 5** Radiation pattern of
15° scanned 50-element array

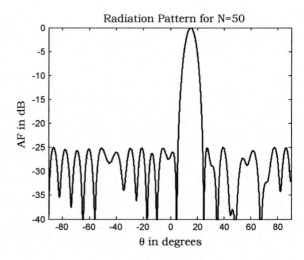

**Fig. 6** Convergence plot of 15° scanned 50-element array

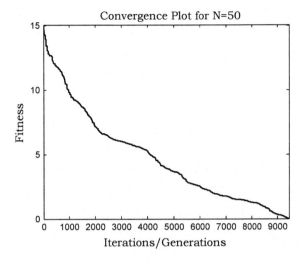

**Fig. 7** Amplitude distribution of 15° scanned 50-element array

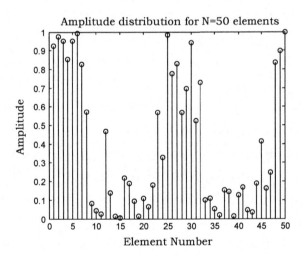

## 5 Conclusion

Performance of the cuckoo search algorithm in circular array synthesis is evaluated in terms of convergence plots and suppressed sidelobe level objectives using amplitude-only technique. Implementation of CSA is evaluated and validated for circular arrays. A very low sidelobe level of −25 dB under both scanned and unscanned cases is reported in this work with no constraint imposed on it. Imposing beamwidth constraint and synthesizing circular arrays with lowest sidelobe levels would be a good scope of future work.

# References

1. C. A. Balanis, Antenna Theory: Analysis and Design, John Wiley & Sons, Singapore, 2nd edition, 2003.
2. P. S. Chowdary, A. M. Prasad, P. M. Rao, and J. Anguera, "Design and performance study of sierpinski fractal based patch antennas for multiband and miniaturization characteristics," Wireless Personal Communications, vol. 83, no. 3, pp. 1713–1730, 2015.
3. K. D. Cheng, "Optimization techniques for antenna arrays," Proceedings of the IEEE, vol. 59, no. 12, pp. 1664–1674, 1971.
4. M. A. Panduro, A. L. Mendez, R. Dominguez, and G. Romero, "Design of non-uniform circular antenna arrays for side lobe reduction using the method of genetic algorithms," AEU —International Journal of Electronics and Communications, vol. 60, no. 10, pp. 713–717, 2006.
5. D. Durbadal Mandal, Md. A. Iqbal Ansari, R. Kar, and S. P. Ghoshal, "Non-uniform concentric circular antenna array design using IPSO technique for side lobe reduction," Procedia Technology, vol. 6, pp. 856–863, 2012.
6. V. V. S. S. S. Chakravarthy and P. M. Rao, "On the convergence characteristics of flower pollination algorithm for circular array synthesis," in Proceedings of the 2nd International Conference on Electronics and Communication Systems (ICECS '15), pp. 485–489, IEEE, Feburary 2015.
7. V. S. S. S. Chakravarthy Vedula, S. R. Chowdary Paladuga, and M. Rao Prithvi, "Synthesis of Circular Array Antenna for Sidelobe Level and Aperture Size Control Using Flower Pollination Algorithm," International Journal of Antennas and Propagation, vol. 2015, Article ID 819712, 9 pages, 2015. https://doi.org/10.1155/2015/819712.
8. V. V. S. S. S. Chakravarthy, P. S. R. Chowdary, Ganapati Panda, Jaume Anguera, Aurora Andújar, and Babita Majhi: On the Linear Antenna Array Synthesis Techniques for Sum and Difference Patterns Using Flower Pollination Algorithm, Arabian Journal for Science and Engineering, Aug (2017).
9. Suraya Mubeen, A. M. Prasad, and A. Jhansi Rani, "On The Beam Forming Characteristics of Linear Array Using Nature Inspired Computing Techniques", ACES Express Journal, 1(6), 181–184, June 2016.
10. V. V. S. S. S. Chakravarthy, K. Naveen Babu, S. Suresh, P. Chaya Devi, and P. Mallikarjuna Rao, "Linear array optimization using teaching learning based optimization," Advances in Intelligent Systems and Computing, vol. 338, pp. 183–187, 2015.
11. X.-S. Yang and S. Deb, "Cuckoo search: recent advances and applications," Neural Computing and Applications, vol. 24, no. 1, pp. 169–174, 2014.

# Conformal Antennas—A Short Survey

N. V. K. Maha Lakshmi, P. V. Subbaiah and A. M. Prasad

**Abstract** Conformal antennas (CA) have wide applications in several civil, commercial, and defence systems. They are the need hour and most essential in aircrafts and ships. Patch antennas are often considered as the better candidate for such CA. In this paper, a consolidated report on several conformal antenna types is presented. General study on the CA with mircrostrips is presented. The singly curved and doubly curved surfaces are considered for discussion.

**Keywords** Conformal antenna · Singly curved · Doubly curved
Microstrip patch antenna

## 1 Introduction

As per the International Electrotechnical Commission (IEC), CA are a radiating system, whose shape is not determined by its electromagnetic features but by the surface of the system where it has to be intake with the advancement in technology; novel techniques and approaches to the system design are must. A typical radiating system refers to a system which acts as interface between the transmitter and receiver in free space. It is possible to modify the characteristics in order to improve the overall system performance. This technique is often responsible for reducing several aspects that affect the image metrics severely and give a better accuracy along with excellent aerodynamics as well as less in volume. This leads to a challenging task for

N. V. K. Maha Lakshmi (✉)
Department of ECE, SRK Institute of Technology, Vijayawada, AP, India
e-mail: mahalakshminvk.nvk@gmail.com

P. V. Subbaiah
Department of ECE, VR Siddhartha College of Engineering, Vijayawada, AP, India
e-mail: drpvsubbaiah99@gmail.com

A. M. Prasad
Department of ECE, University College of Engineering JNTU, Kakinada, AP, India
e-mail: a_malli65@yahoo.com

© Springer Nature Singapore Pte Ltd. 2018
J. Anguera et al. (eds.), *Microelectronics, Electromagnetics and Telecommunications*, Lecture Notes in Electrical Engineering 471,
https://doi.org/10.1007/978-981-10-7329-8_91

the antenna engineers to designing the worthiness of utilizes conformal structures instead of their planar. This antenna can be installed on several aerodynamic systems. They have to posses multiband, broadband, and miniaturization characteristics along with conformal nature.

CA belongs to the class of phased array antenna. They constitute an array of several similar but small flat antennas like dipoles and patches covering the surface. Every antenna equipped with phase shifter device. Which are controlled by a microprocessor, by simply manipulating the current excitation phase of every element. It is possible to sum up all the radiation using the process of interference, forming a strong beam directed in a particular direction. However, in the receiving case, every element combines all these waves in phase independence signals in that direction. This way, the antenna can be made responsive to the signal from one transmitter and at the same time reflecting the interfering signals from other directions.

In the case of conventional phased array, the elements are distributed on flat surface, whereas in the case of conformal antenna, they are distributed on a curved surface. The corresponding phase shifters are used to compensate the phase difference emerging due to difference in path lengths. As the corresponding elements in the CA are very small, these applications are limited to high frequencies and microwave range. However, they readily express miniaturization.

In the present discussion, the deliberations are limited to conformal antenna study and comparison. These CA are featured with less visibility due to inherent miniaturization and can be integrated on the structure. This feature is essential in military environments. The CA can take with any geometry. However, so far investigated geometries are cylindrical, spherical, and conical. Several examples of the CA are shown in Fig. 1.

## 2   Conformal Antenna Arrays

Considering the recent advancements in the wireless technology, the planar microstrip antenna could no more requirements systems. As a result, several CA structures are studied [1–3]. However, the CA can be termed as a planar antenna array on a curved surface. This makes the interconnection in antenna arrays more complex. However, as a result, the performance is affected by its salient features; it has become the topic of the current research to choose the feeding system structure. This makes the design of CA simpler whose performance is superior. Generally, several techniques are used to reduce the antenna size and subsequently enhance its bandwidth, using the Chebyshev polynomial method [4]. Another method is to introduce a large rectangular slot and modify the ground plane [5]. Cylindrical structures have inherently possessed the conformal characteristics [6–10]. Following the above, another studied structure is the sphere [11–13]. Similarly, the other CA which are available in literature are conical, elliptical, or other geometries. The main

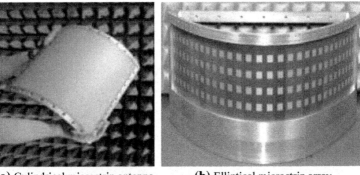

(a) Cylindrical microstrip antenna          (b) Elliptical microstrip array

(c) Aperture array on paraboloid

**Fig. 1** Examples of conformal antennas from Ball Aerospace and EWCA

disadvantage of planar array is that they do not have the desired aerodynamic shape. Also, in addition, it needs aerodynamic radome. Inter-element spacing has degrading effect on the performance closer; they are greater. The mutual coupling degrades the performance. Similarly, they lead to grating lobes. However, in CA, some section is brought over the surface, thereby increasing the scan angle, which reduces the grating lobes. Furthermore, unlike planar array, the corresponding radar cross section appears to be diminished. This is due to the diffraction of the plane when it is incident on a curved surface. Moreover, the reflected energy will be defocused and have lower intensity compared with the reflection from a planar surface. In this basic form, the CA are classified into two categories. They are singly and doubly curved. In this section, these two types are discussed.

## 2.1   Conformal Antennas on Singly Curved Surfaces (SCS)

Antennas on SCS are considered as a basic form of CA, mainly contributing to the improvement in the azimuth coverage (wide coverage). However, it is possible in some cases with omnidirectional also. The circular cylindrical structure is a

**Fig. 2** Structure of the
reference cylindrical
microstrip antenna

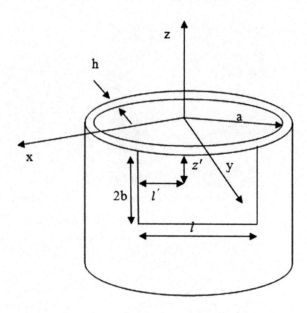

common conformal antenna application. It is considered as one of the 3D
geometries. This SCS is also referred to as developable surfaces as they can be
straightened and further analyzed as a flat surface. As a result, several parameters
are computed using this planar analysis. With this technique, it is possible to
compute. Hence, it can be concluded that the SCS can be studied without involving
geodesic differential equation [14]. The cylindrical patch antenna is shown in Fig. 2
for reference. A thorough parametric analysis is possible by manipulating different
techniques for analyzing conformal physical parameters like radius, permittivity,
and substrate height.

## 2.2    Conformal Antennas on Doubly Curved Surfaces

The complicated CA array is a "smart skin" installed on the surface of the aircraft
body. However, relative to the surface curvature, if small, then the design follows
that of planar with some phase corrections. The radiating elements need to be
highly directional and always projected with the outward from the surface.
Increasing the curvature, the respective elements may not radiate in all directions.
An active section has to be manipulated for scanning the DCS antennas that provide
additional degrees of freedom in their structure. They follow the surface variation in
other directions unlike a cylinder and often produce more elaborate solutions. With
high-degree complexity, CA on DCS is difficult to analyze. As a result, a few
literature is available [1, 9, 10]. The radiation characteristics of CA on DCS cover

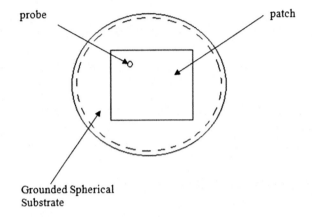

**Fig. 3** Geometry of the quasi-square spherical microstrip antenna

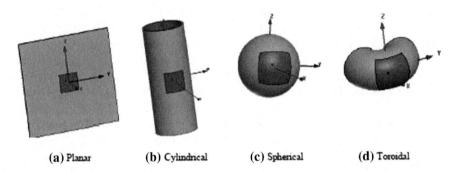

(a) Planar          (b) Cylindrical          (c) Spherical          (d) Toroidal

**Fig. 4** Conformal antennas and their planar counterpart designed using HFSS [23]

nearly full hemisphere and also responsible for aerodynamic structures. The spherical DCS-based CA is shown in Fig. 3.

The planar counterpart realization of CA is of high importance in the analysis of CA. Some designs corresponding to this strategy are shown in Fig. 4.

## 3   Advantages and Disadvantages

There are several advantages as well as disadvantages of CA when compared with planar antennas. When compared with planar 2D antennas, the bandwidth as well as the beamwidth of the CA is very high. Similarly, a single-element antenna is capable of constituting the omnidirectional pattern without any parasitic elements. Moreover, it is also possible to manipulate the gain by controlling the phase errors and the angle of inclination in the CA being conformal; these CA are capable of exhibiting aerodynamic structural features. As a result, such CA does not require additional radome structures to protect the antenna system. This feature typically reduces the radar cross section.

In spite of several advantages, these CA often pulled out due to several drawbacks. One of the prime disadvantages is that no specific dedicated tool is yet bought for design and analysis of simulated CA. As a result, the computation involves in several complex calculations. Moreover, due to its conformal nature, it is not an easy task to derive circular polarization features from CA.

## 4  Conformal Microstrip Antenna Array

The CA generally use the slotted patch antenna for their low-profile structure. They are meant for their features like lightweight, low cost, and their efficiency. They have specific applications in flying high-speed aircrafts with their aerodynamic structure. In missile applications, these microstrip CA are coated with two DCS like cylinder and cone that are used to enhance directivity and SNR. High-degree performance is possible using arrays. The corresponding radiation pattern is manipulated accordingly due to the placement of affected arrays on a conformal surface. This produces omnidirectional patterns which are very much needed in the aerospace systems. The mathematical analysis of CMA on a cylinder is presented by Knghou and Menglin [15] in which $8 \times 8$ array is considered. With the change in curvature Ka $>> 1$, results less effect to pattern at plane of $\theta = 90°$. However, they have more effect at plane $\theta = 90°$. Discarding coupling between elements, the corresponding total electric field strength for an array of N elements is given as

$$E = \sum_{i=1}^{N} E_i e^{-j\varphi_i}$$

Here, $E_i$ refers to field strength of ith element and $\varphi_i$ refers to the phase N radiators.

The radiation characteristics and the parametric analysis of CA on cylindrical structures are discussed in [16]. It is observed that the radiation features are strongly dependent on the cylindrical curvature. Moreover, it is concluded that the array exhibits high sidelobes and wider beamwidth. It is also suggested that the radius of curvature has the capability to define the parameter that must be considered while mechanical and electrical characteristics of CA. A CA array is discussed in [17] where different parameters are of concern during the UWB phased array antenna design. Some of them are physical size of the antenna, increment spacing, and mutual coupling. For conformal placement of an antenna either as an individual antenna or as in an array configuration on any arbitrary surface, it may require very thin antenna. They should be processed preferably on flexible substrates so that they will conform to the surfaces without changing the surface geometry. Conformal microstrip antenna on a conical surface is given in [18]. In this analysis, a thin substrate thickness is considered. Also, the distance of the patch to the cone apex and the curvature radius of the cone surface are large when compared the corresponding wavelength. This analysis is applicable to any arbitrary conical surface that degenerates into a plane surface. Moreover, the analysis is also applicable to any higher order resonant mode. The simulation of 36 GHz conformal rectangular patch antenna on a cone is designed in [19]. In this work, CMA is non-resonant side feeding. This causes high cross polarization. High cross polarization can be restrained by choosing proper ratio of width and length (W/L).

## 5 Conclusion

A short note on the analysis of CA, the operation of CA and comparison between planar and CA enlisting both advantages and disadvantages have been studied. It is observed that the CA is suggested only for applications, where broad beam radiation patterns are desired. Similarly, main drawback of CA is the complexity in analysis, and costs in fabrication are discussed in both SCS and DCS designs of microstrip that are discussed. Further, the design analyzing the conformal antennas using any 3D EM simulator can be considered as a good scope of future work.

Fundamentally, the use of conformal antennas is suggested in case very broad beam radiation patterns are desired. The main drawback of conformal antennas is the increased complexity and cost in manufacturing. There are mainly two types of conformal antennas that are discussed in this work. They are singly curved and doubly curved. The design of the microstrip-based singly curved surfaces and doubly curved surface analysis is also discussed. Designing and analyzing the conformal antennas using any 3D EM simulator are carried out further.

## References

1. V. V. S. S. S. Chakravarthy, P. S. R. Chowdary, Ganapati Panda, Jaume Anguera, Aurora Andújar, and Babita Majhi: On the Linear Antenna Array Synthesis Techniques for Sum and Difference Patterns Using Flower Pollination Algorithm, Arabian Journal for Science and Engineering, Aug (2017).
2. B. Thors and L. Josefsson, "Radiation and Scattering Tradeoff Design for Conformal Arrays", IEEE Trans. on Antennas and Propagation, vol. 51, no. 5, pp. 1069–1076, May 2003.
3. B. D. Braaten, S. Roy, I. Irfanullah, S. Nariyal and D. E. Anagnostou, "Phase Compensated Conformal Antennas for Changing Spherical Surfaces", IEEE Trans. on Antennas and Propagation, vol. 62, no. 4, pp. 1880–1887, April 2014.
4. D.G. Babas and J.N. Sahalos, "Synthesis Method of Series-Fed Microstrip Antenna Arrays", Electronics Letters, vol. 43, no. 2, 18th January 2007.
5. R. Bayderkhani and H.R. Hassani, "Wideband and Low-Sidelobe Slot Antenna Fed by Series-Fed Printed Array", IEEE Trans. on Antennas and propagation, vol. 58, no. 12, pp. 3898–3904, December 2010.

6.  M. V. T. Heckler, "Circularly Polarized Microstrip Antenna Arrays Conformed on Cylindrical Surfaces", MSc. Thesis, Instituto Tecnologico de Aeronautica, Brazil, 2003. (in Portuguese).

7.  G. Gottwald and W. Wiesbeck, "Radiation Efficiency of Conformal Microstrip Antennas on Cylindrical Surfaces", Antennas and Prop. Socety International Symposium, Vol. 4, pp: 1780–1783, June 1995.

8.  K.-L. Wong and S.-Y. KE, "Cylindrical-Rectangular Microstrip Patch Antenna for Circular Polarization", IEEE Trans. On Antennas and Prop., Vol. 41, No. 2, February 1993.

9.  S. Raffaeli, Z. Sipus, and P.-S. Kildal, "Input Impedance And Radiation Pattern of Patch on Multilayer Cylinder", Proceedings of the Millennium Conference on Antenna and Prop., AP2000, April 2000, Switzerland.

10. Q. Jinghui, Z. Lingling, Du Hailong and Li Wei, "Analysis and Simulation of Cylindrical Conformal Omnidirectional Antenna", Microwave conference Proceedings, 2005. Asia-Pacific Conference Proceedings. Vol. 4, Dec. 2005.

11. K.-M. Luk And W.-Y. Tam, "Patch Antennas on A Spherical Body", IEEE Proceedings, Vol. 138, No. 1, February 1991.

12. Z. Sipus, N. Burum and J. Bartolic, "Moment Method Analysis of Rectangular Microstrip Antennas on Spherical Structures", Antennas and Prop. Society International Symposium, Vol. 3a, pp. 126–129, July 2005.

13. K.-M. Luk And W.-Y. Tam, "Analysis of Spherical-Wraparound Microstrip Antennas by Spectral Domain Method", China 1991 International Conference on Circuits and Systems, June 1991.

14. Lars Josefsson, Patrik Persson "Conformal Array Antennas". Handbook of Antenna Technologies. Springer science & Business Media Singapore 2015.

15. M. KNGHOU, and X. Menglin, "A Study of Conformal Microstrip Antenna Array on a Cylinder", IEEE5th International Symposium on Antennas, Propagation and EM Theory, pp 18–21, Aug. 2002.

16. A. M. Ferendeci, "Conformal Wide Bandwidth Antennas and Arrays", IEEE 3$^{rd}$ IEEE International Symposium on Microwave, Antenna, Propagation and EMC Technologies for Wieless Communication, pp. 27–29, Oct. 2009.

17. J. R. Descardeci, And A. J. Giarola, "Microstrip Antenna on a Conical Surface", IEEE Trans. on Antenna and Propagation, Vol. AP-40, pp 460–463,April 1992.

18. Z. Peng, L. Chaowei, and W. Qiang, "The Design and Simulation of Millimeter Wave Conformal Microstrip Antenna Patch on Conical Surface", IEEE Global Symposium on Millimeter Waves, pp. 1772–1775, June 1995.

19. B.R. Piper, and N. V. Shuley, "The Effects of Spherical Conformity on A Wideband L-Probe Circular Patch Antenna". IEEE International Symposium on Antennas and Propagation Society, pp. 289–292, July 2005.

# Internet Security—A Brief Review

Subha Sree Mallela and Sravan Kumar Jonnalagadda

**Abstract** There is a huge increase in the online users and so as the problems faced by them; a typical online user is exposed to virus, worms, bugs, Trojan horses, etc.; in addition, the user is also exposed to sniffers, spooling, and phishing. As a result, the users are constantly prone to cons of privacy due to spyware which monitors the online users. There is even a possible destruction of the personal machine that dies due to malware. Due to these issues, there is an indication that the Internet is not a safe place for online activities. This problem is not only limited to personal suffers but also extended to several corporations and government sector organizations. Several times the confidential and government information are prone to security risks. These attacks may be due to inherent weakness in the networks. Similarly, the carelessness of users also termed reason. Whatever be the reason, the Internet security study has become the demand of the hour. In this paper, a short study on such Internet security issue is considered.

**Keywords** Security · Protocols · Online activities

## 1 Introduction

In one way or the other, Internet usage has become a part of daily life. The usage may be in the form of information to entertainment, financial services, product purchase, and even socializing. As a result, there is a possibility of information collection [1–6]. Internet usage from all kind of activities. The Internet is emerged as a gateway for personal, home, and office convenience.

S. S. Mallela (✉)
Department of CSE, Andhra Loyola Institute of Engineering and Technology,
Vijayawada, AP, India
e-mail: subhasree87@gmail.com

S. K. Jonnalagadda
Department of IT, SRK Institute of Technology, Vijayawada, AP, India
e-mail: jnvsravankumar@gmail.com

© Springer Nature Singapore Pte Ltd. 2018
J. Anguera et al. (eds.), *Microelectronics, Electromagnetics
and Telecommunications*, Lecture Notes in Electrical Engineering 471,
https://doi.org/10.1007/978-981-10-7329-8_92

Generally, the Internet facilitates the data exchange through hosts and servers. There are a wide variety of hosts. Some of them are personal computers and the other can be a supercomputer. These hosts typically have several combinations of hardware and software. However, the protocol governing the exchange of commands, requests, and data like transport control protocol/Internet protocol (TCP/IP). This is the underlying technique in every Internet activity which is often termed as an open technology [7, 8]. As a result, there is always a possibility to express. The user is into quite insecure and risky environment. In many cases, the attachments are through this channel of activity only. The process of authentication can be claimed by a packet of data regarding its origination. This is basically due to the fact that the dominant layers do not perform any authentication procedure.

There is a huge scope of proposing technique to ensure the security of the data through proper authentication. This attracted the interest of several computer engineers with the scope of research.

## 2   Internet Security

Every Internet user is subject to several types of risks associated with the Internet usage. These attacks have become a more common element and most significantly damaging issue in corporate offices, where there is a wide usage of computers. When it comes to the public agencies, any small issue arises as a major panic as the destruction is widely propagating. Most importantly, in most of the cases, the corresponding traces of attacks and their sources are not known until the damage is actually assumed. It is also estimated that the basic data steering and business attacks are through Internet usage only. Seminal of service attacks and unauthorized access attacks have become quite prominent in such cases. All the above issues are responsible for the need for Internet security.

The threats to safe browsing can be of different forms like IP spooling browser session hijacking, denial of services, data stealing, spyware, malware, virus, and sniffing.

Denial of service is often considered as a security breach due to software failure. The random cause can be firewall failure security software. This cannot contribute to complete network security. This becomes a clear weak point for the network hackness. Through the weak point, the threat can be in the form of a small bug, viruses, or a spyware where there is a possibility of address spoofing. The process is often considered as a chain mechanism in which the victims' address book is accessed and the information is sent. Further, the same process of hacking the address of the other and distributing the message is a continuous chain process. Once if the victim opens, the information is recieved which is actually a worm. The spreading of the worms weird behavior starts and destruction of confidential information arouses. Some of such worms which are recently identified are Blaster and Welchia worms.

Encryption and data encoding are techniques to scramble the confidential data and make it understandable. By this way, even when the data is stolen or subjected to any of the abovedescribed issues of Internet security, the confidential data remains safe and undisclosed. Later, the encrypted data is decrypted by the owner to owner and used. During the process of encryption and decryption, a private key is used as an algorithm to make the data unreadable to readable during decryption as well as readable to unreadable in encryption. Several third-party firms have emerged for accomplishing the task and provide high-security keys. However, several Internet security threats had computed intelligence which is sufficient for decoding the keys. This makes the system vulnerable.

## 3 Issues and Solutions

A gathering of intellectuals considers that the nuclear power plants have a ton of basic information to be sent to the administrator workstations. A plant-wide incorporated correspondence organizes, with high throughput, determinism and repetition, which is required between the workstations and the field. Exchanged ethernet setup is a promising prospect for such a coordinated communication. In atomic power plants, the plant information is essential and information misfortune cannot be ignored without serious consequences. Switched ethernet might be a prominent innovation.

System assaults have been found to be as shifted as the framework that they endeavor to enter. Assaults are known to be either deliberate or unexpected and actually capable of gatecrashers which have occupied with focus on the conventions utilized for secure correspondence between organizing gadgets. This audit tends to how exceptionally modern interlopers are infiltrating web systems in spite of large amounts of security. Be that as it may, as the gatecrashers increase, the organized specialists are inferring numerous strategies in keeping assailants from getting to organization systems.

According to the investigation on Internet of Things, which made wake up calls to the enterprises which are organizations, it is required to get ready for the new era of Internet-empowered gadgets that might be found in any place in the world [2]. Agreeing Jericho Forum board part, Andrew Yeomans, the Directive, serves to center security experts on information security over frameworks. "From a Jericho Forum perspective", any fortifying of directions is an impetus to execute unavoidable information-driven security, so the information is ensured wherever it is. The Jericho Forum has featured that the "perimeterized" [that is, traditional] show misses numerous conceivable breaks, particularly information that has been deliberately passed to different associations, which in this manner endure a break.

When the web servers fail, the respective websites also fail to operative. Assailants are abusing any weakness they can to trade-off sites and secure their host servers. The convenience and wide accessibility of web assault toolbox are bolstering the quantity of web assaults, which multiplied in 2015. Site proprietors still are not fixing and refreshing their sites furthermore, servers as regularly as maybe they should. This resembles taking off a window open through which cybercriminals can move through furthermore, exploit whatever they find. In the course of recent years, more than 75% of sites filtered contained unpatched vulnerabilities, one out of seven (15%) of which were regarded basic in 2015.

It is not just modules for web programs that are powerless and misused. Take Word Press, which now controls the fourth of world's sites, for instance. Anybody can compose a Word Press plugin—and they frequently do. Modules go from the helpful to the totally crazy, for example, Logout Roulette: "on each administrator page stack, there's a 1 of every 10 chance you'll be logged out." The issue is that some modules are shockingly unreliable. Windows pulls in many adventures as a result of its huge client base, and the same applies to Word Press modules. Defenseless modules found on Word Press locales can and will be misused. Modules, regardless of whether for programs or servers, should be refreshed frequently as they are defenseless against security blemishes, and out-of date renditions ought to be maintained a strategic distance from where conceivable.

## 4 Integrated Solution

Security arrangements ought to be completely incorporated with rights and the open. Internet security arrangements ought to be completely coordinated with the imperative goals of protecting the central properties of the Internet (open norms, willful cooperation, reusable building squares, respectability, consent-free development and worldwide reach (otherwise called the Internet Invariants [4]) and principal human rights, and qualities and desires (e.g., protection and opportunity of articulation)). Any security arrangement is probably going to affect the Internet's operation and improvement, and also client's rights and desires. Such impacts might be sure or negative. From our viewpoint, it is critical to discover arrangements that help the Internet invariants and key rights and qualities. Security arrangements should be grounded in understanding, created by accord and developmental in viewpoint. Security arrangements should be sufficiently adaptable to develop after some time. We realize that innovation will change and dangers will adjust to exploit new stages and conventions. Accordingly, arrangements should be receptive to new difficulties. Like a human body that may experience the ill effects of infections, yet gets more grounded and stronger subsequently, new advancements, arrangements, and helpful endeavors that expand on "lessons-learned" make the Internet stronger to dangers. Experience demonstrates to us that, in a rapidly developing framework, for example, the Internet, an open agreement-based participatory approach, is the most hearty, adaptable, and lithe. Fractional arrangements and organized sending are vital and ought to be considered important. An accumulation of incremental arrangements might be more successful practically speaking than a fabulous plan.

Regardless of the possibility that an approach does not take care of the issue totally, it may contain it, or to change the monetary condition altogether enough, in order to make the helplessness significantly less alluring to malignant performing artists. The concentrate should be put on characterizing the concurred issue and finding the arrangement. We additionally need to make space for the new, the creative, and the odd. We should be set up to test problematic or non-customary thoughts. At last a procedure, which draws upon the interests and ability of a wide arrangement of partners, is probably going to be the surest way to progress. Focusing on the purpose of greatest effect corresponds to think all inclusive demonstration locally. Security is not accomplished by a solitary arrangement or bit of enactment; it is not unraveled by a solitary specialized fix or would it be able to come to fruition since one organization, government, or on-screen character chooses the security to be vital. Making security and trust in the Internet requires distinctive players (inside their diverse duties and parts) to make a move, nearest to where the issues are happening.

Ordinarily, for more noteworthy viability and proficiency, arrangements ought to be characterized and executed by the littlest, least, or slightest incorporated skilled group [5] at the point in the framework where they can have the most effect. Such people group is much of the time suddenly framed in a base up, self-sorting out a form around particular issues (e.g., spam or directing security), or a territory (e.g., insurance of basic national foundation or security of an Internet trade). However, much as could reasonably be expected, arrangements ought to be founded on interoperable building pieces—e.g., industry-acknowledged benchmarks, best practices, and methodologies.

# 5  Conclusion

Internet security, its issues, and the breaches responsible for theft of valuable personal and confidential information is a sever alarming issue for the Internet users. This is more than a simple insecurity in the case of corporate and company Internet setup. Any confidential information disclosed leads to very serious downfall of the corporate. A short survey on the issues, its face and possible simple solutions is discussed here. Statistical and thorough prediction model of the behavior of the threats and security issues are a better solution to the assumed Internet security problems in future.

# References

1. "Internet Invariants: What Really Matters", http://www.internetsociety.org/internet-invariants-what-really-matters.
2. http://news.cnet.com/8301-1023_3-57525797-93/facebook-hits-1-billion-active-user-milestone/.
3. Ushahidi is an open source project which allows users to crowdsource crisis information to be sent via mobile, http://www.ushahidi.com/.
4. For example, a technique used in the attack against http://www.spamhaus.org in March 2013.
5. For example, a compromise of a Dutch Certificate Authority Diginotar, full report http://www.rijksoverheid.nl/bestanden/documenten-en-publicaties/rapporten/2012/08/13/black-tulip-update/black-tulipupdate.pdf.
6. For more information see RFC 2827 Network Ingress Filtering: Defeating Denial of Service Attacks which employ IP Source Address Spoofing (http://tools.ietf.org/html/rfc2827).
7. Hardin, G. "The Tragedy of the Commons". Science 162 (3859): 1243–1248, 1968.
8. Conficker Working group, http://www.confickerworkinggroup.org/wiki/.

# Antenna Array Synthesis Using Social Group Optimization

V. V. S. S. Sameer Chakravarthy, P. S. R. Chowdary,
Suresh Chandra Satpathy, Sudheer Kumar Terlapu
and Jaume Anguera

**Abstract** Circular array antenna (CAA) design has become a complex and most explored research problem with the advancement in the wireless personal and commercial communication systems. In this paper, the circular array synthesis is performed using novel social group optimization algorithm (SGOA). The synthesis technique employs both nonuniform amplitudes and nonuniform spacing between the elements. The array synthesis problem is translated as an optimization problem with amplitudes and inter-element spacing as two different design variable sets with suppressed sidelobe level (SLL) along with beamwidth constraint as objectives. The SGOA synthesized 30 and 60 element CAA, produced a very low SLL when compared with uniform CAA maintaining the same BW.

**Keywords** Antenna array · Social group optimization · Fitness function
Sidelobe level

V. V. S. S. Chakravarthy (✉) · P. S. R. Chowdary
Department of ECE, Raghu Institute of Technology, Visakhapatnam, India
e-mail: sameervedula@ieee.org

P. S. R. Chowdary
e-mail: satishchowdary@ieee.org

S. C. Satpathy
PVP Siddhartha Institute of Technology, Vijayawada, India
e-mail: sureshsatpathy@ieee.org

S. K. Terlapu
Shri Vishnu Engineering College for Women (Autonomous), Bhimavaram, India
e-mail: skterlapu@gmail.com

J. Anguera
Department of Electronics and Telecommunications, University Ramon Llull,
Barcelona, Spain
e-mail: jaume.anguera@fractus.com

J. Anguera
Technology Department, Fractus Antennas, Barcelona, Spain

© Springer Nature Singapore Pte Ltd. 2018                                   895
J. Anguera et al. (eds.), *Microelectronics, Electromagnetics*
*and Telecommunications*, Lecture Notes in Electrical Engineering 471,
https://doi.org/10.1007/978-981-10-7329-8_93

# 1  Introduction

An array antenna is a collection of multiple stationary and similar radiating elements. These arrays play a significant role in point-to-point communications like wireless, mobile, and radar applications. All the elements in the array collectively operate as a single-element antenna [1, 2], thereby concentrating the radiation to one direction, which is desirable by the abovementioned applications. Generally, communication system performance can be enhanced by employing high directive radiation elements. Earlier, single-element antennas with directivity much less than the required level are used for these applications. Later, antenna arrays with excellent directivity characteristics have replaced these single-element antennas. Radiating elements for modern wireless communications need to posses certain features like enhanced directivity and capability of controlling the SLL and BW [3–5]. Also, in some cases, it is required to steer the main beam in the certain direction. These features are a hard task to achieve in the case of single-element antennas. Hence, the most possible solution is the design of antenna arrays for such applications as they inherently posses the abovementioned characteristics.

The inter-element spacing (d), amplitudes of current excitation (I), and phase excitation ($\alpha$) are often considered as the key parameters of array design. Determining these parameters which produce the desired radiation pattern is known as array synthesis problem. The choice of considering number of properties for synthesis depends on the type of synthesis problem. However, in this paper, both the amplitude and spacing between the elements are considered for the synthesis of circular array. Many conventional numerical techniques which are derivative-based are proposed for such array synthesis. These conventional techniques have a tendency to stuck in the local minima as most of them are local search methods. Moreover, the final solution is dependent on the initial solution. If the initial value is chosen such that its solution lies in the region of solution space that is close to local minima, then the local search gives the best of poor local solution that is available. These conventional techniques have the drawback of long computational time and complex mathematical steps.

For an efficient array design, several evolutionary computing tools are proposed. Certain algorithms like genetic algorithm [6], particle swarm optimization [7], teaching learning-based optimization [8], firefly [9], and flower pollination algorithm [10, 11] have produced excellent results in solving array synthesis problems. In this work, another novel algorithm known as social group optimization algorithm (SGOA) proposed by Satapathy et al. [12] is used for the synthesis of circular array using nonuniform amplitude and nonuniform spacing technique with beamwidth constraint. Further, the paper is organized into five sections. Array optimization problem is discussed in Sect. 2 and formulation of the design problem is given in Sect. 3. Brief discussion on the SGOA and its implementation to the design problem is explained in Sect. 4. Overall conclusions are given in Sect. 5.

## 2  Array Optimization Problem

Array synthesis generally involves in determining appropriate values of I and d which produce desired radiation pattern with desired SLL and BW. Both SLL and BW are two conflicting parameters. Suppressing one of this leads to enhancement of the other. Obtaining lowest SLL with constrained or fixed BW can be considered

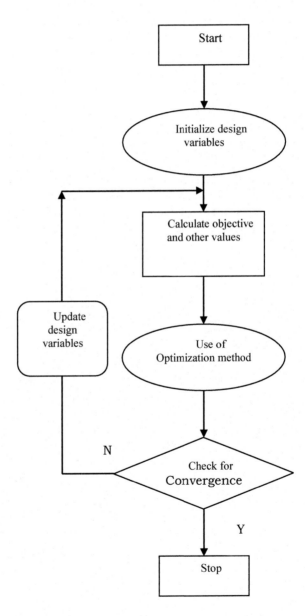

**Fig. 1** Flowchart of antenna optimization

as optimizing the array parameters. Uniform CAA is the simplest form of array in which all the elements are uniformly excited in terms of amplitude, phase, and spacing. However, the resultant SLL is around −7 dB which is very high according to the modern wireless systems. In amplitude only technique where only the amplitudes of excitation current are determined for designing CAA, the corresponding BW gets enhanced severely. In order to control the BW while suppressing, the corresponding inter-element spacing also needs to be controlled. This is possible with inclusion of additional degree of freedom in the design parameter set. A brief description of the antenna array optimization is given in the flowchart in Fig. 1. According to the flowchart, the design variables are modified until the desired pattern is achieved which is also known as convergence.

## 3 Formulation

### 3.1 Array Factor Formulation

The geometry of the circular array is shown in Fig. 2 in which all the elements are arranged on the circumference of the circle whose radius is r.

The circular array is confined to the x-y plane. Also, in the array, all the elements are considered as isotropic elements. The corresponding array factor is given as [1–3]

**Fig. 2** Circular array

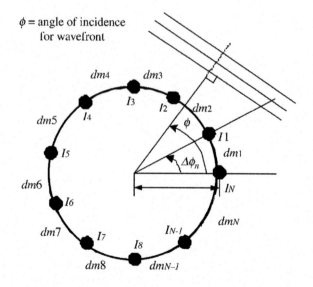

$$AF(\phi) = \sum_{n=1}^{N} I_n \cdot \exp(j.(kr.\cos(\phi - \phi_n) + \beta_n)), \tag{1}$$

where the element index is referred as and N refers to total number of elements distributed in the array. The element excitation phase is given as '$\beta_n$'. Similarly, '$kr$' and $\emptyset_n$ are given as

$$kr = \frac{2\pi r}{\lambda} = \sum_{i=1}^{N} d_i$$

$$\phi_n = \frac{2\pi}{kr} \sum_{i=1}^{n} d_i \tag{2}$$

The example of uniform circular array antenna (UCAA) is considered to interpret its radiation characteristics like SLL and BW. In UCAA, the element excitation is uniform and equals to unity with uniform spacing and no phase difference.

In amplitude–space synthesis, technique has amplitude and space, nonuniformly distributed with phase maintained at constant value. The corresponding amplitude and spacing in the antenna array are represented as a vector given as

$$I = [I_1, I_2, I_3, \ \ldots \ I_N, \ d_1, d_2, d_3, \ \ldots \ d_N]$$
$$\text{and phase} = 0 \tag{3}$$

## 3.2 Fitness Function Formulation

The formulation of fitness function considers both the SLL reduction along with BW constraint. The radiation pattern is the distribution of the computed AF values for all the azimuthal angle ($\theta$) in the range ($-90°$, $90°$). The corresponding fitness evaluation expression is given as follows:

$$SLL_{diff} = SLL_{des} - \max[|AF(\theta)|_{-90}^{\theta_0 - \frac{BW_{obt}}{2}}] \tag{4}$$

$$BW_{diff} = |BWuni - BW_{obt}| \tag{5}$$

$$
\begin{aligned}
f_1 &= SLL_{diff} && \text{if } SLL_{diff} > 0 \\
&= 0 && \text{otherwise}
\end{aligned}
\tag{6}
$$

$$
\begin{aligned}
f_2 &= BW_{diff} && \text{if } BW_{diff} > 0 \\
&= 0 && \text{otherwise}
\end{aligned}
\tag{7}
$$

$$f = c_1 f_1 + c_2 f_2 \tag{8}$$

Here, $SLL_{diff}$ is the difference between the desired SLL ($SLL_{des}$) and the obtained SLL ($SLL_{obt}$) $BW_{diff}$ is the difference between the desired uniform CAA BW (*BWuni*) and the obtained beamwidth (*$BW_{obt}$*).

In this case, $f_1$ is responsible for SLL reduction and $f_2$ controls the BW of the array.

The final fitness $f$ value is calculated as summation of $f_1$ and $f_2$, where $c_1$ and $c_2$ are the two constant biasing weighting factors.

## 4   Sociall Group Optimization

### *4.1   Brief Introduction*

A brief introduction to SGOA is given in this section. The SGOA typically mimics the social behavior of the human beings. A human being inherently possesses several characteristics which are mutually conflicting in oneself. These characteristics are expressed according to the situation prevailing around him. The situation can be a complex problem or a moment of joy. These behavioral traits are sometimes not only useful in handling one's own problem but also can be extended to influence other individual or as a group. Group solving capability can be termed as a more effective means of solving the problem as it sums up every individuals capability in the group. Similarly, the influence can be inherited by an individual from individual or a group and vice versa. Every individual in the society in other words has the capability to solve the problem and hence treated as a possible solution.

The algorithm typically has to phases, namely, improving phase and acquiring phase. During the improving phase, the highly knowledgeable person in the group propagates the knowledge to all the other individuals in the group. Hence, every individual solution is updated using the following expression:

$$Inew_{ij} = c * Iold_{ij} + r * (gbest(j) - Iold_{i,j}) \tag{9}$$

Similarly, during the acquiring stage, the knowledge update or the solution updating procedure takes a decision-based step. Every individual interacts with randomly selected individuals in the social group. This way, knowledge exchange with the randomly selected candidate takes place if it assumes more knowledge than this individual. However, the highly knowledgeable person in the group will have more impact and influence in this strategy. This is explained in the following two equations.

If the individual is less knowledgeable than the randomly selected one,

$$\text{Inew}_{i,j} = \text{Iold}_{i,j} + r1 * (I_{i,j} - I_{r,j}) + r2 * (\text{gbest}_j - I_{i,j}) \tag{10}$$

otherwise

$$\text{Inew}_{i,j} = \text{Iold}_{i,j} + r1 * (I_{r,j} - I_{i,j}) + r2 * (\text{gbest}_j - I_{i,j}) \tag{11}$$

## 4.2 Implementation of the SGOA for CAA Synthesis

Like every population-based optimization algorithm, our SGOA also starts with population initialization. In the array synthesis problem, every individual is an randomly generated array of elements with excitation currents and inter-element spacing as given below:

$$\text{Initial population:} pop = [x_1(k), x_2(k) \dots x_M(k)] \tag{12}$$

Each individual is characterized as a vector and the whole population in the society is a matrix of these vectors as below:

$$\begin{bmatrix} x_1 \\ x_2 \\ . \\ . \\ x_M \end{bmatrix} = \begin{bmatrix} I_1^1, I_1^2, \dots, I_1^N, d_1^1, d_1^2, \dots d_1^N \\ I_2^1, I_2^2, \dots, I_2^N, d_2^1, d_2^2, \dots d_2^N \\ \dots\dots\dots\dots\dots\dots\dots\dots\dots \\ \dots\dots\dots\dots\dots\dots\dots\dots\dots \\ I_M^1, I_M^2, \dots, I_M^N, d_M^1, d_M^2, \dots d_M^N \end{bmatrix} \tag{13}$$

In this paper, the CAA synthesis is transformed into a minimization problem and hence can be formulated as

$$x^*(k) = \arg \mathop{\min}_{m=1, \dots M} f(x_m(k)) \tag{14}$$

## 5  Results

Results pertaining synthesis of CAA using SGOA with amplitude–spacing technique is presented in this section. Two circular arrays with different sizes, each consisting of 30 elements and 60 elements, are considered. Array design using SGOA is performed with the objective of suppressing the SLL to as low as possible

**Table 1** Nonuniform amplitude and inter-element spacing obtained using SGOA

| Number of elements | Parameter | Distribution |
|---|---|---|
| 30 | Amps (normalized) | 0.746, 0.886, 0.37, 0.913, 0.83, 0.753, 0.868, 0.293, 0.439, 0.482, 0.376, 0.619, 0.69, 0.829, 0.999, 0.586, 0.529, 0.878, 0.51, 0.058, 0.897, 0.379, 0.406, 0.01, 0.526, 0.653, 0.95, 0.718, 0.834, 0.736 |
| | Spacing (in λ) | 0.72, 0.52, 0.356, 1.257, 1.181, 1.961, 0.999, 1.98, 1.396, 0.299, 1.418, 1.704, 1.174, 0.761, 1.4, 0.437, 1.712, 0.679, 0.991, 1.99, 1.998, 1.622, 1.535, 1.24, 0.609, 0.54, 1.121, 0.882, 0.937, 0.988 |
| 60 | Amps (normalized) | 0.786, 0.806, 0.967, 0.996, 0.658, 0.473, 0.9, 0.313, 0.661, 0.408, 0.537, 0.95, 0.222, 0.302, 0.151, 0.659, 0.026, 0.433, 0.148, 0.049, 0.338, 0.227, 0.179, 0.503, 0.486, 0.344, 0.603, 0.295, 0.473, 0.859, 0.806, 0.703, 0.794, 0.402, 0.831, 0.734, 0.579, 0.849, 0.449, 0.059, 0.017, 0.057, 0.012, 0.354, 0.237, 0.193, 0.152, 0.302, 0.086, 0.377, 0.201, 0.869, 0.533, 0.501, 0.564, 0.645, 0.863, 0.582, 0.489, 0.574 |
| | Spacing (in λ) | 0.5, 0.495, 0.599, 0.73, 0.803, 0.777, 0.613, 0.608, 0.387, 0.527, 0.669, 0.503, 1.959, 1.091, 1.49, 1.758, 0.595, 1.398, 0.594, 0.44, 1.728, 0.518, 1.956, 0.987, 1.399, 1.154, 1.093, 0.846, 0.355, 1.2, 0.496, 0.487, 0.561, 0.587, 1.63, 0.817, 1.063, 0.618, 1.154, 1.997, 1.236, 1.571, 0.939, 0.937, 1.192, 1.092, 0.835, 1.134, 1.329, 1.128, 1.867, 0.637, 1.523, 1.776, 0.367, 1.565, 0.624, 0.46, 0.632, 0.522 |

while keeping the BW not more than that of the uniform CAA. Simple uniform CAA produces an SLL of −7 dB with a BW dynamically changing with the number of elements in the array. It is obvious that the more the number of elements, the lesser the BW.

In the first example, the CAA with N = 30 is considered. The corresponding amplitudes of current excitation along with inter-elements spacing are determined using SGOA to produce a radiation pattern with a low SLL of −15.83 dB keeping the uniform BW of 18.2°. The obtained nonuniform distribution of amplitudes and spacing are given in Table 1. The corresponding radiation pattern plot of CAA for N = 30 is shown in Fig. 3. Similarly, in the next example, the number of elements in the CAA is increased to N = 60. For uniform CAA of 60 elements, the corresponding BW is 9.1° with an SLL of −7 dB. Using SGOA, the SLL is suppressed to −18.87 dB keeping the BW as minimum as that of uniform CAA as shown in Fig. 4. The corresponding nonuniform amplitude and spacing obtained using SGOA are given in Table 1.

When N = 30, there is an improvement in the SLL suppression to −15.83 which is quite less than the uniform distribution by 8.83 dB. This is significantly a great improvement. The corresponding BW is 10.10 which is much smaller than the uniform distribution CAA BW of 18.20. This certainly refers to the efficiency of the

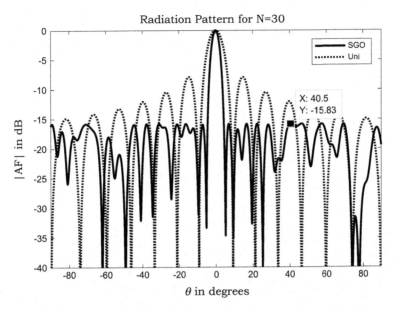

**Fig. 3** Comparison of radiation pattern plots for N = 30 using nonuniform amplitude and spacing obtained by SGOA and uniform distribution

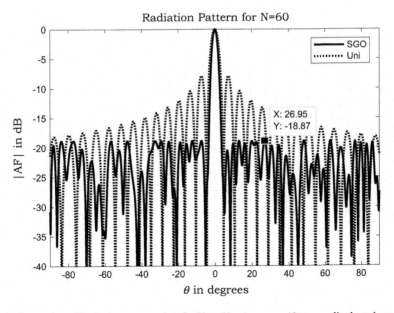

**Fig. 4** Comparison of radiation pattern plots for N = 60 using nonuniform amplitude and spacing obtained by SGOA and uniform distribution

SGOA and its impact in controlling the constrained parameters in array design. Similarly, in the second example where the number of elements is increased to 60, the corresponding SLL is reduced to −15.87 dB almost maintaining the same as that of the previous example. Also, the corresponding BW is less than that of the uniform BW.

# 6  Conclusion

The social group optimization algorithm is effectively implemented for the complex problem of synthesis of circular antenna array. Two examples are demonstrated in which the number of elements is varied from 30 and 60 with the objective of suppressing SLL with the constraint of uniform CAA BW efficiently achieved. There is an improvement in the SLL suppression by more than 100% in both the cases. Also, the corresponding BW in both the examples is much lesser than that of uniform distribution. Further, extending the simulation-based experiment on isotropic elements considered in this work to practical antennas would be a very good scope for this work.

# References

1. C.A. Balanis: Antenna theory: Analysis and design, John Wiley and Sons, Hoboken, NJ, USA, 2005.
2. V. Rabinovich and N. Alexandrov, Antenna Arrays and Automotive Applications, Springer Science + Bussiness Media New York 2013.
3. Monzingo RA and Miller TW, Introduction to adaptive arrays, SciTech Publishing, 2005.
4. Sudheer Kumar Terlapu, G. R. L. V. N. S. Raju and G. S. N. Raju, "Array pattern synthesis using flower pollination algorithm," 2016 IEEE International Conference on ElectroMagnetic Interference & Compatibility (INCEMIC), Bengaluru, India, 2016, pp. 1–4, December 2016.
5. Randy L Haupt, Antenna arrays: a computational approach, Wiley New York, 2010.
6. Goldberg, D. "Genetic Algorithms in Search, Optimization, and Machine Learning." (1989).
7. Kennedy, J.; Eberhart, R., "Particle swarm optimization," Neural Networks, 1995. Proceedings, IEEE International Conference on, vol. 4, no., pp. 1942,1948 vol. 4, Nov/Dec 1995 https://doi.org/10.1109/ICNN.1995.488968.
8. Chakravarthy, V.V.S.S.S.; et al.: Linear array optimization using teaching learning based optimization. In: Advances in Intelligent Systems and Computing, pp. 183–187. Springer, Berlin (2015). https://doi.org/10.1007/978-3-319-13731-5_21.
9. Md. Javeed Ahammed, A. Swathi, Deepika Sanku, V.V.S.S.S. Chakravarthy and H. Ramesh, Performance of Firefly Algorithm for Null Positioning in Linear Arrays, In: Proceedings of 2nd International Conference on Micro-Electronics, Electromagnetics and Telecommunications, Springer, 383–391 (2017).
10. Chakravarthy, V.S.S.S.; Rao, P.M., "On the convergence characteristics of flower pollination algorithm for circular array synthesis," in Electronics and Communication Systems (ICECS), 2015 2nd International Conference on, vol., no., pp. 485–489, 26–27 Feb (2015).

11. VVSSS Chakravarthy, PSR Chowdary, Ganapati Panda, Jaume Anguera, Aurora Andujar, Babita Majhi: On the Linear Antenna Array Synthesis Techniques for Sum and Difference Patterns Using Flower Pollination Algorithm, Arabian Journal for Science and Engineering, Aug (2017).
12. Suresh Satapathy and Anima Naik: "Social group optimization (SGO): a new population evolutionary optimization technique", Complex Intell. Syst., (2) 173–203 (2016).

# On the Design of Fractal UWB Wide-Slot Antenna with Notch Band Characteristics

**Sudheer Kumar Terlapu, P. S. R. Chowdary, Ch Jaya, V. V. S. S. Sameer Chakravarthy and Suresh Chandra Satpathy**

**Abstract** A coplanar waveguide-fed ultra-wideband (UWB) fractal wide-slot antenna with notch band characteristics is proposed. The radiation patch of proposed UWB antenna is designed using cantor set fractals by introducing triangular fractals. The bandwidth is enhanced by introducing symmetrical triangular-tapered corners at the bottom of wide slot. The proposed antenna has a size of 26 × 21 mm$^2$ and has operating frequency over the UWB range (2.8–10.3 GHz) except at the notch band frequency 5–6.3 GHz. The proposed cantor set of fractal wide-slot UWB antenna is designed and the performance of the antenna is verified by observing the antenna parameters such as return loss, gain, VSWR, and radiation characteristics. The results show that the designed antenna with compact size has good impedance bandwidth over the UWB range (2.8–10.3 GHz) and improved radiation characteristics with required notch band.

**Keywords** Antenna · Ultra-wideband · Notch characteristics and fractals

S. K. Terlapu (✉) · C. Jaya
Shri Vishnu Engineering College for Women (Autonomous), Bhimavaram, India
e-mail: skterlapu@gmail.com

C. Jaya
e-mail: jaya.chinni25@gmail.com

P. S. R. Chowdary · V. V. S. S. Sameer Chakravarthy
Department of ECE, Raghu Institute of Technology, Visakhapatnam, India
e-mail: sameersree@gmail.com

S. C. Satpathy
PVP Siddhartha Institute of Technology, Vijayawada, India
e-mail: sureshsatpathy@gmail.com

© Springer Nature Singapore Pte Ltd. 2018
J. Anguera et al. (eds.), *Microelectronics, Electromagnetics and Telecommunications*, Lecture Notes in Electrical Engineering 471,
https://doi.org/10.1007/978-981-10-7329-8_94

# 1 Introduction

A simple and compact UWB-printed monopole antenna with filtering characteristic is presented by A. Nouri and G. R. Dadashzadeh [1]. It consists of a radiating patch with defected ground structure (DGS). A modified shovel-shaped DGS is proposed so as to operate the antenna not only for DSRC systems but also for wireless LAN systems. Hamid Moghadas and AhadTavakoli [2] proposed a dumb-bell DGS cell that is sandwiched between two microstrip patch antennas fed by coaxial probe. DGS dimensions are determined by semi-numerical methods. It also reduces the mutual coupling so that scan blindness can be eliminated.

A novel multi-frequency printed monopole antenna for Wimax and Wireless LAN applications is proposed by Xiaoliang Zhang et al. [3]. It mainly consists of a fork-shaped strip, etched on a modified rectangular ring defected ground plane. By etching a rectangular slot, the proposed antenna can produce three resonant modes. The fabricated antenna parameters are experimentally analyzed which has good antenna performances. A dual-band antenna with a circular patch is designed at 2.5 GHz frequency [4] with a fractional bandwidth of 4.5%. By introducing a circular slot into the ground plane, it radiates by capacitive coupling between the patch and the ground plane. The slot radiates with a fractional bandwidth of 5% at a frequency of 1.95 GHz.

Anil Kumar Gautam et al. [5] presented small, low-profile planar triple-band microstrip antenna for WLAN/WiMAX applications. The designed antenna has a compact size and consists of F-shaped slot radiators with defected ground plane. It exhibits three distinct frequency bands, i.e., 2.0–2.76, 3.04–4.0, and 5.2–6.0 GHz, which covers the complete wireless LAN and WiMAX bands. In order to overcome the radio frequency interference (RFI) that occurs in digital circuits coupled to differentially fed antennas, a wideband balanced filter is proposed by [6]. The notched band characteristics are attained by introducing a T-shaped tuning stub at top of the wide slot [7]. The length and width of T-shaped tuning stub decide the notch band characteristics. Further, these fractals have inherent characteristics like miniaturization and multi-resonance [8, 9]. Therefore, in the present work, an attempt is made to design a UWB antenna using fractals with better notch band characteristics by introducing triangular slot radiating patch.

# 2 Antenna Geometry and Design

A CPW-fed cantor set fractal wide-slot antenna with a band notch characteristic is designed for UWB applications. The antenna is designed using cantor set fractals by introducing triangular fractal slots. Ultra-wideband is attained by implementing the cantor set fractals in three iterations. In the first two iterations, fractal slots are introduced in the radiation patch. The third iteration includes implementing of fractals along with placing of a T-shaped tuning stub in the radiation patch [7]. The

**Table 1** Dimensions of triangular fractal antenna

| Description | Dimensions |
| --- | --- |
| Substrate length L | 26 mm |
| Substrate width W | 21 mm |
| Substrate height H | 1.6 mm |
| Relative permittivity $\epsilon_r$ | 2.55 |
| Length of the slot etched from the ground L1 | 12.5 mm |
| Width of the slot etched from the ground W1 | 19 mm |
| Length of the radiation patch x | 12 mm |
| Width of the radiation patch y | 8.4 mm |
| Length of the stub L2 | 15.4 mm |
| Width of the stub W2 | 0.25 mm |
| Gap between the ground and the stub A | 0.4 mm |
| Gap between ground and the radiation patch g | 0.6 mm |
| Length of the feedline L3 | 6.15 mm |
| Width of the feedline W3 | 3.6 mm |
| Width of slot etched from ground after introducing tapered corners W4 | 16.4 mm |

**Fig. 1** The proposed antenna with triangular fractal first cantor set triangular fractal radiation patch and second cantor set triangular fractal radiation patch

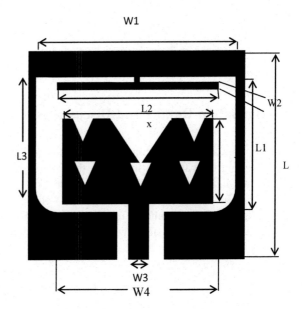

dimensions of the designed antenna are summarized in Table 1. Figures 1 and 2 depict the step-by-step design of the proposed antenna with triangular fractal with first cantor set triangular fractal radiation patch and second cantor set triangular fractal radiation patch.

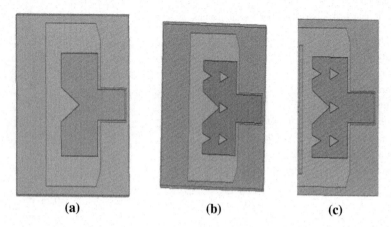

**Fig. 2** **a** Design of first-order cantor set fractal antenna, **b** Design of second-order cantor set fractal antenna, **c** Design of second-order cantor set fractal antenna with T-shaped tuning stub

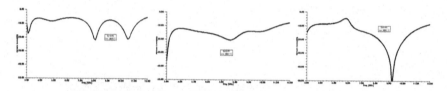

**Fig. 3** Return loss for order triangular cantor set triangular fractal radiation patch in all iterations

# 3 Simulation Results and Discussions

The proposed antenna with triangular fractals is designed. The UWB is attained by implementing the cantor set fractals in three iterations. In the first two iterations, fractals are introduced in the radiation patch. The third iteration includes implementing of fractals along with placing of a T-shaped tuning stub in the radiation patch. The simulation results for return loss, gain, VSWR, and radiation patterns are presented in Figs. 3, 4, 5, and 6 which depicts the attainment of ultra-wideband and notch band characteristics in the consecutive iterations.

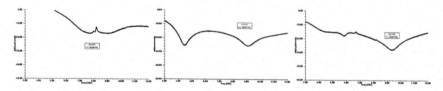

**Fig. 4** Gain plots for triangular cantor set fractal radiation patch for all iterations

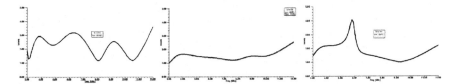

**Fig. 5** VSWR for triangular cantor set fractal radiation patch for all iterations

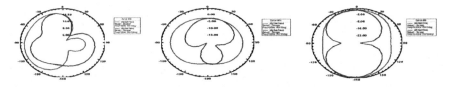

**Fig. 6** Radiation pattern for triangular cantor set fractal radiation patch for all iterations

# 4   Conclusion

The proposed antenna is designed by introducing triangular fractal slots. The zeroth iteration has an impedance bandwidth of 5.7 GHz with VSWR approximately equals to 1.2 over this range. The antenna has resonated only at 3.1, 7.9, and 10.5 GHz, in first iteration with VSWR above two. The second iteration has an impedance bandwidth over 8.3 GHz (2.7–11 GHz) with VSWR approximately 1.21 over this range. The notch band is obtained from 3.4 to 6.3 GHz.

# References

1. A. Nouri and G. R. Dadashzadeh "A compact UWB band-notched printed monopole antenna with defected ground structure." IEEE antennas and wireless propagation letters, Vol. 10, 2011.
2. Hamid Moghadas and AhadTavakoli, "A semi-numerical design algorithm for defected ground structure in microstrip antenna arrays." IEEE Trans. Microw. Theory Tech., Vol. 54, No. 5, 2012.
3. Li Li, Xiaoliang Zhang, Xiaoli Yin, and Le Zhou, "A compact triple-band printed monopole antenna for WLAN/WiMAX applications." IEEE antennas and wireless propagation letters, Vol. 15, 2016.
4. Mohamed Aboualalaa, Adel B. Abdel-Rahman, Ahmed Allam, HalaElsadek, and Ramesh K. Pokharel, "Design of dual band microstrip antenna with enhanced gain for energy harvesting applications." IEEE transactions on antennas and propagation, Vol. 60, No. 1, January, 2016.
5. Anil Kumar Gautam, Lalit Kumar, Binod Kumar Kanaujia, and Karumudi Rambabu, "Mutual coupling reduction by novel fractal defected ground structure bandgap filter." IEEE transactions on antennas and propagation, Vol. 64, No. 3, March 2016.
6. Ying-Cheng Tseng, Pei-Yang Weng, and Tzong-Lin W, "compact wideband balanced filter for eliminating radio-frequency interference on differentially-fed antennas." IEEE transactions on antennas and propagation, Vol. 64, No. 3, March, 2015.

7. Y.-S. Li, X.-D. Yang, C.-Y. Liu, and T. Jiang, "Analysis and investigation of a cantor set fractal UWB antenna with a notch- band characteristic," Progress In Electromagnetics Research B, Vol. 33, 99–114, 2011.
8. P. S. R. Chowdary, A. Mallikarjuna Prasad, P. Mallikarjuna Rao, and Jaume Anguera, "Simulation of Radiation Characteristics of Sierpinski Fractal Geometry for Multiband Applications," International Journal of Information and Electronics Engineering vol. 3, no. 6, pp. 618–621, 2013.
9. Chowdary, P.S.R.; Prasad, A.M.; Rao, P.M.; Jaume Anguera: "Design and Performance Study of Sierpinski Fractal Based Patch Antennas for Multiband and Miniaturization Characteristics", Wireless Pers Commun (2015) 83: 1713.

# Author Index

© Springer Nature Singapore Pte Ltd. 2018                                  913
J. Anguera et al. (eds.), *Microelectronics, Electromagnetics
and Telecommunications*, Lecture Notes in Electrical Engineering 471,
https://doi.org/10.1007/978-981-10-7329-8